SCHAUM'S OUTLINE OF

THEORY AND PROBLEMS

of

ADVANCED
CALCULUS

•

BY

MURRAY R. SPIEGEL, Ph.D.

Professor of Mathematics
Rensselaer Polytechnic Institute

•

SCHAUM'S OUTLINE SERIES

McGRAW-HILL BOOK COMPANY

New York, St. Louis, San Francisco, Toronto, Sydney

ISBN 07-060229-8

25 26 27 28 29 30 SH SH 8 7 6

Preface

The subject commonly called "Advanced Calculus" means different things to different people. To some it essentially represents elementary calculus from an advanced viewpoint, i.e. with rigorous statements and proofs of theorems. To others it represents a variety of special advanced topics which are considered important but which cannot be covered in an elementary course.

In this book an effort has been made to adopt a reasonable compromise between these extreme approaches which, it is believed, will serve a variety of individuals. The early chapters of the book serve in general to review and extend fundamental concepts already presented in elementary calculus. This should be valuable to those who have forgotten some of the calculus studied previously and who need "a bit of refreshing". It may also serve to provide a common background for students who have been given different types of courses in elementary calculus. Later chapters serve to present special advanced topics which are fundamental to the scientist, engineer and mathematician if he is to become proficient in his intended field.

This book has been designed for use either as a supplement to all current standard text-books or as a textbook for a formal course in advanced calculus. It should also prove useful to students taking courses in physics, engineering or any of the numerous other fields in which advanced mathematical methods are employed.

Each chapter begins with a clear statement of pertinent definitions, principles and theorems together with illustrative and other descriptive material. This is followed by graded sets of solved and supplementary problems. The solved problems serve to illustrate and amplify the theory, bring into sharp focus those fine points without which the student continually feels himself on unsafe ground, and provide the repetition of basic principles so vital to effective learning. Numerous proofs of theorems and derivations of basic results are included among the solved problems. The large number of supplementary problems with answers serve as a complete review of the material of each chapter.

Topics covered include the differential and integral calculus of functions of one or more variables and their applications. Vector methods, which lend themselves so readily to concise notation and to geometric and physical interpretations, are introduced early and used whenever they can contribute to motivation and understanding. Special topics include line and surface integrals and integral theorems, infinite series, improper integrals, gamma and beta functions, and Fourier series. Added features are the chapters on Fourier integrals, elliptic integrals and functions of a complex variable which should prove extremely useful in the study of advanced engineering, physics and mathematics.

Considerably more material has been included here than can be covered in most courses. This has been done to make the book more flexible, to provide a more useful book of reference and to stimulate further interest in the topics.

I wish to take this opportunity to thank the staff of the Schaum Publishing Company for their splendid cooperation in meeting the seemingly endless attempts at perfection by the author.

M. R. Spiegel

Rensselaer Polytechnic Institute
December, 1962

CONTENTS

CONTENTS

CONTENTS

Chapter 1

Numbers

SETS

Fundamental in mathematics is the concept of a *set, class* or *collection* of objects having specified characteristics. For example we speak of the set of all university professors, the set of all letters $A, B, C, D, \ldots, Z$ of the English alphabet, etc. The individual objects of the set are called *members* or *elements*. Any part of a set is called a *subset* of the given set, e.g. A, B, C is a subset of $A, B, C, D, \ldots, Z$. The set consisting of no elements is called the *empty set* or *null set*.

REAL NUMBERS

The following types of numbers are already familiar to the student.

1. **Natural numbers** $1, 2, 3, 4, \ldots$, also called *positive integers*, are used in counting members of a set. The symbols varied with the times, e.g. the Romans used $I, II, III, IV, \ldots$. The *sum* $a + b$ and *product* $a \cdot b$ or ab of any two natural numbers a and b is also a natural number. This is often expressed by saying that the set of natural numbers is *closed* under the operations of *addition* and *multiplication*, or satisfies the *closure property* with respect to these operations.

2. **Negative integers and zero** denoted by $-1, -2, -3, \ldots$ and 0 respectively, arose to permit solutions of equations such as $x + b = a$ where a and b are any natural numbers. This leads to the operation of *subtraction*, or *inverse of addition*, and we write $x = a - b$.

   The set of positive and negative integers and zero is called the set of *integers*.

3. **Rational numbers** or *fractions* such as $\frac{2}{3}, -\frac{5}{4}, \ldots$ arose to permit solutions of equations such as $bx = a$ for all integers a and b where $b \neq 0$. This leads to the operation of *division*, or *inverse of multiplication*, and we write $x = a/b$ or $a \div b$ where a is the *numerator* and b the *denominator*.

   The set of integers is a subset of the rational numbers, since integers correspond to rational numbers where $b = 1$.

4. **Irrational numbers** such as $\sqrt{2}$ and π are numbers which are not rational, i.e. cannot be expressed as $\frac{a}{b}$ (called the *quotient* of a and b) where a and b are integers and $b \neq 0$.

   The set of rational and irrational numbers is called the set of *real numbers*.

DECIMAL REPRESENTATION of REAL NUMBERS

Any real number can be expressed in *decimal form*, e.g. $17/10 = 1.7$, $9/100 = 0.09$, $1/6 = 0.16666\ldots$. In the case of a rational number the decimal expansion either terminates or, if it does not terminate, one or a group of digits in the expansion will ultimately repeat as, for example, in $\frac{1}{7} = 0.142857\,142857\,142\ldots$. In the case of an irrational number such as $\sqrt{2} = 1.41423\ldots$ or $\pi = 3.14159\ldots$ no such repetition can occur. We can always consider a decimal expansion as unending, e.g. 1.375 is the same as $1.37500000\ldots$ or $1.3749999\ldots$. To indicate recurring decimals we sometimes place dots over the repeating cycle of digits, e.g. $\frac{1}{7} = 0.\overset{\cdot}{1}4285\overset{\cdot}{7}$, $\frac{19}{6} = 3.1\overset{\cdot}{6}$.

The decimal system uses the ten digits $0, 1, 2, \ldots, 9$. It is possible to design number systems with fewer or more digits, e.g. the *binary system* uses only two digits 0 and 1 (see Problems 32 and 33).

1

GEOMETRIC REPRESENTATION of REAL NUMBERS

The geometric representation of real numbers as points on a line called the *real axis*, as in the figure below, is also well known to the student. For each real number there corresponds one and only one point on the line and conversely, i.e. there is a *one to one* (1-1) *correspondence* between the set of real numbers and the set of points on the line. Because of this we often use point and number interchangeably.

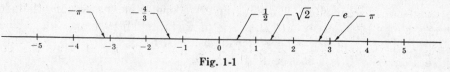

Fig. 1-1

The set of real numbers to the right of 0 is called the set of *positive numbers*; the set to the left of 0 is the set of *negative numbers*, while 0 itself is neither positive nor negative.

Between any two rational numbers (or irrational numbers) on the line there are infinitely many rational (and irrational) numbers. This leads us to call the set of rational (or irrational) numbers an *everywhere dense* set.

OPERATIONS with REAL NUMBERS

If a, b, c belong to the set R of real numbers, then:

1. $a + b$ and ab belong to R Closure law
2. $a + b = b + a$ Commutative law of addition
3. $a + (b + c) = (a + b) + c$ Associative law of addition
4. $ab = ba$ Commutative law of multiplication
5. $a(bc) = (ab)c$ Associative law of multiplication
6. $a(b + c) = ab + ac$ Distributive law
7. $a + 0 = 0 + a = a, \quad 1 \cdot a = a \cdot 1 = a$

   0 is called the *identity with respect to addition*, 1 is called the *identity with respect to multiplication*.

8. For any a there is a number x in R such that $x + a = 0$.

   x is called the *inverse of a with respect to addition* and is denoted by $-a$.

9. For any $a \neq 0$ there is a number x in R such that $ax = 1$.

   x is called the *inverse of a with respect to multiplication* and is denoted by a^{-1} or $1/a$.

These enable us to operate according to the usual rules of algebra. In general any set, such as R, whose members satisfy the above is called a *field*.

INEQUALITIES

If $a - b$ is a nonnegative number we say that a is *greater than or equal to b* or b is *less than or equal to a*, and write respectively $a \geqq b$ or $b \leqq a$. If there is no possibility that $a = b$, we write $a > b$ or $b < a$. Geometrically, $a > b$ if the point on the real axis corresponding to a lies to the right of the point corresponding to b.

> **Examples:** $3 < 5$ or $5 > 3$; $-2 < -1$ or $-1 > -2$; $x \leqq 3$ means that x is a real number which may be 3 or less than 3.

If a, b and c are any given real numbers, then:

1. Either $a > b$, $a = b$ or $a < b$ Law of trichotomy
2. If $a > b$ and $b > c$, then $a > c$ Law of transitivity
3. If $a > b$, then $a + c > b + c$
4. If $a > b$ and $c > 0$, then $ac > bc$
5. If $a > b$ and $c < 0$, then $ac < bc$

ABSOLUTE VALUE of REAL NUMBERS

The absolute value of a real number a, denoted by $|a|$, is defined as a if $a > 0$, $-a$ if $a < 0$, and 0 if $a = 0$.

Examples: $|-5| = 5$, $|+2| = 2$, $|-\frac{3}{4}| = \frac{3}{4}$, $|-\sqrt{2}| = \sqrt{2}$, $|0| = 0$.

1. $|ab| = |a|\,|b|$ or $|abc\ldots m| = |a|\,|b|\,|c|\cdots|m|$
2. $|a + b| \leq |a| + |b|$ or $|a + b + c + \cdots + m| \leq |a| + |b| + |c| + \cdots + |m|$
3. $|a - b| \geq |a| - |b|$

The distance between any two points (real numbers) a and b on the real axis is $|a - b| = |b - a|$.

EXPONENTS and ROOTS

The product $a \cdot a \ldots a$ of a real number a by itself p times is denoted by a^p where p is called the *exponent* and a is called the *base*. The following rules hold.

1. $a^p \cdot a^q = a^{p+q}$ 3. $(a^p)^r = a^{pr}$

2. $\dfrac{a^p}{a^q} = a^{p-q}$ 4. $\left(\dfrac{a}{b}\right)^p = \dfrac{a^p}{b^p}$

These and extensions to any real numbers are possible so long as division by zero is excluded. In particular by using 2, with $p = q$ and $p = 0$ respectively, we are led to the definitions $a^0 = 1$, $a^{-q} = 1/a^q$.

If $a^p = N$, where p is a positive integer, we call a a pth *root* of N, written $\sqrt[p]{N}$. There may be more than one real pth root of N. For example since $2^2 = 4$ and $(-2)^2 = 4$, there are two real square roots of 4, namely 2 and -2. It is customary to denote the positive square root by $\sqrt{4} = 2$ and the negative one by $-\sqrt{4} = -2$.

If p and q are positive integers, we define $a^{p/q} = \sqrt[q]{a^p}$.

LOGARITHMS

If $a^p = N$, p is called the *logarithm* of N to the base a, written $p = \log_a N$. If a and N are positive and $a \neq 1$, there is only one real value for p. The following rules hold.

1. $\log_a MN = \log_a M + \log_a N$ 2. $\log_a \dfrac{M}{N} = \log_a M - \log_a N$

3. $\log_a M^r = r \log_a M$

In practice two bases are used, the *Briggsian system* uses base $a = 10$, the *Napierian system* uses the *natural base* $a = e = 2.71828\ldots$.

AXIOMATIC FOUNDATIONS of the REAL NUMBER SYSTEM

The number system can be built up logically, starting from a basic set of *axioms* or "self evident" truths, usually taken from experience, such as statements 1-9, Page 2.

If we assume as given the natural numbers and the operations of addition and multiplication (although it is possible to start even further back with the concept of sets), we find that statements 1-6, Page 2, with R as the set of natural numbers, hold while 7-9 do not hold.

Taking 7 and 8 as additional requirements, we introduce the numbers $-1, -2, -3, \ldots$ and 0. Then by taking 9 we introduce the rational numbers.

Operations with these newly obtained numbers can be defined by adopting axioms 1-6, where R is now the set of integers. These lead to *proofs* of statements such as $(-2)(-3) = 6$, $-(-4) = 4$, $(0)(5) = 0$, etc., which are usually taken for granted in elementary mathematics.

We can also introduce the concept of order or inequality for integers, and from these inequalities for rational numbers. For example if a, b, c, d are positive integers we define $a/b > c/d$ if and only if $ad > bc$, with similar extensions to negative integers.

Once we have the set of rational numbers and the rules of inequality concerning them, we can order them geometrically as points on the real axis, as already indicated. We can then show that there are points on the line which do not represent rational numbers (such as $\sqrt{2}, \pi$, etc.). These irrational numbers can be defined in various ways one of which uses the idea of *Dedekind cuts* (see Problem 34). From this we can show that the usual rules of algebra apply to irrational numbers and that no further real numbers are possible.

POINT SETS, INTERVALS

A set of points (real numbers) located on the real axis is called a *one-dimensional point set*.

The set of points x such that $a \leqq x \leqq b$ is called a *closed interval* and is denoted by $[a, b]$. The set $a < x < b$ is called an *open interval*, denoted by (a, b). The sets $a < x \leqq b$ and $a \leqq x < b$, denoted by $(a, b]$ and $[a, b)$ respectively, are called *half open* or *half closed* intervals.

The symbol x, which can represent any number or point of a set, is called a *variable*. The given numbers a or b are called *constants*.

> **Example:** The set of all x such that $|x| < 4$, i.e. $-4 < x < 4$, is represented by $(-4, 4)$, an open interval.

The set $x > a$ can also be represented by $a < x < \infty$. Such a set is called an *infinite* or *unbounded interval*. Similarly $-\infty < x < \infty$ represents all real numbers x.

COUNTABILITY

A set is called *countable* or *denumerable* if its elements can be placed in 1-1 correspondence with the natural numbers.

> **Example:** The even natural numbers $2, 4, 6, 8, \ldots$ is a countable set because of the 1-1 correspondence shown.
>
> | Given set | 2 | 4 | 6 | 8 | $\ldots$ |
> | --- | --- | --- | --- | --- | --- |
> | | $\updownarrow$ | $\updownarrow$ | $\updownarrow$ | $\updownarrow$ | |
> | Natural numbers | 1 | 2 | 3 | 4 | $\ldots$ |

A set is *infinite* if it can be placed in 1-1 correspondence with a subset of itself. An infinite set which is countable is called *countably infinite*.

The set of rational numbers is countably infinite while the set of irrational numbers or all real numbers is non-countably infinite (see Problems 17-20).

The number of elements in a set is called its *cardinal number*. A set which is countably infinite is assigned the cardinal number $\aleph_0$ (the Hebrew letter *aleph-null*). The set of real numbers (or any sets which can be placed into 1-1 correspondence with this set) is given the cardinal number C, called the *cardinality of the continuum*.

NEIGHBORHOODS

The set of all points x such that $|x-a| < \delta$ where $\delta > 0$, is called a δ *neighborhood* of the point a. The set of all points x such that $0 < |x-a| < \delta$ in which $x = a$ is excluded, is called a *deleted* δ *neighborhood* of a.

LIMIT POINTS

A *limit point, point of accumulation* or *cluster point* of a set of numbers is a number l such that every deleted δ neighborhood of l contains members of the set. In other words for any $\delta > 0$, however small, we can always find a member x of the set which is not equal to l but which is such that $|x-l| < \delta$. By considering smaller and smaller values of δ we see that there must be infinitely many such values of x.

A finite set cannot have a limit point. An infinite set may or may not have a limit point. Thus the natural numbers have no limit point while the set of rational numbers has infinitely many limit points.

A set containing all its limit points is called a *closed set*. The set of rational numbers is not a closed set since, for example, the limit point $\sqrt{2}$ is not a member of the set (Problem 5). However, the set $0 \leqq x \leqq 1$ is a closed set.

BOUNDS

If for all numbers x of a set there is a number M such that $x \leqq M$, the set is *bounded above* and M is called an *upper bound*. Similarly if $x \geqq m$, the set is *bounded below* and m is called a *lower bound*. If for all x we have $m \leqq x \leqq M$, the set is called *bounded*.

If $\underline{M}$ is a number such that no member of the set is greater than $\underline{M}$ but there is at least one member which exceeds $\underline{M} - \epsilon$ for every $\epsilon > 0$, then $\underline{M}$ is called the *least upper bound* (l.u.b.) of the set. Similarly if no member of the set is smaller than $\bar{m}$ but at least one member is smaller than $\bar{m} + \epsilon$ for every $\epsilon > 0$, then $\bar{m}$ is called the *greatest lower bound* (g.l.b.) of the set.

WEIERSTRASS-BOLZANO THEOREM

The Weierstrass-Bolzano theorem states that every bounded infinite set has at least one limit point. A proof of this is given in Problem 23, Chapter 3.

ALGEBRAIC and TRANSCENDENTAL NUMBERS

A number x which is a solution to the *polynomial equation*

$$a_0 x^n + a_1 x^{n-1} + a_2 x^{n-2} + \cdots + a_{n-1} x + a_n = 0 \tag{1}$$

where $a_0 \neq 0$, $a_1, a_2, \ldots, a_n$ are integers and n is a positive integer, called the *degree* of the equation, is called an *algebraic number*. A number which cannot be expressed as a solution of any polynomial equation with integer coefficients is called a *transcendental number*.

> **Examples:** $\frac{2}{3}$ and $\sqrt{2}$ which are solutions of $3x - 2 = 0$ and $x^2 - 2 = 0$ respectively, are algebraic numbers.

The numbers π and e can be shown to be transcendental numbers. We still cannot determine whether some numbers such as $e\pi$ or $e + \pi$ are algebraic or not.

The set of algebraic numbers is a countably infinite set (see Problem 23) but the set of transcendental numbers is non-countably infinite.

The COMPLEX NUMBER SYSTEM

Since there is no real number x which satisfies the polynomial equation $x^2 + 1 = 0$ or similar equations, the set of complex numbers is introduced.

We can consider a complex number as having the form $a + bi$ where a and b are real numbers called the *real* and *imaginary parts*, and $i = \sqrt{-1}$ is called the *imaginary unit*. Two complex numbers $a + bi$ and $c + di$ are *equal* if and only if $a = c$ and $b = d$. We can consider real numbers as a subset of the set of complex numbers with $b = 0$. The complex number $0 + 0i$ corresponds to the real number 0.

The *absolute value* or *modulus* of $a + bi$ is defined as $|a + bi| = \sqrt{a^2 + b^2}$. The *complex conjugate* of $a + bi$ is defined as $a - bi$. The complex conjugate of the complex number z is often indicated by $\bar{z}$ or z^*.

The set of complex numbers obeys rules 1-9 of Page 2, and thus constitutes a field. In performing operations with complex numbers we can operate as in the algebra of real numbers, replacing i^2 by -1 when it occurs. Inequalities for complex numbers are not defined.

From the point of view of an axiomatic foundation of complex numbers, it is desirable to treat a complex number as an ordered pair (a, b) of real numbers a and b subject to certain operational rules which turn out to be equivalent to those above. For example, we define $(a, b) + (c, d) = (a + c, \ b + d)$, $(a, b)(c, d) = (ac - bd, \ ad + bc)$, $m(a, b) = (ma, mb)$, etc. We then find that $(a, b) = a(1, 0) + b(0, 1)$ and we associate this with $a + bi$, where i is the symbol for $(0, 1)$.

POLAR FORM of COMPLEX NUMBERS

If real scales are chosen on two mutually perpendicular axes $X'OX$ and $Y'OY$ (the x and y axes) as in Fig. 1-2 below, we can locate any point in the plane determined by these lines by the ordered pair of numbers (x, y) called *rectangular coordinates* of the point. Examples of the location of such points are indicated by P, Q, R, S and T in Fig. 1-2.

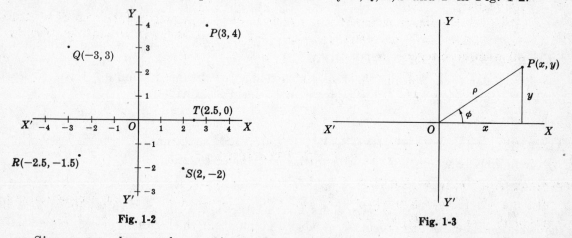

Fig. 1-2   Fig. 1-3

Since a complex number $x + iy$ can be considered as an ordered pair (x, y), we can represent such numbers by points in an xy plane called the *complex plane* or *Argand diagram*. Referring to Fig. 1-3 above we see that $x = \rho \cos \phi$, $y = \rho \sin \phi$ where $\rho = \sqrt{x^2 + y^2} = |x + iy|$ and ϕ, called the *amplitude* or *argument*, is the angle which line OP makes with the positive x axis OX. It follows that

$$z = x + iy = \rho(\cos \phi + i \sin \phi) \tag{2}$$

called the *polar form* of the complex number, where ρ and ϕ are called *polar coordinates*. It is sometimes convenient to write cis ϕ instead of $\cos \phi + i \sin \phi$.

If $z_1 = x_1 + iy_1 = \rho_1(\cos\phi_1 + i\sin\phi_1)$ and $z_2 = x_2 + iy_2 = \rho_2(\cos\phi_2 + i\sin\phi_2)$ we can show that

$$z_1 z_2 = \rho_1\rho_2\{\cos(\phi_1 + \phi_2) + i\sin(\phi_1 + \phi_2)\} \qquad (3)$$

$$\frac{z_1}{z_2} = \frac{\rho_1}{\rho_2}\{\cos(\phi_1 - \phi_2) + i\sin(\phi_1 - \phi_2)\} \qquad (4)$$

$$z^n = \{\rho(\cos\phi + i\sin\phi)\}^n = \rho^n(\cos n\phi + i\sin n\phi) \qquad (5)$$

where n is any real number. Equation (5) is sometimes called *De Moivre's theorem*. We can use this to determine roots of complex numbers. For example if n is a positive integer,

$$z^{1/n} = \{\rho(\cos\phi + i\sin\phi)\}^{1/n} \qquad (6)$$

$$= \rho^{1/n}\left\{\cos\left(\frac{\phi + 2k\pi}{n}\right) + i\sin\left(\frac{\phi + 2k\pi}{n}\right)\right\} \qquad k = 0, 1, 2, 3, \ldots, n-1$$

from which it follows that there are in general n different values for $z^{1/n}$. Later (Chap. 11) we will show that $e^{i\phi} = \cos\phi + i\sin\phi$ where $e = 2.71828\ldots$. This is called *Euler's formula*.

MATHEMATICAL INDUCTION

The principle of *mathematical induction* is an important property of the positive integers. It is especially useful in proving statements involving all positive integers when it is known for example that the statements are valid for $n = 1, 2, 3$ but it is *suspected* or *conjectured* that they hold for all positive integers. The method of proof consists of the following steps.

1. Prove the statement for $n = 1$ (or some other positive integer).
2. Assume the statement true for $n = k$ where k is any positive integer.
3. From the assumption in 2 prove that the statement must be true for $n = k+1$. This is the part of the proof establishing the induction and may be difficult or impossible.
4. Since the statement is true for $n = 1$ [from step 1] it must [from step 3] be true for $n = 1 + 1 = 2$ and from this for $n = 2 + 1 = 3$, etc., and so must be true for all positive integers.

Solved Problems

OPERATIONS with NUMBERS

1. If $x = 4$, $y = 15$, $z = -3$, $p = \frac{2}{3}$, $q = -\frac{1}{6}$, and $r = \frac{3}{4}$, evaluate (a) $x + (y + z)$, (b) $(x + y) + z$, (c) $p(qr)$, (d) $(pq)r$, (e) $x(p + q)$.

(a) $x + (y + z) = 4 + [15 + (-3)] = 4 + 12 = 16$

(b) $(x + y) + z = (4 + 15) + (-3) = 19 - 3 = 16$

The fact that (a) and (b) are equal illustrates the *associative law of addition*.

(c) $p(qr) = \frac{2}{3}\{(-\frac{1}{6})(\frac{3}{4})\} = (\frac{2}{3})(-\frac{3}{24}) = (\frac{2}{3})(-\frac{1}{8}) = -\frac{2}{24} = -\frac{1}{12}$

(d) $(pq)r = \{(\frac{2}{3})(-\frac{1}{6})\}(\frac{3}{4}) = (-\frac{2}{18})(\frac{3}{4}) = (-\frac{1}{9})(\frac{3}{4}) = -\frac{3}{36} = -\frac{1}{12}$

The fact that (c) and (d) are equal illustrates the *associative law of multiplication*.

(e) $x(p + q) = 4(\frac{2}{3} - \frac{1}{6}) = 4(\frac{4}{6} - \frac{1}{6}) = 4(\frac{3}{6}) = \frac{12}{6} = 2$

Another method: $x(p + q) = xp + xq = (4)(\frac{2}{3}) + (4)(-\frac{1}{6}) = \frac{8}{3} - \frac{4}{6} = \frac{8}{3} - \frac{2}{3} = \frac{6}{3} = 2$ using the *distributive law*.

2. Explain why we do not consider (a) $\dfrac{0}{0}$ (b) $\dfrac{1}{0}$ as numbers.

(a) If we define a/b as that number (if it exists) such that $bx = a$, then $0/0$ is that number x such that $0x = 0$. However, this is true for all numbers. Since there is no unique number which $0/0$ can represent, we consider it undefined.

(b) As in (a), if we define $1/0$ as that number x (if it exists) such that $0x = 1$, we conclude that there is no such number.

   Because of these facts we must look upon division by zero as meaningless.

3. Simplify $\dfrac{x^2 - 5x + 6}{x^2 - 2x - 3}$.

$\dfrac{x^2 - 5x + 6}{x^2 - 2x - 3} = \dfrac{(x-3)(x-2)}{(x-3)(x+1)} = \dfrac{x-2}{x+1}$ provided that the cancelled factor $(x-3)$ is not zero, i.e. $x \neq 3$. For $x = 3$ the given fraction is undefined.

RATIONAL and IRRATIONAL NUMBERS

4. Prove that the square of any odd integer is odd.

   Any odd integer has the form $2m + 1$. Since $(2m+1)^2 = 4m^2 + 4m + 1$ is 1 more than the even integer $4m^2 + 4m = 2(2m^2 + 2m)$, the result follows.

5. Prove that there is no rational number whose square is 2.

   Let p/q be a rational number whose square is 2, where we assume that p/q is in lowest terms, i.e. p and q have no common integer factors except ± 1 (we sometimes call such integers *relatively prime*).

   Then $(p/q)^2 = 2$, $p^2 = 2q^2$ and p^2 is even. From Problem 4, p is even since if p were odd, p^2 would be odd. Thus $p = 2m$.

   Substituting $p = 2m$ in $p^2 = 2q^2$ yields $q^2 = 2m^2$, so that q^2 is even and q is even.

   Thus p and q have the common factor 2, contradicting the original assumption that they had no common factors other than ± 1. By virtue of this contradiction there can be no rational number whose square is 2.

6. Show how to find rational numbers whose squares can be made arbitrarily close to 2.

   We restrict ourselves to positive rational numbers. Since $(1)^2 = 1$ and $(2)^2 = 4$, we are led to choose rational numbers between 1 and 2, e.g. $1.1, 1.2, 1.3, \ldots, 1.9$.

   Since $(1.4)^2 = 1.96$ and $(1.5)^2 = 2.25$, we consider rational numbers between 1.4 and 1.5, e.g. $1.41, 1.42, \ldots, 1.49$.

   Continuing in this manner we can obtain closer and closer rational approximations, e.g. $(1.414213562)^2$ is less than 2 while $(1.414213563)^2$ is greater than 2.

7. Given the equation $a_0 x^n + a_1 x^{n-1} + \cdots + a_n = 0$ where $a_0, a_1, \ldots, a_n$ are integers and a_0 and $a_n \neq 0$. Show that if the equation is to have a rational root p/q, then p must divide a_n and q must divide a_0 exactly.

   Since p/q is a root we have, on substituting in the given equation and multiplying by q^n, the result

$$a_0 p^n + a_1 p^{n-1} q + a_2 p^{n-2} q^2 + \cdots + a_{n-1} p q^{n-1} + a_n q^n = 0 \qquad (1)$$

or dividing by p,

$$a_0 p^{n-1} + a_1 p^{n-2} q + \cdots + a_{n-1} q^{n-1} = -\frac{a_n q^n}{p} \qquad (2)$$

Since the left side of (2) is an integer the right side must also be an integer. Then since p and q are relatively prime, p does not divide q^n exactly and so must divide a_n.

   In a similar manner, by transposing the first term of (1) and dividing by q, we can show that q must divide a_0.

8. Prove that $\sqrt{2} + \sqrt{3}$ cannot be a rational number.

   If $x = \sqrt{2} + \sqrt{3}$ then $x^2 = 5 + 2\sqrt{6}$, $x^2 - 5 = 2\sqrt{6}$ and squaring, $x^4 - 10x^2 + 1 = 0$. The only possible rational roots of this equation are ± 1 by Problem 7, and these do not satisfy the equation. It follows that $\sqrt{2} + \sqrt{3}$, which satisfies the equation, cannot be a rational number.

9. Prove that between any two rational numbers there is another rational number.

If a and b are rational numbers, then $\dfrac{a+b}{2}$ is a rational number between a and b.

To prove this assume $a < b$. Then by adding a to both sides, $2a < a+b$ and $a < \dfrac{a+b}{2}$.

Similarly adding b to both sides, $a+b < 2b$ and $\dfrac{a+b}{2} < b$.

Thus $a < \dfrac{a+b}{2} < b$.

To prove that $\dfrac{a+b}{2}$ is a rational number, let $a = \dfrac{p}{q}$ and $b = \dfrac{r}{s}$ where p, q, r, s are integers and $q \neq 0$, $s \neq 0$.

Then $\dfrac{a+b}{2} = \dfrac{1}{2}\left(\dfrac{p}{q} + \dfrac{r}{s}\right) = \dfrac{1}{2}\left(\dfrac{ps}{qs} + \dfrac{qr}{qs}\right) = \dfrac{ps+qr}{2qs}$ is a rational number.

INEQUALITIES

10. For what values of x is $x + 3(2-x) \geqq 4 - x$?

$x + 3(2-x) \geqq 4 - x$ when $x + 6 - 3x \geqq 4 - x$, $6 - 2x \geqq 4 - x$, $6 - 4 \geqq 2x - x$, $2 \geqq x$, i.e. $x \leqq 2$.

11. For what values of x is $x^2 - 3x - 2 < 10 - 2x$?

The required inequality holds when

$$x^2 - 3x - 2 - 10 + 2x < 0, \quad x^2 - x - 12 < 0 \quad \text{or} \quad (x-4)(x+3) < 0$$

This last inequality holds only in the following cases.

Case 1: $x - 4 > 0$ *and* $x + 3 < 0$, i.e. $x > 4$ and $x < -3$. This is *impossible* since x cannot be both greater than 4 and less than -3.

Case 2: $x - 4 < 0$ *and* $x + 3 > 0$, i.e. $x < 4$ and $x > -3$. This is possible when $-3 < x < 4$.

Thus the inequality holds for the set of all x such that $-3 < x < 4$.

12. If $a \geqq 0$ and $b \geqq 0$, prove that $\frac{1}{2}(a+b) \geqq \sqrt{ab}$.

A method of proof is often arrived at by *assuming* the required result to be true and performing valid operations until a result is obtained which is *known* to be true. By reversing the steps (assuming this possible) the proof follows.

In this problem we start with the required result to obtain successively $a + b \geqq 2\sqrt{ab}$, $(a+b)^2 \geqq 4ab$ or $a^2 - 2ab + b^2 \geqq 0$, i.e. $(a-b)^2 \geqq 0$, which is known to be true. Retracing the steps, the result follows.

Another method: Since $(\sqrt{a} - \sqrt{b})^2 \geqq 0$ we have $a - 2\sqrt{ab} + b \geqq 0$ or $\frac{1}{2}(a+b) \geqq \sqrt{ab}$.

This result can be generalized to $\dfrac{a_1 + a_2 + \cdots + a_n}{n} \geqq \sqrt[n]{a_1 a_2 \cdots a_n}$ where $a_1, \ldots, a_n$ are non-negative. The left and right sides are called respectively the *arithmetic mean* and *geometric mean* of the numbers $a_1, \ldots, a_n$.

13. If $a_1, a_2, \ldots, a_n$ and $b_1, b_2, \ldots, b_n$ are any real numbers, prove *Schwarz's inequality*

$$(a_1 b_1 + a_2 b_2 + \cdots + a_n b_n)^2 \leqq (a_1^2 + a_2^2 + \cdots + a_n^2)(b_1^2 + b_2^2 + \cdots + b_n^2)$$

For all real numbers λ, we have

$$(a_1 \lambda + b_1)^2 + (a_2 \lambda + b_2)^2 + \cdots + (a_n \lambda + b_n)^2 \geqq 0$$

Expanding and collecting terms yields

$$A^2 \lambda^2 + 2C\lambda + B^2 \geqq 0 \tag{1}$$

where

$$A^2 = a_1^2 + a_2^2 + \cdots + a_n^2, \quad B^2 = b_1^2 + b_2^2 + \cdots + b_n^2, \quad C = a_1 b_1 + a_2 b_2 + \cdots + a_n b_n \tag{2}$$

Now (1) can be written

$$\lambda^2 + \frac{2C}{A^2}\lambda + \frac{B^2}{A^2} \geqq 0 \quad \text{or} \quad \left(\lambda + \frac{C}{A^2}\right)^2 + \frac{B^2}{A^2} - \frac{C^2}{A^4} \geqq 0 \tag{3}$$

But this last inequality is true for all real λ if and only if $\dfrac{B^2}{A^2} - \dfrac{C^2}{A^4} \geqq 0$ or $C^2 \leqq A^2 B^2$ which gives the required inequality upon using (2).

14. Prove that $\dfrac{1}{2} + \dfrac{1}{4} + \dfrac{1}{8} + \cdots + \dfrac{1}{2^{n-1}} < 1$ for all positive integers $n > 1$.

Let $\qquad\qquad S_n = \frac{1}{2} + \frac{1}{4} + \frac{1}{8} + \cdots + \dfrac{1}{2^{n-1}}$

Then $\qquad\qquad \frac{1}{2}S_n = \qquad \frac{1}{4} + \frac{1}{8} + \cdots + \dfrac{1}{2^{n-1}} + \dfrac{1}{2^n}$

Subtracting, $\quad \frac{1}{2}S_n = \frac{1}{2} - \dfrac{1}{2^n}.$ Thus $S_n = 1 - \dfrac{1}{2^{n-1}} < 1$ for all n.

EXPONENTS, ROOTS and LOGARITHMS

15. Evaluate each of the following.

(a) $\dfrac{3^4 \cdot 3^8}{3^{14}} = \dfrac{3^{4+8}}{3^{14}} = 3^{4+8-14} = 3^{-2} = \dfrac{1}{3^2} = \dfrac{1}{9}$

(b) $\sqrt{\dfrac{(5 \cdot 10^{-6})(4 \cdot 10^2)}{8 \cdot 10^5}} = \sqrt{\dfrac{5 \cdot 4}{8} \cdot \dfrac{10^{-6} \cdot 10^2}{10^5}} = \sqrt{2.5 \cdot 10^{-9}} = \sqrt{25 \cdot 10^{-10}} = 5 \cdot 10^{-5}$ or 0.00005

(c) $\log_{2/3}\left(\frac{27}{8}\right) = x.$ Then $\left(\frac{2}{3}\right)^x = \frac{27}{8} = \left(\frac{3}{2}\right)^3 = \left(\frac{2}{3}\right)^{-3}$ or $x = -3.$

(d) $(\log_a b)(\log_b a) = u.$ Let $\log_a b = x,\ \log_b a = y$ assuming $a, b > 0$ and $a, b \neq 1.$

     Then $a^x = b,\ b^y = a$ and $u = xy.$

     Since $(a^x)^y = a^{xy} = b^y = a$ we have $a^{xy} = a^1$ or $xy = 1$ the required value.

16. If $M > 0,\ N > 0$ and $a > 0$ but $a \neq 1,$ prove that $\log_a \dfrac{M}{N} = \log_a M - \log_a N.$

Let $\log_a M = x,\ \log_a N = y.$ Then $a^x = M,\ a^y = N$ and so

$$\dfrac{M}{N} = \dfrac{a^x}{a^y} = a^{x-y} \qquad \text{or} \qquad \log_a \dfrac{M}{N} = x - y = \log_a M - \log_a N$$

COUNTABILITY

17. Prove that the set of all rational numbers between 0 and 1 inclusive is countable.

Write all fractions with denominator 2, then 3, $\ldots$ considering equivalent fractions such as $\frac{1}{2}, \frac{2}{4}, \frac{3}{6}, \ldots$ no more than once. Then the 1-1 correspondence with the natural numbers can be accomplished as follows.

Rational numbers $0 \quad 1 \quad \frac{1}{2} \quad \frac{1}{3} \quad \frac{2}{3} \quad \frac{1}{4} \quad \frac{3}{4} \quad \frac{1}{5} \quad \frac{2}{5} \quad \cdots$
                             $\updownarrow \quad \updownarrow \quad \updownarrow \quad \updownarrow \quad \updownarrow \quad \updownarrow \quad \updownarrow \quad \updownarrow \quad \updownarrow$
Natural numbers $1 \quad 2 \quad 3 \quad 4 \quad 5 \quad 6 \quad 7 \quad 8 \quad 9 \quad \cdots$

Thus the set of all rational numbers between 0 and 1 inclusive is countable and has cardinal number $\aleph_0$ (see Page 4).

18. If A and B are two countable sets, prove that the set consisting of all elements from A or B (or both) is also countable.

Since A is countable, there is a 1-1 correspondence between elements of A and the natural numbers so that we can denote these elements by $a_1, a_2, a_3, \ldots.$

Similarly we can denote the elements of B by $b_1, b_2, b_3, \ldots.$

Case 1: Suppose elements of A are all distinct from elements of B. Then the set consisting of elements from A or B is countable since we can establish the following 1-1 correspondence.

A or B $a_1 \quad b_1 \quad a_2 \quad b_2 \quad a_3 \quad b_3 \quad \cdots$
                      $\updownarrow \quad \updownarrow \quad \updownarrow \quad \updownarrow \quad \updownarrow \quad \updownarrow$
Natural numbers $1 \quad 2 \quad 3 \quad 4 \quad 5 \quad 6 \quad \cdots$

Case 2: If some elements of A and B are the same, we count them only once as in Problem 17. Then the set of elements belonging to A or B (or both) is countable.

The set consisting of all elements which belong to *A or B* (or both) is often called the *union* of *A* and *B*, denoted by $A \cup B$ or $A + B$.

The set consisting of all elements which are contained in both *A and B* is called the *intersection* of *A* and *B*, denoted by $A \cap B$ or AB. If *A* and *B* are countable, so is $A \cap B$.

The set consisting of all elements in *A* but *not* in *B* is written $A - B$. If we let $\bar{B}$ be the set of elements which are not in *B*, we can also write $A - B = A\bar{B}$. If *A* and *B* are countable, so is $A - B$.

19. Prove that the set of all positive rational numbers is countable.

Consider all rational numbers $x > 1$. With each such rational number we can associate one and only one rational number $1/x$ in $(0, 1)$, i.e. there is a *one to one correspondence* between all rational numbers > 1 and all rational numbers in $(0, 1)$. Since these last are countable by Problem 17, it follows that the set of all rational numbers > 1 are also countable.

From Problem 18 it then follows that the set consisting of all positive rational numbers is countable, since this is composed of the two countable sets of rationals between 0 and 1 and those greater than or equal to 1.

From this we can show that the set of all rational numbers is countable (see Problem 59).

20. Prove that the set of all real numbers in $[0, 1]$ is non-countable.

Every real number in $[0, 1]$ has a decimal expansion $.a_1 a_2 a_3 \ldots$ where $a_1, a_2, \ldots$ are any of the digits $0, 1, 2, \ldots, 9$.

We assume that numbers whose decimal expansions terminate such as 0.7324 are written $0.73240000\ldots$ and that this is the same as $0.73239999\ldots$.

If all real numbers in $[0, 1]$ are countable we can place them in 1-1 correspondence with the natural numbers as in the following list.

$$1 \quad \leftrightarrow \quad 0.a_{11} a_{12} a_{13} a_{14} \ldots$$
$$2 \quad \leftrightarrow \quad 0.a_{21} a_{22} a_{23} a_{24} \ldots$$
$$3 \quad \leftrightarrow \quad 0.a_{31} a_{32} a_{33} a_{34} \ldots$$

We now form a number

$$0.b_1 b_2 b_3 b_4 \ldots$$

where $b_1 \neq a_{11}$, $b_2 \neq a_{22}$, $b_3 \neq a_{33}$, $b_4 \neq a_{44}$, $\ldots$ and where all b's beyond some position are not all 9's.

This number, which is in $[0, 1]$, is different from all numbers in the above list and is thus not in the list, contradicting the assumption that all numbers in $[0, 1]$ were included.

Because of this contradiction it follows that the real numbers in $[0, 1]$ cannot be placed in 1-1 correspondence with the natural numbers, i.e. the set of real numbers in $[0, 1]$ is non-countable.

LIMIT POINTS, BOUNDS, WEIERSTRASS-BOLZANO THEOREM

21. (*a*) Prove that the infinite set of numbers $1, \frac{1}{2}, \frac{1}{3}, \frac{1}{4}, \ldots$ is bounded. (*b*) Determine the least upper bound (l.u.b.) and greatest lower bound (g.l.b.) of the set. (*c*) Prove that 0 is a limit point of the set. (*d*) Is the set a closed set? (*e*) How does this set illustrate the Weierstrass-Bolzano theorem?

(*a*) Since all members of the set are less than 2 and greater than -1 (for example), the set is bounded; 2 is an upper bound, -1 is a lower bound.

We can find smaller upper bounds (e.g. $\frac{3}{2}$) and larger lower bounds (e.g. $-\frac{1}{2}$).

(*b*) Since no member of the set is greater than 1 and since there is at least one member of the set (namely 1) which exceeds $1 - \epsilon$ for every positive number ϵ, we see that 1 is the l.u.b. of the set.

Since no member of the set is less than 0 and since there is at least one member of the set which is less than $0 + \epsilon$ for every positive ϵ (we can always choose for this purpose the number $1/n$ where n is a positive integer greater than $1/\epsilon$), we see that 0 is the g.l.b. of the set.

(c) Let x be any member of the set. Since we can always find a number x such that $0 < |x| < \delta$ for any positive number δ (e.g. we can always pick x to be the number $1/n$ where n is a positive integer greater than $1/\delta$), we see that 0 is a limit point of the set. To put this another way, we see that any deleted δ neighborhood of 0 always includes members of the set, no matter how small we take $\delta > 0$.

(d) The set is not a closed set since the limit point 0 does not belong to the given set.

(e) Since the set is bounded and infinite it must, by the Weierstrass-Bolzano theorem, have at least one limit point. We have found this to be the case, so that the theorem is illustrated.

ALGEBRAIC and TRANSCENDENTAL NUMBERS

22. Prove that $\sqrt[3]{2} + \sqrt{3}$ is an algebraic number.

Let $x = \sqrt[3]{2} + \sqrt{3}$. Then $x - \sqrt{3} = \sqrt[3]{2}$. Cubing both sides and simplifying, we find $x^3 + 9x - 2 = 3\sqrt{3}(x^2 + 1)$. Then squaring both sides and simplifying we find $x^6 - 9x^4 - 4x^3 + 27x^2 + 36x - 23 = 0$.

Since this is a polynomial equation with integral coefficients it follows that $\sqrt[3]{2} + \sqrt{3}$, which is a solution, is an algebraic number.

23. Prove that the set of all algebraic numbers is a countable set.

Algebraic numbers are solutions to polynomial equations of the form $a_0 x^n + a_1 x^{n-1} + \ldots + a_n = 0$ where $a_0, a_1, \ldots, a_n$ are integers.

Let $P = |a_0| + |a_1| + \ldots + |a_n| + n$. For any given value of P there are only a finite number of possible polynomial equations and thus only a finite number of possible algebraic numbers.

Write all algebraic numbers corresponding to $P = 1, 2, 3, 4, \ldots$ avoiding repetitions. Thus all algebraic numbers can be placed into 1-1 correspondence with the natural numbers and so are countable.

COMPLEX NUMBERS

24. Perform the indicated operations.

(a) $(4 - 2i) + (-6 + 5i) = 4 - 2i - 6 + 5i = 4 - 6 + (-2 + 5)i = -2 + 3i$

(b) $(-7 + 3i) - (2 - 4i) = -7 + 3i - 2 + 4i = -9 + 7i$

(c) $(3 - 2i)(1 + 3i) = 3(1 + 3i) - 2i(1 + 3i) = 3 + 9i - 2i - 6i^2 = 3 + 9i - 2i + 6 = 9 + 7i$

(d) $\dfrac{-5 + 5i}{4 - 3i} = \dfrac{-5 + 5i}{4 - 3i} \cdot \dfrac{4 + 3i}{4 + 3i} = \dfrac{(-5 + 5i)(4 + 3i)}{16 - 9i^2} = \dfrac{-20 - 15i + 20i + 15i^2}{16 + 9}$

$= \dfrac{-35 + 5i}{25} = \dfrac{5(-7 + i)}{25} = \dfrac{-7}{5} + \dfrac{1}{5}i$

(e) $\dfrac{i + i^2 + i^3 + i^4 + i^5}{1 + i} = \dfrac{i - 1 + (i^2)(i) + (i^2)^2 + (i^2)^2 i}{1 + i} = \dfrac{i - 1 - i + 1 + i}{1 + i}$

$= \dfrac{i}{1 + i} \cdot \dfrac{1 - i}{1 - i} = \dfrac{i - i^2}{1 - i^2} = \dfrac{i + 1}{2} = \dfrac{1}{2} + \dfrac{1}{2}i$

(f) $|3 - 4i|\,|4 + 3i| = \sqrt{(3)^2 + (-4)^2}\,\sqrt{(4)^2 + (3)^2} = (5)(5) = 25$

(g) $\left|\dfrac{1}{1 + 3i} - \dfrac{1}{1 - 3i}\right| = \left|\dfrac{1 - 3i}{1 - 9i^2} - \dfrac{1 + 3i}{1 - 9i^2}\right| = \left|\dfrac{-6i}{10}\right| = \sqrt{(0)^2 + (-\tfrac{6}{10})^2} = \tfrac{3}{5}$

25. If z_1 and z_2 are two complex numbers, prove that $|z_1 z_2| = |z_1|\,|z_2|$.

Let $z_1 = x_1 + iy_1$, $z_2 = x_2 + iy_2$. Then

$|z_1 z_2| = |(x_1 + iy_1)(x_2 + iy_2)| = |x_1 x_2 - y_1 y_2 + i(x_1 y_2 + x_2 y_1)|$

$= \sqrt{(x_1 x_2 - y_1 y_2)^2 + (x_1 y_2 + x_2 y_1)^2} = \sqrt{x_1^2 x_2^2 + y_1^2 y_2^2 + x_1^2 y_2^2 + x_2^2 y_1^2}$

$= \sqrt{(x_1^2 + y_1^2)(x_2^2 + y_2^2)} = \sqrt{x_1^2 + y^2}\,\sqrt{x_2^2 + y_2^2} = |x_1 + iy_1|\,|x_2 + iy_2| = |z_1|\,|z_2|$.

26. Solve $x^3 - 2x - 4 = 0$.

The possible rational roots using Problem 7 are $\pm 1, \pm 2, \pm 4$. By trial we find $x = 2$ is a root. Then the given equation can be written $(x - 2)(x^2 + 2x + 2) = 0$. The solutions to the *quadratic equation* $ax^2 + bx + c = 0$ are $x = \dfrac{-b \pm \sqrt{b^2 - 4ac}}{2a}$. For $a = 1$, $b = 2$, $c = 2$ this gives $x = \dfrac{-2 \pm \sqrt{4 - 8}}{2} = \dfrac{-2 \pm \sqrt{-4}}{2} = \dfrac{-2 \pm 2i}{2} = -1 \pm i$.

The set of solutions is $2, -1 + i, -1 - i$.

POLAR FORM of COMPLEX NUMBERS

27. Express in polar form (a) $3 + 3i$, (b) $-1 + \sqrt{3}i$, (c) -1, (d) $-2 - 2\sqrt{3}\,i$.

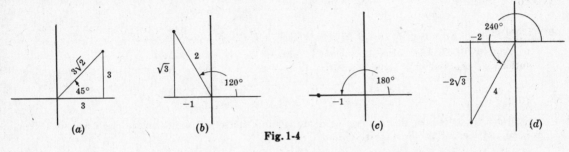

Fig. 1-4

(a)   Amplitude $\phi = 45° = \pi/4$ radians. Modulus $\rho = \sqrt{3^2 + 3^2} = 3\sqrt{2}$. Then
$$3 + 3i = \rho(\cos\phi + i\sin\phi) = 3\sqrt{2}(\cos\pi/4 + i\sin\pi/4) = 3\sqrt{2}\operatorname{cis}\pi/4 = 3\sqrt{2}\,e^{\pi i/4}$$

(b)   Amplitude $\phi = 120° = 2\pi/3$ radians. Modulus $\rho = \sqrt{(-1)^2 + (\sqrt{3})^2} = \sqrt{4} = 2$. Then
$$-1 + \sqrt{3}\,i = 2(\cos 2\pi/3 + i\sin 2\pi/3) = 2\operatorname{cis}2\pi/3 = 2e^{2\pi i/3}$$

(c)   Amplitude $\phi = 180° = \pi$ radians. Modulus $\rho = \sqrt{(-1)^2 + (0)^2} = 1$. Then
$$-1 = 1(\cos\pi + i\sin\pi) = \operatorname{cis}\pi = e^{\pi i}$$

(d)   Amplitude $\phi = 240° = 4\pi/3$ radians. Modulus $\rho = \sqrt{(-2)^2 + (-2\sqrt{3})^2} = 4$. Then
$$-2 - 2\sqrt{3} = 4(\cos 4\pi/3 + i\sin 4\pi/3) = 4\operatorname{cis}4\pi/3 = 4e^{4\pi i/3}$$

28. Evaluate (a) $(-1 + \sqrt{3}\,i)^{10}$, (b) $(-1 + i)^{1/3}$.

(a)   By Problem 27(b) and De Moivre's theorem,
$$(-1 + \sqrt{3}\,i)^{10} = [2(\cos 2\pi/3 + i\sin 2\pi/3)]^{10} = 2^{10}(\cos 20\pi/3 + i\sin 20\pi/3)$$
$$= 1024[\cos(2\pi/3 + 6\pi) + i\sin(2\pi/3 + 6\pi)] = 1024(\cos 2\pi/3 + i\sin 2\pi/3)$$
$$= 1024(-\tfrac{1}{2} + \tfrac{1}{2}\sqrt{3}\,i) = -512 + 512\sqrt{3}\,i$$

(b) $-1 + i = \sqrt{2}(\cos 135° + i\sin 135°) = \sqrt{2}[\cos(135° + k\cdot 360°) + i\sin(135° + k\cdot 360°)]$

Then
$$(-1 + i)^{1/3} = (\sqrt{2})^{1/3}\left[\cos\left(\frac{135° + k\cdot 360°}{3}\right) + i\sin\left(\frac{135° + k\cdot 360°}{3}\right)\right]$$

The results for $k = 0, 1, 2$ are
$$\sqrt[6]{2}(\cos 45° + i\sin 45°),$$
$$\sqrt[6]{2}(\cos 165° + i\sin 165°),$$
$$\sqrt[6]{2}(\cos 285° + i\sin 285°)$$

The results for $k = 3, 4, 5, 6, 7, \ldots$ give repetitions of these. These complex roots are represented geometrically in the complex plane by points P_1, P_2, P_3 on the circle of Fig. 1-5.

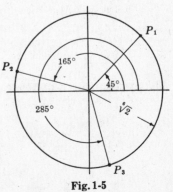

Fig. 1-5

MATHEMATICAL INDUCTION

29. Prove that $1^2 + 2^2 + 3^2 + 4^2 + \cdots + n^2 = \frac{1}{6}n(n+1)(2n+1)$.

The statement is true for $n = 1$ since $1^2 = \frac{1}{6}(1)(1+1)(2 \cdot 1 + 1) = 1$.

Assume the statement true for $n = k$. Then

$$1^2 + 2^2 + 3^2 + \cdots + k^2 = \frac{1}{6}k(k+1)(2k+1)$$

Adding $(k+1)^2$ to both sides,

$$1^2 + 2^2 + 3^2 + \cdots + k^2 + (k+1)^2 = \frac{1}{6}k(k+1)(2k+1) + (k+1)^2 = (k+1)[\frac{1}{6}k(2k+1) + k + 1]$$
$$= \frac{1}{6}(k+1)(2k^2 + 7k + 6) = \frac{1}{6}(k+1)(k+2)(2k+3)$$

which shows that the statement is true for $n = k+1$ *if* it is true for $n = k$. But since it is true for $n = 1$, it follows that it is true for $n = 1 + 1 = 2$ and for $n = 2 + 1 = 3$, $\ldots$, i.e. it is true for all positive integers n.

30. Prove that $x^n - y^n$ has $x - y$ as a factor for all positive integers n.

The statement is true for $n = 1$ since $x^1 - y^1 = x - y$.

Assume the statement true for $n = k$, i.e. assume that $x^k - y^k$ has $x - y$ as a factor. Consider

$$x^{k+1} - y^{k+1} = x^{k+1} - x^k y + x^k y - y^{k+1}$$
$$= x^k(x - y) + y(x^k - y^k)$$

The first term on the right has $x - y$ as a factor, and the second term on the right also has $x - y$ as a factor because of the above assumption.

Thus $x^{k+1} - y^{k+1}$ has $x - y$ as a factor if $x^k - y^k$ does.

Then since $x^1 - y^1$ has $x - y$ as factor, it follows that $x^2 - y^2$ has $x - y$ as a factor, $x^3 - y^3$ has $x - y$ as a factor, etc.

31. Prove *Bernoulli's inequality* $(1+x)^n > 1 + nx$ for $n = 2, 3, \ldots$ if $x > -1$, $x \neq 0$.

The statement is true for $n = 2$ since $(1+x)^2 = 1 + 2x + x^2 > 1 + 2x$.

Assume the statement true for $n = k$, i.e., $(1+x)^k > 1 + kx$.

Multiply both sides by $1 + x$ (which is positive since $x > -1$). Then we have

$$(1+x)^{k+1} > (1+x)(1+kx) = 1 + (k+1)x + kx^2 > 1 + (k+1)x$$

Thus the statement is true for $n = k+1$ if it is true for $n = k$.

But since the statement is true for $n = 2$, it must be true for $n = 2 + 1 = 3$, $\ldots$ and is thus true for all integers greater than or equal to 2.

Note that the result is not true for $n = 1$. However, the modified result $(1+x)^n \geq 1 + nx$ is true for $n = 1, 2, 3, \ldots$.

MISCELLANEOUS PROBLEMS

32. Prove that every positive integer P can be expressed uniquely in the form $P = a_0 2^n + a_1 2^{n-1} + a_2 2^{n-2} + \cdots + a_n$ where the a's are 0's or 1's.

Dividing P by 2, we have $P/2 = a_0 2^{n-1} + a_1 2^{n-2} + \cdots + a_{n-1} + a_n/2$.

Then a_n is the remainder, 0 or 1, obtained when P is divided by 2 and is unique.

Let P_1 be the integer part of $P/2$. Then $P_1 = a_0 2^{n-1} + a_1 2^{n-2} + \cdots + a_{n-1}$.

Dividing P_1 by 2 we see that a_{n-1} is the remainder, 0 or 1, obtained when P_1 is divided by 2 and is unique.

By continuing in this manner, all the a's can be determined as 0's or 1's and are unique.

33. Express the number 23 in the form of Problem 32.

The determination of the coefficients can be arranged as follows.

$$
\begin{array}{r l}
2\,)\,\underline{23} & \\
2\,)\,\underline{11} & \text{Remainder } 1 \\
2\,)\,\underline{5} & \text{Remainder } 1 \\
2\,)\,\underline{2} & \text{Remainder } 1 \\
2\,)\,\underline{1} & \text{Remainder } 0 \\
0 & \text{Remainder } 1
\end{array}
$$

The coefficients are $1\ 0\ 1\ 1\ 1$.   **Check:** $23 = 1 \cdot 2^4 + 0 \cdot 2^3 + 1 \cdot 2^2 + 1 \cdot 2 + 1$.

The number 10111 is said to represent 23 in the *scale of two* or *binary scale*.

34. Dedekind defined a *cut, section* or *partition* in the rational number system as a separation of *all* rational numbers into two classes or sets called L (the left hand class) and R (the right hand class) having the following properties:

   I. The classes are non-empty (i.e. at least one number belongs to each class).

   II. Every rational number is in one class or the other.

   III. Every number in L is less than every number in R.

Prove each of the following statements:

(a) There cannot be a largest number in L and a smallest number in R.

(b) It is possible for L to have a largest number and for R to have no smallest number. What type of number does the cut define in this case?

(c) It is possible for L to have no largest number and for R to have a smallest number. What type of number does the cut define in this case?

(d) It is possible for L to have no largest number and for R to have no smallest number. What type of number does the cut define in this case?

(a)   Let a be the largest rational number in L, and b the smallest rational number in R. Then either $a = b$ or $a < b$.

   We cannot have $a = b$ since by definition of the cut every number in L is *less* than every number in R.

   We cannot have $a < b$ since by Problem 9, $\frac{1}{2}(a + b)$ is a rational number which would be greater than a (and so would have to be in R) but less than b (and so would have to be in L), and by definition a rational number cannot belong to *both* L and R.

(b)   As an indication of the possibility let L contain the number $\frac{2}{3}$ and all rational numbers less than $\frac{2}{3}$, while R contains all rational numbers greater than $\frac{2}{3}$. In this case the cut defines the rational number $\frac{2}{3}$. A similar argument replacing $\frac{2}{3}$ by any other rational number shows that in such case the cut defines a rational number.

(c)   As an indication of the possibility let L contain all rational numbers less than $\frac{2}{3}$ while R contains all rational numbers greater than $\frac{2}{3}$. This cut also defines the rational number $\frac{2}{3}$. A similar argument shows that this cut always defines a rational number.

(d)   As an indication of the possibility let L consist of all negative rational numbers and all positive rational numbers whose squares are less than 2, while R consists of all positive numbers whose squares are greater than 2. We can show that if a is any number of the L class there is always a larger number of the L class, while if b is any number of the R class there is always a smaller number of the R class (see Problem 106). A cut of this type defines an irrational number.

   From $(b), (c), (d)$ it follows that every cut in the rational number system, called a *Dedekind cut*, defines either a rational or an irrational number. By use of Dedekind cuts we can define operations (such as addition, multiplication, etc.) with irrational numbers.

Supplementary Problems

OPERATIONS with NUMBERS

35. Given $x = -3$, $y = 2$, $z = 5$, $a = \frac{3}{2}$ and $b = -\frac{1}{4}$, evaluate:

    (a) $(2x - y)(3y + z)(5x - 2z)$, (b) $\dfrac{xy - 2z^2}{2ab - 1}$, (c) $\dfrac{3a^2b + ab^2}{2a^2b^2 + 1}$, (d) $\dfrac{(ax + by)^2 + (ay - bx)^2}{(ay + bx)^2 + (ax - by)^2}$.

    Ans. (a) 2200, (b) 32, (c) $-51/41$, (d) 1

36. Find the set of values of x for which the following equations are true. Justify all steps in each case.

    (a) $4\{(x - 2) + 3(2x - 1)\} + 2(2x + 1) = 12(x + 2) - 2$ (c) $\sqrt{x^2 + 8x + 7} - \sqrt{2x + 2} = x + 1$

    (b) $\dfrac{1}{8 - x} - \dfrac{1}{x - 2} = \dfrac{1}{4}$ (d) $\dfrac{1 - x}{\sqrt{x^2 - 2x + 5}} = \dfrac{3}{5}$

    Ans. (a) 2, (b) $6, -4$, (c) $-1, 1$, (d) $-\frac{1}{2}$

37. Prove that $\dfrac{x}{(z - x)(x - y)} + \dfrac{y}{(x - y)(y - z)} + \dfrac{z}{(y - z)(z - x)} = 0$ giving restrictions if any.

RATIONAL and IRRATIONAL NUMBERS

38. Find decimal expansions for (a) $\frac{3}{7}$, (b) $\sqrt{5}$.   *Ans.* (a) $0.\overline{428571}$, (b) $2.2360679\ldots$

39. Show that a fraction with denominator 17 and with numerator $1, 2, 3, \ldots, 16$ has 16 digits in the repeating portion of its decimal expansion. Is there any relation between the orders of the digits in these expansions?

40. Prove that (a) $\sqrt{3}$, (b) $\sqrt[3]{2}$ are irrational numbers.

41. Prove that (a) $\sqrt[3]{5} - \sqrt[4]{3}$, (b) $\sqrt{2} + \sqrt{3} + \sqrt{5}$ are irrational numbers.

42. Determine a positive rational number whose square differs from 7 by less than .000001.

43. Prove that every rational number can be expressed as a repeating decimal.

44. Find the values of x for which
    (a) $2x^3 - 5x^2 - 9x + 18 = 0$, (b) $3x^3 + 4x^2 - 35x + 8 = 0$, (c) $x^4 - 21x^2 + 4 = 0$.
    Ans. (a) $3, -2, 3/2$ (b) $8/3, -2 \pm \sqrt{5}$ (c) $\frac{1}{2}(5 \pm \sqrt{17})$, $\frac{1}{2}(-5 \pm \sqrt{17})$

45. If a, b, c, d are rational and m is not a perfect square, prove that $a + b\sqrt{m} = c + d\sqrt{m}$ if and only if $a = c$ and $b = d$.

46. Prove that $\dfrac{1 + \sqrt{3} + \sqrt{5}}{1 - \sqrt{3} + \sqrt{5}} = \dfrac{12\sqrt{5} - 2\sqrt{15} + 14\sqrt{3} - 7}{11}$.

INEQUALITIES

47. Find the set of values of x for which each of the following inequalities holds.

    (a) $\dfrac{1}{x} + \dfrac{3}{2x} \geqq 5$, (b) $x(x + 2) \leqq 24$, (c) $|x + 2| < |x - 5|$, (d) $\dfrac{x}{x + 2} > \dfrac{x + 3}{3x + 1}$.

    Ans. (a) $0 < x \leqq \frac{1}{2}$, (b) $-6 \leqq x \leqq 4$, (c) $x < 3/2$, (d) $x > 3$, $-1 < x < -\frac{1}{3}$, or $x < -2$

48. Prove (a) $|x + y| \leqq |x| + |y|$, (b) $|x + y + z| \leqq |x| + |y| + |z|$, (c) $|x - y| \geqq |x| - |y|$.

49. Prove that for all real x, y, z, $x^2 + y^2 + z^2 \geqq xy + yz + zx$.

50. If $a^2 + b^2 = 1$ and $c^2 + d^2 = 1$, prove that $ac + bd \leqq 1$.

51. If $x > 0$, prove that $x^{n+1} + \dfrac{1}{x^{n+1}} > x^n + \dfrac{1}{x^n}$ where n is any positive integer.

52. Prove that for all real $a \neq 0$, $|a + 1/a| \geqq 2$.

53. Show that in Schwarz's inequality (Problem 13) the equality holds if and only if $a_p = kb_p$, $p = 1, 2, 3, \ldots, n$ where k is any constant.

54. If a_1, a_2, a_3 are positive, prove that $\frac{1}{3}(a_1 + a_2 + a_3) \geqq \sqrt[3]{a_1 a_2 a_3}$.

EXPONENTS, ROOTS and LOGARITHMS

55. Evaluate   (a) $4^{\log_2 8}$,   (b) $\frac{3}{4}\log_{1/8}\left(\frac{1}{128}\right)$,   (c) $\sqrt{\dfrac{(0.00004)(25,000)}{(0.02)^5(0.125)}}$,   (d) $3^{-2\log_3 5}$,   (e) $\left(-\frac{1}{8}\right)^{4/3} - (-27)^{-2/3}$.
    Ans. (a) 64, (b) 7/4, (c) 50,000, (d) 1/25, (e) $-7/144$

56. Prove   (a) $\log_a MN = \log_a M + \log_a N$,   (b) $\log_a M^r = r\log_a M$ indicating restrictions if any.

57. Prove $b^{\log_b a} = a$ giving restrictions if any.

COUNTABILITY

58. (a) Prove that there is a one to one correspondence between the points of the interval $0 \leqq x \leqq 1$ and $-5 \leqq x \leqq -3$. (b) What is the cardinal number of the sets in (a)?
    Ans. (b) C, the cardinal number of the continuum.

59. (a) Prove that the set of all rational numbers is countable. (b) What is the cardinal number of the set in (a)?   Ans. (b) $\aleph_0$

60. Prove that the set of (a) all real numbers, (b) all irrational numbers is non-countable.

61. The *intersection* of two sets A and B, denoted by $A \cap B$ or AB, is the set consisting of all elements belonging to both A and B. Prove that if A and B are countable, so is their intersection.

62. Prove that a countable set of countable sets is countable.

63. Prove that the cardinal number of the set of points inside a square is equal to the cardinal number of the set of points on (a) one side, (b) all four sides. (c) What is the cardinal number in this case? (d) Does a corresponding result hold for a cube?   Ans. (c) C

LIMIT POINTS. BOUNDS. WEIERSTRASS-BOLZANO THEOREM

64. Given the set of numbers $1, 1.1, .9, 1.01, .99, 1.001, .999, \ldots$. (a) Is the set bounded? (b) Does the set have a l.u.b. and g.l.b.? If so, determine them. (c) Does the set have any limit points? If so, determine them. (d) Is the set a closed set?
    Ans. (a) Yes (b) l.u.b. = 1.1, g.l.b. = .9 (c) 1 (d) Yes

65. Given the set $-.9, .9, -.99, .99, -.999, .999$ answer the questions of Problem 64.
    Ans. (a) Yes (b) l.u.b. = 1, g.l.b. = -1 (c) 1, -1 (d) No

66. Give an example of a set which has (a) 3 limit points, (b) no limit points.

67. (a) Prove that every point of the interval $0 < x < 1$ is a limit point.
    (b) Are there any limit points which do not belong to the set in (a)? Justify your answer.

68. Let S be the set of all rational numbers in $(0,1)$ having denominator 2^n, $n = 1, 2, 3, \ldots$. (a) Does S have any limit points? (b) Is S closed?

69. (a) Give an example of a set which has limit points but which is not bounded. (b) Does this contradict the Weierstrass-Bolzano theorem? Explain.

ALGEBRAIC and TRANSCENDENTAL NUMBERS

70. Prove that (a) $\dfrac{\sqrt{3}-\sqrt{2}}{\sqrt{3}+\sqrt{2}}$, (b) $\sqrt{2}+\sqrt{3}+\sqrt{5}$ are algebraic numbers.

71. Prove that the set of transcendental numbers in $(0,1)$ is not countable.

72. Prove that every rational number is algebraic but every irrational number is not necessarily algebraic.

COMPLEX NUMBERS. POLAR FORM

73. Perform each of the indicated operations: (a) $2(5-3i) - 3(-2+i) + 5(i-3)$, (b) $(3-2i)^3$, (c) $\dfrac{5}{3-4i}$
    $+ \dfrac{10}{4+3i}$, (d) $\left(\dfrac{1-i}{1+i}\right)^{10}$, (e) $\left|\dfrac{2-4i}{5+7i}\right|^2$, (f) $\dfrac{(1+i)(2+3i)(4-2i)}{(1+2i)^2(1-i)}$.
    Ans. (a) $1-4i$, (b) $-9-46i$, (c) $\frac{11}{5} - \frac{2}{5}i$, (d) -1, (e) $\frac{10}{37}$, (f) $\frac{16}{5} - \frac{2}{5}i$

74. If z_1 and z_2 are complex numbers, prove (a) $\left|\dfrac{z_1}{z_2}\right| = \dfrac{|z_1|}{|z_2|}$, (b) $|z_1^2| = |z_1|^2$ giving any restrictions.

75. Prove (a) $|z_1 + z_2| \leqq |z_1| + |z_2|$, (b) $|z_1 + z_2 + z_3| \leqq |z_1| + |z_2| + |z_3|$, (c) $|z_1 - z_2| \geqq |z_1| - |z_2|$.

76. Find all solutions of $2x^4 - 3x^3 - 7x^2 - 8x + 6 = 0$.   *Ans.* $3, \frac{1}{2}, -1 \pm i$

77. Let z_1 and z_2 be represented by points P_1 and P_2 in the Argand diagram. Construct lines OP_1 and OP_2, where O is the origin. Show that $z_1 + z_2$ can be represented by the point P_3, where OP_3 is the diagonal of a parallelogram having sides OP_1 and OP_2. This is called the *parallelogram law* of addition of complex numbers. Because of this and other properties, complex numbers can be considered as *vectors* in two dimensions.

78. Interpret geometrically the inequalities of Problem 75.

79. Express in polar form (a) $3\sqrt{3} + 3i$, (b) $-2 - 2i$, (c) $1 - \sqrt{3}\,i$, (d) 5, (e) $-5i$.
Ans. (a) $6 \operatorname{cis} \pi/6$ (b) $2\sqrt{2} \operatorname{cis} 5\pi/4$ (c) $2 \operatorname{cis} 5\pi/3$ (d) $5 \operatorname{cis} 0$ (e) $5 \operatorname{cis} 3\pi/2$

80. Evaluate (a) $[2(\cos 25° + i \sin 25°)][5(\cos 110° + i \sin 110°)]$, (b) $\dfrac{12 \operatorname{cis} 16°}{(3 \operatorname{cis} 44°)(2 \operatorname{cis} 62°)}$.
Ans. (a) $-5\sqrt{2} + 5\sqrt{2}\,i$, (b) $-2i$

81. Determine all the indicated roots and represent them graphically:
(a) $(4\sqrt{2} + 4\sqrt{2}\,i)^{1/3}$, (b) $(-1)^{1/5}$, (c) $(\sqrt{3} - i)^{1/3}$, (d) $i^{1/4}$.
Ans. (a) $2 \operatorname{cis} 15°$, $2 \operatorname{cis} 135°$, $2 \operatorname{cis} 255°$
(b) $\operatorname{cis} 36°$, $\operatorname{cis} 108°$, $\operatorname{cis} 180° = -1$, $\operatorname{cis} 252°$, $\operatorname{cis} 324°$
(c) $\sqrt[3]{2} \operatorname{cis} 110°$, $\sqrt[3]{2} \operatorname{cis} 230°$, $\sqrt[3]{2} \operatorname{cis} 350°$
(d) $\operatorname{cis} 22.5°$, $\operatorname{cis} 112.5°$, $\operatorname{cis} 202.5°$, $\operatorname{cis} 292.5°$

82. Prove that $-1 + \sqrt{3}\,i$ is an algebraic number.

83. If $z_1 = \rho_1 \operatorname{cis} \phi_1$ and $z_2 = \rho_2 \operatorname{cis} \phi_2$, prove (a) $z_1 z_2 = \rho_1 \rho_2 \operatorname{cis}(\phi_1 + \phi_2)$, (b) $z_1/z_2 = (\rho_1/\rho_2) \operatorname{cis}(\phi_1 - \phi_2)$. Interpret geometrically.

MATHEMATICAL INDUCTION

Prove each of the following.

84. $1 + 3 + 5 + \cdots + (2n - 1) = n^2$

85. $\dfrac{1}{1 \cdot 3} + \dfrac{1}{3 \cdot 5} + \dfrac{1}{5 \cdot 7} + \cdots + \dfrac{1}{(2n-1)(2n+1)} = \dfrac{n}{2n+1}$

86. $a + (a + d) + (a + 2d) + \cdots + [a + (n-1)d] = \frac{1}{2} n[2a + (n-1)d]$

87. $\dfrac{1}{1 \cdot 2 \cdot 3} + \dfrac{1}{2 \cdot 3 \cdot 4} + \dfrac{1}{3 \cdot 4 \cdot 5} + \cdots + \dfrac{1}{n(n+1)(n+2)} = \dfrac{n(n+3)}{4(n+1)(n+2)}$

88. $a + ar + ar^2 + \cdots + ar^{n-1} = \dfrac{a(r^n - 1)}{r - 1}$, $r \neq 1$

89. $1^3 + 2^3 + 3^3 + \cdots + n^3 = \frac{1}{4} n^2 (n+1)^2$

90. $1(5) + 2(5)^2 + 3(5)^3 + \cdots + n(5)^{n-1} = \dfrac{5 + (4n-1)5^{n+1}}{16}$

91. $x^{2n-1} + y^{2n-1}$ is divisible by $x + y$ for $n = 1, 2, 3, \ldots$.

92. $(\cos \phi + i \sin \phi)^n = \cos n\phi + i \sin n\phi$. Can this be proved if n is a rational number?

93. $\frac{1}{2} + \cos x + \cos 2x + \cdots + \cos nx = \dfrac{\sin(n + \frac{1}{2})x}{2 \sin \frac{1}{2} x}$, $x \neq 0, \pm 2\pi, \pm 4\pi, \ldots$

94. $\sin x + \sin 2x + \cdots + \sin nx = \dfrac{\cos \frac{1}{2} x - \cos(n + \frac{1}{2})x}{2 \sin \frac{1}{2} x}$, $x \neq 0, \pm 2\pi, \pm 4\pi, \ldots$

95. $(a + b)^n = a^n + {}_nC_1 a^{n-1} b + {}_nC_2 a^{n-2} b^2 + \cdots + {}_nC_{n-1} a b^{n-1} + b^n$
where ${}_nC_r = \dfrac{n(n-1)(n-2)\cdots(n-r+1)}{r!} = \dfrac{n!}{r!\,(n-r)!} = {}_nC_{n-r}$. Here $p! = p(p-1)\cdots 1$ and $0!$
is defined as 1. This is called the *binomial theorem*. The coefficients ${}_nC_0 = 1$, ${}_nC_1 = n$, ${}_nC_2 = \dfrac{n(n-1)}{2!}$,
$\ldots$, ${}_nC_n = 1$ are called the *binomial coefficients*. ${}_nC_r$ is also written $\dbinom{n}{r}$.

MISCELLANEOUS PROBLEMS

96. Express each of the following integers (scale of 10) in the scale of notation indicated: (a) 87 (two), (b) 64 (three), (c) 1736 (nine). Check each answer.   *Ans.* (a) 1010111, (b) 2101, (c) 2338

97. If a number is 144 in the scale of 5, what is the number in the scale of (a) 2, (b) 8? *Ans.* (a) 110001, (b) 61

98. Prove that every rational number p/q between 0 and 1 can be expressed in the form

$$\frac{p}{q} \;=\; \frac{a_1}{2} + \frac{a_2}{2^2} + \cdots + \frac{a_n}{2^n} + \cdots$$

where the a's can be determined uniquely as 0's or 1's and where the process may or may not terminate. The representation $0.a_1 a_2 \ldots a_n \ldots$ is then called the *binary form* of the rational number. [Hint: Multiply both sides successively by 2 and consider remainders.]

99. Express $\frac{2}{3}$ in the scale of (a) 2, (b) 3, (c) 8, (d) 10. *Ans.* (a) 0.1010101..., (b) 0.2 or 0.2000..., (c) 0.5252..., (d) 0.6666...

100. A number in the scale of 2 is 11.01001. What is the number in the scale of 10.   *Ans.* 3.28125

101. In what scale of notation is $3 + 4 = 12$?   *Ans.* 5

102. In the scale of 12, two additional symbols t and e must be used to designate the "digits" ten and eleven respectively. Using these symbols, represent the integer 5110 (scale of ten) in the scale of 12. *Ans.* $2\,e\,5\,t$

103. Find a rational number whose decimal expansion is 1.636363....   *Ans.* 18/11

104. A number in the scale of 10 consists of six digits. If the last digit is removed and placed before the first digit, the new number is one-third as large. Find the original number.   *Ans.* 428571

105. Show that the rational numbers form a field.

106. Using as axioms the relations 1-9 on Page 2, prove that
(a) $(-3)(0) = 0$, (b) $(-2)(+3) = -6$, (c) $(-2)(-3) = 6$.

107. (a) If x is a rational number whose square is less than 2, show that $x + (2 - x^2)/10$ is a larger such number. (b) If x is a rational number whose square is greater than 2, find in terms of x a smaller rational number whose square is greater than 2.

108. Illustrate how you would use Dedekind cuts to define
(a) $\sqrt{5} + \sqrt{3}$, (b) $\sqrt{3} - \sqrt{2}$, (c) $(\sqrt{3})(\sqrt{2})$, (d) $\sqrt{2}/\sqrt{3}$.

Chapter 2

Functions, Limits and Continuity

FUNCTIONS

A function is a rule which establishes a correspondence between two sets. For our present purposes we consider sets of real numbers. If to each value which a variable x can assume there corresponds one or more values of a variable y, we call y a *function* of x and write $y = f(x)$, $y = G(x)$, ... the letters $f, G, \ldots$ symbolizing the function while $f(a), G(a), \ldots$ denote the *value of the function* at $x = a$.

The set of values which x can assume is called the *domain of definition* or simply *domain* of the function; x is called the *independent variable* and y the *dependent variable*.

If only one value of y corresponds to each value of x in the domain of definition, the function is called *single-valued*. If more than one value of y corresponds to some values of x, the function is called *multiple-valued*. Since a multiple-valued function can be considered as a collection of single-valued functions, we shall assume functions to be single-valued unless otherwise indicated.

> **Examples:**
> 1. If to each number in $-1 \leqq x \leqq 1$ we associate a number y given by x^2, then the correspondence between x and x^2 defines a function f which is single-valued.
>
>   The domain of f is $-1 \leqq x \leqq 1$. The value of f at x is given by $y = f(x) = x^2$. For example, $f(-1) = (-1)^2 = 1$ is the value of the function at $x = -1$.
>
> 2. With each time t after the year 1800 we can associate a value P for the population of the United States. The correspondence between P and t defines a single-valued function, say F, and we can write $P = F(t)$.
>
> 3. If $y^2 = x$ where $x > 0$ then to each x there correspond two values of y. Hence y is a double-valued function of x. We can consider this as two single-valued functions f and g where $f(x) = \sqrt{x}$ and $g(x) = -\sqrt{x}$.

Note that although a function is often defined by means of a formula as in Examples 1 and 3, it does not have to be, as seen in Example 2.

For convenience we shall often speak of the function $f(x)$ rather than the function f whose value at x is $f(x)$. The distinction should however be kept in mind.

GRAPH of a FUNCTION

The graph of a function defined by $y = f(x)$ is a pictorial representation of the function and can be obtained by locating on a rectangular coordinate system the points defined by the number pairs (x, y) or $[x, f(x)]$.

BOUNDED FUNCTIONS

If there is a constant M such that $f(x) \leqq M$ for all x in an interval (or other set of numbers), we say that $f(x)$ is *bounded above* in the interval (or the set) and call M an *upper bound* of the function.

If a constant m exists such that $f(x) \geqq m$ for all x in an interval, we say that $f(x)$ is *bounded below* in the interval and call m a *lower bound*.

If $m \leqq f(x) \leqq M$ in an interval, we call $f(x)$ *bounded*. Frequently, when we wish to indicate that a function is bounded we shall write $|f(x)| < P$.

> **Examples: 1.** $f(x) = 3 + x$ is bounded in $-1 \leqq x \leqq 1$. An upper bound is 4 (or any number greater than 4). A lower bound is 2 (or any number less than 2).
>
> **2.** $f(x) = 1/x$ is not bounded in $0 < x < 4$ since by choosing x sufficiently close to zero, $f(x)$ can be made as large as we wish, so that there is no upper bound. However, a lower bound is given by $\frac{1}{4}$ (or any number less than $\frac{1}{4}$).

If $f(x)$ has an upper bound it has a *least upper bound* (l.u.b.); if it has a lower bound it has a *greatest lower bound* (g.l.b.). (See Chapter 1 for these definitions.)

MONOTONIC FUNCTIONS

A function is called *monotonic increasing* in an interval if for any two points x_1 and x_2 in the interval such that $x_1 < x_2$, $f(x_1) \leqq f(x_2)$. If $f(x_1) < f(x_2)$ the function is called *strictly increasing*.

Similarly if $f(x_1) \geqq f(x_2)$ whenever $x_1 < x_2$, then $f(x)$ is *monotonic decreasing*; while if $f(x_1) > f(x_2)$, it is *strictly decreasing*.

INVERSE FUNCTIONS. PRINCIPAL VALUES

If y is a function of x, denoted by $f(x)$, then x is a function of y, denoted by $x = f^{-1}(y)$, called the *inverse function*. Interchange of x and y leads to consideration of $y = f^{-1}(x)$.

If $f(x)$ is single-valued, $f^{-1}(x)$ may be multiple-valued in which case it can be considered as a collection of single-valued functions each of which is called a *branch*. It is often convenient to choose one of these branches, called the *principal branch*, and denote it by $f^{-1}(x)$. In such case the value of the inverse function is called the *principal value*.

> **Example:** The function $y = \sin x$ leads to consideration of $y = \sin^{-1} x$ which is multiple-valued, since for each x in $-1 \leqq x \leqq 1$ there are many values of y. By restricting $\sin^{-1} x$ to be such that $-\pi/2 \leqq \sin^{-1} x \leqq \pi/2$, for example, the function becomes single-valued. In such case the principal value of $\sin^{-1}(-\frac{1}{2}) = -\pi/6$.

MAXIMA and MINIMA

If x_0 is a point of an interval such that $f(x) < f(x_0)$ [or $f(x) > f(x_0)$] for all other x in the interval, then $f(x)$ is said to have an *absolute maximum* [or *absolute minimum*] in the interval at $x = x_0$ of magnitude $f(x_0)$. If this is true only for x in some deleted δ neighborhood of x_0 where $\delta > 0$ [i.e. for all x such that $0 < |x - x_0| < \delta$], then $f(x)$ is said to have a *relative maximum* (or *relative minimum*) at x_0.

TYPES of FUNCTIONS

1. Polynomial functions have the form

$$f(x) = a_0 x^n + a_1 x^{n-1} + \cdots + a_{n-1} x + a_n \tag{1}$$

where $a_0, \ldots, a_n$ are constants and n is a positive integer called the *degree* of the polynomial if $a_0 \neq 0$.

The *fundamental theorem of algebra* states that every polynomial equation $f(x) = 0$ has at least one root. From this we can show that if the degree is n the equation has exactly n roots (counting a repeated root of multiplicity r as r roots).

2. **Algebraic functions** are functions $y = f(x)$ satisfying an equation of the form

$$p_0(x)\,y^n + p_1(x)\,y^{n-1} + \cdots + p_{n-1}(x)\,y + p_n(x) = 0 \qquad (2)$$

where $p_0(x), \ldots, p_n(x)$ are polynomials in x.

      If the function can be expressed as the quotient of two polynomials, i.e. $P(x)/Q(x)$ where $P(x)$ and $Q(x)$ are polynomials, it is called a *rational algebraic function;* otherwise it is an *irrational algebraic function*.

3. **Transcendental functions** are functions which are not algebraic, i.e. do not satisfy equations of the form (2).

Note the analogy with real numbers, polynomials corresponding to integers, rational functions to rational numbers, etc.

SPECIAL TRANSCENDENTAL FUNCTIONS

The following are sometimes called *elementary transcendental functions*.

1. **Exponential function:** $f(x) = a^x$, $a \neq 0, 1$. For properties, see Page 3.

2. **Logarithmic function:** $f(x) = \log_a x$, $a \neq 0, 1$. This and the exponential function are inverse functions. If $a = e = 2.71828\ldots$, called the *natural base of logarithms*, we write $f(x) = \log_e x = \ln x$, called the *natural logarithm* of x. For properties, see Page 3.

3. **Trigonometric functions:**

$$\sin x, \ \cos x, \ \tan x = \frac{\sin x}{\cos x}, \ \csc x = \frac{1}{\sin x}, \ \sec x = \frac{1}{\cos x}, \ \cot x = \frac{1}{\tan x} = \frac{\cos x}{\sin x}$$

The variable x is generally expressed in radians (π radians $= 180°$). For real values of x, $\sin x$ and $\cos x$ lie between -1 and 1 inclusive.

The following are some properties of these functions.

$$\sin^2 x + \cos^2 x = 1 \qquad 1 + \tan^2 x = \sec^2 x \qquad 1 + \cot^2 x = \csc^2 x$$
$$\sin (x \pm y) = \sin x \cos y \pm \cos x \sin y \qquad\qquad \sin (-x) = -\sin x$$
$$\cos (x \pm y) = \cos x \cos y \mp \sin x \sin y \qquad\qquad \cos (-x) = \cos x$$
$$\tan (x \pm y) = \frac{\tan x \pm \tan y}{1 \mp \tan x \tan y} \qquad\qquad\qquad \tan (-x) = -\tan x$$

4. **Inverse trigonometric functions.** The following is a list of the inverse trigonometric functions and their principal values.

(a) $y = \sin^{-1} x$, $(-\pi/2 \leq y \leq \pi/2)$ (d) $y = \csc^{-1} x = \sin^{-1} 1/x$, $(-\pi/2 \leq y \leq \pi/2)$

(b) $y = \cos^{-1} x$, $(0 \leq y \leq \pi)$ (e) $y = \sec^{-1} x = \cos^{-1} 1/x$, $(0 \leq y \leq \pi)$

(c) $y = \tan^{-1} x$, $(-\pi/2 < y < \pi/2)$ (f) $y = \cot^{-1} x = \pi/2 - \tan^{-1} x$, $(0 < y < \pi)$

5. **Hyperbolic functions** are defined in terms of exponential functions as follows.

(a) $\sinh x = \dfrac{e^x - e^{-x}}{2}$ (d) $\csch x = \dfrac{1}{\sinh x} = \dfrac{2}{e^x - e^{-x}}$

(b) $\cosh x = \dfrac{e^x + e^{-x}}{2}$ (e) $\sech x = \dfrac{1}{\cosh x} = \dfrac{2}{e^x + e^{-x}}$

(c) $\tanh x = \dfrac{\sinh x}{\cosh x} = \dfrac{e^x - e^{-x}}{e^x + e^{-x}}$ (f) $\coth x = \dfrac{\cosh x}{\sinh x} = \dfrac{e^x + e^{-x}}{e^x - e^{-x}}$

The following are some properties of these functions.

$$\cosh^2 x - \sinh^2 x = 1 \qquad 1 - \tanh^2 x = \sech^2 x \qquad \coth^2 x - 1 = \csch^2 x$$

$$\sinh(x \pm y) = \sinh x \cosh y \pm \cosh x \sinh y \qquad \sinh(-x) = -\sinh x$$
$$\cosh(x \pm y) = \cosh x \cosh y \pm \sinh x \sinh y \qquad \cosh(-x) = \cosh x$$
$$\tanh(x \pm y) = \frac{\tanh x \pm \tanh y}{1 \pm \tanh x \tanh y} \qquad \tanh(-x) = -\tanh x$$

6. Inverse hyperbolic functions.   If $x = \sinh y$ then $y = \sinh^{-1} x$ is the *inverse hyperbolic sine* of x. The following list gives the principal values of the inverse hyperbolic functions in terms of natural logarithms and the domains for which they are real.

(a) $\sinh^{-1} x = \ln(x + \sqrt{x^2 + 1})$, all x (d) $\operatorname{csch}^{-1} x = \ln\left(\dfrac{1}{x} + \dfrac{\sqrt{x^2 + 1}}{|x|}\right)$, $x \neq 0$

(b) $\cosh^{-1} x = \ln(x + \sqrt{x^2 - 1})$, $x \geqq 1$ (e) $\operatorname{sech}^{-1} x = \ln\left(\dfrac{1 + \sqrt{1 - x^2}}{x}\right)$, $0 < x \leqq 1$

(c) $\tanh^{-1} x = \frac{1}{2}\ln\left(\dfrac{1 + x}{1 - x}\right)$, $|x| < 1$ (f) $\coth^{-1} x = \frac{1}{2}\ln\left(\dfrac{x + 1}{x - 1}\right)$, $|x| > 1$

LIMITS of FUNCTIONS

Let $f(x)$ be defined and single-valued for all values of x near $x = x_0$ with the possible exception of $x = x_0$ itself (i.e. in a deleted δ neighborhood of x_0). We say that the number l is the *limit of $f(x)$ as x approaches x_0* and write $\lim_{x \to x_0} f(x) = l$ if for any positive number ϵ (however small) we can find some positive number δ (usually depending on ϵ) such that $|f(x) - l| < \epsilon$ whenever $0 < |x - x_0| < \delta$. In such case we also say that $f(x)$ approaches l as x approaches x_0 and write $f(x) \to l$ as $x \to x_0$.

In words this means essentially that we can make the absolute value of the difference between $f(x)$ and l as small as we wish by choosing x sufficiently close to x_0, i.e. by choosing the difference in absolute value between x and x_0 sufficiently small (but not zero, i.e. we exclude $x = x_0$).

> **Example:** Let $f(x) = \begin{cases} x^2 & \text{if } x \neq 2 \\ 0 & \text{if } x = 2 \end{cases}$.   Then as x gets closer to 2 (i.e. x approaches 2), $f(x)$ gets closer to 4. We thus *suspect* that $\lim_{x \to 2} f(x) = 4$. To *prove* this we must see whether the above definition of limit (with $l = 4$) is satisfied. For this proof see Problem 10.
>
> Note that $\lim_{x \to 2} f(x) \neq f(2)$, i.e. the limit of $f(x)$ as $x \to 2$ is not the same as the value of $f(x)$ at $x = 2$ since $f(2) = 0$ by definition. The limit would in fact be 4 even if $f(x)$ were not defined at $x = 2$.

When the limit of a function exists it is unique, i.e. it is the only one (see Prob. 17).

RIGHT and LEFT HAND LIMITS

In the definition of limit no restriction was made as to how x should approach x_0. It is sometimes found convenient to restrict this approach. Considering x and x_0 as points on the real axis where x_0 is fixed and x is moving, then x can approach x_0 from the right or from the left. We indicate these respective approaches by writing $x \to x_0+$ and $x \to x_0-$.

If $\lim_{x \to x_0+} f(x) = l_1$ and $\lim_{x \to x_0-} f(x) = l_2$, we call l_1 and l_2 respectively the *right and left hand limits* of $f(x)$ at x_0 and denote them by $f(x_0+)$ or $f(x_0 + 0)$ and $f(x_0-)$ or $f(x_0 - 0)$. The ϵ, δ definitions of limit of $f(x)$ as $x \to x_0+$ or $x \to x_0-$ are the same as those for $x \to x_0$ except for the fact that values of x are restricted to $x > x_0$ or $x < x_0$ respectively.

We have $\lim_{x \to x_0} f(x) = l$ if and only if $\lim_{x \to x_0+} f(x) = \lim_{x \to x_0-} f(x) = l$.

THEOREMS on LIMITS

If $\lim\limits_{x \to x_0} f(x) = A$ and $\lim\limits_{x \to x_0} g(x) = B$, then

1. $\lim\limits_{x \to x_0} \Big(f(x) + g(x) \Big) = \lim\limits_{x \to x_0} f(x) + \lim\limits_{x \to x_0} g(x) = A + B$

2. $\lim\limits_{x \to x_0} \Big(f(x) - g(x) \Big) = \lim\limits_{x \to x_0} f(x) - \lim\limits_{x \to x_0} g(x) = A - B$

3. $\lim\limits_{x \to x_0} \Big(f(x)\, g(x) \Big) = \Big(\lim\limits_{x \to x_0} f(x) \Big)\Big(\lim\limits_{x \to x_0} g(x) \Big) = A B$

4. $\lim\limits_{x \to x_0} \dfrac{f(x)}{g(x)} = \dfrac{\lim\limits_{x \to x_0} f(x)}{\lim\limits_{x \to x_0} g(x)} = \dfrac{A}{B}$ if $B \neq 0$

Similar results hold for right and left hand limits.

INFINITY

It sometimes happens that as $x \to x_0$, $f(x)$ increases or decreases without bound. In such case it is customary to write $\lim\limits_{x \to x_0} f(x) = +\infty$ or $\lim\limits_{x \to x_0} f(x) = -\infty$ respectively. The symbols $+\infty$ (also written ∞) and $-\infty$ are read *plus infinity* (or *infinity*) and *minus infinity* respectively, but it must be emphasized that they are not numbers.

In precise language, we say that $\lim\limits_{x \to x_0} f(x) = \infty$ if for each positive number M we can find a positive number δ (depending on M in general) such that $f(x) > M$ whenever $0 < |x - x_0| < \delta$. Similarly we say that $\lim\limits_{x \to x_0} f(x) = -\infty$ if for each positive number M we can find a positive number δ such that $f(x) < -M$ whenever $0 < |x - x_0| < \delta$. Analogous remarks apply in case $x \to x_0+$ or $x \to x_0-$.

Frequently we wish to examine the behavior of a function as x increases or decreases without bound. In such cases it is customary to write $x \to +\infty$ (or ∞) or $x \to -\infty$ respectively.

We say that $\lim\limits_{x \to +\infty} f(x) = l$, or $f(x) \to l$ as $x \to +\infty$, if for any positive number ϵ we can find a positive number N (depending on ϵ in general) such that $|f(x) - l| < \epsilon$ whenever $x > N$. A similar definition can be formulated for $\lim\limits_{x \to -\infty} f(x)$.

SPECIAL LIMITS

1. $\lim\limits_{x \to 0} \dfrac{\sin x}{x} = 1,$ $\qquad\qquad$ $\lim\limits_{x \to 0} \dfrac{1 - \cos x}{x} = 0$

2. $\lim\limits_{x \to \infty} \Big(1 + \dfrac{1}{x} \Big)^x = e,$ $\qquad$ $\lim\limits_{x \to 0+} (1 + x)^{1/x} = e$

3. $\lim\limits_{x \to 0} \dfrac{e^x - 1}{x} = 1,$ $\qquad\qquad$ $\lim\limits_{x \to 1} \dfrac{x - 1}{\ln x} = 1$

CONTINUITY

Let $f(x)$ be defined and single-valued for all values of x near $x = x_0$ as well as at $x = x_0$ (i.e. in a δ neighborhood of x_0). The function $f(x)$ is called *continuous* at $x = x_0$ if $\lim\limits_{x \to x_0} f(x) = f(x_0)$. Note that this implies three conditions which must be met in order that $f(x)$ be continuous at $x = x_0$.

1. $\lim\limits_{x \to x_0} f(x) = l$ must exist.

2. $f(x_0)$ must exist, i.e. $f(x)$ is defined *at* x_0.

3. $l = f(x_0)$

Equivalently if $f(x)$ is continuous at x_0, we can write this in the suggestive form $\lim\limits_{x \to x_0} f(x) = f\big(\lim\limits_{x \to x_0} x\big)$.

Examples: 1. If $f(x) = \begin{cases} x^2, & x \neq 2 \\ 0, & x = 2 \end{cases}$ then from the example on Page 23, $\lim\limits_{x \to 2} f(x) = 4$. But $f(2) = 0$.

Hence $\lim\limits_{x \to 2} f(x) \neq f(2)$ and the function is not continuous at $x = 2$.

2. If $f(x) = x^2$ for all x, then $\lim\limits_{x \to 2} f(x) = f(2) = 4$ and $f(x)$ is continuous at $x = 2$.

Points where $f(x)$ fails to be continuous are called *discontinuities* of $f(x)$ and $f(x)$ is said to be *discontinuous* at these points.

In constructing a graph of a continuous function the pencil need never leave the paper, while for a discontinuous function this is not true since there is generally a jump taking place. This is of course merely a characteristic property and not a definition of continuity or discontinuity.

Alternative to the above definition of continuity, we can define $f(x)$ as continuous at $x = x_0$ if for any $\epsilon > 0$ we can find $\delta > 0$ such that $|f(x) - f(x_0)| < \epsilon$ whenever $|x - x_0| < \delta$. Note that this is simply the definition of limit with $l = f(x_0)$ and removal of the restriction that $x \neq x_0$.

RIGHT and LEFT HAND CONTINUITY

If $f(x)$ is defined only for $x \geq x_0$, the above definition does not apply. In such case we call $f(x)$ *continuous (on the right)* at $x = x_0$ if $\lim\limits_{x \to x_0+} f(x) = f(x_0)$, i.e. if $f(x_0+) = f(x_0)$.

Similarly, $f(x)$ is *continuous (on the left)* at $x = x_0$ if $\lim\limits_{x \to x_0-} f(x) = f(x_0)$, i.e. $f(x_0-) = f(x_0)$. Definitions in terms of ϵ and δ can be given.

CONTINUITY in an INTERVAL

A function $f(x)$ is said to be *continuous in an interval* if it is continuous at all points of the interval. In particular, if $f(x)$ is defined in the closed interval $a \leq x \leq b$ or $[a, b]$, then $f(x)$ is continuous in the interval if and only if $\lim\limits_{x \to x_0} f(x) = f(x_0)$ for $a < x_0 < b$, $\lim\limits_{x \to a+} f(x) = f(a)$ and $\lim\limits_{x \to b-} f(x) = f(b)$.

THEOREMS on CONTINUITY

Theorem 1. If $f(x)$ and $g(x)$ are continuous at $x = x_0$, so also are the functions $f(x) + g(x)$, $f(x) - g(x)$, $f(x)\, g(x)$ and $\dfrac{f(x)}{g(x)}$, the last only if $g(x_0) \neq 0$. Similar results hold for continuity in an interval.

Theorem 2. The following functions are continuous in every finite interval: (a) all polynomials; (b) $\sin x$ and $\cos x$; (c) a^x, $a > 0$.

Theorem 3. If $y = f(x)$ is continuous at $x = x_0$ and $z = g(y)$ is continuous at $y = y_0$ and if $y_0 = f(x_0)$, then the function $z = g[f(x)]$, called a *function of a function* or *composite function*, is continuous at $x = x_0$. This is sometimes briefly stated as: *A continuous function of a continuous function is continuous.*

Theorem 4. If $f(x)$ is continuous in a closed interval, it is bounded in the interval.

Theorem 5. If $f(x)$ is continuous at $x = x_0$ and $f(x_0) > 0$ [or $f(x_0) < 0$], there exists an interval about $x = x_0$ in which $f(x) > 0$ [or $f(x) < 0$].

Theorem 6. If a function $f(x)$ is continuous in an interval and either strictly increasing or strictly decreasing, the inverse function $f^{-1}(x)$ is single-valued, continuous and either strictly increasing or strictly decreasing.

Theorem 7. If $f(x)$ is continuous in $[a, b]$ and if $f(a) = A$ and $f(b) = B$, then corresponding to any number C between A and B there exists at least one number c in $[a, b]$ such that $f(c) = C$. This is sometimes called the *intermediate value theorem*.

Theorem 8. If $f(x)$ is continuous in $[a, b]$ and if $f(a)$ and $f(b)$ have opposite signs, there is at least one number c for which $f(c) = 0$ where $a < c < b$. This is related to Theorem 7.

Theorem 9. If $f(x)$ is continuous in a closed interval, then $f(x)$ has a maximum value M for at least one value of x in the interval and a minimum value m for at least one value of x in the interval. Furthermore, $f(x)$ assumes all values between m and M for one or more values of x in the interval.

Theorem 10. If $f(x)$ is continuous in a closed interval and if M and m are respectively the least upper bound (l.u.b.) and greatest lower bound (g.l.b.) of $f(x)$, there exists at least one value of x in the interval for which $f(x) = M$ or $f(x) = m$. This is related to Theorem 9.

SECTIONAL CONTINUITY

A function is called *sectionally continuous* or *piecewise continuous* in an interval $a \leqq x \leqq b$ if the interval can be subdivided into a finite number of intervals in each of which the function is continuous and has finite right and left hand limits. Such a function has only a finite number of discontinuities. An example of a function which is sectionally continuous in $a \leqq x \leqq b$ is shown graphically in Fig. 2-1 below. This function has discontinuities at x_1, x_2, x_3 and x_4.

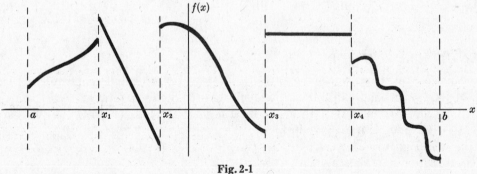

Fig. 2-1

UNIFORM CONTINUITY

Let $f(x)$ be continuous in an interval. Then by definition at each point x_0 of the interval and for any $\epsilon > 0$, we can find $\delta > 0$ (which will in general depend on both ϵ and the particular point x_0) such that $|f(x) - f(x_0)| < \epsilon$ whenever $|x - x_0| < \delta$. If we can find δ for each ϵ which holds for all points of the interval (i.e. if δ depends *only* on ϵ and *not* on x_0), we say that $f(x)$ is *uniformly continuous* in the interval.

Alternatively, $f(x)$ is uniformly continuous in an interval if for any $\epsilon > 0$ we can find $\delta > 0$ such that $|f(x_1) - f(x_2)| < \epsilon$ whenever $|x_1 - x_2| < \delta$ where x_1 and x_2 are any two points in the interval.

Theorem. If $f(x)$ is continuous in a *closed* interval, it is uniformly continuous in the interval.

Solved Problems

FUNCTIONS

1. Let $f(x) = (x-2)(8-x)$ for $2 \leqq x \leqq 8$. (a) Find $f(6)$ and $f(-1)$. (b) What is the domain of definition of $f(x)$? (c) Find $f(1-2t)$ and give the domain of definition. (d) Find $f[f(3)]$, $f[f(5)]$. (e) Graph $f(x)$.

(a) $f(6) = (6-2)(8-6) = 4 \cdot 2 = 8$
     $f(-1)$ is not defined since $f(x)$ is defined only for $2 \leqq x \leqq 8$.

(b) The set of all x such that $2 \leqq x \leqq 8$.

(c) $f(1-2t) = \{(1-2t)-2\}\{8-(1-2t)\} = -(1+2t)(7+2t)$ where t is such that $2 \leqq 1-2t \leqq 8$, i.e. $-7/2 \leqq t \leqq -1/2$.

(d) $f(3) = (3-2)(8-3) = 5$, $f[f(3)] = f(5) = (5-2)(8-5) = 9$.
     $f(5) = 9$ so that $f[f(5)] = f(9)$ is not defined.

(e) The following table shows $f(x)$ for various values of x.

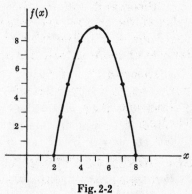

x	2	3	4	5	6	7	8	2.5	7.5
$f(x)$	0	5	8	9	8	5	0	2.75	2.75

Plot points $(2,0)$, $(3,5)$, $(4,8)$, $(5,9)$, $(6,8)$, $(7,5)$, $(8,0)$, $(2.5, 2.75)$, $(7.5, 2.75)$.

These points are only a few of the infinitely many points on the required graph shown in the adjoining Fig. 2-2. This set of points defines a curve which is part of a *parabola*.

Fig. 2-2

2. Let $g(x) = (x-2)(8-x)$ for $2 < x < 8$. (a) Discuss the difference between the graph of $g(x)$ and that of $f(x)$ in Problem 1. (b) What is the l.u.b. and g.l.b. of $g(x)$? (c) Does $g(x)$ attain its l.u.b. and g.l.b. for any value of x in the domain of definition? (d) Answer parts (b) and (c) for the function $f(x)$ of Problem 1.

(a) The graph of $g(x)$ is the same as that in Problem 1 except that the two points $(2,0)$ and $(8,0)$ are missing, since $g(x)$ is not defined at $x=2$ and $x=8$.

(b) The l.u.b. of $g(x)$ is 9. The g.l.b. of $g(x)$ is 0.

(c) The l.u.b. of $g(x)$ is attained for the value $x=5$. The g.l.b. of $g(x)$ is not attained, since there is no value of x in the domain of definition such that $g(x)=0$.

(d) As in (b), the l.u.b. of $f(x)$ is 9 and the g.l.b. of $f(x)$ is 0. The l.u.b. of $f(x)$ is attained for the value $x=5$ and the g.l.b. of $f(x)$ is attained at $x=2$ and $x=8$.

> Note that a function, such as $f(x)$, which is *continuous* in a closed interval attains its l.u.b. and g.l.b. at some point of the interval. However a function, such as $g(x)$, which is not continuous in a closed interval need not attain its l.u.b. and g.l.b. See Problem 34.

3. Let $f(x) = \begin{cases} 1, & \text{if } x \text{ is a rational number} \\ 0, & \text{if } x \text{ is an irrational number} \end{cases}$. (a) Find $f(\tfrac{2}{3})$, $f(-5)$, $f(1.41423)$, $f(\sqrt{2})$,

(b) Construct a graph of $f(x)$ and explain why it is misleading by itself.

(a) $f(\tfrac{2}{3})$ $= 1$ since $\tfrac{2}{3}$ is a rational number
     $f(-5)$ $= 1$ since -5 is a rational number
     $f(1.41423) = 1$ since 1.41423 is a rational number
     $f(\sqrt{2})$ $= 0$ since $\sqrt{2}$ is an irrational number

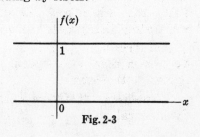

(b) The graph is shown in the adjoining Fig. 2-3. From its appearance it would seem that there are two functional values 0 and 1 corresponding to each value of x, i.e. that $f(x)$ is multiple-valued, whereas it is actually single-valued.

Fig. 2-3

4. Referring to Problem 1, (a) construct the graph of $f^{-1}(x)$, (b) find an expression for $f^{-1}(x)$ and show that $f^{-1}(x)$ is not single-valued.

(a) The graph of $y = f(x)$ or $x = f^{-1}(y)$ is shown in Fig. 2-2 of Problem 1(e). To obtain the graph of $y = f^{-1}(x)$, we have only to interchange the x and y axes. We obtain the graph shown in the adjoining Fig. 2-4 after orienting the axes in the usual manner.

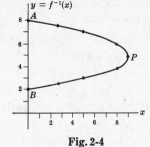

Fig. 2-4

(b) We have $y = (x-2)(8-x)$ or $x^2 - 10x + 16 + y = 0$. Using the quadratic formula,

$$x = f^{-1}(y) = \frac{10 \pm \sqrt{100 - 4(16 + y)}}{2} = 5 \pm \sqrt{9 - y}.$$

Then, $y = f^{-1}(x) = 5 \pm \sqrt{9 - x}$.

In the graph, AP represents $y = 5 + \sqrt{9 - x}$, BP represents $y = 5 - \sqrt{9 - x}$. Thus for each value of x in $0 \leqq x < 9$, $f^{-1}(x)$ is double-valued. This is seen graphically from the fact that every vertical line to the left of P and the right of AB meets the graph in two points.

The functions $5 + \sqrt{9 - x}$ and $5 - \sqrt{9 - x}$ represent the two *branches* of $f^{-1}(x)$. The point where the two branches meet (or have the same value) is sometimes called a *branch point*, in this case at $x = 9$, $y = 5$.

5. (a) Prove that $g(x) = 5 + \sqrt{9 - x}$ is strictly decreasing in $0 \leqq x \leqq 9$. (b) Is it monotonic decreasing in this interval? (c) Does $g(x)$ have a single-valued inverse?

(a) $g(x)$ is strictly decreasing if $g(x_1) > g(x_2)$ whenever $x_1 < x_2$. If $x_1 < x_2$ then $9 - x_1 > 9 - x_2$, $\sqrt{9 - x_1} > \sqrt{9 - x_2}$, $5 + \sqrt{9 - x_1} > 5 + \sqrt{9 - x_2}$ showing that $g(x)$ is strictly decreasing.

(b) Yes, any strictly decreasing function is also monotonic decreasing, since if $g(x_1) > g(x_2)$ it is also true that $g(x_1) \geqq g(x_2)$. However if $g(x)$ is monotonic decreasing, it is not necessarily strictly decreasing.

(c) If $y = 5 + \sqrt{9 - x}$ then $y - 5 = \sqrt{9 - x}$ or squaring, $x = -16 + 10y - y^2 = (y - 2)(8 - y)$ and x is a single-valued function of y, i.e. the inverse function is single-valued.

In general, any strictly decreasing (or increasing) function has a single-valued inverse (see Theorem 6, Page 26).

The results of this problem can be interpreted graphically using the figure of Problem 4.

6. Construct graphs for the functions (a) $f(x) = \begin{cases} x \sin 1/x, & x > 0 \\ 0, & x = 0 \end{cases}$, (b) $f(x) = [x] =$ greatest integer $\leqq x$.

(a) The required graph is shown in Fig. 2-5 below. Since $|x \sin 1/x| \leqq |x|$, the graph is included between $y = x$ and $y = -x$. Note that $f(x) = 0$ when $\sin 1/x = 0$ or $1/x = m\pi$, $m = 1, 2, 3, 4, \ldots$, i.e. where $x = 1/\pi, 1/2\pi, 1/3\pi, \ldots$. The curve oscillates infinitely often between $x = 1/\pi$ and $x = 0$.

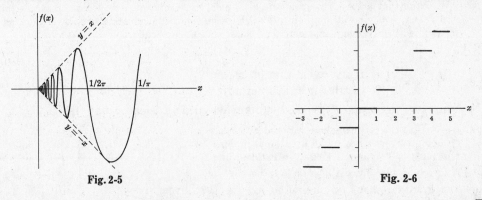

Fig. 2-5   **Fig. 2-6**

(b) The required graph is shown in Fig. 2-6 above. If $1 \leqq x < 2$, then $[x] = 1$. Thus $[1.8] = 1$, $[\sqrt{2}] = 1$, $[1.99999] = 1$. However, $[2] = 2$. Similarly for $2 \leqq x < 3$, $[x] = 2$, etc. Thus there are *jumps* at the integers. The function is sometimes called the *staircase function* or *step function*.

7. (a) Construct the graph of $f(x) = \tan x$. (b) Construct the graph of $\tan^{-1} x$. (c) Show graphically why $\tan^{-1} x$ is a multiple-valued function. (d) Indicate possible principal values for $\tan^{-1} x$. (e) Using your choice, evaluate $\tan^{-1}(-1)$.

(a) The graph of $f(x) = \tan x$ appears in Fig. 2-7 below.

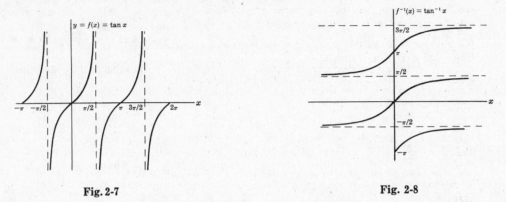

Fig. 2-7   **Fig. 2-8**

(b) If $y = f(x) = \tan x$, then $x = f^{-1}(y) = \tan^{-1} y$. Then the graph of $f^{-1}(x) = \tan^{-1} x$ is obtained by interchanging the x and y axes in the graph of (a). The result, with axes oriented as usual, appears in Fig. 2-8 above.

(c) In Fig. 2-8 of (b), any vertical line meets the graph in infinitely many points. Thus $\tan^{-1} x$ is a multiple-valued function with infinitely many branches.

(d) To define $\tan^{-1} x$ as a single-valued function, it is clear from the graph that we can only do so by restricting its value to any of the following: $-\pi/2 < \tan^{-1} x < \pi/2$, $\pi/2 < \tan^{-1} x < 3\pi/2$, etc. We shall agree to take the first as defining the principal value.

Note that on any of these branches, $\tan^{-1} x$ is a strictly increasing function with a single-valued inverse.

(e) $\tan^{-1}(-1) = -\pi/4$ is the only value lying between $-\pi/2$ and $\pi/2$, i.e. it is the principal value according to our choice in (d).

8. Show that $f(x) = \dfrac{\sqrt{x}+1}{x+1}$, $x \neq -1$, is an irrational algebraic function.

If $y = \dfrac{\sqrt{x}+1}{x+1}$ then $(x+1)y - 1 = \sqrt{x}$ or squaring, $(x+1)^2 y^2 - 2(x+1)y + 1 - x = 0$, a polynomial equation in y whose coefficients are polynomials in x. Thus $f(x)$ is an algebraic function. However, it is not the quotient of two polynomials, so that it is an irrational algebraic function.

9. If $f(x) = \cosh x = \frac{1}{2}(e^x + e^{-x})$, prove that we can choose as the principal value of the inverse function, $\cosh^{-1} x = \ln(x + \sqrt{x^2-1})$, $x \geqq 1$.

If $y = \frac{1}{2}(e^x + e^{-x})$, $e^{2x} - 2ye^x + 1 = 0$. Then using the quadratic formula, $e^x = \dfrac{2y \pm \sqrt{4y^2-4}}{2} = y \pm \sqrt{y^2-1}$. Thus $x = \ln(y \pm \sqrt{y^2-1})$.

Since $y - \sqrt{y^2-1} = (y - \sqrt{y^2-1})\left(\dfrac{y + \sqrt{y^2-1}}{y + \sqrt{y^2-1}}\right) = \dfrac{1}{y + \sqrt{y^2-1}}$, we can also write

$$x = \pm \ln(y + \sqrt{y^2-1}) \quad \text{or} \quad \cosh^{-1} y = \pm \ln(y + \sqrt{y^2-1})$$

Choosing the $+$ sign as defining the principal value and replacing y by x, we have $\cosh^{-1} x = \ln(x + \sqrt{x^2-1})$. The choice $x \geqq 1$ is made so that the inverse function is real.

LIMITS

10. If (a) $f(x) = x^2$, (b) $f(x) = \begin{cases} x^2, & x \neq 2 \\ 0, & x = 2 \end{cases}$, prove that $\lim\limits_{x \to 2} f(x) = 4$.

(a) We must show that given any $\epsilon > 0$ we can find $\delta > 0$ (depending on ϵ in general) such that $|x^2 - 4| < \epsilon$ when $0 < |x - 2| < \delta$.

Choose $\delta \leqq 1$ so that $\quad 0 < |x - 2| < 1$ or $1 < x < 3$, $x \neq 2$.

Then $\quad |x^2 - 4| = |(x-2)(x+2)| = |x-2|\,|x+2| < \delta|x+2| < 5\delta$.

Take δ as 1 or $\epsilon/5$, whichever is smaller. Then we have $|x^2 - 4| < \epsilon$ whenever $0 < |x-2| < \delta$ and the required result is proved.

It is of interest to consider some numerical values. If for example we wish to make $|x^2 - 4| < .05$, we can choose $\delta = \epsilon/5 = .05/5 = .01$. To see that this is actually the case, note that if $0 < |x-2| < .01$ then $1.99 < x < 2.01$ $(x \neq 2)$ and so $3.9601 < x^2 < 4.0401$, $-.0399 < x^2 - 4 < .0401$ and certainly $|x^2 - 4| < .05$ $(x^2 \neq 4)$. The fact that these inequalities also happen to hold at $x = 2$ is merely coincidental.

If we wish to make $|x^2 - 4| < 6$, we can choose $\delta = 1$ and this will be satisfied.

(b) There is no difference between the proof for this case and the proof in (a), since in both cases we exclude $x = 2$.

11. Prove that $\quad \lim\limits_{x \to 1} \dfrac{2x^4 - 6x^3 + x^2 + 3}{x - 1} = -8$.

We must show that for any $\epsilon > 0$ we can find $\delta > 0$ such that $\quad \left| \dfrac{2x^4 - 6x^3 + x^2 + 3}{x - 1} - (-8) \right| < \epsilon$

when $0 < |x-1| < \delta$. Since $x \neq 1$, we can write $\dfrac{2x^4 - 6x^3 + x^2 + 3}{x - 1} = \dfrac{(2x^3 - 4x^2 - 3x - 3)(x - 1)}{x - 1} =$

$2x^3 - 4x^2 - 3x - 3$ on cancelling the common factor $x - 1 \neq 0$.

Then we must show that for any $\epsilon > 0$, we can find $\delta > 0$ such that $|2x^3 - 4x^2 - 3x + 5| < \epsilon$ when $0 < |x-1| < \delta$. Choosing $\delta \leqq 1$, we have $0 < x < 2$, $x \neq 1$.

Now $\quad |2x^3 - 4x^2 - 3x + 5| = |x - 1|\,|2x^2 - 2x - 5| < \delta|2x^2 - 2x - 5| < \delta(|2x^2| + |2x| + 5) < (8 + 4 + 5)\delta = 17\delta$. Taking δ as the smaller of 1 and $\epsilon/17$, the required result follows.

12. Let $\quad f(x) = \begin{cases} \dfrac{|x - 3|}{x - 3}, & x \neq 3 \\ 0, & x = 3 \end{cases}$,   (a) Graph the function.   (b) Find $\lim\limits_{x \to 3+} f(x)$.   (c) Find

$\lim\limits_{x \to 3-} f(x)$.   (d) Find $\lim\limits_{x \to 3} f(x)$.

(a) For $x > 3$, $\dfrac{|x - 3|}{x - 3} = \dfrac{x - 3}{x - 3} = 1$.

For $x < 3$, $\dfrac{|x - 3|}{x - 3} = \dfrac{-(x - 3)}{x - 3} = -1$.

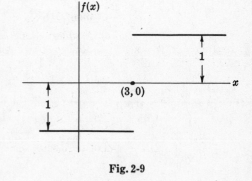

Then the graph, shown in the adjoining Fig. 2-9, consists of the lines $y = 1$, $x > 3$; $y = -1$, $x < 3$ and the point $(3, 0)$.

(b) As $x \to 3$ from the right, $f(x) \to 1$, i.e. $\lim\limits_{x \to 3+} f(x) = 1$, as seems clear from the graph. To prove this we must show that given any $\epsilon > 0$, we can find $\delta > 0$ such that $|f(x) - 1| < \epsilon$ whenever $0 < x - 1 < \delta$.

Now since $x > 1$, $f(x) = 1$ and so the proof consists in the triviality that $|1 - 1| < \epsilon$ whenever $0 < x - 1 < \delta$.

Fig. 2-9

(c) As $x \to 3$ from the left, $f(x) \to -1$, i.e. $\lim\limits_{x \to 3-} f(x) = -1$. A proof can be formulated as in (b).

(d) Since $\lim\limits_{x \to 3+} f(x) \neq \lim\limits_{x \to 3-} f(x)$, $\lim\limits_{x \to 3} f(x)$ does not exist.

13. Prove that $\lim\limits_{x \to 0} x \sin 1/x = 0$.

We must show that given any $\epsilon > 0$, we can find $\delta > 0$ such that $|x \sin 1/x - 0| < \epsilon$ when $0 < |x - 0| < \delta$.

If $0 < |x| < \delta$, then $|x \sin 1/x| = |x|\,|\sin 1/x| \leqq |x| < \delta$ since $|\sin 1/x| \leqq 1$ for all $x \neq 0$.

Making the choice $\delta = \epsilon$, we see that $|x \sin 1/x| < \epsilon$ when $0 < |x| < \delta$, completing the proof.

14. Evaluate $\lim\limits_{x \to 0+} \dfrac{2}{1 + e^{-1/x}}$.

As $x \to 0+$ we *suspect* that $1/x$ increases indefinitely, $e^{1/x}$ increases indefinitely, $e^{-1/x}$ approaches 0, $1 + e^{-1/x}$ approaches 1; thus the required limit is 2.

To *prove* this conjecture we must show that, given $\epsilon > 0$, we can find $\delta > 0$ such that

$$\left| \frac{2}{1 + e^{-1/x}} - 2 \right| < \epsilon \quad \text{when} \quad 0 < x < \delta$$

Now $\qquad \left| \dfrac{2}{1 + e^{-1/x}} - 2 \right| = \left| \dfrac{2 - 2 - 2e^{-1/x}}{1 + e^{-1/x}} \right| = \dfrac{2}{e^{1/x} + 1}$

Since the function on the right is smaller than 1 for all $x > 0$, any $\delta > 0$ will work when $\epsilon \geq 1$. If

$0 < \epsilon < 1$, then $\dfrac{2}{e^{1/x} + 1} < \epsilon$ when $\dfrac{e^{1/x} + 1}{2} > \dfrac{1}{\epsilon}$, $e^{1/x} > \dfrac{2}{\epsilon} - 1$, $\dfrac{1}{x} > \ln\left(\dfrac{2}{\epsilon} - 1\right)$; or $0 < x < \dfrac{1}{\ln(2/\epsilon - 1)} = \delta$

15. Explain exactly what is meant by the statement $\lim\limits_{x \to 1} \dfrac{1}{(x-1)^4} = \infty$ and prove the validity of this statement.

The statement means that for each positive number M, we can find a positive number δ (depending on M in general) such that

$$\frac{1}{(x-1)^4} > M \quad \text{when} \quad 0 < |x - 1| < \delta$$

To prove this note that $\dfrac{1}{(x-1)^4} > M$ when $0 < (x-1)^4 < \dfrac{1}{M}$ or $0 < |x-1| < \dfrac{1}{\sqrt[4]{M}}$.

Choosing $\delta = 1/\sqrt[4]{M}$, the required result follows.

16. Present a geometric proof that $\lim\limits_{\theta \to 0} \dfrac{\sin \theta}{\theta} = 1$.

Construct a circle with center at O and radius $OA = OD = 1$, as in Fig. 2-10 below. Choose point B on OA extended and point C on OD so that lines BD and AC are perpendicular to OD.

It is geometrically evident that

         Area of triangle OAC < Area of sector OAD < Area of triangle OBD

i.e. $\qquad \tfrac{1}{2} \sin \theta \cos \theta \ < \ \tfrac{1}{2}\theta \ < \ \tfrac{1}{2}\tan \theta$

Dividing by $\tfrac{1}{2} \sin \theta$,

$$\cos \theta \ < \ \frac{\theta}{\sin \theta} \ < \ \frac{1}{\cos \theta}$$

or $\qquad\qquad \cos \theta \ < \ \dfrac{\sin \theta}{\theta} \ < \ \dfrac{1}{\cos \theta}$

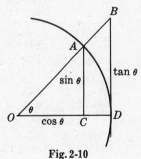

Fig. 2-10

As $\theta \to 0$, $\cos \theta \to 1$ and it follows that $\lim\limits_{\theta \to 0} \dfrac{\sin \theta}{\theta} = 1$.

THEOREMS on LIMITS

17. If $\lim\limits_{x \to x_0} f(x)$ exists, prove that it must be unique.

We must show that if $\lim\limits_{x \to x_0} f(x) = l_1$ and $\lim\limits_{x \to x_0} f(x) = l_2$, then $l_1 = l_2$.

By hypothesis, given any $\epsilon > 0$ we can find $\delta > 0$ such that

$$|f(x) - l_1| < \epsilon/2 \quad \text{when} \quad 0 < |x - x_0| < \delta$$
$$|f(x) - l_2| < \epsilon/2 \quad \text{when} \quad 0 < |x - x_0| < \delta$$

Then by inequality 2, Page 3,

$$|l_1 - l_2| \;=\; |l_1 - f(x) + f(x) - l_2| \;\leqq\; |l_1 - f(x)| + |f(x) - l_2| \;<\; \epsilon/2 + \epsilon/2 \;=\; \epsilon$$

i.e. $|l_1 - l_2|$ is less than any positive number ϵ (however small) and so must be zero. Thus $l_1 = l_2$.

18. If $\lim\limits_{x \to x_0} g(x) = B \neq 0$, prove that there exists $\delta > 0$ such that

$$|g(x)| \;>\; \tfrac{1}{2}|B| \quad \text{for} \quad 0 < |x - x_0| < \delta$$

Since $\lim\limits_{x \to x_0} g(x) = B$, we can find $\delta > 0$ such that $|g(x) - B| < \tfrac{1}{2}|B|$ for $0 < |x - x_0| < \delta$.

Writing $B = B - g(x) + g(x)$, we have

$$|B| \;\leqq\; |B - g(x)| + |g(x)| \;<\; \tfrac{1}{2}|B| + |g(x)|$$

i.e. $|B| < \tfrac{1}{2}|B| + |g(x)|$, from which $|g(x)| > \tfrac{1}{2}|B|$.

19. Given $\lim\limits_{x \to x_0} f(x) = A$ and $\lim\limits_{x \to x_0} g(x) = B$, prove   (a) $\lim\limits_{x \to x_0} [f(x) + g(x)] = A + B$,

(b) $\lim\limits_{x \to x_0} f(x)\, g(x) = AB$,   (c) $\lim\limits_{x \to x_0} \dfrac{1}{g(x)} = \dfrac{1}{B}$ if $B \neq 0$,   (d) $\lim\limits_{x \to x_0} \dfrac{f(x)}{g(x)} = \dfrac{A}{B}$ if $B \neq 0$.

(a)   We must show that for any $\epsilon > 0$ we can find $\delta > 0$ such that

$$|[f(x) + g(x)] - (A + B)| \;<\; \epsilon \quad \text{when} \quad 0 < |x - x_0| < \delta$$

Using inequality 2, Page 3, we have

$$|[f(x) + g(x)] - (A + B)| \;=\; |[f(x) - A] + [g(x) - B]| \;\leqq\; |f(x) - A| + |g(x) - B| \qquad (1)$$

By hypothesis, given $\epsilon > 0$ we can find $\delta_1 > 0$ and $\delta_2 > 0$ such that

$$|f(x) - A| \;<\; \epsilon/2 \quad \text{when} \quad 0 < |x - x_0| < \delta_1 \qquad (2)$$

$$|g(x) - B| \;<\; \epsilon/2 \quad \text{when} \quad 0 < |x - x_0| < \delta_2 \qquad (3)$$

Then from (1), (2) and (3),

$$|[f(x) + g(x)] - (A + B)| \;<\; \epsilon/2 + \epsilon/2 \;=\; \epsilon \quad \text{when} \quad 0 < |x - x_0| < \delta$$

where δ is chosen as the smaller of δ_1 and δ_2.

(b)   We have

$$\begin{aligned}
|f(x)\, g(x) - AB| \;&=\; |\, f(x)\,[g(x) - B] + B[f(x) - A]\,| \qquad (4)\\
&\leqq\; |f(x)|\,|g(x) - B| + |B|\,|f(x) - A| \\
&\leqq\; |f(x)|\,|g(x) - B| + (|B| + 1)\,|f(x) - A|
\end{aligned}$$

Since $\lim\limits_{x \to x_0} f(x) = A$, we can find δ_1 such that $|f(x) - A| < 1$ for $0 < |x - x_0| < \delta_1$, i.e. $A - 1 < f(x) < A + 1$, so that $f(x)$ is bounded, i.e. $|f(x)| < P$ where P is a positive constant.

Since $\lim\limits_{x \to x_0} g(x) = B$, given $\epsilon > 0$ we can find $\delta_2 > 0$ such that $|g(x) - B| < \epsilon/2P$ for $0 < |x - x_0| < \delta_2$.

Since $\lim\limits_{x \to x_0} f(x) = A$, given $\epsilon > 0$ we can find $\delta_3 > 0$ such that $|f(x) - A| < \dfrac{\epsilon}{2(|B| + 1)}$ for $0 < |x - x_0| < \delta_3$.

Using these in (4), we have

$$|f(x)\, g(x) - AB| \;<\; P \cdot \frac{\epsilon}{2P} + (|B| + 1) \cdot \frac{\epsilon}{2(|B| + 1)} \;=\; \epsilon$$

for $0 < |x - x_0| < \delta$ where δ is the smaller of $\delta_1, \delta_2, \delta_3$ and the proof is complete.

(c)   We must show that for any $\epsilon > 0$ we can find $\delta > 0$ such that

$$\left| \frac{1}{g(x)} - \frac{1}{B} \right| \;=\; \frac{|g(x) - B|}{|B|\,|g(x)|} \;<\; \epsilon \quad \text{when} \quad 0 < |x - x_0| < \delta \qquad (5)$$

By hypothesis, given $\epsilon > 0$ we can find $\delta_1 > 0$ such that

$$|g(x) - B| \;<\; \tfrac{1}{2}B^2 \epsilon \quad \text{when} \quad 0 < |x - x_0| < \delta_1$$

By Problem 18, since $\lim\limits_{x \to x_0} g(x) = B \neq 0$, we can find $\delta_2 > 0$ such that
$$|g(x)| > \tfrac{1}{2}|B| \quad \text{when} \quad 0 < |x - x_0| < \delta_2$$

Then if δ is the smaller of δ_1 and δ_2, we can write
$$\left| \frac{1}{g(x)} - \frac{1}{B} \right| = \frac{|g(x) - B|}{|B|\,|g(x)|} < \frac{\tfrac{1}{2}B^2 \epsilon}{|B| \cdot \tfrac{1}{2}|B|} = \epsilon \quad \text{whenever} \quad 0 < |x - x_0| < \delta$$

and the required result is proved.

(d) From parts (b) and (c),
$$\lim_{x \to x_0} \frac{f(x)}{g(x)} = \lim_{x \to x_0} f(x) \cdot \frac{1}{g(x)} = \lim_{x \to x_0} f(x) \cdot \lim_{x \to x_0} \frac{1}{g(x)} = A \cdot \frac{1}{B} = \frac{A}{B}$$

This can also be proved directly (see Problem 69).

The above results can also be proved in the cases $x \to x_0+$, $x \to x_0-$, $x \to \infty$, $x \to -\infty$.

Note: In the proof of (a) we have used the results $|f(x) - A| < \epsilon/2$ and $|g(x) - B| < \epsilon/2$, so that the final result would come out to be $|f(x) + g(x) - (A + B)| < \epsilon$. Of course the proof would be *just as valid* if we had used 2ϵ (or any other positive multiple of ϵ) in place of ϵ. A similar remark holds for the proofs of (b), (c) and (d).

20. Evaluate each of the following, using theorems on limits.

(a) $\lim\limits_{x \to 2} (x^2 - 6x + 4) = \lim\limits_{x \to 2} x^2 + \lim\limits_{x \to 2} (-6x) + \lim\limits_{x \to 2} 4$

$\qquad\qquad = \left(\lim\limits_{x \to 2} x\right)\left(\lim\limits_{x \to 2} x\right) + \left(\lim\limits_{x \to 2} -6\right)\left(\lim\limits_{x \to 2} x\right) + \lim\limits_{x \to 2} 4$

$\qquad\qquad = (2)(2) + (-6)(2) + 4 = -4$

In practice the intermediate steps are omitted.

(b) $\lim\limits_{x \to -1} \dfrac{(x + 3)(2x - 1)}{x^2 + 3x - 2} = \dfrac{\lim\limits_{x \to -1} (x + 3)\ \lim\limits_{x \to -1} (2x - 1)}{\lim\limits_{x \to -1} (x^2 + 3x - 2)} = \dfrac{2 \cdot (-3)}{-4} = \dfrac{3}{2}$

(c) $\lim\limits_{x \to \infty} \dfrac{2x^4 - 3x^2 + 1}{6x^4 + x^3 - 3x} = \lim\limits_{x \to \infty} \dfrac{2 - \dfrac{3}{x^2} + \dfrac{1}{x^4}}{6 + \dfrac{1}{x} - \dfrac{3}{x^3}}$

$\qquad\qquad = \dfrac{\lim\limits_{x \to \infty} 2 + \lim\limits_{x \to \infty} \dfrac{-3}{x^2} + \lim\limits_{x \to \infty} \dfrac{1}{x^4}}{\lim\limits_{x \to \infty} 6 + \lim\limits_{x \to \infty} \dfrac{1}{x} + \lim\limits_{x \to \infty} \dfrac{-3}{x^3}} = \dfrac{2}{6} = \dfrac{1}{3}$

by Problem 19.

(d) $\lim\limits_{h \to 0} \dfrac{\sqrt{4 + h} - 2}{h} = \lim\limits_{h \to 0} \dfrac{\sqrt{4 + h} - 2}{h} \cdot \dfrac{\sqrt{4 + h} + 2}{\sqrt{4 + h} + 2}$

$\qquad\qquad = \lim\limits_{h \to 0} \dfrac{4 + h - 4}{h(\sqrt{4 + h} + 2)} = \lim\limits_{h \to 0} \dfrac{1}{\sqrt{4 + h} + 2} = \dfrac{1}{2 + 2} = \dfrac{1}{4}$

(e) $\lim\limits_{x \to 0+} \dfrac{\sin x}{\sqrt{x}} = \lim\limits_{x \to 0+} \dfrac{\sin x}{x} \cdot \sqrt{x} = \lim\limits_{x \to 0+} \dfrac{\sin x}{x} \cdot \lim\limits_{x \to 0+} \sqrt{x} = 1 \cdot 0 = 0.$

Note that in (c), (d) and (e) if we use the theorems on limits indiscriminately we obtain the so called *indeterminate forms* ∞/∞ and $0/0$. To avoid such predicaments, note that in each case the form of the limit is suitably modified. For other methods of evaluating limits, see Chapter 4.

CONTINUITY

21. Prove that $f(x) = x^2$ is continuous at $x = 2$.

Method 1:

By Problem 10, $\lim\limits_{x \to 2} f(x) = f(2) = 4$ and so $f(x)$ is continuous at $x = 2$.

Method 2:

We must show that given any $\epsilon > 0$, we can find $\delta > 0$ (depending on ϵ) such that $|f(x) - f(2)| = |x^2 - 4| < \epsilon$ when $|x - 2| < \delta$. The proof patterns that given in Problem 10.

22. (a) Prove that $f(x) = \begin{cases} x\sin 1/x, & x \neq 0 \\ 5, & x = 0 \end{cases}$ is not continuous at $x = 0$. (b) Can one redefine $f(0)$ so that $f(x)$ is continuous at $x = 0$?

(a) From Problem 13, $\lim\limits_{x \to 0} f(x) = 0$. But this limit is not equal to $f(0) = 5$, so that $f(x)$ is discontinuous at $x = 0$.

(b) By redefining $f(x)$ so that $f(0) = 0$, the function becomes continuous. Because the function can be made continuous at a point simply by redefining the function at the point, we call the point a *removable discontinuity*.

23. Is the function $f(x) = \dfrac{2x^4 - 6x^3 + x^2 + 3}{x - 1}$ continuous at $x = 1$?

$f(1)$ does not exist, so that $f(x)$ is not continuous at $x = 1$. By redefining $f(x)$ so that $f(1) = \lim\limits_{x \to 1} f(x) = -8$ (see Problem 11), it becomes continuous at $x = 1$, i.e. $x = 1$ is a removable discontinuity.

24. Prove that if $f(x)$ and $g(x)$ are continuous at $x = x_0$, so also are (a) $f(x) + g(x)$, (b) $f(x)\,g(x)$, (c) $\dfrac{f(x)}{g(x)}$ if $f(x_0) \neq 0$.

These results follow at once from the proofs given in Problem 19 by taking $A = f(x_0)$ and $B = g(x_0)$ and rewriting $0 < |x - x_0| < \delta$ as $|x - x_0| < \delta$, i.e. *including* $x = x_0$.

25. Prove that $f(x) = x$ is continuous at any point $x = x_0$.

We must show that, given any $\epsilon > 0$, we can find $\delta > 0$ such that $|f(x) - f(x_0)| = |x - x_0| < \epsilon$ when $|x - x_0| < \delta$. By choosing $\delta = \epsilon$ the result follows at once.

26. Prove that $f(x) = 2x^3 + x$ is continuous at any point $x = x_0$.

Since x is continuous at any point $x = x_0$ (Problem 25) so also is $x \cdot x = x^2$, $x^2 \cdot x = x^3$, $2x^3$ and finally $2x^3 + x$, using the theorem (Problem 24) that sums and products of continuous functions are continuous.

27. Prove that if $f(x) = \sqrt{x - 5}$ for $5 \leqq x \leqq 9$, then $f(x)$ is continuous in this interval.

If x_0 is any point such that $5 < x_0 < 9$, then $\lim\limits_{x \to x_0} f(x) = \lim\limits_{x \to x_0} \sqrt{x - 5} = \sqrt{x_0 - 5} = f(x_0)$. Also, $\lim\limits_{x \to 5+} \sqrt{x - 5} = 0 = f(5)$ and $\lim\limits_{x \to 9-} \sqrt{x - 5} = 2 = f(9)$. Thus the result follows.

Here we have used the result that $\lim\limits_{x \to x_0} \sqrt{f(x)} = \sqrt{\lim\limits_{x \to x_0} f(x)} = \sqrt{f(x_0)}$ if $f(x)$ is continuous at x_0. An ϵ, δ proof, directly from the definition, can also be employed.

28. For what values of x in the domain of definition is each of the following functions continuous.

(a) $f(x) = \dfrac{x}{x^2 - 1}$ *Ans.* all x except $x = \pm 1$ (where the denominator is zero)

(b) $f(x) = \dfrac{1 + \cos x}{3 + \sin x}$ *Ans.* all x

(c) $f(x) = \dfrac{1}{\sqrt[4]{10 + x}}$ *Ans.* all $x > -10$

(d) $f(x) = 10^{-1/(x-3)^2}$ *Ans.* all $x \neq 3$ (see Problem 55)

(e) $f(x) = \begin{cases} 10^{-1/(x-3)^2}, & x \neq 3 \\ 0, & x = 3 \end{cases}$ *Ans.* all x, since $\lim\limits_{x \to 3} f(x) = f(3)$

(f) $f(x) = \dfrac{x - |x|}{x}$

If $x > 0$, $f(x) = \dfrac{x - x}{x} = 0$. If $x < 0$, $f(x) = \dfrac{x + x}{x} = 2$. At $x = 0$, $f(x)$ is undefined.

Then $f(x)$ is continuous for all x except $x = 0$.

(g) $f(x) = \begin{cases} \dfrac{x - |x|}{x}, & x < 0 \\[2mm] 2, & x = 0 \end{cases}$

As in (f), $f(x)$ is continuous for $x < 0$. Then since

$$\lim_{x \to 0-} \frac{x - |x|}{x} = \lim_{x \to 0-} \frac{x + x}{x} = \lim_{x \to 0-} 2 = 2 = f(0)$$

it follows that $f(x)$ is continuous (from the left) at $x = 0$.

Thus $f(x)$ is continuous for all $x \leqq 0$, i.e. everywhere in its domain of definition.

(h) $f(x) = x \csc x = \dfrac{x}{\sin x}$.   *Ans.* all x except $0, \pm\pi, \pm2\pi, \pm3\pi, \ldots$.

(i) $f(x) = x \csc x$, $f(0) = 1$.   Since $\lim\limits_{x \to 0} x \csc x = \lim\limits_{x \to 0} \dfrac{x}{\sin x} = 1 = f(0)$, we see that $f(x)$ is continuous for all x except $\pm\pi, \pm2\pi, \pm3\pi, \ldots$ [compare (h)].

UNIFORM CONTINUITY

29. Prove that $f(x) = x^2$ is uniformly continuous in $0 < x < 1$.

Method 1, using definition.

We must show that given any $\epsilon > 0$ we can find $\delta > 0$ such that $|x^2 - x_0^2| < \epsilon$ when $|x - x_0| < \delta$, where δ depends *only* on ϵ and *not* on x_0 where $0 < x_0 < 1$.

If x and x_0 are any points in $0 < x < 1$, then

$$|x^2 - x_0^2| = |x + x_0|\,|x - x_0| < |1 + 1|\,|x - x_0| = 2\,|x - x_0|$$

Thus if $|x - x_0| < \delta$ it follows that $|x^2 - x_0^2| < 2\delta$. Choosing $\delta = \epsilon/2$, we see that $|x^2 - x_0^2| < \epsilon$ when $|x - x_0| < \delta$, where δ depends only on ϵ and not on x_0. Hence $f(x) = x^2$ is uniformly continuous in $0 < x < 1$.

The above can also be used to prove that $f(x) = x^2$ is uniformly continuous in $0 \leqq x \leqq 1$.

Method 2:

The function $f(x) = x^2$ is continuous in the closed interval $0 \leqq x \leqq 1$. Hence by the theorem on Page 26 it is uniformly continuous in $0 \leqq x \leqq 1$ and thus in $0 < x < 1$.

30. Prove that $f(x) = 1/x$ is not uniformly continuous in $0 < x < 1$.

Method 1:

Suppose that $f(x)$ is uniformly continuous in the given interval. Then for any $\epsilon > 0$ we should be able to find δ, say between 0 and 1, such that $|f(x) - f(x_0)| < \epsilon$ when $|x - x_0| < \delta$ for all x and x_0 in the interval.

Let $x = \delta$ and $x_0 = \dfrac{\delta}{1 + \epsilon}$.   Then $|x - x_0| = \left| \delta - \dfrac{\delta}{1 + \epsilon} \right| = \dfrac{\epsilon}{1 + \epsilon} \delta < \delta$.

However, $\left| \dfrac{1}{x} - \dfrac{1}{x_0} \right| = \left| \dfrac{1}{\delta} - \dfrac{1 + \epsilon}{\delta} \right| = \dfrac{\epsilon}{\delta} > \epsilon$ (since $0 < \delta < 1$).

Thus we have a contradiction and it follows that $f(x) = 1/x$ cannot be uniformly continuous in $0 < x < 1$.

Method 2:

Let x_0 and $x_0 + \delta$ be any two points in $(0, 1)$. Then

$$\left| f(x_0) - f(x_0 + \delta) \right| = \left| \dfrac{1}{x_0} - \dfrac{1}{x_0 + \delta} \right| = \dfrac{\delta}{x_0 (x_0 + \delta)}$$

can be made larger than any positive number by choosing x_0 sufficiently close to 0. Hence the function cannot be uniformly continuous.

MISCELLANEOUS PROBLEMS

31. If $y = f(x)$ is continuous at $x = x_0$, and $z = g(y)$ is continuous at $y = y_0$ where $y_0 = f(x_0)$, prove that $z = g\{f(x)\}$ is continuous at $x = x_0$.

Let $h(x) = g\{f(x)\}$. Since by hypothesis $f(x)$ and $g(y)$ are continuous at x_0 and y_0 respectively, we have

$$\lim_{x \to x_0} f(x) \;=\; f\Big(\lim_{x \to x_0} x\Big) \;=\; f(x_0)$$

$$\lim_{y \to y_0} g(y) \;=\; g\Big(\lim_{y \to y_0} y\Big) \;=\; g(y_0) \;=\; g\{f(x_0)\}$$

Then

$$\lim_{x \to x_0} h(x) \;=\; \lim_{x \to x_0} g\{f(x)\} \;=\; g\Big\{\lim_{x \to x_0} f(x)\Big\} \;=\; g\{f(x_0)\} \;=\; h(x_0)$$

which proves that $h(x) = g\{f(x)\}$ is continuous at $x = x_0$.

32. Prove Theorem 8, Page 26.

Suppose that $f(a) < 0$ and $f(b) > 0$. Since $f(x)$ is continuous there must be an interval $(a, a+h)$, $h > 0$, for which $f(x) < 0$. The set of points $(a, a+h)$ has an upper bound and so has a least upper bound which we call c. Then $f(c) \le 0$. Now we cannot have $f(c) < 0$, because if $f(c)$ were negative we would be able to find an interval about c (including values greater than c) for which $f(x) < 0$; but since c is the least upper bound this is impossible, and so we must have $f(c) = 0$ as required.

If $f(a) > 0$ and $f(b) < 0$, a similar argument can be used.

33. (a) Given $f(x) = 2x^3 - 3x^2 + 7x - 10$, evaluate $f(1)$ and $f(2)$.   (b) Prove that $f(x) = 0$ for some real number x such that $1 < x < 2$.   (c) Show how to calculate the value of x in (b).

(a) $f(1) = 2(1)^3 - 3(1)^2 + 7(1) - 10 = -4$, $f(2) = 2(2)^3 - 3(2)^2 + 7(2) - 10 = 8$.

(b)   If $f(x)$ is continuous in $a \le x \le b$ and if $f(a)$ and $f(b)$ have opposite signs, then there is a value of x between a and b such that $f(x) = 0$ (Problem 32).

To apply this theorem we need only realize that the given polynomial is continuous in $1 \le x \le 2$, since we have already shown in (a) that $f(1) < 0$ and $f(2) > 0$. Thus there *exists* a number c between 1 and 2 such that $f(c) = 0$.

(c) $f(1.5) = 2(1.5)^3 - 3(1.5)^2 + 7(1.5) - 10 = 0.5$. Then applying the theorem of (b) again, we see that the required root lies between 1 and 1.5 and is "most likely" closer to 1.5 than to 1, since $f(1.5) = 0.5$ has a value closer to 0 than $f(1) = -4$ (this is not always a valid conclusion but is worth pursuing in practice).

Thus we consider $x = 1.4$. Since $f(1.4) = 2(1.4)^3 - 3(1.4)^2 + 7(1.4) - 10 = -0.592$, we conclude that there is a root between 1.4 and 1.5 which is most likely closer to 1.5 than to 1.4.

Continuing in this manner, we find that the root is 1.46 to 2 decimal places.

34. Prove Theorem 10, Page 26.

Given any $\epsilon > 0$, we can find x such that $M - f(x) < \epsilon$ by definition of the l.u.b. M.

Then $\dfrac{1}{M - f(x)} > \dfrac{1}{\epsilon}$, so that $\dfrac{1}{M - f(x)}$ is not bounded and hence cannot be continuous in view of Theorem 4, Page 26. However if we suppose that $f(x) \ne M$, then since $M - f(x)$ is continuous, by hypothesis, we must have $\dfrac{1}{M - f(x)}$ also continuous. In view of this contradiction, we must have $f(x) = M$ for at least one value of x in the interval.

Similarly we can show that there exists an x in the interval such that $f(x) = m$ (Problem 93).

Supplementary Problems

FUNCTIONS

35. Give the domain of definition for which each of the following functions is real and single-valued:

   (a) $\sqrt{(3-x)(2x+4)}$,　　(b) $(x-2)/(x^2-4)$,　　(c) $\sqrt{\sin 3x}$,　　(d) $\log_{10}(x^3-3x^2-4x+12)$.

   Ans. (a) $-2 \leqq x \leqq 3$,　　(b) all $x \neq \pm 2$,　　(c) $2m\pi/3 \leqq x \leqq (2m+1)\pi/3$, $m = 0, \pm 1, \pm 2, \ldots$,

       (d) $x > 3$, $-2 < x < 2$.

36. If $f(x) = \dfrac{3x+1}{x-2}$, $x \neq 2$, find: (a) $\dfrac{5f(-1) - 2f(0) + 3f(5)}{6}$;　(b) $\{f(-\frac{1}{2})\}^2$;　(c) $f(2x-3)$;

   (d) $f(x) + f(4/x)$, $x \neq 0$;　(e) $\dfrac{f(h) - f(0)}{h}$, $h \neq 0$;　(f) $f\{f(x)\}$.

   Ans. (a) $\frac{61}{18}$　(b) $\frac{1}{25}$　(c) $\dfrac{6x-8}{2x-5}$, $x \neq 0, \frac{5}{2}, 2$　(d) $\frac{5}{2}$, $x \neq 0, 2$　(e) $\dfrac{7}{2h-4}$, $h \neq 0, 2$

       (f) $\dfrac{10x+1}{x+5}$, $x \neq -5, 2$

37. If $f(x) = 2x^2$, $0 < x \leqq 2$, find (a) the l.u.b. and (b) the g.l.b. of $f(x)$.
   Determine whether $f(x)$ attains its l.u.b. and g.l.b.　　*Ans.* (a) 8, (b) 0

38. Construct a graph for each of the following functions.

   (a) $f(x) = |x|$, $-3 \leqq x \leqq 3$　　　　　(f) $\dfrac{x-[x]}{x}$ where $[x]$ = greatest integer $\leqq x$

   (b) $f(x) = 2 - \dfrac{|x|}{x}$, $-2 \leqq x \leqq 2$　　(g) $f(x) = \cosh x$

   (c) $f(x) = \begin{cases} 0, & x < 0 \\ \frac{1}{2}, & x = 0 \\ 1, & x > 0 \end{cases}$　　　　　(h) $f(x) = \dfrac{\sin x}{x}$

   (d) $f(x) = \begin{cases} -x, & -2 \leqq x \leqq 0 \\ x, & 0 \leqq x \leqq 2 \end{cases}$　　(i) $f(x) = \dfrac{x}{(x-1)(x-2)(x-3)}$

   (e) $f(x) = x^2 \sin 1/x$, $x \neq 0$　　　(j) $f(x) = \dfrac{\sin^2 x}{x^2}$

39. Construct graphs for (a) $x^2/a^2 + y^2/b^2 = 1$, (b) $x^2/a^2 - y^2/b^2 = 1$, (c) $y^2 = 2px$, and (d) $y = 2ax - x^2$, where a, b, p are given constants. If $y = f(x)$ in each of these cases, is $f(x)$ single-valued?

40. (a) From the graph of $y = \cos x$ construct the graph of $y = \cos^{-1} x$. (b) Show graphically why $\cos^{-1} x$ is a multiple-valued function. Indicate possible choices of a principal value of $\cos^{-1} x$. (c) Using the choice in (b), find $\cos^{-1}(1/2) - \cos^{-1}(-1/2)$. Does the value of this depend on the choice? Explain.

41. Work parts (a) and (b) of Problem 40 for (a) $y = \sec^{-1} x$, (b) $y = \cot^{-1} x$.

42. Given the graph for $y = f(x)$, show how to obtain the graph for $y = f(ax+b)$, where a and b are given constants. Illustrate the procedure by obtaining the graphs of
   (a) $y = \cos 3x$, (b) $y = \sin(5x + \pi/3)$, (c) $y = \tan(\pi/6 - 2x)$.

43. Construct graphs for (a) $y = e^{-|x|}$, (b) $y = \ln|x|$, (c) $y = e^{-|x|} \sin x$.

44. Using the conventional principal values on Pages 22 and 23, evaluate:

   (a) $\sin^{-1}(-\sqrt{3}/2)$　　　　　　　　(f) $\sin^{-1} x + \cos^{-1} x$, $-1 \leqq x \leqq 1$

   (b) $\tan^{-1}(1) - \tan^{-1}(-1)$　　　　　(g) $\sin^{-1}(\cos 2x)$, $0 \leqq x \leqq \pi/2$

   (c) $\cot^{-1}(1/\sqrt{3}) - \cot^{-1}(-1/\sqrt{3})$　(h) $\sin^{-1}(\cos 2x)$, $\pi/2 \leqq x \leqq 3\pi/2$

   (d) $\cosh^{-1}\sqrt{2}$　　　　　　　　　(i) $\tanh(\operatorname{csch}^{-1} 3x)$, $x \neq 0$

   (e) $e^{-\coth^{-1}(25/7)}$　　　　　　　　(j) $\cos(2\tan^{-1} x^2)$

   Ans. (a) $-\pi/3$　(c) $-\pi/3$　(e) $\frac{3}{4}$　(g) $\pi/2 - 2x$　(i) $\dfrac{|x|}{x\sqrt{9x^2+1}}$　(j) $\dfrac{1-x^4}{1+x^4}$

       (b) $\pi/2$　(d) $\ln(1+\sqrt{2})$　(f) $\pi/2$　(h) $2x - 3\pi/2$

45. Evaluate (a) $\cos\{\pi \sinh(\ln 2)\}$, (b) $\cosh^{-1}\{\coth(\ln 3)\}$.　　*Ans.* (a) $-\sqrt{2}/2$, (b) $\ln 2$

46. (a) Prove that $\tan^{-1} x + \cot^{-1} x = \pi/2$ if the conventional principal values on Page 22 are taken. (b) Is $\tan^{-1} x + \tan^{-1}(1/x) = \pi/2$ also? Explain.

47. If $f(x) = \tan^{-1} x$, prove that $f(x) + f(y) = f\left(\dfrac{x+y}{1-xy}\right)$, discussing the case $xy = 1$.

48. Prove that $\tan^{-1} a - \tan^{-1} b = \cot^{-1} b - \cot^{-1} a$.

49. Prove the identities:
(a) $1 - \tanh^2 x = \operatorname{sech}^2 x$,   (b) $\sin 3x = 3 \sin x - 4 \sin^3 x$,   (c) $\cos 3x = 4 \cos^3 x - 3 \cos x$,
(d) $\tanh \frac{1}{2} x = (\sinh x)/(1 + \cosh x)$,   (e) $\ln|\csc x - \cot x| = \ln|\tan \frac{1}{2} x|$.

50. Find the relative and absolute maxima and minima of: (a) $f(x) = (\sin x)/x$, $f(0) = 1$;   (b) $f(x) = (\sin^2 x)/x^2$, $f(0) = 1$. Discuss the cases when $f(0)$ is undefined or $f(0)$ is defined but $\ne 1$.

LIMITS

51. Evaluate the following limits, first by using the definition and then using theorems on limits.
(a) $\lim\limits_{x \to 3} (x^2 - 3x + 2)$,   (b) $\lim\limits_{x \to -1} \dfrac{1}{2x - 5}$,   (c) $\lim\limits_{x \to 2} \dfrac{x^2 - 4}{x - 2}$,   (d) $\lim\limits_{x \to 4} \dfrac{\sqrt{x} - 2}{4 - x}$,   (e) $\lim\limits_{h \to 0} \dfrac{(2+h)^4 - 16}{h}$,
(f) $\lim\limits_{x \to 1} \dfrac{\sqrt{x}}{x + 1}$.   *Ans.* (a) 2,   (b) $-\frac{1}{7}$,   (c) 4,   (d) $-\frac{1}{4}$,   (e) 32,   (f) $\frac{1}{2}$

52. Let $f(x) = \begin{cases} 3x - 1, & x < 0 \\ 0, & x = 0 \\ 2x + 5, & x > 0 \end{cases}$.   (a) Construct a graph of $f(x)$.

Evaluate (b) $\lim\limits_{x \to 2} f(x)$, (c) $\lim\limits_{x \to -3} f(x)$, (d) $\lim\limits_{x \to 0+} f(x)$, (e) $\lim\limits_{x \to 0-} f(x)$, (f) $\lim\limits_{x \to 0} f(x)$,   justifying your answer in each case.   *Ans.* (b) 9,   (c) -10,   (d) 5,   (e) -1,   (f) does not exist

53. Evaluate (a) $\lim\limits_{h \to 0+} \dfrac{f(h) - f(0+)}{h}$ and   (b) $\lim\limits_{h \to 0-} \dfrac{f(h) - f(0-)}{h}$, where $f(x)$ is the function of Prob. 52.
Ans. (a) 2,   (b) 3

54. (a) If $f(x) = x^2 \cos 1/x$, evaluate $\lim\limits_{x \to 0} f(x)$, justifying your answer.   (b) Does your answer to (a) still remain the same if we consider $f(x) = x^2 \cos 1/x$, $x \ne 0$, $f(0) = 2$? Explain.

55. Prove that $\lim\limits_{x \to 3} 10^{-1/(x-3)^2} = 0$ using the definition.

56. Let $f(x) = \dfrac{1 + 10^{-1/x}}{2 - 10^{-1/x}}$, $x \ne 0$, $f(0) = \frac{1}{2}$.   Evaluate (a) $\lim\limits_{x \to 0+} f(x)$, (b) $\lim\limits_{x \to 0-} f(x)$, (c) $\lim\limits_{x \to 0} f(x)$, justifying answers in all cases.   *Ans.* (a) $\frac{1}{2}$,   (b) -1,   (c) does not exist.

57. Find (a) $\lim\limits_{x \to 0+} \dfrac{|x|}{x}$, (b) $\lim\limits_{x \to 0-} \dfrac{|x|}{x}$.   Illustrate your answers graphically.   *Ans.* (a) 1,   (b) -1

58. If $f(x)$ is the function defined in Problem 56, does $\lim\limits_{x \to 0} f(|x|)$ exist? Explain.

59. Explain *exactly* what is meant when one writes:
(a) $\lim\limits_{x \to 3} \dfrac{2 - x}{(x - 3)^2} = -\infty$,   (b) $\lim\limits_{x \to 0+} (1 - e^{1/x}) = -\infty$,   (c) $\lim\limits_{x \to \infty} \dfrac{2x + 5}{3x - 2} = \dfrac{2}{3}$.

60. Prove that   (a) $\lim\limits_{x \to \infty} 10^{-x} = 0$,   (b) $\lim\limits_{x \to -\infty} \dfrac{\cos x}{x + \pi} = 0$.

61. Explain why   (a) $\lim\limits_{x \to \infty} \sin x$ does not exist,   (b) $\lim\limits_{x \to \infty} e^{-x} \sin x$ does exist.

62. If $f(x) = \dfrac{3x + |x|}{7x - 5|x|}$, evaluate (a) $\lim\limits_{x \to \infty} f(x)$, (b) $\lim\limits_{x \to -\infty} f(x)$, (c) $\lim\limits_{x \to 0+} f(x)$, (d) $\lim\limits_{x \to 0-} f(x)$, (e) $\lim\limits_{x \to 0} f(x)$.
Ans. (a) 2,   (b) 1/6,   (c) 2,   (d) 1/6,   (e) does not exist

63. If $[x] = $ largest integer $\le x$, evaluate (a) $\lim\limits_{x \to 2+} \{x - [x]\}$, (b) $\lim\limits_{x \to 2-} \{x - [x]\}$.   *Ans.* (a) 0,   (b) 1

64. If $\lim\limits_{x \to x_0} f(x) = A$, prove that (a) $\lim\limits_{x \to x_0} \{f(x)\}^2 = A^2$, (b) $\lim\limits_{x \to x_0} \sqrt[3]{f(x)} = \sqrt[3]{A}$.
What generalizations of these do you suspect are true? Can you prove them?

65. If $\lim\limits_{x \to x_0} f(x) = A$ and $\lim\limits_{x \to x_0} g(x) = B$, prove that
(a) $\lim\limits_{x \to x_0} \{f(x) - g(x)\} = A - B$,   (b) $\lim\limits_{x \to x_0} \{a f(x) + b g(x)\} = aA + bB$ where $a, b = $ any constants.

66. If the limits of $f(x)$, $g(x)$ and $h(x)$ are A, B and C respectively, prove that:
   (a) $\lim\limits_{x \to x_0} \{f(x) + g(x) + h(x)\} = A + B + C$,   (b) $\lim\limits_{x \to x_0} f(x)\, g(x)\, h(x) = ABC$.   Generalize these results.

67. Evaluate each of the following using the theorems on limits.

   (a) $\lim\limits_{x \to 1/2} \left\{ \dfrac{2x^2 - 1}{(3x + 2)(5x - 3)} - \dfrac{2 - 3x}{x^2 - 5x + 3} \right\}$ Ans. (a) $-8/21$

   (b) $\lim\limits_{x \to \infty} \dfrac{(3x - 1)(2x + 3)}{(5x - 3)(4x + 5)}$ (b) $3/10$

   (c) $\lim\limits_{x \to -\infty} \left(\dfrac{3x}{x - 1} - \dfrac{2x}{x + 1} \right)$ (c) 1

   (d) $\lim\limits_{x \to 1} \dfrac{1}{x - 1} \left(\dfrac{1}{x + 3} - \dfrac{2}{3x + 5} \right)$ (d) $1/32$

68. Evaluate $\lim\limits_{h \to 0} \dfrac{\sqrt[3]{8 + h} - 2}{h}$.   (Hint: Let $8 + h = x^3$).   Ans. $1/12$

69. If $\lim\limits_{x \to x_0} f(x) = A$ and $\lim\limits_{x \to x_0} g(x) = B \ne 0$, prove directly that $\lim\limits_{x \to x_0} \dfrac{f(x)}{g(x)} = \dfrac{A}{B}$.

70. Given $\lim\limits_{x \to 0} \dfrac{\sin x}{x} = 1$, evaluate:

   (a) $\lim\limits_{x \to 0} \dfrac{\sin 3x}{x}$ (c) $\lim\limits_{x \to 0} \dfrac{1 - \cos x}{x^2}$ (e) $\lim\limits_{x \to 0} \dfrac{6x - \sin 2x}{2x + 3 \sin 4x}$ (g) $\lim\limits_{x \to 0} \dfrac{1 - 2\cos x + \cos 2x}{x^2}$

   (b) $\lim\limits_{x \to 0} \dfrac{1 - \cos x}{x}$ (d) $\lim\limits_{x \to 3} (x - 3) \csc \pi x$ (f) $\lim\limits_{x \to 0} \dfrac{\cos ax - \cos bx}{x^2}$ (h) $\lim\limits_{x \to 1} \dfrac{3 \sin \pi x - \sin 3 \pi x}{x^3}$

   Ans. (a) 3, (b) 0, (c) 1/2, (d) $-1/\pi$, (e) 2/7, (f) $\frac{1}{2}(b^2 - a^2)$, (g) -1, (h) $4\pi^3$

71. If $\lim\limits_{x \to 0} \dfrac{e^x - 1}{x} = 1$, prove that:

   (a) $\lim\limits_{x \to 0} \dfrac{e^{-ax} - e^{-bx}}{x} = b - a$;   (b) $\lim\limits_{x \to 0} \dfrac{a^x - b^x}{x} = \ln \dfrac{a}{b}$, $a, b > 0$;   (c) $\lim\limits_{x \to 0} \dfrac{\tanh ax}{x} = a$.

72. Prove that $\lim\limits_{x \to x_0} f(x) = l$ if and only if $\lim\limits_{x \to x_0+} f(x) = \lim\limits_{x \to x_0-} f(x) = l$.

CONTINUITY

73. Prove that $f(x) = x^2 - 3x + 2$ is continuous at $x = 4$.

74. Prove that $f(x) = 1/x$ is continuous (a) at $x = 2$, (b) in $1 \le x \le 3$.

75. Investigate the continuity of each of the following functions at the indicated points:

   (a) $f(x) = \dfrac{\sin x}{x}$; $x \ne 0$, $f(0) = 0$; $x = 0$ (c) $f(x) = \dfrac{x^3 - 8}{x^2 - 4}$; $x \ne 2$, $f(2) = 3$; $x = 2$

   (b) $f(x) = x - |x|$; $x = 0$ (d) $f(x) = \begin{cases} \sin \pi x, & 0 < x < 1 \\ \ln x, & 1 < x < 2 \end{cases}$; $x = 1$.

   Ans. (a) discontinuous, (b) continuous, (c) continuous, (d) discontinuous

76. If $[x] = $ greatest integer $\le x$, investigate the continuity of $f(x) = x - [x]$ in the interval (a) $1 < x < 2$, (b) $1 \le x \le 2$.

77. Prove that $f(x) = x^3$ is continuous in every finite interval.

78. If $f(x)/g(x)$ and $g(x)$ are continuous at $x = x_0$, prove that $f(x)$ must be continuous at $x = x_0$.

79. Prove that $f(x) = (\tan^{-1} x)/x$, $f(0) = 1$ is continuous at $x = 0$.

80. Prove that a polynomial is continuous in every finite interval.

81. If $f(x)$ and $g(x)$ are polynomials, prove that $f(x)/g(x)$ is continuous at each point $x = x_0$ for which $g(x_0) \ne 0$.

82. Give the points of discontinuity of each of the following functions.

    (a) $f(x) = \dfrac{x}{(x-2)(x-4)}$ (c) $f(x) = \sqrt{(x-3)(6-x)}, \;\; 3 \leqq x \leqq 6$

    (b) $f(x) = x^2 \sin 1/x, \;\; x \neq 0, \; f(0) = 0$ (d) $f(x) = \dfrac{1}{1 + 2\sin x}.$

    Ans. (a) $x = 2, 4$, (b) none, (c) none, (d) $x = 7\pi/6 \pm 2m\pi, \; 11\pi/6 \pm 2m\pi, \; m = 0, 1, 2, \ldots$

UNIFORM CONTINUITY

83. Prove that $f(x) = x^3$ is uniformly continuous in (a) $0 < x < 2$, (b) $0 \leqq x \leqq 2$, (c) any finite interval.

84. Prove that $f(x) = x^2$ is not uniformly continuous in $0 < x < \infty$.

85. If a is a constant, prove that $f(x) = 1/x^2$ is (a) continuous in $a < x < \infty$ if $a \geqq 0$, (b) uniformly continuous in $a < x < \infty$ if $a > 0$, (c) not uniformly continuous in $0 < x < 1$.

86. If $f(x)$ and $g(x)$ are uniformly continuous in the same interval, prove that (a) $f(x) \pm g(x)$ and (b) $f(x)\, g(x)$ are uniformly continuous in the interval. State and prove an analogous theorem for $f(x)/g(x)$.

MISCELLANEOUS PROBLEMS

87. Give an "ϵ, δ" proof of the theorem of Problem 31.

88. (a) Prove that the equation $\tan x = x$ has a real positive root in each of the intervals $\pi/2 < x < 3\pi/2$, $3\pi/2 < x < 5\pi/2$, $5\pi/2 < x < 7\pi/2$, $\ldots$.

    (b) Illustrate the result in (a) graphically by constructing the graphs of $y = \tan x$ and $y = x$ and locating their points of intersection.

    (c) Determine the value of the smallest positive root of $\tan x = x$.   *Ans.* (c) 4.49 approximately

89. Prove that the only real solution of $\sin x = x$ is $x = 0$.

90. (a) Prove that $\cos x \cosh x + 1 = 0$ has infinitely many real roots.

    (b) Prove that for large values of x the roots approximate those of $\cos x = 0$.

91. Prove that $\displaystyle\lim_{x \to 0} \frac{x^2 \sin (1/x)}{\sin x} = 0$.

92. Suppose $f(x)$ is continuous at $x = x_0$ and assume $f(x_0) > 0$. Prove that there exists an interval $(x_0 - h, x_0 + h)$, where $h > 0$, in which $f(x) > 0$. (See Theorem 5, Page 26.) [Hint: Show that we can make $|f(x) - f(x_0)| < \frac{1}{2} f(x_0)$. Then show that $f(x) \geqq f(x_0) - |f(x) - f(x_0)| > \frac{1}{2} f(x_0) > 0$.]

93. (a) Prove Theorem 10, Page 26, for the greatest lower bound m (see Problem 34). (b) Prove Theorem 9, Page 26, and explain its relationship to Theorem 10.

Chapter 3

Sequences

DEFINITION of a SEQUENCE

A function of a positive integral variable, designated by $f(n)$ or u_n, where $n = 1, 2, 3,$ $\ldots$, is called a *sequence*. Thus a sequence is a set of numbers $u_1, u_2, u_3, \ldots$ in a definite order of arrangement (i.e. a *correspondence* with the natural numbers) and formed according to a definite rule. Each number in the sequence is called a *term*; u_n is called the *n*th *term*. The sequence is called *finite* or *infinite* according as there are or are not a finite number of terms. The sequence $u_1, u_2, u_3, \ldots$ is also designated briefly by $\{u_n\}$.

> **Examples:**
> 1. The set of numbers $2, 7, 12, 17, \ldots, 32$ is a finite sequence; the *n*th term is given by $u_n = f(n) = 2 + 5(n-1) = 5n - 3, \ n = 1, 2, \ldots, 7.$
>
> 2. The set of numbers $1, 1/3, 1/5, 1/7, \ldots$ is an infinite sequence with *n*th term $u_n = 1/(2n-1), \ n = 1, 2, 3, \ldots.$

Unless otherwise specified, we shall consider infinite sequences only.

LIMIT of a SEQUENCE

A number l is called the *limit* of an infinite sequence $u_1, u_2, u_3, \ldots$ if for any positive number ϵ we can find a positive number N depending on ϵ such that $|u_n - l| < \epsilon$ for all integers $n > N$. In such case we write $\lim\limits_{n \to \infty} u_n = l$.

> **Example:** If $u_n = 3 + 1/n = (3n+1)/n$, the sequence is $4, 7/2, 10/3, \ldots$ and we can show that $\lim\limits_{n \to \infty} u_n = 3$.

If the limit of a sequence exists, the sequence is called *convergent*; otherwise it is called *divergent*. A sequence can converge to only one limit, i.e. if a limit exists it is unique. See Problem 8.

A more intuitive but unrigorous way of expressing this concept of limit is to say that a sequence $u_1, u_2, u_3, \ldots$ has a limit l if the successive terms get "closer and closer" to l. This is often used to provide a "guess" as to the value of the limit, after which the definition is applied to see if the guess is really correct.

One should observe the similarities and differences between limits of functions and sequences. In defining $\lim\limits_{x \to \infty} f(x) = l$, the limit l is attained for *all possible approaches* to infinity. In defining $\lim\limits_{n \to \infty} f(n) = l$, the limit l need exist only along a certain approach to infinity, namely along the positive integers. Other possibilities present themselves. For example, in some cases it may be important to consider the limit of $f(x)$ as x approaches ∞ (or in fact any number x_0) along a sequence of rational numbers.

THEOREMS on LIMITS of SEQUENCES

If $\lim\limits_{n \to \infty} a_n = A$ and $\lim\limits_{n \to \infty} b_n = B$, then

1. $\lim\limits_{n \to \infty} (a_n + b_n) = \lim\limits_{n \to \infty} a_n + \lim\limits_{n \to \infty} b_n = A + B$

2. $\lim\limits_{n \to \infty} (a_n - b_n) = \lim\limits_{n \to \infty} a_n - \lim\limits_{n \to \infty} b_n = A - B$

3. $\lim_{n \to \infty} (a_n \cdot b_n) \; = \; \left(\lim_{n \to \infty} a_n\right)\left(\lim_{n \to \infty} b_n\right) \; = \; AB$

4. $\lim_{n \to \infty} \dfrac{a_n}{b_n} \; = \; \dfrac{\lim_{n \to \infty} a_n}{\lim_{n \to \infty} b_n} \; = \; \dfrac{A}{B} \qquad \text{if } \lim_{n \to \infty} b_n = B \neq 0$

> If $B = 0$ and $A \neq 0$, $\lim_{n \to \infty} \dfrac{a_n}{b_n}$ does not exist.

> If $B = 0$ and $A = 0$, $\lim_{n \to \infty} \dfrac{a_n}{b_n}$ may or may not exist.

5. $\lim_{n \to \infty} a_n^p \; = \; \left(\lim_{n \to \infty} a_n\right)^p \; = \; A^p, \qquad \text{for } p = \text{any real number if } A^p \text{ exists.}$

6. $\lim_{n \to \infty} p^{a_n} \; = \; p^{\lim_{n \to \infty} a_n} \; = \; p^A, \qquad \text{for } p = \text{any real number if } p^A \text{ exists.}$

INFINITY

We write $\lim_{n \to \infty} a_n = \infty$ if for each positive number M we can find a positive number N (depending on M) such that $a_n > M$ for all $n > N$. Similarly we write $\lim_{n \to \infty} a_n = -\infty$ if for each positive number M we can find a positive number N such that $a_n < -M$ for all $n > N$. It should be emphasized that ∞ and $-\infty$ are not numbers and the sequences are not convergent. The terminology employed merely indicates that the sequences diverge in a certain manner.

BOUNDED, MONOTONIC SEQUENCES

If $u_n \leq M$ for $n = 1, 2, 3, \ldots$, where M is a constant (independent of n), we say that the sequence $\{u_n\}$ is *bounded above* and M is called an *upper bound*. If $u_n \geq m$, the sequence is *bounded below* and m is called a *lower bound*.

If $m \leq u_n \leq M$ the sequence is called *bounded*, often indicated by $|u_n| \leq P$. Every convergent sequence is bounded, but the converse is not necessarily true.

If $u_{n+1} \geq u_n$ the sequence is called *monotonic increasing*; if $u_{n+1} > u_n$ it is called *strictly increasing*.

Similarly if $u_{n+1} \leq u_n$ the sequence is called *monotonic decreasing*, while if $u_{n+1} < u_n$ it is *strictly decreasing*.

> **Examples:** 1. The sequence $1, 1.1, 1.11, 1.111, \ldots$ is bounded and monotonic increasing. It is also strictly increasing.
>
> 2. The sequence $1, -1, 1, -1, 1, \ldots$ is bounded but not monotonic increasing or decreasing.
>
> 3. The sequence $-1, -1.5, -2, -2.5, -3, \ldots$ is monotonic decreasing and not bounded. However, it is bounded above.

The following theorem is fundamental and is related to the Weierstrass-Bolzano theorem (Chapter 1, Page 5) which is proved in Problem 23.

Theorem. Every bounded monotonic (increasing or decreasing) sequence has a limit.

LEAST UPPER BOUND and GREATEST LOWER BOUND of a SEQUENCE

A number $\underline{M}$ is called the *least upper bound* (l.u.b.) of the sequence $\{u_n\}$ if $u_n \leq \underline{M}$, $n = 1, 2, 3, \ldots$ while at least one term is greater than $\underline{M} - \epsilon$ for any $\epsilon > 0$.

A number $\overline{m}$ is called the *greatest lower bound* (g.l.b.) of the sequence $\{u_n\}$ if $u_n \geq \overline{m}$, $n = 1, 2, 3, \ldots$ while at least one term is less than $\overline{m} + \epsilon$ for any $\epsilon > 0$.

Compare with the definition of l.u.b. and g.l.b. for sets of numbers in general (see Page 5).

LIMIT SUPERIOR, LIMIT INFERIOR

A number $\bar{l}$ is called the *limit superior, greatest limit* or *upper limit* (lim sup or $\overline{\lim}$) of the sequence $\{u_n\}$ if infinitely many terms of the sequence are greater than $\bar{l} - \epsilon$ while only a finite number of terms are greater than $\bar{l} + \epsilon$, where ϵ is any positive number.

A number $\underline{l}$ is called the *limit inferior, least limit* or *lower limit* (lim inf or $\underline{\lim}$) of the sequence $\{u_n\}$ if infinitely many terms of the sequence are less than $\underline{l} + \epsilon$ while only a finite number of terms are less than $\underline{l} - \epsilon$, where ϵ is any positive number.

These correspond to least and greatest limiting points of general sets of numbers.

If infinitely many terms of $\{u_n\}$ exceed any positive number M, we define $\lim \sup \{u_n\} = \infty$. If infinitely many terms are less than $-M$, where M is any positive number, we define $\lim \inf \{u_n\} = -\infty$.

If $\lim_{n \to \infty} u_n = \infty$, we define $\lim \sup \{u_n\} = \lim \inf \{u_n\} = \infty$.

If $\lim_{n \to \infty} u_n = -\infty$, we define $\lim \sup \{u_n\} = \lim \inf \{u_n\} = -\infty$.

Although every bounded sequence is not necessarily convergent, it always has a finite lim sup and lim inf.

A sequence $\{u_n\}$ converges if and only if $\lim \sup u_n = \lim \inf u_n$ is finite.

NESTED INTERVALS

Consider a set of intervals $[a_n, b_n]$, $n = 1, 2, 3, \ldots$, where each interval is contained in the preceding one and $\lim_{n \to \infty} (a_n - b_n) = 0$. Such intervals are called *nested intervals*.

We can prove that to every set of nested intervals there corresponds one and only one real number. This can be used to establish the Weierstrass-Bolzano theorem of Chap. 1. (See Problems 22 and 23.)

CAUCHY'S CONVERGENCE CRITERION

Cauchy's convergence criterion states that a sequence $\{u_n\}$ converges if and only if for each $\epsilon > 0$ we can find a number N such that $|u_p - u_q| < \epsilon$ for all $p, q > N$. This criterion has the advantage that one need not know the limit l in order to demonstrate convergence.

INFINITE SERIES

Let $u_1, u_2, u_3, \ldots$ be a given sequence. Form a new sequence $S_1, S_2, S_3, \ldots$ where

$$S_1 = u_1, \quad S_2 = u_1 + u_2, \quad S_3 = u_1 + u_2 + u_3, \quad \ldots, \quad S_n = u_1 + u_2 + u_3 + \cdots + u_n, \quad \ldots$$

where S_n, called the nth *partial sum*, is the sum of the first n terms of the sequence $\{u_n\}$.

The sequence $S_1, S_2, S_3, \ldots$ is symbolized by

$$u_1 + u_2 + u_3 + \cdots = \sum_{n=1}^{\infty} u_n$$

which is called an *infinite series*. If $\lim_{n \to \infty} S_n = S$ exists, the series is called *convergent* and S is its *sum*, otherwise the series is called *divergent*.

Further discussion of infinite series and other topics related to sequences is given in Chapter 11.

Solved Problems

SEQUENCES

1. Write the first five terms of each of the following sequences.

(a) $\left\{\dfrac{2n-1}{3n+2}\right\}$ Ans. $\dfrac{1}{5}, \dfrac{3}{8}, \dfrac{5}{11}, \dfrac{7}{14}, \dfrac{9}{17}$

(b) $\left\{\dfrac{1-(-1)^n}{n^3}\right\}$ Ans. $\dfrac{2}{1^3}, 0, \dfrac{2}{3^3}, 0, \dfrac{2}{5^3}$

(c) $\left\{\dfrac{(-1)^{n-1}}{2 \cdot 4 \cdot 6 \cdots 2n}\right\}$ Ans. $\dfrac{1}{2}, \dfrac{-1}{2 \cdot 4}, \dfrac{1}{2 \cdot 4 \cdot 6}, \dfrac{-1}{2 \cdot 4 \cdot 6 \cdot 8}, \dfrac{1}{2 \cdot 4 \cdot 6 \cdot 8 \cdot 10}$

(d) $\left\{\dfrac{1}{2} + \dfrac{1}{4} + \dfrac{1}{8} + \cdots + \dfrac{1}{2^n}\right\}$ Ans. $\frac{1}{2}, \frac{1}{2}+\frac{1}{4}, \frac{1}{2}+\frac{1}{4}+\frac{1}{8}, \frac{1}{2}+\frac{1}{4}+\frac{1}{8}+\frac{1}{16}, \frac{1}{2}+\frac{1}{4}+\frac{1}{8}+\frac{1}{16}+\frac{1}{32}$

(e) $\left\{\dfrac{(-1)^{n-1} x^{2n-1}}{(2n-1)!}\right\}$ Ans. $\dfrac{x}{1!}, \dfrac{-x^3}{3!}, \dfrac{x^5}{5!}, \dfrac{-x^7}{7!}, \dfrac{x^9}{9!}$

   Note that $n! = 1 \cdot 2 \cdot 3 \cdot 4 \cdots n$. Thus $1! = 1$, $3! = 1 \cdot 2 \cdot 3 = 6$, $5! = 1 \cdot 2 \cdot 3 \cdot 4 \cdot 5 = 120$, etc. We define $0! = 1$.

2. Two students were asked to write an nth term for the sequence $1, 16, 81, 256, \ldots$ and to write the 5th term of the sequence. One student gave the nth term as $u_n = n^4$. The other student, who did not recognize this simple law of formation, wrote $u_n = 10n^3 - 35n^2 + 50n - 24$. Which student gave the correct 5th term?

   If $u_n = n^4$, then $u_1 = 1^4 = 1$, $u_2 = 2^4 = 16$, $u_3 = 3^4 = 81$, $u_4 = 4^4 = 256$ which agrees with the first four terms of the sequence. Hence the first student gave the 5th term as $u_5 = 5^4 = 625$.

   If $u_n = 10n^3 - 35n^2 + 50n - 24$, then $u_1 = 1$, $u_2 = 16$, $u_3 = 81$, $u_4 = 256$ which also agrees with the first four terms given. Hence the second student gave the 5th term as $u_5 = 601$.

   Both students were correct. Merely giving a finite number of terms of a sequence does not define a unique nth term. In fact an infinite number of nth terms is possible.

LIMIT of a SEQUENCE

3. A sequence has its nth term given by $u_n = \dfrac{3n-1}{4n+5}$. (a) Write the 1st, 5th, 10th, 100th, 1000th, 10,000th and 100,000th terms of the sequence in decimal form. Make a *guess* as to the limit of this sequence as $n \to \infty$. (b) Using the definition of limit verify that the guess in (a) is actually correct.

(a)

$n=1$	$n=5$	$n=10$	$n=100$	$n=1000$	$n=10,000$	$n=100,000$
.22222...	.56000...	.64444...	.73827...	.74881...	.74988...	.74998...

   A good guess is that the limit is $.75000\ldots = \frac{3}{4}$. Note that it is only for *large enough* values of n that a possible limit may become apparent.

(b) We must show that for any given $\epsilon > 0$ (no matter how small) there is a number N (depending on ϵ) such that $|u_n - \frac{3}{4}| < \epsilon$ for all $n > N$.

   Now $\left|\dfrac{3n-1}{4n+5} - \dfrac{3}{4}\right| = \left|\dfrac{-19}{4(4n+5)}\right| < \epsilon$ when $\dfrac{19}{4(4n+5)} < \epsilon$ or

   $$\dfrac{4(4n+5)}{19} > \dfrac{1}{\epsilon}, \qquad 4n+5 > \dfrac{19}{4\epsilon}, \qquad n > \dfrac{1}{4}\left(\dfrac{19}{4\epsilon} - 5\right)$$

   Choosing $N = \frac{1}{4}(19/4\epsilon - 5)$, we see that $|u_n - \frac{3}{4}| < \epsilon$ for all $n > N$, so that $\lim\limits_{n \to \infty} u_n = \frac{3}{4}$ and the proof is complete.

   Note that if $\epsilon = .001$ (for example), $N = \frac{1}{4}(19000/4 - 5) = 1186\frac{1}{4}$. This means that all terms of the sequence beyond the 1186th term differ from $\frac{3}{4}$ in absolute value by less than .001.

4. Prove that $\lim\limits_{n\to\infty} \dfrac{c}{n^p} = 0$ where $c \neq 0$ and $p > 0$ are constants (independent of n).

We must show that for any $\epsilon > 0$ there is a number N such that $|c/n^p - 0| < \epsilon$ for all $n > N$.

Now $\left|\dfrac{c}{n^p}\right| < \epsilon$ when $\dfrac{|c|}{n^p} < \epsilon$, i.e. $n^p > \dfrac{|c|}{\epsilon}$ or $n > \left(\dfrac{|c|}{\epsilon}\right)^{1/p}$. Choosing $N = \left(\dfrac{|c|}{\epsilon}\right)^{1/p}$ (depending on ϵ), we see that $|c/n^p| < \epsilon$ for all $n > N$, proving that $\lim\limits_{n\to\infty} (c/n^p) = 0$.

5. Prove that $\lim\limits_{n\to\infty} \dfrac{1 + 2\cdot 10^n}{5 + 3\cdot 10^n} = \dfrac{2}{3}$.

We must show that for any $\epsilon > 0$ there is a number N such that $\left|\dfrac{1 + 2\cdot 10^n}{5 + 3\cdot 10^n} - \dfrac{2}{3}\right| < \epsilon$ for all $n > N$.

Now $\left|\dfrac{1 + 2\cdot 10^n}{5 + 3\cdot 10^n} - \dfrac{2}{3}\right| = \left|\dfrac{-7}{3(5 + 3\cdot 10^n)}\right| < \epsilon$ when $\dfrac{7}{3(5 + 3\cdot 10^n)} < \epsilon$, i.e. when

$\dfrac{3}{7}(5 + 3\cdot 10^n) > 1/\epsilon$, $\quad 3\cdot 10^n > 7/3\epsilon - 5$, $\quad 10^n > \tfrac{1}{3}(7/3\epsilon - 5)$ or $\quad n > \log_{10}\{\tfrac{1}{3}(7/3\epsilon - 5)\} = N$,

proving the existence of N and thus establishing the required result.

Note that the above value of N is real only if $7/3\epsilon - 5 > 0$, i.e. $0 < \epsilon < 7/15$. If $\epsilon \geq 7/15$, we see that $\left|\dfrac{1 + 2\cdot 10^n}{5 + 3\cdot 10^n} - \dfrac{2}{3}\right| < \epsilon$ for *all* $n > 0$.

6. Explain exactly what is meant by the statements (a) $\lim\limits_{n\to\infty} 3^{2n-1} = \infty$, (b) $\lim\limits_{n\to\infty}(1 - 2n) = -\infty$.

(a) If for each positive number M we can find a positive number N (depending on M) such that $a_n > M$ for all $n > N$, then we write $\lim\limits_{n\to\infty} a_n = \infty$.

In this case, $3^{2n-1} > M$ when $(2n - 1)\log 3 > \log M$, i.e. $n > \tfrac{1}{2}\left(\dfrac{\log M}{\log 3} + 1\right) = N$.

(b) If for each positive number M we can find a positive number N (depending on M) such that $a_n < -M$ for all $n > N$, then we write $\lim\limits_{n\to\infty} a_n = -\infty$.

In this case, $1 - 2n < -M$ when $2n - 1 > M$ or $n > \tfrac{1}{2}(M + 1) = N$.

It should be emphasized that the use of the notations ∞ and $-\infty$ for limits does not in any way imply convergence of the given sequences, since ∞ and $-\infty$ are *not* numbers. Instead, these are notations used to describe that the sequences diverge in specific ways.

7. Prove that $\lim\limits_{n\to\infty} x^n = 0$ if $|x| < 1$.

Method 1:

We can restrict ourselves to $x \neq 0$ since if $x = 0$ the result is clearly true. Given $\epsilon > 0$, we must show that there exists N such that $|x^n| < \epsilon$ for $n > N$. Now $|x^n| = |x|^n < \epsilon$ when $n \log_{10}|x| < \log_{10}\epsilon$.

Dividing by $\log_{10}|x|$, which is negative, yields $n > \dfrac{\log_{10}\epsilon}{\log_{10}|x|} = N$, proving the required result.

Method 2:

Let $|x| = 1/(1 + p)$, where $p > 0$. By Bernoulli's inequality (Prob. 31, Chap. 1), we have

$$|x^n| = |x|^n = 1/(1 + p)^n < 1/(1 + np) < \epsilon \quad \text{for all } n > N. \quad \text{Thus } \lim\limits_{n\to\infty} x^n = 0.$$

THEOREMS on LIMITS of SEQUENCES

8. Prove that if $\lim\limits_{n\to\infty} u_n$ exists, it must be unique.

We must show that if $\lim\limits_{n\to\infty} u_n = l_1$ and $\lim\limits_{n\to\infty} u_n = l_2$, then $l_1 = l_2$.

By hypothesis, given any $\epsilon > 0$ we can find N such that

$$|u_n - l_1| < \tfrac{1}{2}\epsilon \text{ when } n > N, \quad |u_n - l_2| < \tfrac{1}{2}\epsilon \text{ when } n > N$$

Then

$$|l_1 - l_2| = |l_1 - u_n + u_n - l_2| \leqq |l_1 - u_n| + |u_n - l_2| < \tfrac{1}{2}\epsilon + \tfrac{1}{2}\epsilon = \epsilon$$

i.e. $|l_1 - l_2|$ is less than any positive ϵ (however small) and so must be zero. Thus $l_1 = l_2$.

9. If $\lim\limits_{n \to \infty} a_n = A$ and $\lim\limits_{n \to \infty} b_n = B$, prove that $\lim\limits_{n \to \infty} (a_n + b_n) = A + B$.

We must show that for any $\epsilon > 0$, we can find $N > 0$ such that $|(a_n + b_n) - (A + B)| < \epsilon$ for all $n > N$. From inequality 2, Page 3, we have

$$|(a_n + b_n) - (A + B)| = |(a_n - A) + (b_n - B)| \leqq |a_n - A| + |b_n - B| \tag{1}$$

By hypothesis, given $\epsilon > 0$ we can find N_1 and N_2 such that

$$|a_n - A| < \tfrac{1}{2}\epsilon \quad \text{for all } n > N_1 \tag{2}$$

$$|b_n - B| < \tfrac{1}{2}\epsilon \quad \text{for all } n > N_2 \tag{3}$$

Then from (1), (2) and (3),

$$|(a_n + b_n) - (A + B)| < \tfrac{1}{2}\epsilon + \tfrac{1}{2}\epsilon = \epsilon \quad \text{for all } n > N$$

where N is chosen as the larger of N_1 and N_2. Thus the required result follows.

10. Prove that a convergent sequence is bounded.

Given $\lim\limits_{n \to \infty} a_n = A$, we must show that there exists a positive number P such that $|a_n| < P$ for all n. Now

$$|a_n| = |a_n - A + A| \leqq |a_n - A| + |A|$$

But by hypothesis we can find N such that $|a_n - A| < \epsilon$ for all $n > N$, i.e.,

$$|a_n| < \epsilon + |A| \quad \text{for all } n > N$$

It follows that $|a_n| < P$ for all n if we choose P as the largest one of the numbers $a_1, a_2, \ldots, a_N$, $\epsilon + |A|$.

11. If $\lim\limits_{n \to \infty} b_n = B \neq 0$, prove there exists a number N such that $|b_n| > \tfrac{1}{2}|B|$ for all $n > N$.

Since $B = B - b_n + b_n$, we have: $\quad$ (1) $\quad |B| \leqq |B - b_n| + |b_n|$.

Now we can choose N so that $|B - b_n| = |b_n - B| < \tfrac{1}{2}|B|$ for all $n > N$, since $\lim\limits_{n \to \infty} b_n = B$ by hypothesis.

Hence from (1), $\quad |B| < \tfrac{1}{2}|B| + |b_n| \quad$ or $\quad |b_n| > \tfrac{1}{2}|B| \quad$ for all $n > N$.

12. If $\lim\limits_{n \to \infty} a_n = A$ and $\lim\limits_{n \to \infty} b_n = B$, prove that $\lim\limits_{n \to \infty} a_n b_n = AB$.

We have, using Problem 10,

$$|a_n b_n - AB| = |a_n(b_n - B) + B(a_n - A)| \leqq |a_n||b_n - B| + |B||a_n - A| \tag{1}$$

$$\leqq P|b_n - B| + (|B| + 1)|a_n - A|$$

But since $\lim\limits_{n \to \infty} a_n = A$ and $\lim\limits_{n \to \infty} b_n = B$, given any $\epsilon > 0$ we can find N_1 and N_2 such that

$$|b_n - B| < \frac{\epsilon}{2P} \text{ for all } n > N_1 \qquad |a_n - A| < \frac{\epsilon}{2(|B| + 1)} \text{ for all } n > N_2$$

Hence from (1), $|a_n b_n - AB| < \tfrac{1}{2}\epsilon + \tfrac{1}{2}\epsilon = \epsilon$ for all $n > N$, where N is the larger of N_1 and N_2. Thus the result is proved.

13. If $\lim_{n \to \infty} a_n = A$ and $\lim_{n \to \infty} b_n = B \neq 0$, prove *(a)* $\lim_{n \to \infty} \dfrac{1}{b_n} = \dfrac{1}{B}$, *(b)* $\lim_{n \to \infty} \dfrac{a_n}{b_n} = \dfrac{A}{B}$.

(a) We must show that for any given $\epsilon > 0$, we can find N such that

$$\left| \frac{1}{b_n} - \frac{1}{B} \right| = \frac{|B - b_n|}{|B||b_n|} < \epsilon \qquad \text{for all } n > N \tag{1}$$

By hypothesis, given any $\epsilon > 0$, we can find N_1 such that $|b_n - B| < \frac{1}{2}B^2\epsilon$ for all $n > N_1$. Also, since $\lim_{n \to \infty} b_n = B \neq 0$, we can find N_2 such that $|b_n| > \frac{1}{2}|B|$ for all $n > N_2$ (see Problem 11).

Then if N is the larger of N_1 and N_2, we can write *(1)* as

$$\left| \frac{1}{b_n} - \frac{1}{B} \right| = \frac{|b_n - B|}{|B||b_n|} < \frac{\frac{1}{2}B^2\epsilon}{|B| \cdot \frac{1}{2}|B|} = \epsilon \qquad \text{for all } n > N$$

and the proof is complete.

(b) From part *(a)* and Problem 12, we have

$$\lim_{n \to \infty} \frac{a_n}{b_n} = \lim_{n \to \infty} \left(a_n \cdot \frac{1}{b_n} \right) = \lim_{n \to \infty} a_n \cdot \lim_{n \to \infty} \frac{1}{b_n} = A \cdot \frac{1}{B} = \frac{A}{B}$$

This can also be proved directly (see Problem 41).

14. Evaluate each of the following, using theorems on limits.

(a) $\lim_{n \to \infty} \dfrac{3n^2 - 5n}{5n^2 + 2n - 6} = \lim_{n \to \infty} \dfrac{3 - 5/n}{5 + 2/n - 6/n^2} = \dfrac{3 + 0}{5 + 0 + 0} = \dfrac{3}{5}$

(b) $\lim_{n \to \infty} \left\{ \dfrac{n(n+2)}{n+1} - \dfrac{n^3}{n^2+1} \right\} = \lim_{n \to \infty} \left\{ \dfrac{n^3 + n^2 + 2n}{(n+1)(n^2+1)} \right\} = \lim_{n \to \infty} \left\{ \dfrac{1 + 1/n + 2/n^2}{(1 + 1/n)(1 + 1/n^2)} \right\}$

$$= \frac{1 + 0 + 0}{(1+0) \cdot (1+0)} = 1$$

(c) $\lim_{n \to \infty} (\sqrt{n+1} - \sqrt{n}) = \lim_{n \to \infty} (\sqrt{n+1} - \sqrt{n}) \dfrac{\sqrt{n+1} + \sqrt{n}}{\sqrt{n+1} + \sqrt{n}} = \lim_{n \to \infty} \dfrac{1}{\sqrt{n+1} + \sqrt{n}} = 0$

(d) $\lim_{n \to \infty} \dfrac{3n^2 + 4n}{2n - 1} = \lim_{n \to \infty} \dfrac{3 + 4/n}{2/n - 1/n^2}$

Since the limits of the numerator and denominator are 3 and 0 respectively, the limit does not exist.

Since $\dfrac{3n^2 + 4n}{2n - 1} > \dfrac{3n^2}{2n} = \dfrac{3n}{2}$ can be made larger than any positive number M by choosing $n > N$, we can write, if desired, $\lim_{n \to \infty} \dfrac{3n^2 + 4n}{2n - 1} = \infty$.

(e) $\lim_{n \to \infty} \left(\dfrac{2n - 3}{3n + 7} \right)^4 = \left(\lim_{n \to \infty} \dfrac{2 - 3/n}{3 + 7/n} \right)^4 = \left(\dfrac{2}{3} \right)^4 = \dfrac{16}{81}$

(f) $\lim_{n \to \infty} \dfrac{2n^5 - 4n^2}{3n^7 + n^3 - 10} = \lim_{n \to \infty} \dfrac{2/n^2 - 4/n^5}{3 + 1/n^4 - 10/n^7} = \dfrac{0}{3} = 0$

(g) $\lim_{n \to \infty} \dfrac{1 + 2 \cdot 10^n}{5 + 3 \cdot 10^n} = \lim_{n \to \infty} \dfrac{10^{-n} + 2}{5 \cdot 10^{-n} + 3} = \dfrac{2}{3}$　　(Compare with Prob. 5.)

BOUNDED MONOTONIC SEQUENCES

15. Prove that the sequence with nth term $u_n = \dfrac{2n - 7}{3n + 2}$ *(a)* is monotonic increasing, *(b)* is bounded above, *(c)* is bounded below, *(d)* is bounded, *(e)* has a limit.

(a) $\{u_n\}$ is monotonic increasing if $u_{n+1} \geqq u_n$, $n = 1, 2, 3, \ldots$. Now

$$\frac{2(n+1)-7}{3(n+1)+2} \geqq \frac{2n-7}{3n+2} \qquad \text{if and only if} \qquad \frac{2n-5}{3n+5} \geqq \frac{2n-7}{3n+2}$$

or $(2n-5)(3n+2) \geqq (2n-7)(3n+5)$, $6n^2-11n-10 \geqq 6n^2-11n-35$, i.e. $-10 \geqq -35$, which is true. Thus by reversal of steps in the inequalities, we see that $\{u_n\}$ is monotonic increasing. Actually, since $-10 > -35$, the sequence is strictly increasing.

(b) By writing some terms of the sequence, we may *guess* that an upper bound is 2 (for example). To *prove* this we must show that $u_n \leqq 2$. If $(2n-7)/(3n+2) \leqq 2$ then $2n-7 \leqq 6n+4$ or $-4n < 11$, which *is* true. Reversal of steps proves that 2 is an upper bound.

(c) Since this particular sequence is monotonic increasing, the first term -1 is a lower bound, i.e. $u_n \geqq -1$, $n = 1, 2, 3, \ldots$. Any number less than -1 is also a lower bound.

(d) Since the sequence has an upper and lower bound, it is bounded. Thus for example we can write $|u_n| \leqq 2$ for all n.

(e) Since every bounded monotonic (increasing or decreasing) sequence has a limit, the given sequence has a limit. In fact, $\displaystyle\lim_{n\to\infty} \frac{2n-7}{3n+2} = \lim_{n\to\infty} \frac{2-7/n}{3+2/n} = \frac{2}{3}$.

16. A sequence $\{u_n\}$ is defined by the recursion formula $u_{n+1} = \sqrt{3u_n}$, $u_1 = 1$.

(a) Prove that $\displaystyle\lim_{n\to\infty} u_n$ exists. (b) Find the limit in (a).

(a) The terms of the sequence are $u_1 = 1$, $u_2 = \sqrt{3u_1} = 3^{1/2}$, $u_3 = \sqrt{3u_2} = 3^{1/2+1/4}$, ...

The nth term is given by $u_n = 3^{1/2+1/4+\cdots+1/2^{n-1}}$ as can be proved by mathematical induction (Chapter 1).

Clearly, $u_{n+1} \geqq u_n$. Then the sequence is monotone increasing.

By Problem 14, Chapter 1, $u_n \leqq 3^1 = 3$, i.e. u_n is bounded above. Hence u_n is bounded (since a lower bound is zero).

Thus a limit exists, since the sequence is bounded and monotonic increasing.

(b) Let x = required limit. Since $\displaystyle\lim_{n\to\infty} u_{n+1} = \lim_{n\to\infty} \sqrt{3u_n}$, we have $x = \sqrt{3x}$ and $x = 3$. (The other possibility, $x = 0$, is excluded since $u_n \geqq 1$).

Another method: $\displaystyle\lim_{n\to\infty} 3^{1/2+1/4+\cdots+1/2^{n-1}} = \lim_{n\to\infty} 3^{1-1/2^n} = 3^{\lim_{n\to\infty}(1-1/2^n)} = 3^1 = 3$

17. Verify the validity of the entries in the following table.

Sequence	Bounded	Monotonic Increasing	Monotonic Decreasing	Limit Exists
$2, 1.9, 1.8, 1.7, \ldots, 2-(n-1)/10 \ldots$	No	No	Yes	No
$1, -1, 1, -1, \ldots, (-1)^{n-1}, \ldots$	Yes	No	No	No
$\frac{1}{2}, -\frac{1}{3}, \frac{1}{4}, -\frac{1}{5}, \ldots, (-1)^{n-1}/(n+1), \ldots$	Yes	No	No	Yes (0)
$.6, .66, .666, \ldots, \frac{2}{3}(1-1/10^n), \ldots$	Yes	Yes	No	Yes $(\frac{2}{3})$
$-1, +2, -3, +4, -5, \ldots, (-1)^n n, \ldots$	No	No	No	No

18. Prove that $\displaystyle\lim_{n\to\infty}\left(1+\frac{1}{n}\right)^n = e$.

By the binomial theorem, if n is a positive integer (see Problem 95, Chapter 1),

$$(1+x)^n = 1 + nx + \frac{n(n-1)}{2!}x^2 + \frac{n(n-1)(n-2)}{3!}x^3 + \cdots + \frac{n(n-1)\cdots(n-n+1)}{n!}x^n$$

Letting $x = 1/n$,

$$u_n = \left(1+\frac{1}{n}\right)^n = 1 + n\frac{1}{n} + \frac{n(n-1)}{2!}\frac{1}{n^2} + \cdots + \frac{n(n-1)\cdots(n-n+1)}{n!}\frac{1}{n^n}$$

$$= 1 + 1 + \frac{1}{2!}\left(1-\frac{1}{n}\right) + \frac{1}{3!}\left(1-\frac{1}{n}\right)\left(1-\frac{2}{n}\right)$$

$$+ \cdots + \frac{1}{n!}\left(1-\frac{1}{n}\right)\left(1-\frac{2}{n}\right)\cdots\left(1-\frac{n-1}{n}\right)$$

Since each term beyond the first two terms in the last expression is an increasing function of n, it follows that the sequence u_n is a monotonic increasing sequence.

It is also clear that

$$\left(1+\frac{1}{n}\right)^n < 1+1+\frac{1}{2!}+\frac{1}{3!}+\cdots+\frac{1}{n!} < 1+1+\frac{1}{2}+\frac{1}{2^2}+\cdots+\frac{1}{2^{n-1}} < 3$$

by Problem 14, Chapter 1.

Thus u_n is bounded and monotonic increasing, and so has a limit which we denote by e. The value of $e = 2.71828\ldots$.

19. Prove that $\lim\limits_{x\to\infty}\left(1+\dfrac{1}{x}\right)^x = e$, where $x\to\infty$ in any manner whatsoever (i.e. not necessarily along the positive integers, as in Problem 18).

If $n=$ largest integer $\leqq x$, then $n \leqq x \leqq n+1$ and $\left(1+\dfrac{1}{n+1}\right)^n \leqq \left(1+\dfrac{1}{x}\right)^x \leqq \left(1+\dfrac{1}{n}\right)^{n+1}$.

Since $\lim\limits_{n\to\infty}\left(1+\dfrac{1}{n+1}\right)^n = \lim\limits_{n\to\infty}\left(1+\dfrac{1}{n+1}\right)^{n+1} \Big/ \left(1+\dfrac{1}{n+1}\right) = e$

and $\lim\limits_{n\to\infty}\left(1+\dfrac{1}{n}\right)^{n+1} = \lim\limits_{n\to\infty}\left(1+\dfrac{1}{n}\right)^n\left(1+\dfrac{1}{n}\right) = e$

it follows that $\lim\limits_{x\to\infty}\left(1+\dfrac{1}{x}\right)^x = e$.

LEAST UPPER BOUND, GREATEST LOWER BOUND, LIMIT SUPERIOR, LIMIT INFERIOR

20. Find the (a) l.u.b., (b) g.l.b., (c) lim sup ($\overline{\lim}$), and (d) lim inf ($\underline{\lim}$) for the sequence $2, -2, 1, -1, 1, -1, 1, -1, \ldots$.

(a) l.u.b. $= 2$, since all terms are less than or equal to 2 while at least one term (the 1st) is greater than $2-\epsilon$ for any $\epsilon > 0$.

(b) g.l.b. $= -2$, since all terms are greater than or equal to -2 while at least one term (the 2nd) is less than $-2+\epsilon$ for any $\epsilon > 0$.

(c) lim sup or $\overline{\lim} = 1$, since infinitely many terms of the sequence are greater than $1-\epsilon$ for any $\epsilon > 0$ (namely all 1's in the sequence) while only a finite number of terms are greater than $1+\epsilon$ for any $\epsilon > 0$ (namely the 1st term).

(d) lim inf or $\underline{\lim} = -1$, since infinitely many terms of the sequence are less than $-1+\epsilon$ for any $\epsilon > 0$ (namely all -1's in the sequence) while only a finite number of terms are less than $-1-\epsilon$ for any $\epsilon > 0$ (namely the 2nd term).

21. Find the (a) l.u.b., (b) g.l.b., (c) lim sup ($\overline{\lim}$), and (d) lim inf ($\underline{\lim}$) for the sequences in Problem 17.

The results are shown in the following table.

Sequence	l.u.b.	g.l.b.	lim sup or $\overline{\lim}$	lim inf or $\underline{\lim}$
$2, 1.9, 1.8, 1.7, \ldots, 2-(n-1)/10 \ldots$	2	none	$-\infty$	$-\infty$
$1, -1, 1, -1, \ldots, (-1)^{n-1}, \ldots$	1	-1	1	-1
$\frac{1}{2}, -\frac{1}{3}, \frac{1}{4}, -\frac{1}{5}, \ldots, (-1)^{n-1}/(n+1), \ldots$	$\frac{1}{2}$	$-\frac{1}{3}$	0	0
$.6, .66, .666, \ldots, \frac{2}{3}(1-1/10^n), \ldots$	$\frac{2}{3}$	6	$\frac{2}{3}$	$\frac{2}{3}$
$-1, +2, -3, +4, -5, \ldots, (-1)^n n, \ldots$	none	none	$+\infty$	$-\infty$

NESTED INTERVALS

22. Prove that to every set of nested intervals $[a_n, b_n]$, $n = 1, 2, 3, \ldots$, there corresponds one and only one real number.

By definition of nested intervals, $a_{n+1} \geqq a_n$, $b_{n+1} \leqq b_n$, $n = 1, 2, 3, \ldots$ and $\lim\limits_{n \to \infty} (a_n - b_n) = 0$.

Then $a_1 \leqq a_n \leqq b_n \leqq b_1$, and the sequences $\{a_n\}$ and $\{b_n\}$ are bounded and respectively monotonic increasing and decreasing sequences and so converge to a and b.

To show that $a = b$ and thus prove the required result, we note that

$$b - a = (b - b_n) + (b_n - a_n) + (a_n - a) \qquad (1)$$
$$|b - a| \leqq |b - b_n| + |b_n - a_n| + |a_n - a| \qquad (2)$$

Now given any $\epsilon > 0$, we can find N such that for all $n > N$

$$|b - b_n| < \epsilon/3, \qquad |b_n - a_n| < \epsilon/3, \qquad |a_n - a| < \epsilon/3 \qquad (3)$$

so that from (2), $|b - a| < \epsilon$. Since ϵ is any positive number, we must have $b - a = 0$ or $a = b$.

23. Prove the Weierstrass-Bolzano theorem (see Page 5).

Suppose the given bounded infinite set is contained in the finite interval $[a, b]$. Divide this interval into two equal intervals. Then at least one of these, denoted by $[a_1, b_1]$ contains infinitely many points. Dividing $[a_1, b_1]$ into two equal intervals we obtain another interval, say $[a_2, b_2]$, containing infinitely many points. Continuing this process we obtain a set of intervals $[a_n, b_n]$, $n = 1, 2, 3, \ldots$, each interval contained in the preceding one and such that

$$b_1 - a_1 = (b - a)/2, \quad b_2 - a_2 = (b_1 - a_1)/2 = (b - a)/2^2, \quad \ldots, \quad b_n - a_n = (b - a)/2^n$$

from which we see that $\lim\limits_{n \to \infty} (b_n - a_n) = 0$.

This set of nested intervals, by Problem 22, corresponds to a real number which represents a limit point and so proves the theorem.

CAUCHY'S CONVERGENCE CRITERION

24. Prove Cauchy's convergence criterion as stated on Page 43.

Necessity. Suppose the sequence $\{u_n\}$ converges to l. Then given any $\epsilon > 0$, we can find N such that

$$|u_p - l| < \epsilon/2 \text{ for all } p > N \qquad \text{and} \qquad |u_q - l| < \epsilon/2 \text{ for all } q > N$$

Then for both $p > N$ and $q > N$, we have

$$|u_p - u_q| = |(u_p - l) + (l - u_q)| \leqq |u_p - l| + |l - u_q| < \epsilon/2 + \epsilon/2 = \epsilon$$

Sufficiency. Suppose $|u_p - u_q| < \epsilon$ for all $p, q > N$ and any $\epsilon > 0$. Then all the numbers u_N, $u_{N+1}, \ldots$ lie in a finite interval, i.e. the set is bounded and infinite. Hence by the Weierstrass-Bolzano theorem there is at least one limit point, say a.

If a is the only limit point, we have the desired proof and $\lim\limits_{n \to \infty} u_n = a$.

Suppose there are two distinct limit points, say a and b, and suppose $b > a$ (see Fig. 3-1). By definition of limit points, we have

$|u_p - a| < (b - a)/3$ for infinitely many values of p (1)
$|u_q - b| < (b - a)/3$ for infinitely many values of q (2)

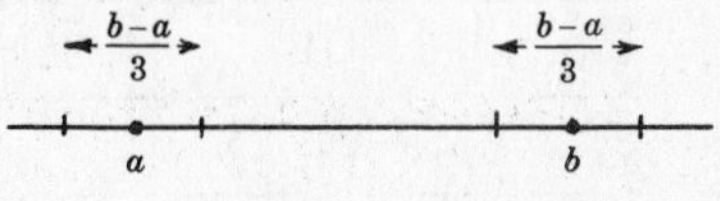

Fig. 3-1

Then since $b - a = (b - u_q) + (u_q - u_p) + (u_p - a)$, we have

$$|b - a| = b - a \leqq |b - u_q| + |u_p - u_q| + |u_p - a| \qquad (3)$$

Using (1) and (2) in (3), we see that $|u_p - u_q| > (b - a)/3$ for infinitely many values of p and q, thus contradicting the hypothesis that $|u_p - u_q| < \epsilon$ for $p, q > N$ and any $\epsilon > 0$. Hence there is only one limit point and the theorem is proved.

INFINITE SERIES

25. Prove that the infinite series (sometimes called the *geometric series*)

$$a + ar + ar^2 + \cdots = \sum_{n=1}^{\infty} ar^{n-1}$$

(a) converges to $a/(1-r)$ if $|r| < 1$, *(b)* diverges if $|r| \geqq 1$.

Let
$$S_n = a + ar + ar^2 + \cdots + ar^{n-1}$$

Then
$$rS_n = ar + ar^2 + \cdots + ar^{n-1} + ar^n$$

Subtract,
$$(1-r)S_n = a \qquad\qquad\qquad - ar^n$$

or
$$S_n = \frac{a(1-r^n)}{1-r}$$

(a) If $|r| < 1$, $\lim_{n \to \infty} S_n = \lim_{n \to \infty} \dfrac{a(1-r^n)}{1-r} = \dfrac{a}{1-r}$ by Problem 7.

(b) If $|r| > 1$, $\lim_{n \to \infty} S_n$ does not exist (see Problem 44).

26. Prove that if a series converges, its nth term must necessarily approach zero.

Since $S_n = u_1 + u_2 + \cdots + u_n$, $S_{n-1} = u_1 + u_2 + \cdots + u_{n-1}$ we have $u_n = S_n - S_{n-1}$.
If the series converges to S, then

$$\lim_{n \to \infty} u_n = \lim_{n \to \infty} (S_n - S_{n-1}) = \lim_{n \to \infty} S_n - \lim_{n \to \infty} S_{n-1} = S - S = 0$$

27. Prove that the series $1 - 1 + 1 - 1 + 1 - 1 + \ldots = \sum_{n=1}^{\infty} (-1)^{n-1}$ diverges.

Method 1:

$\lim_{n \to \infty} (-1)^n \neq 0$, in fact it doesn't exist. Then by Problem 26 the series cannot converge, i.e. it diverges.

Method 2:

The sequence of partial sums is $1, 1-1, 1-1+1, 1-1+1-1, \ldots$ i.e. $1, 0, 1, 0, 1, 0, 1, \ldots$. Since this sequence has no limit, the series diverges.

MISCELLANEOUS PROBLEMS

28. If $\lim_{n \to \infty} u_n = l$, prove that $\lim_{n \to \infty} \dfrac{u_1 + u_2 + \cdots + u_n}{n} = l$.

Let $u_n = v_n + l$. We must show that $\lim_{n \to \infty} \dfrac{v_1 + v_2 + \cdots + v_n}{n} = 0$ if $\lim_{n \to \infty} v_n = 0$. Now

$$\frac{v_1 + v_2 + \cdots + v_n}{n} = \frac{v_1 + v_2 + \cdots + v_P}{n} + \frac{v_{P+1} + v_{P+2} + \cdots + v_n}{n}.$$

so that

$$\left| \frac{v_1 + v_2 + \cdots + v_n}{n} \right| \leqq \frac{|v_1 + v_2 + \cdots + v_P|}{n} + \frac{|v_{P+1}| + |v_{P+2}| + \cdots + |v_n|}{n} \tag{1}$$

Since $\lim_{n \to \infty} v_n = 0$, we can choose P so that $|v_n| < \epsilon/2$ for $n > P$. Then

$$\frac{|v_{P+1}| + |v_{P+2}| + \cdots + |v_n|}{n} < \frac{\epsilon/2 + \epsilon/2 + \cdots + \epsilon/2}{n} = \frac{(n-P)\epsilon/2}{n} < \frac{\epsilon}{2} \tag{2}$$

After choosing P we can choose N so that for $n > N > P$,

$$\frac{|v_1 + v_2 + \cdots + v_P|}{n} < \frac{\epsilon}{2} \tag{3}$$

Then using *(2)* and *(3)*, *(1)* becomes

$$\left| \frac{v_1 + v_2 + \cdots + v_n}{n} \right| < \frac{\epsilon}{2} + \frac{\epsilon}{2} = \epsilon \quad \text{for } n > N$$

thus proving the required result.

29. Prove that $\lim\limits_{n \to \infty} (1 + n + n^2)^{1/n} = 1$.

Let $(1 + n + n^2)^{1/n} = 1 + u_n$ where $u_n \geqq 0$. Now by the binomial theorem,

$$1 + n + n^2 = (1 + u_n)^n = 1 + nu_n + \frac{n(n-1)}{2!} u_n^2 + \frac{n(n-1)(n-2)}{3!} u_n^3 + \cdots + u_n^n$$

Then $1 + n + n^2 > 1 + \dfrac{n(n-1)(n-2)}{3!} u_n^3$ or $0 < u_n^3 < \dfrac{6(n^2 + n)}{n(n-1)(n-2)}$.

Hence $\lim\limits_{n \to \infty} u_n^3 = 0$ and $\lim\limits_{n \to \infty} u_n = 0$. Thus $\lim\limits_{n \to \infty} (1 + n + n^2)^{1/n} = \lim\limits_{n \to \infty} (1 + u_n) = 1$.

30. Prove that $\lim\limits_{n \to \infty} \dfrac{a^n}{n!} = 0$ for all constants a.

The result follows if we can prove that $\lim\limits_{n \to \infty} \dfrac{|a|^n}{n!} = 0$ (see Problem 39). We can assume $a \neq 0$.

Let $u_n = \dfrac{|a|^n}{n!}$. Then $\dfrac{u_n}{u_{n-1}} = \dfrac{|a|}{n}$. If n is large enough, say $n > 2|a|$, and if we call $N = [2|a| + 1]$, i.e. the greatest integer $\leqq 2|a| + 1$, then

$$\frac{u_{N+1}}{u_N} < \frac{1}{2}, \quad \frac{u_{N+2}}{u_{N+1}} < \frac{1}{2}, \quad \cdots, \quad \frac{u_n}{u_{n-1}} < \frac{1}{2}$$

Multiplying these inequalities yields $\dfrac{u_n}{u_N} < (\tfrac{1}{2})^{n-N}$ or $u_n < (\tfrac{1}{2})^{n-N} u_N$.

Since $\lim\limits_{n \to \infty} (\tfrac{1}{2})^{n-N} = 0$ (using Problem 7), it follows that $\lim\limits_{n \to \infty} u_n = 0$.

31. The expression $a_1 + \cfrac{1}{a_2 + \cfrac{1}{a_3 + \cdots}}$ indicated briefly by $a_1 + \dfrac{1}{a_2 +}\dfrac{1}{a_3 +} \cdots$, where

$a_1, a_2, \ldots$ are positive integers, is an example of a *continued fraction*. Its value is defined as the limit of the sequence $a_1, \quad a_1 + \dfrac{1}{a_2}, \quad a_1 + \cfrac{1}{a_2 + \cfrac{1}{a_3}}, \quad \ldots$ when this limit

exists, and the continued fraction is said to *converge* to this limit. The successive terms of the sequence are called the successive *convergents* of the continued fraction. In case the constants $a_1, a_2, \ldots$ repeat after some point, the continued fraction is called *recurring*. Given the recurring continued fraction

$$2 + \frac{1}{2+}\frac{1}{2+}\frac{1}{2+} \cdots$$

(*a*) Find the first ten convergents and guess at a possible limit. (*b*) Assuming that the limit exists, find its value.

(*a*)   The first convergent $= 2$

The second convergent $= 2 + 1/2 = 5/2 = 2.5$

The third convergent $= 2 + \dfrac{1}{2 + 1/2} = 2 + \dfrac{1}{5/2} = \dfrac{12}{5} = 2.4$

The fourth convergent $= 2 + \dfrac{1}{12/5} = \dfrac{29}{12} = 2.4166\ldots$

The fifth convergent $= 2 + \dfrac{1}{29/12} = \dfrac{70}{29} = 2.4137\ldots$

Similarly, we find for the sixth through tenth convergents respectively the values

$\dfrac{169}{70} = 2.4140\ldots, \quad \dfrac{408}{169} = 2.4142\ldots, \quad \dfrac{985}{408} = 2.4142\ldots, \quad \dfrac{2378}{985} = 2.4142\ldots, \quad \dfrac{5741}{2378} = 2.4142\ldots$

From the results it is reasonable to guess that the required limit accurate to four decimal places is 2.4142.

It is of interest to note that if P_n/Q_n and P_{n+1}/Q_{n+1} are the nth and $(n+1)$st convergents respectively, then the $(n+2)$nd convergent is

$$\frac{P_{n+2}}{Q_{n+2}} = \frac{2P_{n+1} + P_n}{2Q_{n+1} + Q_n}$$

For the general result in the case of any continued fraction, see Problem 75(a).

(b) Assume the limit to be given by x. Then clearly we must have $x = 2 + 1/x$. Thus $x^2 - 2x - 1 = 0$ or $x = 1 \pm \sqrt{2}$. Since the limit cannot be negative, it must be $1 + \sqrt{2}$. This agrees with the guess in (a), since $\sqrt{2} = 1.4142$ approximately.

Note that this continued fraction can be defined by the recursion formula

$$u_{n+1} = 2 + 1/u_n, \quad u_1 = 2$$

and if $\lim_{n \to \infty} u_n = x$ exists, this yields $x = 2 + 1/x$ as above.

Supplementary Problems

SEQUENCES

32. Write the first four terms of each of the following sequences:

(a) $\left\{\dfrac{\sqrt{n}}{n+1}\right\}$, (b) $\left\{\dfrac{(-1)^{n+1}}{n!}\right\}$, (c) $\left\{\dfrac{(2x)^{n-1}}{(2n-1)^5}\right\}$, (d) $\left\{\dfrac{(-1)^n x^{2n-1}}{1 \cdot 3 \cdot 5 \cdots (2n-1)}\right\}$, (e) $\left\{\dfrac{\cos nx}{x^2 + n^2}\right\}$.

Ans. (a) $\dfrac{\sqrt{1}}{2}, \dfrac{\sqrt{2}}{3}, \dfrac{\sqrt{3}}{4}, \dfrac{\sqrt{4}}{5}$ (c) $\dfrac{1}{1^5}, \dfrac{2x}{3^5}, \dfrac{4x^2}{5^5}, \dfrac{8x^3}{7^5}$ (e) $\dfrac{\cos x}{x^2 + 1^2}, \dfrac{\cos 2x}{x^2 + 2^2}, \dfrac{\cos 3x}{x^2 + 3^2}, \dfrac{\cos 4x}{x^2 + 4^2}$

(b) $\dfrac{1}{1!}, -\dfrac{1}{2!}, \dfrac{1}{3!}, -\dfrac{1}{4!}$ (d) $\dfrac{-x}{1}, \dfrac{x^3}{1 \cdot 3}, \dfrac{-x^5}{1 \cdot 3 \cdot 5}, \dfrac{x^7}{1 \cdot 3 \cdot 5 \cdot 7}$

33. Find a possible nth term for the sequences whose first 5 terms are indicated and find the 6th term:

(a) $\dfrac{-1}{5}, \dfrac{3}{8}, \dfrac{-5}{11}, \dfrac{7}{14}, \dfrac{-9}{17}, \ldots$ (b) $1, 0, 1, 0, 1, \ldots$ (c) $\dfrac{2}{3}, 0, \dfrac{3}{4}, 0, \dfrac{4}{5}, \ldots$

Ans. (a) $\dfrac{(-1)^n(2n-1)}{(3n+2)}$ (b) $\dfrac{1-(-1)^n}{2}$ (c) $\dfrac{(n+3)}{(n+5)} \cdot \dfrac{1-(-1)^n}{2}$

34. The *Fibonacci sequence* is the sequence $\{u_n\}$ where $u_{n+2} = u_{n+1} + u_n$ and $u_1 = 1$, $u_2 = 1$. (a) Find the first 6 terms of the sequence. (b) Show that the nth term is given by $u_n = (a^n - b^n)/\sqrt{5}$ where $a = \frac{1}{2}(1 + \sqrt{5})$, $b = \frac{1}{2}(1 - \sqrt{5})$. *Ans.* (a) $1, 1, 2, 3, 5, 8$

LIMITS of SEQUENCES

35. Using the definition of limit, prove that:

(a) $\lim_{n \to \infty} \dfrac{4 - 2n}{3n + 2} = \dfrac{-2}{3}$, (b) $\lim_{n \to \infty} 2^{-1/\sqrt{n}} = 1$, (c) $\lim_{n \to \infty} \dfrac{n^4 + 1}{n^2} = \infty$, (d) $\lim_{n \to \infty} \dfrac{\sin n}{n} = 0$.

36. Find the least positive integer N such that $|(3n+2)/(n-1) - 3| < \epsilon$ for all $n > N$ if (a) $\epsilon = .01$, (b) $\epsilon = .001$, (c) $\epsilon = .0001$. $Ans.$ (a) 502, (b) 5002, (c) 50,002

37. Using the definition of limit, prove that $\lim_{n \to \infty} (2n-1)/(3n+4)$ cannot be $\frac{1}{2}$.

38. Prove that $\lim_{n \to \infty} (-1)^n n$ does not exist.

39. Prove that if $\lim_{n \to \infty} |u_n| = 0$ then $\lim_{n \to \infty} u_n = 0$. Is the converse true?

40. If $\lim_{n \to \infty} u_n = l$, prove that (a) $\lim_{n \to \infty} cu_n = cl$ where c is any constant, (b) $\lim_{n \to \infty} u_n^2 = l^2$, (c) $\lim_{n \to \infty} u_n^p = l^p$ where p is a positive integer, (d) $\lim_{n \to \infty} \sqrt{u_n} = \sqrt{l}$, $l \geqq 0$.

41. Give a direct proof that $\lim_{n \to \infty} a_n/b_n = A/B$ if $\lim_{n \to \infty} a_n = A$ and $\lim_{n \to \infty} b_n = B \neq 0$.

42. Prove that (a) $\lim_{n \to \infty} 3^{1/n} = 1$, (b) $\lim_{n \to \infty} (\frac{2}{3})^{1/n} = 1$, (c) $\lim_{n \to \infty} (\frac{3}{4})^n = 0$.

43. If $r > 1$, prove that $\lim_{n \to \infty} r^n = \infty$, carefully explaining the significance of this statement.

44. If $|r| > 1$, prove that $\lim_{n \to \infty} r^n$ does not exist.

45. Evaluate each of the following, using theorems on limits.

(a) $\lim_{n \to \infty} \dfrac{4 - 2n - 3n^2}{2n^2 + n}$ (c) $\lim_{n \to \infty} \dfrac{\sqrt{3n^2 - 5n + 4}}{2n - 7}$ (e) $\lim_{n \to \infty} (\sqrt{n^2 + n} - n)$

(b) $\lim_{n \to \infty} \sqrt[3]{\dfrac{(3 - \sqrt{n})(\sqrt{n} + 2)}{8n - 4}}$ (d) $\lim_{n \to \infty} \dfrac{4 \cdot 10^n - 3 \cdot 10^{2n}}{3 \cdot 10^{n-1} + 2 \cdot 10^{2n-1}}$ (f) $\lim_{n \to \infty} (2^n + 3^n)^{1/n}$

$Ans.$ (a) $-3/2$, (b) $-1/2$, (c) $\sqrt{3}/2$, (d) -15, (e) $1/2$, (f) 3

BOUNDED MONOTONIC SEQUENCES

46. Prove that the sequence with nth term $u_n = \sqrt{n}/(n+1)$ (a) is monotonic decreasing, (b) is bounded below, (c) is bounded above, (d) has a limit.

47. If $u_n = \dfrac{1}{1+n} + \dfrac{1}{2+n} + \dfrac{1}{3+n} + \cdots + \dfrac{1}{n+n}$, prove that $\lim_{n \to \infty} u_n$ exists and lies between 0 and 1.

48. If $u_{n+1} = \sqrt{u_n + 1}$, $u_1 = 1$, prove that $\lim_{n \to \infty} u_n = \frac{1}{2}(1 + \sqrt{5})$.

49. If $u_{n+1} = \frac{1}{2}(u_n + p/u_n)$ where $p > 0$ and $u_1 > 0$, prove that $\lim_{n \to \infty} u_n = \sqrt{p}$. Show how this can be used to determine $\sqrt{2}$.

50. If u_n is monotonic increasing (or monotonic decreasing) prove that S_n/n, where $S_n = u_1 + u_2 + \cdots + u_n$, is also monotonic increasing (or monotonic decreasing).

LEAST UPPER BOUND, GREATEST LOWER BOUND, LIMIT SUPERIOR, LIMIT INFERIOR

51. Find the l.u.b., g.l.b., lim sup $(\overline{\lim})$, lim inf $(\underline{\lim})$ for each sequence:

(a) $-1, \frac{1}{3}, -\frac{1}{5}, \frac{1}{7}, \ldots, (-1)^n/(2n-1), \ldots$ (c) $1, -3, 5, -7, \ldots, (-1)^{n-1}(2n-1), \ldots$

(b) $\frac{2}{3}, -\frac{3}{4}, \frac{4}{5}, -\frac{5}{6}, \ldots, (-1)^{n+1}(n+1)/(n+2), \ldots$ (d) $1, 4, 1, 16, 1, 36, \ldots, n^{1+(-1)^n}, \ldots$

$Ans.$ (a) $\frac{1}{3}, -1, 0, 0$ (b) $1, -1, 1, -1$ (c) none, none, $+\infty, -\infty$ (d) none, 1, $+\infty$, 1

52. Prove that a bounded sequence $\{u_n\}$ is convergent if and only if $\overline{\lim}\, u_n = \underline{\lim}\, u_n$.

INFINITE SERIES

53. Find the sum of the series $\sum_{n=1}^{\infty} (\frac{2}{3})^n$.　　　*Ans.* 2

54. Evaluate $\sum_{n=1}^{\infty} (-1)^{n-1}/5^n$.　　　*Ans.* $\frac{1}{6}$

55. Prove that $\dfrac{1}{1 \cdot 2} + \dfrac{1}{2 \cdot 3} + \dfrac{1}{3 \cdot 4} + \dfrac{1}{4 \cdot 5} + \cdots = \sum_{n=1}^{\infty} \dfrac{1}{n(n+1)} = 1$.　[Hint: $\dfrac{1}{n(n+1)} = \dfrac{1}{n} - \dfrac{1}{n+1}$]

56. Prove that multiplication of each term of an infinite series by a constant (not zero) does not affect the convergence or divergence.

57. Prove that the series $1 + \dfrac{1}{2} + \dfrac{1}{3} + \cdots + \dfrac{1}{n} + \ldots$ diverges.　[Hint: Let $S_n = 1 + \dfrac{1}{2} + \dfrac{1}{3} + \cdots + \dfrac{1}{n}$. Then prove that $|S_{2n} - S_n| > \frac{1}{2}$, giving a contradiction with Cauchy's convergence criterion.]

MISCELLANEOUS PROBLEMS

58. If $a_n \leqq u_n \leqq b_n$ for all $n > N$, and $\lim_{n \to \infty} a_n = \lim_{n \to \infty} b_n = l$, prove that $\lim_{n \to \infty} u_n = l$.

59. If $\lim_{n \to \infty} a_n = \lim_{n \to \infty} b_n = 0$, and θ is independent of n, prove that $\lim_{n \to \infty} (a_n \cos n\theta + b_n \sin n\theta) = 0$. Is the result true when θ depends on n?

60. Let $u_n = \frac{1}{2}\{1 + (-1)^n\}$, $n = 1, 2, 3, \ldots$. If $S_n = u_1 + u_2 + \cdots + u_n$, prove that $\lim_{n \to \infty} S_n/n = \frac{1}{2}$.

61. Prove that (a) $\lim_{n \to \infty} n^{1/n} = 1$, (b) $\lim_{n \to \infty} (a+n)^{p/n} = 1$ where a and p are constants.

62. If $\lim_{n \to \infty} |u_{n+1}/u_n| = |a| < 1$, prove that $\lim_{n \to \infty} u_n = 0$.

63. If $|a| < 1$, prove that $\lim_{n \to \infty} n^p a^n = 0$ where the constant $p > 0$.

64. Prove that $\lim \dfrac{2^n n!}{n^n} = 0$.

65. Prove that $\lim_{n \to \infty} n \sin 1/n = 1$.

66. If $\{u_n\}$ is the Fibonacci sequence (Problem 34), prove that $\lim_{n \to \infty} u_{n+1}/u_n = \frac{1}{2}(1 + \sqrt{5})$.

67. Prove that the sequence $u_n = (1 + 1/n)^{n+1}$, $n = 1, 2, 3, \ldots$ is a monotonic decreasing sequence whose limit is e. [Hint: Show that $u_n/u_{n-1} \leqq 1$.]

68. If $a_n \geqq b_n$ for all $n > N$ and $\lim_{n \to \infty} a_n = A$, $\lim_{n \to \infty} b_n = B$, prove that $A \geqq B$.

69. If $|u_n| \leqq |v_n|$ and $\lim_{n \to \infty} v_n = 0$, prove that $\lim_{n \to \infty} u_n = 0$.

70. Prove that $\lim_{n \to \infty} \dfrac{1}{n}\left(1 + \dfrac{1}{2} + \dfrac{1}{3} + \cdots + \dfrac{1}{n}\right) = 0$.

71. Prove that $[a_n, b_n]$, where $a_n = (1 + 1/n)^n$ and $b_n = (1 + 1/n)^{n+1}$, is a set of nested intervals defining the number e.

72. Prove that every bounded monotonic (increasing or decreasing) sequence has a limit.

73. Verify the values of each of the following continued fractions.

(a) $3 + \dfrac{1}{2+} \dfrac{1}{3+} + \dfrac{1}{2+} \cdots = \frac{1}{2}(3 + \sqrt{15})$　　　(c) $a + \dfrac{1}{b+} \dfrac{1}{a+} \dfrac{1}{b+} \cdots = \dfrac{a}{2} + \sqrt{\dfrac{a^2}{4} + \dfrac{a}{b}}$

(b) $a + \dfrac{1}{a+} \dfrac{1}{a+} \dfrac{1}{a+} \cdots = \frac{1}{2}(a + \sqrt{a^2 + 4})$　　　(d) $\dfrac{1}{2-} \dfrac{1}{2-} \dfrac{1}{2-} \dfrac{1}{2-} \cdots = 1$

74. Express (a) 174/251, (b) $\sqrt{3}$, (c) $\sqrt{6}$, and (d) 3.14159 as continued fractions.

Ans. (a) $\dfrac{1}{1+}\ \dfrac{1}{2+}\ \dfrac{1}{3+}\ \dfrac{1}{1+}\ \dfrac{1}{5+}\ \dfrac{1}{1+}\ \dfrac{1}{2}$ ⠀⠀⠀⠀⠀⠀⠀⠀ (c) $2\ +\ \dfrac{1}{2+}\ \dfrac{1}{4+}\ \dfrac{1}{2+}\ \dfrac{1}{4+}\cdots$

⠀⠀⠀ (b) $1\ +\ \dfrac{1}{1+}\ \dfrac{1}{2+}\ \dfrac{1}{1+}\ \dfrac{1}{2+}\cdots$ ⠀⠀⠀⠀⠀⠀⠀ (d) $3\ +\ \dfrac{1}{7+}\ \dfrac{1}{15+}\ \dfrac{1}{1+}\ \dfrac{1}{25+}\ \dfrac{1}{7+}\ \dfrac{1}{4}$

[Hint: In (b) add and subtract the greatest integer less than $\sqrt{3}$ (namely 1) to obtain

$$\sqrt{3}\ =\ 1 + (\sqrt{3}-1)\ =\ 1 + \frac{1}{1/(\sqrt{3}-1)}\ =\ 1 + \frac{1}{(\sqrt{3}+1)/2}$$

Then add and subtract the greatest integer in $(\sqrt{3}+1)/2$ (namely 1) to obtain

$$(\sqrt{3}+1)/2\ =\ 1 + (\sqrt{3}-1)/2\ =\ 1 + \frac{1}{2/(\sqrt{3}-1)}\ =\ 1 + \frac{1}{\sqrt{3}+1}$$

Then add and subtract the greatest integer in $\sqrt{3}+1$ (namely 2) to obtain

$$\sqrt{3}\ +\ 1\ =\ 2 + (\sqrt{3}-1)\ =\ 2 + \frac{1}{1/(\sqrt{3}-1)}\ =\ 2 + \frac{1}{(\sqrt{3}+1)/2}$$

after which repetition occurs.]

75. Given the continued fraction $a_1 + \dfrac{1}{a_2+}\ \dfrac{1}{a_3+}\ \dfrac{1}{a_4+}\ \cdots,\quad a_n > 0,$ whose nth convergent is P_n/Q_n, prove each of the following and illustrate by means of examples.

(a) $P_n\ =\ a_n P_{n-1} + P_{n-2},\quad Q_n\ =\ a_n Q_{n-1} + Q_{n-2}$

(b) $P_n Q_{n-1} - P_{n-1} Q_n\ =\ (-1)^{n-1}$

(c) The successive convergents are alternately less than and greater than the continued fraction.

(d) The convergents of odd order are less than the continued fraction but are increasing; the convergents of even order are greater than the continued fraction but are decreasing.

(e) The continued fraction always converges.

76. (a) Prove that if P_n/Q_n and P_{n+1}/Q_{n+1} are two successive convergents to the continued fraction in Problem 75, then $\left| \dfrac{P_{n+1}}{Q_{n+1}} - \dfrac{P_n}{Q_n} \right| \leqq \dfrac{1}{a_{n+1}Q_n^2} \leqq \dfrac{1}{Q_n^2}.$ (b) Find the first convergent to $\sqrt{3}$ which is accurate to two decimal places.   Ans. (b) 26/15

77. Let $\{u_n\}$ be a sequence such that $u_{n+2}\ =\ a\,u_{n+1} + b\,u_n$ where a and b are constants. This is called a second order difference equation for u_n. (a) Assuming a solution of the form $u_n = r^n$ where r is a constant, prove that r must satisfy the equation $r^2 - ar - b = 0$. (b) Use (a) to show that a solution of the difference equation (called a general solution) is $u_n = Ar_1^n + Br_2^n$, where A and B are arbitrary constants and r_1 and r_2 are the two solutions of $r^2 - ar - b = 0$ assumed different. (c) In case $r_1 = r_2$ in (b), show that a (general) solution is $u_n = (A + Bn)r_1^n$.

78. Solve the following difference equations subject to the given conditions: (a) $u_{n+2} = u_{n+1} + u_n$, $u_1 = 1$, $u_2 = 1$ (compare Prob. 34); (b) $u_{n+2} = 2u_{n+1} + 3u_n$, $u_1 = 3$, $u_2 = 5$; (c) $u_{n+2} = 4u_{n+1} - 4u_n$, $u_1 = 2$, $u_2 = 8$. Ans. (a) Same as in Prob. 34,   (b) $u_n = 2(3)^{n-1} + (-1)^{n-1}$ (c) $u_n = n \cdot 2^n$

79. (a) Prove that the nth convergent to the continued fraction $1 + \dfrac{1}{1+}\ \dfrac{1}{1+}\ \cdots$ is

$$\frac{1}{2}\left\{ \frac{(1+\sqrt{5})^{n+1} - (1-\sqrt{5})^{n+1}}{(1+\sqrt{5})^n - (1-\sqrt{5})^n} \right\}$$

⠀⠀⠀ [Hint: Use Prob. 34.]

(b) By taking the limits as $n \to \infty$ in (a), find the value of the continued fraction.

80. Work Problems 73(a) − (d) by first finding the nth convergent.

Chapter 4

Derivatives

DEFINITION of a DERIVATIVE

Let $f(x)$ be defined at any point x_0 in (a, b). The derivative of $f(x)$ at $x = x_0$ is defined as

$$f'(x_0) = \lim_{h \to 0} \frac{f(x_0 + h) - f(x_0)}{h} \tag{1}$$

if this limit exists.

The derivative can also be defined in various other equivalent ways; for example,

$$f'(x_0) = \lim_{x \to x_0} \frac{f(x) - f(x_0)}{x - x_0} = \lim_{\Delta x \to 0} \frac{f(x_0 + \Delta x) - f(x_0)}{\Delta x} \tag{2}$$

A function is called *differentiable* at a point $x = x_0$ if it has a derivative at this point, i.e. if $f'(x_0)$ exists. If $f(x)$ is differentiable at $x = x_0$ it must be continuous there. However, the converse is not necessarily true (see Problems 3 and 4).

RIGHT and LEFT HAND DERIVATIVES

The *right hand derivative* of $f(x)$ at $x = x_0$ is defined as

$$f'_+(x_0) = \lim_{h \to 0+} \frac{f(x_0 + h) - f(x_0)}{h} \tag{3}$$

if this limit exists. Note that in this case $h (= \Delta x)$ is restricted only to positive values as it approaches zero.

Similarly, the *left hand derivative* of $f(x)$ at $x = x_0$ is defined as

$$f'_-(x_0) = \lim_{h \to 0-} \frac{f(x_0 + h) - f(x_0)}{h} \tag{4}$$

if this limit exists. In this case h is restricted to negative values as it approaches zero.

A function $f(x)$ has a derivative at $x = x_0$ if and only if $f'_+(x_0) = f'_-(x_0)$.

DIFFERENTIABILITY in an INTERVAL

If a function has a derivative at all points of an interval, it is said to be *differentiable in the interval*. In particular if $f(x)$ is defined in the closed interval $a \leq x \leq b$, i.e. $[a, b]$, then $f(x)$ is differentiable in the interval if and only if $f'(x_0)$ exists for each x_0 such that $a < x_0 < b$ and if $f'_+(a)$ and $f'_-(b)$ both exist.

If a function has a continuous derivative, it is sometimes called *continuously differentiable*.

SECTIONAL DIFFERENTIABILITY

A function is called *sectionally or piecewise differentiable* or *sectionally or piecewise smooth* in an interval $a \leq x \leq b$ if $f'(x)$ is sectionally continuous. An example of a sectionally continuous function is shown graphically on Page 26.

GRAPHICAL INTERPRETATION of the DERIVATIVE

Let the graph of $y = f(x)$ be represented by the curve $APQB$ shown in Fig. 4-1 below. The difference quotient

$$\frac{QR}{PR} \;=\; \frac{f(x_0 + \Delta x) - f(x_0)}{\Delta x} \;=\; \tan\theta \tag{5}$$

is the *slope* of the *secant line* joining points P and Q of the curve. As $\Delta x \to 0$, this secant line approaches the tangent line PS to the curve at the point P. Then

$$\lim_{\Delta x \to 0} \frac{f(x_0 + \Delta x) - f(x_0)}{\Delta x} \;=\; \frac{SR}{PR} \;=\; \tan\alpha \tag{6}$$

is the slope of the tangent line to the curve at the point P.

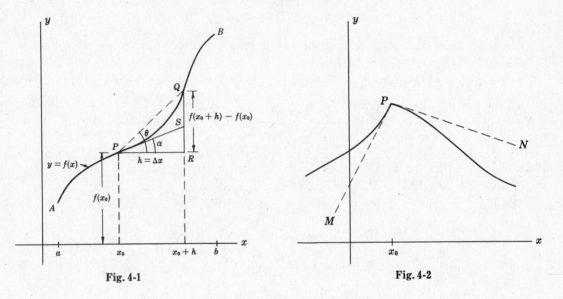

Fig. 4-1   Fig. 4-2

An equation for the tangent line to the curve $y = f(x)$ at the point where $x = x_0$ is given by

$$y - f(x_0) \;=\; f'(x_0)(x - x_0) \tag{7}$$

The fact that a function can be continuous at a point and yet not be differentiable there is shown graphically in Fig. 4-2. In this case there are two tangent lines at P represented by PM and PN. The slopes of these tangent lines are $f'_-(x_0)$ and $f'_+(x_0)$ respectively.

DIFFERENTIALS

Let $\Delta x = dx$ be an increment given to x. Then

$$\Delta y \;=\; f(x + \Delta x) - f(x) \tag{8}$$

is called the *increment* in $y = f(x)$. If $f(x)$ is continuous and has a continuous first derivative in an interval, then

$$\Delta y \;=\; f'(x)\,\Delta x + \epsilon\,\Delta x \;=\; f'(x)\,dx + \epsilon\,dx \tag{9}$$

where $\epsilon \to 0$ as $\Delta x \to 0$. The expression

$$dy \;\; = \;\; f'(x)\, dx \tag{10}$$

is called the *differential of y or f(x)* or the *principal part* of Δy. Note that $\Delta y \neq dy$ in general. However if $\Delta x = dx$ is small, then dy is a close approximation of Δy (see Problem 11). The quantity dx, called the *differential of x*, and dy need not be small.

Because of the definitions (8) and (10), we often write

$$\frac{dy}{dx} \;\; = \;\; f'(x) \;\; = \;\; \lim_{\Delta x \to 0} \frac{f(x + \Delta x) - f(x)}{\Delta x} \;\; = \;\; \lim_{\Delta x \to 0} \frac{\Delta y}{\Delta x} \tag{11}$$

It is emphasized that dx and dy are *not* the limits of Δx and Δy as $\Delta x \to 0$, since these limits are zero whereas dx and dy are not necessarily zero. Instead, given dx we determine dy from (10), i.e. dy is a dependent variable determined from the independent variable dx for a given x.

Geometrically, dy is represented in Fig. 4-1 above, for the particular value $x = x_0$, by the line segment SR, whereas Δy is represented by QR.

RULES for DIFFERENTIATION

If f, g and h are differentiable functions, the following differentiation rules are valid.

1. $\dfrac{d}{dx}\{f(x) + g(x)\} \;\; = \;\; \dfrac{d}{dx}f(x) + \dfrac{d}{dx}g(x) \;\; = \;\; f'(x) + g'(x)$

2. $\dfrac{d}{dx}\{f(x) - g(x)\} \;\; = \;\; \dfrac{d}{dx}f(x) - \dfrac{d}{dx}g(x) \;\; = \;\; f'(x) - g'(x)$

3. $\dfrac{d}{dx}\{C f(x)\} \;\; = \;\; C\dfrac{d}{dx}f(x) \;\; = \;\; C f'(x)$ where C is any constant

4. $\dfrac{d}{dx}\{f(x)\, g(x)\} \;\; = \;\; f(x)\dfrac{d}{dx}g(x) + g(x)\dfrac{d}{dx}f(x) \;\; = \;\; f(x)\, g'(x) + g(x)\, f'(x)$

5. $\dfrac{d}{dx}\left\{\dfrac{f(x)}{g(x)}\right\} \;\; = \;\; \dfrac{g(x)\dfrac{d}{dx}f(x) - f(x)\dfrac{d}{dx}g(x)}{[g(x)]^2} \;\; = \;\; \dfrac{g(x)\, f'(x) - f(x)\, g'(x)}{[g(x)]^2}$ if $g(x) \neq 0$

6. If $y = f(u)$ where $u = g(x)$, then

$$\frac{dy}{dx} \;\; = \;\; \frac{dy}{du} \cdot \frac{du}{dx} \;\; = \;\; f'(u)\frac{du}{dx} \;\; = \;\; f'\{g(x)\}\, g'(x) \tag{12}$$

Similarly if $y = f(u)$ where $u = g(v)$ and $v = h(x)$, then

$$\frac{dy}{dx} \;\; = \;\; \frac{dy}{du} \cdot \frac{du}{dv} \cdot \frac{dv}{dx} \tag{13}$$

The results (12) and (13) are often called *chain rules* for differentiation of composite functions.

7. If $y = f(x)$, then $x = f^{-1}(y)$; and dy/dx and dx/dy are related by

$$\frac{dy}{dx} \;\; = \;\; \frac{1}{dx/dy} \tag{14}$$

8. If $x = f(t)$ and $y = g(t)$, then

$$\frac{dy}{dx} \;\; = \;\; \frac{dy/dt}{dx/dt} \;\; = \;\; \frac{f'(t)}{g'(t)} \tag{15}$$

Similar rules can be formulated for differentials. For example,

$$d\{f(x) + g(x)\} \;\; = \;\; d f(x) + d g(x) \;\; = \;\; f'(x)\, dx + g'(x)\, dx \;\; = \;\; \{f'(x) + g'(x)\}dx$$
$$d\{f(x)\, g(x)\} \;\; = \;\; f(x)\, d g(x) + g(x)\, d f(x) \;\; = \;\; \{f(x)\, g'(x) + g(x)\, f'(x)\}dx$$

DERIVATIVES of SPECIAL FUNCTIONS

In the following we assume that u is a differentiable function of x; if $u = x$, $du/dx = 1$. The inverse functions are defined according to the principal values given in Chapter 2.

1. $\dfrac{d}{dx}(C) = 0$

2. $\dfrac{d}{dx}u^n = nu^{n-1}\dfrac{du}{dx}$

3. $\dfrac{d}{dx}\sin u = \cos u\dfrac{du}{dx}$

4. $\dfrac{d}{dx}\cos u = -\sin u\dfrac{du}{dx}$

5. $\dfrac{d}{dx}\tan u = \sec^2 u\dfrac{du}{dx}$

6. $\dfrac{d}{dx}\cot u = -\csc^2 u\dfrac{du}{dx}$

7. $\dfrac{d}{dx}\sec u = \sec u\tan u\dfrac{du}{dx}$

8. $\dfrac{d}{dx}\csc u = -\csc u\cot u\dfrac{du}{dx}$

9. $\dfrac{d}{dx}\log_a u = \dfrac{\log_a e}{u}\dfrac{du}{dx} \quad a > 0,\ a \neq 1$

10. $\dfrac{d}{dx}\log_e u = \dfrac{d}{dx}\ln u = \dfrac{1}{u}\dfrac{du}{dx}$

11. $\dfrac{d}{dx}a^u = a^u\ln a\dfrac{du}{dx}$

12. $\dfrac{d}{dx}e^u = e^u\dfrac{du}{dx}$

13. $\dfrac{d}{dx}\sin^{-1}u = \dfrac{1}{\sqrt{1-u^2}}\dfrac{du}{dx}$

14. $\dfrac{d}{dx}\cos^{-1}u = -\dfrac{1}{\sqrt{1-u^2}}\dfrac{du}{dx}$

15. $\dfrac{d}{dx}\tan^{-1}u = \dfrac{1}{1+u^2}\dfrac{du}{dx}$

16. $\dfrac{d}{dx}\cot^{-1}u = -\dfrac{1}{1+u^2}\dfrac{du}{dx}$

17. $\dfrac{d}{dx}\sec^{-1}u = \pm\dfrac{1}{u\sqrt{u^2-1}}\dfrac{du}{dx} \quad \begin{cases} +\text{ if } u>1 \\ -\text{ if } u<-1 \end{cases}$

18. $\dfrac{d}{dx}\csc^{-1}u = \mp\dfrac{1}{u\sqrt{u^2-1}}\dfrac{du}{dx} \quad \begin{cases} -\text{ if } u>1 \\ +\text{ if } u<-1 \end{cases}$

19. $\dfrac{d}{dx}\sinh u = \cosh u\dfrac{du}{dx}$

20. $\dfrac{d}{dx}\cosh u = \sinh u\dfrac{du}{dx}$

21. $\dfrac{d}{dx}\tanh u = \operatorname{sech}^2 u\dfrac{du}{dx}$

22. $\dfrac{d}{dx}\coth u = -\operatorname{csch}^2 u\dfrac{du}{dx}$

23. $\dfrac{d}{dx}\operatorname{sech} u = -\operatorname{sech} u\tanh u\dfrac{du}{dx}$

24. $\dfrac{d}{dx}\operatorname{csch} u = -\operatorname{csch} u\coth u\dfrac{du}{dx}$

25. $\dfrac{d}{dx}\sinh^{-1}u = \dfrac{1}{\sqrt{1+u^2}}\dfrac{du}{dx}$

26. $\dfrac{d}{dx}\cosh^{-1}u = \dfrac{1}{\sqrt{u^2-1}}\dfrac{du}{dx}$

27. $\dfrac{d}{dx}\tanh^{-1}u = \dfrac{1}{1-u^2}\dfrac{du}{dx}, \quad |u| < 1$

28. $\dfrac{d}{dx}\coth^{-1}u = \dfrac{1}{1-u^2}\dfrac{du}{dx}, \quad |u| > 1$

29. $\dfrac{d}{dx}\operatorname{sech}^{-1}u = -\dfrac{1}{u\sqrt{u^2-1}}\dfrac{du}{dx}$

30. $\dfrac{d}{dx}\operatorname{csch}^{-1}u = -\dfrac{1}{u\sqrt{u^2+1}}\dfrac{du}{dx}$

HIGHER ORDER DERIVATIVES

If $f(x)$ is differentiable in an interval, its derivative is given by $f'(x)$, y' or dy/dx, where $y = f(x)$. If $f'(x)$ is also differentiable in the interval, its derivative is denoted by $f''(x)$, y'' or $\dfrac{d}{dx}\left(\dfrac{dy}{dx}\right) = \dfrac{d^2y}{dx^2}$. Similarly the nth derivative of $f(x)$, if it exists, is denoted by $f^{(n)}(x)$, $y^{(n)}$ or $\dfrac{d^n y}{dx^n}$, where n is called the *order* of the derivative. Thus derivatives of the first, second, third, ... orders are given by $f'(x)$, $f''(x)$, $f'''(x)$, ...

Computation of higher order derivatives follows by repeated application of the differentiation rules given above.

MEAN VALUE THEOREMS

1. **Rolle's theorem.** If $f(x)$ is continuous in $[a, b]$ and differentiable in (a, b) and if $f(a) = f(b) = 0$, then there exists a point ξ in (a, b) such that $f'(\xi) = 0$.

2. **The theorem of the mean.** If $f(x)$ is continuous in $[a, b]$ and differentiable in (a, b), then there exists a point ξ in (a, b) such that

$$\frac{f(b) - f(a)}{b - a} = f'(\xi) \qquad a < \xi < b \qquad (16)$$

Rolle's theorem is a special case of this where $f(a) = f(b) = 0$.

The result (16) can be written in various alternative forms; for example, if x and x_0 are in (a, b), then

$$f(x) = f(x_0) + f'(\xi)(x - x_0) \qquad \xi \text{ between } x_0 \text{ and } x \qquad (17)$$

We can also write (16) with $b = a + h$, in which case $\xi = a + \theta h$, where $0 < \theta < 1$.

The theorem of the mean is also called the *law of the mean*.

3. **Cauchy's generalized theorem of the mean.** If $f(x)$ and $g(x)$ are continuous in $[a, b]$ and differentiable in (a, b), then there exists a point ξ in (a, b) such that

$$\frac{f(b) - f(a)}{g(b) - g(a)} = \frac{f'(\xi)}{g'(\xi)} \qquad a < \xi < b \qquad (18)$$

where we assume $g(a) \neq g(b)$ and $f'(x), g'(x)$ are not simultaneously zero. Note that the special case $g(x) = x$ yields (16).

4. **Taylor's theorem of the mean.** If $f^{(n)}(x)$ is continuous in $[a, b]$ and differentiable in (a, b), then there exists a point ξ in (a, b) such that

$$f(b) = f(a) + f'(a)(b - a) + \frac{f''(a)(b - a)^2}{2!} + \cdots + \frac{f^{(n)}(a)(b - a)^n}{n!} + R_n \qquad (19)$$

where R_n, called the *remainder*, can be written in either of the following forms:

Lagrange's form: $\qquad R_n = \dfrac{f^{(n+1)}(\xi)(b - a)^{n+1}}{(n + 1)!} \qquad a < \xi < b \quad (20)$

Cauchy's form: $\qquad R_n = \dfrac{f^{(n+1)}(\xi)(b - \xi)^n(b - a)}{n!} \qquad a < \xi < b \quad (21)$

See Problems 26, 81-84. In both forms the values of ξ are different in general.

The result (19) can be written in various alternative forms; for example, if x and x_0 are in (a, b), then using Lagrange's form of the remainder R_n, we have for ξ between x_0 and x,

$$f(x) = f(x_0) + f'(x_0)(x - x_0) + \cdots + \frac{f^{(n)}(x_0)(x - x_0)^n}{n!} + \frac{f^{(n+1)}(\xi)(x - x_0)^{n+1}}{(n + 1)!} \qquad (22)$$

This is often called a *Taylor series for $f(x)$ with a remainder* and is used to approximate $f(x)$ by a polynomial, in which case R_n is the *error term*.

If $\lim\limits_{n \to \infty} R_n = 0$ in (19), the infinite series obtained is called a *Taylor series* for $f(x)$ about $x = x_0$. If $x_0 = 0$, the series is called a *Maclaurin series*. Such series, called *power series*, generally converge for all values of x in some interval, called the *interval of convergence*, and diverge for all x outside this interval (see Chapter 11 for further discussion).

In referring to Taylor's theorem of the mean, the Lagrange form of the remainder will be assumed unless otherwise stated.

SPECIAL EXPANSIONS

The following are some important expansions. The remainder R_n in each case can be obtained by using either (20) or (21).

$$1. \qquad e^x \;=\; 1 + x + \frac{x^2}{2!} + \frac{x^3}{3!} + \cdots + \frac{x^n}{n!} + R_n$$

$$2. \qquad \sin x \;=\; x - \frac{x^3}{3!} + \frac{x^5}{5!} - \frac{x^7}{7!} + \cdots + \frac{(-1)^{n-1}x^{2n-1}}{(2n-1)!} + R_n$$

$$3. \qquad \cos x \;=\; 1 - \frac{x^2}{2!} + \frac{x^4}{4!} - \frac{x^6}{6!} + \cdots + \frac{(-1)^{n-1}x^{2n-2}}{(2n-2)!} + R_n$$

$$4. \quad \ln(1+x) \;=\; x - \frac{x^2}{2} + \frac{x^3}{3} - \frac{x^4}{4} + \cdots + \frac{(-1)^{n-1}x^n}{n} + R_n$$

$$5. \qquad \tan^{-1}x \;=\; x - \frac{x^3}{3} + \frac{x^5}{5} - \frac{x^7}{7} + \cdots + \frac{(-1)^{n-1}x^{2n-1}}{2n-1} + R_n$$

In 1-3, $\lim_{n\to\infty} R_n = 0$ for all x. In 4, $\lim_{n\to\infty} R_n = 0$ for $-1 < x \leqq 1$. In 5, $\lim_{n\to\infty} R_n = 0$ for $-1 \leqq x \leqq 1$. Further discussion of such expansions is given in Chapter 11.

L'HOSPITAL'S RULES

If $\lim_{x\to x_0} f(x) = A$ and $\lim_{x\to x_0} g(x) = B$ where A and B are either both zero or both infinite, $\lim_{x\to x_0} \dfrac{f(x)}{g(x)}$ is often called an *indeterminate* of the form 0/0 or ∞/∞ respectively, although such terminology is somewhat misleading since there is usually nothing indeterminate involved. The following theorems, called *L'Hospital's rules*, facilitate evaluation of such limits.

1. If $f(x)$ and $g(x)$ are differentiable in the interval (a, b) except possibly at a point x_0 in this interval, and if $g'(x) \neq 0$ for $x \neq x_0$, then

$$\lim_{x\to x_0} \frac{f(x)}{g(x)} \;=\; \lim_{x\to x_0} \frac{f'(x)}{g'(x)} \qquad (23)$$

whenever the limit on the right can be found. In case $f'(x)$ and $g'(x)$ satisfy the same conditions as $f(x)$ and $g(x)$ given above, the process can be repeated.

2. If $\lim_{x\to x_0} f(x) = \infty$ and $\lim_{x\to x_0} g(x) = \infty$, the result (23) is also valid.

These can be extended to cases where $x \to \infty$ or $-\infty$, and to cases where $x_0 = a$ or $x_0 = b$ in which only one sided limits, such as $x \to a+$ or $x \to b-$, are involved.

Limits represented by the so-called *indeterminate forms* $0 \cdot \infty$, ∞^0, 0^0, 1^∞ and $\infty - \infty$ can be evaluated on replacing them by equivalent limits for which the above rules are applicable (see Problems 33-36).

Sometimes evaluation of such limits is facilitated by using Taylor's theorem of the mean, as in Problems 32 and 36.

APPLICATIONS

1. **Maxima and minima.** Suppose that at $x = x_0$, $f(x)$ satisfies the conditions

$$f'(x_0) \;=\; f''(x_0) \;=\; \cdots \;=\; f^{(2p-1)}(x_0) \;=\; 0 \quad \text{and} \quad f^{(2p)}(x_0) \neq 0 \qquad (24)$$

for some positive integer p (usually $p = 1$). Then

   (a) $f(x)$ has a relative maximum at $x = x_0$ if $f^{(2p)}(x_0) < 0$

   (b) $f(x)$ has a relative minimum at $x = x_0$ if $f^{(2p)}(x_0) > 0$

See Problem 39. In practice, to find the relative maxima and minima of a differentiable function $f(x)$ we solve the equation $f'(x) = 0$ to obtain the *critical points* x_0 and then use (24). Graphically the necessary condition $f'(x_0) = 0$ follows, since at a relative maximum or minimum point $x = x_0$ the tangent line to $y = f(x)$ must be parallel to the x axis.

· 2. **Rates of change.** We can interpret $dy/dx = f'(x)$ as the rate of change of $y = f(x)$ with respect to x. If $f'(x_0) > 0$, then y is increasing at $x = x_0$; if $f'(x_0) < 0$, then y is decreasing at $x = x_0$.

3. **Velocity and acceleration.** If s is the instantaneous *displacement* of a particle from a point O on a line at time t then ds/dt is its instantaneous *velocity* and d^2s/dt^2 is its instantaneous *acceleration* at time t.

Solved Problems

DERIVATIVES

1. Let $f(x) = \dfrac{3+x}{3-x}$, $x \neq 3$. Evaluate $f'(2)$ from the definition.

$$f'(2) = \lim_{h \to 0} \frac{f(2+h) - f(2)}{h} = \lim_{h \to 0} \frac{1}{h}\left(\frac{5+h}{1-h} - 5\right) = \lim_{h \to 0} \frac{1}{h} \cdot \frac{6h}{1-h} = \lim_{h \to 0} \frac{6}{1-h} = 6$$

Note: By using rules of elementary calculus, we find

$$f'(x) = \frac{(3-x)\dfrac{d}{dx}(3+x) - (3+x)\dfrac{d}{dx}(3-x)}{(3-x)^2} = \frac{(3-x)(1) - (3+x)(-1)}{(3-x)^2} = \frac{6}{(3-x)^2}$$

at all points x where the derivative exists. Putting $x = 2$, we find $f'(2) = 6$. Although such rules are often useful, one must be careful not to apply them indiscriminately (see Problem 5).

2. Let $f(x) = \sqrt{2x-1}$. Evaluate $f'(5)$ from the definition.

$$f'(5) = \lim_{h \to 0} \frac{f(5+h) - f(5)}{h} = \lim_{h \to 0} \frac{\sqrt{9+2h} - 3}{h}$$

$$= \lim_{h \to 0} \frac{\sqrt{9+2h} - 3}{h} \cdot \frac{\sqrt{9+2h} + 3}{\sqrt{9+2h} + 3} = \lim_{h \to 0} \frac{9+2h-9}{h(\sqrt{9+2h} + 3)} = \lim_{h \to 0} \frac{2}{\sqrt{9+2h} + 3} = \frac{1}{3}$$

By using rules of elementary calculus, we find $f'(x) = \dfrac{d}{dx}(2x-1)^{1/2} = \frac{1}{2}(2x-1)^{-1/2}\dfrac{d}{dx}(2x-1) = (2x-1)^{-1/2}$. Then $f'(5) = 9^{-1/2} = \frac{1}{3}$.

3. If $f(x)$ has a derivative at $x = x_0$, prove that $f(x)$ must be continuous at $x = x_0$.

$$f(x_0 + h) - f(x_0) = \frac{f(x_0 + h) - f(x_0)}{h} \cdot h, \quad h \neq 0$$

Then

$$\lim_{h \to 0} f(x_0 + h) - f(x_0) = \lim_{h \to 0} \frac{f(x_0 + h) - f(x_0)}{h} \cdot \lim_{h \to 0} h = f'(x_0) \cdot 0 = 0$$

since $f'(x_0)$ exists by hypothesis. Thus

$$\lim_{h \to 0} f(x_0 + h) - f(x_0) = 0 \quad \text{or} \quad \lim_{h \to 0} f(x_0 + h) = f(x_0)$$

showing that $f(x)$ is continuous at $x = x_0$.

4. Let $f(x) = \begin{cases} x \sin 1/x, & x \neq 0 \\ 0, & x = 0 \end{cases}$.

   (*a*) Is $f(x)$ continuous at $x = 0$?   (*b*) Does $f(x)$ have a derivative at $x = 0$?

   (*a*)   By Problem 22(*b*) of Chapter 2, $f(x)$ is continuous at $x = 0$.

   (*b*) $f'(0) = \lim\limits_{h \to 0} \dfrac{f(0+h) - f(0)}{h} = \lim\limits_{h \to 0} \dfrac{f(h) - f(0)}{h} = \lim\limits_{h \to 0} \dfrac{h \sin 1/h - 0}{h} = \lim\limits_{h \to 0} \sin \dfrac{1}{h}$

   which does not exist.

   This example shows that even though a function is continuous at a point, it need not have a derivative at the point, i.e. the converse of the theorem in Problem 3 is not necessarily true.

   It is possible to construct a function which is continuous at every point of an interval but has a derivative nowhere.

5. Let $f(x) = \begin{cases} x^2 \sin 1/x, & x \neq 0 \\ 0, & x = 0 \end{cases}$.

   (*a*) Is $f(x)$ differentiable at $x = 0$?   (*b*) Is $f'(x)$ continuous at $x = 0$?

   (*a*) $f'(0) = \lim\limits_{h \to 0} \dfrac{f(h) - f(0)}{h} = \lim\limits_{h \to 0} \dfrac{h^2 \sin 1/h - 0}{h} = \lim\limits_{h \to 0} h \sin \dfrac{1}{h} = 0$

   by Problem 13, Chapter 2. Then $f(x)$ has a derivative (is differentiable) at $x = 0$ and its value is 0.

   (*b*)   From elementary calculus differentiation rules, if $x \neq 0$,

   $$f'(x) = \frac{d}{dx}\left(x^2 \sin \frac{1}{x} \right) = x^2 \frac{d}{dx}\left(\sin \frac{1}{x} \right) + \left(\sin \frac{1}{x} \right) \frac{d}{dx}(x^2)$$

   $$= x^2 \left(\cos \frac{1}{x} \right)\left(-\frac{1}{x^2} \right) + \left(\sin \frac{1}{x} \right)(2x) = -\cos \frac{1}{x} + 2x \sin \frac{1}{x}$$

   Since $\lim\limits_{x \to 0} f'(x) = \lim\limits_{x \to 0}\left(-\cos \dfrac{1}{x} + 2x \sin \dfrac{1}{x} \right)$ does not exist (because $\lim\limits_{x \to 0} \cos 1/x$ does not exist), $f'(x)$ cannot be continuous at $x = 0$ in spite of the fact that $f'(0)$ exists.

   This shows that we cannot calculate $f'(0)$ in this case by simply calculating $f'(x)$ and putting $x = 0$, as is frequently supposed in elementary calculus. It is only when the derivative of a function is *continuous* at a point that this procedure gives the right answer. This happens to be true for most functions arising in elementary calculus.

6. Present an "ϵ, δ" definition of the derivative of $f(x)$ at $x = x_0$.

   $f(x)$ has a derivative $f'(x_0)$ at $x = x_0$ if, given any $\epsilon > 0$, we can find $\delta > 0$ such that

   $$\left| \frac{f(x_0 + h) - f(x_0)}{h} - f'(x_0) \right| < \epsilon \quad \text{when} \quad 0 < |h| < \delta$$

RIGHT and LEFT HAND DERIVATIVES

7. Let $f(x) = |x|$. (*a*) Calculate the right hand derivative of $f(x)$ at $x = 0$. (*b*) Calculate the left hand derivative of $f(x)$ at $x = 0$. (*c*) Does $f(x)$ have a derivative at $x = 0$? (*d*) Illustrate the conclusions in (*a*), (*b*) and (*c*) from a graph.

   (*a*) $f'_+(0) = \lim\limits_{h \to 0+} \dfrac{f(h) - f(0)}{h} = \lim\limits_{h \to 0+} \dfrac{|h| - 0}{h} = \lim\limits_{h \to 0+} \dfrac{h}{h} = 1$

   since $|h| = h$ for $h > 0$.

   (*b*) $f'_-(0) = \lim\limits_{h \to 0-} \dfrac{f(h) - f(0)}{h} = \lim\limits_{h \to 0-} \dfrac{|h| - 0}{h} = \lim\limits_{h \to 0-} \dfrac{-h}{h} = -1$

   since $|h| = -h$ for $h < 0$.

(c) No. The derivative at 0 does not exist if the right and left hand derivatives are unequal.

(d) The required graph is shown in the adjoining Fig. 4-3. Note that the slopes of the lines $y = x$ and $y = -x$ are 1 and -1 respectively, representing the right and left hand derivatives at $x = 0$. However, the derivative at $x = 0$ does not exist.

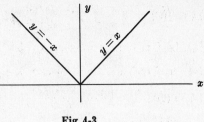

Fig. 4-3

8. Prove that $f(x) = x^2$ is differentiable in $0 \leqq x \leqq 1$.

Let x_0 be any value such that $0 < x_0 < 1$. Then

$$f'(x_0) = \lim_{h \to 0} \frac{f(x_0 + h) - f(x_0)}{h} = \lim_{h \to 0} \frac{(x_0 + h)^2 - x_0^2}{h} = \lim_{h \to 0} (2x_0 + h) = 2x_0$$

At the end point $x = 0$,

$$f'_+(0) = \lim_{h \to 0+} \frac{f(0 + h) - f(0)}{h} = \lim_{h \to 0+} \frac{h^2 - 0}{h} = \lim_{h \to 0+} h = 0$$

At the end point $x = 1$,

$$f'_-(1) = \lim_{h \to 0-} \frac{f(1 + h) - f(1)}{h} = \lim_{h \to 0-} \frac{(1 + h)^2 - 1}{h} = \lim_{h \to 0-} (2 + h) = 2$$

Then $f(x)$ is differentiable in $0 \leqq x \leqq 1$. We may write $f'(x) = 2x$ for any x in this interval. It is customary to write $f'_+(0) = f'(0)$ and $f'_-(1) = f'(1)$ in this case.

9. Find an equation for the tangent line to $y = x^2$ at the point where (a) $x = 1/3$, (b) $x = 1$.

(a) From Problem 8, $f'(x_0) = 2x_0$ so that $f'(1/3) = 2/3$. Then the equation of the tangent line is

$$y - f(x_0) = f'(x_0)(x - x_0) \quad \text{or} \quad y - \tfrac{1}{9} = \tfrac{2}{3}(x - \tfrac{1}{3}), \quad \text{i.e.} \quad y = \tfrac{2}{3}x - \tfrac{1}{9}$$

(b) As in part (a), $y - f(1) = f'(1)(x - 1)$ or $y - 1 = 2(x - 1)$, i.e. $y = 2x - 1$.

DIFFERENTIALS

10. If $y = f(x) = x^3 - 6x$, find (a) Δy, (b) dy, (c) $\Delta y - dy$.

(a)
$$\begin{aligned}
\Delta y &= f(x + \Delta x) - f(x) = \{(x + \Delta x)^3 - 6(x + \Delta x)\} - \{x^3 - 6x\} \\
&= x^3 + 3x^2 \Delta x + 3x(\Delta x)^2 + (\Delta x)^3 - 6x - 6\Delta x - x^3 + 6x \\
&= (3x^2 - 6)\Delta x + 3x(\Delta x)^2 + (\Delta x)^3
\end{aligned}$$

(b) $dy = $ principal part of $\Delta y = (3x^2 - 6)\Delta x = (3x^2 - 6)dx$, since by definition $\Delta x = dx$.

Note that $f'(x) = 3x^2 - 6$ and $dy = (3x^2 - 6)dx$, i.e. $dy/dx = 3x^2 - 6$. It must be emphasized that dy and dx are not necessarily small.

(c) From (a) and (b), $\Delta y - dy = 3x(\Delta x)^2 + (\Delta x)^3 = \epsilon \Delta x$, where $\epsilon = 3x \Delta x + (\Delta x)^2$.

Note that $\epsilon \to 0$ as $\Delta x \to 0$, i.e. $\dfrac{\Delta y - dy}{\Delta x} \to 0$ as $\Delta x \to 0$. Hence $\Delta y - dy$ is an infinitesimal of higher order than Δx (see Problem 92).

In case Δx is small, dy and Δy are approximately equal.

11. Evaluate $\sqrt[3]{25}$ approximately by use of differentials.

If Δx is small, $\Delta y = f(x + \Delta x) - f(x) = f'(x)\Delta x$ approximately.

Let $f(x) = \sqrt[3]{x}$. Then $\sqrt[3]{x + \Delta x} - \sqrt[3]{x} \approx \tfrac{1}{3}x^{-2/3}\Delta x$ (where $\approx$ denotes *approximately equal to*).

If $x = 27$ and $\Delta x = -2$, we have

$$\sqrt[3]{27-2} - \sqrt[3]{27} \approx \tfrac{1}{3}(27)^{-2/3}(-2), \qquad \text{i.e.} \qquad \sqrt[3]{25} - 3 \approx -2/27$$

Then $\sqrt[3]{25} \approx 3 - 2/27$ or 2.926.

It is interesting to observe that $(2.926)^3 = 25.05$, so that the approximation is fairly good.

DIFFERENTIATION RULES. DIFFERENTIATION of SPECIAL FUNCTIONS

12. Prove the formula $\dfrac{d}{dx}\{f(x)\,g(x)\} = f(x)\dfrac{d}{dx}g(x) + g(x)\dfrac{d}{dx}f(x)$, assuming f and g are differentiable.

By definition,

$$\frac{d}{dx}\{f(x)\,g(x)\} = \lim_{\Delta x \to 0} \frac{f(x+\Delta x)\,g(x+\Delta x) - f(x)\,g(x)}{\Delta x}$$

$$= \lim_{\Delta x \to 0} \frac{f(x+\Delta x)\,\{g(x+\Delta x) - g(x)\} + g(x)\,\{f(x+\Delta x) - f(x)\}}{\Delta x}$$

$$= \lim_{\Delta x \to 0} f(x+\Delta x)\left\{\frac{g(x+\Delta x) - g(x)}{\Delta x}\right\} + \lim_{\Delta x \to 0} g(x)\left\{\frac{f(x+\Delta x) - f(x)}{\Delta x}\right\}$$

$$= f(x)\,\frac{d}{dx}g(x) + g(x)\,\frac{d}{dx}f(x)$$

Another method:

Let $u = f(x)$, $v = g(x)$. Then $\Delta u = f(x+\Delta x) - f(x)$ and $\Delta v = g(x+\Delta x) - g(x)$, i.e. $f(x+\Delta x) = u + \Delta u$, $g(x+\Delta x) = v + \Delta v$. Thus

$$\frac{d}{dx}uv = \lim_{\Delta x \to 0}\frac{(u+\Delta u)(v+\Delta v) - uv}{\Delta x} = \lim_{\Delta x \to 0}\frac{u\Delta v + v\Delta u + \Delta u\Delta v}{\Delta x}$$

$$= \lim_{\Delta x \to 0}\left(u\frac{\Delta v}{\Delta x} + v\frac{\Delta u}{\Delta x} + \frac{\Delta u}{\Delta x}\Delta v\right) = u\frac{dv}{dx} + v\frac{du}{dx}$$

where it is noted that $\Delta v \to 0$ as $\Delta x \to 0$, since v is supposed differentiable and thus continuous.

13. If $y = f(u)$ where $u = g(x)$, prove that $\dfrac{dy}{dx} = \dfrac{dy}{du}\cdot\dfrac{du}{dx}$ assuming that f and g are differentiable.

Let x be given an increment $\Delta x \neq 0$. Then as a consequence u and y take on increments Δu and Δy respectively, where

$$\Delta y = f(u+\Delta u) - f(u), \qquad \Delta u = g(x+\Delta x) - g(x) \tag{1}$$

Note that as $\Delta x \to 0$, $\Delta y \to 0$ and $\Delta u \to 0$.

If $\Delta u \neq 0$, let us write $\epsilon = \dfrac{\Delta y}{\Delta u} - \dfrac{dy}{du}$ so that $\epsilon \to 0$ as $\Delta u \to 0$ and

$$\Delta y = \frac{dy}{du}\Delta u + \epsilon\,\Delta u \tag{2}$$

If $\Delta u = 0$ for values of Δx, then (1) shows that $\Delta y = 0$ for these values of Δx. For such cases, we define $\epsilon = 0$.

It follows that in both cases, $\Delta u \neq 0$ or $\Delta u = 0$, (2) holds. Dividing (2) by $\Delta x \neq 0$ and taking the limit as $\Delta x \to 0$, we have

$$\frac{dy}{dx} = \lim_{\Delta x \to 0}\frac{\Delta y}{\Delta x} = \lim_{\Delta x \to 0}\left(\frac{dy}{du}\frac{\Delta u}{\Delta x} + \epsilon\frac{\Delta u}{\Delta x}\right) = \frac{dy}{du}\cdot\lim_{\Delta x \to 0}\frac{\Delta u}{\Delta x} + \lim_{\Delta x \to 0}\epsilon\cdot\lim_{\Delta x \to 0}\frac{\Delta u}{\Delta x}$$

$$= \frac{dy}{du}\frac{du}{dx} + 0\cdot\frac{du}{dx} = \frac{dy}{du}\cdot\frac{du}{dx} \tag{3}$$

14. Given $\dfrac{d}{dx}(\sin x) = \cos x$ and $\dfrac{d}{dx}(\cos x) = -\sin x$, derive the formulas

(a) $\dfrac{d}{dx}(\tan x) = \sec^2 x$, (b) $\dfrac{d}{dx}(\sin^{-1} x) = \dfrac{1}{\sqrt{1-x^2}}$.

(a) $\dfrac{d}{dx}(\tan x) = \dfrac{d}{dx}\left(\dfrac{\sin x}{\cos x}\right) = \dfrac{\cos x \dfrac{d}{dx}(\sin x) - \sin x \dfrac{d}{dx}(\cos x)}{\cos^2 x}$

$\qquad = \dfrac{(\cos x)(\cos x) - (\sin x)(-\sin x)}{\cos^2 x} = \dfrac{1}{\cos^2 x} = \sec^2 x$

(b) If $y = \sin^{-1} x$, then $x = \sin y$. Taking the derivative with respect to x,

$$1 = \cos y \dfrac{dy}{dx} \quad \text{or} \quad \dfrac{dy}{dx} = \dfrac{1}{\cos y} = \dfrac{1}{\sqrt{1 - \sin^2 y}} = \dfrac{1}{\sqrt{1-x^2}}$$

We have supposed here that the principal value $-\pi/2 \leqq \sin^{-1} x \leqq \pi/2$, is chosen so that $\cos y$ is positive, thus accounting for our writing $\cos y = \sqrt{1 - \sin^2 y}$ rather than $\cos y = \pm\sqrt{1 - \sin^2 y}$.

15. Derive the formula $\dfrac{d}{dx}(\log_a u) = \dfrac{\log_a e}{u}\dfrac{du}{dx}$ $(a > 0, \ a \neq 1)$, where u is a differentiable function of x.

Consider $y = f(u) = \log_a u$. By definition,

$$\dfrac{dy}{du} = \lim_{\Delta u \to 0} \dfrac{f(u + \Delta u) - f(u)}{\Delta u} = \lim_{\Delta u \to 0} \dfrac{\log_a (u + \Delta u) - \log_a u}{\Delta u}$$

$$= \lim_{\Delta u \to 0} \dfrac{1}{\Delta u} \log_a\left(\dfrac{u + \Delta u}{u}\right) = \lim_{\Delta u \to 0} \dfrac{1}{u} \log_a\left(1 + \dfrac{\Delta u}{u}\right)^{u/\Delta u}$$

Since the logarithm is a continuous function, this can be written

$$\dfrac{1}{u} \log_a\left\{ \lim_{\Delta u \to 0} \left(1 + \dfrac{\Delta u}{u}\right)^{u/\Delta u} \right\} = \dfrac{1}{u} \log_a e$$

by Problem 19, Chapter 3, with $x = u/\Delta u$.

Then by Problem 13, $\dfrac{d}{dx}(\log_a u) = \dfrac{\log_a e}{u}\dfrac{du}{dx}$.

16. Calculate dy/dx if (a) $xy^3 - 3x^2 = xy + 5$, (b) $e^{xy} + y \ln x = \cos 2x$.

(a) Differentiate with respect to x, considering y as a function of x. (We sometimes say that y is an *implicit function of x*, since we cannot solve explicitly for y in terms of x.) Then

$$\dfrac{d}{dx}(xy^3) - \dfrac{d}{dx}(3x^2) = \dfrac{d}{dx}(xy) + \dfrac{d}{dx}(5) \quad \text{or} \quad (x)(3y^2 y') + (y^3)(1) - 6x = (x)(y') + (y)(1) + 0$$

where $y' = dy/dx$. Solving, $y' = (6x - y^3 + y)/(3xy^2 - x)$.

(b) $\dfrac{d}{dx}(e^{xy}) + \dfrac{d}{dx}(y \ln x) = \dfrac{d}{dx}(\cos 2x), \quad e^{xy}(xy' + y) + \dfrac{y}{x} + (\ln x)y' = -2 \sin 2x$.

Solving, $\qquad\qquad\qquad y' = -\dfrac{2x \sin 2x + xye^{xy} + y}{x^2 e^{xy} + x \ln x}$

17. If $y = \cosh(x^2 - 3x + 1)$, find (a) dy/dx, (b) d^2y/dx^2.

(a) Let $y = \cosh u$, where $u = x^2 - 3x + 1$. Then $dy/du = \sinh u$, $du/dx = 2x - 3$, and

$$\dfrac{dy}{dx} = \dfrac{dy}{du} \cdot \dfrac{du}{dx} = (\sinh u)(2x - 3) = (2x - 3) \sinh(x^2 - 3x + 1)$$

(b) $\dfrac{d^2y}{dx^2} = \dfrac{d}{dx}\left(\dfrac{dy}{dx}\right) = \dfrac{d}{dx}\left(\sinh u \dfrac{du}{dx}\right) = \sinh u \dfrac{d^2u}{dx^2} + \cosh u \left(\dfrac{du}{dx}\right)^2$

$\qquad = (\sinh u)(2) + (\cosh u)(2x - 3)^2 = 2 \sinh(x^2 - 3x + 1) + (2x - 3)^2 \cosh(x^2 - 3x + 1)$

18. If $x^2y + y^3 = 2$, find (a) y', (b) y'' at the point $(1, 1)$.

(a) Differentiating with respect to x, $x^2y' + 2xy + 3y^2y' = 0$ and

$$y' = \frac{-2xy}{x^2 + 3y^2} = -\frac{1}{2} \text{ at } (1, 1)$$

(b) $y'' = \dfrac{d}{dx}(y') = \dfrac{d}{dx}\left(\dfrac{-2xy}{x^2 + 3y^2}\right) = -\dfrac{(x^2 + 3y^2)(2xy' + 2y) - (2xy)(2x + 6yy')}{(x^2 + 3y^2)^2}$

Substituting $x = 1$, $y = 1$ and $y' = -\frac{1}{2}$, we find $y'' = -\frac{3}{8}$.

MEAN VALUE THEOREMS

19. Prove Rolle's theorem.

Case 1:

$f(x) \equiv 0$ in $[a, b]$. Then $f'(x) = 0$ for all x in (a, b).

Case 2:

$f(x) \not\equiv 0$ in $[a, b]$. Since $f(x)$ is continuous there are points at which $f(x)$ attains its maximum and minimum values, denoted by M and m respectively (see Problem 34, Chapter 2).

Since $f(x) \not\equiv 0$, at least one of the values M, m is not zero. Suppose, for example, $M \neq 0$ and that $f(\xi) = M$ (see Fig. 4-4). For this case, $f(\xi + h) \leqq f(\xi)$.

If $h > 0$, then $\dfrac{f(\xi + h) - f(\xi)}{h} \leqq 0$ and

$(1) \quad \lim_{h \to 0+} \dfrac{f(\xi + h) - f(\xi)}{h} \leqq 0$

If $h < 0$, then $\dfrac{f(\xi + h) - f(\xi)}{h} \geqq 0$ and

$(2) \quad \lim_{h \to 0-} \dfrac{f(\xi + h) - f(\xi)}{h} \geqq 0$

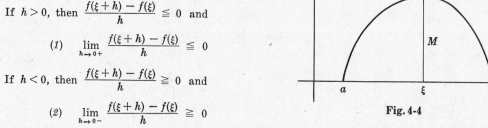

Fig. 4-4

But by hypothesis $f(x)$ has a derivative at all points in (a, b). Then the right hand derivative (1) must be equal to the left hand derivative (2). This can happen only if they are both equal to zero, in which case $f'(\xi) = 0$ as required.

A similar argument can be used in case $M = 0$ and $m \neq 0$.

20. (a) Prove the theorem of the mean. (b) Give a geometric interpretation of this theorem.

(a) Define $F(x) = f(x) - f(a) - (x - a)\dfrac{f(b) - f(a)}{b - a}$.

Then $F(a) = 0$ and $F(b) = 0$.

Also, if $f(x)$ satisfies the conditions on continuity and differentiability specified in Rolle's theorem, then $F(x)$ satisfies them also.

Then applying Rolle's theorem to the function $F(x)$, we obtain

$$F'(\xi) = f'(\xi) - \frac{f(b) - f(a)}{b - a} = 0, \quad a < \xi < b \quad \text{or} \quad f'(\xi) = \frac{f(b) - f(a)}{b - a}, \quad a < \xi < b$$

(b) Let curve ACB in Fig. 4-5 represent the graph of $f(x)$. Geometrically it appears that there is a point ξ between a and b where the tangent line to this curve at point C is parallel to the chord AB.

Slope of tangent line $= f'(\xi)$.

Slope of chord $AB = \dfrac{f(b) - f(a)}{b - a}$.

Then ξ is such that $f'(\xi) = \dfrac{f(b) - f(a)}{b - a}$.

It is interesting to note that the function $F(x)$ of part (a) represents the difference in ordinates of curve ACB and line AB at any point x in (a, b).

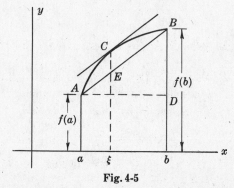

Fig. 4-5

21. Verify the theorem of the mean for $f(x) = 2x^2 - 7x + 10$, $a = 2$, $b = 5$.

$f(2) = 4$, $f(5) = 25$, $f'(\xi) = 4\xi - 7$.   Then the theorem of the mean states that $4\xi - 7 = (25 - 4)/(5 - 2)$ or $\xi = 3.5$.   Since $2 < \xi < 5$, the theorem is verified.

22. If $f'(x) = 0$ at all points of the interval (a, b), prove that $f(x)$ must be a constant in the interval.

Let $x_1 < x_2$ be any two different points in (a, b). By the theorem of the mean, for $x_1 < \xi < x_2$,

$$\frac{f(x_2) - f(x_1)}{x_2 - x_1} \; = \; f'(\xi) \; = \; 0$$

Thus $f(x_1) = f(x_2) = $ constant. From this it follows that if two functions have the same derivative at all points of (a, b), the functions can only differ by a constant.

23. If $f'(x) > 0$ at all points of the interval (a, b), prove that $f(x)$ is strictly increasing.

Let $x_1 < x_2$ be any two different points in (a, b). By the theorem of the mean, for $x_1 < \xi < x_2$,

$$\frac{f(x_2) - f(x_1)}{x_2 - x_1} \; = \; f'(\xi) \; > \; 0$$

Then $f(x_2) > f(x_1)$ for $x_2 > x_1$, and so $f(x)$ is strictly increasing.

24. (a) Prove that $\dfrac{b-a}{1+b^2} < \tan^{-1} b - \tan^{-1} a < \dfrac{b-a}{1+a^2}$ if $a < b$.

(b) Show that $\dfrac{\pi}{4} + \dfrac{3}{25} < \tan^{-1} \dfrac{4}{3} < \dfrac{\pi}{4} + \dfrac{1}{6}$.

(a) Let $f(x) = \tan^{-1} x$. Since $f'(x) = 1/(1 + x^2)$ and $f'(\xi) = 1/(1 + \xi^2)$, we have by the theorem of the mean,

$$\frac{\tan^{-1} b - \tan^{-1} a}{b - a} \; = \; \frac{1}{1 + \xi^2} \qquad a < \xi < b$$

Since $\xi > a$, $1/(1 + \xi^2) < 1/(1 + a^2)$. Since $\xi < b$, $1/(1 + \xi^2) > 1/(1 + b^2)$. Then

$$\frac{1}{1 + b^2} \; < \; \frac{\tan^{-1} b - \tan^{-1} a}{b - a} \; < \; \frac{1}{1 + a^2}$$

and the required result follows on multiplying by $b - a$.

(b) Let $b = 4/3$ and $a = 1$ in the result of part (a). Then since $\tan^{-1} 1 = \pi/4$, we have

$$\frac{3}{25} < \tan^{-1}\frac{4}{3} - \tan^{-1} 1 < \frac{1}{6} \qquad \text{or} \qquad \frac{\pi}{4} + \frac{3}{25} < \tan^{-1}\frac{4}{3} < \frac{\pi}{4} + \frac{1}{6}$$

25. Prove Cauchy's generalized theorem of the mean.

Consider $G(x) = f(x) - f(a) - \alpha\{g(x) - g(a)\}$,   where α is a constant.   Then $G(x)$ satisfies the conditions of Rolle's theorem, provided $f(x)$ and $g(x)$ satisfy the continuity and differentiability conditions of Rolle's theorem and if $G(a) = G(b) = 0$. Both latter conditions are satisfied if the constant $\alpha = \dfrac{f(b) - f(a)}{g(b) - g(a)}$.

Applying Rolle's theorem, $G'(\xi) = 0$ for $a < \xi < b$, we have

$$f'(\xi) - \alpha g'(\xi) \; = \; 0 \qquad \text{or} \qquad \frac{f'(\xi)}{g'(\xi)} \; = \; \frac{f(b) - f(a)}{g(b) - g(a)}, \qquad a < \xi < b$$

as required.

TAYLOR'S THEOREM of the MEAN

26. Prove Taylor's theorem of the mean with the Lagrange remainder for the case $n = 1$.

We must show that

$$f(b) \;=\; f(a) + f'(a)(b - a) + \frac{f''(\xi)}{2!}(b - a)^2 \qquad a < \xi < b \tag{1}$$

Consider the function

$$H(x) \;=\; f(b) - f(x) - f'(x)(b - x) - (b - x)^2 A \tag{2}$$

where A is an undetermined constant. Motivation for forming $H(x)$ is attained by replacing a by x in (1) and transposing all terms to the right.

From (2), $H(b) = 0$. To obtain $H(a) = 0$, we must choose

$$A \;=\; \frac{f(b) - f(a) - f'(a)(b - a)}{(b - a)^2} \tag{3}$$

Assuming $f(x)$ and $f'(x)$ satisfy the continuity and differentiability conditions of Rolle's theorem, then $H(x)$ also satisfies the conditions of Rolle's theorem; hence there is a value ξ between a and b such that $H'(\xi) = 0$.

From (2), $H'(x) = -f''(x)(b - x) + 2(b - x)A$ and $H'(\xi) = -f''(\xi)(b - \xi) + 2(b - \xi)A = 0$ for $A = f''(\xi)/2!$ (since $\xi \neq b$). Substituting this value of A into (3) and solving for $f(b)$ yields the required result (1).

Generalizations to $n > 1$ follow by similar reasoning.

27. (a) Prove that

$$\sin x \;=\; \sin a + (\cos a)(x - a) - \frac{(\sin a)(x - a)^2}{2!} - \frac{(\cos \xi)(x - a)^3}{3!}$$

where ξ is between a and x.

(b) Use part (a) to evaluate $\sin 51°$ and estimate the error made.

(a) Let $f(x) = \sin x$. Then $f'(x) = \cos x$, $f''(x) = -\sin x$, $f'''(x) = -\cos x$, so that $f(a) = \sin a$, $f'(a) = \cos a$, $f''(a) = -\sin a$, $f'''(\xi) = -\cos \xi$.

Then substituting into Taylor's formula with $n = 2$, i.e.,

$$f(x) \;=\; f(a) + f'(a)(x - a) + \frac{f''(a)(x - a)^2}{2!} + \frac{f'''(\xi)(x - a)^3}{3!}$$

where ξ is between a and x, the required result is obtained.

(b) Let $x = 51° = 51\pi/180$ radians and $a = 45° = 45\pi/180$ radians. Then $x - a = \pi/30$ radians. Thus from part (a), since $\sin 45° = \cos 45° = \sqrt{2}/2$,

$$\sin 51° \;=\; \frac{\sqrt{2}}{2} + \frac{\sqrt{2}}{2}\left(\frac{\pi}{30}\right) - \frac{(\sqrt{2}/2)(\pi/30)^2}{2!} - \frac{(\cos \xi)(\pi/30)^3}{3!}$$

The absolute value of the error term $= \left| \dfrac{-(\cos \xi)(\pi/30)^3}{3!} \right| < \dfrac{1}{3!}\left(\dfrac{\pi}{30}\right)^3 < .0002$.

Hence the sum of the first three terms, namely 0.777, is accurate to 3 decimal places.

If more accuracy is desired, more terms in Taylor's formula should be taken.

28. (a) Prove that $e^x \;=\; 1 + x + x^2/2! + x^3/3! + \cdots$.

(b) Prove that e is an irrational number.

(a) Let $f(x) = e^x$. Then all derivatives of $f(x)$ are equal to e^x. If $a = 0$, then $f(0) = f'(0) = \cdots = f^{(n)}(0) = e^0 = 1$ while $f^{(n+1)}(\xi) = e^\xi$, so that Taylor's formula becomes

$$e^x \;=\; 1 + x + \frac{x^2}{2!} + \frac{x^3}{3!} + \cdots + \frac{x^n}{n!} + \frac{x^{n+1} e^\xi}{(n+1)!} \tag{1}$$

where ξ is between 0 and x.

Let $R_n = \dfrac{x^{n+1}}{(n+1)!} e^\xi$. If $x > 0$, $|R_n| < \dfrac{x^{n+1}}{(n+1)!} e^x$ and $\lim\limits_{n \to \infty} |R_n| = 0$ (see Prob. 30, Chap. 3).

If $x < 0$, $|R_n| < \dfrac{|x|^{n+1}}{(n+1)!}$ and $\lim\limits_{n \to \infty} |R_n| = 0$ also. If $x = 0$, $|R_n| = 0$ and $\lim\limits_{n \to \infty} |R_n| = 0$.

Thus for all x, i.e. $-\infty < x < \infty$, $\lim\limits_{n \to \infty} |R_n| = 0$ and we may write

$$e^x = 1 + x + x^2/2! + x^3/3! + \cdots \tag{2}$$

i.e. the series converges for all x.

(b) From (1), letting $x = 1$ and assuming that e is a rational number $\dfrac{p}{q}$ in lowest terms where p and q are positive integers, we have

$$e = \frac{p}{q} = 1 + 1 + \frac{1}{2!} + \frac{1}{3!} + \cdots + \frac{1}{n!} + \frac{e^\xi}{(n+1)!} \qquad 0 < \xi < 1 \tag{3}$$

Choose $n > q$ and multiply both sides of (1) by $n!$ Then

$$n!\,e = n!\frac{p}{q} = n! + n! + \frac{n!}{2!} + \frac{n!}{3!} + \cdots + 1 + \frac{e^\xi}{n+1} \tag{4}$$

Now $e^\xi/(n+1)$ is a number between 0 and 1, while every other term in (4) is a positive integer. Thus assuming that e is rational, we arrive at the contradiction that an integer is equal to a non-integer. Hence e must be irrational.

L'HOSPITAL'S RULE

29. Prove L'Hospital's rule for the case of the "indeterminate forms" (a) 0/0, (b) ∞/∞.

(a) We shall suppose that $f(x)$ and $g(x)$ are differentiable in $a < x < b$ and $f(x_0) = 0$, $g(x_0) = 0$, where $a < x_0 < b$.

By Cauchy's generalized theorem of the mean (Problem 25),

$$\frac{f(x)}{g(x)} = \frac{f(x) - f(x_0)}{g(x) - g(x_0)} = \frac{f'(\xi)}{g'(\xi)} \qquad x_0 < \xi < x$$

Then

$$\lim_{x \to x_0+} \frac{f(x)}{g(x)} = \lim_{x \to x_0+} \frac{f'(\xi)}{g'(\xi)} = \lim_{x \to x_0+} \frac{f'(x)}{g'(x)} = L$$

since as $x \to x_0+$, $\xi \to x_0+$.

Modification of the above procedure can be used to establish the result if $x \to x_0-$, $x \to x_0$, $x \to \infty$, $x \to -\infty$.

(b) We suppose that $f(x)$ and $g(x)$ are differentiable in $a < x < b$, and $\lim\limits_{x \to x_0+} f(x) = \infty$, $\lim\limits_{x \to x_0+} g(x) = \infty$ where $a < x_0 < b$.

Assume x_1 is such that $a < x_0 < x < x_1 < b$. By Cauchy's generalized theorem of the mean,

$$\frac{f(x) - f(x_1)}{g(x) - g(x_1)} = \frac{f'(\xi)}{g'(\xi)} \qquad x < \xi < x_1$$

Hence

$$\frac{f(x) - f(x_1)}{g(x) - g(x_1)} = \frac{f(x)}{g(x)} \cdot \frac{1 - f(x_1)/f(x)}{1 - g(x_1)/g(x)} = \frac{f'(\xi)}{g'(\xi)}$$

from which we see that

$$\frac{f(x)}{g(x)} = \frac{f'(\xi)}{g'(\xi)} \cdot \frac{1 - g(x_1)/g(x)}{1 - f(x_1)/f(x)} \tag{1}$$

Let us now suppose that $\lim\limits_{x \to x_0+} \dfrac{f'(x)}{g'(x)} = L$ and write (1) as

$$\frac{f(x)}{g(x)} = \left(\frac{f'(\xi)}{g'(\xi)} - L\right)\left(\frac{1 - g(x_1)/g(x)}{1 - f(x_1)/f(x)}\right) + L\left(\frac{1 - g(x_1)/g(x)}{1 - f(x_1)/f(x)}\right) \tag{2}$$

We can choose x_1 so close to x_0 that $|f'(\xi)/g'(\xi) - L| < \epsilon$. Keeping x_1 fixed, we see that

$$\lim_{x \to x_0+} \left(\frac{1 - g(x_1)/g(x)}{1 - f(x_1)/f(x)} \right) = 1 \qquad \text{since } \lim_{x \to x_0+} f(x) = \infty \text{ and } \lim_{x \to x_0} g(x) = \infty$$

Then taking the limit as $x \to x_0+$ on both sides of (2), we see that, as required,

$$\lim_{x \to x_0+} \frac{f(x)}{g(x)} = L = \lim_{x \to x_0+} \frac{f'(x)}{g'(x)}$$

Appropriate modifications of the above procedure establish the result if $x \to x_0-$, $x \to x_0$, $x \to \infty$, $x \to -\infty$.

30. Evaluate   (a) $\lim\limits_{x \to 0} \dfrac{e^{2x} - 1}{x}$,   (b) $\lim\limits_{x \to 1} \dfrac{1 + \cos \pi x}{x^2 - 2x + 1}$,   (c) $\lim\limits_{x \to 0+} \dfrac{\ln \cos 3x}{\ln \cos 2x}$.

All of these have the "indeterminate form" 0/0.

(a) $\lim\limits_{x \to 0} \dfrac{e^{2x} - 1}{x} = \lim\limits_{x \to 0} \dfrac{2e^{2x}}{1} = 2$

(b) $\lim\limits_{x \to 1} \dfrac{1 + \cos \pi x}{x^2 - 2x + 1} = \lim\limits_{x \to 1} \dfrac{-\pi \sin \pi x}{2x - 2} = \lim\limits_{x \to 1} \dfrac{-\pi^2 \cos \pi x}{2} = \dfrac{\pi^2}{2}$

Note: Here L'Hospital's rule is applied twice, since the first application again yields the "indeterminate form" 0/0 and the conditions for L'Hospital's rule are satisfied once more.

(c) $\lim\limits_{x \to 0+} \dfrac{\ln \cos 3x}{\ln \cos 2x} = \lim\limits_{x \to 0+} \dfrac{(-3 \sin 3x)/(\cos 3x)}{(-2 \sin 2x)/(\cos 2x)} = \lim\limits_{x \to 0+} \dfrac{3 \sin 3x \cos 2x}{2 \sin 2x \cos 3x}$

$\qquad = \left(\lim\limits_{x \to 0+} \dfrac{\sin 3x}{\sin 2x} \right) \cdot \left(\lim\limits_{x \to 0+} \dfrac{3 \cos 2x}{2 \cos 3x} \right) = \left(\lim\limits_{x \to 0+} \dfrac{3 \cos 3x}{2 \cos 2x} \right) \cdot \left(\dfrac{3}{2} \right)$

$\qquad = \left(\dfrac{3}{2} \right) \cdot \left(\dfrac{3}{2} \right) = \dfrac{9}{4}$

Note that in the fourth step we have taken advantage of a theorem on limits to simplify further calculations.

31. Evaluate   (a) $\lim\limits_{x \to \infty} \dfrac{3x^2 - x + 5}{5x^2 + 6x - 3}$,   (b) $\lim\limits_{x \to \infty} x^2 e^{-x}$,   (c) $\lim\limits_{x \to 0+} \dfrac{\ln \tan 2x}{\ln \tan 3x}$.

All of these have or can be arranged to have the "indeterminate form" ∞/∞.

(a) $\lim\limits_{x \to \infty} \dfrac{3x^2 - x + 5}{5x^2 + 6x - 3} = \lim\limits_{x \to \infty} \dfrac{6x - 1}{10x + 6} = \lim\limits_{x \to \infty} \dfrac{6}{10} = \dfrac{3}{5}$

(b) $\lim\limits_{x \to \infty} x^2 e^{-x} = \lim\limits_{x \to \infty} \dfrac{x^2}{e^x} = \lim\limits_{x \to \infty} \dfrac{2x}{e^x} = \lim\limits_{x \to \infty} \dfrac{2}{e^x} = 0$

(c) $\lim\limits_{x \to 0+} \dfrac{\ln \tan 2x}{\ln \tan 3x} = \lim\limits_{x \to 0+} \dfrac{(2 \sec^2 2x)/(\tan 2x)}{(3 \sec^2 3x)/(\tan 3x)} = \lim\limits_{x \to 0+} \dfrac{2 \sec^2 2x \tan 3x}{3 \sec^2 3x \tan 2x}$

$\qquad = \left(\lim\limits_{x \to 0+} \dfrac{2 \sec^2 2x}{3 \sec^2 3x} \right) \left(\lim\limits_{x \to 0+} \dfrac{\tan 3x}{\tan 2x} \right) = \left(\dfrac{2}{3} \right) \left(\lim\limits_{x \to 0+} \dfrac{3 \sec^2 3x}{2 \sec^2 2x} \right) = 1$

32. Evaluate $\lim\limits_{x \to 0} \dfrac{\sin x - \tan^{-1} x}{x^2 \ln (1 + x)}$.

Although L'Hospital's rule is applicable here, its successive use becomes involved. Using Taylor's theorem of the mean, however, the limit is obtained quickly and easily. We use the results (see Page 62)

$$\sin x = x - \frac{x^3}{3!} + \frac{Px^5}{5!}, \quad \tan^{-1} x = x - \frac{x^3}{3} + Qx^5, \quad \ln(1 + x) = x - \frac{x^2}{2} + Rx^3$$

Then the required limit equals

$$\lim_{x \to 0+} \frac{(x - x^3/3! + Px^5) - (x - x^3/3 + Qx^5)}{x^3 - x^4/2 + Rx^5} = \lim_{x \to 0} \frac{\frac{1}{6} + (P - Q)x^2}{1 - x/2 \pm Rx^2} = \frac{1}{6}$$

33. Evaluate $\lim_{x \to 0+} x^2 \ln x$.

$$\lim_{x \to 0+} x^2 \ln x = \lim_{x \to 0+} \frac{\ln x}{1/x^2} = \lim_{x \to 0+} \frac{1/x}{-2/x^3} = \lim_{x \to 0+} \frac{-x^2}{2} = 0$$

The given limit has the "indeterminate form" $0 \cdot \infty$. In the second step the form is altered so as to give the indeterminate form ∞/∞ and L'Hospital's rule is then applied.

34. Find $\lim_{x \to 0} (\cos x)^{1/x^2}$.

Since $\lim_{x \to 0} \cos x = 1$ and $\lim_{x \to 0} 1/x^2 = \infty$, the limit takes the "indeterminate form" 1^∞.

Let $F(x) = (\cos x)^{1/x^2}$. Then $\ln F(x) = (\ln \cos x)/x^2$ to which L'Hospital's rule can be applied. We have

$$\lim_{x \to 0} \frac{\ln \cos x}{x^2} = \lim_{x \to 0} \frac{(-\sin x)/(\cos x)}{2x} = \lim_{x \to 0} \frac{-\sin x}{2x \cos x} = \lim_{x \to 0} \frac{-\cos x}{-2x \sin x + 2 \cos x} = -\frac{1}{2}$$

Thus $\lim_{x \to 0} \ln F(x) = -\frac{1}{2}$. But since the logarithm is a continuous function, $\lim_{x \to 0} \ln F(x) = \ln \left(\lim_{x \to 0} F(x) \right)$. Then

$$\ln \left(\lim_{x \to 0} F(x) \right) = -\frac{1}{2} \quad \text{or} \quad \lim_{x \to 0} F(x) = \lim_{x \to 0} (\cos x)^{1/x^2} = e^{-1/2}$$

35. If $F(x) = (e^{3x} - 5x)^{1/x}$, find (a) $\lim_{x \to \infty} F(x)$ and (b) $\lim_{x \to 0} F(x)$.

The respective indeterminate forms in (a) and (b) are ∞^0 and 1^∞.

Let $G(x) = \ln F(x) = \frac{\ln (e^{3x} - 5x)}{x}$. Then $\lim_{x \to \infty} G(x)$ and $\lim_{x \to 0} G(x)$ assume the indeterminate forms ∞/∞ and $0/0$ respectively, and L'Hospital's rule applies. We have

(a) $\lim_{x \to \infty} \frac{\ln (e^{3x} - 5x)}{x} = \lim_{x \to \infty} \frac{3e^{3x} - 5}{e^{3x} - 5x} = \lim_{x \to \infty} \frac{9e^{3x}}{3e^{3x} - 5} = \lim_{x \to \infty} \frac{27e^{3x}}{9e^{3x}} = 3$

Then, as in Problem 34, $\lim_{x \to \infty} (e^{3x} - 5x)^{1/x} = e^3$.

(b) $\lim_{x \to 0} \frac{\ln (e^{3x} - 5x)}{x} = \lim_{x \to 0} \frac{3e^{3x} - 5}{e^{3x} - 5x} = -2 \quad \text{and} \quad \lim_{x \to 0} (e^{3x} - 5x)^{1/x} = e^{-2}$

36. Evaluate $\lim_{x \to 0} \left(\frac{1}{\sin^2 x} - \frac{1}{x^2} \right)$.

This has the indeterminate form $\infty - \infty$. By writing the limit as $\lim_{x \to 0} \frac{x^2 - \sin^2 x}{x^2 \sin^2 x}$, it is seen that L'Hospital's rule is applicable. However, this process proves laborious. Two methods of procedure are possible.

Method 1: The required limit can be written

$$\lim_{x \to 0} \left(\frac{x^2 - \sin^2 x}{x^4} \right) \left(\frac{x^2}{\sin^2 x} \right) = \lim_{x \to 0} \frac{x^2 - \sin^2 x}{x^4}$$

since $\lim_{x \to 0} \frac{x^2}{\sin^2 x} = \left(\lim_{x \to 0} \frac{x}{\sin x} \right)^2 = 1$. Now by successive applications of L'Hospital's rule,

$$\lim_{x \to 0} \frac{x^2 - \sin^2 x}{x^4} = \lim_{x \to 0} \frac{2x - 2 \sin x \cos x}{4x^3} = \lim_{x \to 0} \frac{2x - \sin 2x}{4x^3}$$

$$= \lim_{x \to 0} \frac{2 - 2 \cos 2x}{12x^2} = \lim_{x \to 0} \frac{4 \sin 2x}{24x} = \lim_{x \to 0} \frac{8 \cos 2x}{24} = \frac{1}{3}$$

Method 2: Using Taylor's theorem, we have

$$\lim_{x \to 0} \frac{x^2 - \sin^2 x}{x^2 \sin^2 x} = \lim_{x \to 0} \frac{x^2 - (x - x^3/6 + Px^5)^2}{x^2(x - x^3/6 + Px^5)^2}$$

$$= \lim_{x \to 0} \frac{x^4/3 + \text{terms involving } x^6 \text{ and higher}}{x^4 + \text{terms involving } x^6 \text{ and higher}}$$

$$= \lim_{x \to 0} \frac{1/3 + \text{terms involving } x^2 \text{ and higher}}{1 + \text{terms involving } x^2 \text{ and higher}} = \frac{1}{3}$$

MISCELLANEOUS PROBLEMS

37. If $x = g(t)$ and $y = f(t)$ are twice differentiable, find (a) dy/dx, (b) d^2y/dx^2.

(a) Letting primes denote derivatives with respect to t, we have

$$\frac{dy}{dx} = \frac{dy/dt}{dx/dt} = \frac{f'(t)}{g'(t)} \qquad \text{if } g'(t) \neq 0$$

(b) $\dfrac{d^2y}{dx^2} = \dfrac{d}{dx}\left(\dfrac{dy}{dx}\right) = \dfrac{d}{dx}\left(\dfrac{f'(t)}{g'(t)}\right) = \dfrac{\dfrac{d}{dt}\left(\dfrac{f'(t)}{g'(t)}\right)}{dx/dt} = \dfrac{\dfrac{d}{dt}\left(\dfrac{f'(t)}{g'(t)}\right)}{g'(t)}$

$$= \frac{1}{g'(t)} \left\{ \frac{g'(t) f''(t) - f'(t) g''(t)}{[g'(t)]^2} \right\} = \frac{g'(t) f''(t) - f'(t) g''(t)}{[g'(t)]^3} \qquad \text{if } g'(t) \neq 0$$

38. Let $f(x) = \begin{cases} e^{-1/x^2}, & x \neq 0 \\ 0, & x = 0 \end{cases}$. Prove that (a) $f'(0) = 0$, (b) $f''(0) = 0$.

(a) $f'_+(0) = \lim_{h \to 0+} \dfrac{f(h) - f(0)}{h} = \lim_{h \to 0+} \dfrac{e^{-1/h^2} - 0}{h} = \lim_{h \to 0+} \dfrac{e^{-1/h^2}}{h}$

If $h = 1/u$, using L'Hospital's rule this limit equals

$$\lim_{u \to \infty} u e^{-u^2} = \lim_{u \to \infty} u/e^{u^2} = \lim_{u \to \infty} 1/2u e^{u^2} = 0$$

Similarly, replacing $h \to 0+$ by $h \to 0-$ and $u \to \infty$ by $u \to -\infty$, we find $f'_-(0) = 0$. Thus $f'_+(0) = f'_-(0) = 0$, and so $f'(0) = 0$.

(b) $f''_+(0) = \lim_{h \to 0+} \dfrac{f'(h) - f'(0)}{h} = \lim_{h \to 0+} \dfrac{e^{-1/h^2} \cdot 2h^{-3} - 0}{h} = \lim_{h \to 0+} \dfrac{2 e^{-1/h^2}}{h^4} = \lim_{u \to \infty} \dfrac{2u^4}{e^{u^2}} = 0$

by successive applications of L'Hospital's rule.

Similarly, $f''_-(0) = 0$ and so $f''(0) = 0$.

In general, $f^{(n)}(0) = 0$ for $n = 1, 2, 3, \ldots$ (see Problem 89).

39. Let $f(x)$ be such that $f^{(IV)}(x)$ exists in $a \leq x \leq b$, and suppose that $f'(x_0) = f''(x_0) = f'''(x_0) = 0$ where $a < x_0 < b$. Prove that $f(x)$ has a relative maximum or minimum at x_0 according as $f^{(IV)}(x_0) < 0$ or > 0 respectively.

By Taylor's theorem of the mean, if ξ is between x_0 and x, then

$$f(x) \; = \; f(x_0) \, + \, f'(x_0) \, (x - x_0) \, + \, \frac{f''(x_0) \, (x - x_0)^2}{2!} \, + \, \frac{f'''(x_0) \, (x - x_0)^3}{3!} \, + \, \frac{f^{(\mathrm{IV})}(\xi) \, (x - x_0)^4}{4!}$$

$$= \; f(x_0) \, + \, \frac{f^{(\mathrm{IV})}(\xi) \, (x - x_0)^4}{4!}$$

If $f^{(\mathrm{IV})}(x_0) > 0$, then for all x in a deleted δ neighborhood of x_0 we have $f(x) > f(x_0)$, so that $f(x)$ has a relative minimum value at $x = x_0$. Similarly if $f^{(\mathrm{IV})}(x_0) < 0$, then for all x in a deleted δ neighborhood of x_0 we have $f(x) < f(x_0)$, so that $f(x)$ has a relative maximum value at $x = x_0$.

40. Find the length of the longest ladder which can be carried around the corner of a corridor, whose dimensions are indicated in the figure below, if it is assumed that the ladder is carried parallel to the floor.

The length of the *longest* ladder is the same as the *shortest* straight line segment AB [Fig. 4-6] which touches both outer walls and the corner formed by the inner walls.

As seen from Fig. 4-6, the length of the ladder AB is

$$L \; = \; a \sec \theta \, + \, b \csc \theta$$

L is a minimum when

$$dL/d\theta \; = \; a \sec \theta \tan \theta \, - \, b \csc \theta \cot \theta \; = \; 0$$

i.e., $\qquad a \sin^3 \theta \, = \, b \cos^3 \theta \quad$ or $\quad \tan \theta \, = \, \sqrt[3]{b/a}$

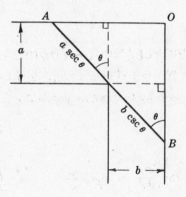

Fig. 4-6

Then $\qquad \sec \theta \; = \; \dfrac{\sqrt{a^{2/3} + b^{2/3}}}{a^{1/3}}, \quad \csc \theta \; = \; \dfrac{\sqrt{a^{2/3} + b^{2/3}}}{b^{1/3}}$

so that $\qquad L \; = \; a \sec \theta \, + \, b \csc \theta \; = \; (a^{2/3} + b^{2/3})^{3/2}$

Although it is geometrically evident that this gives the minimum length, we can prove this analytically by showing that $d^2L/d\theta^2$ for $\theta = \tan^{-1} \sqrt[3]{b/a}$ is positive (see Problem 88).

Supplementary Problems

DERIVATIVES

41. Use the definition to compute the derivatives of each of the following functions at the indicated point:
   (a) $(3x - 4)/(2x + 3)$, $x = 1$; (b) $x^3 - 3x^2 + 2x - 5$, $x = 2$; (c) $\sqrt{x}$, $x = 4$; (d) $\sqrt[3]{6x - 4}$, $x = 2$.
   Ans. (a) $17/25$, (b) 2, (c) $\tfrac{1}{4}$, (d) $\tfrac{1}{2}$

42. Let $f(x) = \begin{cases} x^3 \sin 1/x, & x \neq 0 \\ 0, & x = 0 \end{cases}$. Prove that (a) $f(x)$ is continuous at $x = 0$, (b) $f(x)$ has a derivative at $x = 0$, (c) $f'(x)$ is continuous at $x = 0$.

43. Let $f(x) = \begin{cases} xe^{-1/x^2}, & x \neq 0 \\ 0, & x = 0 \end{cases}$. Determine whether $f(x)$ (a) is continuous at $x = 0$, (b) has a derivative at $x = 0$. *Ans.* (a) Yes; (b) Yes, 0

44. Give an alternative proof of the theorem in Problem 3, Page 63, using "ϵ, δ definitions".

45. If $f(x) = e^x$, show that $f'(x_0) = e^{x_0}$ depends on the result $\lim_{h \to 0} (e^h - 1)/h = 1$.

46. Use the results $\lim_{h \to 0} (\sin h)/h = 1$, $\lim_{h \to 0} (1 - \cos h)/h = 0$ to prove that if $f(x) = \sin x$, $f'(x_0) = \cos x_0$.

RIGHT and LEFT HAND DERIVATIVES

47. Let $f(x) = x|x|$. (a) Calculate the right hand derivative of $f(x)$ at $x = 0$. (b) Calculate the left hand derivative of $f(x)$ at $x = 0$. (c) Does $f(x)$ have a derivative at $x = 0$? (d) Illustrate the conclusions in (a), (b) and (c) from a graph.   *Ans.* (a) 0; (b) 0; (c) Yes, 0

48. Discuss the (a) continuity and (b) differentiability of $f(x) = x^p \sin 1/x$, $f(0) = 0$, where p is any positive number. What happens in case p is any real number?

49. Let $f(x) = \begin{cases} 2x - 3, & 0 \le x \le 2 \\ x^2 - 3, & 2 < x \le 4 \end{cases}$.   Discuss the (a) continuity and (b) differentiability of $f(x)$ in $0 \le x \le 4$.

50. Prove that the derivative of $f(x)$ at $x = x_0$ exists if and only if $f'_+(x_0) = f'_-(x_0)$.

51. (a) Prove that $f(x) = x^3 - x^2 + 5x - 6$ is differentiable in $a \le x \le b$, where a and b are any constants. (b) Find equations for the tangent lines to the curve $y = x^3 - x^2 + 5x - 6$ at $x = 0$ and $x = 1$. Illustrate by means of a graph. (c) Determine the point of intersection of the tangent lines in (b). (d) Find $f'(x)$, $f''(x)$, $f'''(x)$, $f^{(IV)}(x)$, ...
    Ans. (b) $y = 5x - 6$, $y = 6x - 7$; (c) $(1, -1)$; (d) $3x^2 - 2x + 5$, $6x - 2$, 6, 0, 0, 0, ...

52. Explain clearly the difference between (a) $f'_+(x_0)$ and $f'(x_0+)$, (b) $f'_-(x_0)$ and $f'(x_0-)$.

53. If $f(x) = x^2|x|$, discuss the existence of successive derivatives of $f(x)$ at $x = 0$.

DIFFERENTIALS

54. If $y = f(x) = x + 1/x$, find (a) Δy, (b) dy, (c) $\Delta y - dy$, (d) $(\Delta y - dy)/\Delta x$, (e) dy/dx.
    Ans. (a) $\Delta x - \dfrac{\Delta x}{x(x + \Delta x)}$, (b) $\left(1 - \dfrac{1}{x^2}\right)\Delta x$, (c) $\dfrac{(\Delta x)^2}{x^2(x + \Delta x)}$, (d) $\dfrac{\Delta x}{x^2(x + \Delta x)}$, (e) $1 - \dfrac{1}{x^2}$. Note: $\Delta x = dx$.

55. If $f(x) = x^2 + 3x$, find (a) Δy, (b) dy, (c) $\Delta y/\Delta x$, (d) dy/dx and (e) $(\Delta y - dy)/\Delta x$, if $x = 1$ and $\Delta x = .01$.
    Ans. (a) .0501, (b) .05, (c) 5.01, (d) 5, (e) .01

56. Using differentials, compute approximate values for each of the following: (a) $\sin 31°$, (b) $\ln(1.12)$, (c) $\sqrt[5]{36}$.   *Ans.* (a) 0.515, (b) 0.12, (c) 2.0125

57. If $y = \sin x$, evaluate (a) Δy, (b) dy. (c) Prove that $(\Delta y - dy)/\Delta x \to 0$ as $\Delta x \to 0$.

DIFFERENTIATION RULES and SPECIAL FUNCTIONS

58. Prove: (a) $\dfrac{d}{dx}\{f(x) + g(x)\} = \dfrac{d}{dx}f(x) + \dfrac{d}{dx}g(x)$,   (b) $\dfrac{d}{dx}\{f(x) - g(x)\} = \dfrac{d}{dx}f(x) - \dfrac{d}{dx}g(x)$,

    (c) $\dfrac{d}{dx}\left\{\dfrac{f(x)}{g(x)}\right\} = \dfrac{g(x)f'(x) - f(x)g'(x)}{[g(x)]^2}$, $g(x) \ne 0$.

59. Evaluate (a) $\dfrac{d}{dx}\{x^3 \ln(x^2 - 2x + 5)\}$ at $x = 1$, (b) $\dfrac{d}{dx}\{\sin^2(3x + \pi/6)\}$ at $x = 0$.
    Ans. (a) $3 \ln 4$, (b) $\frac{3}{2}\sqrt{3}$

60. Derive the formulas: (a) $\dfrac{d}{dx}a^u = a^u \ln a\,\dfrac{du}{dx}$, $a > 0$, $a \ne 1$; (b) $\dfrac{d}{dx}\csc u = -\csc u \cot u\,\dfrac{du}{dx}$;

    (c) $\dfrac{d}{dx}\tanh u = \operatorname{sech}^2 u\,\dfrac{du}{dx}$ where u is a differentiable function of x.

61. Compute (a) $\dfrac{d}{dx}\tan^{-1} x$, (b) $\dfrac{d}{dx}\csc^{-1} x$, (c) $\dfrac{d}{dx}\sinh^{-1} x$, (d) $\dfrac{d}{dx}\coth^{-1} x$, paying attention to the use of principal values.

62. If $y = x^x$, compute dy/dx. [Hint: Take logarithms before differentiating]. *Ans.* $x^x(1 + \ln x)$.

63. If $y = \{\ln(3x+2)\}^{\sin^{-1}(2x+.5)}$, find dy/dx at $x = 0$. *Ans.* $\left(\dfrac{\pi}{4\ln 2} + \dfrac{2\ln\ln 2}{\sqrt{3}}\right)(\ln 2)^{\pi/6}$

64. If $y = f(u)$, where $u = g(v)$ and $v = h(x)$, prove that $\dfrac{dy}{dx} = \dfrac{dy}{du} \cdot \dfrac{du}{dv} \cdot \dfrac{dv}{dx}$ assuming f, g and h are differentiable.

65. Calculate (a) dy/dx and (b) d^2y/dx^2 if $xy - \ln y = 1$.
    Ans. (a) $y^2/(1-xy)$, (b) $(3y^3 - 2xy^4)/(1-xy)^3$ provided $xy \neq 1$

66. If $y = \tan x$, prove that $y''' = 2(1+y^2)(1+3y^2)$.

67. If $x = \sec t$ and $y = \tan t$, evaluate (a) dy/dx, (b) d^2y/dx^2, (c) d^3y/dx^3, at $t = \pi/4$.
    Ans. (a) $\sqrt{2}$, (b) -1, (c) $3\sqrt{2}$

68. Prove that $\dfrac{d^2y}{dx^2} = -\dfrac{d^2x}{dy^2}\Big/\left(\dfrac{dx}{dy}\right)^3$, stating precise conditions under which it holds.

69. Establish formulas (a) 7, (b) 18 and (c) 27, on Page 60.

MEAN VALUE THEOREMS

70. Let $f(x) = 1 - (x-1)^{2/3}$, $0 \le x \le 2$. (a) Construct the graph of $f(x)$. (b) Explain why Rolle's theorem is not applicable to this function, i.e. there is no value ξ for which $f'(\xi) = 0$, $0 < \xi < 2$.

71. Verify Rolle's theorem for $f(x) = x^2(1-x)^2$, $0 \le x \le 1$.

72. Prove that between any two real roots of $e^x \sin x = 1$ there is at least one real root of $e^x \cos x = -1$.
    [Hint: Apply Rolle's theorem to the function $e^{-x} - \sin x$.]

73. (a) If $0 < a < b$, prove that $(1 - a/b) < \ln b/a < (b/a - 1)$
    (b) Use the result of (a) to show that $\frac{1}{6} < \ln 1.2 < \frac{1}{5}$.

74. Prove that $(\pi/6 + \sqrt{3}/15) < \sin^{-1}.6 < (\pi/6 + 1/8)$ by using the theorem of the mean.

75. Prove the statement in the last paragraph of Problem 20(b).

76. (a) If $f'(x) \le 0$ at all points of (a, b), prove that $f(x)$ is monotonic decreasing in (a, b).
    (b) Under what conditions is $f(x)$ strictly decreasing in (a, b)?

77. (a) Prove that $(\sin x)/x$ is strictly decreasing in $(0, \pi/2)$. (b) Prove that $0 \le \sin x \le 2x/\pi$ for $0 \le x \le \pi/2$.

78. (a) Prove that $\dfrac{\sin b - \sin a}{\cos a - \cos b} = \cot \xi$, where ξ is between a and b.
    (b) By placing $a = 0$ and $b = x$ in (a), show that $\xi = x/2$. Does the result hold if $x < 0$?

TAYLOR'S THEOREM of the MEAN

79. (a) Prove that $\ln(1+x) = x - \dfrac{x^2}{2} + \dfrac{x^3}{3} - \dfrac{x^4}{4} + \dfrac{x^5}{5(1+\xi)^5}$, $0 < \xi < x$. (b) Use (a) to evaluate $\ln(1.1)$ and estimate the accuracy. *Ans.* (b) 0.09531 with an error less than $2 \cdot 10^{-6}$.

80. Evaluate (a) $\cos 64°$, (b) $\tan^{-1} 0.2$, (c) $\cosh 1$, (d) $e^{-0.3}$ to 3 decimal places.
    Ans. (a) 0.438, (b) 0.197, (c) 1.543, (d) 0.741

81. Prove Taylor's theorem of the mean with the Lagrange remainder for (a) $n = 2$, (b) $n = 3$, (c) $n =$ any positive integer.

82. Derive the result (21), Page 61, for the Cauchy form of the remainder.
[Hint: Apply Rolle's theorem to

$$H(x) = f(b) - f(x) - f'(x)(b-x) - \frac{f''(x)(b-x)^2}{2!} - \cdots - \frac{f^{(n)}(x)(b-x)^n}{n!} - (b-x)A$$

where A is chosen so that $H(a) = 0$.]

83. Prove that the Lagrange and Cauchy forms of the remainder in Taylor's theorem can be written in the forms

$$\frac{h^{n+1}f^{n+1}(a+\theta h)}{(n+1)!} \quad \text{and} \quad \frac{h^{n+1}(1-\theta)^n f^{(n+1)}(a+\theta h)}{n!}$$

respectively, where $h = b - a$ and $0 < \theta < 1$.

84. By writing $(b-x)^p A$ in place of the last term $(b-x)A$ in the hint of Problem 82, obtain Taylor's theorem with (a) the Lagrange remainder, (b) the Cauchy remainder, using suitable special values for p.

L'HOSPITAL'S RULE

85. Evaluate each of the following limits.

(a) $\lim\limits_{x \to 0} \dfrac{x - \sin x}{x^3}$

(e) $\lim\limits_{x \to 0+} x^3 \ln x$

(i) $\lim\limits_{x \to 0} (1/x - \csc x)$

(m) $\lim\limits_{x \to \infty} x \ln\left(\dfrac{x+3}{x-3}\right)$

(b) $\lim\limits_{x \to 0} \dfrac{e^{2x} - 2e^x + 1}{\cos 3x - 2\cos 2x + \cos x}$

(f) $\lim\limits_{x \to 0} (3^x - 2^x)/x$

(j) $\lim\limits_{x \to 0} x^{\sin x}$

(n) $\lim\limits_{x \to 0} \left(\dfrac{\sin x}{x}\right)^{1/x^2}$

(c) $\lim\limits_{x \to 1+} (x^2 - 1)\tan \pi x/2$

(g) $\lim\limits_{x \to \infty} (1 - 3/x)^{2x}$

(k) $\lim\limits_{x \to 0} (1/x^2 - \cot^2 x)$

(o) $\lim\limits_{x \to \infty} (x + e^x + e^{2x})^{1/x}$

(d) $\lim\limits_{x \to \infty} x^3 e^{-2x}$

(h) $\lim\limits_{x \to \infty} (1 + 2x)^{1/3x}$

(l) $\lim\limits_{x \to 0} \dfrac{\tan^{-1} x - \sin^{-1} x}{x(1 - \cos x)}$

(p) $\lim\limits_{x \to 0+} (\sin x)^{1/\ln x}$

Ans. (a) $\frac{1}{6}$, (b) -1, (c) $-4/\pi$, (d) 0, (e) 0, (f) $\ln 3/2$, (g) e^{-6}, (h) 1, (i) 0, (j) 1, (k) $\frac{2}{3}$, (l) $\frac{1}{3}$, (m) 6, (n) $e^{-1/6}$, (o) e^2, (p) e

MISCELLANEOUS PROBLEMS

86. Prove that $\sqrt{\dfrac{1-x}{1+x}} < \dfrac{\ln(1+x)}{\sin^{-1} x} < 1$ if $0 < x < 1$.

87. If $\Delta f(x) = f(x + \Delta x) - f(x)$,   (a) prove that $\Delta\{\Delta f(x)\} = \Delta^2 f(x) = f(x + 2\Delta x) - 2f(x + \Delta x) + f(x)$,

(b) derive an expression for $\Delta^n f(x)$ where n is any positive integer,   (c) show that $\lim\limits_{\Delta x \to 0} \dfrac{\Delta^n f(x)}{(\Delta x)^n} = f^{(n)}(x)$ if this limit exists.

88. Complete the analytic proof mentioned at the end of Problem 40.

89. (a) If $f(x)$ is the function of Problem 38, prove that $f^{(n)}(0) = 0$ for $n = 1, 2, 3, \ldots$.   (b) Write Taylor's series with a remainder for this function and prove that $f(x) = R_n$.   (c) Explain why R_n cannot approach zero as $n \to \infty$ and discuss the consequences of this.

90. Find the relative maxima and minima of $f(x) = x^x$, $x > 0$.
Ans. $f(x)$ has a relative minimum when $x = e^{-1}$.

91. A particle travels with constant velocities v_1 and v_2 in mediums I and II respectively (see adjoining Fig. 4-7). Show that in order to go from point P to point Q in the least time, it must follow path PAQ where A is such that

$$(\sin \theta_1)/(\sin \theta_2) = v_1/v_2$$

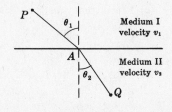

Fig. 4-7

92. A variable α is called an *infinitesimal* if it has zero as a limit. Given two infinitesimals α and β, we say that α is an infinitesimal of *higher order* (or the *same order*) if $\lim \alpha/\beta = 0$ (or $\lim \alpha/\beta = l \neq 0$). Prove that as $x \to 0$, (a) $\sin^2 2x$ and $(1 - \cos 3x)$ are infinitesimals of the same order, (b) $(x^3 - \sin^3 x)$ is an infinitesimal of higher order than $\{x - \ln(1+x) - 1 + \cos x\}$.

93. Why can we not use L'Hospital's rule to prove that $\lim\limits_{x \to 0} \dfrac{x^2 \sin 1/x}{\sin x} = 0$ (see Prob. 91, Chap. 2)?

94. Can we use L'Hospital's rule to evaluate the limit of the sequence $u_n = n^3 e^{-n^2}$, $n = 1, 2, 3, \ldots$? Explain.

95. If a is an approximate root of $f(x) = 0$, show that a better approximation, in general, is given by $a - f(a)/f'(a)$ (*Newton's method*).

[Hint: Assume the actual root is $a + h$, so that $f(a + h) = 0$. Then use the fact that for small h, $f(a + h) = f(a) + h f'(a)$ approximately.]

96. Using successive applications of Problem 95, obtain the positive root of (a) $x^3 - 2x^2 - 2x - 7 = 0$, (b) $5 \sin x = 4x$ to 3 decimal places.   *Ans.* (a) 3.268, (b) 1.131

97. If D denotes the operator d/dx so that $Dy \equiv dy/dx$ while $D^k y \equiv d^k y/dx^k$, prove *Leibnitz's formula*

$$D^n(uv) = (D^n u)v + {}_nC_1(D^{n-1}u)(Dv) + {}_nC_2(D^{n-2}u)(D^2v) + \cdots + {}_nC_r(D^{n-r}u)(D^r v) + \cdots + uD^n v$$

where ${}_nC_r = \binom{n}{r}$ are the binomial coefficients (see Problem 95, Chapter 1).

98. Prove that $\dfrac{d^n}{dx^n}(x^2 \sin x) = \{x^2 - n(n-1)\} \sin(x + n\pi/2) - 2nx \cos(x + n\pi/2)$.

99. If $f'(x_0) = f''(x_0) = \cdots = f^{(2n)}(x_0) = 0$ but $f^{(2n+1)}(x_0) \neq 0$, discuss the behavior of $f(x)$ in the neighborhood of $x = x_0$. The point x_0 in such case is often called a *point of inflection*.

100. Let $f(x)$ be twice differentiable in (a, b) and suppose that $f'(a) = f'(b) = 0$. Prove that there exists at least one point ξ in (a, b) such that $|f''(\xi)| \geq \dfrac{4}{(b-a)^2}\{f(b) - f(a)\}$. Give a physical interpretation involving velocity and acceleration of a particle.

Chapter 5

Integrals

DEFINITION of a DEFINITE INTEGRAL

The concept of a definite integral is often motivated by consideration of the area bounded by the curve $y = f(x)$, the x axis and the ordinates erected at $x = a$ and $x = b$ (see Fig. 5-1). However, the definition can be given without appealing to geometry.

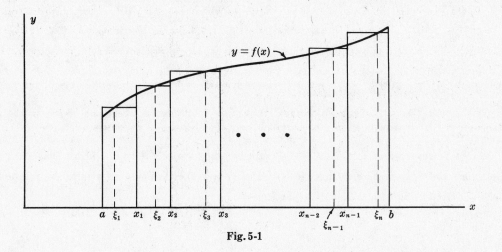

Fig. 5-1

Subdivide the interval $a \leqq x \leqq b$ into n sub-intervals by means of the points $x_1, x_2, \ldots, x_{n-1}$ chosen arbitrarily. In each of the new intervals $(a, x_1), (x_1, x_2), \ldots, (x_{n-1}, b)$ choose points $\xi_1, \xi_2, \ldots, \xi_n$ arbitrarily. Form the sum

$$f(\xi_1)(x_1 - a) + f(\xi_2)(x_2 - x_1) + f(\xi_3)(x_3 - x_2) + \cdots + f(\xi_n)(b - x_{n-1}) \tag{1}$$

By writing $x_0 = a$, $x_n = b$ and $x_k - x_{k-1} = \Delta x_k$, this can be written

$$\sum_{k=1}^{n} f(\xi_k)(x_k - x_{k-1}) = \sum_{k=1}^{n} f(\xi_k) \Delta x_k \tag{2}$$

Geometrically this sum represents the total area of all rectangles in the above figure.

We now let the number of subdivisions n increase in such a way that each $\Delta x_k \to 0$. If as a result the sum (1) or (2) approaches a limit which does not depend on the mode of subdivision, we denote this limit by

$$\int_a^b f(x)\,dx \tag{3}$$

which is called the *definite integral of $f(x)$ between a and b*. In this symbol $f(x)\,dx$ is often called the *integrand*, and $[a, b]$ is called the *range of integration*. We sometimes call a and b the limits of integration, a being the lower limit of integration and b the upper limit.

80

The limit (*3*) exists whenever $f(x)$ is continuous (or sectionally continuous) in $a \leq x \leq b$ (see Problem 35). When this limit exists we say that $f(x)$ is *Riemann integrable* or simply *integrable* in $[a, b]$.

Geometrically the value of this definite integral represents the area bounded by the curve $y = f(x)$, the x axis and the ordinates at $x = a$ and $x = b$ only if $f(x) \geq 0$. If $f(x)$ is sometimes positive and sometimes negative, the definite integral represents the algebraic sum of the areas above and below the x axis, treating areas above the x axis as positive and areas below the x axis as negative.

MEASURE ZERO

A set of points on the x axis is said to have *measure zero* if the sum of the lengths of intervals enclosing all the points can be made arbitrary small (less than any given positive number ϵ). We can show (see Problem 6) that any countable set of points on the real axis has measure zero. In particular, the set of rational numbers which is countable (see Problems 17 and 59, Chapter 1), has measure zero.

An important theorem in the theory of Riemann integration is the following:

Theorem. If $f(x)$ is bounded in $[a, b]$, then a necessary and sufficient condition for the existence of $\displaystyle\int_a^b f(x)\,dx$ is that the set of discontinuities of $f(x)$ have measure zero.

PROPERTIES of DEFINITE INTEGRALS

If $f(x)$ and $g(x)$ are integrable in $[a, b]$ then

1. $\displaystyle\int_a^b \{f(x) \pm g(x)\}\,dx \;=\; \int_a^b f(x)\,dx \pm \int_a^b g(x)\,dx$

2. $\displaystyle\int_a^b A\,f(x)\,dx \;=\; A\int_a^b f(x)\,dx \qquad$ where A is any constant

3. $\displaystyle\int_a^b f(x)\,dx \;=\; \int_a^c f(x)\,dx \;+\; \int_c^b f(x)\,dx$

   provided $f(x)$ is integrable in $[a, c]$ and $[c, b]$.

4. $\displaystyle\int_a^b f(x)\,dx \;=\; -\int_b^a f(x)\,dx$

5. $\displaystyle\int_a^a f(x)\,dx \;=\; 0$

6. If in $a \leq x \leq b$, $m \leq f(x) \leq M$ where m and M are constants, then
$$m(b-a) \;\leq\; \int_a^b f(x)\,dx \;\leq\; M(b-a)$$

7. If in $a \leq x \leq b$, $f(x) \leq g(x)$ then
$$\int_a^b f(x)\,dx \;\leq\; \int_a^b g(x)\,dx$$

8. $\displaystyle\left|\int_a^b f(x)\,dx\right| \;\leq\; \int_a^b |f(x)|\,dx \qquad$ if $a < b$

MEAN VALUE THEOREMS for INTEGRALS

1. First mean value theorem. If $f(x)$ is continuous in $[a, b]$, there is a point ξ in (a, b) such that
$$\int_a^b f(x)\,dx \;=\; (b-a)\,f(\xi) \tag{4}$$

2. Generalized first mean value theorem. If $f(x)$ and $g(x)$ are continuous in $[a, b]$, and $g(x)$ does not change sign in the interval, then there is a point ξ in (a, b) such that

$$\int_a^b f(x)\, g(x)\, dx \;=\; f(\xi) \int_a^b g(x)\, dx \tag{5}$$

This reduces to (4) if $g(x) = 1$.

3. Bonnet's second mean value theorem. If $f(x)$ and $g(x)$ are continuous in $[a, b]$ and if $g(x)$ is a positive monotonic decreasing function, then there is a point ξ in (a, b) such that

$$\int_a^b f(x)\, g(x)\, dx \;=\; g(a) \int_a^\xi f(x)\, dx \tag{6}$$

If $g(x)$ is a positive monotonic increasing function, then there is a point ξ in (a, b) such that

$$\int_a^b f(x)\, g(x)\, dx \;=\; g(b) \int_\xi^b f(x)\, dx \tag{7}$$

4. Generalized second mean value theorem. If $f(x)$ and $g(x)$ are continuous in $[a, b]$ and if $g(x)$ is monotonic increasing or monotonic decreasing and is not necessarily always positive as in 3, there is a point ξ in (a, b) such that

$$\int_a^b f(x)\, g(x)\, dx \;=\; g(a) \int_a^\xi f(x)\, dx \;+\; g(b) \int_\xi^b f(x)\, dx \tag{8}$$

This result holds also if we replace continuity by integrability.

INDEFINITE INTEGRALS

If $f(x)$ is given, then any function $F(x)$ such that $F'(x) = f(x)$ is called an *indefinite integral* or *anti-derivative* of $f(x)$. Clearly if $F(x)$ is an indefinite integral of $f(x)$, so also is $F(x) + c$ where c is any constant since $[F(x) + c]' = F'(x) = f(x)$. Thus all indefinite integrals differ by a constant. We often use the symbol $\int f(x)\, dx$ to denote any indefinite integral of $f(x)$.

> **Example:** If $F'(x) = x^2$, then $F(x) = \int x^2\, dx = x^3/3 + c$ is an indefinite integral or anti-derivative of x^2.

FUNDAMENTAL THEOREM of INTEGRAL CALCULUS

If $f(x)$ is continuous in $[a, b]$ and $F(x)$ is any function such that $F'(x) = f(x)$ [i.e. $F(x)$ is an indefinite integral or antiderivative of $f(x)$], then

$$\int_a^b f(x)\, dx \;=\; F(b) - F(a) \tag{9}$$

This important theorem enables us to calculate definite integrals without direct use of the definition whenever an indefinite integral is known.

> **Example:** To calculate $\int_1^2 x^2\, dx$, we note that $F'(x) = x^2$, $F(x) = x^3/3 + c$, and we have
>
> $$\int_1^2 x^2\, dx \;=\; F(2) - F(1) \;=\; \left(\frac{2^3}{3} + c\right) - \left(\frac{1^3}{3} + c\right) \;=\; \frac{7}{3}$$
>
> Since c disappears anyway, it is convenient to write more simply
>
> $$\int_1^2 x^2\, dx \;=\; \frac{x^3}{3}\Big|_1^2 \;=\; \frac{2^3}{3} - \frac{1^3}{3} \;=\; \frac{7}{3}$$

DEFINITE INTEGRALS with VARIABLE LIMITS of INTEGRATION

An indefinite integral can be expressed as a definite integral with variable upper limit by writing

$$\int f(x)\, dx \;=\; \int_a^x f(x)\, dx \;+\; c \tag{10}$$

It follows that

$$\frac{d}{dx} \int_a^x f(x)\, dx \;=\; f(x) \tag{11}$$

Since a definite integral depends only on the limits of integration, we can use any variable as the symbol of integration. For example, $\int_a^b f(x)\, dx = \int_a^b f(t)\, dt = \int_a^b f(u)\, du$, etc. For this reason the variable is often called a *dummy variable*. We can write (11), for instance, as

$$\frac{d}{dx} \int_a^x f(t)\, dt \;=\; f(x) \tag{12}$$

The result can be generalized to the case where the lower and upper integration limits are variable. Thus we have

$$\frac{d}{dx} \int_{u(x)}^{v(x)} f(t)\, dt \;=\; f\{v(x)\}\frac{dv}{dx} \;-\; f\{u(x)\}\frac{du}{dx} \tag{13}$$

Example: $\qquad \dfrac{d}{dx}\displaystyle\int_x^{x^2} \dfrac{\sin t}{t}\, dt \;=\; \dfrac{\sin x^2}{x^2}\dfrac{d(x^2)}{dx} \;-\; \dfrac{\sin x}{x}\dfrac{d(x)}{dx} \;=\; \dfrac{2\sin x^2 - \sin x}{x}$

CHANGE of VARIABLE of INTEGRATION

If a determination of $\int f(x)\, dx$ is not immediately obvious in terms of elementary functions, useful results may be obtained by changing the variable form x to t according to the transformation $x = g(t)$. The fundamental theorem enabling us to do this is summarized in the statement

$$\int f(x)\, dx \;=\; \int f\{g(t)\}\, g'(t)\, dt \tag{14}$$

where after obtaining the indefinite integral on the right we replace t by its value in terms of x, i.e. $t = g^{-1}(x)$, assumed single-valued. This result is analogous to the chain rule for differentiation (see Page 59).

The corresponding theorem for definite integrals is

$$\int_a^b f(x)\, dx \;=\; \int_\alpha^\beta f\{g(t)\}\, g'(t)\, dt \tag{15}$$

where $g(\alpha) = a$ and $g(\beta) = b$, i.e. $\alpha = g^{-1}(a)$, $\beta = g^{-1}(b)$. This result is certainly valid if $f(x)$ is continuous in $[a, b]$ and if $g(t)$ is continuous and has a continuous derivative in $\alpha \leqq t \leqq \beta$.

INTEGRALS of SPECIAL FUNCTIONS

The following results can be demonstrated by differentiating both sides to produce an identity. In each case an arbitrary constant c (which has been omitted here) should be added.

1. $\displaystyle\int u^n\, du \;=\; \frac{u^{n+1}}{n+1} \quad n \neq -1$

2. $\displaystyle\int \frac{du}{u} \;=\; \ln|u|$

3. $\displaystyle\int \sin u\, du \;=\; -\cos u$

4. $\displaystyle\int \cos u\, du \;=\; \sin u$

5. $\displaystyle\int \tan u\, du \;=\; \ln|\sec u|$
   $\qquad\qquad\qquad\; =\; -\ln|\cos u|$

6. $\displaystyle\int \cot u\, du \;=\; \ln|\sin u|$

7. $\displaystyle\int \sec u \, du = \ln |\sec u + \tan u|$
$= \ln |\tan (u/2 + \pi/4)|$

8. $\displaystyle\int \csc u \, du = \ln |\csc u - \cot u|$
$= \ln |\tan u/2|$

9. $\displaystyle\int \sec^2 u \, du = \tan u$

10. $\displaystyle\int \csc^2 u \, du = -\cot u$

11. $\displaystyle\int \sec u \tan u \, du = \sec u$

12. $\displaystyle\int \csc u \cot u \, du = -\csc u$

13. $\displaystyle\int a^u \, du = \frac{a^u}{\ln a} \quad a > 0, \; a \neq 1$

14. $\displaystyle\int e^u \, du = e^u$

15. $\displaystyle\int \sinh u \, du = \cosh u$

16. $\displaystyle\int \cosh u \, du = \sinh u$

17. $\displaystyle\int \tanh u \, du = \ln \cosh u$

18. $\displaystyle\int \coth u \, du = \ln |\sinh u|$

19. $\displaystyle\int \operatorname{sech} u \, du = \tan^{-1}(\sinh u)$

20. $\displaystyle\int \operatorname{csch} u \, du = -\coth^{-1}(\cosh u)$

21. $\displaystyle\int \operatorname{sech}^2 u \, du = \tanh u$

22. $\displaystyle\int \operatorname{csch}^2 u \, du = -\coth u$

23. $\displaystyle\int \operatorname{sech} u \tanh u \, du = -\operatorname{sech} u$

24. $\displaystyle\int \operatorname{csch} u \coth u \, du = -\operatorname{csch} u$

25. $\displaystyle\int \frac{du}{\sqrt{a^2 - u^2}} = \sin^{-1}\frac{u}{a} \;\text{ or }\; -\cos^{-1}\frac{u}{a}$

26. $\displaystyle\int \frac{du}{\sqrt{u^2 \pm a^2}} = \ln |u + \sqrt{u^2 \pm a^2}|$

27. $\displaystyle\int \frac{du}{u^2 + a^2} = \frac{1}{a}\tan^{-1}\frac{u}{a} \;\text{ or }\; -\frac{1}{a}\cot^{-1}\frac{u}{a}$

28. $\displaystyle\int \frac{du}{u^2 - a^2} = \frac{1}{2a}\ln\left|\frac{u-a}{u+a}\right|$

29. $\displaystyle\int \frac{du}{u\sqrt{a^2 \pm u^2}} = \frac{1}{a}\ln\left|\frac{u}{a + \sqrt{a^2 \pm u^2}}\right|$

30. $\displaystyle\int \frac{du}{u\sqrt{u^2 - a^2}} = \frac{1}{a}\cos^{-1}\frac{a}{u} \;\text{ or }\; \frac{1}{a}\sec^{-1}\frac{u}{a}$

31. $\displaystyle\int \sqrt{u^2 \pm a^2} \, du = \frac{u}{2}\sqrt{u^2 \pm a^2}$
$\pm \frac{a^2}{2}\ln |u + \sqrt{u^2 \pm a^2}|$

32. $\displaystyle\int \sqrt{a^2 - u^2} \, du = \frac{u}{2}\sqrt{a^2 - u^2} + \frac{a^2}{2}\sin^{-1}\frac{u}{a}$

33. $\displaystyle\int e^{au} \sin bu \, du = \frac{e^{au}(a \sin bu - b \cos bu)}{a^2 + b^2}$

34. $\displaystyle\int e^{au} \cos bu \, du = \frac{e^{au}(a \cos bu + b \sin bu)}{a^2 + b^2}$

SPECIAL METHODS of INTEGRATION

1. Integration by parts.

$$\int u \, dv = uv - \int v \, du \quad \text{or} \quad \int f(x) \, g'(x) \, dx = f(x) \, g(x) - \int f'(x) \, g(x) \, dx$$

where $u = f(x)$ and $v = g(x)$. The corresponding result for definite integrals over the interval $[a, b]$ is certainly valid if $f(x)$ and $g(x)$ are continuous and have continuous derivatives in $[a, b]$. See Problems 18-20.

2. Partial fractions.

Any rational function $\dfrac{P(x)}{Q(x)}$ where $P(x)$ and $Q(x)$ are polynomials, with the degree of $P(x)$ less than that of $Q(x)$, can be written as the sum of rational functions having the form $\dfrac{A}{(ax+b)^r}$, $\dfrac{Ax+B}{(ax^2+bx+c)^r}$ where $r = 1, 2, 3, \ldots$ which can always be integrated in terms of elementary functions.

Example 1: $\dfrac{3x-2}{(4x-3)(2x+5)^3} = \dfrac{A}{4x-3} + \dfrac{B}{(2x+5)^3} + \dfrac{C}{(2x+5)^2} + \dfrac{D}{2x+5}$

Example 2: $\dfrac{5x^2-x+2}{(x^2+2x+4)^2(x-1)} = \dfrac{Ax+B}{(x^2+2x+4)^2} + \dfrac{Cx+D}{x^2+2x+4} + \dfrac{E}{x-1}$

The constants, A, B, C, etc., can be found by clearing of fractions and equating coefficients of like powers of x on both sides of the equation or by using special methods (see Problem 21).

3. **Rational functions of sin x and cos x** can always be integrated in terms of elementary functions by the substitution $\tan x/2 = u$ (see Problem 22).

4. **Special devices** depending on the particular form of the integrand are often employed (see Problems 23 and 24).

IMPROPER INTEGRALS

If the range of integration $[a, b]$ is not finite or if $f(x)$ is not defined or not bounded at one or more points of $[a, b]$, then the integral of $f(x)$ over this range is called an *improper integral*. By use of appropriate limiting operations, we may define the integrals in such cases.

Example 1: $\displaystyle\int_0^\infty \frac{dx}{1+x^2} = \lim_{M\to\infty} \int_0^M \frac{dx}{1+x^2} = \lim_{M\to\infty} \tan^{-1} x \Big|_0^M = \lim_{M\to\infty} \tan^{-1} M = \pi/2$

Example 2: $\displaystyle\int_0^1 \frac{dx}{\sqrt{x}} = \lim_{\epsilon\to 0+} \int_\epsilon^1 \frac{dx}{\sqrt{x}} = \lim_{\epsilon\to 0+} 2\sqrt{x}\Big|_\epsilon^1 = \lim_{\epsilon\to 0+} (2 - 2\sqrt{\epsilon}) = 2$

Example 3: $\displaystyle\int_0^1 \frac{dx}{x} = \lim_{\epsilon\to 0+} \int_\epsilon^1 \frac{dx}{x} = \lim_{\epsilon\to 0+} \ln x \Big|_\epsilon^1 = \lim_{\epsilon\to 0+} (-\ln \epsilon)$.

Since this limit does not exist we say that the integral diverges (i.e. does not converge).

For further examples, see Problems 33, 78-80. For further discussion of improper integrals, see Chapter 12.

NUMERICAL METHODS for EVALUATING DEFINITE INTEGRALS

Numerical methods for evaluating definite integrals are available in case the integrals cannot be evaluated exactly. The following special numerical methods are based on subdividing the interval $[a, b]$ into n equal parts of length $\Delta x = (b - a)/n$. For simplicity we denote $f(a + k\,\Delta x) = f(x_k)$ by y_k, where $k = 0, 1, 2, \ldots, n$. The symbol $\approx$ means "approximately equal". In general, the approximation improves as n increases.

1. **Rectangular rule.**

$$\int_a^b f(x)\,dx \;\approx\; \Delta x\,\{y_0 + y_1 + y_2 + \cdots + y_{n-1}\} \quad \text{or} \quad \Delta x\,\{y_1 + y_2 + y_3 + \cdots + y_n\} \tag{16}$$

The geometric interpretation is evident from the figure on Page 80.

2. **Trapezoidal rule.**

$$\int_a^b f(x)\,dx \;\approx\; \frac{\Delta x}{2}\{y_0 + 2y_1 + 2y_2 + \cdots + 2y_{n-1} + y_n\} \tag{17}$$

This is obtained by taking the mean of the approximations in (16). Geometrically this replaces the curve $y = f(x)$ by a set of approximating line segments.

3. **Simpson's rule.**

$$\int_a^b f(x)\,dx \;\approx\; \frac{\Delta x}{3}\{y_0 + 4y_1 + 2y_2 + 4y_3 + 2y_4 + 4y_5 + \cdots + 2y_{n-2} + 4y_{n-1} + y_n\} \tag{18}$$

This is obtained by dividing $[a, b]$ into an even number of equal intervals, (i.e. n is even) and approximating $f(x)$ by a quadratic through 3 successive points corresponding to $x_0, x_1, x_2;\ x_1, x_2, x_3;\ \ldots;\ x_{n-2}, x_{n-1}, x_n$. Geometrically this replaces the curve $y = f(x)$ by a set of approximating parabolic arcs.

4. **Taylor's theorem of the mean** can sometimes be used, as in Problem 26.

APPLICATIONS

The use of the integral as a limit of a sum enables us to solve many physical or geometrical problems such as determination of areas, volumes, arc lengths, moments of inertia, centroids, etc.

Solved Problems

DEFINITION of a DEFINITE INTEGRAL

1. If $f(x)$ is continuous in $[a, b]$ prove that

$$\lim_{n \to \infty} \frac{b-a}{n} \sum_{k=1}^{n} f\left(a + \frac{k(b-a)}{n}\right) = \int_{a}^{b} f(x)\, dx$$

Since $f(x)$ is continuous, the limit exists independent of the mode of subdivision (see Problem 35). Choose the subdivision of $[a, b]$ into n equal parts of equal length $\Delta x = (b-a)/n$ (see Fig. 5-1, Page 80). Let $\xi_k = a + k(b-a)/n$, $k = 1, 2, \ldots, n$. Then

$$\lim_{n \to \infty} \sum_{k=1}^{n} f(\xi_k)\, \Delta x_k = \lim_{n \to \infty} \frac{b-a}{n} \sum_{k=1}^{n} f\left(a + \frac{k(b-a)}{n}\right) = \int_{a}^{b} f(x)\, dx$$

2. Express $\displaystyle \lim_{n \to \infty} \frac{1}{n} \sum_{k=1}^{n} f\left(\frac{k}{n}\right)$ as a definite integral.

Let $a = 0$, $b = 1$ in Problem 1. Then

$$\lim_{n \to \infty} \frac{1}{n} \sum_{k=1}^{n} f\left(\frac{k}{n}\right) = \int_{0}^{1} f(x)\, dx$$

3. (a) Express $\displaystyle \int_{0}^{1} x^2\, dx$ as a limit of a sum, and use the result to evaluate the given definite integral. (b) Interpret the result geometrically.

(a) If $f(x) = x^2$, then $f(k/n) = (k/n)^2 = k^2/n^2$. Thus by Problem 2,

$$\lim_{n \to \infty} \frac{1}{n} \sum_{k=1}^{n} \frac{k^2}{n^2} = \int_{0}^{1} x^2\, dx$$

This can be written, using Problem 29 of Chapter 1,

$$\int_{0}^{1} x^2\, dx = \lim_{n \to \infty} \frac{1}{n}\left(\frac{1^2}{n^2} + \frac{2^2}{n^2} + \cdots + \frac{n^2}{n^2}\right) = \lim_{n \to \infty} \frac{1^2 + 2^2 + \cdots + n^2}{n^3}$$

$$= \lim_{n \to \infty} \frac{n(n+1)(2n+1)}{6n^3} = \lim_{n \to \infty} \frac{(1 + 1/n)(2 + 1/n)}{6} = \frac{1}{3}$$

which is the required limit.

Note: By using the fundamental theorem of the calculus, we observe that $\displaystyle \int_{0}^{1} x^2\, dx = (x^3/3)\big|_{0}^{1} = 1^3/3 - 0^3/3 = 1/3$.

(b) The area bounded by the curve $y = x^2$, the x axis and the line $x = 1$ is equal to $\frac{1}{3}$.

4. Evaluate $\displaystyle \lim_{n \to \infty} \left\{ \frac{1}{n+1} + \frac{1}{n+2} + \cdots + \frac{1}{n+n} \right\}$.

The required limit can be written

$$\lim_{n \to \infty} \frac{1}{n}\left\{ \frac{1}{1 + 1/n} + \frac{1}{1 + 2/n} + \cdots + \frac{1}{1 + n/n} \right\} = \lim_{n \to \infty} \frac{1}{n} \sum_{k=1}^{n} \frac{1}{1 + k/n}$$

$$= \int_{0}^{1} \frac{dx}{1 + x} = \ln(1 + x)\big|_{0}^{1} = \ln 2$$

using Problem 2 and the fundamental theorem of the calculus.

5. Prove that $\lim\limits_{n\to\infty} \dfrac{1}{n}\left\{\sin\dfrac{t}{n} + \sin\dfrac{2t}{n} + \cdots + \sin\dfrac{(n-1)t}{n}\right\} = \dfrac{1-\cos t}{t}$.

Let $a = 0$, $b = t$, $f(x) = \sin x$ in Problem 1.   Then

$$\lim_{n\to\infty} \frac{t}{n}\sum_{k=1}^{n}\sin\frac{kt}{n} = \int_0^t \sin x\,dx = 1 - \cos t$$

and so

$$\lim_{n\to\infty}\frac{1}{n}\sum_{k=1}^{n-1}\sin\frac{kt}{n} = \frac{1-\cos t}{t}$$

using the fact that $\lim\limits_{n\to\infty}\dfrac{\sin t}{n} = 0$.

MEASURE ZERO

6. Prove that a countable point set has measure zero.

Let the point set be denoted by $x_1, x_2, x_3, x_4, \ldots$ and suppose that intervals of lengths less than $\epsilon/2,\ \epsilon/4,\ \epsilon/8,\ \epsilon/16,\ \ldots$ respectively enclose the points, where ϵ is any positive number. Then the sum of the lengths of the intervals is less than $\epsilon/2 + \epsilon/4 + \epsilon/8 + \cdots = \epsilon$ (let $a = \epsilon/2$ and $r = \frac{1}{2}$ in Problem 25(a) of Chapter 3), showing that the set has measure zero.

PROPERTIES of DEFINITE INTEGRALS

7. (a) If $f(x)$ is continuous in $[a, b]$ and $m \leqq f(x) \leqq M$ where m and M are constants, prove that

$$m(b-a) \;\leqq\; \int_a^b f(x)\,dx \;\leqq\; M(b-a)$$

(b) Interpret the result of (a) geometrically.

(a) We have

$$m\,\Delta x_k \;\leqq\; f(\xi_k)\,\Delta x_k \;\leqq\; M\,\Delta x_k \qquad k = 1, 2, \ldots, n$$

Summing from $k = 1$ to n and using the fact that

$$\sum_{k=1}^{n}\Delta x_k = (x_1 - a) + (x_2 - x_1) + \cdots + (b - x_{n-1}) = b - a$$

it follows that

$$m(b-a) \;\leqq\; \sum_{k=1}^{n} f(\xi_k)\,\Delta x_k \;\leqq\; M(b-a)$$

Taking the limit as $n \to \infty$ and each $\Delta x_k \to 0$ yields the required result.

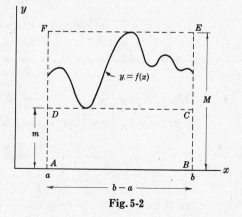

(b) Assume $f(x) \geqq 0$ and continuous in $[a, b]$ with graph shown in the adjoining Fig. 5-2. It is geometrically evident that

Area $ABCD \;\leqq\;$ Area under $y = f(x) \;\leqq\;$ Area $ABEF$
i.e.,

$$m(b-a) \;\leqq\; \int_a^b f(x)\,dx \;\leqq\; M(b-a)$$

A similar interpretation can be made if the restriction $f(x) \geqq 0$ is removed. The result also holds if $f(x)$ is sectionally continuous in $[a, b]$.

Fig. 5-2

8. Prove that $\left|\displaystyle\int_a^b f(x)\,dx\right| \leqq \displaystyle\int_a^b |f(x)|\,dx \quad$ if $\ a < b$.

By inequality 2, Page 3,

$$|\sum_{k=1}^{n} f(\xi_k)\,\Delta x_k| \ \leqq \ \sum_{k=1}^{n} |f(\xi_k)\,\Delta x_k| \ = \ \sum_{k=1}^{n} |f(\xi_k)|\,\Delta x_k$$

Taking the limit as $n \to \infty$ and each $\Delta x_k \to 0$, we have the required result.

9. Prove that $\displaystyle\lim_{n\to\infty} \int_0^{2\pi} \frac{\sin nx}{x^2 + n^2}\,dx \ = \ 0$.

$$\left|\int_0^{2\pi} \frac{\sin nx}{x^2 + n^2}\,dx\right| \ \leqq \ \int_0^{2\pi}\left|\frac{\sin nx}{x^2 + n^2}\right|\,dx \ \leqq \ \int_0^{2\pi} \frac{dx}{n^2} \ = \ \frac{2\pi}{n^2}$$

Then $\displaystyle\lim_{n\to\infty}\left|\int_0^{2\pi} \frac{\sin nx}{x^2 + n^2}\,dx\right| \ = \ 0$, and so the required result follows.

MEAN VALUE THEOREMS for INTEGRALS

10. (a) If $f(x)$ is continuous in $[a, b]$, prove that there is a point ξ in (a, b) such that

$$\int_a^b f(x)\,dx \ = \ (b - a)\,f(\xi)$$

(b) Interpret the result of (a) geometrically.

(a) Since $f(x)$ is continuous in $[a, b]$, we can find constants m and M such that $m \leqq f(x) \leqq M$. Then by Problem 7,

$$m \ \leqq \ \frac{\int_a^b f(x)\,dx}{b - a} \ \leqq \ M$$

Since $f(x)$ is continuous it takes on all values between m and M (see Chapter 2, Problems 34, 93). In particular there must be a value ξ such that

$$f(\xi) \ = \ \frac{\int_a^b f(x)\,dx}{b - a} \qquad a < \xi < b$$

The required result follows on multiplying by $b - a$.

(b) If $f(x) \geqq 0$ with graph as shown in the figure of Problem 7(b), we can interpret $\displaystyle\int_a^b f(x)\,dx$ as the shaded area under the curve $y = f(x)$. Geometrically this area should equal that of a rectangle with base $b - a$ and height $f(\xi)$ for some value ξ between a and b.

FUNDAMENTAL THEOREM of INTEGRAL CALCULUS

11. If $F(x) = \displaystyle\int_a^x f(t)\,dt$ where $f(x)$ is continuous in $[a, b]$, prove that $F'(x) = f(x)$.

$$\frac{F(x+h) - F(x)}{h} \ = \ \frac{1}{h}\left\{\int_a^{x+h} f(t)\,dt \ - \ \int_a^x f(t)\,dt\right\} \ = \ \frac{1}{h}\int_x^{x+h} f(t)\,dt$$

$$= \ f(\xi) \qquad \xi \text{ between } x \text{ and } x + h$$

by the first mean value theorem for integrals (Problem 10).

Then if x is any point interior to $[a, b]$,

$$F'(x) \ = \ \lim_{h\to 0} \frac{F(x+h) - F(x)}{h} \ = \ \lim_{h\to 0} f(\xi) \ = \ f(x)$$

since f is continuous.

If $x = a$ or $x = b$, we use right or left hand limits respectively and the result holds in these cases as well.

12. Prove the fundamental theorem of the integral calculus.

By Problem 11, if $F(x)$ is *any* function whose derivative is $f(x)$, we can write

$$F(x) \;=\; \int_a^x f(t)\,dt \;+\; c$$

where c is any constant (see last line of Problem 22, Chapter 4).

Since $F(a) = c$, it follows that $\quad F(b) \;=\; \int_a^b f(t)\,dt + F(a) \quad$ or $\quad \int_a^b f(t)\,dt \;=\; F(b) - F(a)$.

13. If $f(x)$ is continuous in $[a, b]$, prove that $\quad F(x) \;=\; \int_a^x f(t)\,dt \quad$ is continuous in $[a, b]$.

If x is any point interior to $[a, b]$, then as in Problem 11,

$$\lim_{h \to 0} F(x + h) \,-\, F(x) \;=\; \lim_{h \to 0} h\, f(\xi) \;=\; 0$$

and $F(x)$ is continuous.

If $x = a$ and $x = b$, we use right and left hand limits respectively to show that $F(x)$ is continuous at $x = a$ and $x = b$.

Another method:

By Problem 11 and Problem 3, Chapter 4, it follows that $F'(x)$ exists and so $F(x)$ must be continuous.

CHANGE of VARIABLES and SPECIAL METHODS of INTEGRATION

14. Prove the result (*14*), Page 83, for changing the variable of integration.

Let $\quad F(x) \;=\; \int_a^x f(x)\,dx \quad$ and $\quad G(t) \;=\; \int_\alpha^t f\{g(t)\}\, g'(t)\,dt, \quad$ where $\quad x = g(t)$.

Then $\quad dF = f(x)\,dx, \quad dG = f\{g(t)\}\, g'(t)\,dt$.

Since $dx = g'(t)\,dt$, it follows that $f(x)\,dx = f\{g(t)\}\, g'(t)\,dt$ so that $dF(x) = dG(t)$, from which $F(x) = G(t) + c$.

Now when $x = a$, $t = \alpha$ or $F(a) = G(\alpha) + c$. But $F(a) = G(\alpha) = 0$, so that $c = 0$. Hence $F(x) = G(t)$. Since $x = b$ when $t = \beta$, we have

$$\int_a^b f(x)\,dx \;=\; \int_\alpha^\beta f\{g(t)\}\, g'(t)\,dt$$

as required.

15. Evaluate:

(*a*) $\displaystyle\int (x + 2) \sin (x^2 + 4x - 6)\,dx$ (*c*) $\displaystyle\int_{-1}^{1} \frac{dx}{\sqrt{(x + 2)(3 - x)}}$ (*e*) $\displaystyle\int_0^{1/\sqrt{2}} \frac{x \sin^{-1} x^2}{\sqrt{1 - x^4}}\,dx$

(*b*) $\displaystyle\int \frac{\cot (\ln x)}{x}\,dx$ (*d*) $\displaystyle\int 2^{-x} \tanh 2^{1-x}\,dx$ (*f*) $\displaystyle\int \frac{x\,dx}{\sqrt{x^2 + x + 1}}$

(*a*) **Method 1:**

Let $x^2 + 4x - 6 = u$. Then $(2x + 4)\,dx = du$, $(x + 2)\,dx = \tfrac{1}{2}\,du$ and the integral becomes

$$\tfrac{1}{2} \int \sin u\,du \;=\; -\tfrac{1}{2} \cos u + c \;=\; -\tfrac{1}{2} \cos (x^2 + 4x - 6) + c$$

Method 2:

$$\int (x + 2) \sin (x^2 + 4x - 6)\,dx \;=\; \tfrac{1}{2} \int \sin (x^2 + 4x - 6)\, d(x^2 + 4x - 6) \;=\; -\tfrac{1}{2} \cos (x^2 + 4x - 6) + c$$

(*b*) Let $\ln x = u$. Then $(dx)/x = du$ and the integral becomes

$$\int \cot u\,du \;=\; \ln |\sin u| + c \;=\; \ln |\sin (\ln x)| + c$$

(c) **Method 1:**

$$\int \frac{dx}{\sqrt{(x+2)(3-x)}} = \int \frac{dx}{\sqrt{6+x-x^2}} = \int \frac{dx}{\sqrt{6-(x^2-x)}} = \int \frac{dx}{\sqrt{25/4-(x-\frac{1}{2})^2}}$$

Letting $x - \frac{1}{2} = u$, this becomes $\displaystyle\int \frac{du}{\sqrt{25/4 - u^2}} = \sin^{-1}\frac{u}{5/2} + c = \sin^{-1}\left(\frac{2x-1}{5}\right) + c.$

Then $\displaystyle\int_{-1}^{1} \frac{dx}{\sqrt{(x+2)(3-x)}} = \sin^{-1}\left(\frac{2x-1}{5}\right)\Big|_{-1}^{1} = \sin^{-1}\left(\frac{1}{5}\right) - \sin^{-1}\left(-\frac{3}{5}\right)$

$$= \sin^{-1}.2 + \sin^{-1}.6$$

Method 2:

Let $x - \frac{1}{2} = u$ as in Method 1. Now when $x = -1$, $u = -\frac{3}{2}$; and when $x = 1$, $u = \frac{1}{2}$. Thus by Problem 14,

$$\int_{-1}^{1} \frac{dx}{\sqrt{(x+2)(3-x)}} = \int_{-1}^{1} \frac{dx}{\sqrt{25/4-(x-\frac{1}{2})^2}} = \int_{-3/2}^{1/2} \frac{du}{\sqrt{25/4-u^2}} = \sin^{-1}\frac{u}{5/2}\Big|_{-3/2}^{1/2}$$

$$= \sin^{-1}.2 + \sin^{-1}.6$$

(d) Let $2^{1-x} = u$. Then $-2^{1-x}(\ln 2)\,dx = du$ and $2^{-x}\,dx = -\dfrac{du}{2\ln 2}$, so that the integral becomes

$$-\frac{1}{2\ln 2}\int \tanh u \, du = -\frac{1}{2\ln 2}\ln\cosh 2^{1-x} + c$$

(e) Let $\sin^{-1} x^2 = u$. Then $du = \dfrac{1}{\sqrt{1-(x^2)^2}}\,2x\,dx = \dfrac{2x\,dx}{\sqrt{1-x^4}}$ and the integral becomes

$$\frac{1}{2}\int u \, du = \frac{1}{4}u^2 + c = \frac{1}{4}(\sin^{-1} x^2)^2 + c$$

Thus $\displaystyle\int_0^{1/\sqrt{2}} \frac{x\sin^{-1} x^2}{\sqrt{1-x^4}}\,dx = \frac{1}{4}(\sin^{-1} x^2)^2\Big|_0^{1/\sqrt{2}} = \frac{1}{4}(\sin^{-1}\tfrac{1}{2})^2 = \frac{\pi^2}{144}.$

(f)
$$\int \frac{x\,dx}{\sqrt{x^2+x+1}} = \frac{1}{2}\int \frac{2x+1-1}{\sqrt{x^2+x+1}}dx = \frac{1}{2}\int \frac{2x+1}{\sqrt{x^2+x+1}}dx - \frac{1}{2}\int \frac{dx}{\sqrt{x^2+x+1}}$$

$$= \frac{1}{2}\int (x^2+x+1)^{-1/2}\,d(x^2+x+1) - \frac{1}{2}\int \frac{dx}{\sqrt{(x+\frac{1}{2})^2+\frac{3}{4}}}$$

$$= \sqrt{x^2+x+1} - \frac{1}{2}\ln|x+\tfrac{1}{2}+\sqrt{(x+\tfrac{1}{2})^2+\tfrac{3}{4}}| + c$$

16. Show that $\displaystyle\int_1^2 \frac{dx}{(x^2-2x+4)^{3/2}} = \frac{1}{6}.$

Write the integral as $\displaystyle\int_1^2 \frac{dx}{[(x-1)^2+3]^{3/2}}$. Let $x - 1 = \sqrt{3}\tan u$, $dx = \sqrt{3}\sec^2 u\,du$. When $x = 1$, $u = \tan^{-1} 0 = 0$; when $x = 2$, $u = \tan^{-1} 1/\sqrt{3} = \pi/6$. Then the integral becomes

$$\int_0^{\pi/6} \frac{\sqrt{3}\sec^2 u\,du}{[3+3\tan^2 u]^{3/2}} = \int_0^{\pi/6} \frac{\sqrt{3}\sec^2 u\,du}{[3\sec^2 u]^{3/2}} = \frac{1}{3}\int_0^{\pi/6}\cos u\,du = \frac{1}{3}\sin u\Big|_0^{\pi/6} = \frac{1}{6}.$$

17. Determine $\displaystyle\int_e^{e^2} \frac{dx}{x(\ln x)^3}.$

Let $\ln x = y$, $(dx)/x = dy$. When $x = e$, $y = 1$; when $x = e^2$, $y = 2$. Then the integral becomes

$$\int_1^2 \frac{dy}{y^3} = \frac{y^{-2}}{-2}\Big|_1^2 = \frac{3}{8}$$

18. Find $\int x^n \ln x \, dx$ if (a) $n \neq -1$, (b) $n = -1$.

(a) Use integration by parts, letting $u = \ln x$, $dv = x^n \, dx$, so that $du = (dx)/x$, $v = x^{n+1}/(n+1)$. Then

$$\int x^n \ln x \, dx = \int u \, dv = uv - \int v \, du = \frac{x^{n+1}}{n+1} \ln x - \int \frac{x^{n+1}}{n+1} \cdot \frac{dx}{x}$$

$$= \frac{x^{n+1}}{n+1} \ln x - \frac{x^{n+1}}{(n+1)^2} + c$$

(b) $\int x^{-1} \ln x \, dx = \int \ln x \, d(\ln x) = \frac{1}{2}(\ln x)^2 + c.$

19. Find $\int 3^{\sqrt{2x+1}} \, dx$.

Let $\sqrt{2x+1} = y$, $2x+1 = y^2$. Then $dx = y \, dy$ and the integral becomes $\int 3^y \cdot y \, dy$.

Integrate by parts, letting $u = y$, $dv = 3^y \, dy$; then $du = dy$, $v = 3^y/(\ln 3)$, and we have

$$\int 3^y \cdot y \, dy = \int u \, dv = uv - \int v \, du = \frac{y \cdot 3^y}{\ln 3} - \int \frac{3^y}{\ln 3} \, dy = \frac{y \cdot 3^y}{\ln 3} - \frac{3^y}{(\ln 3)^2} + c$$

20. Find $\int_0^1 x \ln (x+3) \, dx$.

Let $u = \ln (x+3)$, $dv = x \, dx$. Then $du = \dfrac{dx}{x+3}$, $v = \dfrac{x^2}{2}$. Hence on integrating by parts,

$$\int x \ln (x+3) \, dx = \frac{x^2}{2} \ln (x+3) - \frac{1}{2} \int \frac{x^2 \, dx}{x+3} = \frac{x^2}{2} \ln (x+3) - \frac{1}{2} \int \left(x - 3 + \frac{9}{x+3} \right) dx$$

$$= \frac{x^2}{2} \ln (x+3) - \frac{1}{2} \left\{ \frac{x^2}{2} - 3x + 9 \ln (x+3) \right\} + c$$

Then

$$\int_0^1 x \ln (x+3) \, dx = \frac{5}{4} - 4 \ln 4 + \frac{9}{2} \ln 3$$

21. Determine $\int \dfrac{6-x}{(x-3)(2x+5)} \, dx$.

Use the method of *partial fractions*. Let $\dfrac{6-x}{(x-3)(2x+5)} = \dfrac{A}{x-3} + \dfrac{B}{2x+5}$.

Method 1:

To determine the constants A and B, multiply both sides by $(x-3)(2x+5)$ to obtain

$$6-x = A(2x+5) + B(x-3) \quad \text{or} \quad 6-x = 5A - 3B + (2A+B)x \tag{1}$$

Since this is an identity, $5A - 3B = 6$, $2A + B = -1$ and $A = 3/11$, $B = -17/11$. Then

$$\int \frac{6-x}{(x-3)(2x+5)} \, dx = \int \frac{3/11}{x-3} \, dx + \int \frac{-17/11}{2x+5} \, dx = \frac{3}{11} \ln |x-3| - \frac{17}{22} \ln |2x+5| + c$$

Method 2:

Substitute suitable values for x in the identity (1). For example, letting $x = 3$ and $x = -5/2$ in (1), we find at once $A = 3/11$, $B = -17/11$.

22. Evaluate $\int \dfrac{dx}{5 + 3 \cos x}$ by using the substitution $\tan x/2 = u$.

From Fig. 5-3 we see that

$$\sin x/2 = \frac{u}{\sqrt{1+u^2}}, \qquad \cos x/2 = \frac{1}{\sqrt{1+u^2}}$$

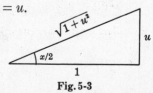

Fig. 5-3

Then $\cos x = \cos^2 x/2 - \sin^2 x/2 = \dfrac{1-u^2}{1+u^2}.$

Also $du = \tfrac{1}{2}\sec^2 x/2\,dx$ or $dx = 2\cos^2 x/2\,du = \dfrac{2\,du}{1+u^2}.$

Thus the integral becomes $\displaystyle\int \frac{du}{u^2+4} = \tfrac{1}{2}\tan^{-1}u/2 + c = \tfrac{1}{2}\tan^{-1}(\tfrac{1}{2}\tan x/2) + c.$

23. Evaluate $\displaystyle\int_0^\pi \frac{x\sin x}{1+\cos^2 x}\,dx.$

Let $x = \pi - y$.   Then

$$I = \int_0^\pi \frac{x\sin x}{1+\cos^2 x}\,dx = \int_0^\pi \frac{(\pi-y)\sin y}{1+\cos^2 y}\,dy = \pi\int_0^\pi \frac{\sin y}{1+\cos^2 y}\,dy - \int_0^\pi \frac{y\sin y}{1+\cos^2 y}\,dy$$

$$= -\pi\int_0^\pi \frac{d(\cos y)}{1+\cos^2 y} - I = -\pi\tan^{-1}(\cos y)\Big|_0^\pi - I = \pi^2/2 - I$$

i.e. $I = \pi^2/2 - I$ or $I = \pi^2/4$.

24. Prove that $\displaystyle\int_0^{\pi/2} \frac{\sqrt{\sin x}}{\sqrt{\sin x}+\sqrt{\cos x}}\,dx = \frac{\pi}{4}.$

Letting $x = \pi/2 - y$, we have

$$I = \int_0^{\pi/2} \frac{\sqrt{\sin x}}{\sqrt{\sin x}+\sqrt{\cos x}}\,dx = \int_0^{\pi/2} \frac{\sqrt{\cos y}}{\sqrt{\cos y}+\sqrt{\sin y}}\,dy = \int_0^{\pi/2} \frac{\sqrt{\cos x}}{\sqrt{\cos x}+\sqrt{\sin x}}\,dx$$

Then

$$I + I = \int_0^{\pi/2} \frac{\sqrt{\sin x}}{\sqrt{\sin x}+\sqrt{\cos x}}\,dx + \int_0^{\pi/2} \frac{\sqrt{\cos x}}{\sqrt{\cos x}+\sqrt{\sin x}}\,dx$$

$$= \int_0^{\pi/2} \frac{\sqrt{\sin x}+\sqrt{\cos x}}{\sqrt{\sin x}+\sqrt{\cos x}}\,dx = \int_0^{\pi/2} dx = \frac{\pi}{2}$$

from which $2I = \pi/2$ and $I = \pi/4$.

The same method can be used to prove that for all real values of m,

$$\int_0^{\pi/2} \frac{\sin^m x}{\sin^m x + \cos^m x}\,dx = \frac{\pi}{4}$$

(see Problem 94).

Note: This problem and Problem 23 show that some definite integrals can be evaluated without first finding the corresponding indefinite integrals.

NUMERICAL METHODS for EVALUATING DEFINITE INTEGRALS

25. Evaluate $\displaystyle\int_0^1 \frac{dx}{1+x^2}$ approximately, using (*a*) the trapezoidal rule, (*b*) Simpson's rule, where the interval $[0,1]$ is divided into $n=4$ equal parts.

Let $f(x) = 1/(1+x^2)$. Using the notation on Page 85, we find $\Delta x = (b-a)/n = (1-0)/4 = 0.25$. Then keeping 4 decimal places, we have: $y_0 = f(0) = 1.0000$, $y_1 = f(0.25) = 0.9412$, $y_2 = f(0.50) = 0.8000$, $y_3 = f(0.75) = 0.6400$, $y_4 = f(1) = 0.5000$.

(*a*) The trapezoidal rule gives

$$\frac{\Delta x}{2}\{y_0 + 2y_1 + 2y_2 + 2y_3 + y_4\} = \frac{0.25}{2}\{1.0000 + 2(0.9412) + 2(0.8000) + 2(0.6400) + 0.500\} = 0.7828.$$

(b) Simpson's rule gives

$$\frac{\Delta x}{3}\{y_0 + 4y_1 + 2y_2 + 4y_3 + y_4\} = \frac{0.25}{3}\{1.0000 + 4(0.9412) + 2(0.8000) + 4(0.6400) + 0.5000\} = 0.7854.$$

The true value is $\pi/4 \approx 0.7854$.

26. (a) Evaluate $\displaystyle\int_0^1 e^{x^2}\,dx$ approximately by using Taylor's theorem of the mean and (b) estimate the maximum error.

As in Problem 28, Chapter 4, we find

$$e^x = 1 + x + \frac{x^2}{2!} + \frac{x^3}{3!} + \frac{x^4}{4!} + \frac{x^5 e^\xi}{5!} \qquad 0 < \xi < x$$

Then replacing x by x^2,

$$e^{x^2} = 1 + x^2 + \frac{x^4}{2!} + \frac{x^6}{3!} + \frac{x^8}{4!} + \frac{x^{10} e^\xi}{5!} \qquad 0 < \xi < x^2$$

Integrating from 0 to 1,

$$\int_0^1 e^{x^2}\,dx = \int_0^1 \left(1 + x^2 + \frac{x^4}{2!} + \frac{x^6}{3!} + \frac{x^8}{4!}\right)dx + E \qquad \left(\text{where the error } E = \int_0^1 \frac{x^{10}}{5!}\,e^\xi\,dx\right)$$

$$= 1 + \frac{1}{3} + \frac{1}{5\cdot 2!} + \frac{1}{7\cdot 3!} + \frac{1}{9\cdot 4!} + E = 1.4618 + E$$

Now $\displaystyle |E| = \left|\int_0^1 \frac{x^{10}}{5!}\,e^\xi\,dx\right| \leqq \int_0^1 \left|\frac{x^{10}}{5!}\,e^\xi\right|dx \leqq e\int_0^1 \frac{x^{10}}{5!}\,dx = \frac{e}{11\cdot 5!} < 0.0021.$

Thus the maximum error is less than 0.0021, and so the value of the integral is 1.46 accurate to two decimal places. By using more terms in Taylor's theorem, better accuracy can be attained.

APPLICATIONS

27. Find the (a) area and (b) moment of inertia about the y axis of the region in the xy plane bounded by $y = 4 - x^2$ and the x axis.

(a) Subdivide the region into rectangles as in the figure on Page 80. A typical rectangle is shown in the adjoining Fig. 5-4. Then

$$\text{Required area} = \lim_{n \to \infty} \sum_{k=1}^n f(\xi_k)\,\Delta x_k$$

$$= \lim_{n \to \infty} \sum_{k=1}^n (4 - \xi_k^2)\,\Delta x_k$$

$$= \int_{-2}^2 (4 - x^2)\,dx = \frac{32}{3}$$

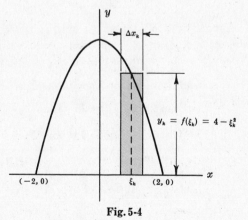

$$y_k = f(\xi_k) = 4 - \xi_k^2$$

Fig. 5-4

(b) Assuming unit density, the moment of inertia about the y axis of the typical rectangle shown above is $\xi_k^2 f(\xi_k)\,\Delta x_k$. Then

$$\text{Required moment of inertia} = \lim_{n \to \infty} \sum_{k=1}^n \xi_k^2 f(\xi_k)\,\Delta x_k = \lim_{n \to \infty} \sum_{k=1}^n \xi_k^2(4 - \xi_k^2)\,\Delta x_k$$

$$= \int_{-2}^2 x^2(4 - x^2)\,dx = \frac{128}{15}$$

28. Find the length of arc of the parabola $y = x^2$ from $x = 0$ to $x = 1$.

$$\text{Required arc length} \;=\; \int_0^1 \sqrt{1 + (dy/dx)^2}\,dx \;=\; \int_0^1 \sqrt{1 + (2x)^2}\,dx$$

$$=\; \int_0^1 \sqrt{1 + 4x^2}\,dx \;=\; \tfrac{1}{2}\int_0^2 \sqrt{1 + u^2}\,du$$

$$=\; \tfrac{1}{2}\{\tfrac{1}{2}u\sqrt{1+u^2} + \tfrac{1}{2}\ln(u + \sqrt{1+u^2})\}\,\big|_0^2 \;=\; \tfrac{1}{2}\sqrt{5} + \tfrac{1}{4}\ln(2 + \sqrt{5})$$

29. Find the volume generated by revolving the region of Problem 27 about the x axis.

$$\text{Required volume} \;=\; \lim_{n \to \infty} \sum_{k=1}^{n} \pi\, y_k^2\, \Delta x_k \;=\; \pi \int_{-2}^{2} (4 - x^2)^2\,dx \;=\; 512\pi/15.$$

MISCELLANEOUS PROBLEMS

30. If $f(x)$ and $g(x)$ are continuous in $[a, b]$, prove *Schwarz's inequality for integrals:*

$$\left(\int_a^b f(x)\,g(x)\,dx \right)^2 \;\leqq\; \int_a^b \{f(x)\}^2\,dx \int_a^b \{g(x)\}^2\,dx$$

We have

$$\int_a^b \{f(x) + \lambda\,g(x)\}^2\,dx \;=\; \int_a^b \{f(x)\}^2\,dx + 2\lambda \int_a^b f(x)\,g(x)\,dx + \lambda^2 \int_a^b \{g(x)\}^2\,dx \;\geqq\; 0$$

for all real values of λ.   Hence by Problem 13 of Chapter 1, using *(1)* with

$$A^2 \;=\; \int_a^b \{g(x)\}^2\,dx, \qquad B^2 \;=\; \int_a^b \{f(x)\}^2\,dx, \qquad C \;=\; \int_a^b f(x)\,g(x)\,dx$$

we find $C^2 \leqq A^2 B^2$, which gives the required result.

31. Prove the second mean value theorem of equation *(8)*, Page 82, under the assumption of the existence and continuity of $g'(x)$ in $[a, b]$ in addition to the other assumptions.

Let $F(x) = \displaystyle\int_a^x f(t)\,dt$.   Then on integrating by parts,

$$\int_a^b f(x)\,g(x)\,dx \;=\; \int_a^b g(x)\,F'(x)\,dx$$

$$=\; g(x)\,F(x)\,\big|_a^b - \int_a^b g'(x)\,F(x)\,dx$$

$$=\; g(b)\,F(b) - \int_a^b g'(x)\,F(x)\,dx$$

Case 1: $g(x)$ is monotonic increasing, i.e. $g'(x) \geqq 0$.

Then by the generalized first mean value theorem for integrals (see Page 82) we have,

$$\int_a^b g'(x)\,F(x)\,dx \;=\; F(\xi) \int_a^b g'(x)\,dx \;=\; F(\xi)\,[g(b) - g(a)]$$

where ξ is in (a, b) so that

$$\int_a^b f(x)\,g(x)\,dx \;=\; g(b)\,F(b) - F(\xi)\,[g(b) - g(a)]$$

$$=\; g(a)\,F(\xi) + g(b)\,[F(b) - F(\xi)]$$

$$=\; g(a) \int_a^\xi f(x)\,dx + g(b) \int_\xi^b f(x)\,dx$$

Case 2: $g(x)$ is monotonic decreasing, i.e. $g'(x) \leqq 0$.

The proof is similar to that in Case 1.

32. (a) If $f^{(n+1)}(x)$ is continuous in $[a, b]$, prove that for x in $[a, b]$,

$$f(x) = f(a) + f'(a)(x-a) + \frac{f''(a)(x-a)^2}{2!} + \ldots + \frac{f^{(n)}(a)(x-a)^n}{n!} + \frac{1}{n!}\int_a^x (x-t)^n f^{(n+1)}(t)\,dt$$

(b) Use (a) to obtain the Lagrange and Cauchy forms of the remainder in Taylor's theorem (see Page 61).

(a) We use mathematical induction (see Chapter 1). The result holds for $n = 0$ since

$$f(x) = f(a) + \int_a^x f'(t)\,dt = f(a) + f(t)\big|_a^x = f(a) + f(x) - f(a) \tag{1}$$

Assume that it holds for $n = k$. Then integrating by parts, using

$$\frac{(x-t)^k}{k!}\,dt = dv, \quad f^{(k+1)}(t) = u \quad \text{so that} \quad v = -\frac{(x-t)^{k+1}}{(k+1)!}, \quad du = f^{(k+2)}(t)\,dt$$

we find

$$\frac{1}{k!}\int_a^x (x-t)^k f^{(k+1)}(t)\,dt = -\frac{f^{(k+1)}(t)(x-t)^{k+1}}{(k+1)!}\bigg|_a^x + \frac{1}{(k+1)!}\int_a^x (x-t)^{k+1} f^{(k+2)}(t)\,dt$$

$$= \frac{f^{(k+1)}(a)(x-a)^{k+1}}{(k+1)!} + \frac{1}{(k+1)!}\int_a^x (x-t)^{k+1} f^{(k+2)}(t)\,dt$$

which shows that the result holds for $n = k+1$. Thus it holds for all integers $n \geqq 0$.

(b) We have by the generalized first mean value theorem for integrals (see Page 82),

$$\int_a^x F(t)\,G(t)\,dt = F(\xi)\int_a^x G(t)\,dt \qquad \xi \text{ between } a \text{ and } x$$

Letting $F(t) = f^{(n+1)}(t)$, $G(t) = \frac{(x-t)^n}{n!}$, we obtain

$$\frac{1}{n!}\int_a^x (x-t)^n f^{(n+1)}(t)\,dt = \frac{f^{(n+1)}(\xi)}{n!}\int_a^x (x-t)^n\,dt = \frac{f^{(n+1)}(\xi)(x-a)^{n+1}}{(n+1)!}$$

giving the *Lagrange form* of the remainder, [equation (20), Page 61], with b replaced by x.

Letting $F(t) = \frac{f^{(n+1)}(t)(x-t)^n}{n!}$, $G(t) = 1$, we have

$$\frac{1}{n!}\int_a^x (x-t)^n f^{(n+1)}(t)\,dt = \frac{f^{(n+1)}(\xi)(x-\xi)^n}{n!}\int_a^x 1\,dt = \frac{f^{(n+1)}(\xi)(x-\xi)^n(x-a)}{n!}$$

giving the *Cauchy form* of the remainder, [equation (21), Page 61], with b replaced by x.

33. Prove that $\displaystyle\lim_{M\to\infty}\int_0^M \frac{dx}{x^4+4} = \frac{\pi}{8}$.

We have $x^4 + 4 = x^4 + 4x^2 + 4 - 4x^2 = (x^2+2)^2 - (2x)^2 = (x^2+2+2x)(x^2+2-2x)$.

According to the method of partial fractions, assume

$$\frac{1}{x^4+4} = \frac{Ax+B}{x^2+2x+2} + \frac{Cx+D}{x^2-2x+2}$$

Then $\quad 1 = (A+C)x^3 + (B-2A+2C+D)x^2 + (2A-2B+2C+2D)x + 2B + 2D$

so that $\quad A+C = 0, \ B-2A+2C+D = 0, \ 2A-2B+2C+2D = 0, \ 2B+2D = 1$

Solving simultaneously, $A = \frac{1}{8}$, $B = \frac{1}{4}$, $C = -\frac{1}{8}$, $D = \frac{1}{4}$. Thus

$$\int \frac{dx}{x^4+4} = \frac{1}{8}\int \frac{x+2}{x^2+2x+2}\,dx - \frac{1}{8}\int \frac{x-2}{x^2-2x+2}\,dx$$

$$= \frac{1}{8}\int \frac{x+1}{(x+1)^2+1}\,dx + \frac{1}{8}\int \frac{dx}{(x+1)^2+1} - \frac{1}{8}\int \frac{x-1}{(x-1)^2+1}\,dx + \frac{1}{8}\int \frac{dx}{(x-1)^2+1}$$

$$= \frac{1}{16}\ln(x^2+2x+2) + \frac{1}{8}\tan^{-1}(x+1) - \frac{1}{16}\ln(x^2-2x+2) + \frac{1}{8}\tan^{-1}(x-1) + C$$

Then
$$\lim_{M \to \infty} \int_0^M \frac{dx}{x^4 + 4} = \lim_{M \to \infty} \left\{ \frac{1}{16} \ln \left(\frac{M^2 + 2M + 2}{M^2 - 2M + 2} \right) + \frac{1}{8} \tan^{-1}(M+1) + \frac{1}{8} \tan^{-1}(M-1) \right\} = \frac{\pi}{8}$$

We denote this limit by $\int_0^\infty \frac{dx}{x^4 + 4}$, called an *improper integral of the first kind*. Such integrals are considered further in Chapter 12. See also Problem 78.

34. Evaluate $\lim_{x \to 0} \dfrac{\int_0^x \sin t^3 \, dt}{x^4}$.

The conditions of L'Hospital's rule are satisfied, so that the required limit is

$$\lim_{x \to 0} \frac{\dfrac{d}{dx} \displaystyle\int_0^x \sin t^3 \, dt}{\dfrac{d}{dx} (x^4)} = \lim_{x \to 0} \frac{\sin x^3}{4x^3} = \lim_{x \to 0} \frac{\dfrac{d}{dx} (\sin x^3)}{\dfrac{d}{dx} (4x^3)} = \lim_{x \to 0} \frac{3x^2 \cos x^3}{12x^2} = \frac{1}{4}$$

35. Prove that if $f(x)$ is continuous in $[a, b]$ then $\displaystyle\int_a^b f(x) \, dx$ exists.

Let $\sigma = \displaystyle\sum_{k=1}^n f(\xi_k) \Delta x_k$, using the notation of Page 80. Since $f(x)$ is continuous we can find numbers M_k and m_k representing the l.u.b. and g.l.b. of $f(x)$ in the interval $[x_{k-1}, x_k]$, i.e. such that $m_k \leqq f(x) \leqq M_k$. We then have

$$m(b-a) \leqq s = \sum_{k=1}^n m_k \Delta x_k \leqq \sigma \leqq \sum_{k=1}^n M_k \Delta x_k = S \leqq M(b-a) \tag{1}$$

where m and M are the g.l.b. and l.u.b. of $f(x)$ in $[a, b]$. The sums s and S are sometimes called the *lower* and *upper sums* respectively.

Now choose a second mode of subdivision of $[a, b]$ and consider the corresponding lower and upper sums denoted by s' and S' respectively. We must have

$$s' \leqq S \qquad \text{and} \qquad S' \geqq s \tag{2}$$

To prove this we choose a third mode of subdivision obtained by using the division points of both the first and second modes of subdivision and consider the corresponding lower and upper sums, denoted by t and T respectively. By Problem 89, we have

$$s \leqq t \leqq T \leqq S' \qquad \text{and} \qquad s' \leqq t \leqq T \leqq S \tag{3}$$

which proves (2).

From (2) it is also clear that as the number of subdivisions is increased, the upper sums are monotonic decreasing and the lower sums are monotonic increasing. Since according to (1) these sums are also bounded, it follows that they have limiting values which we shall call $\bar{s}$ and $\underline{S}$ respectively. By Problem 90, $\bar{s} \leqq \underline{S}$. In order to prove that the integral exists, we must show that $\bar{s} = \underline{S}$.

Since $f(x)$ is continuous in the closed interval $[a, b]$, it is uniformly continuous. Then given any $\epsilon > 0$, we can take each Δx_k so small that $M_k - m_k < \epsilon/(b-a)$. It follows that

$$S - s = \sum_{k=1}^n (M_k - m_k) \Delta x_k < \frac{\epsilon}{b-a} \sum_{k=1}^n \Delta x_k = \epsilon \tag{4}$$

Now $S - s = (S - \underline{S}) + (\underline{S} - \bar{s}) + (\bar{s} - s)$ and it follows that each term in parentheses is positive and so is less than ϵ by (4). In particular, since $\underline{S} - \bar{s}$ is a definite number it must be zero, i.e. $\underline{S} = \bar{s}$. Thus the limits of the upper and lower sums are equal and the proof is complete.

Supplementary Problems

DEFINITION of a DEFINITE INTEGRAL

36. (a) Express $\int_0^1 x^3\,dx$ as a limit of a sum. (b) Use the result of (a) to evaluate the given definite integral. (c) Interpret the result geometrically. *Ans.* (b) $\frac{1}{4}$

37. Using the definition, evaluate (a) $\int_0^2 (3x+1)\,dx$, (b) $\int_3^6 (x^2-4x)\,dx$. *Ans.* (a) 8, (b) 9

38. Prove that $\displaystyle\lim_{n\to\infty}\left\{\frac{n}{n^2+1^2}+\frac{n}{n^2+2^2}+\cdots+\frac{n}{n^2+n^2}\right\}=\frac{\pi}{4}$.

39. Prove that $\displaystyle\lim_{n\to\infty}\left\{\frac{1^p+2^p+3^p+\cdots+n^p}{n^{p+1}}=\frac{1}{p+1}\right\}$ if $p>-1$.

40. Using the definition, prove that $\int_a^b e^x\,dx=e^b-e^a$.

41. Work Problem 5 directly, using Problem 94 of Chapter 1.

42. Prove that $\displaystyle\lim_{n\to\infty}\left\{\frac{1}{\sqrt{n^2+1^2}}+\frac{1}{\sqrt{n^2+2^2}}+\cdots+\frac{1}{\sqrt{n^2+n^2}}\right\}=\ln(1+\sqrt{2})$.

43. Prove that $\displaystyle\lim_{n\to\infty}\sum_{k=1}^n \frac{n}{n^2+k^2x^2}=\frac{\tan^{-1}x}{x}$ if $x\neq 0$.

PROPERTIES of DEFINITE INTEGRALS

44. Prove (a) Property 2, (b) Property 3 on Page 81.

45. If $f(x)$ is integrable in (a,c) and (c,b), prove that $\int_a^b f(x)\,dx=\int_a^c f(x)\,dx+\int_c^b f(x)\,dx$.

46. If $f(x)$ and $g(x)$ are integrable in $[a,b]$ and $f(x)\leqq g(x)$, prove that $\int_a^b f(x)\,dx\leqq\int_a^b g(x)\,dx$.

47. Prove that $1-\cos x\geqq x^2/\pi$ for $0\leqq x\leqq\pi/2$.

48. Prove that $\left|\int_0^1 \frac{\cos nx}{x+1}\,dx\right|\leqq\ln 2$ for all n.

49. Prove that $\left|\int_1^{\sqrt{3}} \frac{e^{-x}\sin x}{x^2+1}\,dx\right|\leqq\frac{\pi}{12e}$.

MEAN VALUE THEOREMS for INTEGRALS

50. Prove the result (5), Page 82. [Hint: If $m\leqq f(x)\leqq M$, then $m\,g(x)\leqq f(x)\,g(x)\leqq M\,g(x)$. Now integrate and apply Property 7, Page 82.]

51. Prove that there exist values ξ_1 and ξ_2 in $0\leqq x\leqq 1$ such that
$$\int_0^1 \frac{\sin \pi x}{x^2+1}\,dx=\frac{2}{\pi(\xi_1^2+1)}=\frac{\pi}{4}\sin\pi\xi_2$$

52. Prove that there is a value ξ in $0\leqq x\leqq\pi$ such that $\int_0^\pi e^{-x}\cos x\,dx=\sin\xi$.

CHANGE of VARIABLES and SPECIAL METHODS of INTEGRATION

53. Evaluate: (a) $\int x^2 e^{\sin x^3}\cos x^3\,dx$, (b) $\int_0^1 \frac{\tan^{-1}t}{1+t^2}\,dt$, (c) $\int_1^3 \frac{dx}{\sqrt{4x-x^2}}$, (d) $\int \frac{\operatorname{csch}^2\sqrt{u}}{\sqrt{u}}\,du$,

(e) $\int_{-2}^2 \frac{dx}{16-x^2}$. *Ans.* (a) $\frac{1}{3}e^{\sin x^3}+c$, (b) $\pi^2/32$, (c) $\pi/3$, (d) $-2\coth\sqrt{u}+c$, (e) $\frac{1}{4}\ln 3$.

54. Show that   (a) $\int_0^1 \dfrac{dx}{(3 + 2x - x^2)^{3/2}} = \dfrac{\sqrt{3}}{12}$,   (b) $\int \dfrac{dx}{x^2\sqrt{x^2 - 1}} = \dfrac{\sqrt{x^2 - 1}}{x} + c$.

55. Prove that   (a) $\int \sqrt{u^2 \pm a^2}\, du = \tfrac{1}{2}u\sqrt{u^2 \pm a^2} \pm \tfrac{1}{2}a^2 \ln|u + \sqrt{u^2 \pm a^2}|$

(b) $\int \sqrt{a^2 - u^2}\, du = \tfrac{1}{2}u\sqrt{a^2 - u^2} + \tfrac{1}{2}a^2 \sin^{-1} u/a + c$, $a > 0$.

56. Find $\int \dfrac{x\, dx}{\sqrt{x^2 + 2x + 5}}$.   *Ans.* $\sqrt{x^2 + 2x + 5} - \ln|x + 1 + \sqrt{x^2 + 2x + 5}| + c$.

57. Establish the validity of the method of integration by parts.

58. Evaluate   (a) $\int_0^\pi x \cos 3x\, dx$,   (b) $\int x^3 e^{-2x}\, dx$.   *Ans.* (a) $-2/9$,   (b) $-\tfrac{1}{8}e^{-2x}(4x^3 + 6x^2 + 6x + 3) + c$

59. Show that   (a) $\int_0^1 x^2 \tan^{-1} x\, dx = \tfrac{1}{12}\pi - \tfrac{1}{6} + \tfrac{1}{6}\ln 2$

(b) $\int_{-2}^2 \sqrt{x^2 + x + 1}\, dx = \dfrac{5\sqrt{7}}{4} + \dfrac{3\sqrt{3}}{4} + \dfrac{3}{8}\ln\!\left(\dfrac{5 + 2\sqrt{7}}{2\sqrt{3} - 3}\right)$.

60. (a) If $u = f(x)$ and $v = g(x)$ have continuous nth derivatives, prove that

$$\int uv^{(n)}\, dx = uv^{(n-1)} - u'v^{(n-2)} + u''v^{(n-3)} - \cdots (-1)^n \int u^{(n)} v\, dx$$

called *generalized integration by parts*.   (b) What simplifications occur if $u^{(n)} = 0$?   Discuss.   (c) Use (a) to evaluate $\int_0^\pi x^4 \sin x\, dx$.   *Ans.* (c) $\pi^4 - 12\pi^2 + 48$

61. Show that $\int_0^1 \dfrac{x\, dx}{(x + 1)^2(x^2 + 1)} = \dfrac{\pi - 2}{8}$.

[Hint: Use partial fractions, i.e. assume $\dfrac{x}{(x + 1)^2(x^2 + 1)} = \dfrac{A}{(x + 1)^2} + \dfrac{B}{x + 1} + \dfrac{Cx + D}{x^2 + 1}$ and find A, B, C, D.]

62. Prove that $\int_0^\pi \dfrac{dx}{\alpha - \cos x} = \dfrac{\pi}{\sqrt{\alpha^2 - 1}}$, $\alpha > 1$.

NUMERICAL METHODS for EVALUATING DEFINITE INTEGRALS

63. Evaluate $\int_0^1 \dfrac{dx}{1 + x}$ approximately, using (a) the trapezoidal rule, (b) Simpson's rule, taking $n = 4$. Compare with the exact value, $\ln 2 = 0.6931$.

64. Using (a) the trapezoidal rule, (b) Simpson's rule evaluate $\int_0^{\pi/2} \sin^2 x\, dx$ by obtaining the values of $\sin^2 x$ at $x = 0°, 10°, \ldots, 90°$ and compare with the exact value $\pi/4$.

65. Prove the (a) rectangular rule, (b) trapezoidal rule, i.e. (16) and (17) of Page 85.

66. Prove Simpson's rule.

67. Evaluate to 3 decimal places using numerical integration: (a) $\int_1^2 \dfrac{dx}{1 + x^2}$,   (b) $\int_0^1 \cosh x^2\, dx$. *Ans.* (a) 0.322, (b) 1.105

APPLICATIONS

68. Find the (a) area and (b) moment of inertia about the y axis of the region in the xy plane bounded by $y = \sin x$, $0 \leqq x \leqq \pi$ and the x axis, assuming unit density.   *Ans.* (a) 2, (b) $\pi^2 - 4$

69. Find the moment of inertia about the x axis of the region bounded by $y = x^2$ and $y = x$, if the density is proportional to the distance from the x axis.   *Ans.* $\tfrac{1}{3}M$, where M = mass of the region.

70. Show that the arc length of the *catenary* $y = \cosh x$ from $x = 0$ to $x = \ln 2$ is $\tfrac{3}{4}$.

71. Show that the length of one arch of the *cycloid* $x = a(\theta - \sin \theta)$, $y = a(1 - \cos \theta)$, $(0 \leqq \theta \leqq 2\pi)$ is $8a$.

72. Prove that the area bounded by the ellipse $x^2/a^2 + y^2/b^2 = 1$ is πab.

73. Find the volume of the region obtained by revolving the curve $y = \sin x$, $0 \leqq x \leqq \pi$, about the x axis. *Ans.* $\pi^2/2$

74. Prove that the centroid of the region bounded by $y = \sqrt{a^2 - x^2}$, $-a \leqq x \leqq a$ and the x axis is located at $(0, 4a/3\pi)$.

75. (a) If $\rho = f(\phi)$ is the equation of a curve in polar coordinates, show that the area bounded by this curve and the lines $\phi = \phi_1$ and $\phi = \phi_2$ is $\dfrac{1}{2} \displaystyle\int_{\phi_1}^{\phi_2} \rho^2 d\phi$. (b) Find the area bounded by one loop of the *lemniscate* $\rho^2 = a^2 \cos 2\phi$.   *Ans.* (b) a^2

76. (a) Prove that the arc length of the curve in Problem 75(a) is $\displaystyle\int_{\phi_1}^{\phi_2} \sqrt{\rho^2 + (d\rho/d\phi)^2}\, d\phi$. (b) Find the length of arc of the *cardioid* $\rho = a(1 - \cos \phi)$.   *Ans.* (b) $8a$

MISCELLANEOUS PROBLEMS

77. Establish the theorem of the mean for derivatives from the first mean value theorem for integrals. [Hint: Let $f(x) = F'(x)$ in (4), Page 81.]

78. Prove that   (a) $\displaystyle\lim_{\epsilon \to 0+} \int_0^{4-\epsilon} \frac{dx}{\sqrt{4-x}} = 4$,   (b) $\displaystyle\lim_{\epsilon \to 0+} \int_\epsilon^8 \frac{dx}{\sqrt[3]{x}} = 6$,   (c) $\displaystyle\lim_{\epsilon \to 0+} \int_0^{1-\epsilon} \frac{dx}{\sqrt{1-x^2}} = \frac{\pi}{2}$

and give a geometric interpretation of the results.

[These limits, denoted usually by $\displaystyle\int_0^4 \frac{dx}{\sqrt{4-x}}$, $\displaystyle\int_0^8 \frac{dx}{\sqrt[3]{x}}$ and $\displaystyle\int_0^1 \frac{dx}{\sqrt{1-x^2}}$ respectively, are called *improper integrals of the second kind* (see Problem 33) since the integrands are not bounded in the range of integration. For further discussion of improper integrals, see Chapter 12.]

79. Prove that   (a) $\displaystyle\lim_{M \to \infty} \int_0^M x^5 e^{-x}\, dx = 4! = 24$,   (b) $\displaystyle\lim_{\epsilon \to 0+} \int_1^{2-\epsilon} \frac{dx}{\sqrt{x(2-x)}} = \frac{\pi}{2}$.

80. Evaluate   (a) $\displaystyle\int_0^\infty \frac{dx}{1+x^3}$,   (b) $\displaystyle\int_0^{\pi/2} \frac{\sin 2x}{(\sin x)^{4/3}}\, dx$,   (c) $\displaystyle\int_0^\infty \frac{dx}{x + \sqrt{x^2+1}}$.

Ans. (a) $\dfrac{2\pi}{3\sqrt{3}}$ (b) 3 (c) does not exist

81. Evaluate $\displaystyle\lim_{x \to \pi/2} \frac{e x^2/\pi - e\pi/4 + \int_x^{\pi/2} e^{\sin t}\, dt}{1 + \cos 2x}$.   *Ans.* $e/2\pi$

82. Prove: (a) $\dfrac{d}{dx} \displaystyle\int_{x^2}^{x^3} (t^2 + t + 1)\, dt = 3x^8 + x^5 - 2x^3 + 3x^2 - 2x$, (b) $\dfrac{d}{dx} \displaystyle\int_x^{x^2} \cos t^2\, dt = 2x \cos x^4 - \cos x^2$.

83. Prove the result (13) on Page 83.

84. Prove that   (a) $\displaystyle\int_0^\pi \sqrt{1 + \sin x}\, dx = 4$,   (b) $\displaystyle\int_0^{\pi/2} \frac{dx}{\sin x + \cos x} = \sqrt{2} \ln (\sqrt{2} + 1)$.

85. Explain the fallacy: $I = \displaystyle\int_{-1}^1 \frac{dx}{1+x^2} = -\int_{-1}^1 \frac{dy}{1+y^2} = -I$, using the transformation $x = 1/y$. Hence $I = 0$. But $I = \tan^{-1}(1) - \tan^{-1}(-1) = \pi/4 - (-\pi/4) = \pi/2$. Thus $\pi/2 = 0$.

86. Prove that $\displaystyle\int_0^{1/2} \frac{\cos \pi x}{\sqrt{1+x^2}}\, dx \leqq \frac{1}{4} \tan^{-1} \frac{1}{2}$.

87. Evaluate $\displaystyle\lim_{n \to \infty} \left\{ \frac{\sqrt{n+1} + \sqrt{n+2} + \cdots + \sqrt{2n-1}}{n^{3/2}} \right\}$.   *Ans.* $\frac{2}{3}(2\sqrt{2} - 1)$

88. Prove that $\quad f(x) = \begin{cases} 1 & \text{if } x \text{ is irrational} \\ 0 & \text{if } x \text{ is rational} \end{cases}$ is not Riemann integrable in $[0, 1]$.

   [Hint: In (2), Page 80, let ξ_k, $k = 1, 2, 3, \ldots, n$ be first rational and then irrational points of subdivision and examine the lower and upper sums of Problem 35.]

89. Prove the result (3) of Problem 35. [Hint: First consider the effect of only one additional point of subdivision.]

90. In Problem 35, prove that $\bar{s} \leq \underline{S}$. [Hint: Assume the contrary and obtain a contradiction.]

91. If $f(x)$ is sectionally continuous in $[a, b]$, prove that $\displaystyle\int_a^b f(x)\, dx$ exists. [Hint: Enclose each point of discontinuity in an interval, noting that the sum of the lengths of such intervals can be made arbitrarily small. Then consider the difference between the upper and lower sums.]

92. If $\quad f(x) = \begin{cases} 2x & 0 < x < 1 \\ 3 & x = 1 \\ 6x - 1 & 1 < x < 2 \end{cases}$, find $\displaystyle\int_0^2 f(x)\, dx$. Interpret the result graphically.   *Ans.* 9

93. Evaluate $\displaystyle\int_0^3 \{x - [x] + \tfrac{1}{2}\}\, dx$ where $[x]$ denotes the greatest integer less than or equal to x. Interpret the result graphically.   *Ans.* 3

94. (a) Prove that $\displaystyle\int_0^{\pi/2} \frac{\sin^m x}{\sin^m x + \cos^m x}\, dx = \frac{\pi}{4}$ for all real values of m.

   (b) Prove that $\displaystyle\int_0^{2\pi} \frac{dx}{1 + \tan^4 x} = \pi$.

95. Prove that $\displaystyle\int_0^{\pi/2} \frac{\sin x}{x}\, dx$ exists.

96. Show that $\displaystyle\int_0^{0.5} \frac{\tan^{-1} x}{x}\, dx = 0.4872$ approximately.

97. Show that $\displaystyle\int_0^{\pi} \frac{x\, dx}{1 + \cos^2 x} = \frac{\pi^2}{2\sqrt{2}}$.

Chapter 6

Partial Derivatives

FUNCTIONS of TWO or MORE VARIABLES

A variable z is said to be a *function* of two variables x and y if for each given pair (x, y) we can determine one or more values of z. This definition is in keeping with the general definition of function as a correspondence between two sets (see Page 20). Here the two sets are (1) a set of number pairs (x, y) [represented geometrically by a two dimensional point set in the xy plane] and (2) a set of real numbers represented by the variable z.

We use the notation $f(x, y)$, $F(x, y)$, etc., to denote the value of the function at (x, y) and write $z = f(x, y)$, $z = F(x, y)$, etc. We shall also sometimes use the notation $z = z(x, y)$ although it should be understood that in this case z is used in two senses, namely as a function and as a variable.

> **Example:** If $f(x, y) = x^2 + 2y^3$, then $f(3, -1) = (3)^2 + 2(-1)^3 = 7$.

The concept is easily extended. Thus $w = F(x, y, z)$ denotes the value of a function at (x, y, z) [a point in 3 dimensional space], etc.

DEPENDENT and INDEPENDENT VARIABLES. DOMAIN of a FUNCTION

If $z = F(x, y)$, we call z a *dependent variable* and x and y the *independent variables*. The function is called *single-valued* if only one value of z corresponds to each pair (x, y) for which the function is defined. If there is more than one value of z, the function is *multiple-valued* and can be considered as a collection of single-valued functions. Hence we shall restrict ourselves to single-valued functions, unless otherwise indicated.

The set of values (points), (x, y) for which a function is defined is called the *domain of definition* or simply *domain* of the function.

> **Example:** If $z = \sqrt{1 - (x^2 + y^2)}$, the domain for which z is real consists of the set of points (x, y) such that $x^2 + y^2 \leq 1$, i.e. the set of points inside and on a circle in the xy plane having center at $(0, 0)$ and radius 1.

THREE DIMENSIONAL RECTANGULAR COORDINATE SYSTEMS

A three dimensional rectangular coordinate system obtained by constructing 3 mutually perpendicular axes (the x, y *and* z *axes*) intersecting in point O (the origin) forms a natural extension of the usual xy plane for representing functions of two variables graphically. A point in 3 dimensions is represented by the triplet (x, y, z) called *coordinates* of the point. In this coordinate system $z = f(x, y)$ [or $F(x, y, z) = 0$] represents a surface, in general.

> **Example:** The set of points (x, y, z) such that $z = \sqrt{1 - (x^2 + y^2)}$ comprises the surface of a hemisphere of radius 1 and center at $(0, 0, 0)$.

For functions of more than two variables such geometric interpretation fails, although the terminology is still employed. For example, (x, y, z, w) is a point in 4 dimensional space, and $w = f(x, y, z)$ [or $F(x, y, z, w) = 0$] represents a *hypersurface* in 4 dimensions; thus $x^2 + y^2 + z^2 + w^2 = a^2$ represents a *hypersphere* in 4 dimensions with radius $a > 0$ and center at $(0, 0, 0, 0)$.

101

NEIGHBORHOODS

The set of all points (x, y) such that $|x - x_0| < \delta$, $|y - y_0| < \delta$ where $\delta > 0$, is called a *rectangular δ neighborhood* of (x_0, y_0); the set $0 < |x - x_0| < \delta$, $0 < |y - y_0| < \delta$ which excludes (x_0, y_0) is called a *rectangular deleted δ neighborhood* of (x_0, y_0). Similar remarks can be made for other neighborhoods, e.g. $(x - x_0)^2 + (y - y_0)^2 < \delta^2$ is a *circular δ neighborhood* of (x_0, y_0).

A point (x_0, y_0) is called a *limit point, accumulation point or cluster point* of a point set S if every deleted δ neighborhood of (x_0, y_0) contains points of S. As in the case of one dimensional point sets, every bounded infinite set has at least one limit point (the Weirstrass-Bolzano theorem, see Pages 5 and 50). A set containing all its limit points is called a *closed set*.

REGIONS

A point P belonging to a point set S is called an *interior point* of S if there exists a deleted δ neighborhood of P all of whose points belong to S. A point P not belonging to S is called an *exterior point* of S if there exists a deleted δ neighborhood of P all of whose points do not belong to S. A point P is called a *boundary point* of S if every deleted δ neighborhood of P contains points belonging to S and also points not belonging to S.

If any two points of a set S can be joined by a path consisting of a finite number of broken line segments all of whose points belong to S, then S is called a *connected set*. A *region* is a connected set which consists of interior points or interior and boundary points. A *closed region* is a region containing all its boundary points. An *open region* consists only of interior points.

Examples of some regions are shown graphically in Figures 6-1(a), (b) and (c) below. The rectangular region of Fig. 6-1(a), including the boundary, represents the set of points $a \leq x \leq b$, $c \leq y \leq d$ which is a natural extension of the closed interval $a \leq x \leq b$ for one dimension. The set $a < x < b$, $c < y < d$ corresponds to the boundary being excluded.

In the regions of Figures 6-1(a) and 6-1(b), any *simple closed curve* (one which does not intersect itself anywhere) lying inside the region can be shrunk to a point which also lies in the region. Such regions are called *simply-connected regions*. In Fig. 6-1(c) however, a simple closed curve $ABCD$ surrounding one of the "holes" in the region cannot be shrunk to a point without leaving the region. Such regions are called *multiply-connected regions*.

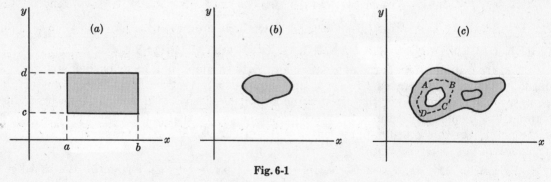

Fig. 6-1

LIMITS

Let $f(x, y)$ be defined in a deleted δ neighborhood of (x_0, y_0) [i.e. $f(x, y)$ may be undefined at (x_0, y_0)]. We say that l is the *limit* of $f(x, y)$ as x approaches x_0 and y approaches y_0 [or (x, y) approaches (x_0, y_0)] and write $\lim\limits_{\substack{x \to x_0 \\ y \to y_0}} f(x, y) = l$ [or $\lim\limits_{(x, y) \to (x_0, y_0)} f(x, y) = l$] if for any positive number ϵ we can find some positive number δ [depending on ϵ and (x_0, y_0), in general] such that $|f(x, y) - l| < \epsilon$ whenever $0 < |x - x_0| < \delta$ and $0 < |y - y_0| < \delta$.

If desired we can use the deleted circular neighborhood $0 < (x - x_0)^2 + (y - y_0)^2 < \delta^2$ instead of the deleted rectangular neighborhood.

Example: Let $f(x, y) = \begin{cases} 3xy & \text{if } (x, y) \neq (1, 2) \\ 0 & \text{if } (x, y) = (1, 2) \end{cases}$. As $x \to 1$ and $y \to 2$ [or $(x, y) \to (1, 2)$], $f(x, y)$ gets closer to $3(1)(2) = 6$ and we *suspect* that $\lim_{\substack{x \to 1 \\ y \to 2}} f(x, y) = 6$. To *prove* this we must show that the above definition of limit with $l = 6$ is satisfied. Such a proof can be supplied by a method similar to that of Problem 4.

Note that $\lim_{\substack{x \to 1 \\ y \to 2}} f(x, y) \neq f(1, 2)$ since $f(1, 2) = 0$. The limit would in fact be 6 even if $f(x, y)$ were not defined at $(1, 2)$. Thus the existence of the limit of $f(x, y)$ as $(x, y) \to (x_0, y_0)$ is in no way dependent on the existence of a value of $f(x, y)$ at (x_0, y_0).

Note that in order for $\lim_{(x, y) \to (x_0, y_0)} f(x, y)$ to exist, it must have the same value regardless of the approach of (x, y) to (x_0, y_0). It follows that if two different approaches give different values, the limit cannot exist (see Problem 7). This implies, as in the case of functions of one variable, that if a limit exists it is unique.

The concept of one-sided limits for functions of one variable is easily extended to functions of more than one variable.

Example 1: $\lim_{\substack{x \to 0+ \\ y \to 1}} \tan^{-1}(y/x) = \pi/2, \quad \lim_{\substack{x \to 0- \\ y \to 1}} \tan^{-1}(y/x) = -\pi/2$

Example 2: $\lim_{\substack{x \to 0 \\ y \to 1}} \tan^{-1}(y/x)$ does not exist, as is clear from the fact that the two different approaches of Example 1 give different results.

In general the theorems on limits, concepts of infinity, etc., for functions of one variable (see Page 24) apply as well, with appropriate modifications, to functions of two or more variables.

ITERATED LIMITS

The *iterated limits* $\lim_{x \to x_0} \left\{ \lim_{y \to y_0} f(x, y) \right\}$ and $\lim_{y \to y_0} \left\{ \lim_{x \to x_0} f(x, y) \right\}$, [also denoted by $\lim_{x \to x_0} \lim_{y \to y_0} f(x, y)$ and $\lim_{y \to y_0} \lim_{x \to x_0} f(x, y)$ respectively] are not necessarily equal. Although they must be equal if $\lim_{\substack{x \to x_0 \\ y \to y_0}} f(x, y)$ is to exist, their equality does not guarantee the existence of this last limit.

Example: If $f(x, y) = \dfrac{x - y}{x + y}$, then $\lim_{x \to 0} \left(\lim_{y \to 0} \dfrac{x - y}{x + y} \right) = \lim_{x \to 0} (1) = 1$ and $\lim_{y \to 0} \left(\lim_{x \to 0} \dfrac{x - y}{x + y} \right) = \lim_{y \to 0} (-1) = -1$. Thus the iterated limits are not equal and so $\lim_{\substack{x \to 0 \\ y \to 0}} f(x, y)$ cannot exist.

CONTINUITY

Let $f(x, y)$ be defined in a δ neighborhood of (x_0, y_0) [i.e. $f(x, y)$ must be defined *at* (x_0, y_0) as well as near it]. We say that $f(x, y)$ is *continuous* at (x_0, y_0) if for any positive number ϵ we can find some positive number δ [depending on ϵ and (x_0, y_0) in general] such that $|f(x, y) - f(x_0, y_0)| < \epsilon$ whenever $|x - x_0| < \delta$ and $|y - y_0| < \delta$. Note that three conditions must be satisfied in order that $f(x, y)$ be continuous at (x_0, y_0).

1. $\lim\limits_{(x,y) \to (x_0,y_0)} f(x,y) = l$, i.e. the limit exists as $(x,y) \to (x_0,y_0)$

2. $f(x_0,y_0)$ must exist, i.e. $f(x,y)$ is defined at (x_0,y_0)

3. $l = f(x_0,y_0)$

If desired we can write this in the suggestive form $\lim\limits_{\substack{x \to x_0 \\ y \to y_0}} f(x,y) = f(\lim\limits_{x \to x_0} x, \lim\limits_{y \to y_0} y)$.

Example: If $f(x,y) = \begin{cases} 3xy & (x,y) \neq (1,2) \\ 0 & (x,y) = (1,2) \end{cases}$, then $\lim\limits_{(x,y) \to (1,2)} f(x,y) = 6 \neq f(1,2)$. Hence $f(x,y)$ is not continuous at $(1,2)$. If we redefine the function so that $f(x,y) = 6$ for $(x,y) = (1,2)$, then the function is continuous at $(1,2)$.

If a function is not continuous at a point (x_0,y_0), it is said to be *discontinuous* at (x_0,y_0) which is then called a *point of discontinuity*. If, as in the above example, it is possible so to redefine the value of a function at a point of discontinuity that the new function is continuous, we say that the point is a *removable discontinuity* of the old function. A function is said to be *continuous in a region* $\mathcal{R}$ of the xy plane if it is continuous at every point of $\mathcal{R}$.

Many of the theorems on continuity for functions of a single variable can, with suitable modification, be extended to functions of two or more variables.

UNIFORM CONTINUITY

In the definition of continuity of $f(x,y)$ at (x_0,y_0), δ depends on ϵ and also (x_0,y_0) in general. If in a region $\mathcal{R}$ we can find a δ which depends only on ϵ but not on any particular point (x_0,y_0) in $\mathcal{R}$ [i.e. the same δ will work for *all* points in $\mathcal{R}$], then $f(x,y)$ is said to be *uniformly continuous* in $\mathcal{R}$. As in the case of functions of one variable, it can be proved that a function which is continuous in a closed and bounded region is uniformly continuous in the region.

PARTIAL DERIVATIVES

The ordinary derivative of a function of several variables with respect to one of the independent variables, keeping all other independent variables constant, is called the partial derivative of the function with respect to the variable. Partial derivatives of $f(x,y)$ with respect to x and y are denoted by $\dfrac{\partial f}{\partial x}$ $\left[\text{or } f_x, f_x(x,y), \dfrac{\partial f}{\partial x}\Big|_y \right]$ and $\dfrac{\partial f}{\partial y}$ $\left[\text{or } f_y, f_y(x,y), \dfrac{\partial f}{\partial y}\Big|_x \right]$ respectively, the latter notations being used when it is needed to emphasize which variables are held constant.

By definition,

$$\frac{\partial f}{\partial x} = \lim_{\Delta x \to 0} \frac{f(x+\Delta x, y) - f(x,y)}{\Delta x}, \qquad \frac{\partial f}{\partial y} = \lim_{\Delta y \to 0} \frac{f(x, y+\Delta y) - f(x,y)}{\Delta y} \qquad (1)$$

when these limits exist. The derivatives evaluated at the particular point (x_0,y_0) are often indicated by $\dfrac{\partial f}{\partial x}\Big|_{(x_0,y_0)} = f_x(x_0,y_0)$ and $\dfrac{\partial f}{\partial y}\Big|_{(x_0,y_0)} = f_y(x_0,y_0)$ respectively.

Example: If $f(x,y) = 2x^3 + 3xy^2$, then $f_x = \partial f/\partial x = 6x^2 + 3y^2$ and $f_y = \partial f/\partial y = 6xy$. Also, $f_x(1,2) = 6(1)^2 + 3(2)^2 = 18$, $f_y(1,2) = 6(1)(2) = 12$.

If a function f has continuous partial derivatives $\partial f/\partial x$, $\partial f/\partial y$ in a region, then f must be continuous in the region. However, the existence of these partial derivatives alone is not enough to guarantee the continuity of f (see Problem 9).

HIGHER ORDER PARTIAL DERIVATIVES

If $f(x, y)$ has partial derivatives at each point (x, y) in a region, then $\partial f/\partial x$ and $\partial f/\partial y$ are themselves functions of x and y which may also have partial derivatives. These second derivatives are denoted by

$$\frac{\partial}{\partial x}\left(\frac{\partial f}{\partial x}\right) = \frac{\partial^2 f}{\partial x^2} = f_{xx}, \quad \frac{\partial}{\partial y}\left(\frac{\partial f}{\partial y}\right) = \frac{\partial^2 f}{\partial y^2} = f_{yy}, \quad \frac{\partial}{\partial x}\left(\frac{\partial f}{\partial y}\right) = \frac{\partial^2 f}{\partial x\,\partial y} = f_{yx}, \quad \frac{\partial}{\partial y}\left(\frac{\partial f}{\partial x}\right) = \frac{\partial^2 f}{\partial y\,\partial x} = f_{xy} \quad (2)$$

If f_{xy} and f_{yx} are continuous, then $f_{xy} = f_{yx}$ and the order of differentiation is immaterial; otherwise they may not be equal (see Problems 13 and 43).

> **Example:** If $f(x, y) = 2x^3 + 3xy^2$ (see preceding example), then $f_{xx} = 12x$, $f_{yy} = 6x$, $f_{xy} = 6y = f_{yx}$. In such case $f_{xx}(1, 2) = 12$, $f_{yy}(1, 2) = 6$, $f_{xy}(1, 2) = f_{yx}(1, 2) = 12$.

In a similar manner higher order derivatives are defined. For example, $\dfrac{\partial^3 f}{\partial x^2\,\partial y} = f_{yxx}$ is the derivative of f taken once with respect to y and twice with respect to x.

DIFFERENTIALS

Let $\Delta x = dx$ and $\Delta y = dy$ be increments given to x and y respectively. Then

$$\Delta z = f(x + \Delta x, y + \Delta y) - f(x, y) = \Delta f \quad (3)$$

is called the *increment* in $z = f(x, y)$. If $f(x, y)$ has continuous first partial derivatives in a region, then

$$\Delta z = \frac{\partial f}{\partial x}\Delta x + \frac{\partial f}{\partial y}\Delta y + \epsilon_1 \Delta x + \epsilon_2 \Delta y = \frac{\partial z}{\partial x}dx + \frac{\partial z}{\partial y}dy + \epsilon_1 dx + \epsilon_2 dy = \Delta f \quad (4)$$

where ϵ_1 and ϵ_2 approach zero as Δx and Δy approach zero (see Problem 14). The expression

$$dz = \frac{\partial z}{\partial x}dx + \frac{\partial z}{\partial y}dy \quad \text{or} \quad df = \frac{\partial f}{\partial x}dx + \frac{\partial f}{\partial y}dy \quad (5)$$

is called the *total differential* or simply *differential* of z or f, or the *principal part* of Δz or Δf. Note that $\Delta z \neq dz$ in general. However, if $\Delta x = dx$ and $\Delta y = dy$ are "small", then dz is a close approximation of Δz (see Problem 15). The quantities dx and dy, called *differentials* of x and y respectively, need not be small.

If f is such that Δf (or Δz) can be expressed in the form (4) where ϵ_1 and ϵ_2 approach zero as Δx and Δy approach zero, we call f *differentiable* at (x, y). The mere existence of f_x and f_y does not in itself guarantee differentiability; however, continuity of f_x and f_y does (although this condition happens to be slightly stronger than necessary). In case f_x and f_y are continuous in a region $\mathcal{R}$, we shall say that f is *continuously differentiable* in $\mathcal{R}$.

THEOREMS on DIFFERENTIALS

In the following we shall assume that all functions have continuous first partial derivatives in a region $\mathcal{R}$, i.e. the functions are continuously differentiable in $\mathcal{R}$.

1. If $z = f(x_1, x_2, \ldots, x_n)$, then

$$df = \frac{\partial f}{\partial x_1}dx_1 + \frac{\partial f}{\partial x_2}dx_2 + \cdots + \frac{\partial f}{\partial x_n}dx_n \quad (6)$$

regardless of whether the variables $x_1, x_2, \ldots, x_n$ are independent or dependent on other variables (see Problem 20). This is a generalization of the result (5). In (6) we often use z in place of f.

2. If $f(x_1, x_2, \ldots, x_n) = c$, a constant, then $df = 0$. Note that in this case $x_1, x_2, \ldots, x_n$ cannot all be independent variables.

3. The expression $P(x, y)\, dx + Q(x, y)\, dy$ or briefly $P\, dx + Q\, dy$ is the differential of $f(x, y)$ if and only if $\dfrac{\partial P}{\partial y} = \dfrac{\partial Q}{\partial x}$. In such case $P\, dx + Q\, dy$ is called an *exact differential*.

4. The expression $P(x, y, z)\, dx + Q(x, y, z)\, dy + R(x, y, z)\, dz$ or briefly $P\, dx + Q\, dy + R\, dz$ is the differential of $f(x, y, z)$ if and only if $\dfrac{\partial P}{\partial y} = \dfrac{\partial Q}{\partial x}$, $\dfrac{\partial Q}{\partial z} = \dfrac{\partial R}{\partial y}$, $\dfrac{\partial R}{\partial x} = \dfrac{\partial P}{\partial z}$. In such case $P\, dx + Q\, dy + R\, dz$ is called an *exact differential*.

Proofs of Theorems 3 and 4 are best supplied by methods of later chapters (see Chapter 10, Problems 13 and 30).

DIFFERENTIATION of COMPOSITE FUNCTIONS

Let $z = f(x, y)$ where $x = g(r, s)$, $y = h(r, s)$ so that z is a function of r and s. Then

$$\frac{\partial z}{\partial r} = \frac{\partial z}{\partial x}\frac{\partial x}{\partial r} + \frac{\partial z}{\partial y}\frac{\partial y}{\partial r}, \qquad \frac{\partial z}{\partial s} = \frac{\partial z}{\partial x}\frac{\partial x}{\partial s} + \frac{\partial z}{\partial y}\frac{\partial y}{\partial s} \tag{7}$$

In general, if $u = F(x_1, \ldots, x_n)$ where $x_1 = f_1(r_1, \ldots, r_p), \ldots, x_n = f_n(r_1, \ldots, r_p)$, then

$$\frac{\partial u}{\partial r_k} = \frac{\partial u}{\partial x_1}\frac{\partial x_1}{\partial r_k} + \frac{\partial u}{\partial x_2}\frac{\partial x_2}{\partial r_k} + \cdots + \frac{\partial u}{\partial x_n}\frac{\partial x_n}{\partial r_k} \qquad k = 1, 2, \ldots, p \tag{8}$$

If in particular $x_1, x_2, \ldots, x_n$ depend on only one variable s, then

$$\frac{du}{ds} = \frac{\partial u}{\partial x_1}\frac{dx_1}{ds} + \frac{\partial u}{\partial x_2}\frac{dx_2}{ds} + \cdots + \frac{\partial u}{\partial x_n}\frac{dx_n}{ds} \tag{9}$$

These results, often called *chain rules*, are useful in transforming derivatives from one set of variables to another.

Higher derivatives are obtained by repeated application of the chain rules.

EULER'S THEOREM on HOMOGENEOUS FUNCTIONS

A function $F(x_1, x_2, \ldots, x_n)$ is called *homogeneous of degree p* if, for all values of the parameter λ and some constant p, we have the identity

$$F(\lambda x_1, \lambda x_2, \ldots, \lambda x_n) = \lambda^p F(x_1, x_2, \ldots, x_n) \tag{10}$$

Example: $F(x, y) = x^4 + 2xy^3 - 5y^4$ is homogeneous of degree 4, since
$$F(\lambda x, \lambda y) = (\lambda x)^4 + 2(\lambda x)(\lambda y)^3 - 5(\lambda y)^4 = \lambda^4(x^4 + 2xy^3 - 5y^4) = \lambda^4 F(x, y)$$

Euler's theorem on homogeneous functions states that if $F(x_1, x_2, \ldots, x_n)$ is homogeneous of degree p then (see Problem 25)

$$x_1\frac{\partial F}{\partial x_1} + x_2\frac{\partial F}{\partial x_2} + \cdots + x_n\frac{\partial F}{\partial x_n} = pF \tag{11}$$

IMPLICIT FUNCTIONS

In general, an equation such as $F(x, y, z) = 0$ defines one variable, say z, as a function of the other two variables x and y. Then z is sometimes called an *implicit function* of x and y, as distinguished from a so-called *explicit function* f, where $z = f(x, y)$, which is such that $F[x, y, f(x, y)] \equiv 0$.

Differentiation of implicit functions is not difficult provided the dependent and independent variables are kept clearly in mind.

JACOBIANS

If $F(u, v)$ and $G(u, v)$ are differentiable in a region, the *Jacobian determinant*, or briefly the *Jacobian*, of F and G with respect to u and v is the second order functional determinant defined by

$$\frac{\partial(F, G)}{\partial(u, v)} = \begin{vmatrix} \dfrac{\partial F}{\partial u} & \dfrac{\partial F}{\partial v} \\[2mm] \dfrac{\partial G}{\partial u} & \dfrac{\partial G}{\partial v} \end{vmatrix} = \begin{vmatrix} F_u & F_v \\ G_u & G_v \end{vmatrix} \tag{7}$$

Similarly, the third order determinant

$$\frac{\partial(F, G, H)}{\partial(u, v, w)} = \begin{vmatrix} F_u & F_v & F_w \\ G_u & G_v & G_w \\ H_u & H_v & H_w \end{vmatrix}$$

is called the Jacobian of F, G and H with respect to u, v and w. Extensions are easily made.

PARTIAL DERIVATIVES USING JACOBIANS

Jacobians often prove useful in obtaining partial derivatives of implicit functions. Thus for example, given the simultaneous equations

$$F(x, y, u, v) = 0, \quad G(x, y, u, v) = 0$$

we may, in general, consider u and v as functions of x and y. In this case, we have (see Problem 31)

$$\frac{\partial u}{\partial x} = -\frac{\dfrac{\partial(F, G)}{\partial(x, v)}}{\dfrac{\partial(F, G)}{\partial(u, v)}}, \quad \frac{\partial u}{\partial y} = -\frac{\dfrac{\partial(F, G)}{\partial(y, v)}}{\dfrac{\partial(F, G)}{\partial(u, v)}}, \quad \frac{\partial v}{\partial x} = -\frac{\dfrac{\partial(F, G)}{\partial(u, x)}}{\dfrac{\partial(F, G)}{\partial(u, v)}}, \quad \frac{\partial v}{\partial y} = -\frac{\dfrac{\partial(F, G)}{\partial(u, y)}}{\dfrac{\partial(F, G)}{\partial(u, v)}}$$

The ideas are easily extended. Thus if we consider the simultaneous equations

$$F(u, v, w, x, y) = 0, \quad G(u, v, w, x, y) = 0, \quad H(u, v, w, x, y) = 0$$

we may, for example, consider u, v and w as functions of x and y. In this case,

$$\frac{\partial u}{\partial x} = -\frac{\dfrac{\partial(F, G, H)}{\partial(x, v, w)}}{\dfrac{\partial(F, G, H)}{\partial(u, v, w)}}, \quad \frac{\partial w}{\partial y} = -\frac{\dfrac{\partial(F, G, H)}{\partial(u, v, y)}}{\dfrac{\partial(F, G, H)}{\partial(u, v, w)}}$$

with similar results for the remaining partial derivatives (see Problem 33).

THEOREMS on JACOBIANS

In the following we assume that all functions are continuously differentiable.

1. A necessary and sufficient condition that the equations $F(u, v, x, y, z) = 0$, $G(u, v, x, y, z) = 0$ can be solved for u and v (for example) is that $\dfrac{\partial(F, G)}{\partial(u, v)}$ is not identically zero in a region $\mathcal{R}$.

Similar results are valid for m equations in n variables, where $m < n$.

2. If x and y are functions of u and v while u and v are functions of r and s, then (see Problem 45)

$$\frac{\partial(x, y)}{\partial(r, s)} = \frac{\partial(x, y)}{\partial(u, v)} \frac{\partial(u, v)}{\partial(r, s)} \tag{9}$$

This is an example of a *chain rule* for Jacobians. These ideas are capable of generalization (see Problems 114 and 116, for example).

3. If $u = f(x, y)$ and $v = g(x, y)$, then a necessary and sufficient condition that a functional relation of the form $\phi(u, v) = 0$ exists between u and v is that $\dfrac{\partial(u, v)}{\partial(x, y)}$ be identically zero. Similar results hold for n functions of n variables.

Further discussion of Jacobians appears in Chapter 7 where vector interpretations are employed.

TRANSFORMATIONS

The set of equations

$$\begin{cases} x = F(u, v) \\ y = G(u, v) \end{cases} \tag{10}$$

defines, in general, a *transformation* or *mapping* which establishes a correspondence between points in the uv and xy planes. If to each point in the uv plane there corresponds one and only one point in the xy plane, and conversely, we speak of a *one to one transformation* or *mapping*. This will be so if F and G are continuously differentiable with Jacobian not identically zero in a region. In such case (which we shall assume unless otherwise stated) equations (10) are said to define a *continuously differentiable transformation* or *mapping*.

Under the transformation (10) a closed region $\mathcal{R}$ of the xy plane is, in general, mapped into a closed region $\mathcal{R}'$ of the uv plane. Then if ΔA_{xy} and ΔA_{uv} denote respectively the areas of these regions, we can show that

$$\lim \frac{\Delta A_{xy}}{\Delta A_{uv}} = \left| \frac{\partial(x, y)}{\partial(u, v)} \right| \tag{11}$$

where lim denotes the limit as ΔA_{xy} (or ΔA_{uv}) approaches zero. The Jacobian on the right of (11) is often called the *Jacobian of the transformation* (10).

If we solve (10) for u and v in terms of x and y, we obtain the transformation $u = f(x, y)$, $v = g(x, y)$ often called the *inverse transformation* corresponding to (10). The Jacobians $\dfrac{\partial(u, v)}{\partial(x, y)}$ and $\dfrac{\partial(x, y)}{\partial(u, v)}$ of these transformations are reciprocals of each other (see Problem 45). Hence if one Jacobian is different from zero in a region, so also is the other.

The above ideas can be extended to transformations in three or higher dimensions. We shall deal further with these topics in Chapter 7, where use is made of the simplicity of vector notation and interpretation.

CURVILINEAR COORDINATES

If (x, y) are the rectangular coordinates of a point in the xy plane, we can think of (u, v) as also specifying coordinates of the same point, since by knowing (u, v) we can determine (x, y) from (10). The coordinates (u, v) are called *curvilinear coordinates* of the point.

> **Example:** The polar coordinates (ρ, ϕ) of a point correspond to the case $u = \rho$, $v = \phi$. In this case the transformation equations (10) are $x = \rho \cos \phi$, $y = \rho \sin \phi$.

For curvilinear coordinates in higher dimensional spaces, see Chapter 7.

MEAN VALUE THEOREMS

1. **First mean value theorem.** If $f(x, y)$ is continuous in a closed region and if the first partial derivatives exist in the open region (i.e. excluding boundary points), then

$$f(x_0+h, y_0+k) - f(x_0, y_0) = h f_x(x_0+\theta h, y_0+\theta k) + k f_y(x_0+\theta h, y_0+\theta k) \quad 0 < \theta < 1 \quad (12)$$

This is sometimes written in a form in which $h = \Delta x = x - x_0$ and $k = \Delta y = y - y_0$.

2. **Taylor's theorem of the mean.** If all the nth partial derivatives of $f(x, y)$ are continuous in a closed region and if the $(n+1)$st partial derivatives exist in the open region, then

$$f(x_0+h, y_0+k) = f(x_0, y_0) + \left(h \frac{\partial}{\partial x} + k \frac{\partial}{\partial y} \right) f(x_0, y_0) + \frac{1}{2!} \left(h \frac{\partial}{\partial x} + k \frac{\partial}{\partial y} \right)^2 f(x_0, y_0) + \cdots$$

$$+ \frac{1}{n!} \left(h \frac{\partial}{\partial x} + k \frac{\partial}{\partial y} \right)^n f(x_0, y_0) + R_n \quad (13)$$

where R_n, the remainder after n terms, is given by

$$R_n = \frac{1}{(n+1)!} \left(h \frac{\partial}{\partial x} + k \frac{\partial}{\partial y} \right)^{n+1} f(x_0+\theta h, y_0+\theta k) \quad 0 < \theta < 1 \quad (14)$$

and where we use the operator notation

$$\left(h \frac{\partial}{\partial x} + k \frac{\partial}{\partial y} \right) f(x_0, y_0) \equiv h f_x(x_0, y_0) + k f_y(x_0, y_0)$$

$$\left(h \frac{\partial}{\partial x} + k \frac{\partial}{\partial y} \right)^2 f(x_0, y_0) \equiv \left(h^2 \frac{\partial^2}{\partial x^2} + 2hk \frac{\partial^2}{\partial x \, \partial y} + k^2 \frac{\partial^2}{\partial y^2} \right) f(x_0, y_0)$$

$$\equiv h^2 f_{xx}(x_0, y_0) + 2hk f_{xy}(x_0, y_0) + k^2 f_{yy}(x_0, y_0) \quad (15)$$

etc., where we expand $\left(h \frac{\partial}{\partial x} + k \frac{\partial}{\partial y} \right)^n$ formally by the binomial theorem.

Equation (13) is sometimes written in a form where $h = \Delta x = x - x_0$ and $k = \Delta y = y - y_0$. Note that (12) is a special case of (13) where $n = 0$.

In case $\lim\limits_{n \to \infty} R_n = 0$ for all (x, y) in a region, the result can be used to obtain an infinite series expansion of $f(x, y)$ in powers of $x - x_0$ and $y - y_0$ convergent in this region, called the *region of convergence*. This series is called a *Taylor series* in 2 variables. Extensions to 3 or more variables can be made.

Solved Problems

FUNCTIONS and GRAPHS

1. If $f(x, y) = x^3 - 2xy + 3y^2$, find: *(a)* $f(-2, 3)$; *(b)* $f\left(\dfrac{1}{x}, \dfrac{2}{y}\right)$; *(c)* $\dfrac{f(x, y+k) - f(x, y)}{k}$, $k \neq 0$.

(a) $f(-2, 3) = (-2)^3 - 2(-2)(3) + 3(3)^2 = -8 + 12 + 27 = 31$

(b) $f\left(\dfrac{1}{x}, \dfrac{2}{y}\right) = \left(\dfrac{1}{x}\right)^3 - 2\left(\dfrac{1}{x}\right)\left(\dfrac{2}{y}\right) + 3\left(\dfrac{2}{y}\right)^2 = \dfrac{1}{x^3} - \dfrac{4}{xy} + \dfrac{12}{y^2}$

(c) $\dfrac{f(x, y+k) - f(x, y)}{k} = \dfrac{1}{k}\{[x^3 - 2x(y+k) + 3(y+k)^2] - [x^3 - 2xy + 3y^2]\}$

$\qquad\qquad\qquad\qquad = \dfrac{1}{k}(x^3 - 2xy - 2kx + 3y^2 + 6ky + 3k^2 - x^3 + 2xy - 3y^2)$

$\qquad\qquad\qquad\qquad = \dfrac{1}{k}(-2kx + 6ky + 3k^2) = -2x + 6y + 3k.$

2. Give the domain of definition for which each of the following functions are defined and real, and indicate this domain graphically.

(a) $f(x, y) = \ln\{(16 - x^2 - y^2)(x^2 + y^2 - 4)\}$

The function is defined and real for all points (x, y) such that

$$(16 - x^2 - y^2)(x^2 + y^2 - 4) > 0, \quad \text{i.e.} \quad 4 < x^2 + y^2 < 16$$

which is the required domain of definition. This point set consists of all points *interior* to the circle of radius 4 with center at the origin and *exterior* to the circle of radius 2 with center at the origin, as in the figure. The corresponding region, shown shaded in Fig. 6-2 below, is an *open region*.

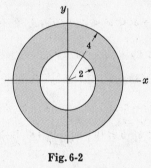

Fig. 6-2

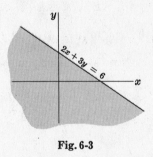

Fig. 6-3

(b) $f(x, y) = \sqrt{6 - (2x + 3y)}$

The function is defined and real for all points (x, y) such that $2x + 3y \leqq 6$, which is the required domain of definition.

The corresponding (unbounded) region of the xy plane is shown shaded in Fig. 6-3 above.

3. Sketch and name the surface in 3 dimensional space represented by each of the following.

(a) $2x + 4y + 3z = 12$.

Trace on xy plane ($z = 0$) is the straight line $x + 2y = 6$, $z = 0$.

Trace on yz plane ($x = 0$) is the straight line $4y + 3z = 12$, $x = 0$.

Trace on xz plane ($y = 0$) is the straight line $2x + 3z = 12$, $y = 0$.

These are represented by AB, BC and AC in Fig. 6-4.

The surface is a plane intersecting the x, y and z axes in the points $A(6, 0, 0)$, $B(0, 3, 0)$, $C(0, 0, 4)$. The lengths $\overline{OA} = 6$, $\overline{OB} = 3$, $\overline{OC} = 4$ are called the x, y and z *intercepts* respectively.

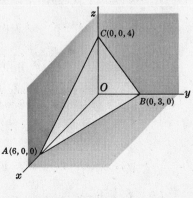

Fig. 6-4

(b) $\dfrac{x^2}{a^2} + \dfrac{y^2}{b^2} - \dfrac{z^2}{c^2} = 1$

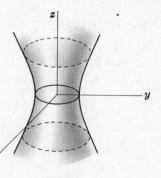

Trace on xy plane ($z = 0$) is the ellipse $\dfrac{x^2}{a^2} + \dfrac{y^2}{b^2} = 1$, $z = 0$.

Trace on yz plane ($x = 0$) is the hyperbola $\dfrac{y^2}{b^2} - \dfrac{z^2}{c^2} = 1$, $x = 0$.

Trace on xz plane ($y = 0$) is the hyperbola $\dfrac{x^2}{a^2} - \dfrac{z^2}{c^2} = 1$, $y = 0$.

Trace on any plane $z = p$ parallel to the xy plane is the ellipse

$$\frac{x^2}{a^2(1 + p^2/c^2)} + \frac{y^2}{b^2(1 + p^2/c^2)} = 1$$

Fig. 6-5

As $|p|$ increases from zero, the elliptic cross-section increases in size.

The surface is a *hyperboloid of one sheet* (see Fig. 6-5).

LIMITS and CONTINUITY

4. Prove that $\lim\limits_{\substack{x \to 1 \\ y \to 2}} (x^2 + 2y) = 5$.

Method 1, using definition of limit.

We must show that given any $\epsilon > 0$, we can find $\delta > 0$ such that $|x^2 + 2y - 5| < \epsilon$ when $0 < |x - 1| < \delta$, $0 < |y - 2| < \delta$.

If $0 < |x - 1| < \delta$ and $0 < |y - 2| < \delta$, then $1 - \delta < x < 1 + \delta$ and $2 - \delta < y < 2 + \delta$, excluding $x = 1$, $y = 2$.

Thus $1 - 2\delta + \delta^2 < x^2 < 1 + 2\delta + \delta^2$ and $4 - 2\delta < 2y < 4 + 2\delta$. Adding,

$$5 - 4\delta + \delta^2 < x^2 + 2y < 5 + 4\delta + \delta^2 \qquad \text{or} \qquad -4\delta + \delta^2 < x^2 + 2y - 5 < 4\delta + \delta^2$$

Now if $\delta \leqq 1$, it certainly follows that $-5\delta < x^2 + 2y - 5 < 5\delta$, i.e. $|x^2 + 2y - 5| < 5\delta$ whenever $0 < |x - 1| < \delta$, $0 < |y - 2| < \delta$. Then choosing $5\delta = \epsilon$, i.e. $\delta = \epsilon/5$ (or $\delta = 1$, whichever is smaller), it follows that $|x^2 + 2y - 5| < \epsilon$ when $0 < |x - 1| < \delta$, $0 < |y - 2| < \delta$, i.e. $\lim\limits_{\substack{x \to 1 \\ y \to 2}} (x^2 + 2y) = 5$.

Method 2, using theorems on limits.

$$\lim_{\substack{x \to 1 \\ y \to 2}} (x^2 + 2y) = \lim_{\substack{x \to 1 \\ y \to 2}} x^2 + \lim_{\substack{x \to 1 \\ y \to 2}} 2y = 1 + 4 = 5$$

5. Prove that $f(x, y) = x^2 + 2y$ is continuous at $(1, 2)$.

By Problem 4, $\lim\limits_{\substack{x \to 1 \\ y \to 2}} f(x, y) = 5$. Also, $f(1, 2) = 1^2 + 2(2) = 5$.

Then $\lim\limits_{\substack{x \to 1 \\ y \to 2}} f(x, y) = f(1, 2)$ and the function is continuous at $(1, 2)$.

Alternatively we can show, in much the same manner as in the first method of Prob. 4, that given any $\epsilon > 0$ we can find $\delta > 0$ such that $|f(x, y) - f(1, 2)| < \epsilon$ when $|x - 1| < \delta$, $|y - 2| < \delta$.

6. Determine whether $f(x, y) = \begin{cases} x^2 + 2y, & (x, y) \neq (1, 2) \\ 0, & (x, y) = (1, 2) \end{cases}$

(a) has a limit as $x \to 1$ and $y \to 2$, (b) is continuous at $(1, 2)$.

(a) By Problem 4, it follows that $\lim\limits_{\substack{x \to 1 \\ y \to 2}} f(x, y) = 5$, since the *limit* has nothing to do with the *value* at $(1, 2)$.

(b) Since $\lim\limits_{\substack{x \to 1 \\ y \to 2}} f(x, y) = 5$ and $f(1, 2) = 0$, it follows that $\lim\limits_{\substack{x \to 1 \\ y \to 2}} f(x, y) \neq f(1, 2)$. Hence the function is *discontinuous* at $(1, 2)$.

7. Investigate the continuity of $f(x, y) = \begin{cases} \dfrac{x^2 - y^2}{x^2 + y^2} & (x, y) \neq (0, 0) \\ 0 & (x, y) = (0, 0) \end{cases}$ at $(0, 0)$.

Let $x \to 0$ and $y \to 0$ in such a way that $y = mx$ (a line in the xy plane). Then along this line,

$$\lim_{\substack{x \to 0 \\ y \to 0}} \frac{x^2 - y^2}{x^2 + y^2} = \lim_{x \to 0} \frac{x^2 - m^2 x^2}{x^2 + m^2 x^2} = \lim_{x \to 0} \frac{x^2(1 - m^2)}{x^2(1 + m^2)} = \frac{1 - m^2}{1 + m^2}$$

Since the limit of the function depends on the manner of approach to $(0, 0)$ (i.e. the slope m of the line), the function cannot be continuous at $(0, 0)$.

Another method:

Since $\lim\limits_{x \to 0} \left\{ \lim\limits_{y \to 0} \dfrac{x^2 - y^2}{x^2 + y^2} \right\} = \lim\limits_{x \to 0} \dfrac{x^2}{x^2} = 1$ and $\lim\limits_{y \to 0} \left\{ \lim\limits_{x \to 0} \dfrac{x^2 - y^2}{x^2 + y^2} \right\} = -1$ are not equal, $\lim\limits_{\substack{x \to 0 \\ y \to 0}} f(x, y)$

cannot exist. Hence $f(x, y)$ cannot be continuous at $(0, 0)$.

PARTIAL DERIVATIVES

8. If $f(x, y) = 2x^2 - xy + y^2$, find (a) $\partial f / \partial x$, and (b) $\partial f / \partial y$ at (x_0, y_0) directly from the definition.

(a) $\dfrac{\partial f}{\partial x}\Big|_{(x_0, y_0)} = f_x(x_0, y_0) = \lim\limits_{h \to 0} \dfrac{f(x_0 + h, y_0) - f(x_0, y_0)}{h}$

$= \lim\limits_{h \to 0} \dfrac{[2(x_0 + h)^2 - (x_0 + h)y_0 + y_0^2] - [2x_0^2 - x_0 y_0 + y_0^2]}{h}$

$= \lim\limits_{h \to 0} \dfrac{4hx_0 + 2h^2 - hy_0}{h} = \lim\limits_{h \to 0} (4x_0 + 2h - y_0) = 4x_0 - y_0$

(b) $\dfrac{\partial f}{\partial y}\Big|_{(x_0, y_0)} = f_y(x_0, y_0) = \lim\limits_{k \to 0} \dfrac{f(x_0, y_0 + k) - f(x_0, y_0)}{k}$

$= \lim\limits_{k \to 0} \dfrac{[2x_0^2 - x_0(y_0 + k) + (y_0 + k)^2] - [2x_0^2 - x_0 y_0 + y_0^2]}{k}$

$= \lim\limits_{k \to 0} \dfrac{-kx_0 + 2ky_0 + k^2}{k} = \lim\limits_{k \to 0} (-x_0 + 2y_0 + k) = -x_0 + 2y_0$

Since the limits exist for all points (x_0, y_0), we can write $f_x(x, y) = f_x = 4x - y$, $f_y(x, y) = f_y = -x + 2y$ which are themselves functions of x and y.

Note that *formally* $f_x(x_0, y_0)$ is obtained from $f(x, y)$ by differentiating with respect to x, keeping y constant and then putting $x = x_0$, $y = y_0$. Similarly $f_y(x_0, y_0)$ is obtained by differentiating f with respect to y, keeping x constant. This procedure, while often lucrative in practice, need not always yield correct results (see Problem 9). It will work if the partial derivatives are continuous.

9. Let $f(x, y) = \begin{cases} xy/(x^2 + y^2) & (x, y) \neq (0, 0) \\ 0 & \text{otherwise} \end{cases}$.   Prove that (a) $f_x(0, 0)$ and $f_y(0, 0)$ both exist but that (b) $f(x, y)$ is discontinuous at $(0, 0)$.

(a) $f_x(0, 0) = \lim_{h \to 0} \dfrac{f(h, 0) - f(0, 0)}{h} = \lim_{h \to 0} \dfrac{0}{h} = 0$

$f_y(0, 0) = \lim_{k \to 0} \dfrac{f(0, k) - f(0, 0)}{k} = \lim_{k \to 0} \dfrac{0}{k} = 0$

(b) Let $x \to 0$ and $y \to 0$ along the line $y = mx$ in the xy plane. Then $\lim_{\substack{x \to 0 \\ y \to 0}} f(x, y) = \lim_{x \to 0} \dfrac{mx^2}{x^2 + m^2 x^2} = \dfrac{m}{1 + m^2}$ so that the limit depends on the approach and therefore does not exist. Hence $f(x, y)$ is not continuous at $(0, 0)$.

Note that unlike the situation for functions of one variable, the existence of the first partial derivatives at a point *does not* imply continuity at the point.

Note also that if $(x, y) \neq (0, 0)$, $f_x = \dfrac{y^3 - x^2 y}{(x^2 + y^2)^2}$, $f_y = \dfrac{x^3 - xy^2}{(x^2 + y^2)^2}$ and $f_x(0, 0)$, $f_y(0, 0)$ cannot be computed from them by merely letting $x = 0$ and $y = 0$. See remark at the end of Problem 5(b), Chapter 4.

10. If $\phi(x, y) = x^3 y + e^{xy^2}$, find (a) ϕ_x, (b) ϕ_y, (c) ϕ_{xx}, (d) ϕ_{yy}, (e) ϕ_{xy}, (f) ϕ_{yx}.

(a) $\phi_x = \dfrac{\partial \phi}{\partial x} = \dfrac{\partial}{\partial x}(x^3 y + e^{xy^2}) = 3x^2 y + e^{xy^2} \cdot y^2 = 3x^2 y + y^2 e^{xy^2}$

(b) $\phi_y = \dfrac{\partial \phi}{\partial y} = \dfrac{\partial}{\partial y}(x^3 y + e^{xy^2}) = x^3 + e^{xy^2} \cdot 2xy = x^3 + 2xy\, e^{xy^2}$

(c) $\phi_{xx} = \dfrac{\partial^2 \phi}{\partial x^2} = \dfrac{\partial}{\partial x}\left(\dfrac{\partial \phi}{\partial x}\right) = \dfrac{\partial}{\partial x}(3x^2 y + y^2 e^{xy^2}) = 6xy + y^2(e^{xy^2} \cdot y^2) = 6xy + y^4 e^{xy^2}$

(d) $\phi_{yy} = \dfrac{\partial^2 \phi}{\partial y^2} = \dfrac{\partial}{\partial y}(x^3 + 2xy\, e^{xy^2}) = 0 + 2xy \cdot \dfrac{\partial}{\partial y}(e^{xy^2}) + e^{xy^2} \dfrac{\partial}{\partial y}(2xy)$

$= 2xy \cdot e^{xy^2} \cdot 2xy + e^{xy^2} \cdot 2x = 4x^2 y^2 e^{xy^2} + 2x\, e^{xy^2}$

(e) $\phi_{xy} = \dfrac{\partial^2 \phi}{\partial y\, \partial x} = \dfrac{\partial}{\partial y}\left(\dfrac{\partial \phi}{\partial x}\right) = \dfrac{\partial}{\partial y}(3x^2 y + y^2 e^{xy^2}) = 3x^2 + y^2 \cdot e^{xy^2} \cdot 2xy + e^{xy^2} \cdot 2y$

$= 3x^2 + 2xy^3 e^{xy^2} + 2y e^{xy^2}$

(f) $\phi_{yx} = \dfrac{\partial^2 \phi}{\partial x\, \partial y} = \dfrac{\partial}{\partial x}\left(\dfrac{\partial \phi}{\partial y}\right) = \dfrac{\partial}{\partial x}(x^3 + 2xy\, e^{xy^2}) = 3x^2 + 2xy \cdot e^{xy^2} \cdot y^2 + e^{xy^2} \cdot 2y$

$= 3x^2 + 2xy^3 e^{xy^2} + 2y\, e^{xy^2}$

Note that $\phi_{xy} = \phi_{yx}$ in this case. This is because the second partial derivatives exist and are continuous for all (x, y) in a region $\mathcal{R}$. When this is not true we may have $\phi_{xy} \neq \phi_{yx}$ (see Problem 43, for example).

11. Show that $U(x, y, z) = (x^2 + y^2 + z^2)^{-1/2}$ satisfies Laplace's partial differential equation $\dfrac{\partial^2 U}{\partial x^2} + \dfrac{\partial^2 U}{\partial y^2} + \dfrac{\partial^2 U}{\partial z^2} = 0$.

We assume here that $(x, y, z) \neq (0, 0, 0)$. Then

$\dfrac{\partial U}{\partial x} = -\tfrac{1}{2}(x^2 + y^2 + z^2)^{-3/2} \cdot 2x = -x(x^2 + y^2 + z^2)^{-3/2}$

$\dfrac{\partial^2 U}{\partial x^2} = \dfrac{\partial}{\partial x}[-x(x^2 + y^2 + z^2)^{-3/2}] = (-x)[-\tfrac{3}{2}(x^2 + y^2 + z^2)^{-5/2} \cdot 2x] + (x^2 + y^2 + z^2)^{-3/2} \cdot (-1)$

$= \dfrac{3x^2}{(x^2 + y^2 + z^2)^{5/2}} - \dfrac{(x^2 + y^2 + z^2)}{(x^2 + y^2 + z^2)^{5/2}} = \dfrac{2x^2 - y^2 - z^2}{(x^2 + y^2 + z^2)^{5/2}}$

Similarly, $\dfrac{\partial^2 U}{\partial y^2} = \dfrac{2y^2 - x^2 - z^2}{(x^2 + y^2 + z^2)^{5/2}}, \quad \dfrac{\partial^2 U}{\partial z^2} = \dfrac{2z^2 - x^2 - y^2}{(x^2 + y^2 + z^2)^{5/2}}.$

Adding, $\dfrac{\partial^2 U}{\partial x^2} + \dfrac{\partial^2 U}{\partial y^2} + \dfrac{\partial^2 U}{\partial z^2} = 0.$

12. If $z = x^2 \tan^{-1} \dfrac{y}{x}$, find $\dfrac{\partial^2 z}{\partial x\, \partial y}$ at $(1,1)$.

$$\frac{\partial z}{\partial y} = x^2 \cdot \frac{1}{1 + (y/x)^2} \frac{\partial}{\partial y}\left(\frac{y}{x}\right) = x^2 \cdot \frac{x^2}{x^2 + y^2} \cdot \frac{1}{x} = \frac{x^3}{x^2 + y^2}$$

$$\frac{\partial^2 z}{\partial x\, \partial y} = \frac{\partial}{\partial x}\left(\frac{\partial z}{\partial y}\right) = \frac{\partial}{\partial x}\left(\frac{x^3}{x^2 + y^2}\right) = \frac{(x^2 + y^2)(3x^2) - (x^3)(2x)}{(x^2 + y^2)^2} = \frac{2 \cdot 3 - 1 \cdot 2}{2^2} = 1 \quad \text{at } (1,1).$$

The result can be written $z_{xy}(1,1) = 1$.

Note: In this calculation we are using the fact that z_{xy} is continuous at $(1,1)$ (see remark at the end of Problem 9).

13. If $f(x, y)$ is defined in a region $\mathcal{R}$ and if f_{xy} and f_{yx} exist and are continuous at a point of $\mathcal{R}$, prove that $f_{xy} = f_{yx}$ at this point.

Let (x_0, y_0) be the point of $\mathcal{R}$. Consider

$$G = f(x_0 + h, y_0 + k) - f(x_0, y_0 + k) - f(x_0 + h, y_0) + f(x_0, y_0)$$

Define   *(1)* $\phi(x, y) = f(x + h, y) - f(x, y)$ *(2)* $\psi(x, y) = f(x, y + k) - f(x, y)$

Then   *(3)* $G = \phi(x_0, y_0 + k) - \phi(x_0, y_0)$ *(4)* $G = \psi(x_0 + h, y_0) - \psi(x_0, y_0)$

Applying the theorem of the mean for functions of one variable (see Page 61) to *(3)* and *(4)*, we have

(5) $G = k\phi_y(x_0, y_0 + \theta_1 k) = k\{f_y(x_0 + h, y_0 + \theta_1 k) - f_y(x_0, y_0 + \theta_1 k)\}$ $0 < \theta_1 < 1$

(6) $G = h\psi_x(x_0 + \theta_2 h, y_0) = h\{f_x(x_0 + \theta_2 h, y_0 + k) - f_x(x_0 + \theta_2 h, y_0)\}$ $0 < \theta_2 < 1$

Applying the theorem of the mean again to *(5)* and *(6)*, we have

(7) $G = hk\, f_{yx}(x_0 + \theta_3 h, y_0 + \theta_1 k)$ $0 < \theta_1 < 1,\ 0 < \theta_3 < 1$

(8) $G = hk\, f_{xy}(x_0 + \theta_2 h, y_0 + \theta_4 k)$ $0 < \theta_2 < 1,\ 0 < \theta_4 < 1$

From *(7)* and *(8)* we have

(9) $f_{yx}(x_0 + \theta_3 h, y_0 + \theta_1 k) = f_{xy}(x_0 + \theta_2 h, y_0 + \theta_4 k)$

Letting $h \to 0$ and $k \to 0$ in *(9)* we have, since f_{xy} and f_{yx} are assumed continuous at (x_0, y_0),

$$f_{yx}(x_0, y_0) = f_{xy}(x_0, y_0)$$

as required. For an example where this fails to hold, see Problem 43.

DIFFERENTIALS

14. Let $f(x, y)$ have continuous first partial derivatives in a region $\mathcal{R}$ of the xy plane. Prove that

$$\Delta f = f(x + \Delta x,\ y + \Delta y) - f(x, y) = f_x \Delta x + f_y \Delta y + \epsilon_1 \Delta x + \epsilon_2 \Delta y$$

where ϵ_1 and ϵ_2 approach zero as Δx and Δy approach zero.

Applying the theorem of the mean for functions of one variable (see Page 61), we have

(1) $\Delta f = \{f(x + \Delta x,\ y + \Delta y) - f(x,\ y + \Delta y)\} + \{f(x,\ y + \Delta y) - f(x, y)\}$

$\qquad = \Delta x\, f_x(x + \theta_1 \Delta x,\ y + \Delta y) + \Delta y\, f_y(x,\ y + \theta_2 \Delta y)$ $0 < \theta_1 < 1,\ 0 < \theta_2 < 1$

Since, by hypothesis, f_x and f_y are continuous, it follows that

$$f_x(x + \theta_1 \Delta x, \, y + \Delta y) \;=\; f_x(x, y) + \epsilon_1, \qquad f_y(x, \, y + \theta_2 \Delta y) \;=\; f_y(x, y) + \epsilon_2$$

where $\epsilon_1 \to 0$, $\epsilon_2 \to 0$ as $\Delta x \to 0$ and $\Delta y \to 0$.

Thus $\;\; \Delta f \;=\; f_x \Delta x + f_y \Delta y + \epsilon_1 \Delta x + \epsilon_2 \Delta y \;\;$ as required.

Defining $\Delta x = dx$, $\Delta y = dy$, we have $\;\; \Delta f \;=\; f_x \, dx + f_y \, dy + \epsilon_1 \, dx + \epsilon_2 \, dy$.

We call $\;\; df \;=\; f_x \, dx + f_y \, dy \;\;$ the *differential* of f (or z) or the *principal part* of Δf (or Δz).

---

15. If $\;\; z = f(x, y) = x^2 y - 3y$, find (*a*) Δz, (*b*) dz. (*c*) Determine Δz and dz if $x = 4$, $y = 3$, $\Delta x = -0.01$, $\Delta y = 0.02$. (*d*) How might you determine $f(5.12, 6.85)$ without direct computation?

Solution:

(*a*) $\;\; \Delta z \;=\; f(x + \Delta x, \, y + \Delta y) - f(x, y)$

$\qquad\quad =\; \{(x + \Delta x)^2 (y + \Delta y) - 3(y + \Delta y)\} - \{x^2 y - 3y\}$

$\qquad\quad =\; \underbrace{2xy \, \Delta x + (x^2 - 3) \, \Delta y}_{(A)} \;+\; \underbrace{(\Delta x)^2 y + 2x \, \Delta x \, \Delta y + (\Delta x)^2 \, \Delta y}_{(B)}$

The sum (A) is the *principal part* of Δz and is the differential of z, i.e. dz. Thus

(*b*) $\qquad\qquad\qquad dz \;=\; 2xy \, \Delta x + (x^2 - 3) \, \Delta y \;=\; 2xy \, dx + (x^2 - 3) \, dy$

Another method: $\quad dz \;=\; \dfrac{\partial z}{\partial x} \, dx + \dfrac{\partial z}{\partial y} \, dy \;=\; 2xy \, dx + (x^2 - 3) \, dy$

(*c*) $\;\; \Delta z \;=\; f(x + \Delta x, \, y + \Delta y) - f(x, y) \;=\; f(4 - 0.01, \, 3 + 0.02) - f(4, 3)$

$\qquad\quad =\; \{(3.99)^2 (3.02) - 3(3.02)\} - \{(4)^2(3) - 3(3)\} \;=\; 0.018702$

$\;\; dz \;=\; 2xy \, dx + (x^2 - 3) \, dy \;=\; 2(4)(3)(-0.01) + (4^2 - 3)(0.02) \;=\; 0.02$

Note that in this case Δz and dz are approximately equal, due to the fact that $\Delta x = dx$ and $\Delta y = dy$ are sufficiently small.

(*d*) We must find $f(x + \Delta x, \, y + \Delta y)$ when $x + \Delta x = 5.12$ and $y + \Delta y = 6.85$. We can accomplish this by choosing $x = 5$, $\Delta x = 0.12$, $y = 7$, $\Delta y = -0.15$. Since Δx and Δy are small, we use the fact that $f(x + \Delta x, \, y + \Delta y) = f(x, y) + \Delta z$ is approximately equal to $f(x, y) + dz$, i.e. $z + dz$.

Now $\qquad z = f(x, y) = f(5, 7) = (5)^2(7) - 3(7) = 154$

$\qquad dz \;=\; 2xy \, dx + (x^2 - 3) \, dy \;=\; 2(5)(7)(0.12) + (5^2 - 3)(-0.15) \;=\; 5.1$

Then the required value is $154 + 5.1 = 159.1$ approximately. The value obtained by direct computation is 159.01864.

---

16. (*a*) Let $U = x^2 e^{y/x}$. Find dU. (*b*) Show that $(3x^2 y - 2y^2) \, dx + (x^3 - 4xy + 6y^2) \, dy$ can be written as an exact differential of a function $\phi(x, y)$ and find this function.

(*a*) **Method 1:**

$$\frac{\partial U}{\partial x} \;=\; x^2 e^{y/x} \left(-\frac{y}{x^2} \right) + 2x e^{y/x}, \qquad \frac{\partial U}{\partial y} \;=\; x^2 e^{y/x} \left(\frac{1}{x} \right)$$

Then $\qquad dU \;=\; \dfrac{\partial U}{\partial x} \, dx + \dfrac{\partial U}{\partial y} \, dy \;=\; (2x e^{y/x} - y e^{y/x}) \, dx + x e^{y/x} \, dy$

Method 2:

$\qquad dU \;=\; x^2 \, d(e^{y/x}) + e^{y/x} \, d(x^2) \;=\; x^2 e^{y/x} \, d(y/x) + 2x e^{y/x} \, dx$

$\qquad\qquad =\; x^2 e^{y/x} \left(\dfrac{x \, dy - y \, dx}{x^2} \right) + 2x e^{y/x} \, dx \;=\; (2x e^{y/x} - y e^{y/x}) \, dx + x e^{y/x} \, dy$

(*b*) **Method 1:**

Suppose that $\;\; (3x^2 y - 2y^2) \, dx + (x^3 - 4xy + 6y^2) \, dy \;=\; d\phi \;=\; \dfrac{\partial \phi}{\partial x} \, dx + \dfrac{\partial \phi}{\partial y} \, dy$.

Then $\qquad\qquad (1) \;\; \dfrac{\partial \phi}{\partial x} \;=\; 3x^2 y - 2y^2, \qquad (2) \;\; \dfrac{\partial \phi}{\partial y} \;=\; x^3 - 4xy + 6y^2$

From (1), integrating with respect to x keeping y constant, we have

$$\phi = x^3y - 2xy^2 + F(y)$$

where $F(y)$ is the "constant" of integration. Substituting this into (2) yields

$$x^3 - 4xy + F'(y) = x^3 - 4xy + 6y^2 \quad \text{from which} \quad F'(y) = 6y^2, \text{ i.e. } F(y) = 2y^3 + c$$

Hence the required function is $\phi = x^3y - 2xy^2 + 2y^3 + c$, where c is an arbitrary constant.

Note that by Theorem 3, Page 106, the existence of such a function is guaranteed, since if $P = 3x^2y - 2y^2$ and $Q = x^3 - 4xy + 6y^2$, then $\partial P/\partial y = 3x^2 - 4y = \partial Q/\partial x$ identically. If $\partial P/\partial y \neq \partial Q/\partial x$ this function would not exist and the given expression would not be an exact differential.

Method 2:

$$
\begin{aligned}
(3x^2y - 2y^2)\,dx + (x^3 - 4xy + 6y^2)\,dy &= (3x^2y\,dx + x^3\,dy) - (2y^2\,dx + 4xy\,dy) + 6y^2\,dy \\
&= d(x^3y) - d(2xy^2) + d(2y^3) = d(x^3y - 2xy^2 + 2y^3) \\
&= d(x^3y - 2xy^2 + 2y^3 + c)
\end{aligned}
$$

Then the required function is $x^3y - 2xy^2 + 2y^3 + c$.

This method, called the *grouping method*, is based on one's ability to recognize exact differential combinations and is less direct than Method 1. Naturally, before attempting to apply any method, one should determine whether the given expression is an exact differential by using Theorem 3, Page 106. See last paragraph of Method 1.

DIFFERENTIATION of COMPOSITE FUNCTIONS

17. Let $z = f(x, y)$ and $x = \phi(t)$, $y = \psi(t)$ where f, ϕ, ψ are assumed differentiable. Prove that

$$\frac{dz}{dt} = \frac{\partial z}{\partial x}\frac{dx}{dt} + \frac{\partial z}{\partial y}\frac{dy}{dt}$$

Using the results of Problem 14, we have

$$\frac{dz}{dt} = \lim_{\Delta t \to 0} \frac{\Delta z}{\Delta t} = \lim_{\Delta t \to 0}\left\{\frac{\partial z}{\partial x}\frac{\Delta x}{\Delta t} + \frac{\partial z}{\partial y}\frac{\Delta y}{\Delta t} + \epsilon_1\frac{\Delta x}{\Delta t} + \epsilon_2\frac{\Delta y}{\Delta t}\right\} = \frac{\partial z}{\partial x}\frac{dx}{dt} + \frac{\partial z}{\partial y}\frac{dy}{dt}$$

since as $\Delta t \to 0$ we have $\Delta x \to 0$, $\Delta y \to 0$, $\epsilon_1 \to 0$, $\epsilon_2 \to 0$, $\dfrac{\Delta x}{\Delta t} \to \dfrac{dx}{dt}$, $\dfrac{\Delta y}{\Delta t} \to \dfrac{dy}{dt}$.

18. If $z = e^{xy^2}$, $x = t\cos t$, $y = t\sin t$, compute dz/dt at $t = \pi/2$.

$$\frac{dz}{dt} = \frac{\partial z}{\partial x}\frac{dx}{dt} + \frac{\partial z}{\partial y}\frac{dy}{dt} = (y^2e^{xy^2})(-t\sin t + \cos t) + (2xye^{xy^2})(t\cos t + \sin t).$$

At $t = \pi/2$, $x = 0$, $y = \pi/2$. Then $\left.\dfrac{dz}{dt}\right|_{t=\pi/2} = (\pi^2/4)(-\pi/2) + (0)(1) = -\pi^3/8$.

Another method. Substitute x and y to obtain $z = e^{t^3\sin^2 t\cos t}$ and then differentiate.

19. If $z = f(x, y)$ where $x = \phi(u, v)$ and $y = \psi(u, v)$, prove that

(a) $\dfrac{\partial z}{\partial u} = \dfrac{\partial z}{\partial x}\dfrac{\partial x}{\partial u} + \dfrac{\partial z}{\partial y}\dfrac{\partial y}{\partial u}$, (b) $\dfrac{\partial z}{\partial v} = \dfrac{\partial z}{\partial x}\dfrac{\partial x}{\partial v} + \dfrac{\partial z}{\partial y}\dfrac{\partial y}{\partial v}$.

(a) From Problem 14, assuming the differentiability of f, ϕ, ψ, we have

$$\frac{\partial z}{\partial u} = \lim_{\Delta u \to 0}\frac{\Delta z}{\Delta u} \doteq \lim_{\Delta u \to 0}\left\{\frac{\partial z}{\partial x}\frac{\Delta x}{\Delta u} + \frac{\partial z}{\partial y}\frac{\Delta y}{\Delta u} + \epsilon_1\frac{\Delta x}{\Delta u} + \epsilon_2\frac{\Delta y}{\Delta u}\right\} = \frac{\partial z}{\partial x}\frac{\partial x}{\partial u} + \frac{\partial z}{\partial y}\frac{\partial y}{\partial u}$$

(b) The result is proved as in (a) by replacing Δu by Δv and letting $\Delta v \to 0$.

20. Prove that $dz = \dfrac{\partial z}{\partial x}dx + \dfrac{\partial z}{\partial y}dy$ even if x and y are dependent variables.

Suppose x and y depend on three variables u, v, w, for example. Then

$$(1)\quad dx = x_u\,du + x_v\,dv + x_w\,dw \qquad (2)\quad dy = y_u\,du + y_v\,dv + y_w\,dw$$

Thus $\quad z_x\,dx + z_y\,dy = (z_x x_u + z_y y_u)\,du + (z_x x_v + z_y y_v)\,dv + (z_x x_w + z_y y_w)\,dw$

$$= z_u\,du + z_v\,dv + z_w\,dw = dz$$

using obvious generalizations of Prob. 19.

21. If $\ T = x^3 - xy + y^3,\ x = \rho\cos\phi,\ y = \rho\sin\phi,\ $ find $\ (a)\ \partial T/\partial\rho,\ (b)\ \partial T/\partial\phi.$

$$\frac{\partial T}{\partial\rho} = \frac{\partial T}{\partial x}\frac{\partial x}{\partial\rho} + \frac{\partial T}{\partial y}\frac{\partial y}{\partial\rho} = (3x^2 - y)(\cos\phi) + (3y^2 - x)(\sin\phi)$$

$$\frac{\partial T}{\partial\phi} = \frac{\partial T}{\partial x}\frac{\partial x}{\partial\phi} + \frac{\partial T}{\partial y}\frac{\partial y}{\partial\phi} = (3x^2 - y)(-\rho\sin\phi) + (3y^2 - x)(\rho\cos\phi)$$

This may also be worked by direct substitution of x and y in T.

22. If $\ U = z\sin y/x\ $ where $\ x = 3r^2 + 2s,\ y = 4r - 2s^3,\ z = 2r^2 - 3s^2,\ $ find $\ (a)\ \partial U/\partial r,$ $(b)\ \partial U/\partial s.$

$(a)\ \dfrac{\partial U}{\partial r} = \dfrac{\partial U}{\partial x}\dfrac{\partial x}{\partial r} + \dfrac{\partial U}{\partial y}\dfrac{\partial y}{\partial r} + \dfrac{\partial U}{\partial z}\dfrac{\partial z}{\partial r}$

$$= \left\{\left(z\cos\frac{y}{x}\right)\left(-\frac{y}{x^2}\right)\right\}(6r) + \left\{\left(z\cos\frac{y}{x}\right)\left(\frac{1}{x}\right)\right\}(4) + \left(\sin\frac{y}{x}\right)(4r)$$

$$= -\frac{6ryz}{x^2}\cos\frac{y}{x} + \frac{4z}{x}\cos\frac{y}{x} + 4r\sin\frac{y}{x}$$

$(b)\ \dfrac{\partial U}{\partial s} = \dfrac{\partial U}{\partial x}\dfrac{\partial x}{\partial s} + \dfrac{\partial U}{\partial y}\dfrac{\partial y}{\partial s} + \dfrac{\partial U}{\partial z}\dfrac{\partial z}{\partial s}$

$$= \left\{\left(z\cos\frac{y}{x}\right)\left(-\frac{y}{x^2}\right)\right\}(2) + \left\{\left(z\cos\frac{y}{x}\right)\left(\frac{1}{x}\right)\right\}(-6s^2) + \left(\sin\frac{y}{x}\right)(-6s)$$

$$= -\frac{2yz}{x^2}\cos\frac{y}{x} - \frac{6s^2 z}{x}\cos\frac{y}{x} - 6s\sin\frac{y}{x}$$

23. If $\ x = \rho\cos\phi,\ y = \rho\sin\phi,\ $ show that $\ \left(\dfrac{\partial V}{\partial x}\right)^2 + \left(\dfrac{\partial V}{\partial y}\right)^2 = \left(\dfrac{\partial V}{\partial\rho}\right)^2 + \dfrac{1}{\rho^2}\left(\dfrac{\partial V}{\partial\phi}\right)^2.$

Using the subscript notation for partial derivatives, we have

$$V_\rho = V_x x_\rho + V_y y_\rho = V_x\cos\phi + V_y\sin\phi \qquad\qquad (1)$$

$$V_\phi = V_x x_\phi + V_y y_\phi = V_x(-\rho\sin\phi) + V_y(\rho\cos\phi) \qquad\qquad (2)$$

Dividing both sides of (2) by ρ, we have

$$\frac{1}{\rho}V_\phi = -V_x\sin\phi + V_y\cos\phi \qquad\qquad (3)$$

Then from (1) and (3), we have

$$V_\rho^2 + \frac{1}{\rho^2}V_\phi^2 = (V_x\cos\phi + V_y\sin\phi)^2 + (-V_x\sin\phi + V_y\cos\phi)^2 = V_x^2 + V_y^2$$

24. Show that $z = f(x^2 y)$, where f is differentiable, satisfies $x(\partial z/\partial x) = 2y(\partial z/\partial y)$.

Let $x^2 y = u$. Then $z = f(u)$. Thus

$$\frac{\partial z}{\partial x} = \frac{\partial z}{\partial u}\frac{\partial u}{\partial x} = f'(u)\cdot 2xy, \qquad \frac{\partial z}{\partial y} = \frac{\partial z}{\partial u}\frac{\partial u}{\partial y} = f'(u)\cdot x^2$$

Then $\quad x\dfrac{\partial z}{\partial x} = f'(u)\cdot 2x^2 y,\ \ 2y\dfrac{\partial z}{\partial y} = f'(u)\cdot 2x^2 y\ \ $ and so $\ x\dfrac{\partial z}{\partial x} = 2y\dfrac{\partial z}{\partial y}.$

Another method:

We have $\qquad dz = f'(x^2y)\,d(x^2y) = f'(x^2y)(2xy\,dx + x^2\,dy)$.

Also, $\qquad dz = \dfrac{\partial z}{\partial x}\,dx + \dfrac{\partial z}{\partial y}\,dy$.

Then $\qquad \dfrac{\partial z}{\partial x} = 2xy\,f'(x^2y), \quad \dfrac{\partial z}{\partial y} = x^2\,f'(x^2y)$.

Elimination of $f'(x^2y)$ yields $\quad x\dfrac{\partial z}{\partial x} = 2y\dfrac{\partial z}{\partial y}$.

25. If for all values of the parameter λ and for some constant p, $F(\lambda x, \lambda y) = \lambda^p\,F(x, y)$ identically, where F is assumed differentiable, prove that $x(\partial F/\partial x) + y(\partial F/\partial y) = pF$.

Let $\lambda x = u, \ \lambda y = v$. Then

$$F(u, v) = \lambda^p\,F(x, y) \tag{1}$$

The derivative with respect to λ of the left side of (1) is

$$\frac{\partial F}{\partial \lambda} = \frac{\partial F}{\partial u}\frac{\partial u}{\partial \lambda} + \frac{\partial F}{\partial v}\frac{\partial v}{\partial \lambda} = \frac{\partial F}{\partial u}x + \frac{\partial F}{\partial v}y$$

The derivative with respect to λ of the right side of (1) is $p\lambda^{p-1}F$. Then

$$x\frac{\partial F}{\partial u} + y\frac{\partial F}{\partial v} = p\lambda^{p-1}F \tag{2}$$

Letting $\lambda = 1$ in (2), so that $u = x, \ v = y$, we have $x(\partial F/\partial x) + y(\partial F/\partial y) = pF$.

26. If $F(x, y) = x^4y^2\sin^{-1}y/x$, show that $x(\partial F/\partial x) + y(\partial F/\partial y) = 6F$.

Since $F(\lambda x, \lambda y) = (\lambda x)^4(\lambda y)^2\sin^{-1}\lambda y/\lambda x = \lambda^6 x^4y^2\sin^{-1}y/x = \lambda^6\,F(x, y)$, the result follows from Problem 25 with $p = 6$. It can of course also be shown by direct differentiation.

27. Prove that $Y = f(x + at) + g(x - at)$ satisfies $\partial^2 Y/\partial t^2 = a^2(\partial^2 Y/\partial x^2)$, where f and g are assumed to be at least twice differentiable and a is any constant.

Let $u = x + at, \ v = x - at$ so that $Y = f(u) + g(v)$. Then if $f'(u) \equiv df/du, \ g'(v) \equiv dg/dv$,

$$\frac{\partial Y}{\partial t} = \frac{\partial Y}{\partial u}\frac{\partial u}{\partial t} + \frac{\partial Y}{\partial v}\frac{\partial v}{\partial t} = a\,f'(u) - a\,g'(v), \quad \frac{\partial Y}{\partial x} = \frac{\partial Y}{\partial u}\frac{\partial u}{\partial x} + \frac{\partial Y}{\partial v}\frac{\partial v}{\partial x} = f'(u) + g'(v)$$

By further differentiation, using the notation $f''(u) \equiv d^2f/du^2, \ g''(v) \equiv d^2g/dv^2$, we have

$$(1) \quad \frac{\partial^2 Y}{\partial t^2} = \frac{\partial Y_t}{\partial t} = \frac{\partial Y_t}{\partial u}\frac{\partial u}{\partial t} + \frac{\partial Y_t}{\partial v}\frac{\partial v}{\partial t} = \frac{\partial}{\partial u}\{a\,f'(u) - a\,g'(v)\}\,(a) + \frac{\partial}{\partial v}\{a\,f'(u) - a\,g'(v)\}\,(-a)$$

$$= a^2\,f''(u) + a^2\,g''(v)$$

$$(2) \quad \frac{\partial^2 Y}{\partial x^2} = \frac{\partial Y_x}{\partial x} = \frac{\partial Y_x}{\partial u}\frac{\partial u}{\partial x} + \frac{\partial Y_x}{\partial v}\frac{\partial v}{\partial x} = \frac{\partial}{\partial u}\{f'(u) + g'(v)\} + \frac{\partial}{\partial v}\{f'(u) + g'(v)\}$$

$$= f''(u) + g''(v)$$

Then from (1) and (2), $\quad \partial^2 Y/\partial t^2 = a^2(\partial^2 Y/\partial x^2)$.

28. If $x = 2r - s$ and $y = r + 2s$, find $\dfrac{\partial^2 U}{\partial y\, \partial x}$ in terms of derivatives with respect to r and s.

Solving $x = 2r - s$, $y = r + 2s$ for r and s: $r = (2x + y)/5$, $s = (2y - x)/5$.

Then $\partial r/\partial x = 2/5$, $\partial s/\partial x = -1/5$, $\partial r/\partial y = 1/5$, $\partial s/\partial y = 2/5$. Hence we have

$$\frac{\partial U}{\partial x} \;=\; \frac{\partial U}{\partial r}\frac{\partial r}{\partial x} + \frac{\partial U}{\partial s}\frac{\partial s}{\partial x} \;=\; \frac{2}{5}\frac{\partial U}{\partial r} - \frac{1}{5}\frac{\partial U}{\partial s}$$

$$\frac{\partial^2 U}{\partial y\, \partial x} \;=\; \frac{\partial}{\partial y}\left(\frac{\partial U}{\partial x}\right) \;=\; \frac{\partial}{\partial r}\left(\frac{2}{5}\frac{\partial U}{\partial r} - \frac{1}{5}\frac{\partial U}{\partial s}\right)\frac{\partial r}{\partial y} + \frac{\partial}{\partial s}\left(\frac{2}{5}\frac{\partial U}{\partial r} - \frac{1}{5}\frac{\partial U}{\partial s}\right)\frac{\partial s}{\partial y}$$

$$=\; \left(\frac{2}{5}\frac{\partial^2 U}{\partial r^2} - \frac{1}{5}\frac{\partial^2 U}{\partial r\, \partial s}\right)\left(\frac{1}{5}\right) + \left(\frac{2}{5}\frac{\partial^2 U}{\partial s\, \partial r} - \frac{1}{5}\frac{\partial^2 U}{\partial s^2}\right)\left(\frac{2}{5}\right)$$

$$=\; \frac{1}{25}\left(2\frac{\partial^2 U}{\partial r^2} + 3\frac{\partial^2 U}{\partial r\, \partial s} - 2\frac{\partial^2 U}{\partial s^2}\right)$$

assuming U has continuous second partial derivatives.

IMPLICIT FUNCTIONS and JACOBIANS

29. If $U = x^3 y$, find dU/dt if $\;(1)\; x^5 + y = t$, $\;(2)\; x^2 + y^3 = t^2$.

Equations (1) and (2) define x and y as (implicit) functions of t. Then differentiating with respect to t, we have

$$(3)\;\; 5x^4(dx/dt) + dy/dt = 1 \qquad\qquad (4)\;\; 2x(dx/dt) + 3y^2(dy/dt) = 2t$$

Solving (3) and (4) simultaneously for dx/dt and dy/dt,

$$\frac{dx}{dt} \;=\; \frac{\begin{vmatrix} 1 & 1 \\ 2t & 3y^2 \end{vmatrix}}{\begin{vmatrix} 5x^4 & 1 \\ 2x & 3y^2 \end{vmatrix}} \;=\; \frac{3y^2 - 2t}{15x^4y^2 - 2x}\,, \qquad \frac{dy}{dt} \;=\; \frac{\begin{vmatrix} 5x^4 & 1 \\ 2x & 2t \end{vmatrix}}{\begin{vmatrix} 5x^4 & 1 \\ 2x & 3y^2 \end{vmatrix}} \;=\; \frac{10x^4t - 2x}{15x^4y^2 - 2x}$$

Then $\quad \dfrac{dU}{dt} \;=\; \dfrac{\partial U}{\partial x}\dfrac{dx}{dt} + \dfrac{\partial U}{\partial y}\dfrac{dy}{dt} \;=\; (3x^2y)\left(\dfrac{3y^2 - 2t}{15x^4y^2 - 2x}\right) + (x^3)\left(\dfrac{10x^4t - 2x}{15x^4y^2 - 2x}\right)$.

30. If $F(x, y, z) = 0$ defines z as an implicit function of x and y in a region $\mathcal{R}$ of the xy plane, prove that $(a)\; \partial z/\partial x = -F_x/F_z$ and $(b)\; \partial z/\partial y = -F_y/F_z$, where $F_z \neq 0$.

Since z is a function of x and y, $\quad dz \;=\; \dfrac{\partial z}{\partial x}dx + \dfrac{\partial z}{\partial y}dy$.

Then $\quad dF \;=\; \dfrac{\partial F}{\partial x}dx + \dfrac{\partial F}{\partial y}dy + \dfrac{\partial F}{\partial z}dz \;=\; \left(\dfrac{\partial F}{\partial x} + \dfrac{\partial F}{\partial z}\dfrac{\partial z}{\partial x}\right)dx + \left(\dfrac{\partial F}{\partial y} + \dfrac{\partial F}{\partial z}\dfrac{\partial z}{\partial y}\right)dy \;=\; 0$.

Since x and y are independent, we have

$$(1)\;\; \frac{\partial F}{\partial x} + \frac{\partial F}{\partial z}\frac{\partial z}{\partial x} \;=\; 0 \qquad\qquad (2)\;\; \frac{\partial F}{\partial y} + \frac{\partial F}{\partial z}\frac{\partial z}{\partial y} \;=\; 0$$

from which the required results are obtained. If desired, equations (1) and (2) can be written directly.

31. If $F(x, y, u, v) = 0$ and $G(x, y, u, v) = 0$, find $(a)\; \partial u/\partial x$, $(b)\; \partial u/\partial y$, $(c)\; \partial v/\partial x$, $(d)\; \partial v/\partial y$.

The two equations in general define the dependent variables u and v as (implicit) functions of the independent variables x and y. Using the subscript notation, we have

$$(1)\;\; dF \;=\; F_x\, dx + F_y\, dy + F_u\, du + F_v\, dv \;=\; 0$$
$$(2)\;\; dG \;=\; G_x\, dx + G_y\, dy + G_u\, du + G_v\, dv \;=\; 0$$

Also, since u and v are functions of x and y,

$$(3)\;\; du \;=\; u_x\, dx + u_y\, dy \qquad\qquad (4)\;\; dv \;=\; v_x\, dx + v_y\, dy$$

Substituting (3) and (4) in (1) and (2) yields

$$(5) \quad dF = (F_x + F_u u_x + F_v v_x)\, dx + (F_y + F_u u_y + F_v v_y)\, dy = 0$$

$$(6) \quad dG = (G_x + G_u u_x + G_v v_x)\, dx + (G_y + G_u u_y + G_v v_y)\, dy = 0$$

Since x and y are independent, the coefficients of dx and dy in (5) and (6) are zero. Hence we obtain

$$(7) \quad \begin{cases} F_u u_x + F_v v_x = -F_x \\ G_u u_x + G_v v_x = -G_x \end{cases} \qquad (8) \quad \begin{cases} F_u u_y + F_v v_y = -F_y \\ G_u u_y + G_v v_y = -G_y \end{cases}$$

Solving (7) and (8) gives

$$(a) \quad u_x = \frac{\partial u}{\partial x} = \frac{\begin{vmatrix} -F_x & F_v \\ -G_x & G_v \end{vmatrix}}{\begin{vmatrix} F_u & F_v \\ G_u & G_v \end{vmatrix}} = -\frac{\frac{\partial(F, G)}{\partial(x, v)}}{\frac{\partial(F, G)}{\partial(u, v)}} \qquad (b) \quad v_x = \frac{\partial v}{\partial x} = \frac{\begin{vmatrix} F_u & -F_x \\ G_u & -G_x \end{vmatrix}}{\begin{vmatrix} F_u & F_v \\ G_u & G_v \end{vmatrix}} = -\frac{\frac{\partial(F, G)}{\partial(u, x)}}{\frac{\partial(F, G)}{\partial(u, v)}}$$

$$(c) \quad u_y = \frac{\partial u}{\partial y} = \frac{\begin{vmatrix} -F_y & F_v \\ -G_y & G_v \end{vmatrix}}{\begin{vmatrix} F_u & F_v \\ G_u & G_v \end{vmatrix}} = -\frac{\frac{\partial(F, G)}{\partial(y, v)}}{\frac{\partial(F, G)}{\partial(u, v)}} \qquad (d) \quad v_y = \frac{\partial v}{\partial y} = \frac{\begin{vmatrix} F_u & -F_y \\ G_u & -G_y \end{vmatrix}}{\begin{vmatrix} F_u & F_v \\ G_u & G_v \end{vmatrix}} = -\frac{\frac{\partial(F, G)}{\partial(u, y)}}{\frac{\partial(F, G)}{\partial(u, v)}}$$

The functional determinant $\begin{vmatrix} F_u & F_v \\ G_u & G_v \end{vmatrix}$, denoted by $\dfrac{\partial(F, G)}{\partial(u, v)}$ or $J\left(\dfrac{F, G}{u, v}\right)$, is the *Jacobian* of F and G with respect to u and v and is supposed $\neq 0$.

Note that it is possible to devise mnemonic rules for writing at once the required partial derivatives in terms of Jacobians (see also Problem 33).

32. If $u^2 - v = 3x + y$ and $u - 2v^2 = x - 2y$, find (a) $\partial u/\partial x$, (b) $\partial v/\partial x$, (c) $\partial u/\partial y$, (d) $\partial v/\partial y$.

Method 1: Differentiate the given equations with respect to x, considering u and v as functions of x and y. Then

$$(1) \quad 2u\frac{\partial u}{\partial x} - \frac{\partial v}{\partial x} = 3 \qquad\qquad (2) \quad \frac{\partial u}{\partial x} - 4v\frac{\partial v}{\partial x} = 1$$

Solving, $\quad \dfrac{\partial u}{\partial x} = \dfrac{1 - 12v}{1 - 8uv}, \quad \dfrac{\partial v}{\partial x} = \dfrac{2u - 3}{1 - 8uv}.$

Differentiating with respect to y, we have

$$(3) \quad 2u\frac{\partial u}{\partial y} - \frac{\partial v}{\partial y} = 1 \qquad\qquad (4) \quad \frac{\partial u}{\partial y} - 4v\frac{\partial v}{\partial y} = -2$$

Solving, $\quad \dfrac{\partial u}{\partial y} = \dfrac{-2 - 4v}{1 - 8uv}, \quad \dfrac{\partial v}{\partial y} = \dfrac{-4u - 1}{1 - 8uv}.$

We have, of course, assumed that $1 - 8uv \neq 0$.

Method 2: The given equations are $F = u^2 - v - 3x - y = 0$, $G = u - 2v^2 - x + 2y = 0$. Then by Problem 31,

$$\frac{\partial u}{\partial x} = -\frac{\frac{\partial(F, G)}{\partial(x, v)}}{\frac{\partial(F, G)}{\partial(u, v)}} = -\frac{\begin{vmatrix} F_x & F_v \\ G_x & G_v \end{vmatrix}}{\begin{vmatrix} F_u & F_v \\ G_u & G_v \end{vmatrix}} = -\frac{\begin{vmatrix} -3 & -1 \\ -1 & -4v \end{vmatrix}}{\begin{vmatrix} 2u & -1 \\ 1 & -4v \end{vmatrix}} = \frac{1 - 12v}{1 - 8uv}$$

provided $1 - 8uv \neq 0$. Similarly the other partial derivatives are obtained.

33. If $F(u, v, w, x, y) = 0$, $G(u, v, w, x, y) = 0$, $H(u, v, w, x, y) = 0$, find

$$(a) \left.\frac{\partial v}{\partial y}\right|_x, \quad (b) \left.\frac{\partial x}{\partial v}\right|_w, \quad (c) \left.\frac{\partial w}{\partial u}\right|_y.$$

From 3 equations in 5 variables, we can (theoretically at least) determine 3 variables in terms of the remaining 2. Thus 3 variables are dependent and 2 are independent. If we were asked to determine $\partial v/\partial y$, we would know that v is a dependent variable and y is an independent variable, but would not know the remaining independent variable. However, the particular notation $\left.\dfrac{\partial v}{\partial y}\right|_x$ serves to indicate that we are to obtain $\partial v/\partial y$ keeping x constant, i.e. x is the other independent variable.

(a) Differentiating the given equations with respect to y, keeping x constant, gives

$$(1) \quad F_u\,u_y + F_v\,v_y + F_w\,w_y + F_y = 0 \qquad (2) \quad G_u\,u_y + G_v\,v_y + G_w\,w_y + G_y = 0$$
$$(3) \quad H_u\,u_y + H_v\,v_y + H_w\,w_y + H_y = 0$$

Solving simultaneously for v_y, we have

$$v_y = \left.\frac{\partial v}{\partial y}\right|_x = -\frac{\begin{vmatrix} F_u & F_y & F_w \\ G_u & G_y & G_w \\ H_u & H_y & H_w \end{vmatrix}}{\begin{vmatrix} F_u & F_v & F_w \\ G_u & G_v & G_w \\ H_u & H_v & H_w \end{vmatrix}} = -\frac{\dfrac{\partial(F,G,H)}{\partial(u,y,w)}}{\dfrac{\partial(F,G,H)}{\partial(u,v,w)}}$$

Equations (1), (2) and (3) can also be obtained by using differentials as in Problem 31.

The Jacobian method is very suggestive for writing results immediately, as seen in this problem and Problem 31. Thus observe that in calculating $\left.\dfrac{\partial v}{\partial y}\right|_x$ the result is the negative of the quotient of two Jacobians, the numerator containing the independent variable y, the denominator containing the dependent variable v in the same relative positions. Using this scheme, we have

$$(b) \quad \left.\frac{\partial x}{\partial v}\right|_w = -\frac{\dfrac{\partial(F,G,H)}{\partial(v,y,u)}}{\dfrac{\partial(F,G,H)}{\partial(x,y,u)}} \qquad\qquad (c) \quad \left.\frac{\partial w}{\partial u}\right|_y = -\frac{\dfrac{\partial(F,G,H)}{\partial(u,x,v)}}{\dfrac{\partial(F,G,H)}{\partial(w,x,v)}}$$

34. If $z^3 - xz - y = 0$, prove that $\dfrac{\partial^2 z}{\partial x\,\partial y} = -\dfrac{3z^2 + x}{(3z^2 - x)^3}$.

Differentiating with respect to x, keeping y constant and remembering that z is the dependent variable depending on the independent variables x and y, we find

$$3z^2\frac{\partial z}{\partial x} - x\frac{\partial z}{\partial x} - z = 0 \qquad \text{and} \qquad (1) \quad \frac{\partial z}{\partial x} = \frac{z}{3z^2 - x}$$

Differentiating with respect to y, keeping x constant, we find

$$3z^2\frac{\partial z}{\partial y} - x\frac{\partial z}{\partial y} - 1 = 0 \qquad \text{and} \qquad (2) \quad \frac{\partial z}{\partial y} = \frac{1}{3z^2 - x}$$

Differentiating (2) with respect to x and using (1), we have

$$\frac{\partial^2 z}{\partial x\,\partial y} = \frac{-1}{(3z^2 - x)^2}\left(6z\frac{\partial z}{\partial x} - 1\right) = \frac{1 - 6z[z/(3z^2 - x)]}{(3z^2 - x)^2} = -\frac{3z^2 + x}{(3z^2 - x)^3}$$

The result can also be obtained by differentiating (1) with respect to y and using (2).

35. Let $u = f(x, y)$ and $v = g(x, y)$, where f and g are continuously differentiable in some region $\mathcal{R}$. Prove that a necessary and sufficient condition that there exists a functional relation between u and v of the form $\phi(u, v) = 0$ is the vanishing of the Jacobian, i.e. $\dfrac{\partial(u, v)}{\partial(x, y)} = 0$ identically.

Necessity. We have to prove that if the functional relation $\phi(u, v) = 0$ exists, then the Jacobian $\dfrac{\partial(u, v)}{\partial(x, y)} = 0$ identically. To do this, we note that

$$d\phi \;=\; \phi_u\, du + \phi_v\, dv \;=\; \phi_u(u_x\, dx + u_y\, dy) + \phi_v(v_x\, dx + v_y\, dy)$$
$$=\; (\phi_u u_x + \phi_v v_x)\, dx + (\phi_u u_y + \phi_v v_y)\, dy \;=\; 0$$

Then   (1) $\phi_u u_x + \phi_v v_x = 0$ (2) $\phi_u u_y + \phi_v v_y = 0$

Now ϕ_u and ϕ_v cannot be identically zero since if they were, there would be no functional relation, contrary to hypothesis. Hence it follows from (1) and (2) that $\begin{vmatrix} u_x & v_x \\ u_y & v_y \end{vmatrix} = \dfrac{\partial(u, v)}{\partial(x, y)} = 0$ identically.

Sufficiency. We have to prove that if the Jacobian $\dfrac{\partial(u, v)}{\partial(x, y)} = 0$ identically, then there exists a functional relation between u and v, i.e. $\phi(u, v) = 0$.

Let us first suppose that both $u_x = 0$ and $u_y = 0$. In this case the Jacobian is identically zero and u is a constant c_1, so that the trivial functional relation $u = c_1$ is obtained.

Let us now assume that we do not have both $u_x = 0$ and $u_y = 0$; for definiteness, assume $u_x \neq 0$. We may then, according to Theorem 1, Page 108, solve for x in the equation $u = f(x, y)$ to obtain $x = F(u, y)$, from which it follows that

(1) $u = f\{F(u, y), y\}$ (2) $v = g\{F(u, y), y\}$

From these we have respectively,

(3) $du = u_x\, dx + u_y\, dy = u_x(F_u\, du + F_y\, dy) + u_y\, dy = u_x F_u\, du + (u_x F_y + u_y)\, dy$

(4) $dv = v_x\, dx + v_y\, dy = v_x(F_u\, du + F_y\, dy) + v_y\, dy = v_x F_u\, du + (v_x F_y + v_y)\, dy$

From (3), $u_x F_u = 1$ and $u_x F_y + u_y = 0$ or (5) $F_y = -u_y/u_x$. Using this, (4) becomes

(6) $dv = v_x F_u\, du + \{v_x(-u_y/u_x) + v_y\}\, dy = v_x F_u\, du + \left(\dfrac{u_x v_y - u_y v_x}{u_x}\right) dy.$

But by hypothesis $\dfrac{\partial(u, v)}{\partial(x, y)} = \begin{vmatrix} u_x & u_y \\ v_x & v_y \end{vmatrix} = u_x v_y - u_y v_x = 0$ identically, so that (6) becomes

$dv = v_x F_u\, du$. This means essentially that referring to (2), $\partial v/\partial y = 0$ which means that v is not dependent on y but depends only on u, i.e. v is a function of u, which is the same as saying that the functional relation $\phi(u, v) = 0$ exists.

36. (a) If $u = \dfrac{x + y}{1 - xy}$ and $v = \tan^{-1} x + \tan^{-1} y$, find $\dfrac{\partial(u, v)}{\partial(x, y)}$.

(b) Are u and v functionally related? If so, find the relationship.

(a) $\dfrac{\partial(u, v)}{\partial(x, y)} = \begin{vmatrix} u_x & u_y \\ v_x & v_y \end{vmatrix} = \begin{vmatrix} \dfrac{1 + y^2}{(1 - xy)^2} & \dfrac{1 + x^2}{(1 - xy)^2} \\ \dfrac{1}{1 + x^2} & \dfrac{1}{1 + y^2} \end{vmatrix} = 0$ if $xy \neq 1$.

(b) By Problem 35, since the Jacobian is identically zero in a region, there must be a functional relationship between u and v. This is seen to be $\tan v = u$, i.e. $\phi(u, v) = u - \tan v = 0$. We can show this directly by solving for x (say) in one of the equations and then substituting in the other. Thus, for example, from $v = \tan^{-1} x + \tan^{-1} y$ we find $\tan^{-1} x = v - \tan^{-1} y$ and so

$$x \;=\; \tan(v - \tan^{-1} y) \;=\; \frac{\tan v - \tan(\tan^{-1} y)}{1 + \tan v \tan(\tan^{-1} y)} \;=\; \frac{\tan v - y}{1 + y \tan v}$$

Then substituting this in $u = (x + y)/(1 - xy)$ and simplifying, we find $u = \tan v$.

37. (a) If $x = u - v + w$, $y = u^2 - v^2 - w^2$ and $z = u^3 + v$, evaluate the Jacobian $\dfrac{\partial(x, y, z)}{\partial(u, v, w)}$ and (b) explain the significance of the non-vanishing of this Jacobian.

(a) $\dfrac{\partial(x, y, z)}{\partial(u, v, w)} = \begin{vmatrix} x_u & x_v & x_w \\ y_u & y_v & y_w \\ z_u & z_v & z_w \end{vmatrix} = \begin{vmatrix} 1 & -1 & 1 \\ 2u & -2v & -2w \\ 3u^2 & 1 & 0 \end{vmatrix} = 6wu^2 + 2u + 6u^2v + 2w$

(b) The given equations can be solved simultaneously for u, v, w in terms of x, y, z in a region $\mathcal{R}$ if the Jacobian is not zero in $\mathcal{R}$.

TRANSFORMATIONS, CURVILINEAR COORDINATES

38. A region $\mathcal{R}$ in the xy plane is bounded by $x + y = 6$, $x - y = 2$ and $y = 0$. (a) Determine the region $\mathcal{R}'$ in the uv plane into which $\mathcal{R}$ is mapped under the transformation $x = u + v$, $y = u - v$. (b) Compute $\dfrac{\partial(x, y)}{\partial(u, v)}$. (c) Compare the result of (b) with the ratio of the areas of $\mathcal{R}$ and $\mathcal{R}'$.

(a) The region $\mathcal{R}$ shown shaded in Fig. 6-6(a) below is a triangle bounded by the lines $x + y = 6$, $x - y = 2$ and $y = 0$ which for distinguishing purposes are shown dotted, dashed and heavy respectively.

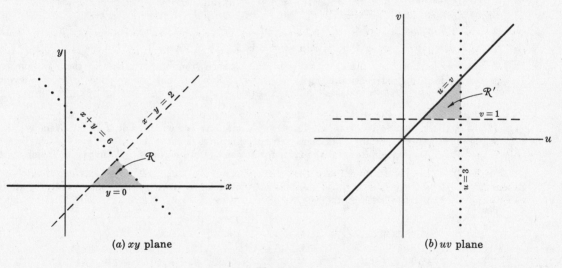

(a) xy plane   (b) uv plane

Fig. 6-6

Under the given transformation the line $x + y = 6$ is transformed into $(u + v) + (u - v) = 6$, i.e. $2u = 6$ or $u = 3$, which is a line (shown dotted) in the uv plane of Fig. 6-6(b) above.

Similarly, $x - y = 2$ becomes $(u + v) - (u - v) = 2$ or $v = 1$, which is a line (shown dashed) in the uv plane. In like manner, $y = 0$ becomes $u - v = 0$ or $u = v$, which is a line shown heavy in the uv plane. Then the required region is bounded by $u = 3$, $v = 1$ and $u = v$, and is shown shaded in Fig. 6-6(b).

(b) $\dfrac{\partial(x, y)}{\partial(u, v)} = \begin{vmatrix} \dfrac{\partial x}{\partial u} & \dfrac{\partial x}{\partial v} \\ \dfrac{\partial y}{\partial u} & \dfrac{\partial y}{\partial v} \end{vmatrix} = \begin{vmatrix} \dfrac{\partial}{\partial u}(u + v) & \dfrac{\partial}{\partial v}(u + v) \\ \dfrac{\partial}{\partial u}(u - v) & \dfrac{\partial}{\partial v}(u - v) \end{vmatrix} = \begin{vmatrix} 1 & 1 \\ 1 & -1 \end{vmatrix} = 2$

(c) The area of triangular region $\mathcal{R}$ is 4, whereas the area of triangular region $\mathcal{R}'$ is 2. Hence the ratio is $4/2 = 2$, agreeing with the value of the Jacobian in (b). Since the Jacobian is constant in this case, the areas of any regions $\mathcal{R}$ in the xy plane are twice the areas of corresponding mapped regions $\mathcal{R}'$ in the uv plane.

39. A region $\mathcal{R}$ in the xy plane is bounded by $x^2 + y^2 = a^2$, $x^2 + y^2 = b^2$, $x = 0$ and $y = 0$, where $0 < a < b$. (a) Determine the region $\mathcal{R}'$ into which $\mathcal{R}$ is mapped under the transformation $x = \rho \cos \phi$, $y = \rho \sin \phi$, where $\rho > 0$, $0 \le \phi < 2\pi$. (b) Discuss what happens when $a = 0$. (c) Compute $\dfrac{\partial(x, y)}{\partial(\rho, \phi)}$. (d) Compute $\dfrac{\partial(\rho, \phi)}{\partial(x, y)}$.

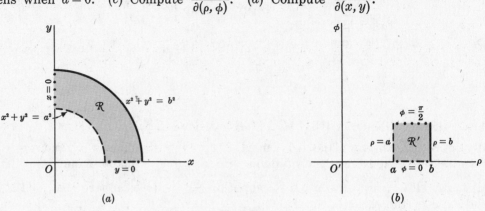

Fig. 6-7

(a) The region $\mathcal{R}$ [shaded in Fig. 6-7(a) above] is bounded by $x = 0$ (dotted), $y = 0$ (dotted and dashed), $x^2 + y^2 = a^2$ (dashed), $x^2 + y^2 = b^2$ (heavy).

  Under the given transformation, $x^2 + y^2 = a^2$ and $x^2 + y^2 = b^2$ become $\rho^2 = a^2$ and $\rho^2 = b^2$ or $\rho = a$ and $\rho = b$ respectively. Also, $x = 0$, $a \le y \le b$ becomes $\phi = \pi/2$, $a \le \rho \le b$; $y = 0$, $a \le x \le b$ becomes $\phi = 0$, $a \le \rho \le b$.

  The required region $\mathcal{R}'$ is shown shaded in Fig. 6-7(b) above.

  Another method: Using the fact that ρ is the distance from the origin O of the xy plane and ϕ is the angle measured from the positive x axis, it is clear that the required region is given by $a \le \rho \le b$, $0 \le \phi \le \pi/2$ as indicated in Fig. 6-7(b).

(b) If $a = 0$, the region $\mathcal{R}$ becomes one-fourth of a circular region of radius b (bounded by 3 sides) while $\mathcal{R}'$ remains a rectangle. The reason for this is that the point $x = 0$, $y = 0$ is mapped into $\rho = 0$, ϕ = an indeterminate and the transformation is not one to one at this point which is sometimes called a *singular point*.

(c) $\dfrac{\partial(x, y)}{\partial(\rho, \phi)} = \begin{vmatrix} \dfrac{\partial}{\partial \rho} (\rho \cos \phi) & \dfrac{\partial}{\partial \phi} (\rho \cos \phi) \\[2mm] \dfrac{\partial}{\partial \rho} (\rho \sin \phi) & \dfrac{\partial}{\partial \phi} (\rho \sin \phi) \end{vmatrix} = \begin{vmatrix} \cos \phi & -\rho \sin \phi \\[2mm] \sin \phi & \rho \cos \phi \end{vmatrix}$

  $= \rho(\cos^2 \phi + \sin^2 \phi) = \rho$

(d) From Problem 45(b) we have, letting $u = \rho$, $v = \phi$,

$$\frac{\partial(x, y)}{\partial(\rho, \phi)} \frac{\partial(\rho, \phi)}{\partial(x, y)} = 1 \qquad \text{so that, using (c),} \qquad \frac{\partial(\rho, \phi)}{\partial(x, y)} = \frac{1}{\rho}$$

This can also be obtained by direct differentiation.

  Note that from the Jacobians of these transformations it is clear why $\rho = 0$ (i.e. $x = 0$, $y = 0$) is a singular point.

MEAN VALUE THEOREMS, TAYLOR'S THEOREM

40. Prove the first mean value theorem for functions of two variables.

  Let $F(t) = f(x_0 + ht, y_0 + kt)$. By the mean value theorem for functions of one variable,

$$F(1) - F(0) = F'(\theta) \qquad 0 < \theta < 1 \tag{1}$$

If $x = x_0 + ht$, $y = y_0 + kt$, then $F(t) = f(x, y)$, so that by Problem 17,

$$F'(t) = f_x(dx/dt) + f_y(dy/dt) = hf_x + kf_y \quad \text{and} \quad F'(\theta) = h f_x(x_0 + \theta h, y_0 + \theta k) + k f_y(x_0 + \theta h, y_0 + \theta k)$$

where $0 < \theta < 1$. Thus (1) becomes

$$f(x_0 + h, y_0 + k) - f(x_0, y_0) = h f_x(x_0 + \theta h, y_0 + \theta k) + k f_y(x_0 + \theta h, y_0 + \theta k) \qquad (2)$$

where $0 < \theta < 1$ as required.

Note that (2), which is analogous to (1) of Problem 14 where $h = \Delta x$, has the advantage of being more symmetric (and also more useful), since only a single number θ is involved.

41. Prove Taylor's theorem of the mean for functions of two variables.

Let $F(t) = f(x_0 + ht, y_0 + kt)$ as in Problem 40. By Taylor's theorem of the mean for functions of one variable,

$$F(t) = F(0) + F'(0)\, t + \frac{F''(0)\, t^2}{2!} + \cdots + \frac{F^{(n)}(0)}{n!} t^n + \frac{F^{(n+1)}(\theta)}{(n+1)!} t^{n+1} \quad 0 < \theta < t \qquad (1)$$

and if $t = 1$,

$$F(1) = F(0) + F'(0) + \frac{F''(0)}{2!} + \cdots + \frac{F^{(n)}(0)}{n!} + \frac{F^{(n+1)}(\theta)}{(n+1)!} \quad 0 < \theta < 1 \qquad (2)$$

From Problem 40,

$$F'(0) = h f_x(x_0, y_0) + k f_y(x_0, y_0) = \left(h \frac{\partial}{\partial x} + k \frac{\partial}{\partial y} \right) f(x_0, y_0)$$

where we have used the symbolic operator notation. Similarly,

$$F''(t) = \frac{d}{dt} F'(t) = \frac{d}{dt} (hf_x + kf_y) = h\left(\frac{\partial f_x}{\partial x} \frac{dx}{dt} + \frac{\partial f_x}{\partial y} \frac{dy}{dt} \right) + k\left(\frac{\partial f_y}{\partial x} \frac{dx}{dt} + \frac{\partial f_y}{\partial y} \frac{dy}{dt} \right)$$

$$= h^2 f_{xx} + 2hk f_{xy} + k^2 f_{yy}$$

from which

$$F''(0) = h^2 f_{xx}(x_0, y_0) + 2hk f_{xy}(x_0, y_0) + k^2 f_{yy}(x_0, y_0) = \left(h \frac{\partial}{\partial x} + k \frac{\partial}{\partial y} \right)^2 f(x_0, y_0)$$

since the second partial derivatives of f are supposed continuous.

In like manner we can verify (by mathematical induction) that for all positive integers n,

$$F^{(n)}(0) = \left(h \frac{\partial}{\partial x} + k \frac{\partial}{\partial y} \right)^n f(x_0, y_0), \qquad F^{(n+1)}(\theta) = \left(h \frac{\partial}{\partial x} + k \frac{\partial}{\partial y} \right)^n f(x_0 + \theta h, y_0 + \theta k)$$

where $0 < \theta < 1$. Substituting these in (2), the required result follows.

42. Expand $x^2 y + 3y - 2$ in powers of $x - 1$ and $y + 2$.

Use Taylor's theorem of the mean with $h = x - x_0$, $k = y - y_0$, where $x_0 = 1$, $y_0 = -2$. Then

$$f(x, y) = x^2 y + 3y - 2, \quad f_x = 2xy, \quad f_y = x^2 + 3, \quad f_{xx} = 2y, \quad f_{xy} = 2x, \quad f_{yy} = 0, \quad f_{xxx} = 0, \quad f_{xxy} = 2, \quad f_{xyy} = 0, \quad f_{yyy} = 0$$

All higher derivatives are zero. Thus

$$f(1, -2) = -10, \quad f_x(1, -2) = -4, \quad f_y(1, -2) = 4, \quad f_{xx}(1, -2) = -4, \quad f_{xy}(1, -2) = 2, \quad f_{yy}(1, -2) = 0$$
$$f_{xxx}(1, -2) = 0, \quad f_{xxy}(1, -2) = 2, \quad f_{xyy}(1, -2) = 0, \quad f_{yyy}(1, -2) = 0$$

By Taylor's theorem,

$$f(x, y) = f(1, -2) + h f_x(1, -2) + k f_y(1, -2) + \frac{1}{2!} \{ h^2 f_{xx}(1, -2) + 2hk f_{xy}(1, -2) + k^2 f_{yy}(1, -2) \}$$
$$+ \frac{1}{3!} \{ h^3 f_{xxx}(1, -2) + 3h^2 k f_{xxy}(1, -2) + 3hk^2 f_{xyy}(1, -2) + k^3 f_{yyy}(1, -2) \} + R_3$$

where R_3 is the remainder and is zero in this case.

Substituting the values of the derivatives obtained above, we find

$$x^2 y + 3y - 2 = -10 - 4(x - 1) + 4(y + 2) - 2(x - 1)^2 + 2(x - 1)(y + 2) + (x - 1)^2(y + 2)$$

as can be verified directly in this case by algebraic processes.

MISCELLANEOUS PROBLEMS

43. Let $f(x, y) = \begin{cases} xy\left(\dfrac{x^2 - y^2}{x^2 + y^2}\right) & (x, y) \neq (0, 0) \\ 0 & (x, y) = (0, 0) \end{cases}$.

Compute (a) $f_x(0,0)$, (b) $f_y(0,0)$, (c) $f_{xx}(0,0)$, (d) $f_{yy}(0,0)$, (e) $f_{xy}(0,0)$, (f) $f_{yx}(0,0)$.

(a) $f_x(0, 0) = \lim_{h \to 0} \dfrac{f(h, 0) - f(0, 0)}{h} = \lim_{h \to 0} \dfrac{0}{h} = 0$

(b) $f_y(0, 0) = \lim_{h \to 0} \dfrac{f(0, k) - f(0, 0)}{k} = \lim_{k \to 0} \dfrac{0}{k} = 0$

If $(x, y) \neq (0, 0)$,

$$f_x(x, y) = \frac{\partial}{\partial x}\left\{xy\left(\frac{x^2 - y^2}{x^2 + y^2}\right)\right\} = xy\left(\frac{4xy^2}{(x^2 + y^2)^2}\right) + y\left(\frac{x^2 - y^2}{x^2 + y^2}\right)$$

$$f_y(x, y) = \frac{\partial}{\partial y}\left\{xy\left(\frac{x^2 - y^2}{x^2 + y^2}\right)\right\} = xy\left(\frac{-4x^2y}{(x^2 + y^2)^2}\right) + x\left(\frac{x^2 - y^2}{x^2 + y^2}\right)$$

Then

(c) $f_{xx}(0, 0) = \lim_{h \to 0} \dfrac{f_x(h, 0) - f_x(0, 0)}{h} = \lim_{h \to 0} \dfrac{0}{h} = 0$

(d) $f_{yy}(0, 0) = \lim_{k \to 0} \dfrac{f_y(0, k) - f_y(0, 0)}{k} = \lim_{k \to 0} \dfrac{0}{k} = 0$

(e) $f_{xy}(0, 0) = \lim_{k \to 0} \dfrac{f_x(0, k) - f_x(0, 0)}{k} = \lim_{k \to 0} \dfrac{-k}{k} = -1$

(f) $f_{yx}(0, 0) = \lim_{h \to 0} \dfrac{f_y(h, 0) - f_y(0, 0)}{h} = \lim_{h \to 0} \dfrac{h}{h} = 1$

Note that $f_{xy} \neq f_{yx}$ at $(0, 0)$. See Problem 13.

44. Show that under the transformation $x = \rho \cos \phi$, $y = \rho \sin \phi$ the equation $\dfrac{\partial^2 V}{\partial x^2} + \dfrac{\partial^2 V}{\partial y^2} = 0$ becomes $\dfrac{\partial^2 V}{\partial \rho^2} + \dfrac{1}{\rho}\dfrac{\partial V}{\partial \phi} + \dfrac{1}{\rho^2}\dfrac{\partial^2 V}{\partial \phi^2} = 0$.

We have

$$(1) \quad \frac{\partial V}{\partial x} = \frac{\partial V}{\partial \rho}\frac{\partial \rho}{\partial x} + \frac{\partial V}{\partial \phi}\frac{\partial \phi}{\partial x} \qquad (2) \quad \frac{\partial V}{\partial y} = \frac{\partial V}{\partial \rho}\frac{\partial \rho}{\partial y} + \frac{\partial V}{\partial \phi}\frac{\partial \phi}{\partial y}$$

Differentiate $x = \rho \cos \phi$, $y = \rho \sin \phi$ with respect to x, remembering that ρ and ϕ are functions of x and y

$$1 = -\rho \sin \phi \frac{\partial \phi}{\partial x} + \cos \phi \frac{\partial \rho}{\partial x}, \qquad 0 = \rho \cos \phi \frac{\partial \phi}{\partial x} + \sin \phi \frac{\partial \rho}{\partial x}$$

Solving simultaneously,

$$\frac{\partial \rho}{\partial x} = \cos \phi, \qquad \frac{\partial \phi}{\partial x} = -\frac{\sin \phi}{\rho} \tag{3}$$

Similarly, differentiate with respect to y. Then

$$0 = -\rho \sin \phi \frac{\partial \phi}{\partial y} + \cos \phi \frac{\partial \rho}{\partial y}, \qquad 1 = \rho \cos \phi \frac{\partial \phi}{\partial y} + \sin \phi \frac{\partial \rho}{\partial y}$$

Solving simultaneously,

$$\frac{\partial \rho}{\partial y} = \sin \phi, \qquad \frac{\partial \phi}{\partial y} = \frac{\cos \phi}{\rho} \tag{4}$$

Then from (1) and (2),

$$(5) \quad \frac{\partial V}{\partial x} = \cos \phi \frac{\partial V}{\partial \rho} - \frac{\sin \phi}{\rho}\frac{\partial V}{\partial \phi} \qquad (6) \quad \frac{\partial V}{\partial y} = \sin \phi \frac{\partial V}{\partial \rho} + \frac{\cos \phi}{\rho}\frac{\partial V}{\partial \phi}$$

Hence

$$\frac{\partial^2 V}{\partial x^2} = \frac{\partial}{\partial x}\left(\frac{\partial V}{\partial x}\right) = \frac{\partial}{\partial \rho}\left(\frac{\partial V}{\partial x}\right)\frac{\partial \rho}{\partial x} + \frac{\partial}{\partial \phi}\left(\frac{\partial V}{\partial x}\right)\frac{\partial \phi}{\partial x}$$

$$= \frac{\partial}{\partial \rho}\left(\cos\phi\,\frac{\partial V}{\partial \rho} - \frac{\sin\phi}{\rho}\frac{\partial V}{\partial \phi}\right)\frac{\partial \rho}{\partial x} + \frac{\partial}{\partial \phi}\left(\cos\phi\,\frac{\partial V}{\partial \rho} - \frac{\sin\phi}{\rho}\frac{\partial V}{\partial \phi}\right)\frac{\partial \phi}{\partial x}$$

$$= \left(\cos\phi\,\frac{\partial^2 V}{\partial \rho^2} + \frac{\sin\phi}{\rho^2}\frac{\partial V}{\partial \phi} - \frac{\sin\phi}{\rho}\frac{\partial^2 V}{\partial \rho\,\partial \phi}\right)(\cos\phi)$$

$$+ \left(-\sin\phi\,\frac{\partial V}{\partial \rho} + \cos\phi\,\frac{\partial^2 V}{\partial \rho\,\partial \phi} - \frac{\cos\phi}{\rho}\frac{\partial V}{\partial \phi} - \frac{\sin\phi}{\rho}\frac{\partial^2 V}{\partial \phi^2}\right)\left(-\frac{\sin\phi}{\rho}\right)$$

which simplifies to

$$\frac{\partial^2 V}{\partial x^2} = \cos^2\phi\,\frac{\partial^2 V}{\partial \rho^2} + \frac{2\sin\phi\cos\phi}{\rho^2}\frac{\partial V}{\partial \phi} - \frac{2\sin\phi\cos\phi}{\rho}\frac{\partial^2 V}{\partial \rho\,\partial \phi} + \frac{\sin^2\phi}{\rho}\frac{\partial V}{\partial \rho} + \frac{\sin^2\phi}{\rho^2}\frac{\partial^2 V}{\partial \phi^2} \qquad (7)$$

Similarly,

$$\frac{\partial^2 V}{\partial y^2} = \sin^2\phi\,\frac{\partial^2 V}{\partial \rho^2} - \frac{2\sin\phi\cos\phi}{\rho^2}\frac{\partial V}{\partial \phi} + \frac{2\sin\phi\cos\phi}{\rho}\frac{\partial^2 V}{\partial \rho\,\partial \phi} + \frac{\cos^2\phi}{\rho}\frac{\partial V}{\partial \rho} + \frac{\cos^2\phi}{\rho^2}\frac{\partial^2 V}{\partial \phi^2} \qquad (8)$$

Adding (7) and (8) we find, as required, $\dfrac{\partial^2 V}{\partial x^2} + \dfrac{\partial^2 V}{\partial y^2} = \dfrac{\partial^2 V}{\partial \rho^2} + \dfrac{1}{\rho}\dfrac{\partial V}{\partial \rho} + \dfrac{1}{\rho^2}\dfrac{\partial^2 V}{\partial \phi^2} = 0.$

45. (a) If $x = f(u, v)$ and $y = g(u, v)$, where $u = \phi(r, s)$ and $v = \psi(r, s)$, prove that $\dfrac{\partial(x, y)}{\partial(r, s)} = \dfrac{\partial(x, y)}{\partial(u, v)}\dfrac{\partial(u, v)}{\partial(r, s)}.$

(b) Prove that $\dfrac{\partial(x, y)}{\partial(u, v)}\dfrac{\partial(u, v)}{\partial(x, y)} = 1$ provided $\dfrac{\partial(x, y)}{\partial(u, v)} \neq 0$, and interpret geometrically.

(a) $\dfrac{\partial(x, y)}{\partial(r, s)} = \begin{vmatrix} x_r & x_s \\ y_r & y_s \end{vmatrix} = \begin{vmatrix} x_u u_r + x_v v_r & x_u u_s + x_v v_s \\ y_u u_r + y_v v_r & y_u u_s + y_v v_s \end{vmatrix}$

$$= \begin{vmatrix} x_u & x_v \\ y_u & y_v \end{vmatrix}\begin{vmatrix} u_r & u_s \\ v_r & v_s \end{vmatrix} = \frac{\partial(x, y)}{\partial(u, v)}\frac{\partial(u, v)}{\partial(r, s)}$$

using a theorem on multiplication of determinants (see Problem 115). We have assumed here, of course, the existence of the partial derivatives involved.

(b) Place $r = x$, $s = y$ in the result of (a). Then $\dfrac{\partial(x, y)}{\partial(u, v)}\dfrac{\partial(u, v)}{\partial(x, y)} = \dfrac{\partial(x, y)}{\partial(x, y)} = 1.$

The equations $x = f(u, v)$, $y = g(u, v)$ defines a transformation between points (x, y) in the xy plane and points (u, v) in the uv plane. The inverse transformation is given by $u = \phi(x, y)$, $v = \psi(x, y)$. The result obtained states that the Jacobians of these transformations are reciprocals of each other.

46. Show that $F(xy, z - 2x) = 0$ satisfies under suitable conditions the equation $x(\partial z/\partial x) - y(\partial z/\partial y) = 2x$. What are these conditions?

Let $u = xy$, $v = z - 2x$. Then $F(u, v) = 0$ and

(1) $\qquad dF = F_u\,du + F_v\,dv = F_u(x\,dy + y\,dx) + F_v(dz - 2\,dx) = 0$

Taking z as dependent variable and x and y as independent variables, we have $dz = z_x\,dx + z_y\,dy$. Then substituting in (1), we find

$$(yF_u + F_v z_x - 2)\,dx + (xF_u + F_v z_y)\,dy = 0$$

Hence we have, since x and y are independent,

(2) $\quad yF_u + F_v z_x - 2 = 0 \qquad\qquad$ (3) $\quad xF_u + F_v z_y = 0$

Solve for F_u in (3) and substitute in (2). Then we obtain the required result $xz_x - yz_y = 2x$ upon dividing by F_v (supposed not equal to zero).

The result will certainly be valid if we assume that $F(u, v)$ is continuously differentiable and that $F_v \neq 0$.

Supplementary Problems

FUNCTIONS and GRAPHS

47. If $f(x,y) = \dfrac{2x+y}{1-xy}$, find (a) $f(1,-3)$, (b) $\dfrac{f(2+h, 3) - f(2, 3)}{h}$, (c) $f(x+y, xy)$.

  Ans. (a) $-\frac{1}{4}$, (b) $\dfrac{11}{5(3h+5)}$, (c) $\dfrac{2x + 2y + xy}{1 - x^2 y - xy^2}$

48. If $g(x, y, z) = x^2 - yz + 3xy$, find (a) $g(1, -2, 2)$, (b) $g(x+1, y-1, z^2)$, (c) $g(xy, xz, x+y)$.
  Ans. (a) -1, (b) $x^2 - x - 2 - yz^2 + z^2 + 3xy + 3y$, (c) $x^2 y^2 - x^2 z - xyz + 3x^2 yz$

49. Give the domain of definition for which each of the following functions are defined and real, and indicate this domain graphically.

  (a) $f(x,y) = \dfrac{1}{x^2 + y^2 - 1}$, (b) $f(x,y) = \ln(x+y)$, (c) $f(x,y) = \sin^{-1}\left(\dfrac{2x-y}{x+y}\right)$.

  Ans. (a) $x^2 + y^2 \neq 1$, (b) $x+y > 0$, (c) $\left|\dfrac{2x-y}{x+y}\right| \leq 1$

50. (a) What is the domain of definition for which $f(x,y,z) = \sqrt{\dfrac{x+y+z-1}{x^2+y^2+z^2-1}}$ is defined and real? (b) Indicate this domain graphically.
  Ans. (a) $x+y+z \leq 1$, $x^2+y^2+z^2 < 1$ and $x+y+z \geq 1$, $x^2+y^2+z^2 > 1$.

51. Sketch and name the surface in 3 dimensional space represented by each of the following.

  (a) $3x + 2z = 12$,   (d) $x^2 + z^2 = y^2$,   (g) $x^2 + y^2 = 2y$,
  (b) $4z = x^2 + y^2$,   (e) $x^2 + y^2 + z^2 = 16$,   (h) $z = x + y$,
  (c) $z = x^2 - 4y^2$,   (f) $x^2 - 4y^2 - 4z^2 = 36$,   (i) $y^2 = 4z$,
  (j) $x^2 + y^2 + z^2 - 4x + 6y + 2z - 2 = 0$.

  Ans. (a) plane, (b) paraboloid of revolution, (c) hyperbolic paraboloid, (d) right circular cone, (e) sphere, (f) hyperboloid of two sheets, (g) right circular cylinder, (h) plane, (i) parabolic cylinder, (j) sphere, center at $(2, -3, -1)$ and radius 4.

52. Construct a graph of the region bounded by $x^2 + y^2 = a^2$ and $x^2 + z^2 = a^2$, where a is a constant.

53. Describe graphically the set of points (x, y, z) such that:
  (a) $x^2 + y^2 + z^2 = 1$, $x^2 + y^2 = z^2$; (b) $x^2 + y^2 < z < x + y$.

54. The *level curves* for a function $z = f(x, y)$ are curves in the xy plane defined by $f(x, y) = c$, where c is any constant. They provide a way of representing the function graphically. Similarly, the *level surfaces* of $w = f(x, y, z)$ are the surfaces in a rectangular (xyz) coordinate system defined by $f(x, y, z) = c$, where c is any constant. Describe and graph the level curves and surfaces for each of the following functions: (a) $f(x, y) = \ln(x^2 + y^2 - 1)$, (b) $f(x, y) = 4xy$, (c) $f(x, y) = \tan^{-1} y/(x+1)$, (d) $f(x, y) = x^{2/3} + y^{2/3}$, (e) $f(x, y, z) = x^2 + 4y^2 + 16z^2$, (f) $\sin(x+z)/(1-y)$.

LIMITS and CONTINUITY

55. Prove that (a) $\lim\limits_{\substack{x \to 4 \\ y \to -1}} (3x - 2y) = 14$ and (b) $\lim\limits_{(x,y) \to (2,1)} (xy - 3x + 4) = 0$ by using the definition.

56. If $\lim f(x,y) = A$ and $\lim g(x,y) = B$, where $\lim$ denotes *limit as* $(x,y) \to (x_0, y_0)$, prove that:
(a) $\lim \{f(x,y) + g(x,y)\} = A + B$, (b) $\lim \{f(x,y)\, g(x,y)\} = AB$.

57. Under what conditions is the limit of the quotient of two functions equal to the quotient of their limits? Prove your answer.

58. Evaluate each of the following limits where they exist.

(a) $\lim\limits_{\substack{x \to 1 \\ y \to 2}} \dfrac{3 - x + y}{4 + x - 2y}$
  (c) $\lim\limits_{\substack{x \to 4 \\ y \to \pi}} x^2 \sin \dfrac{y}{x}$
  (e) $\lim\limits_{\substack{x \to 0 \\ y \to 1}} e^{-1/x^2(y-1)^2}$
  (g) $\lim\limits_{\substack{x \to 0+ \\ y \to 1-}} \dfrac{x + y - 1}{\sqrt{x} - \sqrt{1 - y}}$

(b) $\lim\limits_{\substack{x \to 0 \\ y \to 0}} \dfrac{3x - 2y}{2x - 3y}$
  (d) $\lim\limits_{\substack{x \to 0 \\ y \to 0}} \dfrac{x \sin (x^2 + y^2)}{x^2 + y^2}$
  (f) $\lim\limits_{\substack{x \to 0 \\ y \to 0}} \dfrac{2x - y}{x^2 + y^2}$
  (h) $\lim\limits_{\substack{x \to 2 \\ y \to 1}} \dfrac{\sin^{-1}(xy - 2)}{\tan^{-1}(3xy - 6)}$

Ans. (a) 4, (b) does not exist, (c) $8\sqrt{2}$, (d) 0, (e) 0, (f) does not exist, (g) 0, (h) 1/3

59. Formulate a definition of limit for functions of (a) 3, (b) n variables.

60. Does $\lim \dfrac{4x + y - 3z}{2x - 5y + 2z}$ as $(x,y,z) \to (0,0,0)$ exist? Justify your answer.

61. Investigate the continuity of each of the following functions at the indicated points:

(a) $x^2 + y^2$; (x_0, y_0). (b) $\dfrac{x}{3x + 5y}$; $(0,0)$. (c) $(x^2 + y^2) \sin \dfrac{1}{x^2 + y^2}$ if $(x,y) \neq (0,0)$, 0 if $(x,y) = (0,0)$; $(0,0)$.

Ans. (a) continuous, (b) discontinuous, (c) continuous

62. Using the definition, prove that $f(x,y) = xy + 6x$ is continuous at (a) $(1,2)$, (b) (x_0, y_0).

63. Prove that the function of Problem 62 is uniformly continuous in the square region defined by $0 \leqq x \leqq 1$, $0 \leqq y \leqq 1$.

PARTIAL DERIVATIVES

64. If $f(x,y) = \dfrac{x - y}{x + y}$, find (a) $\partial f/\partial x$ and (b) $\partial f/\partial y$ at $(2,-1)$ from the definition and verify your answer by differentiation rules.   *Ans.* (a) -2, (b) -4

65. If $f(x,y) = \begin{cases} (x^2 - xy)/(x + y) & \text{for } (x,y) \neq (0,0) \\ 0 & \text{for } (x,y) = (0,0) \end{cases}$, find (a) $f_x(0,0)$, (b) $f_y(0,0)$.
Ans. (a) 1, (b) 0

66. Investigate $\lim\limits_{(x,y) \to (0,0)} f_x(x,y)$ for the function in the preceding problem and explain why this limit (if it exists) is or is not equal to $f_x(0,0)$.

67. If $f(x,y) = (x - y) \sin (3x + 2y)$, compute (a) f_x, (b) f_y, (c) f_{xx}, (d) f_{yy}, (e) f_{xy}, (f) f_{yx} at $(0, \pi/3)$.
Ans. (a) $\frac{1}{2}(\pi + \sqrt{3})$, (b) $\frac{1}{6}(2\pi - 3\sqrt{3})$, (c) $\frac{3}{2}(\pi\sqrt{3} - 2)$, (d) $\frac{2}{3}(\pi\sqrt{3} + 3)$, (e) $\frac{1}{2}(2\pi\sqrt{3} + 1)$, (f) $\frac{1}{2}(2\pi\sqrt{3} + 1)$

68. (a) Prove by direct differentiation that $z = xy \tan (y/x)$ satisfies the equation $x(\partial z/\partial x) + y(\partial z/\partial y) = 2z$ if $(x,y) \neq (0,0)$. (b) Discuss part (a) for all other points (x,y) assuming $z = 0$ at $(0,0)$.

69. Verify that $f_{xy} = f_{yx}$ for the functions (a) $(2x - y)/(x + y)$, (b) $x \tan xy$ and (c) $\cosh (y + \cos x)$, indicating possible exceptional points and investigate these points.

70. Show that $z = \ln \{(x - a)^2 + (y - b)^2\}$ satisfies $\partial^2 z/\partial x^2 + \partial^2 z/\partial y^2 = 0$ except at (a,b).

71. Show that $z = x \cos{(y/x)} + \tan{(y/x)}$ satisfies $x^2 z_{xx} + 2xy z_{xy} + y^2 z_{yy} = 0$ except at points for which $x = 0$.

72. Show that if $w = \left(\dfrac{x - y + z}{x + y - z} \right)^n$, then:

     (a) $x\dfrac{\partial w}{\partial x} + y\dfrac{\partial w}{\partial y} + z\dfrac{\partial w}{\partial z} = 0$,   (b) $x^2\dfrac{\partial^2 w}{\partial x^2} + y^2\dfrac{\partial^2 w}{\partial y^2} + z^2\dfrac{\partial^2 w}{\partial z^2} + 2xy\dfrac{\partial^2 w}{\partial x\,\partial y} + 2xz\dfrac{\partial^2 w}{\partial x\,\partial z} + 2yz\dfrac{\partial^2 w}{\partial y\,\partial z} = 0.$

     Indicate possible exceptional points.

DIFFERENTIALS

73. If $z = x^3 - xy + 3y^2$, compute (a) Δz and (b) dz where $x = 5$, $y = 4$, $\Delta x = -0.2$, $\Delta y = 0.1$. Explain why Δz and dz are approximately equal. (c) Find Δz and dz if $x = 5$, $y = 4$, $\Delta x = -2$, $\Delta y = 1$.
     Ans. (a) -11.658, (b) -12.3, (c) $\Delta z = -66$, $dz = -123$.

74. Compute $\sqrt[5]{(3.8)^2 + 2(2.1)^3}$ approximately, using differentials.
     Ans. 2.01

75. Find dF and dG if (a) $F(x, y) = x^3 y - 4xy^2 + 8y^3$, (b) $G(x, y, z) = 8xy^2 z^3 - 3x^2 yz$, (c) $F(x, y) = xy^2 \ln{(y/x)}$.
     Ans. (a) $(3x^2 y - 4y^2)\,dx + (x^3 - 8xy + 24y^2)\,dy$
           (b) $(8y^2 z^3 - 6xyz)\,dx + (16xyz^3 - 3x^2 z)\,dy + (24xy^2 z^2 - 3x^2 y)\,dz$
           (c) $\{y^2 \ln{(y/x)} - y^2\}\,dx + \{2xy \ln{(y/x)} + xy\}\,dy$

76. Prove that (a) $d(UV) = U\,dV + V\,dU$, (b) $d(U/V) = (V\,dU - U\,dV)/V^2$, (c) $d(\ln U) = (dU)/U$, (d) $d(\tan^{-1} V) = (dV)/(1 + V^2)$ where U and V are differentiable functions of two or more variables.

77. Determine whether each of the following are exact differentials of a function and if so, find the function.
     (a) $(2xy^2 + 3y \cos 3x)\,dx + (2x^2 y + \sin 3x)\,dy$
     (b) $(6xy - y^2)\,dx + (2xe^y - x^2)\,dy$
     (c) $(z^3 - 3y)\,dx + (12y^2 - 3x)\,dy + 3xz^2\,dz$
     Ans. (a) $x^2 y^2 + y \sin 3x + c$, (b) not exact, (c) $xz^3 + 4y^3 - 3xy + c$

DIFFERENTIATION of COMPOSITE FUNCTIONS

78. (a) If $U(x, y, z) = 2x^2 - yz + xz^2$, $x = 2 \sin t$, $y = t^2 - t + 1$, $z = 3e^{-t}$, find dU/dt at $t = 0$.
     (b) If $H(x, y) = \sin{(3x - y)}$, $x^3 + 2y = 2t^3$, $x - y^2 = t^2 + 3t$, find dH/dt.

     Ans. (a) 24, (b) $\left(\dfrac{36t^2 y + 12t + 9x^2 - 6t^2 + 6x^2 t + 18}{6x^2 y + 2} \right) \cos{(3x - y)}$

79. If $F(x, y) = (2x + y)/(y - 2x)$, $x = 2u - 3v$, $y = u + 2v$, find (a) $\partial F/\partial u$, (b) $\partial F/\partial v$, (c) $\partial^2 F/\partial u^2$, (d) $\partial^2 F/\partial v^2$, (e) $\partial^2 F/\partial u\,\partial v$, where $u = 2$, $v = 1$.   Ans. (a) 7, (b) -14, (c) 21, (d) 112, (e) -49

80. If $U = x^2 F(y/x)$, show that under suitable restrictions on F, $x(\partial U/\partial x) + y(\partial U/\partial y) = 2U$.

81. If $x = u \cos\alpha - v \sin\alpha$ and $y = u \sin\alpha + v \cos\alpha$, where α is a constant, show that
$$(\partial V/\partial x)^2 + (\partial V/\partial y)^2 = (\partial V/\partial u)^2 + (\partial V/\partial v)^2$$

82. Show that if $x = \rho \cos\phi$, $y = \rho \sin\phi$, the equations
$$\frac{\partial u}{\partial x} = \frac{\partial v}{\partial y}, \quad \frac{\partial u}{\partial y} = -\frac{\partial v}{\partial x} \qquad \text{become} \qquad \frac{\partial u}{\partial \rho} = \frac{1}{\rho}\frac{\partial v}{\partial \phi}, \quad \frac{\partial v}{\partial \rho} = -\frac{1}{\rho}\frac{\partial u}{\partial \phi}$$

83. Use Problem 82 to show that under the transformation $x = \rho \cos\phi$, $y = \rho \sin\phi$, the equation
$$\frac{\partial^2 u}{\partial x^2} + \frac{\partial^2 u}{\partial y^2} = 0 \qquad \text{becomes} \qquad \frac{\partial^2 u}{\partial \rho^2} + \frac{1}{\rho}\frac{\partial u}{\partial \rho} + \frac{1}{\rho^2}\frac{\partial^2 u}{\partial \phi^2} = 0$$

IMPLICIT FUNCTIONS and JACOBIANS

84. If $F(x, y) = 0$, prove that $dy/dx = -F_x/F_y$.

85. Find (a) dy/dx and (b) d^2y/dx^2 if $x^3 + y^3 - 3xy = 0$.
    Ans. (a) $(y - x^2)/(y^2 - x)$, (b) $-2xy/(y^2 - x)^3$

86. If $xu^2 + v = y^3$, $2yu - xv^3 = 4x$, find (a) $\dfrac{\partial u}{\partial x}$ (b) $\dfrac{\partial v}{\partial y}$.   Ans. (a) $\dfrac{v^3 - 3xu^2v^2 + 4}{6x^2uv^2 + 2y}$, (b) $\dfrac{2xu^2 + 3y^3}{3x^2uv^2 + y}$

87. If $u = f(x, y)$, $v = g(x, y)$ are differentiable, prove that $\dfrac{\partial u}{\partial x} \dfrac{\partial x}{\partial u} + \dfrac{\partial v}{\partial x} \dfrac{\partial x}{\partial v} = 1$. Explain clearly which variables are considered independent in each partial derivative.

88. If $f(x, y, r, s) = 0$, $g(x, y, r, s) = 0$, prove that $\dfrac{\partial y}{\partial r} \dfrac{\partial r}{\partial x} + \dfrac{\partial y}{\partial s} \dfrac{\partial s}{\partial x} = 0$, explaining which variables are independent. What notation could you use to indicate the independent variables considered?

89. If $F(x, y) = 0$, show that $\dfrac{d^2y}{dx^2} = -\dfrac{F_{xx}F_y^2 - 2F_{xy}F_xF_y + F_{yy}F_x^2}{F_y^3}$.

90. Evaluate $\dfrac{\partial(F, G)}{\partial(u, v)}$ if $F(u, v) = 3u^2 - uv$, $G(u, v) = 2uv^2 + v^3$.   Ans. $24u^2v + 16uv^2 - 3v^3$

91. If $F = x + 3y^2 - z^3$, $G = 2x^2yz$, and $H = 2z^2 - xy$, evaluate $\dfrac{\partial(F, G, H)}{\partial(x, y, z)}$ at $(1, -1, 0)$.   Ans. 10

92. If $u = \sin^{-1} x + \sin^{-1} y$ and $v = x\sqrt{1 - y^2} + y\sqrt{1 - x^2}$, determine whether there is a functional relationship between u and v, and if so find it.

93. If $F = xy + yz + zx$, $G = x^2 + y^2 + z^2$, and $H = x + y + z$, determine whether there is a functional relationship connecting F, G, and H, and if so find it.   Ans. $H^2 - G - 2F = 0$

94. (a) If $x = f(u, v, w)$, $y = g(u, v, w)$, and $z = h(u, v, w)$, prove that $\dfrac{\partial(x, y, z)}{\partial(u, v, w)} \dfrac{\partial(u, v, w)}{\partial(x, y, w)} = 1$ provided $\dfrac{\partial(x, y, z)}{\partial(u, v, w)} \neq 0$. (b) Give an interpretation of the result of (a) in terms of transformations.

95. If $f(x, y, z) = 0$ and $g(x, y, z) = 0$, show that
$$\frac{dx}{\dfrac{\partial(f, g)}{\partial(y, z)}} = \frac{dy}{\dfrac{\partial(f, g)}{\partial(z, x)}} = \frac{dz}{\dfrac{\partial(f, g)}{\partial(x, y)}}$$
giving conditions under which the result is valid.

96. If $x + y^2 = u$, $y + z^2 = v$, $z + x^2 = w$, find (a) $\dfrac{\partial x}{\partial u}$, (b) $\dfrac{\partial^2 x}{\partial u^2}$, (c) $\dfrac{\partial^2 x}{\partial u\, \partial v}$ assuming that the equations define x, y and z as twice differentiable functions of u, v and w.
    Ans. (a) $\dfrac{1}{1 + 8xyz}$, (b) $\dfrac{16x^2y - 8yz - 32x^2z^2}{(1 + 8xyz)^3}$, (c) $\dfrac{16y^2z - 8xz - 32x^2y^2}{(1 + 8xyz)^3}$

97. State and prove a theorem similar to that in Problem 35, for the case where $u = f(x, y, z)$, $v = g(x, y, z)$, $w = h(x, y, z)$.

TRANSFORMATIONS, CURVILINEAR COORDINATES

98. Given the transformation $x = 2u + v$, $y = u - 3v$. (a) Sketch the region $\mathcal{R}'$ of the uv plane into which the region $\mathcal{R}$ of the xy plane bounded by $x = 0, x = 1, y = 0, y = 1$ is mapped under the transformation. (b) Compute $\dfrac{\partial(x, y)}{\partial(u, v)}$. (c) Compare the result of (b) with the ratios of the areas of $\mathcal{R}$ and $\mathcal{R}'$.
    Ans. (b) -7

99. (a) Prove that under a *linear transformation* $x = a_1u + a_2v$, $y = b_1u + b_2v$ $(a_1b_2 - a_2b_1 \neq 0)$ lines and circles in the xy plane are mapped respectively into lines and circles in the uv plane. (b) Compute the Jacobian J of the transformation and discuss the significance of $J = 0$.

100. Given $x = \cos u \cosh v$, $y = \sin u \sinh v$. (a) Show that in general the coordinate curves $u = a$ and $v = b$ in the uv plane are mapped into hyperbolas and ellipses, respectively, in the xy plane. (b) Compute $\left|\dfrac{\partial(x, y)}{\partial(u, v)}\right|$. (c) Compute $\left|\dfrac{\partial(u, v)}{\partial(x, y)}\right|$.

Ans. (b) $\sin^2 u \cosh^2 v + \cos^2 u \sinh^2 v$, (c) $(\sin^2 u \cosh^2 v + \cos^2 u \sinh^2 v)^{-1}$

101. Given the transformation $x = 2u + 3v - w$, $y = u - 2v + w$, $z = 2u - 2v + w$. (a) Sketch the region $\mathcal{R}'$ of the uvw space into which the region $\mathcal{R}$ of the xyz space bounded by $x = 0, x = 8, y = 0, y = 4, z = 0$, $z = 6$ is mapped. (b) Compute $\dfrac{\partial(x, y, z)}{\partial(u, v, w)}$. (c) Compare the result of (b) with the ratios of the volumes of $\mathcal{R}$ and $\mathcal{R}'$. *Ans.* (b) 1

102. Given the spherical coordinate transformation $x = r \sin\theta \cos\phi$, $y = r \sin\theta \sin\phi$, $z = r \cos\theta$, where $r \geq 0$, $0 \leq \theta \leq \pi$, $0 \leq \phi < 2\pi$. Describe the coordinate surfaces (a) $r = a$, (b) $\theta = b$, and (c) $\phi = c$, where a, b, c are any constants. *Ans.* (a) spheres, (b) cones, (c) planes

103. (a) Verify that for the spherical coordinate transformation of Problem 102, $J = \dfrac{\partial(x, y, z)}{\partial(r, \theta, \phi)} = r^2 \sin\theta$. (b) Discuss the case where $J = 0$.

MEAN VALUE THEOREMS

104. Prove that $\ln \dfrac{x + y}{2} = \dfrac{x + y - 2}{2 + \theta(x + y - 2)}$, $0 < \theta < 1$, where $x > 0, y > 0$.

105. Expand $f(x, y) = \sin xy$ in powers of $x - 1$ and $y - \tfrac{1}{2}\pi$, up to and including second degree terms. *Ans.* $1 - \tfrac{1}{8}\pi^2(x - 1)^2 - \tfrac{1}{2}\pi(x - 1)(y - \tfrac{1}{2}\pi) - \tfrac{1}{2}(y - \tfrac{1}{2}\pi)^2$

106. Expand $f(x, y) = y^2/x^3$ in powers of $x - 1$ and $y + 1$, up to and including second degree terms and write the remainder.

Ans. $1 - 3(x - 1) - 2(y + 1) + 6(x - 1)^2 + 6(x - 1)(y + 1) + (y + 1)^2$

$$- \frac{10(x - 1)^3[1 - \theta(y + 1)]^2 + 12(x - 1)^2(y + 1)[1 + \theta(x - 1)][1 - \theta(y + 1)] + 3(x - 1)(y + 1)^2[1 + \theta(x - 1)]^2}{[1 + \theta(x - 1)]^6}$$

where $0 < \theta < 1$.

107. Prove the first mean value theorem for functions of 3 variables.

108. Generalize and prove Taylor's theorem of the mean for functions of 3 variables.

MISCELLANEOUS PROBLEMS

109. If $F(P, V, T) = 0$, prove that (a) $\dfrac{\partial P}{\partial T}\bigg|_V \dfrac{\partial T}{\partial V}\bigg|_P = -\dfrac{\partial P}{\partial V}\bigg|_T$, (b) $\dfrac{\partial P}{\partial T}\bigg|_V \dfrac{\partial T}{\partial V}\bigg|_P \dfrac{\partial V}{\partial P}\bigg|_T = -1$.

These results are useful in *thermodynamics*, where P, V, T correspond to pressure, volume and temperature of a physical system.

110. Show that $F(x/y, z/y) = 0$ satisfies $x(\partial z/\partial x) + y(\partial z/\partial y) = z$.

111. Show that $F(x + y - z, x^2 + y^2) = 0$ satisfies $x(\partial z/\partial y) - y(\partial z/\partial x) = x - y$.

112. If $x = f(u, v)$ and $y = g(u, v)$, prove that $\dfrac{\partial v}{\partial x} = -\dfrac{1}{J}\dfrac{\partial y}{\partial u}$ where $J = \dfrac{\partial(x, y)}{\partial(u, v)}$.

113. If $x = f(u, v)$, $y = g(u, v)$, $z = h(u, v)$ and $F(x, y, z) = 0$, prove that

$$\frac{\partial(y, z)}{\partial(u, v)} dx + \frac{\partial(z, x)}{\partial(u, v)} dy + \frac{\partial(x, y)}{\partial(u, v)} dz = 0$$

114. If $x = \phi(u, v, w)$, $y = \psi(u, v, w)$ and $u = f(r, s)$, $v = g(r, s)$, $w = h(r, s)$, prove that

$$\frac{\partial(x, y)}{\partial(r, s)} = \frac{\partial(x, y)}{\partial(u, v)}\frac{\partial(u, v)}{\partial(r, s)} + \frac{\partial(x, y)}{\partial(v, w)}\frac{\partial(v, w)}{\partial(r, s)} + \frac{\partial(x, y)}{\partial(w, u)}\frac{\partial(w, u)}{\partial(r, s)}$$

115. (a) Prove that $\begin{vmatrix} a & b \\ c & d \end{vmatrix} \cdot \begin{vmatrix} e & f \\ g & h \end{vmatrix} = \begin{vmatrix} ae+bg & af+bh \\ ce+dg & cf+dh \end{vmatrix}$, thus establishing the rule for the

product of two second order determinants referred to in Problem 45. (b) Generalize the result of (a) to determinants of order $3, 4, \ldots$.

116. If x, y and z are functions of u, v and w, while u, v and w are functions of r, s and t, prove that

$$\frac{\partial(x,y,z)}{\partial(r,s,t)} = \frac{\partial(x,y,z)}{\partial(u,v,w)} \cdot \frac{\partial(u,v,w)}{\partial(r,s,t)}$$

117. If D_x and D_y denote the operators $\partial/\partial x$ and $\partial/\partial y$ respectively, show that if the Taylor series for $f(x+h, y+k)$ exists it can be written in the form

$$f(x+h, y+k) = e^{hD_x + kD_y} f(x,y)$$

118. Given the equations $F_j(x_1, \ldots, x_m, y_1, \ldots, y_n) = 0$ where $j = 1, 2, \ldots, n$. Prove that under suitable conditions on F_j,

$$\frac{\partial y_r}{\partial x_s} = -\frac{\partial(F_1, F_2, \ldots, F_r, \ldots, F_n)}{\partial(y_1, y_2, \ldots, x_s, \ldots, y_n)} \Bigg/ \frac{\partial(F_1, F_2, \ldots, F_n)}{\partial(y_1, y_2, \ldots, y_n)}$$

119. (a) If $F(x,y)$ is homogeneous of degree 2, prove that $x^2 \dfrac{\partial^2 F}{\partial x^2} + 2xy \dfrac{\partial^2 F}{\partial x\,\partial y} + y^2 \dfrac{\partial^2 F}{\partial y^2} = 2F$.

(b) Illustrate by using the special case $F(x,y) = x^2 \ln(y/x)$.

Note that the result can be written in operator form, using $D_x \equiv \partial/\partial x$ and $D_y \equiv \partial/\partial y$, as $(x D_x + y D_y)^2 F = 2F$. [Hint: Differentiate both sides of equation (1), Problem 25, twice with respect to λ.]

120. Generalize the result of Problem 119 as follows. If $F(x_1, x_2, \ldots, x_n)$ is homogeneous of degree p, then for any positive integer r, if $D_{x_j} \equiv \partial/\partial x_j$,

$$(x_1 D_{x_1} + x_2 D_{x_2} + \cdots + x_n D_{x_n})^r F = p(p-1) \cdots (p-r+1)F$$

121. (a) Let x and y be determined from u and v according to $x + iy = (u+iv)^3$. Prove that under this transformation the equation

$$\frac{\partial^2 \phi}{\partial x^2} + \frac{\partial^2 \phi}{\partial y^2} = 0 \quad \text{is transformed into} \quad \frac{\partial^2 \phi}{\partial u^2} + \frac{\partial^2 \phi}{\partial v^2} = 0$$

(b) Is the result in (a) true if $x + iy = F(u+iv)$? Prove your statements.

Chapter 7

Vectors

VECTORS and SCALARS

There are quantities in physics characterized by both magnitude and direction, such as displacement, velocity, force and acceleration. To describe such quantities, we introduce the concept of a *vector* as a directed line segment $\overrightarrow{PQ}$ from one point P called the *initial point* to another point Q called the *terminal point*. We denote vectors by bold faced letters or letters with an arrow over them. Thus $\overrightarrow{PQ}$ is denoted by $\mathbf{A}$ or $\vec{A}$ as in Fig. 7-1. The *magnitude* or *length* of the vector is then denoted by $|\overrightarrow{PQ}|$, $\overrightarrow{PQ}$, $|\mathbf{A}|$ or $|\vec{A}|$.

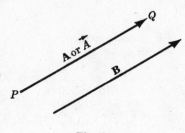

Fig. 7-1

Other quantities in physics are characterized by magnitude only, such as mass, length and temperature. Such quantities are often called *scalars* to distinguish them from vectors, but it must be emphasized that apart from units such as feet, degrees, etc., they are nothing more than real numbers. We can thus denote them by ordinary letters as usual.

VECTOR ALGEBRA

The operations of addition, subtraction and multiplication familiar in the algebra of numbers are, with suitable definition, capable of extension to an algebra of vectors. The following definitions are fundamental.

1. Two vectors $\mathbf{A}$ and $\mathbf{B}$ are *equal* if they have the same magnitude and direction regardless of their initial points. Thus $\mathbf{A} = \mathbf{B}$ in Fig. 7-1 above.

2. A vector having direction opposite to that of vector $\mathbf{A}$ but with the same magnitude is denoted by $-\mathbf{A}$ [see Fig. 7-2].

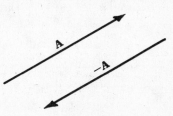

Fig. 7-2

3. The *sum* or *resultant* of vectors $\mathbf{A}$ and $\mathbf{B}$ of Fig. 7-3(a) below is a vector $\mathbf{C}$ formed by placing the initial point of $\mathbf{B}$ on the terminal point of $\mathbf{A}$ and joining the initial point of $\mathbf{A}$ to the terminal point of $\mathbf{B}$ [see Fig. 7-3(b) below]. The sum $\mathbf{C}$ is written $\mathbf{C} = \mathbf{A} + \mathbf{B}$. The definition here is equivalent to the *parallelogram law* for vector addition as indicated in Fig. 7-3(c) below.

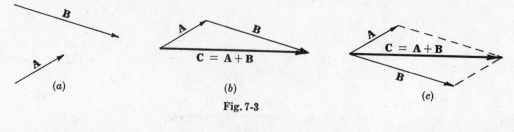

Fig. 7-3

134

Extensions to sums of more than two vectors are immediate. For example, Fig. 7-4 below shows how to obtain the sum or resultant **E** of the vectors **A**, **B**, **C** and **D**.

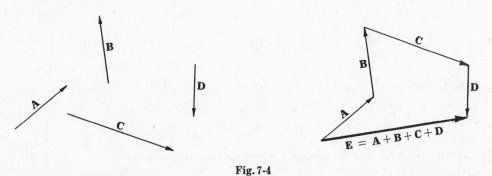

Fig. 7-4

4. The *difference* of vectors **A** and **B**, represented by **A** − **B**, is that vector **C** which added to **B** gives **A**. Equivalently, **A** − **B** may be defined as **A** + (−**B**). If **A** = **B**, then **A** − **B** is defined as the *null* or *zero vector* and is represented by the symbol **0**. This has a magnitude of zero but its direction is not defined.

5. Multiplication of a vector **A** by a scalar m produces a vector m**A** with magnitude $|m|$ times the magnitude of **A** and direction the same as or opposite to that of **A** according as m is positive or negative. If $m = 0$, m**A** = **0**, the null vector.

LAWS of VECTOR ALGEBRA

If **A**, **B** and **C** are vectors, and m and n are scalars, then

1. $\mathbf{A} + \mathbf{B} = \mathbf{B} + \mathbf{A}$ Commutative Law for Addition
2. $\mathbf{A} + (\mathbf{B} + \mathbf{C}) = (\mathbf{A} + \mathbf{B}) + \mathbf{C}$ Associative Law for Addition
3. $m(n\mathbf{A}) = (mn)\mathbf{A} = n(m\mathbf{A})$ Associative Law for Multiplication
4. $(m+n)\mathbf{A} = m\mathbf{A} + n\mathbf{A}$ Distributive Law
5. $m(\mathbf{A} + \mathbf{B}) = m\mathbf{A} + m\mathbf{B}$ Distributive Law

Note that in these laws only multiplication of a vector by one or more scalars is defined. On Pages 136 and 137 we define products of vectors.

UNIT VECTORS

Unit vectors are vectors having unit length. If **A** is any vector with length $A > 0$, then **A**/A is a unit vector, denoted by **a**, having the same direction as **A**. Then **A** = A**a**.

RECTANGULAR UNIT VECTORS

The rectangular unit vectors **i**, **j** and **k** are unit vectors having the direction of the positive x, y and z axes of a rectangular coordinate system [see Fig. 7-5]. We use right-handed rectangular coordinate systems unless otherwise specified. Such systems derive their name from the fact that a right threaded screw rotated through 90° from Ox to Oy will advance in the positive z direction. In general,

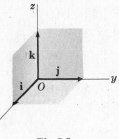

Fig. 7-5

three vectors **A**, **B** and **C** which have coincident initial points and are not coplanar are said to form a *right-handed system* or *dextral system* if a right threaded screw rotated through an angle less than 180° from **A** to **B** will advance in the direction **C** [see Fig. 7-6 below].

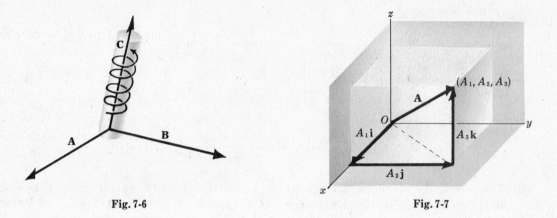

Fig. 7-6   Fig. 7-7

COMPONENTS of a VECTOR

Any vector **A** in 3 dimensions can be represented with initial point at the origin O of a rectangular coordinate system [see Fig. 7-7 above]. Let (A_1, A_2, A_3) be the rectangular coordinates of the terminal point of vector **A** with initial point at O. The vectors $A_1\mathbf{i}$, $A_2\mathbf{j}$ and $A_3\mathbf{k}$ are called the *rectangular component vectors*, or simply component vectors, of **A** in the x, y and z directions respectively. A_1, A_2 and A_3 are called the *rectangular components*, or simply *components*, of **A** in the x, y and z directions respectively.

The sum or resultant of $A_1\mathbf{i}$, $A_2\mathbf{j}$ and $A_3\mathbf{k}$ is the vector **A**, so that we can write

$$\mathbf{A} = A_1\mathbf{i} + A_2\mathbf{j} + A_3\mathbf{k} \tag{1}$$

The magnitude of **A** is

$$A = |\mathbf{A}| = \sqrt{A_1^2 + A_2^2 + A_3^2} \tag{2}$$

In particular, the *position vector* or *radius vector* **r** from O to the point (x, y, z) is written

$$\mathbf{r} = x\mathbf{i} + y\mathbf{j} + z\mathbf{k} \tag{3}$$

and has magnitude $r = |\mathbf{r}| = \sqrt{x^2 + y^2 + z^2}$.

DOT or SCALAR PRODUCT

The dot or scalar product of two vectors **A** and **B**, denoted by **A·B** (read **A** dot **B**) is defined as the product of the magnitudes of **A** and **B** and the cosine of the angle between them. In symbols,

$$\mathbf{A} \cdot \mathbf{B} = AB \cos\theta, \qquad 0 \le \theta \le \pi \tag{4}$$

Note that **A·B** is a scalar and not a vector.

The following laws are valid:

1. $\mathbf{A} \cdot \mathbf{B} = \mathbf{B} \cdot \mathbf{A}$ Commutative Law for Dot Products
2. $\mathbf{A} \cdot (\mathbf{B} + \mathbf{C}) = \mathbf{A} \cdot \mathbf{B} + \mathbf{A} \cdot \mathbf{C}$ Distributive Law
3. $m(\mathbf{A} \cdot \mathbf{B}) = (m\mathbf{A}) \cdot \mathbf{B} = \mathbf{A} \cdot (m\mathbf{B}) = (\mathbf{A} \cdot \mathbf{B})m$,   where m is a scalar.
4. $\mathbf{i} \cdot \mathbf{i} = \mathbf{j} \cdot \mathbf{j} = \mathbf{k} \cdot \mathbf{k} = 1$, $\mathbf{i} \cdot \mathbf{j} = \mathbf{j} \cdot \mathbf{k} = \mathbf{k} \cdot \mathbf{i} = 0$

5. If $\mathbf{A} = A_1\mathbf{i} + A_2\mathbf{j} + A_3\mathbf{k}$ and $\mathbf{B} = B_1\mathbf{i} + B_2\mathbf{j} + B_3\mathbf{k}$, then

$$\mathbf{A}\cdot\mathbf{B} = A_1B_1 + A_2B_2 + A_3B_3$$
$$\mathbf{A}\cdot\mathbf{A} = A^2 = A_1^2 + A_2^2 + A_3^2$$
$$\mathbf{B}\cdot\mathbf{B} = B^2 = B_1^2 + B_2^2 + B_3^2$$

6. If $\mathbf{A}\cdot\mathbf{B} = 0$ and $\mathbf{A}$ and $\mathbf{B}$ are not null vectors, then $\mathbf{A}$ and $\mathbf{B}$ are perpendicular.

CROSS or VECTOR PRODUCT

The cross or vector product of $\mathbf{A}$ and $\mathbf{B}$ is a vector $\mathbf{C} = \mathbf{A}\times\mathbf{B}$ (read $\mathbf{A}$ cross $\mathbf{B}$). The magnitude of $\mathbf{A}\times\mathbf{B}$ is defined as the product of the magnitudes of $\mathbf{A}$ and $\mathbf{B}$ and the sine of the angle between them. The direction of the vector $\mathbf{C} = \mathbf{A}\times\mathbf{B}$ is perpendicular to the plane of $\mathbf{A}$ and $\mathbf{B}$ and such that $\mathbf{A}$, $\mathbf{B}$ and $\mathbf{C}$ form a right-handed system. In symbols,

$$\mathbf{A}\times\mathbf{B} = AB\sin\theta\,\mathbf{u}, \qquad 0 \leqq \theta \leqq \pi \tag{5}$$

where $\mathbf{u}$ is a unit vector indicating the direction of $\mathbf{A}\times\mathbf{B}$. If $\mathbf{A}=\mathbf{B}$ or if $\mathbf{A}$ is parallel to $\mathbf{B}$, then $\sin\theta = 0$ and we define $\mathbf{A}\times\mathbf{B} = \mathbf{0}$.

The following laws are valid:

1. $\mathbf{A}\times\mathbf{B} = -\mathbf{B}\times\mathbf{A}$ (Commutative Law for Cross Products Fails)
2. $\mathbf{A}\times(\mathbf{B}+\mathbf{C}) = \mathbf{A}\times\mathbf{B} + \mathbf{A}\times\mathbf{C}$ Distributive Law
3. $m(\mathbf{A}\times\mathbf{B}) = (m\mathbf{A})\times\mathbf{B} = \mathbf{A}\times(m\mathbf{B}) = (\mathbf{A}\times\mathbf{B})m$,   where m is a scalar.
4. $\mathbf{i}\times\mathbf{i} = \mathbf{j}\times\mathbf{j} = \mathbf{k}\times\mathbf{k} = 0$, $\mathbf{i}\times\mathbf{j} = \mathbf{k}$, $\mathbf{j}\times\mathbf{k} = \mathbf{i}$, $\mathbf{k}\times\mathbf{i} = \mathbf{j}$
5. If $\mathbf{A} = A_1\mathbf{i} + A_2\mathbf{j} + A_3\mathbf{k}$ and $\mathbf{B} = B_1\mathbf{i} + B_2\mathbf{j} + B_3\mathbf{k}$, then

$$\mathbf{A}\times\mathbf{B} = \begin{vmatrix} \mathbf{i} & \mathbf{j} & \mathbf{k} \\ A_1 & A_2 & A_3 \\ B_1 & B_2 & B_3 \end{vmatrix}$$

6. $|\mathbf{A}\times\mathbf{B}| = $ the area of a parallelogram with sides $\mathbf{A}$ and $\mathbf{B}$.
7. If $\mathbf{A}\times\mathbf{B} = 0$ and $\mathbf{A}$ and $\mathbf{B}$ are not null vectors, then $\mathbf{A}$ and $\mathbf{B}$ are parallel.

TRIPLE PRODUCTS

Dot and cross multiplication of three vectors $\mathbf{A}$, $\mathbf{B}$ and $\mathbf{C}$ may produce meaningful products of the form $(\mathbf{A}\cdot\mathbf{B})\mathbf{C}$, $\mathbf{A}\cdot(\mathbf{B}\times\mathbf{C})$ and $\mathbf{A}\times(\mathbf{B}\times\mathbf{C})$. The following laws are valid:

1. $(\mathbf{A}\cdot\mathbf{B})\mathbf{C} \neq \mathbf{A}(\mathbf{B}\cdot\mathbf{C})$ in general
2. $\mathbf{A}\cdot(\mathbf{B}\times\mathbf{C}) = \mathbf{B}\cdot(\mathbf{C}\times\mathbf{A}) = \mathbf{C}\cdot(\mathbf{A}\times\mathbf{B}) = $ volume of a parallelepiped having $\mathbf{A}$, $\mathbf{B}$, and $\mathbf{C}$ as edges, or the negative of this volume according as $\mathbf{A}$, $\mathbf{B}$ and $\mathbf{C}$ do or do not form a right-handed system. If $\mathbf{A} = A_1\mathbf{i} + A_2\mathbf{j} + A_3\mathbf{k}$, $\mathbf{B} = B_1\mathbf{i} + B_2\mathbf{j} + B_3\mathbf{k}$ and $\mathbf{C} = C_1\mathbf{i} + C_2\mathbf{j} + C_3\mathbf{k}$, then

$$\mathbf{A}\cdot(\mathbf{B}\times\mathbf{C}) = \begin{vmatrix} A_1 & A_2 & A_3 \\ B_1 & B_2 & B_3 \\ C_1 & C_2 & C_3 \end{vmatrix} \tag{6}$$

3. $\mathbf{A}\times(\mathbf{B}\times\mathbf{C}) \neq (\mathbf{A}\times\mathbf{B})\times\mathbf{C}$ (Associative Law for Cross Products Fails)
4. $\mathbf{A}\times(\mathbf{B}\times\mathbf{C}) = (\mathbf{A}\cdot\mathbf{C})\mathbf{B} - (\mathbf{A}\cdot\mathbf{B})\mathbf{C}$
    $(\mathbf{A}\times\mathbf{B})\times\mathbf{C} = (\mathbf{A}\cdot\mathbf{C})\mathbf{B} - (\mathbf{B}\cdot\mathbf{C})\mathbf{A}$

The product $\mathbf{A}\cdot(\mathbf{B}\times\mathbf{C})$ is sometimes called the *scalar triple product* or *box product* and may be denoted by $[\mathbf{ABC}]$. The product $\mathbf{A}\times(\mathbf{B}\times\mathbf{C})$ is called the *vector triple product*.

In $\mathbf{A} \cdot (\mathbf{B} \times \mathbf{C})$ parentheses are sometimes omitted and we write $\mathbf{A} \cdot \mathbf{B} \times \mathbf{C}$. However, parentheses must be used in $\mathbf{A} \times (\mathbf{B} \times \mathbf{C})$ (see Problem 29). Note that $\mathbf{A} \cdot (\mathbf{B} \times \mathbf{C}) = (\mathbf{A} \times \mathbf{B}) \cdot \mathbf{C}$. This is often expressed by stating that in a scalar triple product the dot and the cross can be interchanged without affecting the result (see Problem 26).

AXIOMATIC APPROACH to VECTOR ANALYSIS

From the above remarks it is seen that a vector $\mathbf{r} = x\mathbf{i} + y\mathbf{j} + z\mathbf{k}$ is determined when its 3 components (x, y, z) relative to some coordinate system are known. In adopting an axiomatic approach it is thus quite natural for us to make the following

Definition.   A 3 dimensional vector is an *ordered triplet* of real numbers (A_1, A_2, A_3).

With this as starting point we can define equality, vector addition and subtraction, etc. Thus if $\mathbf{A} = (A_1, A_2, A_3)$ and $\mathbf{B} = (B_1, B_2, B_3)$, we define

1. $\mathbf{A} = \mathbf{B}$ if and only if $A_1 = B_1,\ A_2 = B_2,\ A_3 = B_3$
2. $\mathbf{A} + \mathbf{B} = (A_1 + B_1,\ A_2 + B_2,\ A_3 + B_3)$
3. $\mathbf{A} - \mathbf{B} = (A_1 - B_1,\ A_2 - B_2,\ A_3 - B_3)$
4. $\mathbf{0} = (0, 0, 0)$
5. $m\mathbf{A} = m(A_1, A_2, A_3) = (mA_1, mA_2, mA_3)$
6. $\mathbf{A} \cdot \mathbf{B} = A_1 B_1 + A_2 B_2 + A_3 B_3$
7. Length or magnitude of $\mathbf{A} = |\mathbf{A}| = \sqrt{\mathbf{A} \cdot \mathbf{A}} = \sqrt{A_1^2 + A_2^2 + A_3^2}$

From these we obtain other properties of vectors, such as $\mathbf{A} + \mathbf{B} = \mathbf{B} + \mathbf{A}$, $\mathbf{A} + (\mathbf{B} + \mathbf{C}) = (\mathbf{A} + \mathbf{B}) + \mathbf{C}$, $\mathbf{A} \cdot (\mathbf{B} + \mathbf{C}) = \mathbf{A} \cdot \mathbf{B} + \mathbf{A} \cdot \mathbf{C}$, etc. By defining the unit vectors

$$\mathbf{i} = (1, 0, 0), \quad \mathbf{j} = (0, 1, 0), \quad \mathbf{k} = (0, 0, 1) \tag{7}$$

we can then show that

$$\mathbf{A} = A_1 \mathbf{i} + A_2 \mathbf{j} + A_3 \mathbf{k} \tag{8}$$

In like manner we can define $\mathbf{A} \times \mathbf{B} = (A_2 B_3 - A_3 B_2,\ A_3 B_1 - A_1 B_3,\ A_1 B_2 - A_2 B_1)$.

After this axiomatic approach has been developed we can interpret the results geometrically or physically. For example, we can show that $\mathbf{A} \cdot \mathbf{B} = AB \cos \theta$, $|\mathbf{A} \times \mathbf{B}| = AB \sin \theta$, etc.

In the above we have considered three dimensional vectors. It is easy to extend the idea of a vector to higher dimensions. For example, a *four dimensional vector* is defined as an *ordered quadruple* (A_1, A_2, A_3, A_4).

VECTOR FUNCTIONS

If corresponding to each value of a scalar u we associate a vector $\mathbf{A}$, then $\mathbf{A}$ is called a *function* of u denoted by $\mathbf{A}(u)$. In three dimensions we can write $\mathbf{A}(u) = A_1(u)\mathbf{i} + A_2(u)\mathbf{j} + A_3(u)\mathbf{k}$.

The function concept is easily extended. Thus if to each point (x, y, z) there corresponds a vector $\mathbf{A}$, then $\mathbf{A}$ is a function of (x, y, z), indicated by $\mathbf{A}(x, y, z) = A_1(x, y, z)\mathbf{i} + A_2(x, y, z)\mathbf{j} + A_3(x, y, z)\mathbf{k}$.

We sometimes say that a vector function $\mathbf{A}(x, y, z)$ defines a *vector field* since it associates a vector with each point of a region. Similarly $\phi(x, y, z)$ defines a *scalar field* since it associates a scalar with each point of a region.

LIMITS, CONTINUITY and DERIVATIVES of VECTOR FUNCTIONS

Limits, continuity and derivatives of vector functions follow rules similar to those for scalar functions already considered. The following statements show the analogy which exists.

1. The vector function $\mathbf{A}(u)$ is said to be *continuous* at u_0 if given any positive number ϵ, we can find some positive number δ such that $|\mathbf{A}(u) - \mathbf{A}(u_0)| < \epsilon$ whenever $|u - u_0| < \delta$. This is equivalent to the statement $\lim\limits_{u \to u_0} \mathbf{A}(u) = \mathbf{A}(u_0)$.

2. The derivative of $\mathbf{A}(u)$ is defined as

$$\frac{d\mathbf{A}}{du} = \lim_{\Delta u \to 0} \frac{\mathbf{A}(u + \Delta u) - \mathbf{A}(u)}{\Delta u}$$

provided this limit exists. In case $\mathbf{A}(u) = A_1(u)\mathbf{i} + A_2(u)\mathbf{j} + A_3(u)\mathbf{k}$; then

$$\frac{d\mathbf{A}}{du} = \frac{dA_1}{du}\mathbf{i} + \frac{dA_2}{du}\mathbf{j} + \frac{dA_3}{du}\mathbf{k}$$

Higher derivatives such as $d^2\mathbf{A}/du^2$, etc., can be similarly defined.

3. If $\mathbf{A}(x,y,z) = A_1(x,y,z)\mathbf{i} + A_2(x,y,z)\mathbf{j} + A_3(x,y,z)\mathbf{k}$, then

$$d\mathbf{A} = \frac{\partial \mathbf{A}}{\partial x}dx + \frac{\partial \mathbf{A}}{\partial y}dy + \frac{\partial \mathbf{A}}{\partial z}dz$$

is the *differential* of $\mathbf{A}$.

4. Derivatives of products obey rules similar to those for scalar functions. However, when cross products are involved the order may be important. Some examples are:

(a) $\dfrac{d}{du}(\phi \mathbf{A}) = \phi\dfrac{d\mathbf{A}}{du} + \dfrac{d\phi}{du}\mathbf{A}$,

(b) $\dfrac{\partial}{\partial y}(\mathbf{A} \cdot \mathbf{B}) = \mathbf{A} \cdot \dfrac{\partial \mathbf{B}}{\partial y} + \dfrac{\partial \mathbf{A}}{\partial y} \cdot \mathbf{B}$,

(c) $\dfrac{\partial}{\partial z}(\mathbf{A} \times \mathbf{B}) = \mathbf{A} \times \dfrac{\partial \mathbf{B}}{\partial z} + \dfrac{\partial \mathbf{A}}{\partial z} \times \mathbf{B}$

GEOMETRIC INTERPRETATION of a VECTOR DERIVATIVE

If $\mathbf{r}$ is the vector joining the origin O of a coordinate system and the point (x,y,z), then specification of the vector function $\mathbf{r}(u)$ defines x, y and z as functions of u. As u changes, the terminal point of $\mathbf{r}$ describes a *space curve* (see Fig. 7-8) having parametric equations $x = x(u)$, $y = y(u)$, $z = z(u)$. If the parameter u is the arc length s measured from some fixed point on the curve, then

$$\frac{d\mathbf{r}}{ds} = \mathbf{T} \qquad (9)$$

is a unit vector in the direction of the tangent to the curve and is called the *unit tangent vector*. If u is the time t, then

$$\frac{d\mathbf{r}}{dt} = \mathbf{v} \qquad (10)$$

is the *velocity* with which the terminal point

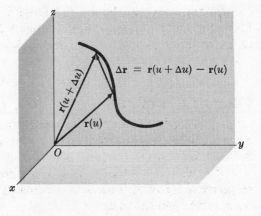

Fig. 7-8

of $\mathbf{r}$ describes the curve. We have

$$\mathbf{v} = \frac{d\mathbf{r}}{dt} = \frac{d\mathbf{r}}{ds}\frac{ds}{dt} = \frac{ds}{dt}\mathbf{T} = v\mathbf{T} \tag{11}$$

from which we see that the magnitude of $\mathbf{v}$ is $v = ds/dt$. Similarly,

$$\frac{d^2\mathbf{r}}{dt^2} = \mathbf{a} \tag{12}$$

is the *acceleration* with which the terminal point of $\mathbf{r}$ describes the curve. These concepts have important applications in *mechanics* and *differential geometry*.

GRADIENT, DIVERGENCE and CURL

Consider the vector operator ∇ (*del*) defined by

$$\nabla \equiv \mathbf{i}\frac{\partial}{\partial x} + \mathbf{j}\frac{\partial}{\partial y} + \mathbf{k}\frac{\partial}{\partial z} \tag{13}$$

Then if $\phi(x,y,z)$ and $\mathbf{A}(x,y,z)$ have continuous first partial derivatives in a region (a condition which is in many cases stronger than necessary), we can define the following.

1. Gradient. The *gradient* of ϕ is defined by

$$\text{grad } \phi = \nabla\phi = \left(\mathbf{i}\frac{\partial}{\partial x} + \mathbf{j}\frac{\partial}{\partial y} + \mathbf{k}\frac{\partial}{\partial z}\right)\phi = \mathbf{i}\frac{\partial\phi}{\partial x} + \mathbf{j}\frac{\partial\phi}{\partial y} + \mathbf{k}\frac{\partial\phi}{\partial z} \tag{14}$$

$$= \frac{\partial\phi}{\partial x}\mathbf{i} + \frac{\partial\phi}{\partial y}\mathbf{j} + \frac{\partial\phi}{\partial z}\mathbf{k}$$

An interesting interpretation is that if $\phi(x,y,z) = c$ is the equation of a surface, then $\nabla\phi$ is a normal to this surface (see Problem 36).

2. Divergence. The *divergence* of $\mathbf{A}$ is defined by

$$\text{div } \mathbf{A} = \nabla\cdot\mathbf{A} = \left(\mathbf{i}\frac{\partial}{\partial x} + \mathbf{j}\frac{\partial}{\partial y} + \mathbf{k}\frac{\partial}{\partial z}\right)\cdot(A_1\mathbf{i} + A_2\mathbf{j} + A_3\mathbf{k}) \tag{15}$$

$$= \frac{\partial A_1}{\partial x} + \frac{\partial A_2}{\partial y} + \frac{\partial A_3}{\partial z}$$

3. Curl. The *curl* of $\mathbf{A}$ is defined by

$$\text{curl } \mathbf{A} = \nabla\times\mathbf{A} = \left(\mathbf{i}\frac{\partial}{\partial x} + \mathbf{j}\frac{\partial}{\partial y} + \mathbf{k}\frac{\partial}{\partial z}\right)\times(A_1\mathbf{i} + A_2\mathbf{j} + A_3\mathbf{k}) \tag{16}$$

$$= \begin{vmatrix} \mathbf{i} & \mathbf{j} & \mathbf{k} \\ \dfrac{\partial}{\partial x} & \dfrac{\partial}{\partial y} & \dfrac{\partial}{\partial z} \\ A_1 & A_2 & A_3 \end{vmatrix}$$

$$= \mathbf{i}\begin{vmatrix} \dfrac{\partial}{\partial y} & \dfrac{\partial}{\partial z} \\ A_2 & A_3 \end{vmatrix} - \mathbf{j}\begin{vmatrix} \dfrac{\partial}{\partial x} & \dfrac{\partial}{\partial z} \\ A_1 & A_2 \end{vmatrix} + \mathbf{k}\begin{vmatrix} \dfrac{\partial}{\partial x} & \dfrac{\partial}{\partial y} \\ A_1 & A_2 \end{vmatrix}$$

$$= \left(\frac{\partial A_3}{\partial y} - \frac{\partial A_2}{\partial z}\right)\mathbf{i} + \left(\frac{\partial A_1}{\partial z} - \frac{\partial A_3}{\partial x}\right)\mathbf{j} + \left(\frac{\partial A_2}{\partial x} - \frac{\partial A_1}{\partial y}\right)\mathbf{k}$$

Note that in the expansion of the determinant, the operators $\partial/\partial x, \partial/\partial y, \partial/\partial z$ must precede A_1, A_2, A_3.

FORMULAS INVOLVING ∇

If the partial derivatives of $\mathbf{A}$, $\mathbf{B}$, U and V are assumed to exist, then

1. $\nabla(U+V) = \nabla U + \nabla V$ or $\operatorname{grad}(U+V) = \operatorname{grad} u + \operatorname{grad} V$

2. $\nabla \cdot (\mathbf{A}+\mathbf{B}) = \nabla \cdot \mathbf{A} + \nabla \cdot \mathbf{B}$ or $\operatorname{div}(\mathbf{A}+\mathbf{B}) = \operatorname{div} \mathbf{A} + \operatorname{div} \mathbf{B}$

3. $\nabla \times (\mathbf{A}+\mathbf{B}) = \nabla \times \mathbf{A} + \nabla \times \mathbf{B}$ or $\operatorname{curl}(\mathbf{A}+\mathbf{B}) = \operatorname{curl} \mathbf{A} + \operatorname{curl} \mathbf{B}$

4. $\nabla \cdot (U\mathbf{A}) = (\nabla U) \cdot \mathbf{A} + U(\nabla \cdot \mathbf{A})$

5. $\nabla \times (U\mathbf{A}) = (\nabla U) \times \mathbf{A} + U(\nabla \times \mathbf{A})$

6. $\nabla \cdot (\mathbf{A} \times \mathbf{B}) = \mathbf{B} \cdot (\nabla \times \mathbf{A}) - \mathbf{A} \cdot (\nabla \times \mathbf{B})$

7. $\nabla \times (\mathbf{A} \times \mathbf{B}) = (\mathbf{B} \cdot \nabla)\mathbf{A} - \mathbf{B}(\nabla \cdot \mathbf{A}) - (\mathbf{A} \cdot \nabla)\mathbf{B} + \mathbf{A}(\nabla \cdot \mathbf{B})$

8. $\nabla(\mathbf{A} \cdot \mathbf{B}) = (\mathbf{B} \cdot \nabla)\mathbf{A} + (\mathbf{A} \cdot \nabla)\mathbf{B} + \mathbf{B} \times (\nabla \times \mathbf{A}) + \mathbf{A} \times (\nabla \times \mathbf{B})$

9. $\nabla \cdot (\nabla U) \equiv \nabla^2 U \equiv \dfrac{\partial^2 U}{\partial x^2} + \dfrac{\partial^2 U}{\partial y^2} + \dfrac{\partial^2 U}{\partial z^2}$ is called the *Laplacian* of U

    and $\nabla^2 \equiv \dfrac{\partial^2}{\partial x^2} + \dfrac{\partial^2}{\partial y^2} + \dfrac{\partial^2}{\partial z^2}$ is called the *Laplacian operator*.

10. $\nabla \times (\nabla U) = 0$.   The curl of the gradient of U is zero.

11. $\nabla \cdot (\nabla \times \mathbf{A}) = 0$.   The divergence of the curl of $\mathbf{A}$ is zero.

12. $\nabla \times (\nabla \times \mathbf{A}) = \nabla(\nabla \cdot \mathbf{A}) - \nabla^2\mathbf{A}$

VECTOR INTERPRETATION of JACOBIANS.
ORTHOGONAL CURVILINEAR COORDINATES.

The transformation equations

$$x = f(u_1, u_2, u_3), \quad y = g(u_1, u_2, u_3), \quad z = h(u_1, u_2, u_3) \tag{17}$$

[where we assume that f, g, h are continuous, have continuous partial derivatives and have a single-valued inverse] establish a one to one correspondence between points in an xyz and $u_1u_2u_3$ rectangular coordinate system.   In vector notation the transformation (*17*) can be written

$$\mathbf{r} = x\mathbf{i} + y\mathbf{j} + z\mathbf{k} = f(u_1, u_2, u_3)\mathbf{i} + g(u_1, u_2, u_3)\mathbf{j} + h(u_1, u_2, u_3)\mathbf{k} \tag{18}$$

A point P in Fig. 7-9 can then be defined not only by *rectangular coordinates* (x, y, z) but by coordinates (u_1, u_2, u_3) as well.   We call (u_1, u_2, u_3) the *curvilinear coordinates* of the point.

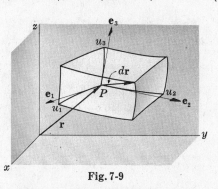

Fig. 7-9

If u_2 and u_3 are constant, then as u_1 varies, $\mathbf{r}$ describes a curve which we call the u_1 *coordinate curve*.   Similarly we define the u_2 and u_3 coordinate curves through P.

From (*18*), we have

$$d\mathbf{r} = \frac{\partial \mathbf{r}}{\partial u_1} du_1 + \frac{\partial \mathbf{r}}{\partial u_2} du_2 + \frac{\partial \mathbf{r}}{\partial u_3} du_3 \quad (19)$$

The vector $\partial \mathbf{r}/\partial u_1$ is tangent to the u_1 coordinate curve at P.   If $\mathbf{e}_1$ is a unit vector at P in this direction, we can write $\partial \mathbf{r}/\partial u_1 = h_1\mathbf{e}_1$ where $h_1 = |\partial \mathbf{r}/\partial u_1|$.   Similarly we can write $\partial \mathbf{r}/\partial u_2 = h_2\mathbf{e}_2$ and $\partial \mathbf{r}/\partial u_3 = h_3\mathbf{e}_3$, where $h_2 = |\partial \mathbf{r}/\partial u_2|$ and $h_3 = |\partial \mathbf{r}/\partial u_3|$ respectively.   Then (*19*) can be written

$$d\mathbf{r} = h_1 du_1 \mathbf{e}_1 + h_2 du_2 \mathbf{e}_2 + h_3 du_3 \mathbf{e}_3 \tag{20}$$

The quantities h_1, h_2, h_3 are sometimes called *scale factors*.

If e_1, e_2, e_3 are mutually perpendicular at any point P, the curvilinear coordinates are called *orthogonal*. In such case the element of arc length ds is given by

$$ds^2 = d\mathbf{r} \cdot d\mathbf{r} = h_1^2\, du_1^2 + h_2^2\, du_2^2 + h_3^2\, du_3^2 \qquad (21)$$

and corresponds to the square of the length of the diagonal in the above parallelepiped.

Also, in the case of orthogonal coordinates the volume of the parallelepiped is given by

$$dV = |(h_1\, du_1\, e_1) \cdot (h_2\, du_2\, e_2) \times (h_3\, du_3\, e_3)| = h_1 h_2 h_3\, du_1\, du_2\, du_3 \qquad (22)$$

which can be written as

$$dV = \left| \frac{\partial \mathbf{r}}{\partial u_1} \cdot \frac{\partial \mathbf{r}}{\partial u_2} \times \frac{\partial \mathbf{r}}{\partial u_3} \right| du_1\, du_2\, du_3 = \left| \frac{\partial(x, y, z)}{\partial(u_1, u_2, u_3)} \right| du_1\, du_2\, du_3 \qquad (23)$$

where $\partial(x, y, z)/\partial(u_1, u_2, u_3)$ is the *Jacobian* of the transformation.

It is clear that when the Jacobian vanishes there is no parallelepiped and explains geometrically the significance of the vanishing of a Jacobian as treated in Chapter 6.

GRADIENT, DIVERGENCE, CURL and LAPLACIAN in ORTHOGONAL CURVILINEAR COORDINATES

If Φ is a scalar function and $\mathbf{A} = A_1 e_1 + A_2 e_2 + A_3 e_3$ a vector function of orthogonal curvilinear coordinates u_1, u_2, u_3, we have the following results.

1. $\nabla \Phi = \text{grad } \Phi = \dfrac{1}{h_1}\dfrac{\partial \Phi}{\partial u_1} e_1 + \dfrac{1}{h_2}\dfrac{\partial \Phi}{\partial u_2} e_2 + \dfrac{1}{h_3}\dfrac{\partial \Phi}{\partial u_3} e_3$

2. $\nabla \cdot \mathbf{A} = \text{div } \mathbf{A} = \dfrac{1}{h_1 h_2 h_3}\left[\dfrac{\partial}{\partial u_1}(h_2 h_3 A_1) + \dfrac{\partial}{\partial u_2}(h_3 h_1 A_2) + \dfrac{\partial}{\partial u_3}(h_1 h_2 A_3) \right]$

3. $\nabla \times \mathbf{A} = \text{curl } \mathbf{A} = \dfrac{1}{h_1 h_2 h_3} \begin{vmatrix} h_1 e_1 & h_2 e_2 & h_3 e_3 \\ \dfrac{\partial}{\partial u_1} & \dfrac{\partial}{\partial u_2} & \dfrac{\partial}{\partial u_3} \\ h_1 A_1 & h_2 A_2 & h_3 A_3 \end{vmatrix}$

4. $\nabla^2 \Phi = \text{Laplacian of } \Phi = \dfrac{1}{h_1 h_2 h_3}\left[\dfrac{\partial}{\partial u_1}\left(\dfrac{h_2 h_3}{h_1}\dfrac{\partial \Phi}{\partial u_1}\right) + \dfrac{\partial}{\partial u_2}\left(\dfrac{h_3 h_1}{h_2}\dfrac{\partial \Phi}{\partial u_2}\right) \right.$
$$\left. + \dfrac{\partial}{\partial u_3}\left(\dfrac{h_1 h_2}{h_3}\dfrac{\partial \Phi}{\partial u_3}\right) \right]$$

These reduce to the usual expressions in rectangular coordinates if we replace (u_1, u_2, u_3) by (x, y, z), in which case e_1, e_2 and e_3 are replaced by $\mathbf{i}, \mathbf{j}$ and $\mathbf{k}$ and $h_1 = h_2 = h_3 = 1$.

SPECIAL CURVILINEAR COORDINATES

1. Cylindrical Coordinates (ρ, ϕ, z). See Fig. 7-10.

Transformation equations:

$$x = \rho \cos\phi, \quad y = \rho \sin\phi, \quad z = z$$

where $\rho \geqq 0,\ 0 \leqq \phi < 2\pi,\ -\infty < z < \infty$.

Scale factors: $h_1 = 1,\ h_2 = \rho,\ h_3 = 1$

Element of arc length: $ds^2 = d\rho^2 + \rho^2\, d\phi^2 + dz^2$

Jacobian: $\dfrac{\partial(x, y, z)}{\partial(\rho, \phi, z)} = \rho$

Element of volume: $dV = \rho\, d\rho\, d\phi\, dz$

Laplacian:

$$\nabla^2 U = \frac{1}{\rho}\frac{\partial}{\partial \rho}\left(\rho \frac{\partial U}{\partial \rho}\right) + \frac{1}{\rho^2}\frac{\partial^2 U}{\partial \phi^2} + \frac{\partial^2 U}{\partial z^2} = \frac{\partial^2 U}{\partial \rho^2} + \frac{1}{\rho}\frac{\partial U}{\partial \rho} + \frac{1}{\rho^2}\frac{\partial^2 U}{\partial \phi^2} + \frac{\partial^2 U}{\partial z^2}$$

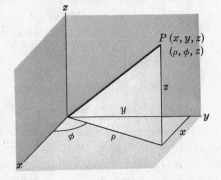

Fig. 7-10

Note that corresponding results can be obtained for polar coordinates in the plane by omitting z dependence. In such case for example, $ds^2 = d\rho^2 + \rho^2\,d\phi^2$, while the element of volume is replaced by the element of area, $dA = \rho\,d\rho\,d\phi$.

2. **Spherical Coordinates** (r, θ, ϕ). See Fig. 7-11.

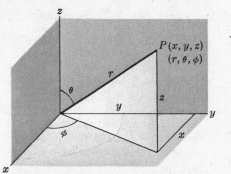

Transformation equations:

$x = r \sin\theta \cos\phi$, $y = r \sin\theta \sin\phi$, $z = r \cos\theta$

where $r \geqq 0$, $0 \leqq \theta \leqq \pi$, $0 \leqq \phi < 2\pi$.

Scale factors: $h_1 = 1$, $h_2 = r$, $h_3 = r \sin\theta$

Element of arc length:

$$ds^2 = dr^2 + r^2\,d\theta^2 + r^2 \sin^2\theta\,d\phi^2$$

Jacobian: $\dfrac{\partial(x, y, z)}{\partial(r, \theta, \phi)} = r^2 \sin\theta$

Fig. 7-11

Element of volume: $dV = r^2 \sin\theta\,dr\,d\theta\,d\phi$

Laplacian: $\nabla^2 U = \dfrac{1}{r^2}\dfrac{\partial}{\partial r}\left(r^2 \dfrac{\partial U}{\partial r}\right) + \dfrac{1}{r^2 \sin\theta}\dfrac{\partial}{\partial \theta}\left(\sin\theta \dfrac{\partial U}{\partial \theta}\right) + \dfrac{1}{r^2 \sin^2\theta}\dfrac{\partial^2 U}{\partial \phi^2}.$

Other types of coordinate systems are possible.

Solved Problems

VECTOR ALGEBRA

1. Show that addition of vectors is commutative, i.e. $\mathbf{A} + \mathbf{B} = \mathbf{B} + \mathbf{A}$. See Fig. 7-12 below.

$$\mathbf{OP} + \mathbf{PQ} = \mathbf{OQ} \quad \text{or} \quad \mathbf{A} + \mathbf{B} = \mathbf{C},$$

and

$$\mathbf{OR} + \mathbf{RQ} = \mathbf{OQ} \quad \text{or} \quad \mathbf{B} + \mathbf{A} = \mathbf{C}.$$

Then $\mathbf{A} + \mathbf{B} = \mathbf{B} + \mathbf{A}$.

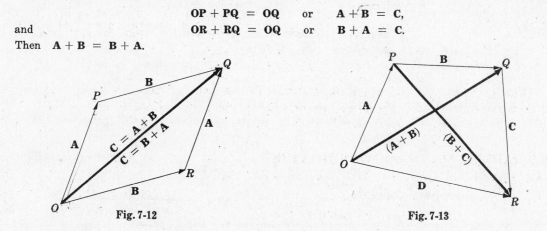

Fig. 7-12   **Fig. 7-13**

2. Show that the addition of vectors is associative, i.e. $\mathbf{A} + (\mathbf{B} + \mathbf{C}) = (\mathbf{A} + \mathbf{B}) + \mathbf{C}$. See Fig. 7-13 above.

$$\mathbf{OP} + \mathbf{PQ} = \mathbf{OQ} = (\mathbf{A} + \mathbf{B}) \quad \text{and} \quad \mathbf{PQ} + \mathbf{QR} = \mathbf{PR} = (\mathbf{B} + \mathbf{C})$$

Since

$$\mathbf{OP} + \mathbf{PR} = \mathbf{OR} = \mathbf{D}, \quad \text{i.e. } \mathbf{A} + (\mathbf{B} + \mathbf{C}) = \mathbf{D}$$

$$\mathbf{OQ} + \mathbf{QR} = \mathbf{OR} = \mathbf{D}, \quad \text{i.e. } (\mathbf{A} + \mathbf{B}) + \mathbf{C} = \mathbf{D}$$

we have $\mathbf{A} + (\mathbf{B} + \mathbf{C}) = (\mathbf{A} + \mathbf{B}) + \mathbf{C}$.

Extensions of the results of Problems 1 and 2 show that the order of addition of any number of vectors is immaterial.

3. An automobile travels 3 miles due north, then 5 miles northeast as shown in Fig. 7-14. Represent these displacements graphically and determine the resultant displacement (*a*) graphically, (*b*) analytically.

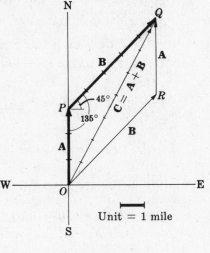

Vector **OP** or **A** represents displacement of 3 mi due north.

Vector **PQ** or **B** represents displacement of 5 mi northeast.

Vector **OQ** or **C** represents the resultant displacement or sum of vectors **A** and **B**, i.e. **C** = **A** + **B**. This is the *triangle law* of vector addition.

The resultant vector **OQ** can also be obtained by constructing the diagonal of the parallelogram *OPQR* having vectors **OP** = **A** and **OR** (equal to vector **PQ** or **B**) as sides. This is the *parallelogram law* of vector addition.

(*a*) *Graphical Determination of Resultant.* Lay off the 1 mile unit on vector **OQ** to find the magnitude 7.4 mi (approximately). Angle *EOQ* = 61.5°, using a protractor. Then vector **OQ** has magnitude 7.4 mi and direction 61.5° north of east.

Unit = 1 mile

Fig. 7-14

(*b*) *Analytical Determination of Resultant.* From triangle *OPQ*, denoting the magnitudes of **A**, **B**, **C** by A, B, C, we have by the law of cosines

$$C^2 = A^2 + B^2 - 2AB \cos \angle OPQ = 3^2 + 5^2 - 2(3)(5) \cos 135° = 34 + 15\sqrt{2} = 55.21$$

and $C = 7.43$ (approximately).

By the law of sines, $\dfrac{A}{\sin \angle OQP} = \dfrac{C}{\sin \angle OPQ}$. Then

$$\sin \angle OQP = \frac{A \sin \angle OPQ}{C} = \frac{3(0.707)}{7.43} = 0.2855 \quad \text{and} \quad \angle OQP \doteq 16°35'$$

Thus vector **OQ** has magnitude 7.43 mi and direction $(45° + 16°35') = 61°35'$ north of east.

4. Prove that if **a** and **b** are non-collinear, then $x\mathbf{a} + y\mathbf{b} = 0$ implies $x = y = 0$.

Suppose $x \neq 0$. Then $x\mathbf{a} + y\mathbf{b} = 0$ implies $x\mathbf{a} = -y\mathbf{b}$ or $\mathbf{a} = -(y/x)\mathbf{b}$, i.e. **a** and **b** must be parallel to the same line (collinear), contrary to hypothesis. Thus $x = 0$; then $y\mathbf{b} = 0$, from which $y = 0$.

5. If $x_1\mathbf{a} + y_1\mathbf{b} = x_2\mathbf{a} + y_2\mathbf{b}$, where **a** and **b** are non-collinear, then $x_1 = x_2$ and $y_1 = y_2$.

$x_1\mathbf{a} + y_1\mathbf{b} = x_2\mathbf{a} + y_2\mathbf{b}$ can be written

$$x_1\mathbf{a} + y_1\mathbf{b} - (x_2\mathbf{a} + y_2\mathbf{b}) = 0 \quad \text{or} \quad (x_1 - x_2)\mathbf{a} + (y_1 - y_2)\mathbf{b} = 0$$

Hence by Problem 4, $x_1 - x_2 = 0$, $y_1 - y_2 = 0$ or $x_1 = x_2$, $y_1 = y_2$.

Extensions are possible (see Problem 49).

6. Prove that the diagonals of a parallelogram bisect each other.

Let *ABCD* be the given parallelogram with diagonals intersecting at *P* as shown in Fig. 7-15.

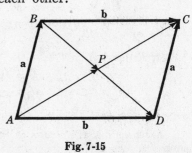

Since **BD** + **a** = **b**, **BD** = **b** − **a**. Then **BP** = x(**b** − **a**).

Since **AC** = **a** + **b**, **AP** = y(**a** + **b**).

But **AB** = **AP** + **PB** = **AP** − **BP**, i.e. **a** = y(**a** + **b**) − x(**b** − **a**) = $(x + y)$**a** + $(y - x)$**b**.

Since **a** and **b** are non-collinear we have by Problem 5, $x + y = 1$ and $y - x = 0$, i.e. $x = y = \frac{1}{2}$ and *P* is the midpoint of both diagonals.

Fig. 7-15

7. Prove that the line joining the midpoints of two sides of a triangle is parallel to the third side and has half its length.

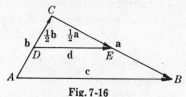

Fig. 7-16

From Fig. 7-16, $\mathbf{AC} + \mathbf{CB} = \mathbf{AB}$ or $\mathbf{b} + \mathbf{a} = \mathbf{c}$.

Let $\mathbf{DE} = \mathbf{d}$ be the line joining the midpoints of sides AC and CB. Then

$$\mathbf{d} = \mathbf{DC} + \mathbf{CE} = \tfrac{1}{2}\mathbf{b} + \tfrac{1}{2}\mathbf{a} = \tfrac{1}{2}(\mathbf{b} + \mathbf{a}) = \tfrac{1}{2}\mathbf{c}$$

Thus $\mathbf{d}$ is parallel to $\mathbf{c}$ and has half its length.

8. Prove that the magnitude A of the vector $\mathbf{A} = A_1\mathbf{i} + A_2\mathbf{j} + A_3\mathbf{k}$ is $A = \sqrt{A_1^2 + A_2^2 + A_3^2}$. See Fig. 7-17.

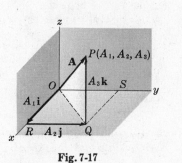

Fig. 7-17

By the Pythagorean theorem,

$$(\overline{OP})^2 = (\overline{OQ})^2 + (\overline{QP})^2$$

where $\overline{OP}$ denotes the magnitude of vector $\mathbf{OP}$, etc. Similarly, $(\overline{OQ})^2 = (\overline{OR})^2 + (\overline{RQ})^2$.

Then $(\overline{OP})^2 = (\overline{OR})^2 + (\overline{RQ})^2 + (\overline{QP})^2$ or $A^2 = A_1^2 + A_2^2 + A_3^2$, i.e. $A = \sqrt{A_1^2 + A_2^2 + A_3^2}$.

9. Determine the vector having initial point $P(x_1, y_1, z_1)$ and terminal point $Q(x_2, y_2, z_2)$ and find its magnitude. See Fig. 7-18.

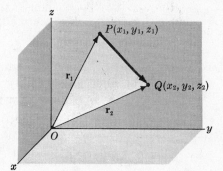

Fig. 7-18

The position vector of P is $\mathbf{r}_1 = x_1\mathbf{i} + y_1\mathbf{j} + z_1\mathbf{k}$.
The position vector of Q is $\mathbf{r}_2 = x_2\mathbf{i} + y_2\mathbf{j} + z_2\mathbf{k}$.

$\mathbf{r}_1 + \mathbf{PQ} = \mathbf{r}_2$ or

$$\begin{aligned}\mathbf{PQ} = \mathbf{r}_2 - \mathbf{r}_1 &= (x_2\mathbf{i} + y_2\mathbf{j} + z_2\mathbf{k}) - (x_1\mathbf{i} + y_1\mathbf{j} + z_1\mathbf{k}) \\ &= (x_2 - x_1)\mathbf{i} + (y_2 - y_1)\mathbf{j} + (z_2 - z_1)\mathbf{k}.\end{aligned}$$

Magnitude of $\mathbf{PQ} = \overline{PQ}$
$$= \sqrt{(x_2 - x_1)^2 + (y_2 - y_1)^2 + (z_2 - z_1)^2}.$$

Note that this is the distance between points P and Q.

The DOT or SCALAR PRODUCT

10. Prove that the projection of $\mathbf{A}$ on $\mathbf{B}$ is equal to $\mathbf{A} \cdot \mathbf{b}$, where $\mathbf{b}$ is a unit vector in the direction of $\mathbf{B}$.

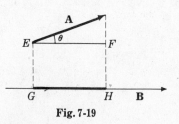

Fig. 7-19

Through the initial and terminal points of $\mathbf{A}$ pass planes perpendicular to $\mathbf{B}$ at G and H respectively as in the adjacent Fig. 7-19; then

Projection of $\mathbf{A}$ on $\mathbf{B} = \overline{GH} = \overline{EF} = A\cos\theta = \mathbf{A} \cdot \mathbf{b}$

unit vector.

11. Prove $\mathbf{A} \cdot (\mathbf{B} + \mathbf{C}) = \mathbf{A} \cdot \mathbf{B} + \mathbf{A} \cdot \mathbf{C}$.

Let $\mathbf{a}$ be a unit vector in the direction of $\mathbf{A}$; then

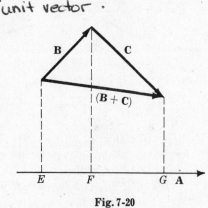

Fig. 7-20

Projection of $(\mathbf{B} + \mathbf{C})$ on $\mathbf{A}$ = projection of $\mathbf{B}$ on $\mathbf{A}$ + projection of $\mathbf{C}$ on $\mathbf{A}$

$$(\mathbf{B} + \mathbf{C}) \cdot \mathbf{a} = \mathbf{B} \cdot \mathbf{a} + \mathbf{C} \cdot \mathbf{a}$$

Multiplying by A,

$$(\mathbf{B} + \mathbf{C}) \cdot A\mathbf{a} = \mathbf{B} \cdot A\mathbf{a} + \mathbf{C} \cdot A\mathbf{a}$$

and $(\mathbf{B} + \mathbf{C}) \cdot \mathbf{A} = \mathbf{B} \cdot \mathbf{A} + \mathbf{C} \cdot \mathbf{A}$

Then by the commutative law for dot products,

$$\mathbf{A} \cdot (\mathbf{B} + \mathbf{C}) = \mathbf{A} \cdot \mathbf{B} + \mathbf{A} \cdot \mathbf{C}$$

and the distributive law is valid.

12. Prove that $(\mathbf{A}+\mathbf{B})\cdot(\mathbf{C}+\mathbf{D}) = \mathbf{A}\cdot\mathbf{C} + \mathbf{A}\cdot\mathbf{D} + \mathbf{B}\cdot\mathbf{C} + \mathbf{B}\cdot\mathbf{D}.$

By Problem 11, $(\mathbf{A}+\mathbf{B})\cdot(\mathbf{C}+\mathbf{D}) = \mathbf{A}\cdot(\mathbf{C}+\mathbf{D}) + \mathbf{B}\cdot(\mathbf{C}+\mathbf{D}) = \mathbf{A}\cdot\mathbf{C} + \mathbf{A}\cdot\mathbf{D} + \mathbf{B}\cdot\mathbf{C} + \mathbf{B}\cdot\mathbf{D}.$

The ordinary laws of algebra are valid for dot products where the operations are defined.

13. Evaluate each of the following.

(a) $\mathbf{i}\cdot\mathbf{i} = |\mathbf{i}|\,|\mathbf{i}|\,\cos 0° = (1)(1)(1) = 1$

(b) $\mathbf{i}\cdot\mathbf{k} = |\mathbf{i}|\,|\mathbf{k}|\,\cos 90° = (1)(1)(0) = 0$

(c) $\mathbf{k}\cdot\mathbf{j} = |\mathbf{k}|\,|\mathbf{j}|\,\cos 90° = (1)(1)(0) = 0$

(d) $\mathbf{j}\cdot(2\mathbf{i}-3\mathbf{j}+\mathbf{k}) = 2\mathbf{j}\cdot\mathbf{i} - 3\mathbf{j}\cdot\mathbf{j} + \mathbf{j}\cdot\mathbf{k} = 0 - 3 + 0 = -3$

(e) $(2\mathbf{i}-\mathbf{j})\cdot(3\mathbf{i}+\mathbf{k}) = 2\mathbf{i}\cdot(3\mathbf{i}+\mathbf{k}) - \mathbf{j}\cdot(3\mathbf{i}+\mathbf{k}) = 6\mathbf{i}\cdot\mathbf{i} + 2\mathbf{i}\cdot\mathbf{k} - 3\mathbf{j}\cdot\mathbf{i} - \mathbf{j}\cdot\mathbf{k} = 6 + 0 - 0 - 0 = 6$

14. If $\mathbf{A} = A_1\mathbf{i}+A_2\mathbf{j}+A_3\mathbf{k}$ and $\mathbf{B} = B_1\mathbf{i}+B_2\mathbf{j}+B_3\mathbf{k}$, prove that $\mathbf{A}\cdot\mathbf{B} = A_1B_1 + A_2B_2 + A_3B_3.$

$$\begin{aligned}
\mathbf{A}\cdot\mathbf{B} &= (A_1\mathbf{i}+A_2\mathbf{j}+A_3\mathbf{k})\cdot(B_1\mathbf{i}+B_2\mathbf{j}+B_3\mathbf{k}) \\
&= A_1\mathbf{i}\cdot(B_1\mathbf{i}+B_2\mathbf{j}+B_3\mathbf{k}) + A_2\mathbf{j}\cdot(B_1\mathbf{i}+B_2\mathbf{j}+B_3\mathbf{k}) + A_3\mathbf{k}\cdot(B_1\mathbf{i}+B_2\mathbf{j}+B_3\mathbf{k}) \\
&= A_1B_1\mathbf{i}\cdot\mathbf{i} + A_1B_2\mathbf{i}\cdot\mathbf{j} + A_1B_3\mathbf{i}\cdot\mathbf{k} + A_2B_1\mathbf{j}\cdot\mathbf{i} + A_2B_2\mathbf{j}\cdot\mathbf{j} + A_2B_3\mathbf{j}\cdot\mathbf{k} \\
&\quad + A_3B_1\mathbf{k}\cdot\mathbf{i} + A_3B_2\mathbf{k}\cdot\mathbf{j} + A_3B_3\mathbf{k}\cdot\mathbf{k} \\
&= A_1B_1 + A_2B_2 + A_3B_3
\end{aligned}$$

since $\mathbf{i}\cdot\mathbf{i} = \mathbf{j}\cdot\mathbf{j} = \mathbf{k}\cdot\mathbf{k} = 1$ and all other dot products are zero.

15. If $\mathbf{A} = A_1\mathbf{i}+A_2\mathbf{j}+A_3\mathbf{k}$, show that $A = \sqrt{\mathbf{A}\cdot\mathbf{A}} = \sqrt{A_1^2 + A_2^2 + A_3^2}.$

$\mathbf{A}\cdot\mathbf{A} = (A)(A)\cos 0° = A^2.$ Then $A = \sqrt{\mathbf{A}\cdot\mathbf{A}}.$

Also, $\mathbf{A}\cdot\mathbf{A} = (A_1\mathbf{i}+A_2\mathbf{j}+A_3\mathbf{k})\cdot(A_1\mathbf{i}+A_2\mathbf{j}+A_3\mathbf{k})$

$= (A_1)(A_1) + (A_2)(A_2) + (A_3)(A_3) = A_1^2 + A_2^2 + A_3^2$

by Problem 14, taking $\mathbf{B}=\mathbf{A}.$

Then $A = \sqrt{\mathbf{A}\cdot\mathbf{A}} = \sqrt{A_1^2 + A_2^2 + A_3^2}$ is the magnitude of $\mathbf{A}$.   Sometimes $\mathbf{A}\cdot\mathbf{A}$ is written $\mathbf{A}^2$.

The CROSS or VECTOR PRODUCT

16. Prove $\mathbf{A}\times\mathbf{B} = -\mathbf{B}\times\mathbf{A}.$

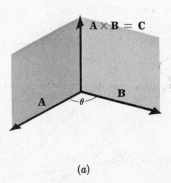

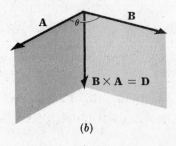

(a)   (b)

Fig. 7-21

$\mathbf{A}\times\mathbf{B} = \mathbf{C}$ has magnitude $AB\sin\theta$ and direction such that $\mathbf{A}$, $\mathbf{B}$ and $\mathbf{C}$ form a right-handed system [Fig. 7-21(a) above].

$\mathbf{B}\times\mathbf{A} = \mathbf{D}$ has magnitude $BA\sin\theta$ and direction such that $\mathbf{B}$, $\mathbf{A}$ and $\mathbf{D}$ form a right-handed system [Fig. 7-21(b) above].

Then $\mathbf{D}$ has the same magnitude as $\mathbf{C}$ but is opposite in direction, i.e. $\mathbf{C}=-\mathbf{D}$ or $\mathbf{A}\times\mathbf{B} = -\mathbf{B}\times\mathbf{A}.$

The commutative law for cross products is not valid.

17. Prove that $\mathbf{A} \times (\mathbf{B} + \mathbf{C}) = \mathbf{A} \times \mathbf{B} + \mathbf{A} \times \mathbf{C}$ for the case where $\mathbf{A}$ is perpendicular to $\mathbf{B}$ and also to $\mathbf{C}$.

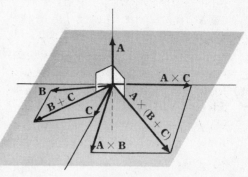

Since $\mathbf{A}$ is perpendicular to $\mathbf{B}$, $\mathbf{A} \times \mathbf{B}$ is a vector perpendicular to the plane of $\mathbf{A}$ and $\mathbf{B}$ and having magnitude $AB \sin 90° = AB$ or magnitude of $A\mathbf{B}$. This is equivalent to multiplying vector $\mathbf{B}$ by A and rotating the resultant vector through $90°$ to the position shown in Fig. 7-22.

Similarly, $\mathbf{A} \times \mathbf{C}$ is the vector obtained by multiplying $\mathbf{C}$ by A and rotating the resultant vector through $90°$ to the position shown.

In like manner, $\mathbf{A} \times (\mathbf{B} + \mathbf{C})$ is the vector obtained by multiplying $\mathbf{B} + \mathbf{C}$ by A and rotating the resultant vector through $90°$ to the position shown.

Since $\mathbf{A} \times (\mathbf{B} + \mathbf{C})$ is the diagonal of the parallelogram with $\mathbf{A} \times \mathbf{B}$ and $\mathbf{A} \times \mathbf{C}$ as sides, we have $\mathbf{A} \times (\mathbf{B} + \mathbf{C}) = \mathbf{A} \times \mathbf{B} + \mathbf{A} \times \mathbf{C}$.

Fig. 7-22

18. Prove that $\mathbf{A} \times (\mathbf{B} + \mathbf{C}) = \mathbf{A} \times \mathbf{B} + \mathbf{A} \times \mathbf{C}$ in the general case where $\mathbf{A}$, $\mathbf{B}$ and $\mathbf{C}$ are non-coplanar. See Fig. 7-23.

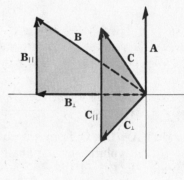

Resolve $\mathbf{B}$ into two component vectors, one perpendicular to $\mathbf{A}$ and the other parallel to $\mathbf{A}$, and denote them by $\mathbf{B}_\perp$ and $\mathbf{B}_{||}$ respectively. Then $\mathbf{B} = \mathbf{B}_\perp + \mathbf{B}_{||}$.

If θ is the angle between $\mathbf{A}$ and $\mathbf{B}$, then $B_\perp = B \sin \theta$. Thus the magnitude of $\mathbf{A} \times \mathbf{B}_\perp$ is $AB \sin \theta$, the same as the magnitude of $\mathbf{A} \times \mathbf{B}$. Also, the direction of $\mathbf{A} \times \mathbf{B}_\perp$ is the same as the direction of $\mathbf{A} \times \mathbf{B}$. Hence $\mathbf{A} \times \mathbf{B}_\perp = \mathbf{A} \times \mathbf{B}$.

Similarly if $\mathbf{C}$ is resolved into two component vectors $\mathbf{C}_{||}$ and $\mathbf{C}_\perp$, parallel and perpendicular respectively to $\mathbf{A}$, then $\mathbf{A} \times \mathbf{C}_\perp = \mathbf{A} \times \mathbf{C}$.

Fig. 7-23

Also, since $\mathbf{B} + \mathbf{C} = \mathbf{B}_\perp + \mathbf{B}_{||} + \mathbf{C}_\perp + \mathbf{C}_{||} = (\mathbf{B}_\perp + \mathbf{C}_\perp) + (\mathbf{B}_{||} + \mathbf{C}_{||})$ it follows that

$$\mathbf{A} \times (\mathbf{B}_\perp + \mathbf{C}_\perp) = \mathbf{A} \times (\mathbf{B} + \mathbf{C})$$

Now $\mathbf{B}_\perp$ and $\mathbf{C}_\perp$ are vectors perpendicular to $\mathbf{A}$ and so by Problem 17,

$$\mathbf{A} \times (\mathbf{B}_\perp + \mathbf{C}_\perp) = \mathbf{A} \times \mathbf{B}_\perp + \mathbf{A} \times \mathbf{C}_\perp$$

Then
$$\mathbf{A} \times (\mathbf{B} + \mathbf{C}) = \mathbf{A} \times \mathbf{B} + \mathbf{A} \times \mathbf{C}$$

and the distributive law holds. Multiplying by -1, using Problem 16, this becomes $(\mathbf{B} + \mathbf{C}) \times \mathbf{A} = \mathbf{B} \times \mathbf{A} + \mathbf{C} \times \mathbf{A}$. Note that the order of factors in cross products is important. The usual laws of algebra apply only if proper order is maintained.

19. If $\mathbf{A} = A_1\mathbf{i} + A_2\mathbf{j} + A_3\mathbf{k}$ and $\mathbf{B} = B_1\mathbf{i} + B_2\mathbf{j} + B_3\mathbf{k}$, prove that $\mathbf{A} \times \mathbf{B} = \begin{vmatrix} \mathbf{i} & \mathbf{j} & \mathbf{k} \\ A_1 & A_2 & A_3 \\ B_1 & B_2 & B_3 \end{vmatrix}$.

$$\begin{aligned}
\mathbf{A} \times \mathbf{B} &= (A_1\mathbf{i} + A_2\mathbf{j} + A_3\mathbf{k}) \times (B_1\mathbf{i} + B_2\mathbf{j} + B_3\mathbf{k}) \\
&= A_1\mathbf{i} \times (B_1\mathbf{i} + B_2\mathbf{j} + B_3\mathbf{k}) + A_2\mathbf{j} \times (B_1\mathbf{i} + B_2\mathbf{j} + B_3\mathbf{k}) + A_3\mathbf{k} \times (B_1\mathbf{i} + B_2\mathbf{j} + B_3\mathbf{k}) \\
&= A_1B_1\mathbf{i} \times \mathbf{i} + A_1B_2\mathbf{i} \times \mathbf{j} + A_1B_3\mathbf{i} \times \mathbf{k} + A_2B_1\mathbf{j} \times \mathbf{i} + A_2B_2\mathbf{j} \times \mathbf{j} + A_2B_3\mathbf{j} \times \mathbf{k} \\
&\quad + A_3B_1\mathbf{k} \times \mathbf{i} + A_3B_2\mathbf{k} \times \mathbf{j} + A_3B_3\mathbf{k} \times \mathbf{k} \\
&= (A_2B_3 - A_3B_2)\mathbf{i} + (A_3B_1 - A_1B_3)\mathbf{j} + (A_1B_2 - A_2B_1)\mathbf{k} = \begin{vmatrix} \mathbf{i} & \mathbf{j} & \mathbf{k} \\ A_1 & A_2 & A_3 \\ B_1 & B_2 & B_3 \end{vmatrix}.
\end{aligned}$$

20. If $\mathbf{A} = 3\mathbf{i} - \mathbf{j} + 2\mathbf{k}$ and $\mathbf{B} = 2\mathbf{i} + 3\mathbf{j} - \mathbf{k}$, find $\mathbf{A} \times \mathbf{B}$.

$$\mathbf{A} \times \mathbf{B} = \begin{vmatrix} \mathbf{i} & \mathbf{j} & \mathbf{k} \\ 3 & -1 & 2 \\ 2 & 3 & -1 \end{vmatrix} = \mathbf{i} \begin{vmatrix} -1 & 2 \\ 3 & -1 \end{vmatrix} - \mathbf{j} \begin{vmatrix} 3 & 2 \\ 2 & -1 \end{vmatrix} + \mathbf{k} \begin{vmatrix} 3 & -1 \\ 2 & 3 \end{vmatrix}$$
$$= -5\mathbf{i} + 7\mathbf{j} + 11\mathbf{k}$$

21. Prove that the area of a parallelogram with sides $\mathbf{A}$ and $\mathbf{B}$ is $|\mathbf{A} \times \mathbf{B}|$.

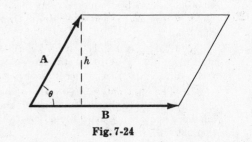

$$\text{Area of parallelogram} = h\,|\mathbf{B}|$$
$$= |\mathbf{A}|\,\sin\theta\,|\mathbf{B}|$$
$$= |\mathbf{A} \times \mathbf{B}|.$$

Note that the area of the triangle with sides $\mathbf{A}$ and $\mathbf{B} = \frac{1}{2}|\mathbf{A} \times \mathbf{B}|$.

Fig. 7-24

22. Find the area of the triangle with vertices at $P(2,3,5)$, $Q(4,2,-1)$, $R(3,6,4)$.

$$\mathbf{PQ} = (4-2)\mathbf{i} + (2-3)\mathbf{j} + (-1-5)\mathbf{k} = 2\mathbf{i} - \mathbf{j} - 6\mathbf{k}$$
$$\mathbf{PR} = (3-2)\mathbf{i} + (6-3)\mathbf{j} + (4-5)\mathbf{k} = \mathbf{i} + 3\mathbf{j} - \mathbf{k}$$

$$\text{Area of triangle} = \frac{1}{2}|\mathbf{PQ} \times \mathbf{PR}| = \frac{1}{2}|(2\mathbf{i} - \mathbf{j} - 6\mathbf{k}) \times (\mathbf{i} + 3\mathbf{j} - \mathbf{k})|$$
$$= \frac{1}{2}\left|\begin{vmatrix} \mathbf{i} & \mathbf{j} & \mathbf{k} \\ 2 & -1 & -6 \\ 1 & 3 & -1 \end{vmatrix}\right| = \frac{1}{2}|19\mathbf{i} - 4\mathbf{j} + 7\mathbf{k}|$$
$$= \frac{1}{2}\sqrt{(19)^2 + (-4)^2 + (7)^2} = \frac{1}{2}\sqrt{426}$$

TRIPLE PRODUCTS

23. Show that $\mathbf{A} \cdot (\mathbf{B} \times \mathbf{C})$ is in absolute value equal to the volume of a parallelepiped with sides $\mathbf{A}$, $\mathbf{B}$ and $\mathbf{C}$.

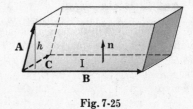

Let $\mathbf{n}$ be a unit normal to parallelogram I, having the direction of $\mathbf{B} \times \mathbf{C}$, and let h be the height of the terminal point of $\mathbf{A}$ above the parallelogram I.

Fig. 7-25

$$\text{Volume of parallelepiped} = (\text{height } h)(\text{area of parallelogram } I)$$
$$= (\mathbf{A} \cdot \mathbf{n})(|\mathbf{B} \times \mathbf{C}|)$$
$$= \mathbf{A} \cdot \{|\mathbf{B} \times \mathbf{C}|\,\mathbf{n}\} = \mathbf{A} \cdot (\mathbf{B} \times \mathbf{C})$$

If $\mathbf{A}$, $\mathbf{B}$ and $\mathbf{C}$ do not form a right-handed system, $\mathbf{A} \cdot \mathbf{n} < 0$ and the volume $= |\mathbf{A} \cdot (\mathbf{B} \times \mathbf{C})|$.

24. If $\mathbf{A} = A_1\mathbf{i} + A_2\mathbf{j} + A_3\mathbf{k}$, $\mathbf{B} = B_1\mathbf{i} + B_2\mathbf{j} + B_3\mathbf{k}$, $\mathbf{C} = C_1\mathbf{i} + C_2\mathbf{j} + C_3\mathbf{k}$ show that

$$\mathbf{A} \cdot (\mathbf{B} \times \mathbf{C}) = \begin{vmatrix} A_1 & A_2 & A_3 \\ B_1 & B_2 & B_3 \\ C_1 & C_2 & C_3 \end{vmatrix}$$

$$\mathbf{A} \cdot (\mathbf{B} \times \mathbf{C}) = \mathbf{A} \cdot \begin{vmatrix} \mathbf{i} & \mathbf{j} & \mathbf{k} \\ B_1 & B_2 & B_3 \\ C_1 & C_2 & C_3 \end{vmatrix}$$
$$= (A_1\mathbf{i} + A_2\mathbf{j} + A_3\mathbf{k}) \cdot [(B_2C_3 - B_3C_2)\mathbf{i} + (B_3C_1 - B_1C_3)\mathbf{j} + (B_1C_2 - B_2C_1)\mathbf{k}]$$
$$= A_1(B_2C_3 - B_3C_2) + A_2(B_3C_1 - B_1C_3) + A_3(B_1C_2 - B_2C_1) = \begin{vmatrix} A_1 & A_2 & A_3 \\ B_1 & B_2 & B_3 \\ C_1 & C_2 & C_3 \end{vmatrix}.$$

25. Find the volume of a parallelepiped with sides $\mathbf{A} = 3\mathbf{i} - \mathbf{j}$, $\mathbf{B} = \mathbf{j} + 2\mathbf{k}$, $\mathbf{C} = \mathbf{i} + 5\mathbf{j} + 4\mathbf{k}$.

By Problems 23 and 24, volume of parallelepiped $= |\mathbf{A} \cdot (\mathbf{B} \times \mathbf{C})| = \left| \begin{vmatrix} 3 & -1 & 0 \\ 0 & 1 & 2 \\ 1 & 5 & 4 \end{vmatrix} \right|$

$$= |-20| = 20.$$

26. Prove that $\mathbf{A} \cdot (\mathbf{B} \times \mathbf{C}) = (\mathbf{A} \times \mathbf{B}) \cdot \mathbf{C}$, i.e. the dot and cross can be interchanged.

By Problem 24: $\mathbf{A} \cdot (\mathbf{B} \times \mathbf{C}) = \begin{vmatrix} A_1 & A_2 & A_3 \\ B_1 & B_2 & B_3 \\ C_1 & C_2 & C_3 \end{vmatrix}$, $(\mathbf{A} \times \mathbf{B}) \cdot \mathbf{C} = \mathbf{C} \cdot (\mathbf{A} \times \mathbf{B}) = \begin{vmatrix} C_1 & C_2 & C_3 \\ A_1 & A_2 & A_3 \\ B_1 & B_2 & B_3 \end{vmatrix}$

Since the two determinants are equal, the required result follows.

27. Let $\mathbf{r}_1 = x_1\mathbf{i} + y_1\mathbf{j} + z_1\mathbf{k}$, $\mathbf{r}_2 = x_2\mathbf{i} + y_2\mathbf{j} + z_2\mathbf{k}$ and $\mathbf{r}_3 = x_3\mathbf{i} + y_3\mathbf{j} + z_3\mathbf{k}$ be the position vectors of points $P_1(x_1, y_1, z_1)$, $P_2(x_2, y_2, z_2)$ and $P_3(x_3, y_3, z_3)$. Find an equation for the plane passing through P_1, P_2 and P_3. See Fig. 7-26.

We assume that P_1, P_2 and P_3 do not lie in the same straight line; hence they determine a plane.

Let $\mathbf{r} = x\mathbf{i} + y\mathbf{j} + z\mathbf{k}$ denote the position vector of any point $P(x, y, z)$ in the plane. Consider vectors $\mathbf{P_1P_2} = \mathbf{r}_2 - \mathbf{r}_1$, $\mathbf{P_1P_3} = \mathbf{r}_3 - \mathbf{r}_1$ and $\mathbf{P_1P} = \mathbf{r} - \mathbf{r}_1$ which all lie in the plane. Then

$$\mathbf{P_1P} \cdot \mathbf{P_1P_2} \times \mathbf{P_1P_3} = 0$$

or $(\mathbf{r} - \mathbf{r}_1) \cdot (\mathbf{r}_2 - \mathbf{r}_1) \times (\mathbf{r}_3 - \mathbf{r}_1) = 0$

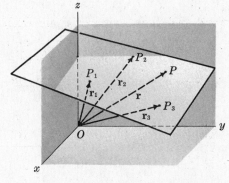

Fig. 7-26

In terms of rectangular coordinates this becomes

$[(x - x_1)\mathbf{i} + (y - y_1)\mathbf{j} + (z - z_1)\mathbf{k}] \cdot [(x_2 - x_1)\mathbf{i} + (y_2 - y_1)\mathbf{j} + (z_2 - z_1)\mathbf{k}]$
$\qquad \times [(x_3 - x_1)\mathbf{i} + (y_3 - y_1)\mathbf{j} + (z_3 - z_1)\mathbf{k}] = 0$

or, using Problem 24, $\begin{vmatrix} x - x_1 & y - y_1 & z - z_1 \\ x_2 - x_1 & y_2 - y_1 & z_2 - z_1 \\ x_3 - x_1 & y_3 - y_1 & z_3 - z_1 \end{vmatrix} = 0$

28. Find an equation for the plane passing through the points $P_1(3, 1, -2)$, $P_2(-1, 2, 4)$, $P_3(2, -1, 1)$.

The position vectors of P_1, P_2, P_3 and any point $P(x, y, z)$ on the plane are respectively

$$\mathbf{r}_1 = 3\mathbf{i} + \mathbf{j} - 2\mathbf{k}, \quad \mathbf{r}_2 = -\mathbf{i} + 2\mathbf{j} + 4\mathbf{k}, \quad \mathbf{r}_3 = 2\mathbf{i} - \mathbf{j} + \mathbf{k}, \quad \mathbf{r} = x\mathbf{i} + y\mathbf{j} + z\mathbf{k}$$

Then $\mathbf{PP_1} = \mathbf{r} - \mathbf{r}_1$, $\mathbf{P_2P_1} = \mathbf{r}_2 - \mathbf{r}_1$, $\mathbf{P_3P_1} = \mathbf{r}_3 - \mathbf{r}_1$, all lie in the required plane and so the required equation is $(\mathbf{r} - \mathbf{r}_1) \cdot (\mathbf{r}_2 - \mathbf{r}_1) \times (\mathbf{r}_3 - \mathbf{r}_1) = 0$, i.e.,

$$\{(x - 3)\mathbf{i} + (y - 1)\mathbf{j} + (z + 2)\mathbf{k}\} \cdot \{-4\mathbf{i} + \mathbf{j} + 6\mathbf{k}\} \times \{-\mathbf{i} - 2\mathbf{j} + 3\mathbf{k}\} = 0$$

$$\{(x - 3)\mathbf{i} + (y - 1)\mathbf{j} + (z + 2)\mathbf{k}\} \cdot \{15\mathbf{i} + 6\mathbf{j} + 9\mathbf{k}\} = 0$$

$$15(x - 3) + 6(y - 1) + 9(z + 2) = 0 \quad \text{or} \quad 5x + 2y + 3z = 11$$

Another method: By Problem 27, the required equation is

$$\begin{vmatrix} x - 3 & y - 1 & z + 2 \\ -1 - 3 & 2 - 1 & 4 + 2 \\ 2 - 3 & -1 - 1 & 1 + 2 \end{vmatrix} = 0 \quad \text{or} \quad 5x + 2y + 3z = 11$$

29. If $\mathbf{A} = \mathbf{i} + \mathbf{j}$, $\mathbf{B} = 2\mathbf{i} - 3\mathbf{j} + \mathbf{k}$, $\mathbf{C} = 4\mathbf{j} - 3\mathbf{k}$, find (a) $(\mathbf{A} \times \mathbf{B}) \times \mathbf{C}$, (b) $\mathbf{A} \times (\mathbf{B} \times \mathbf{C})$.

(a) $\mathbf{A} \times \mathbf{B} = \begin{vmatrix} \mathbf{i} & \mathbf{j} & \mathbf{k} \\ 1 & 1 & 0 \\ 2 & -3 & 1 \end{vmatrix} = \mathbf{i} - \mathbf{j} - 5\mathbf{k}$.   Then $(\mathbf{A} \times \mathbf{B}) \times \mathbf{C} = \begin{vmatrix} \mathbf{i} & \mathbf{j} & \mathbf{k} \\ 1 & -1 & -5 \\ 0 & 4 & -3 \end{vmatrix} = 23\mathbf{i} + 3\mathbf{j} + 4\mathbf{k}$.

(b) $\mathbf{B} \times \mathbf{C} = \begin{vmatrix} \mathbf{i} & \mathbf{j} & \mathbf{k} \\ 2 & -3 & 1 \\ 0 & 4 & -3 \end{vmatrix} = 5\mathbf{i} + 6\mathbf{j} + 8\mathbf{k}$.   Then $\mathbf{A} \times (\mathbf{B} \times \mathbf{C}) = \begin{vmatrix} \mathbf{i} & \mathbf{j} & \mathbf{k} \\ 1 & 1 & 0 \\ 5 & 6 & 8 \end{vmatrix} = 8\mathbf{i} - 8\mathbf{j} + \mathbf{k}$.

It follows that, in general, $(\mathbf{A} \times \mathbf{B}) \times \mathbf{C} \neq \mathbf{A} \times (\mathbf{B} \times \mathbf{C})$.

DERIVATIVES

30. If $\mathbf{r} = (t^3 + 2t)\mathbf{i} - 3e^{-2t}\mathbf{j} + 2\sin 5t\,\mathbf{k}$,   find (a) $\dfrac{d\mathbf{r}}{dt}$, (b) $\left|\dfrac{d\mathbf{r}}{dt}\right|$, (c) $\dfrac{d^2\mathbf{r}}{dt^2}$, (d) $\left|\dfrac{d^2\mathbf{r}}{dt^2}\right|$ at $t = 0$ and give a possible physical significance.

(a) $\dfrac{d\mathbf{r}}{dt} = \dfrac{d}{dt}(t^3 + 2t)\mathbf{i} + \dfrac{d}{dt}(-3e^{-2t})\mathbf{j} + \dfrac{d}{dt}(2\sin 5t)\mathbf{k} = (3t^2 + 2)\mathbf{i} + 6e^{-2t}\mathbf{j} + 10\cos 5t\,\mathbf{k}$

At $t = 0$, $d\mathbf{r}/dt = 2\mathbf{i} + 6\mathbf{j} + 10\mathbf{k}$

(b) From (a), $|d\mathbf{r}/dt| = \sqrt{(2)^2 + (6)^2 + (10)^2} = \sqrt{140} = 2\sqrt{35}$ at $t = 0$.

(c) $\dfrac{d^2\mathbf{r}}{dt^2} = \dfrac{d}{dt}\left(\dfrac{d\mathbf{r}}{dt}\right) = \dfrac{d}{dt}\{(3t^2 + 2)\mathbf{i} + 6e^{-2t}\mathbf{j} + 10\cos 5t\,\mathbf{k}\} = 6t\mathbf{i} - 12e^{-2t}\mathbf{j} - 50\sin 5t\,\mathbf{k}$

At $t = 0$, $d^2\mathbf{r}/dt^2 = -12\mathbf{j}$.

(d) From (c), $|d^2\mathbf{r}/dt^2| = 12$ at $t = 0$.

If t represents time, these represent respectively the velocity, magnitude of the velocity, acceleration and magnitude of the acceleration at $t = 0$ of a particle moving along the space curve $x = t^3 + 2t$, $y = -3e^{-2t}$, $z = 2\sin 5t$.

31. Prove that $\dfrac{d}{du}(\mathbf{A} \cdot \mathbf{B}) = \mathbf{A} \cdot \dfrac{d\mathbf{B}}{du} + \dfrac{d\mathbf{A}}{du} \cdot \mathbf{B}$,   where $\mathbf{A}$ and $\mathbf{B}$ are differentiable functions of u.

Method 1: $\dfrac{d}{du}(\mathbf{A} \cdot \mathbf{B}) = \lim\limits_{\Delta u \to 0} \dfrac{(\mathbf{A} + \Delta\mathbf{A}) \cdot (\mathbf{B} + \Delta\mathbf{B}) - \mathbf{A} \cdot \mathbf{B}}{\Delta u}$

$= \lim\limits_{\Delta u \to 0} \dfrac{\mathbf{A} \cdot \Delta\mathbf{B} + \Delta\mathbf{A} \cdot \mathbf{B} + \Delta\mathbf{A} \cdot \Delta\mathbf{B}}{\Delta u}$

$= \lim\limits_{\Delta u \to 0} \left(\mathbf{A} \cdot \dfrac{\Delta\mathbf{B}}{\Delta u} + \dfrac{\Delta\mathbf{A}}{\Delta u} \cdot \mathbf{B} + \dfrac{\Delta\mathbf{A}}{\Delta u} \cdot \Delta\mathbf{B}\right) = \mathbf{A} \cdot \dfrac{d\mathbf{B}}{du} + \dfrac{d\mathbf{A}}{du} \cdot \mathbf{B}$

Method 2:   Let $\mathbf{A} = A_1\mathbf{i} + A_2\mathbf{j} + A_3\mathbf{k}$, $\mathbf{B} = B_1\mathbf{i} + B_2\mathbf{j} + B_3\mathbf{k}$.   Then

$\dfrac{d}{du}(\mathbf{A} \cdot \mathbf{B}) = \dfrac{d}{du}(A_1 B_1 + A_2 B_2 + A_3 B_3)$

$= \left(A_1 \dfrac{dB_1}{du} + A_2 \dfrac{dB_2}{du} + A_3 \dfrac{dB_3}{du}\right) + \left(\dfrac{dA_1}{du} B_1 + \dfrac{dA_2}{du} B_2 + \dfrac{dA_3}{du} B_3\right)$

$= \mathbf{A} \cdot \dfrac{d\mathbf{B}}{du} + \dfrac{d\mathbf{A}}{du} \cdot \mathbf{B}$

32. If $\phi(x, y, z) = x^2 yz$ and $\mathbf{A} = 3x^2 y\mathbf{i} + yz^2\mathbf{j} - xz\mathbf{k}$, find $\dfrac{\partial^2}{\partial y\,\partial z}(\phi\mathbf{A})$ at the point $(1, -2, -1)$.

$\phi\mathbf{A} = (x^2 yz)(3x^2 y\mathbf{i} + yz^2\mathbf{j} - xz\mathbf{k}) = 3x^4 y^2 z\mathbf{i} + x^2 y^2 z^3\mathbf{j} - x^3 yz^2\mathbf{k}$

$\dfrac{\partial}{\partial z}(\phi\mathbf{A}) = \dfrac{\partial}{\partial z}(3x^4 y^2 z\mathbf{i} + x^2 y^2 z^3\mathbf{j} - x^3 yz^2\mathbf{k}) = 3x^4 y^2\mathbf{i} + 3x^2 y^2 z^2\mathbf{j} - 2x^3 yz\mathbf{k}$

$$\frac{\partial^2}{\partial y\,\partial z}(\phi\mathbf{A}) \;=\; \frac{\partial}{\partial y}\,(3x^4y^2\mathbf{i} + 3x^2y^2z^2\mathbf{j} - 2x^3yz\mathbf{k}) \;=\; 6x^4y\mathbf{i} + 6x^2yz^2\mathbf{j} - 2x^3z\mathbf{k}$$

If $x=1$, $y=-2$, $z=-1$, this becomes $-12\mathbf{i} - 12\mathbf{j} + 2\mathbf{k}$.

33. If $\mathbf{A} = x^2 \sin y\,\mathbf{i} + z^2 \cos y\,\mathbf{j} - xy^2\mathbf{k}$, find $d\mathbf{A}$.

Method 1:

$$\frac{\partial\mathbf{A}}{\partial x} = 2x\sin y\mathbf{i} - y^2\mathbf{k}, \quad \frac{\partial\mathbf{A}}{\partial y} = x^2\cos y\,\mathbf{i} - z^2\sin y\,\mathbf{j} - 2xy\mathbf{k}, \quad \frac{\partial\mathbf{A}}{\partial z} = 2z\cos y\mathbf{j}$$

$$\begin{aligned}
d\mathbf{A} \;=\;& \frac{\partial\mathbf{A}}{\partial x}\,dx + \frac{\partial\mathbf{A}}{\partial y}\,dy + \frac{\partial\mathbf{A}}{\partial z}\,dz \\
\;=\;& (2x\sin y\,\mathbf{i} - y^2\mathbf{k})\,dx + (x^2\cos y\,\mathbf{i} - z^2\sin y\,\mathbf{j} - 2xy\mathbf{k})\,dy + (2z\cos y\,\mathbf{j})\,dz \\
\;=\;& (2x\sin y\,dx + x^2\cos y\,dy)\mathbf{i} + (2z\cos y\,dz - z^2\sin y\,dy)\mathbf{j} - (y^2\,dx + 2xy\,dy)\mathbf{k}
\end{aligned}$$

Method 2:

$$\begin{aligned}
d\mathbf{A} \;=\;& d(x^2\sin y)\mathbf{i} + d(z^2\cos y)\mathbf{j} - d(xy^2)\mathbf{k} \\
\;=\;& (2x\sin y\,dx + x^2\cos y\,dy)\mathbf{i} + (2z\cos y\,dz - z^2\sin y\,dy)\mathbf{j} - (y^2\,dx + 2xy\,dy)\mathbf{k}
\end{aligned}$$

GRADIENT, DIVERGENCE and CURL

34. If $\phi = x^2yz^3$ and $\mathbf{A} = xz\mathbf{i} - y^2\mathbf{j} + 2x^2y\mathbf{k}$, find (a) $\nabla\phi$, (b) $\nabla\cdot\mathbf{A}$, (c) $\nabla\times\mathbf{A}$, (d) div $(\phi\mathbf{A})$, (e) curl $(\phi\mathbf{A})$.

(a) $$\begin{aligned}
\nabla\phi \;=\;& \left(\mathbf{i}\frac{\partial}{\partial x} + \mathbf{j}\frac{\partial}{\partial y} + \mathbf{k}\frac{\partial}{\partial z}\right)\phi \;=\; \frac{\partial\phi}{\partial x}\mathbf{i} + \frac{\partial\phi}{\partial y}\mathbf{j} + \frac{\partial\phi}{\partial z}\mathbf{k} \;=\; \frac{\partial}{\partial x}\,(x^2yz^3)\mathbf{i} + \frac{\partial}{\partial y}\,(x^2yz^3)\mathbf{j} + \frac{\partial}{\partial z}\,(x^2yz^3)\mathbf{k} \\
\;=\;& 2xyz^3\mathbf{i} + x^2z^3\mathbf{j} + 3x^2yz^2\mathbf{k}
\end{aligned}$$

(b) $$\begin{aligned}
\nabla\cdot\mathbf{A} \;=\;& \left(\mathbf{i}\frac{\partial}{\partial x} + \mathbf{j}\frac{\partial}{\partial y} + \mathbf{k}\frac{\partial}{\partial z}\right)\cdot(xz\mathbf{i} - y^2\mathbf{j} + 2x^2y\mathbf{k}) \\
\;=\;& \frac{\partial}{\partial x}\,(xz) + \frac{\partial}{\partial y}\,(-y^2) + \frac{\partial}{\partial z}\,(2x^2y) \;=\; z - 2y
\end{aligned}$$

(c) $$\begin{aligned}
\nabla\times\mathbf{A} \;=\;& \left(\mathbf{i}\frac{\partial}{\partial x} + \mathbf{j}\frac{\partial}{\partial y} + \mathbf{k}\frac{\partial}{\partial z}\right)\times(xz\mathbf{i} - y^2\mathbf{j} + 2x^2y\mathbf{k}) \\
\;=\;& \begin{vmatrix} \mathbf{i} & \mathbf{j} & \mathbf{k} \\ \partial/\partial x & \partial/\partial y & \partial/\partial z \\ xz & -y^2 & 2x^2y \end{vmatrix} \\
\;=\;& \left(\frac{\partial}{\partial y}\,(2x^2y) - \frac{\partial}{\partial z}(-y^2)\right)\mathbf{i} + \left(\frac{\partial}{\partial z}\,(xz) - \frac{\partial}{\partial x}\,(2x^2y)\right)\mathbf{j} + \left(\frac{\partial}{\partial x}\,(-y^2) - \frac{\partial}{\partial y}\,(xz)\right)\mathbf{k} \\
\;=\;& 2x^2\mathbf{i} + (x - 4xy)\mathbf{j}
\end{aligned}$$

(d) $$\begin{aligned}
\text{div }(\phi\mathbf{A}) \;=\;& \nabla\cdot(\phi\mathbf{A}) \;=\; \nabla\cdot(x^3yz^4\mathbf{i} - x^2y^3z^3\mathbf{j} + 2x^4y^2z^3\mathbf{k}) \\
\;=\;& \frac{\partial}{\partial x}\,(x^3yz^4) + \frac{\partial}{\partial y}\,(-x^2y^3z^3) + \frac{\partial}{\partial z}\,(2x^4y^2z^3) \\
\;=\;& 3x^2yz^4 - 3x^2y^2z^3 + 6x^4y^2z^2
\end{aligned}$$

(e) $$\begin{aligned}
\text{curl }(\phi\mathbf{A}) \;=\;& \nabla\times(\phi\mathbf{A}) \;=\; \nabla\times(x^3yz^4\mathbf{i} - x^2y^3z^3\mathbf{j} + 2x^4y^2z^3\mathbf{k}) \\
\;=\;& \begin{vmatrix} \mathbf{i} & \mathbf{j} & \mathbf{k} \\ \partial/\partial x & \partial/\partial y & \partial/\partial z \\ x^3yz^4 & -x^2y^3z^3 & 2x^4y^2z^3 \end{vmatrix} \\
\;=\;& (4x^4yz^3 + 3x^2y^3z^2)\mathbf{i} + (4x^3yz^3 - 8x^3y^2z^3)\mathbf{j} - (2xy^3z^3 + x^3z^4)\mathbf{k}
\end{aligned}$$

35. Prove $\nabla \cdot (\phi \mathbf{A}) = (\nabla \phi) \cdot \mathbf{A} + \phi(\nabla \cdot \mathbf{A})$.

$$\begin{aligned}
\nabla \cdot (\phi \mathbf{A}) &= \nabla \cdot (\phi A_1 \mathbf{i} + \phi A_2 \mathbf{j} + \phi A_3 \mathbf{k}) \\
&= \frac{\partial}{\partial x}(\phi A_1) + \frac{\partial}{\partial y}(\phi A_2) + \frac{\partial}{\partial z}(\phi A_3) \\
&= \frac{\partial \phi}{\partial x} A_1 + \frac{\partial \phi}{\partial y} A_2 + \frac{\partial \phi}{\partial z} A_3 + \phi\left(\frac{\partial A_1}{\partial x} + \frac{\partial A_2}{\partial y} + \frac{\partial A_3}{\partial z}\right) \\
&= \left(\frac{\partial \phi}{\partial x}\mathbf{i} + \frac{\partial \phi}{\partial y}\mathbf{j} + \frac{\partial \phi}{\partial z}\mathbf{k}\right) \cdot (A_1 \mathbf{i} + A_2 \mathbf{j} + A_3 \mathbf{k}) \\
&\qquad + \phi\left(\frac{\partial}{\partial x}\mathbf{i} + \frac{\partial}{\partial y}\mathbf{j} + \frac{\partial}{\partial z}\mathbf{k}\right) \cdot (A_1 \mathbf{i} + A_2 \mathbf{j} + A_3 \mathbf{k}) \\
&= (\nabla \phi) \cdot \mathbf{A} + \phi(\nabla \cdot \mathbf{A})
\end{aligned}$$

36. Prove that $\nabla \phi$ is a vector perpendicular to the surface $\phi(x,y,z) = c$, where c is a constant.

Let $\mathbf{r} = x\mathbf{i} + y\mathbf{j} + z\mathbf{k}$ be the position vector to any point $P(x,y,z)$ on the surface.

Then $d\mathbf{r} = dx\,\mathbf{i} + dy\,\mathbf{j} + dz\,\mathbf{k}$ lies in the plane tangent to the surface at P. But

$$d\phi = \frac{\partial \phi}{\partial x}dx + \frac{\partial \phi}{\partial y}dy + \frac{\partial \phi}{\partial z}dz = 0 \quad \text{or} \quad \left(\frac{\partial \phi}{\partial x}\mathbf{i} + \frac{\partial \phi}{\partial y}\mathbf{j} + \frac{\partial \phi}{\partial z}\mathbf{k}\right) \cdot (dx\,\mathbf{i} + dy\,\mathbf{j} + dz\,\mathbf{k}) = 0$$

i.e. $\nabla \phi \cdot d\mathbf{r} = 0$ so that $\nabla \phi$ is perpendicular to $d\mathbf{r}$ and therefore to the surface.

37. Find a unit normal to the surface $2x^2 + 4yz - 5z^2 = -10$ at the point $P(3,-1,2)$.

By Problem 36, a vector normal to the surface is

$$\nabla(2x^2 + 4yz - 5z^2) = 4x\mathbf{i} + 4z\mathbf{j} + (4y - 10z)\mathbf{k} = 12\mathbf{i} + 8\mathbf{j} - 24\mathbf{k} \quad \text{at } (3,-1,2)$$

Then a unit normal to the surface at P is $\dfrac{12\mathbf{i} + 8\mathbf{j} - 24\mathbf{k}}{\sqrt{(12)^2 + (8)^2 + (-24)^2}} = \dfrac{3\mathbf{i} + 2\mathbf{j} - 6\mathbf{k}}{7}$.

Another unit normal to the surface at P is $-\dfrac{3\mathbf{i} + 2\mathbf{j} - 6\mathbf{k}}{7}$.

38. If $\phi = 2x^2 y - xz^3$, find (a) $\nabla \phi$ and (b) $\nabla^2 \phi$.

(a) $\nabla \phi = \dfrac{\partial \phi}{\partial x}\mathbf{i} + \dfrac{\partial \phi}{\partial y}\mathbf{j} + \dfrac{\partial \phi}{\partial z}\mathbf{k} = (4xy - z^3)\mathbf{i} + 2x^2 \mathbf{j} - 3xz^2 \mathbf{k}$

(b) $\nabla^2 \phi = $ Laplacian of $\phi = \nabla \cdot \nabla \phi = \dfrac{\partial}{\partial x}(4xy - z^3) + \dfrac{\partial}{\partial y}(2x^2) + \dfrac{\partial}{\partial z}(-3xz^2)$

$\qquad = 4y - 6xz$

Another method:

$$\nabla^2 \phi = \frac{\partial^2 \phi}{\partial x^2} + \frac{\partial^2 \phi}{\partial y^2} + \frac{\partial^2 \phi}{\partial z^2} = \frac{\partial^2}{\partial x^2}(2x^2 y - xz^3) + \frac{\partial^2}{\partial y^2}(2x^2 y - xz^3) + \frac{\partial^2}{\partial z^2}(2x^2 y - xz^3)$$

$$= 4y - 6xz$$

39. Prove div curl $\mathbf{A} = 0$.

$$\text{div curl } \mathbf{A} = \nabla \cdot (\nabla \times \mathbf{A}) = \nabla \cdot \begin{vmatrix} \mathbf{i} & \mathbf{j} & \mathbf{k} \\ \partial/\partial x & \partial/\partial y & \partial/\partial z \\ A_1 & A_2 & A_3 \end{vmatrix}$$

$$= \nabla \cdot \left[\left(\frac{\partial A_3}{\partial y} - \frac{\partial A_2}{\partial z}\right)\mathbf{i} + \left(\frac{\partial A_1}{\partial z} - \frac{\partial A_3}{\partial x}\right)\mathbf{j} + \left(\frac{\partial A_2}{\partial x} - \frac{\partial A_1}{\partial y}\right)\mathbf{k}\right]$$

$$= \frac{\partial}{\partial x}\left(\frac{\partial A_3}{\partial y} - \frac{\partial A_2}{\partial z}\right) + \frac{\partial}{\partial y}\left(\frac{\partial A_1}{\partial z} - \frac{\partial A_3}{\partial x}\right) + \frac{\partial}{\partial z}\left(\frac{\partial A_2}{\partial x} - \frac{\partial A_1}{\partial y}\right)$$

$$= \frac{\partial^2 A_3}{\partial x\,\partial y} - \frac{\partial^2 A_2}{\partial x\,\partial z} + \frac{\partial^2 A_1}{\partial y\,\partial z} - \frac{\partial^2 A_3}{\partial y\,\partial x} + \frac{\partial^2 A_2}{\partial z\,\partial x} - \frac{\partial^2 A_1}{\partial z\,\partial y}$$

$$= 0$$

assuming that $\mathbf{A}$ has continuous second partial derivatives so that the order of differentiation is immaterial.

JACOBIANS and CURVILINEAR COORDINATES

40. Find ds^2 in (a) cylindrical and (b) spherical coordinates and determine the scale factors.

(a) **Method 1:**

$$x = \rho \cos\phi, \qquad y = \rho \sin\phi, \qquad z = z$$

$$dx = -\rho \sin\phi \, d\phi + \cos\phi \, d\rho, \qquad dy = \rho \cos\phi \, d\phi + \sin\phi \, d\rho, \qquad dz = dz$$

Then
$$ds^2 = dx^2 + dy^2 + dz^2 = (-\rho \sin\phi \, d\phi + \cos\phi \, d\rho)^2$$
$$+ (\rho \cos\phi \, d\phi + \sin\phi \, d\rho)^2 + (dz)^2$$
$$= (d\rho)^2 + \rho^2 (d\phi)^2 + (dz)^2 = h_1^2 (d\rho)^2 + h_2^2 (d\phi)^2 + h_3^2 (dz)^2$$

and $h_1 = h_\rho = 1$, $h_2 = h_\phi = \rho$, $h_3 = h_z = 1$ are the scale factors.

Method 2: The position vector is $\mathbf{r} = \rho \cos\phi \, \mathbf{i} + \rho \sin\phi \, \mathbf{j} + z\mathbf{k}$. Then

$$dr = \frac{\partial \mathbf{r}}{\partial \rho} d\rho + \frac{\partial \mathbf{r}}{\partial \phi} d\phi + \frac{\partial \mathbf{r}}{\partial z} dz$$
$$= (\cos\phi \, \mathbf{i} + \sin\phi \, \mathbf{j}) \, d\rho + (-\rho \sin\phi \, \mathbf{i} + \rho \cos\phi \, \mathbf{j}) \, d\phi + \mathbf{k} \, dz$$
$$= (\cos\phi \, d\rho - \rho \sin\phi \, d\phi)\mathbf{i} + (\sin\phi \, d\rho + \rho \cos\phi \, d\phi)\mathbf{j} + \mathbf{k} \, dz$$

Thus
$$ds^2 = d\mathbf{r} \cdot d\mathbf{r} = (\cos\phi \, d\rho - \rho \sin\phi \, d\phi)^2 + (\sin\phi \, d\rho + \rho \cos\phi \, d\phi)^2 + (dz)^2$$
$$= (d\rho)^2 + \rho^2 (d\phi)^2 + (dz)^2$$

(b)
$$x = r \sin\theta \cos\phi, \qquad y = r \sin\theta \sin\phi, \qquad z = r \cos\theta$$

Then
$$dx = -r \sin\theta \sin\phi \, d\phi + r \cos\theta \cos\phi \, d\theta + \sin\theta \cos\phi \, dr$$
$$dy = r \sin\theta \cos\phi \, d\phi + r \cos\theta \sin\phi \, d\theta + \sin\theta \sin\phi \, dr$$
$$dz = -r \sin\theta \, d\theta + \cos\theta \, dr$$

and
$$(ds)^2 = (dx)^2 + (dy)^2 + (dz)^2 = (dr)^2 + r^2 (d\theta)^2 + r^2 \sin^2\theta \, (d\phi)^2$$

The scale factors are $h_1 = h_r = 1$, $h_2 = h_\theta = r$, $h_3 = h_\phi = r \sin\theta$.

41. Find the volume element dV in (a) cylindrical and (b) spherical coordinates and sketch.

The volume element in orthogonal curvilinear coordinates u_1, u_2, u_3 is

$$dV = h_1 h_2 h_3 \, du_1 \, du_2 \, du_3 = \left| \frac{\partial(x, y, z)}{\partial(u_1, u_2, u_3)} \right| du_1 \, du_2 \, du_3$$

(a) In cylindrical coordinates, $u_1 = \rho$, $u_2 = \phi$, $u_3 = z$, $h_1 = 1$, $h_2 = \rho$, $h_3 = 1$ [see Problem 40(a)]. **Then**

$$dV = (1)(\rho)(1) \, d\rho \, d\phi \, dz = \rho \, d\rho \, d\phi \, dz$$

This can also be observed directly from Fig. 7-27(a) below.

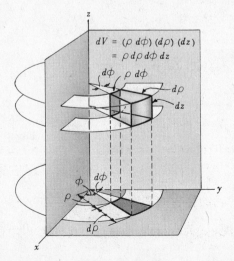

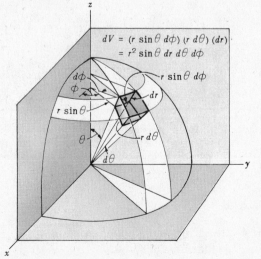

(a) Volume element in cylindrical coordinates.   (b) Volume element in spherical coordinates.

Fig. 7-27

(b) In spherical coordinates, $u_1 = r,\ u_2 = \theta,\ u_3 = \phi,\ h_1 = 1,\ h_2 = r,\ h_3 = r\sin\theta$ [see Problem 40(b)]. **Then**

$$dV \;=\; (1)(r)(r\sin\theta)\,dr\,d\theta\,d\phi \;=\; r^2\sin\theta\,dr\,d\theta\,d\phi$$

This can also be observed directly from Fig. 7-27(b) above.

42. Express in cylindrical coordinates: (a) grad Φ, (b) div $\mathbf{A}$, (c) $\nabla^2\Phi$.

Let $u_1 = \rho,\ u_2 = \phi,\ u_3 = z,\ h_1 = 1,\ h_2 = \rho,\ h_3 = 1$ [see Problem 40(a)] in the results 1, 2 and 4 on Page 142. Then

(a) $\operatorname{grad}\Phi \;=\; \nabla\Phi \;=\; \dfrac{1}{1}\dfrac{\partial\Phi}{\partial\rho}\mathbf{e}_1 + \dfrac{1}{\rho}\dfrac{\partial\Phi}{\partial\phi}\mathbf{e}_2 + \dfrac{1}{1}\dfrac{\partial\Phi}{\partial z}\mathbf{e}_3 \;=\; \dfrac{\partial\Phi}{\partial\rho}\mathbf{e}_1 + \dfrac{1}{\rho}\dfrac{\partial\Phi}{\partial\phi}\mathbf{e}_2 + \dfrac{\partial\Phi}{\partial z}\mathbf{e}_3$

where $\mathbf{e}_1, \mathbf{e}_2, \mathbf{e}_3$ are the unit vectors in the directions of increasing ρ, ϕ, z respectively.

(b) $\operatorname{div}\mathbf{A} \;=\; \nabla\cdot\mathbf{A} \;=\; \dfrac{1}{(1)(\rho)(1)}\left[\dfrac{\partial}{\partial\rho}\Big((\rho)(1)A_1\Big) + \dfrac{\partial}{\partial\phi}\Big((1)(1)A_2\Big) + \dfrac{\partial}{\partial z}\Big((1)(\rho)A_3\Big)\right]$

$\qquad\qquad\qquad\quad=\; \dfrac{1}{\rho}\left[\dfrac{\partial}{\partial\rho}(\rho A_1) + \dfrac{\partial A_2}{\partial\phi} + \dfrac{\partial A_3}{\partial z}\right]$

where $\mathbf{A} = A_1\mathbf{e}_1 + A_2\mathbf{e}_2 + A_3\mathbf{e}_3$.

(c) $\nabla^2\Phi \;=\; \dfrac{1}{(1)(\rho)(1)}\left[\dfrac{\partial}{\partial\rho}\left(\dfrac{(\rho)(1)}{(1)}\dfrac{\partial\Phi}{\partial\rho}\right) + \dfrac{\partial}{\partial\phi}\left(\dfrac{(1)(1)}{(\rho)}\dfrac{\partial\Phi}{\partial\phi}\right) + \dfrac{\partial}{\partial z}\left(\dfrac{(1)(\rho)}{(1)}\dfrac{\partial\Phi}{\partial z}\right)\right]$

$\qquad\qquad=\; \dfrac{1}{\rho}\dfrac{\partial}{\partial\rho}\left(\rho\dfrac{\partial\Phi}{\partial\rho}\right) + \dfrac{1}{\rho^2}\dfrac{\partial^2\Phi}{\partial\phi^2} + \dfrac{\partial^2\Phi}{\partial z^2}$

MISCELLANEOUS PROBLEMS

43. Prove that $\operatorname{grad} f(r) = \dfrac{f'(r)}{r}\mathbf{r}$, where $r = \sqrt{x^2 + y^2 + z^2}$ and $f'(r) = df/dr$ is assumed to exist.

$\operatorname{grad} f(r) \;=\; \nabla f(r) \;=\; \dfrac{\partial}{\partial x}f(r)\,\mathbf{i} + \dfrac{\partial}{\partial y}f(r)\,\mathbf{j} + \dfrac{\partial}{\partial z}f(r)\,\mathbf{k}$

$\qquad\qquad\quad=\; f'(r)\dfrac{\partial r}{\partial x}\mathbf{i} + f'(r)\dfrac{\partial r}{\partial y}\mathbf{j} + f'(r)\dfrac{\partial r}{\partial z}\mathbf{k}$

$\qquad\qquad\quad=\; f'(r)\dfrac{x}{r}\mathbf{i} + f'(r)\dfrac{y}{r}\mathbf{j} + f'(r)\dfrac{z}{r}\mathbf{k} \;=\; \dfrac{f'(r)}{r}(x\mathbf{i} + y\mathbf{j} + z\mathbf{k}) \;=\; \dfrac{f'(r)}{r}\mathbf{r}$

Another method: In orthogonal curvilinear coordinates u_1, u_2, u_3, we have

$$\nabla\Phi \;=\; \dfrac{1}{h_1}\dfrac{\partial\Phi}{\partial u_1}\mathbf{e}_1 + \dfrac{1}{h_2}\dfrac{\partial\Phi}{\partial u_2}\mathbf{e}_2 + \dfrac{1}{h_3}\dfrac{\partial\Phi}{\partial u_3}\mathbf{e}_3 \tag{1}$$

If in particular we use spherical coordinates, we have $u_1 = r,\ u_2 = \theta,\ u_3 = \phi$. Then letting $\Phi = f(r)$, a function of r alone, the last two terms on the right of (1) are zero. Hence we have, on observing that $\mathbf{e}_1 = \mathbf{r}/r$ and $h_1 = 1$, the result

$$\nabla f(r) \;=\; \dfrac{1}{1}\dfrac{\partial f(r)}{\partial r}\dfrac{\mathbf{r}}{r} \;=\; \dfrac{f'(r)}{r}\mathbf{r} \tag{2}$$

44. (a) Find the Laplacian of $\phi = f(r)$. (b) Prove that $\phi = 1/r$ is a solution of Laplace's equation $\nabla^2\phi = 0$.

(a) By Problem 43,

$$\nabla \phi \;=\; \nabla f(r) \;=\; \frac{f'(r)}{r}\mathbf{r}$$

By Problem 35, assuming that $f(r)$ has continuous second partial derivatives, we have

$$\text{Laplacian of } \phi \;=\; \nabla^2 \phi \;=\; \nabla \cdot (\nabla \phi) \;=\; \nabla \cdot \left\{ \frac{f'(r)}{r}\mathbf{r} \right\}$$

$$=\; \nabla \left\{ \frac{f'(r)}{r} \right\} \cdot \mathbf{r} \;+\; \frac{f'(r)}{r}(\nabla \cdot \mathbf{r}) \;=\; \frac{1}{r}\frac{d}{dr}\left\{ \frac{f'(r)}{r} \right\}\mathbf{r} \cdot \mathbf{r} \;+\; \frac{f'(r)}{r}(3)$$

$$=\; \frac{r f''(r) - f'(r)}{r^3}r^2 \;+\; \frac{3 f'(r)}{r} \;=\; f''(r) \;+\; \frac{2}{r}f'(r)$$

Another method:　In spherical coordinates, we have

$$\nabla^2 U \;=\; \frac{1}{r^2}\frac{\partial}{\partial r}\left(r^2 \frac{\partial U}{\partial r} \right) \;+\; \frac{1}{r^2 \sin \theta}\frac{\partial}{\partial \theta}\left(\sin \theta \frac{\partial U}{\partial \theta} \right) \;+\; \frac{1}{r^2 \sin^2 \theta}\frac{\partial^2 U}{\partial \phi^2}$$

If $U = f(r)$, the last two terms on the right are zero and we find

$$\nabla^2 f(r) \;=\; \frac{1}{r^2}\frac{d}{dr}\left(r^2 f'(r) \right) \;=\; f''(r) \;+\; \frac{2}{r}f'(r)$$

(b) From the result in part (a), we have

$$\nabla^2 \left(\frac{1}{r} \right) \;=\; \frac{d^2}{dr^2}\left(\frac{1}{r} \right) \;+\; \frac{2}{r}\frac{d}{dr}\left(\frac{1}{r} \right) \;=\; \frac{2}{r^3} \;-\; \frac{2}{r^3} \;=\; 0$$

showing that $1/r$ is a solution of Laplace's equation.

45. A particle moves along a space curve $\mathbf{r} = \mathbf{r}(t)$, where t is the time measured from some initial time. If $v = |d\mathbf{r}/dt| = ds/dt$ is the magnitude of the velocity of the particle (s is the arc length along the space curve measured from the initial position), prove that the acceleration $\mathbf{a}$ of the particle is given by

$$\mathbf{a} \;=\; \frac{dv}{dt}\mathbf{T} \;+\; \frac{v^2}{\rho}\mathbf{N}$$

where $\mathbf{T}$ and $\mathbf{N}$ are unit tangent and normal vectors to the space curve and

$$\rho \;=\; \left| \frac{d^2\mathbf{r}}{ds^2} \right|^{-1} \;=\; \left\{ \left(\frac{d^2 x}{ds^2} \right)^2 + \left(\frac{d^2 y}{ds^2} \right)^2 + \left(\frac{d^2 z}{ds^2} \right)^2 \right\}^{-1/2}$$

The velocity of the particle is given by $\mathbf{v} = v\mathbf{T}$. Then the acceleration is given by

$$\mathbf{a} \;=\; \frac{d\mathbf{v}}{dt} \;=\; \frac{d}{dt}(v\mathbf{T}) \;=\; \frac{dv}{dt}\mathbf{T} + v\frac{d\mathbf{T}}{dt} \;=\; \frac{dv}{dt}\mathbf{T} + v\frac{d\mathbf{T}}{ds}\frac{ds}{dt} \;=\; \frac{dv}{dt}\mathbf{T} + v^2\frac{d\mathbf{T}}{ds} \qquad (1)$$

Since $\mathbf{T}$ has unit magnitude, we have $\mathbf{T} \cdot \mathbf{T} = 1$. Then differentiating with respect to s,

$$\mathbf{T} \cdot \frac{d\mathbf{T}}{ds} + \frac{d\mathbf{T}}{ds} \cdot \mathbf{T} \;=\; 0, \qquad 2\mathbf{T} \cdot \frac{d\mathbf{T}}{ds} \;=\; 0 \quad \text{or} \quad \mathbf{T} \cdot \frac{d\mathbf{T}}{ds} \;=\; 0$$

from which it follows that $d\mathbf{T}/ds$ is perpendicular to $\mathbf{T}$. Denoting by $\mathbf{N}$ the unit vector in the direction of $d\mathbf{T}/ds$, and called the *principal normal* to the space curve, we have

$$\frac{d\mathbf{T}}{ds} \;=\; \kappa \mathbf{N} \qquad\qquad (2)$$

where κ is the magnitude of $d\mathbf{T}/ds$. Now since $\mathbf{T} = d\mathbf{r}/ds$ [see equation (9), Page 139], we have $d\mathbf{T}/ds = d^2\mathbf{r}/ds^2$. Hence

$$\kappa \;=\; \left| \frac{d^2\mathbf{r}}{ds^2} \right| \;=\; \left\{ \left(\frac{d^2x}{ds^2} \right)^2 + \left(\frac{d^2y}{ds^2} \right)^2 + \left(\frac{d^2z}{ds^2} \right)^2 \right\}^{1/2}$$

Defining $\rho = 1/\kappa$, (2) becomes $d\mathbf{T}/ds = \mathbf{N}/\rho$. Thus from (1) we have, as required,

$$\mathbf{a} \;=\; \frac{dv}{dt}\mathbf{T} + \frac{v^2}{\rho}\mathbf{N}$$

The components dv/dt and v^2/ρ in the direction of $\mathbf{T}$ and $\mathbf{N}$ are called the *tangential* and *normal* components of the acceleration, the latter being sometimes called the *centripetal acceleration*. The quantities ρ and κ are respectively the *radius of curvature* and *curvature* of the space curve.

Supplementary Problems

VECTOR ALGEBRA

46. Given any two vectors $\mathbf{A}$ and $\mathbf{B}$, illustrate geometrically the equality $4\mathbf{A} + 3(\mathbf{B} - \mathbf{A}) = \mathbf{A} + 3\mathbf{B}$.

47. A man travels 25 miles northeast, 15 miles due east and 10 miles due south. By using an appropriate scale determine graphically (a) how far and (b) in what direction he is from his starting position. Is it possible to determine the answer analytically?   *Ans.* 33.6 miles, 13.2° north of east

48. If $\mathbf{A}$ and $\mathbf{B}$ are any two non-zero vectors which do not have the same direction, prove that $m\mathbf{A} + n\mathbf{B}$ is a vector lying in the plane determined by $\mathbf{A}$ and $\mathbf{B}$.

49. If $\mathbf{A}$, $\mathbf{B}$ and $\mathbf{C}$ are non-coplanar vectors (vectors which do not all lie in the same plane) and $x_1\mathbf{A} + y_1\mathbf{B} + z_1\mathbf{C} = x_2\mathbf{A} + y_2\mathbf{B} + z_2\mathbf{C}$, prove that necessarily $x_1 = x_2$, $y_1 = y_2$, $z_1 = z_2$.

50. Let $ABCD$ be any quadrilateral and points P, Q, R and S the midpoints of successive sides. Prove (a) that $PQRS$ is a parallelogram and (b) that the perimeter of $PQRS$ is equal to the sum of the lengths of the diagonals of $ABCD$.

51. Prove that the medians of a triangle intersect at a point which is a trisection point of each median.

52. Find a unit vector in the direction of the resultant of vectors $\mathbf{A} = 2\mathbf{i} - \mathbf{j} + \mathbf{k}$, $\mathbf{B} = \mathbf{i} + \mathbf{j} + 2\mathbf{k}$, $\mathbf{C} = 3\mathbf{i} - 2\mathbf{j} + 4\mathbf{k}$.   *Ans.* $(6\mathbf{i} - 2\mathbf{j} + 7\mathbf{k})/\sqrt{89}$

The DOT or SCALAR PRODUCT

53. Evaluate $|(\mathbf{A} + \mathbf{B}) \cdot (\mathbf{A} - \mathbf{B})|$ if $\mathbf{A} = 2\mathbf{i} - 3\mathbf{j} + 5\mathbf{k}$ and $\mathbf{B} = 3\mathbf{i} + \mathbf{j} - 2\mathbf{k}$.   *Ans.* 24

54. Prove the law of cosines for a triangle. [Hint: Take the sides as $\mathbf{A}, \mathbf{B}, \mathbf{C}$ where $\mathbf{C} = \mathbf{A} - \mathbf{B}$. Then use $\mathbf{C} \cdot \mathbf{C} = (\mathbf{A} - \mathbf{B}) \cdot (\mathbf{A} - \mathbf{B})$.]

55. Find a so that $2\mathbf{i} - 3\mathbf{j} + 5\mathbf{k}$ and $3\mathbf{i} + a\mathbf{j} - 2\mathbf{k}$ are perpendicular.   *Ans.* $a = -4/3$

56. If $\mathbf{A} = 2\mathbf{i} + \mathbf{j} + \mathbf{k}$, $\mathbf{B} = \mathbf{i} - 2\mathbf{j} + 2\mathbf{k}$ and $\mathbf{C} = 3\mathbf{i} - 4\mathbf{j} + 2\mathbf{k}$, find the projection of $\mathbf{A} + \mathbf{C}$ in the direction of $\mathbf{B}$.   *Ans.* 17/3

57. A triangle has vertices at $A(2,3,1)$, $B(-1,1,2)$, $C(1,-2,3)$. Find (a) the length of the median drawn from B to side AC and (b) the acute angle which this median makes with side BC.
Ans. (a) $\frac{1}{2}\sqrt{26}$, (b) $\cos^{-1}\sqrt{91}/14$

58. Prove that the diagonals of a rhombus are perpendicular to each other.

59. Prove that the vector $(A\mathbf{B} + B\mathbf{A})/(A+B)$ represents the bisector of the angle between $\mathbf{A}$ and $\mathbf{B}$.

The CROSS or VECTOR PRODUCT

60. If $\mathbf{A} = 2\mathbf{i} - \mathbf{j} + \mathbf{k}$ and $\mathbf{B} = \mathbf{i} + 2\mathbf{j} - 3\mathbf{k}$, find $|(2\mathbf{A}+\mathbf{B}) \times (\mathbf{A}-2\mathbf{B})|$.   Ans. $25\sqrt{3}$

61. Find a unit vector perpendicular to the plane of the vectors $\mathbf{A} = 3\mathbf{i} - 2\mathbf{j} + 4\mathbf{k}$ and $\mathbf{B} = \mathbf{i} + \mathbf{j} - 2\mathbf{k}$.
Ans. $\pm(2\mathbf{j} + \mathbf{k})/\sqrt{5}$

62. If $\mathbf{A} \times \mathbf{B} = \mathbf{A} \times \mathbf{C}$, does $\mathbf{B} = \mathbf{C}$ necessarily?

63. Find the area of the triangle with vertices $(2,-3,1)$, $(1,-1,2)$, $(-1,2,3)$.   Ans. $\frac{1}{2}\sqrt{3}$

64. Find the shortest distance from the point $(3,2,1)$ to the plane determined by $(1,1,0)$, $(3,-1,1)$, $(-1,0,2)$.   Ans. 2

TRIPLE PRODUCTS

65. If $\mathbf{A} = 2\mathbf{i} + \mathbf{j} - 3\mathbf{k}$, $\mathbf{B} = \mathbf{i} - 2\mathbf{j} + \mathbf{k}$, $\mathbf{C} = -\mathbf{i} + \mathbf{j} - 4\mathbf{k}$, find (a) $\mathbf{A} \cdot (\mathbf{B} \times \mathbf{C})$, (b) $\mathbf{C} \cdot (\mathbf{A} \times \mathbf{B})$, (c) $\mathbf{A} \times (\mathbf{B} \times \mathbf{C})$, (d) $(\mathbf{A} \times \mathbf{B}) \times \mathbf{C}$.   Ans. (a) 20, (b) 20, (c) $8\mathbf{i} - 19\mathbf{j} - \mathbf{k}$, (d) $25\mathbf{i} - 15\mathbf{j} - 10\mathbf{k}$

66. Prove that (a) $\mathbf{A} \cdot (\mathbf{B} \times \mathbf{C}) = \mathbf{B} \cdot (\mathbf{C} \times \mathbf{A}) = \mathbf{C} \cdot (\mathbf{A} \times \mathbf{B})$
(b) $\mathbf{A} \times (\mathbf{B} \times \mathbf{C}) = \mathbf{B}(\mathbf{A} \cdot \mathbf{C}) - \mathbf{C}(\mathbf{A} \cdot \mathbf{B})$.

67. Find an equation for the plane passing through $(2,-1,-2)$, $(-1,2,-3)$, $(4,1,0)$.
Ans. $2x + y - 3z = 9$

68. Find the volume of the tetrahedron with vertices at $(2,1,1)$, $(1,-1,2)$, $(0,1,-1)$, $(1,-2,1)$.   Ans. $\frac{4}{3}$

69. Prove that $(\mathbf{A} \times \mathbf{B}) \cdot (\mathbf{C} \times \mathbf{D}) + (\mathbf{B} \times \mathbf{C}) \cdot (\mathbf{A} \times \mathbf{D}) + (\mathbf{C} \times \mathbf{A}) \cdot (\mathbf{B} \times \mathbf{D}) = 0$.

DERIVATIVES

70. A particle moves along the space curve $\mathbf{r} = e^{-t} \cos t\,\mathbf{i} + e^{-t} \sin t\,\mathbf{j} + e^{-t}\mathbf{k}$. Find the magnitude of the (a) velocity and (b) acceleration at any time t.   Ans. (a) $\sqrt{3}\,e^{-t}$, (b) $\sqrt{5}\,e^{-t}$

71. Prove that $\dfrac{d}{du}(\mathbf{A} \times \mathbf{B}) = \mathbf{A} \times \dfrac{d\mathbf{B}}{du} + \dfrac{d\mathbf{A}}{du} \times \mathbf{B}$ where $\mathbf{A}$ and $\mathbf{B}$ are differentiable functions of u.

72. Find a unit vector tangent to the space curve $x = t$, $y = t^2$, $z = t^3$ at the point where $t = 1$.
Ans. $(\mathbf{i} + 2\mathbf{j} + 3\mathbf{k})/\sqrt{14}$

73. If $\mathbf{r} = \mathbf{a} \cos \omega t + \mathbf{b} \sin \omega t$, where $\mathbf{a}$ and $\mathbf{b}$ are any constant noncollinear vectors and ω is a constant scalar, prove that (a) $\mathbf{r} \times \dfrac{d\mathbf{r}}{dt} = \omega(\mathbf{a} \times \mathbf{b})$, (b) $\dfrac{d^2\mathbf{r}}{dt^2} + \omega^2\mathbf{r} = \mathbf{0}$.

74. If $\mathbf{A} = x^2\mathbf{i} - yj + xz\mathbf{k}$, $\mathbf{B} = yi + xj - xyz\mathbf{k}$ and $\mathbf{C} = \mathbf{i} - yj + x^3z\mathbf{k}$, find (a) $\dfrac{\partial^2}{\partial x\,\partial y}(\mathbf{A} \times \mathbf{B})$ and (b) $d[\mathbf{A} \cdot (\mathbf{B} \times \mathbf{C})]$ at the point $(1,-1,2)$.   Ans. (a) $-4\mathbf{i} + 8\mathbf{j}$, (b) $8\,dx$

75. If $\mathbf{R} = x^2y\mathbf{i} - 2y^2z\mathbf{j} + xy^2z^2\mathbf{k}$, find $\left|\dfrac{\partial^2\mathbf{R}}{\partial x^2} \times \dfrac{\partial^2\mathbf{R}}{\partial y^2}\right|$ at the point $(2,1,-2)$.   Ans. $16\sqrt{5}$

GRADIENT, DIVERGENCE and CURL

76. If $U, V, \mathbf{A}, \mathbf{B}$ have continuous partial derivatives prove that:

(a) $\nabla(U+V) = \nabla U + \nabla V$, (b) $\nabla \cdot (\mathbf{A}+\mathbf{B}) = \nabla \cdot \mathbf{A} + \nabla \cdot \mathbf{B}$, (c) $\nabla \times (\mathbf{A}+\mathbf{B}) = \nabla \times \mathbf{A} + \nabla \times \mathbf{B}$.

77. If $\phi = xy + yz + zx$ and $\mathbf{A} = x^2y\mathbf{i} + y^2z\mathbf{j} + z^2x\mathbf{k}$, find (a) $\mathbf{A} \cdot \nabla\phi$, (b) $\phi \nabla \cdot \mathbf{A}$ and (c) $(\nabla\phi) \times \mathbf{A}$ at the point $(3, -1, 2)$.   *Ans.* (a) 25, (b) 2, (c) $56\mathbf{i} - 30\mathbf{j} + 47\mathbf{k}$

78. Show that $\nabla \times (r^2\mathbf{r}) = \mathbf{0}$ where $\mathbf{r} = x\mathbf{i} + y\mathbf{j} + z\mathbf{k}$ and $r = |\mathbf{r}|$.

79. Prove: (a) $\nabla \times (U\mathbf{A}) = (\nabla U) \times \mathbf{A} + U(\nabla \times \mathbf{A})$, (b) $\nabla \cdot (\mathbf{A} \times \mathbf{B}) = \mathbf{B} \cdot (\nabla \times \mathbf{A}) - \mathbf{A} \cdot (\nabla \times \mathbf{B})$.

80. Prove that curl grad $U = 0$, stating appropriate conditions on U.

81. Find a unit normal to the surface $x^2y - 2xz + 2y^2z^4 = 10$ at the point $(2, 1, -1)$.
Ans. $\pm (3\mathbf{i} + 4\mathbf{j} - 6\mathbf{k})/\sqrt{61}$

82. If $\mathbf{A} = 3xz^2\mathbf{i} - yz\mathbf{j} + (x+2z)\mathbf{k}$, find curl curl $\mathbf{A}$.   *Ans.* $-6x\mathbf{i} + (6z-1)\mathbf{k}$

83. (a) Prove that $\nabla \times (\nabla \times \mathbf{A}) = -\nabla^2\mathbf{A} + \nabla(\nabla \cdot \mathbf{A})$. (b) Verify the result in (a) if $\mathbf{A}$ is given as in Problem 82.

JACOBIANS and CURVILINEAR COORDINATES

84. Prove that $\left| \dfrac{\partial(x,y,z)}{\partial(u_1, u_2, u_3)} \right| = \left| \dfrac{\partial \mathbf{r}}{\partial u_1} \cdot \dfrac{\partial \mathbf{r}}{\partial u_2} \times \dfrac{\partial \mathbf{r}}{\partial u_3} \right|$.

85. Express (a) grad Φ, (b) div $\mathbf{A}$, (c) $\nabla^2\Phi$ in spherical coordinates.

Ans. (a) $\dfrac{\partial \Phi}{\partial r}\mathbf{e}_1 + \dfrac{1}{r}\dfrac{\partial \Phi}{\partial \theta}\mathbf{e}_2 + \dfrac{1}{r \sin \theta}\dfrac{\partial \Phi}{\partial \phi}\mathbf{e}_3$

(b) $\dfrac{1}{r^2}\dfrac{\partial}{\partial r}(r^2 A_1) + \dfrac{1}{r \sin \theta}\dfrac{\partial}{\partial \theta}(\sin \theta\, A_2) + \dfrac{1}{r \sin \theta}\dfrac{\partial A_3}{\partial \phi}$ where $\mathbf{A} = A_1\mathbf{e}_1 + A_2\mathbf{e}_2 + A_3\mathbf{e}_3$

(c) $\dfrac{1}{r^2}\dfrac{\partial}{\partial r}\left(r^2 \dfrac{\partial \Phi}{\partial r}\right) + \dfrac{1}{r^2 \sin \theta}\dfrac{\partial}{\partial \theta}\left(\sin \theta \dfrac{\partial \Phi}{\partial \theta}\right) + \dfrac{1}{r^2 \sin^2 \theta}\dfrac{\partial^2 \Phi}{\partial \phi^2}$

86. The transformation from rectangular to *parabolic cylindrical coordinates* is defined by the equations $x = \frac{1}{2}(u^2 - v^2)$, $y = uv$, $z = z$. (a) Prove that the system is orthogonal. (b) Find ds^2 and the scale factors. (c) Find the Jacobian of the transformation and the volume element.

Ans. (b) $ds^2 = (u^2+v^2)\,du^2 + (u^2+v^2)\,dv^2 + dz^2$, $h_1 = h_2 = \sqrt{u^2+v^2}$, $h_3 = 1$

(c) $u^2 + v^2$, $(u^2+v^2)\,du\,dv\,dz$

87. Write (a) $\nabla^2\Phi$ and (b) div $\mathbf{A}$ in parabolic cylindrical coordinates.

Ans. (a) $\nabla^2\Phi = \dfrac{1}{u^2+v^2}\left(\dfrac{\partial^2 \Phi}{\partial u^2} + \dfrac{\partial^2 \Phi}{\partial v^2}\right) + \dfrac{\partial^2 \Phi}{\partial z^2}$

(b) div $\mathbf{A} = \dfrac{1}{u^2+v^2}\left\{\dfrac{\partial}{\partial u}(\sqrt{u^2+v^2}\,A_1) + \dfrac{\partial}{\partial v}(\sqrt{u^2+v^2}\,A_2)\right\} + \dfrac{\partial A_3}{\partial z}$

88. Prove that for orthogonal curvilinear coordinates,

$$\nabla\Phi = \frac{\mathbf{e}_1}{h_1}\frac{\partial \Phi}{\partial u_1} + \frac{\mathbf{e}_2}{h_2}\frac{\partial \Phi}{\partial u_2} + \frac{\mathbf{e}_3}{h_3}\frac{\partial \Phi}{\partial u_3}$$

[Hint: Let $\nabla\Phi = a_1\mathbf{e}_1 + a_2\mathbf{e}_2 + a_3\mathbf{e}_3$ and use the fact that $d\Phi = \nabla\Phi \cdot d\mathbf{r}$ must be the same in both rectangular and the curvilinear coordinates.]

89. Give a vector interpretation to the theorem in Problem 35 of Chapter 6.

MISCELLANEOUS PROBLEMS

90. If $\mathbf{A}$ is a differentiable function of u and $|\mathbf{A}(u)| = 1$, prove that $d\mathbf{A}/du$ is perpendicular to $\mathbf{A}$.

91. Prove formulas 6, 7 and 8 on Page 141.

92. If ρ and ϕ are polar coordinates and A, B, n are any constants, prove that $U = \rho^n(A \cos n\phi + B \sin n\phi)$ satisfies Laplace's equation.

93. If $V = \dfrac{2 \cos \theta + 3 \sin^3 \theta \cos \phi}{r^2}$, find $\nabla^2 V$. _Ans._ $\dfrac{6 \sin \theta \cos \phi\ (4 - 5 \sin^2 \theta)}{r^4}$

94. Find the most general function of (a) the cylindrical coordinate ρ, (b) the spherical coordinate r, (c) the spherical coordinate θ which satisfies Laplace's equation.
    Ans. (a) $A + B \ln \rho$, (b) $A + B/r$, (c) $A + B \ln (\csc \theta - \cot \theta)$ where A and B are any constants.

95. Let $\mathbf{T}$ and $\mathbf{N}$ denote respectively the unit _tangent vector_ and unit _principal normal_ vector to a space curve $\mathbf{r} = \mathbf{r}(u)$, where $\mathbf{r}(u)$ is assumed differentiable. Define a vector $\mathbf{B} = \mathbf{T} \times \mathbf{N}$ called the unit _binormal vector_ to the space curve. Prove that

$$\frac{d\mathbf{T}}{ds} = \kappa\mathbf{N}, \qquad \frac{d\mathbf{B}}{ds} = -\tau\mathbf{N}, \qquad \frac{d\mathbf{N}}{ds} = \tau\mathbf{B} - \kappa\mathbf{T}$$

These are called the _Frenet-Serret_ formulas and are of fundamental importance in _differential geometry_. In these formulas κ is called the _curvature_, τ is called the _torsion_; and the reciprocals of these, $\rho = 1/\kappa$ and $\sigma = 1/\tau$, are called the _radius of curvature_ and _radius of torsion_ respectively.

96. (a) Prove that the radius of curvature at any point of the plane curve $y = f(x)$, $z = 0$ where $f(x)$ is differentiable, is given by

$$\rho = \left| \frac{(1 + y'^2)^{3/2}}{y''} \right|$$

(b) Find the radius of curvature at the point $(\pi/2, 1, 0)$ of the curve $y = \sin x$, $z = 0$.
Ans. (b) $2\sqrt{2}$

97. Prove that the acceleration of a particle along a space curve is given respectively in (a) cylindrical, (b) spherical coordinates by

$$(\ddot{\rho} - \rho\dot{\phi}^2)\mathbf{e}_\rho + (\rho\ddot{\phi} + 2\dot{\rho}\dot{\phi})\mathbf{e}_\phi + \ddot{z}\mathbf{e}_z$$

$$(\ddot{r} - r\dot{\theta}^2 - r\dot{\phi}^2 \sin^2 \theta)\mathbf{e}_r + (r\ddot{\theta} + 2\dot{r}\dot{\theta} - r\dot{\phi}^2 \sin \theta \cos \theta)\mathbf{e}_\theta + (2\dot{r}\dot{\phi} \sin \theta + 2r\dot{\theta}\dot{\phi} \cos \theta + r\ddot{\phi} \sin \theta)\mathbf{e}_\phi$$

where dots denote time derivatives and $\mathbf{e}_\rho, \mathbf{e}_\phi, \mathbf{e}_z, \mathbf{e}_r, \mathbf{e}_\theta, \mathbf{e}_\phi$ are unit vectors in the directions of increasing $\rho, \phi, z, r, \theta, \phi$ respectively.

98. Let $\mathbf{E}$ and $\mathbf{H}$ be two vectors assumed to have continuous partial derivatives (of second order at least) with respect to position and time. Suppose further that $\mathbf{E}$ and $\mathbf{H}$ satisfy the equations

$$\nabla \cdot \mathbf{E} = 0, \quad \nabla \cdot \mathbf{H} = 0, \quad \nabla \times \mathbf{E} = -\frac{1}{c}\frac{\partial \mathbf{H}}{\partial t}, \quad \nabla \times \mathbf{H} = \frac{1}{c}\frac{\partial \mathbf{E}}{\partial t} \qquad (1)$$

prove that $\mathbf{E}$ and $\mathbf{H}$ satisfy the equation

$$\nabla^2 \psi = \frac{1}{c^2}\frac{\partial^2 \psi}{\partial t^2} \qquad (2)$$

[The vectors $\mathbf{E}$ and $\mathbf{H}$ are called _electric_ and _magnetic field vectors_ in _electromagnetic theory_. Equations (1) are a special case of _Maxwell's equations_. The result (2) led Maxwell to the conclusion that light was an electromagnetic phenomena. The constant c is the velocity of light.]

99. Use the relations in Problem 98 to show that

$$\frac{\partial}{\partial t}\left\{\tfrac{1}{2}(E^2 + H^2)\right\} + c\,\nabla \cdot (\mathbf{E} \times \mathbf{H}) = 0$$

100. Let A_1, A_2, A_3 be the components of vector $\mathbf{A}$ in an xyz rectangular coordinate system with unit vectors $\mathbf{i}_1, \mathbf{i}_2, \mathbf{i}_3$ (the usual $\mathbf{i}, \mathbf{j}, \mathbf{k}$ vectors), and A'_1, A'_2, A'_3 the components of $\mathbf{A}$ in an $x'y'z'$ rectangular coordinate system which has the same origin as the xyz system but is rotated with respect to it and has the unit vectors $\mathbf{i}'_1, \mathbf{i}'_2, \mathbf{i}'_3$. Prove that the following relations (often called _invariance_ relations) must hold:

$$A_n = l_{1n}A'_1 + l_{2n}A'_2 + l_{3n}A'_3 \qquad n = 1, 2, 3$$

where $\mathbf{i}'_m \cdot \mathbf{i}_n = l_{mn}$.

101. If **A** is the vector of Problem 100, prove that the divergence of **A**, i.e. $\nabla \cdot \mathbf{A}$, is an invariant (often called a *scalar invariant*), i.e. prove that

$$\frac{\partial A_1'}{\partial x'} + \frac{\partial A_2'}{\partial y'} + \frac{\partial A_3'}{\partial z'} = \frac{\partial A_1}{\partial x} + \frac{\partial A_2}{\partial y} + \frac{\partial A_3}{\partial z}$$

The results of this and the preceding problem express an obvious requirement that physical quantities must not depend on coordinate systems in which they are observed. Such ideas when generalized lead to an important subject called *tensor analysis* which is basic to the *theory of relativity*.

102. Prove that (*a*) $\mathbf{A} \cdot \mathbf{B}$, (*b*) $\mathbf{A} \times \mathbf{B}$, (*c*) $\nabla \times \mathbf{A}$ are invariant under the transformation of Problem 100.

103. If u_1, u_2, u_3 are orthogonal curvilinear coordinates, prove that

$$(a) \quad \frac{\partial(u_1, u_2, u_3)}{\partial(x, y, z)} = \nabla u_1 \cdot \nabla u_2 \times \nabla u_3 \qquad (b) \quad \left(\frac{\partial \mathbf{r}}{\partial u_1} \cdot \frac{\partial \mathbf{r}}{\partial u_2} \times \frac{\partial \mathbf{r}}{\partial u_3}\right)(\nabla u_1 \cdot \nabla u_2 \times \nabla u_3) = 1$$

and give the significance of these in terms of Jacobians.

104. Use the axiomatic approach to vectors to prove relation (*8*) on Page 138.

105. A set of n vectors $\mathbf{A}_1, \mathbf{A}_2, \ldots, \mathbf{A}_n$ is called *linearly dependent* if there exists a set of scalars $c_1, c_2, \ldots, c_n$ not all zero such that $c_1\mathbf{A}_1 + c_2\mathbf{A}_2 + \cdots + c_n\mathbf{A}_n = 0$ identically, otherwise the set is called *linearly independent*. (*a*) Prove that the vectors $\mathbf{A}_1 = 2\mathbf{i} - 3\mathbf{j} + 5\mathbf{k}$, $\mathbf{A}_2 = \mathbf{i} + \mathbf{j} - 2\mathbf{k}$, $\mathbf{A}_3 = 3\mathbf{i} - 7\mathbf{j} + 12\mathbf{k}$ are linearly dependent. (*b*) Prove that any four 3 dimensional vectors are linearly dependent. (*c*) Prove that a necessary and sufficient condition that the vectors $\mathbf{A}_1 = a_1\mathbf{i} + b_1\mathbf{j} + c_1\mathbf{k}$, $\mathbf{A}_2 = a_2\mathbf{i} + b_2\mathbf{j} + c_2\mathbf{k}$, $\mathbf{A}_3 = a_3\mathbf{i} + b_3\mathbf{j} + c_3\mathbf{k}$ be linearly independent is that $\mathbf{A}_1 \cdot \mathbf{A}_2 \times \mathbf{A}_3 \neq 0$. Give a geometrical interpretation of this.

106. A complex number can be defined as an ordered pair (a, b) of real numbers a and b subject to certain rules of operation for addition and multiplication. (*a*) What are these rules? (*b*) How can the rules in (*a*) be used to define subtraction and division? (*c*) Explain why complex numbers can be considered as two-dimensional vectors. (*d*) Describe similarities and differences between various operations involving complex numbers and the vectors considered in this chapter.

Chapter 8

Applications of Partial Derivatives

APPLICATIONS to GEOMETRY

1. Tangent Plane to a Surface.

Let $F(x, y, z) = 0$ be the equation of a surface S such as shown in Fig. 8-1. We shall assume that F, and all other functions in this chapter, is continuously differentiable unless otherwise indicated. Suppose we wish to find the equation of a tangent plane to S at the point $P(x_0, y_0, z_0)$. A vector normal to S at this point is $\mathbf{N}_0 = \nabla F|_P$, the subscript P indicating that the gradient is to be evaluated at the point $P(x_0, y_0, z_0)$.

If $\mathbf{r}_0$ and $\mathbf{r}$ are the vectors drawn respectively from O to $P(x_0, y_0, z_0)$ and $Q(x, y, z)$ on the plane, the equation of the plane is

$$(\mathbf{r} - \mathbf{r}_0) \cdot \mathbf{N}_0 = (\mathbf{r} - \mathbf{r}_0) \cdot \nabla F|_P = 0 \qquad (1)$$

since $\mathbf{r} - \mathbf{r}_0$ is perpendicular to $\mathbf{N}_0$.

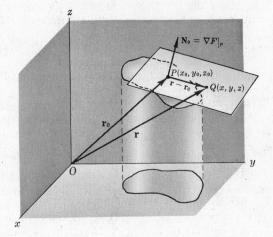

Fig. 8-1

In rectangular form this is

$$\frac{\partial F}{\partial x}\Big|_P (x - x_0) + \frac{\partial F}{\partial y}\Big|_P (y - y_0) + \frac{\partial F}{\partial z}\Big|_P (z - z_0) = 0 \qquad (2)$$

In case the equation of the surface is given in orthogonal curvilinear coordinates in the form $F(u_1, u_2, u_3) = 0$, the equation of the tangent plane can be obtained using the result on Page 142 for the gradient in these coordinates. See Problem 4.

2. Normal Line to a Surface.

Suppose we require equations for the normal line to the surface S at $P(x_0, y_0, z_0)$. If we now let $\mathbf{r}$ be the vector drawn from O in Fig. 8-1 to any point (x, y, z) on the normal $\mathbf{N}_0$, we see that $\mathbf{r} - \mathbf{r}_0$ is collinear with $\mathbf{N}_0$ and so the required condition is

$$(\mathbf{r} - \mathbf{r}_0) \times \mathbf{N}_0 = (\mathbf{r} - \mathbf{r}_0) \times \nabla F|_P = 0 \qquad (3)$$

In rectangular form this is

$$\frac{x - x_0}{\dfrac{\partial F}{\partial x}\Big|_P} = \frac{y - y_0}{\dfrac{\partial F}{\partial y}\Big|_P} = \frac{z - z_0}{\dfrac{\partial F}{\partial z}\Big|_P} \qquad (4)$$

Setting each of these ratios equal to a parameter (such as t or u) and solving for x, y and z yields the *parametric equations* of the normal line.

The equations for the normal line can also be written when the equation of the surface is expressed in orthogonal curvilinear coordinates.

3. Tangent Line to a Curve.

Let the parametric equations of curve C of Fig. 8-2 be $x = f(u)$, $y = g(u)$, $z = h(u)$ where we shall suppose, unless otherwise indicated, that f, g and h are continuously differentiable. We wish to find equations for the tangent line to C at the point $P(x_0, y, z_0)$ where $u = u_0$.

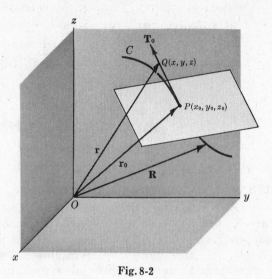

Fig. 8-2

If $\mathbf{R} = f(u)\mathbf{i} + g(u)\mathbf{j} + h(u)\mathbf{k}$, a vector tangent to C at the point P is given by $\mathbf{T_0} = \dfrac{d\mathbf{R}}{du}\Big|_P$.

If $\mathbf{r_0}$ and $\mathbf{r}$ denote the vectors drawn respectively from O to $P(x_0, y_0, z_0)$ and $Q(x, y, z)$ on the tangent line, then since $\mathbf{r} - \mathbf{r_0}$ is collinear with $\mathbf{T_0}$ we have

$$(\mathbf{r} - \mathbf{r_0}) \times \mathbf{T_0} = (\mathbf{r} - \mathbf{r_0}) \times \frac{d\mathbf{R}}{du}\Big|_P = 0 \qquad (5)$$

In rectangular form this becomes

$$\frac{x - x_0}{f'(u_0)} = \frac{y - y_0}{g'(u_0)} = \frac{z - z_0}{h'(u_0)} \qquad (6)$$

The parametric form is obtained by setting each ratio equal to u.

If the curve C is given as the intersection of two surfaces with equations $F(x, y, z) = 0$ and $G(x, y, z) = 0$, the corresponding equations of the tangent line are

$$\frac{x - x_0}{\begin{vmatrix} F_y & F_z \\ G_y & G_z \end{vmatrix}_P} = \frac{y - y_0}{\begin{vmatrix} F_z & F_x \\ G_z & G_x \end{vmatrix}_P} = \frac{z - z_0}{\begin{vmatrix} F_x & F_y \\ G_x & G_y \end{vmatrix}_P} \qquad (7)$$

Note that the determinants in (7) are Jacobians. A similar result can be found when the surfaces are given in terms of orthogonal curvilinear coordinates.

4. Normal Plane to a Curve.

Suppose we wish to find an equation for the normal plane to curve C at $P(x_0, y_0, z_0)$ of Fig. 8-2 (i.e. the plane perpendicular to the tangent line to C at this point). Letting $\mathbf{r}$ be the vector from O to any point (x, y, z) on this plane, it follows that $\mathbf{r} - \mathbf{r_0}$ is perpendicular to $\mathbf{T_0}$. Then the required equation is

$$(\mathbf{r} - \mathbf{r_0}) \cdot \mathbf{T_0} = (\mathbf{r} - \mathbf{r_0}) \cdot \frac{d\mathbf{R}}{du}\Big|_P = 0 \qquad (8)$$

In rectangular form this becomes

$$f'(u_0)(x - x_0) + g'(u_0)(y - y_0) + h'(u_0)(z - z_0) = 0 \qquad (9)$$

when the curve has parametric equations $x = f(u)$, $y = g(u)$, $z = h(u)$, and

$$\begin{vmatrix} F_y & F_z \\ G_y & G_z \end{vmatrix}_P (x - x_0) + \begin{vmatrix} F_z & F_x \\ G_z & G_x \end{vmatrix}_P (y - y_0) + \begin{vmatrix} F_x & F_y \\ G_x & G_y \end{vmatrix}_P (z - z_0) = 0 \qquad (10)$$

when the curve is defined by $F(x, y, z) = 0$, $G(x, y, z) = 0$.

5. Envelopes.

If $\phi(x, y, \alpha) = 0$ is a one parameter family of curves in the xy plane, there may be a curve E which is tangent at each point to some member of the family and such that each member of the family is tangent to E.   If E exists, its equation can be found by solving simultaneously the equations

$$\phi(x, y, \alpha) \;=\; 0, \qquad \phi_\alpha(x, y, \alpha) \;=\; 0 \tag{11}$$

and E is called the *envelope* of the family.

The result can be extended to determine the envelope of a one parameter family of surfaces $\phi(x, y, z, \alpha)$.   This envelope can be found from

$$\phi(x, y, z, \alpha) \;=\; 0, \qquad \phi_\alpha(x, y, z, \alpha) \;=\; 0 \tag{12}$$

Extensions to two (or more) parameter families can be made.

DIRECTIONAL DERIVATIVES

Suppose $F(x, y, z)$ is defined at a point (x, y, z) on a given space curve C.   Let $F(x + \Delta x,\, y + \Delta y,\, z + \Delta z)$ be the value of the function at a neighboring point on C and let Δs denote the length of arc of the curve between those points.   Then

$$\lim_{\Delta s \to 0} \frac{\Delta F}{\Delta s} \;=\; \lim_{\Delta s \to 0} \frac{F(x + \Delta x,\, y + \Delta y,\, z + \Delta z) - F(x, y, z)}{\Delta s} \tag{13}$$

if it exists, is called the *directional derivative* of F at the point (x, y, z) along the curve C and is given by

$$\frac{dF}{ds} \;=\; \frac{\partial F}{\partial x}\frac{dx}{ds} + \frac{\partial F}{\partial y}\frac{dy}{ds} + \frac{\partial F}{\partial z}\frac{dz}{ds} \tag{14}$$

In vector form this can be written

$$\frac{dF}{ds} \;=\; \left(\frac{\partial F}{\partial x}\mathbf{i} + \frac{\partial F}{\partial y}\mathbf{j} + \frac{\partial F}{\partial z}\mathbf{k}\right) \cdot \left(\frac{dx}{ds}\mathbf{i} + \frac{dy}{ds}\mathbf{j} + \frac{dz}{ds}\mathbf{k}\right) \;=\; \nabla F \cdot \frac{d\mathbf{r}}{ds} \;=\; \nabla F \cdot \mathbf{T} \tag{15}$$

from which it follows that the directional derivative is given by the component of ∇F in the direction of the tangent to C.

The maximum value of the directional derivative is given by $|\nabla F|$.

These maxima occur in directions normal to the surfaces $F(x, y, z) = c$ (where c is any constant) which are sometimes called *equipotential surfaces* or *level surfaces*.

DIFFERENTIATION UNDER the INTEGRAL SIGN

Let

$$\phi(\alpha) \;=\; \int_{u_1}^{u_2} f(x, \alpha)\, dx \qquad a \leqq \alpha \leqq b \tag{16}$$

where u_1 and u_2 may depend on the parameter α.   Then

$$\frac{d\phi}{d\alpha} \;=\; \int_{u_1}^{u_2} \frac{\partial f}{\partial \alpha}\, dx \;+\; f(u_2, \alpha)\frac{du_2}{d\alpha} \;-\; f(u_1, \alpha)\frac{du_1}{d\alpha} \tag{17}$$

for $a \leqq \alpha \leqq b$, if $f(x, \alpha)$ and $\partial f / \partial \alpha$ are continuous in both x and α in some region of the $x\alpha$ plane including $u_1 \leqq x \leqq u_2$, $a \leqq \alpha \leqq b$ and if u_1 and u_2 are continuous and have continuous derivatives for $a \leqq \alpha \leqq b$.

In case u_1 and u_2 are constants, the last two terms of (17) are zero.

The result (17), called *Leibnitz's rule*, is often useful in evaluating definite integrals (see Problems 15, 29).

INTEGRATION UNDER the INTEGRAL SIGN

If $\phi(\alpha)$ is defined by (16) and $f(x, \alpha)$ is continuous in x and α in a region including $u_1 \leqq x \leqq u_2$, $a \leqq x \leqq b$, then if u_1 and u_2 are constants,

$$\int_a^b \phi(\alpha)\, d\alpha \;=\; \int_a^b \left\{ \int_{u_1}^{u_2} f(x, \alpha)\, dx \right\} d\alpha \;=\; \int_{u_1}^{u_2} \left\{ \int_a^b f(x, \alpha)\, d\alpha \right\} dx \tag{18}$$

The result is known as *interchange of the order of integration* or *integration under the integral sign*.

MAXIMA and MINIMA

A point (x_0, y_0) is called a *relative maximum point* or *relative minimum point* of $f(x, y)$ respectively according as $f(x_0 + h, y_0 + k) < f(x_0, y_0)$ or $f(x_0 + h, y_0 + k) > f(x_0, y_0)$ for all h and k such that $0 < |h| < \delta$, $0 < |k| < \delta$ where δ is a sufficiently small positive number.

A necessary condition that a differentiable function $f(x, y)$ have a relative maximum or minimum is

$$\frac{\partial f}{\partial x} = 0, \qquad \frac{\partial f}{\partial y} = 0 \tag{19}$$

If (x_0, y_0) is a point (called a *critical point*) satisfying equations (19) and if Δ is defined by

$$\Delta = \left\{ \left(\frac{\partial^2 f}{\partial x^2} \right) \left(\frac{\partial^2 f}{\partial y^2} \right) - \left(\frac{\partial^2 f}{\partial x \, \partial y} \right)^2 \right\} \bigg|_{(x_0, y_0)} \tag{20}$$

then

1. (x_0, y_0) is a relative maximum point if $\Delta > 0$ and $\dfrac{\partial^2 f}{\partial x^2}\bigg|_{(x_0, y_0)} < 0 \left(\text{or} \quad \dfrac{\partial^2 f}{\partial y^2}\bigg|_{(x_0, y_0)} < 0 \right)$

2. (x_0, y_0) is a relative minimum point if $\Delta > 0$ and $\dfrac{\partial^2 f}{\partial x^2}\bigg|_{(x_0, y_0)} > 0 \left(\text{or} \quad \dfrac{\partial^2 f}{\partial y^2}\bigg|_{(x_0, y_0)} > 0 \right)$

3. (x_0, y_0) is neither a relative maximum or minimum point if $\Delta < 0$. If $\Delta < 0$, (x_0, y_0) is sometimes called a *saddle point*.

4. no information is obtained if $\Delta = 0$ (in such case further investigation is necessary).

METHOD of LAGRANGE MULTIPLIERS for MAXIMA and MINIMA

A method for obtaining the relative maximum or minimum values of a function $F(x, y, z)$ subject to a *constraint condition* $\phi(x, y, z) = 0$, consists of the formation of the auxiliary function

$$G(x, y, z) \equiv F(x, y, z) + \lambda \phi(x, y, z) \tag{21}$$

subject to the conditions

$$\frac{\partial G}{\partial x} = 0, \qquad \frac{\partial G}{\partial y} = 0, \qquad \frac{\partial G}{\partial z} = 0 \tag{22}$$

which are necessary conditions for a relative maximum or minimum. The parameter λ, which is independent of x, y, z, is called a *Lagrange multiplier*.

The method can be generalized. If we wish to find the relative maximum or minimum values of a function $F(x_1, x_2, x_3, \ldots, x_n)$ subject to the *constraint conditions* $\phi_1(x_1, \ldots, x_n) = 0$, $\phi_2(x_1, \ldots, x_n) = 0$, $\ldots$, $\phi_k(x_1, \ldots, x_n) = 0$, we form the auxiliary function

$$G(x_1, x_2, \ldots, x_n) \equiv F + \lambda_1 \phi_1 + \lambda_2 \phi_2 + \cdots + \lambda_k \phi_k \tag{23}$$

subject to the (necessary) conditions

$$\frac{\partial G}{\partial x_1} = 0, \quad \frac{\partial G}{\partial x_2} = 0, \quad \ldots, \quad \frac{\partial G}{\partial x_n} \equiv 0 \tag{24}$$

where $\lambda_1, \lambda_2, \ldots, \lambda_k$, which are independent of $x_1, x_2, \ldots, x_n$, are the *Lagrange multipliers*.

APPLICATIONS to ERRORS

The theory of differentials can be applied to obtain errors in a function of x, y, z, etc., when the errors in x, y, z, etc., are known. See Problem 28.

Solved Problems

TANGENT PLANE and NORMAL LINE to a SURFACE

1. Find equations for the (a) tangent plane and (b) normal line to the surface $x^2yz + 3y^2 = 2xz^2 - 8z$ at the point $(1, 2, -1)$.

   (a) The equation of the surface is $F = x^2yz + 3y^2 - 2xz^2 + 8z = 0$. A normal to the surface at $(1, 2, -1)$ is

   $$\mathbf{N}_0 = \nabla F|_{(1,2,-1)} = (2xyz - 2z^2)\mathbf{i} + (x^2z + 6y)\mathbf{j} + (x^2y - 4xz + 8)\mathbf{k}|_{(1,2,-1)}$$
   $$= -6\mathbf{i} + 11\mathbf{j} + 14\mathbf{k}$$

   Referring to Fig. 8-1, Page 161:

   The vector from O to any point (x, y, z) on the tangent plane is $\mathbf{r} = x\mathbf{i} + y\mathbf{j} + z\mathbf{k}$.

   The vector from O to the point $(1, 2, -1)$ on the tangent plane is $\mathbf{r}_0 = \mathbf{i} + 2\mathbf{j} - \mathbf{k}$.

   The vector $\mathbf{r} - \mathbf{r}_0 = (x-1)\mathbf{i} + (y-2)\mathbf{j} + (z+1)\mathbf{k}$ lies in the tangent plane and is thus perpendicular to $\mathbf{N}_0$.

   Then the required equation is

   $$(\mathbf{r} - \mathbf{r}_0) \cdot \mathbf{N}_0 = 0 \quad \text{i.e.} \quad \{(x-1)\mathbf{i} + (y-2)\mathbf{j} + (z+1)\mathbf{k}\} \cdot \{-6\mathbf{i} + 11\mathbf{j} + 14\mathbf{k}\} = 0$$
   $$-6(x-1) + 11(y-2) + 14(z+1) = 0 \quad \text{or} \quad 6x - 11y - 14z + 2 = 0$$

   (b) Let $\mathbf{r} = x\mathbf{i} + y\mathbf{j} + z\mathbf{k}$ be the vector from O to any point (x, y, z) of the normal $\mathbf{N}_0$. The vector from O to the point $(1, 2, -1)$ on the normal is $\mathbf{r}_0 = \mathbf{i} + 2\mathbf{j} - \mathbf{k}$. The vector $\mathbf{r} - \mathbf{r}_0 = (x-1)\mathbf{i} + (y-2)\mathbf{j} + (z+1)\mathbf{k}$ is collinear with $\mathbf{N}_0$. Then

   $$(\mathbf{r} - \mathbf{r}_0) \times \mathbf{N}_0 = 0 \quad \text{i.e.} \quad \begin{vmatrix} \mathbf{i} & \mathbf{j} & \mathbf{k} \\ x-1 & y-2 & z+1 \\ -6 & 11 & 14 \end{vmatrix} = 0$$

   which is equivalent to the equations

   $$11(x-1) = -6(y-2), \quad 14(y-2) = 11(z+1), \quad 14(x-1) = -6(z+1)$$

   These can be written as

   $$\frac{x-1}{-6} = \frac{y-2}{11} = \frac{z+1}{14}$$

   often called the *standard form* for the equations of a line. By setting each of these ratios equal to the parameter t, we have

   $$x = 1 - 6t, \quad y = 2 + 11t, \quad z = 14t - 1$$

   called the *parametric equations* for the line.

2. In what point does the normal line of Problem 1(b) meet the plane $x + 3y - 2z = 10$?

   Substituting the parametric equations of Problem 1(b), we have

   $$1 - 6t + 3(2 + 11t) - 2(14t - 1) = 10 \quad \text{or} \quad t = -1$$

   Then $x = 1 - 6t = 7$, $y = 2 + 11t = -9$, $z = 14t - 1 = -15$ and the required point is $(7, -9, -15)$.

3. Show that the surface $x^2 - 2yz + y^3 = 4$ is perpendicular to any member of the family of surfaces $x^2 + 1 = (2 - 4a)y^2 + az^2$ at the point of intersection $(1, -1, 2)$.

   Let the equations of the two surfaces be written in the form

   $$F = x^2 - 2yz + y^3 - 4 = 0 \quad \text{and} \quad G = x^2 + 1 - (2-4a)y^2 - az^2 = 0$$

   Then

   $$\nabla F = 2x\mathbf{i} + (3y^2 - 2z)\mathbf{j} - 2y\mathbf{k}, \quad \nabla G = 2x\mathbf{i} - 2(2-4a)y\mathbf{j} - 2az\mathbf{k}$$

Thus the normals to the two surfaces at $(1, -1, 2)$ are given by

$$\mathbf{N_1} = 2\mathbf{i} - \mathbf{j} + 2\mathbf{k}, \qquad \mathbf{N_2} = 2\mathbf{i} + 2(2 - 4a)\mathbf{j} - 4a\mathbf{k}$$

Since $\mathbf{N_1} \cdot \mathbf{N_2} = (2)(2) - 2(2 - 4a) - (2)(4a) \equiv 0$, it follows that $\mathbf{N_1}$ and $\mathbf{N_2}$ are perpendicular for all a, and so the required result follows.

4. The equation of a surface is given in spherical coordinates by $F(r, \theta, \phi) = 0$, where we suppose that F is continuously differentiable. (a) Find an equation for the tangent plane to the surface at the point (r_0, θ_0, ϕ_0). (b) Find an equation for the tangent plane to the surface $r = 4 \cos \theta$ at the point $(2\sqrt{2}, \pi/4, 3\pi/4)$. (c) Find a set of equations for the normal line to the surface in (b) at the indicated point.

(a) The gradient of Φ in orthogonal curvilinear coordinates is

$$\nabla \Phi = \frac{1}{h_1} \frac{\partial \Phi}{\partial u_1} \mathbf{e_1} + \frac{1}{h_2} \frac{\partial \Phi}{\partial u_2} \mathbf{e_2} + \frac{1}{h_3} \frac{\partial \Phi}{\partial u_3} \mathbf{e_3}$$

where
$$\mathbf{e_1} = \frac{1}{h_1} \frac{\partial \mathbf{r}}{\partial u_1}, \quad \mathbf{e_2} = \frac{1}{h_2} \frac{\partial \mathbf{r}}{\partial u_2}, \quad \mathbf{e_3} = \frac{1}{h_3} \frac{\partial \mathbf{r}}{\partial u_3}$$

(see Pages 141, 142).

In spherical coordinates $u_1 = r$, $u_2 = \theta$, $u_3 = \phi$, $h_1 = 1$, $h_2 = r$, $h_3 = r \sin \theta$ and $\mathbf{r} = x\mathbf{i} + y\mathbf{j} + z\mathbf{k} = r \sin \theta \cos \phi \mathbf{i} + r \sin \theta \sin \phi \mathbf{j} + r \cos \theta \mathbf{k}$.

Then

$$\begin{cases} \mathbf{e_1} = \sin \theta \cos \phi \, \mathbf{i} + \sin \theta \sin \phi \, \mathbf{j} + \cos \theta \, \mathbf{k} \\ \mathbf{e_2} = \cos \theta \cos \phi \, \mathbf{i} + \cos \theta \sin \phi \, \mathbf{j} - \sin \theta \, \mathbf{k} \\ \mathbf{e_3} = -\sin \phi \, \mathbf{i} + \cos \phi \, \mathbf{j} \end{cases} \tag{1}$$

and

$$\nabla F = \frac{\partial F}{\partial r} \mathbf{e_1} + \frac{1}{r} \frac{\partial F}{\partial \theta} \mathbf{e_2} + \frac{1}{r \sin \theta} \frac{\partial F}{\partial \phi} \mathbf{e_3} \tag{2}$$

As on Page 161, the required equation is $(\mathbf{r} - \mathbf{r_0}) \cdot \nabla F|_P = 0$.

Now substituting (1) in (2), we have

$$\nabla F|_P = \left\{ \frac{\partial F}{\partial r} \Big|_P \sin \theta_0 \cos \phi_0 + \frac{1}{r_0} \frac{\partial F}{\partial \theta} \Big|_P \cos \theta_0 \cos \phi_0 - \frac{\sin \phi_0}{r_0 \sin \theta_0} \frac{\partial F}{\partial \phi} \Big|_P \right\} \mathbf{i}$$

$$+ \left\{ \frac{\partial F}{\partial r} \Big|_P \sin \theta_0 \sin \phi_0 + \frac{1}{r_0} \frac{\partial F}{\partial \theta} \Big|_P \cos \theta_0 \sin \phi_0 + \frac{\cos \phi_0}{r_0 \sin \theta_0} \frac{\partial F}{\partial \phi} \Big|_P \right\} \mathbf{j}$$

$$+ \left\{ \frac{\partial F}{\partial r} \Big|_P \cos \theta_0 - \frac{1}{r_0} \frac{\partial F}{\partial \theta} \Big|_P \sin \theta_0 \right\} \mathbf{k}$$

Denoting the expressions in braces by A, B, C respectively so that $\nabla F|_P = A\mathbf{i} + B\mathbf{j} + C\mathbf{k}$, we see that the required equation is $A(x - x_0) + B(y - y_0) + C(z - z_0) = 0$. This can be written in spherical coordinates by using the transformation equations for x, y and z in these coordinates.

(b) We have $F = r - 4 \cos \theta = 0$. Then $\partial F/\partial r = 1$, $\partial F/\partial \theta = 4 \sin \theta$, $\partial F/\partial \phi = 0$.

Since $r_0 = 2\sqrt{2}$, $\theta_0 = \pi/4$, $\phi_0 = 3\pi/4$, we have from part (a), $\nabla F|_P = A\mathbf{i} + B\mathbf{j} + C\mathbf{k} = -\mathbf{i} + \mathbf{j}$.

From the transformation equations the given point has rectangular coordinates $(-\sqrt{2}, \sqrt{2}, 2)$, and so $\mathbf{r} - \mathbf{r_0} = (x + \sqrt{2})\mathbf{i} + (y - \sqrt{2})\mathbf{j} + (z - 2)\mathbf{k}$.

The required equation of the plane is thus $-(x + \sqrt{2}) + (y - \sqrt{2}) = 0$ or $y - x = 2\sqrt{2}$. In spherical coordinates this becomes $r \sin \theta \sin \phi - r \sin \theta \cos \phi = 2\sqrt{2}$.

In rectangular coordinates the equation $r = 4 \cos \theta$ becomes $x^2 + y^2 + (z - 2)^2 = 4$ and the tangent plane can be determined from this as in Problem 1. In other cases, however, it may not be so easy to obtain the equation in rectangular form, and in such cases the method of part (a) is simpler to use.

(c) The equations of the normal line can be represented by

$$\frac{x+\sqrt{2}}{-1} = \frac{y-\sqrt{2}}{1} = \frac{z-2}{0}$$

the significance of the right hand member being that the line lies in the plane $z = 2$. Thus the required line is given by

$$\frac{x+\sqrt{2}}{-1} = \frac{y-\sqrt{2}}{1}, \quad z = 0 \quad \text{or} \quad x+y = 0, \ z = 0$$

TANGENT LINE and NORMAL PLANE to a CURVE

5. Find equations for the (a) tangent line and (b) normal plane to the curve $x = t - \cos t$, $y = 3 + \sin 2t$, $z = 1 + \cos 3t$ at the point where $t = \frac{1}{2}\pi$.

(a) The vector from origin O (see Fig. 8-2, Page 162) to any point of curve C is $\mathbf{R} = (t - \cos t)\mathbf{i} + (3 + \sin 2t)\mathbf{j} + (1 + \cos 3t)\mathbf{k}$. Then a vector tangent to C at the point where $t = \frac{1}{2}\pi$ is

$$\mathbf{T_0} = \left.\frac{d\mathbf{R}}{dt}\right|_{t=\frac{1}{2}\pi} = (1 + \sin t)\mathbf{i} + 2\cos 2t\,\mathbf{j} - 3\sin 3t\,\mathbf{k}\,|_{t=\frac{1}{2}\pi} = 2\mathbf{i} - 2\mathbf{j} + 3\mathbf{k}$$

The vector from O to the point where $t = \frac{1}{2}\pi$ is $\mathbf{r_0} = \frac{1}{2}\pi\mathbf{i} + 3\mathbf{j} + \mathbf{k}$.

The vector from O to any point (x, y, z) on the tangent line is $\mathbf{r} = x\mathbf{i} + y\mathbf{j} + z\mathbf{k}$.

Then $\mathbf{r} - \mathbf{r_0} = (x - \frac{1}{2}\pi)\mathbf{i} + (y - 3)\mathbf{j} + (z - 1)\mathbf{k}$ is collinear with $\mathbf{T_0}$, so that the required equation is

$$(\mathbf{r} - \mathbf{r_0}) \times \mathbf{T_0} = 0, \quad \text{i.e.} \quad \begin{vmatrix} \mathbf{i} & \mathbf{j} & \mathbf{k} \\ x - \frac{1}{2}\pi & y - 3 & z - 1 \\ 2 & -2 & 3 \end{vmatrix} = 0$$

and the required equations are $\dfrac{x - \frac{1}{2}\pi}{2} = \dfrac{y - 3}{-2} = \dfrac{z - 1}{3}$ or in parametric form $x = 2t + \frac{1}{2}\pi$, $y = 3 - 2t$, $z = 3t + 1$.

(b) Let $r = x\mathbf{i} + y\mathbf{j} + z\mathbf{k}$ be the vector from O to any point (x, y, z) of the normal plane. The vector from O to the point where $t = \frac{1}{2}\pi$ is $\mathbf{r_0} = \frac{1}{2}\pi\mathbf{i} + 3\mathbf{j} + \mathbf{k}$. The vector $\mathbf{r} - \mathbf{r_0} = (x - \frac{1}{2}\pi)\mathbf{i} + (y - 3)\mathbf{j} + (z - 1)\mathbf{k}$ lies in the normal plane and hence is perpendicular to $\mathbf{T_0}$. Then the required equation is $(\mathbf{r} - \mathbf{r_0}) \cdot \mathbf{T_0} = 0$ or $2(x - \frac{1}{2}\pi) - 2(y - 3) + 3(z - 1) = 0$.

6. Find equations for the (a) tangent line and (b) normal plane to the curve $3x^2y + y^2z = -2$, $2xz - x^2y = 3$ at the point $(1, -1, 1)$.

(a) The equations of the surfaces intersecting in the curve are

$$F = 3x^2y + y^2z + 2 = 0, \qquad G = 2xz - x^2y - 3 = 0$$

The normals to each surface at the point $P(1, -1, 1)$ are respectively

$$\mathbf{N_1} = \nabla F|_P = 6xy\mathbf{i} + (3x^2 + 2yz)\mathbf{j} + y^2\mathbf{k} = -6\mathbf{i} + \mathbf{j} + \mathbf{k}$$
$$\mathbf{N_2} = \nabla G|_P = (2z - 2xy)\mathbf{i} - x^2\mathbf{j} + 2x\mathbf{k} = 4\mathbf{i} - \mathbf{j} + 2\mathbf{k}$$

Then a tangent vector to the curve at P is

$$\mathbf{T_0} = \mathbf{N_1} \times \mathbf{N_2} = (-6\mathbf{i} + \mathbf{j} + \mathbf{k}) \times (4\mathbf{i} - \mathbf{j} + 2\mathbf{k}) = 3\mathbf{i} + 16\mathbf{j} + 2\mathbf{k}$$

Thus, as in Problem 5(a), the tangent line is given by

$$(\mathbf{r} - \mathbf{r_0}) \times \mathbf{T_0} = 0 \quad \text{or} \quad \{(x - 1)\mathbf{i} + (y + 1)\mathbf{j} + (z - 1)\mathbf{k}\} \times \{3\mathbf{i} + 16\mathbf{j} + 2\mathbf{k}\} = 0$$

i.e., $\dfrac{x - 1}{3} = \dfrac{y + 1}{16} = \dfrac{z - 1}{2}$ or $x = 1 + 3t$, $y = 16t - 1$, $z = 2t + 1$

(b) As in Problem 5(b) the normal plane is given by

$$(r - \mathbf{r_0}) \cdot \mathbf{T_0} = 0 \quad \text{or} \quad \{(x - 1)\mathbf{i} + (y + 1)\mathbf{j} + (z - 1)\mathbf{k}\} \cdot \{3\mathbf{i} + 16\mathbf{j} + 2\mathbf{k}\} = 0$$

i.e., $3(x - 1) + 16(y + 1) + 2(z - 1) = 0$ or $3x + 16y + 2z = -11$

The results in (a) and (b) can also be obtained by using equations (7) and (10) respectively on Page 162.

7. Establish equation (*10*), Page 162.

Suppose the curve is defined by the intersection of two surfaces whose equations are $F(x, y, z) = 0$, $G(x, y, z) = 0$ where we assume F and G continuously differentiable.

The normals to each surface at point P are given respectively by $\mathbf{N}_1 = \nabla F|_P$ and $\mathbf{N}_2 = \nabla G|_P$. Then a tangent vector to the curve at P is $\mathbf{T}_0 = \mathbf{N}_1 \times \mathbf{N}_2 = \nabla F|_P \times \nabla G|_P$. Thus the equation of the normal plane is $(\mathbf{r} - \mathbf{r}_0) \cdot \mathbf{T}_0 = 0$. Now

$$\mathbf{T}_0 = \nabla F|_P \times \nabla G|_P = \{(F_x\mathbf{i} + F_y\mathbf{j} + F_z\mathbf{k}) \times (G_x\mathbf{i} + G_y\mathbf{j} + G_z\mathbf{k})\}\,|_P$$

$$= \begin{vmatrix} \mathbf{i} & \mathbf{j} & \mathbf{k} \\ F_x & F_y & F_z \\ G_x & G_y & G_z \end{vmatrix}_P = \begin{vmatrix} F_y & F_z \\ G_y & G_z \end{vmatrix}_P \mathbf{i} + \begin{vmatrix} F_z & F_x \\ G_z & G_x \end{vmatrix}_P \mathbf{j} + \begin{vmatrix} F_x & F_y \\ G_x & G_y \end{vmatrix}_P \mathbf{k}$$

and so the required equation is

$$(r - r_0) \cdot \nabla F|_P = 0 \quad \text{or} \quad \begin{vmatrix} F_y & F_z \\ G_y & G_z \end{vmatrix}_P (x - x_0) + \begin{vmatrix} F_z & F_x \\ G_z & G_x \end{vmatrix}_P (y - y_0) + \begin{vmatrix} F_x & F_y \\ G_x & G_y \end{vmatrix}_P (z - z_0) = 0$$

ENVELOPES

8. Prove that the envelope of the family $\phi(x, y, \alpha) = 0$, if it exists, can be obtained by solving simultaneously the equations $\phi = 0$ and $\phi_\alpha = 0$.

Assume parametric equations of the envelope to be $x = f(\alpha)$, $y = g(\alpha)$. Then $\phi(f(\alpha), g(\alpha), \alpha) = 0$ identically, and so upon differentiating with respect to α [assuming that ϕ, f and g have continuous derivatives], we have

$$\phi_x f'(\alpha) + \phi_y g'(\alpha) + \phi_\alpha = 0 \tag{1}$$

The slope of any member of the family $\phi(x, y, \alpha) = 0$ at (x, y) is given by $\phi_x\,dx + \phi_y\,dy = 0$ or $\dfrac{dy}{dx} = -\dfrac{\phi_x}{\phi_y}$. The slope of the envelope at (x, y) is $\dfrac{dy}{dx} = \dfrac{dy/d\alpha}{dx/d\alpha} = \dfrac{g'(\alpha)}{f'(\alpha)}$. Then at any point where the envelope and a member of the family are tangent, we must have

$$-\frac{\phi_x}{\phi_y} = \frac{g'(\alpha)}{f'(\alpha)} \quad \text{or} \quad \phi_x f'(\alpha) + \phi_y g'(\alpha) = 0 \tag{2}$$

Comparing (*2*) with (*1*) we see that $\phi_\alpha = 0$ and the required result follows.

9. (*a*) Find the envelope of the family $x \sin \alpha + y \cos \alpha = 1$. (*b*) Illustrate the results geometrically.

(*a*) By Problem 8 the envelope, if it exists, is obtained by solving simultaneously the equations $\phi(x, y, \alpha) = x \sin \alpha + y \sin \alpha - 1 = 0$ and $\phi_\alpha(x, y, \alpha) = x \cos \alpha - y \cos \alpha = 0$. From these equations we find $x = \sin \alpha$, $y = \cos \alpha$ or $x^2 + y^2 = 1$.

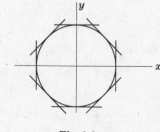

Fig. 8-3

(*b*) The given family is a family of straight lines, some members of which are indicated in Fig. 8-3. The envelope is the circle $x^2 + y^2 = 1$.

10. Find the envelope of the family of surfaces $z = 2\alpha x - \alpha^2 y$.

By a generalization of Problem 8 the required envelope, if it exists, is obtained by solving simultaneously the equations

$$(1) \quad \phi = 2\alpha x - \alpha^2 y - z = 0 \quad \text{and} \quad (2) \quad \phi_\alpha = 2x - 2\alpha y = 0$$

From (*2*) $\alpha = x/y$. Then substitution in (*1*) yields $x^2 = yz$, the required envelope.

11. Find the envelope of the two parameter family of surfaces $z = \alpha x + \beta y - \alpha\beta$.

The envelope of the family $F(x, y, z, \alpha, \beta) = 0$, if it exists, is obtained by eliminating α and β between the equations $F = 0$, $F_\alpha = 0$, $F_\beta = 0$ (see Problem 43). Now

$$F = z - \alpha x - \beta y + \alpha\beta = 0, \quad F_\alpha = -x + \beta = 0, \quad F_\beta = -y + \alpha = 0$$

Then $\beta = x$, $\alpha = y$ and we have $z = xy$.

DIRECTIONAL DERIVATIVES

12. Find the directional derivative of $F = x^2yz^3$ along the curve $x = e^{-u}$, $y = 2\sin u + 1$, $z = u - \cos u$ at the point P where $u = 0$.

The point P corresponding to $u = 0$ is $(1, 1, -1)$. Then

$$\nabla F = 2xyz^3\mathbf{i} + x^2z^3\mathbf{j} + 3x^2yz^2\mathbf{k} = -2\mathbf{i} - \mathbf{j} + 3\mathbf{k} \text{ at } P$$

A tangent vector to the curve is

$$\frac{d\mathbf{r}}{du} = \frac{d}{du}\{e^{-u}\mathbf{i} + (2\sin u + 1)\mathbf{j} + (u - \cos u)\mathbf{k}\}$$

$$= -e^{-u}\mathbf{i} + 2\cos u\,\mathbf{j} + (1 + \sin u)\mathbf{k} = -\mathbf{i} + 2\mathbf{j} + \mathbf{k} \text{ at } P$$

and the unit tangent vector in this direction is $\mathbf{T}_0 = \dfrac{-\mathbf{i} + 2\mathbf{j} + \mathbf{k}}{\sqrt{6}}$.

Then

Directional derivative $= \nabla F \cdot \mathbf{T}_0 = (-2\mathbf{i} - \mathbf{j} + 3\mathbf{k}) \cdot \left(\dfrac{-\mathbf{i} + 2\mathbf{j} + \mathbf{k}}{\sqrt{6}}\right) = \dfrac{3}{\sqrt{6}} = \dfrac{1}{2}\sqrt{6}$.

Since this is positive, F is increasing in this direction.

13. Prove that the greatest rate of change of F, i.e. the maximum directional derivative, takes place in the direction of, and has the magnitude of, the vector ∇F.

$\dfrac{dF}{ds} = \nabla F \cdot \dfrac{d\mathbf{r}}{ds}$ is the projection of ∇F in the direction $\dfrac{d\mathbf{r}}{ds}$. This projection is a maximum when ∇F and $d\mathbf{r}/ds$ have the same direction. Then the maximum value of dF/ds takes place in the direction of ∇F, and the magnitude is $|\nabla F|$.

14. (a) Find the directional derivative of $U = 2x^3y - 3y^2z$ at $P(1, 2, -1)$ in a direction toward $Q(3, -1, 5)$. (b) In what direction from P is the directional derivative a maximum? (c) What is the magnitude of the maximum directional derivative?

(a) $\nabla U = 6x^2y\mathbf{i} + (2x^3 - 6yz)\mathbf{j} - 3y^2\mathbf{k} = 12\mathbf{i} + 14\mathbf{j} - 12\mathbf{k}$ at P.

The vector from P to Q $= (3 - 1)\mathbf{i} + (-1 - 2)\mathbf{j} + [5 - (-1)]\mathbf{k} = 2\mathbf{i} - 3\mathbf{j} + 6\mathbf{k}$.

The unit vector from P to Q $= \mathbf{T} = \dfrac{2\mathbf{i} - 3\mathbf{j} + 6\mathbf{k}}{\sqrt{(2)^2 + (-3)^2 + (6)^2}} = \dfrac{2\mathbf{i} - 3\mathbf{j} + 6\mathbf{k}}{7}$.

Then

Directional derivative at $P = (12\mathbf{i} + 14\mathbf{j} - 12\mathbf{k}) \cdot \left(\dfrac{2\mathbf{i} - 3\mathbf{j} + 6\mathbf{k}}{7}\right) = -\dfrac{90}{7}$

i.e. U is decreasing in this direction.

(b) From Problem 13, the directional derivative is a maximum in the direction $12\mathbf{i} + 14\mathbf{j} - 12\mathbf{k}$.

(c) From Problem 13, the value of the maximum directional derivative is $|12\mathbf{i} + 14\mathbf{j} - 12\mathbf{k}| = \sqrt{144 + 196 + 144} = 22$.

DIFFERENTIATION UNDER the INTEGRAL SIGN

15. Prove Leibnitz's rule for differentiating under the integral sign.

Let $\phi(\alpha) = \int_{u_1(\alpha)}^{u_2(\alpha)} f(x, \alpha)\, dx$. Then

$$\Delta\phi = \phi(\alpha + \Delta\alpha) - \phi(\alpha) = \int_{u_1(\alpha+\Delta\alpha)}^{u_2(\alpha+\Delta\alpha)} f(x, \alpha + \Delta\alpha)\, dx - \int_{u_1(\alpha)}^{u_2(\alpha)} f(x, \alpha)\, dx$$

$$= \int_{u_1(\alpha+\Delta\alpha)}^{u_1(\alpha)} f(x, \alpha + \Delta\alpha)\, dx + \int_{u_1(\alpha)}^{u_2(\alpha)} f(x, \alpha + \Delta\alpha)\, dx + \int_{u_2(\alpha)}^{u_2(\alpha+\Delta\alpha)} f(x, \alpha + \Delta\alpha)\, dx$$

$$- \int_{u_1(\alpha)}^{u_2(\alpha)} f(x, \alpha)\, dx$$

$$= \int_{u_1(\alpha)}^{u_2(\alpha)} [f(x, \alpha + \Delta\alpha) - f(x, \alpha)]\, dx + \int_{u_2(\alpha)}^{u_2(\alpha+\Delta\alpha)} f(x, \alpha + \Delta\alpha)\, dx - \int_{u_1(\alpha)}^{u_1(\alpha+\Delta\alpha)} f(x, \alpha + \Delta\alpha)\, dx$$

By the mean value theorems for integrals, we have

$$\int_{u_1(\alpha)}^{u_2(\alpha)} [f(x, \alpha + \Delta\alpha) - f(x, \alpha)]\, dx = \Delta\alpha \int_{u_1(\alpha)}^{u_2(\alpha)} f_\alpha(x, \xi)\, dx \qquad (1)$$

$$\int_{u_1(\alpha)}^{u_1(\alpha+\Delta\alpha)} f(x, \alpha + \Delta\alpha)\, dx = f(\xi_1, \alpha + \Delta\alpha)[u_1(\alpha + \Delta\alpha) - u_1(\alpha)] \qquad (2)$$

$$\int_{u_2(\alpha)}^{u_2(\alpha+\Delta\alpha)} f(x, \alpha + \Delta\alpha)\, dx = f(\xi_2, \alpha + \Delta\alpha)[u_2(\alpha + \Delta\alpha) - u_2(\alpha)] \qquad (3)$$

where ξ is between α and $\alpha + \Delta\alpha$, ξ_1 is between $u_1(\alpha)$ and $u_1(\alpha + \Delta\alpha)$ and ξ_2 is between $u_2(\alpha)$ and $u_2(\alpha + \Delta\alpha)$.

Then

$$\frac{\Delta\phi}{\Delta\alpha} = \int_{u_1(\alpha)}^{u_2(\alpha)} f_\alpha(x, \xi)\, dx + f(\xi_2, \alpha + \Delta\alpha)\frac{\Delta u_2}{\Delta\alpha} - f(\xi_1, \alpha + \Delta\alpha)\frac{\Delta u_1}{\Delta\alpha}$$

Taking the limit as $\Delta\alpha \to 0$, making use of the fact that the functions are assumed to have continuous derivatives, we obtain

$$\frac{d\phi}{d\alpha} = \int_{u_1(\alpha)}^{u_2(\alpha)} f_\alpha(x, \alpha)\, dx + f[u_2(\alpha), \alpha]\frac{du_2}{d\alpha} - f[u_1(\alpha), \alpha]\frac{du_1}{d\alpha}$$

16. If $\phi(\alpha) = \int_{\alpha}^{\alpha^2} \frac{\sin \alpha x}{x}\, dx$, find $\phi'(\alpha)$ where $\alpha \neq 0$.

By Leibnitz's rule,

$$\phi'(\alpha) = \int_{\alpha}^{\alpha^2} \frac{\partial}{\partial\alpha}\left(\frac{\sin \alpha x}{x}\right) dx + \frac{\sin(\alpha \cdot \alpha^2)}{\alpha^2}\frac{d}{d\alpha}(\alpha^2) - \frac{\sin(\alpha \cdot \alpha)}{\alpha}\frac{d}{d\alpha}(\alpha)$$

$$= \int_{\alpha}^{\alpha^2} \cos \alpha x\, dx + \frac{2 \sin \alpha^3}{\alpha} - \frac{\sin \alpha^2}{\alpha}$$

$$= \left.\frac{\sin \alpha x}{\alpha}\right|_{\alpha}^{\alpha^2} + \frac{2 \sin \alpha^3}{\alpha} - \frac{\sin \alpha^2}{\alpha} = \frac{3 \sin \alpha^3 - 2 \sin \alpha^2}{\alpha}$$

17. If $\int_0^\pi \frac{dx}{\alpha - \cos x} = \frac{\pi}{\sqrt{\alpha^2 - 1}}$, $\alpha > 1$ find $\int_0^\pi \frac{dx}{(2 - \cos x)^2}$. (See Problem 62, Chapter 5.)

By Leibnitz's rule, if $\phi(\alpha) = \int_0^\pi \frac{dx}{\alpha - \cos x} = \pi(\alpha^2 - 1)^{-1/2}$, then

$$\phi'(\alpha) = -\int_0^\pi \frac{dx}{(\alpha - \cos x)^2} = -\tfrac{1}{2}\pi(\alpha^2 - 1)^{-3/2}\, 2\alpha = \frac{-\pi\alpha}{(\alpha^2 - 1)^{3/2}}$$

Thus $\int_0^\pi \frac{dx}{(\alpha - \cos x)^2} = \frac{\pi\alpha}{(\alpha^2 - 1)^{3/2}}$ from which $\int_0^\pi \frac{dx}{(2 - \cos x)^2} = \frac{2\pi}{3\sqrt{3}}$.

INTEGRATION UNDER the INTEGRAL SIGN

18. Prove the result *(18)*, Page 163, for integration under the integral sign.

Consider   (1) $\psi(\alpha) = \int_{u_1}^{u_2}\left\{\int_a^\alpha f(x,\alpha)\,d\alpha\right\}dx$

By Leibnitz's rule,

$$\psi'(\alpha) = \int_{u_1}^{u_2}\frac{\partial}{\partial\alpha}\left\{\int_a^\alpha f(x,\alpha)\,d\alpha\right\}dx = \int_{u_1}^{u_2}f(x,\alpha)\,dx = \phi(\alpha)$$

Then by integration,   (2) $\psi(\alpha) = \int_a^\alpha \phi(\alpha)\,d\alpha + c$

Since $\psi(a) = 0$ from *(1)*, we have $c = 0$ in *(2)*. Thus from *(1)* and *(2)* with $c = 0$, we find

$$\int_{u_1}^{u_2}\left\{\int_a^\alpha f(x,\alpha)\,dx\right\}dx = \int_a^\alpha\left\{\int_{u_1}^{u_2}f(x,\alpha)\,dx\right\}d\alpha$$

Putting $\alpha = b$, the required result follows.

19. Prove that $\int_0^\pi \ln\left(\dfrac{b-\cos x}{a-\cos x}\right)dx = \pi\ln\left(\dfrac{b+\sqrt{b^2-1}}{a+\sqrt{a^2-1}}\right)$ if $a, b > 1$.

From Problem 62, Chapter 5, $\int_0^\pi \dfrac{dx}{\alpha-\cos x} = \dfrac{\pi}{\sqrt{\alpha^2-1}}, \quad \alpha > 1.$

Integrating the left side with respect to α from a to b yields

$$\int_0^\pi\left\{\int_a^b\frac{d\alpha}{\alpha-\cos x}\right\}dx = \int_0^\pi \ln(\alpha-\cos x)\Big|_a^b\,dx = \int_0^\pi \ln\left(\frac{b-\cos x}{a-\cos x}\right)dx$$

Integrating the right side with respect to α from a to b yields

$$\int_0^\pi \frac{\pi\,d\alpha}{\sqrt{\alpha^2-1}} = \pi\ln(\alpha+\sqrt{\alpha^2-1})\Big|_a^b = \pi\ln\left(\frac{b+\sqrt{b^2-1}}{a+\sqrt{a^2-1}}\right)$$

and the required result follows.

MAXIMA and MINIMA

20. Prove that a necessary condition for $f(x,y)$ to have a relative extremum (maximum or minimum) at (x_0, y_0) is that $f_x(x_0, y_0) = 0, \ f_y(x_0, y_0) = 0$.

If $f(x_0, y_0)$ is to be an extreme value for $f(x, y)$, then it must be an extreme value for both $f(x, y_0)$ and $f(x_0, y)$. But a necessary condition that these have extreme values at $x = x_0$ and $y = y_0$ respectively is $f_x(x_0, y_0) = 0, \ f_y(x_0, y_0) = 0$ (using results for functions of one variable).

21. Let $f(x, y)$ be continuous and have continuous partial derivatives, of order two at least, in some region $\mathcal{R}$ including the point (x_0, y_0). Prove that a sufficient condition that $f(x_0, y_0)$ is a relative maximum is that $\Delta = f_{xx}(x_0, y_0)\,f_{yy}(x_0, y_0) - f_{xy}^2(x_0, y_0) > 0$ and $f_{xx}(x_0, y_0) < 0$.

By Taylor's theorem of the mean (see Page 109), using $f_x(x_0, y_0) = 0, \ f_y(x_0, y_0) = 0$, we have

$$f(x_0+h, y_0+k) - f(x_0, y_0) = \tfrac{1}{2}(h^2 f_{xx} + 2hk f_{xy} + k^2 f_{yy}) \tag{1}$$

where the second derivatives on the right are evaluated at $x_0 + \theta h, y_0 + \theta k$ where $0 < \theta < 1$. On completing the square on the right of *(1)* we find

$$f(x_0+h, y_0+k) - f(x_0, y_0) = \tfrac{1}{2}f_{xx}\left\{\left(h + \frac{f_{xy}}{f_{xx}}k\right)^2 + \left(\frac{f_{xx}f_{yy} - f_{xy}^2}{f_{xx}^2}\right)k^2\right\} \tag{2}$$

Now by hypothesis there is a neighborhood of (x_0, y_0) such that $f_{xx} < 0$. Also the sum of the terms in braces must be positive, since $f_{xx}f_{yy} - f_{xy}^2 > 0$ by hypothesis. Thus it follows that

$$f(x_0 + h, y_0 + k) \leqq f(x_0, y_0)$$

for all sufficiently small h and k. But this states that $f(x_0, y_0)$ is a relative maximum.

Similarly we can establish sufficient conditions for a relative minimum.

22. Find the relative maxima and minima of $f(x, y) = x^3 + y^3 - 3x - 12y + 20$.

$f_x = 3x^2 - 3 = 0$ when $x = \pm 1$, $f_y = 3y^2 - 12 = 0$ when $y = \pm 2$. Then critical points are $P(1, 2)$, $Q(-1, 2)$, $R(1, -2)$, $S(-1, -2)$.

$f_{xx} = 6x$, $f_{yy} = 6y$, $f_{xy} = 0$. Then $\Delta = f_{xx}f_{yy} - f_{xy}^2 = 36xy$.

At $P(1, 2)$, $\Delta > 0$ and f_{xx} (or $f_{yy}) > 0$; hence P is a relative minimum point.

At $Q(-1, 2)$, $\Delta < 0$ and Q is neither a relative maximum or minimum point.

At $R(1, -2)$, $\Delta < 0$ and R is neither a relative maximum or minimum point.

At $S(-1, -2)$, $\Delta > 0$ and f_{xx} (or $f_{yy}) < 0$ so S is a relative maximum point.

Thus the relative minimum value of $f(x, y)$ occurring at P is 2, while the relative maximum value occurring at S is 38. Points Q and R are *saddle points*.

23. A rectangular box, open at the top, is to have a volume of 32 cubic feet. What must be the dimensions so that the total surface is a minimum?

If x, y and z are the edges (see Fig. 8-4), then

    (1)   Volume of box $= V = xyz = 32$

    (2)   Surface area of box $= S = xy + 2yz + 2xz$

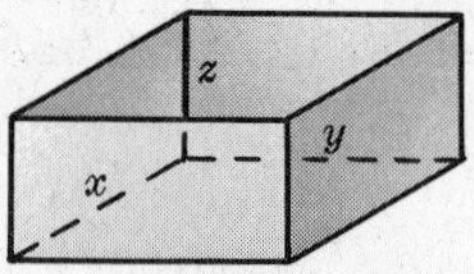

or, since $z = 32/xy$ from *(1)*,

$$S = xy + \frac{64}{x} + \frac{64}{y}$$

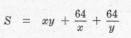

Fig. 8-4

$$\frac{\partial S}{\partial x} = y - \frac{64}{x^2} = 0 \quad \text{when } (3) \ x^2y = 64, \qquad \frac{\partial S}{\partial y} = x - \frac{64}{y^2} = 0 \quad \text{when } (4) \ xy^2 = 64$$

Dividing equations *(3)* and *(4)*, we find $y = x$ so that $x^3 = 64$ or $x = y = 4$ and $z = 2$.

For $x = y = 4$, $\Delta = S_{xx}S_{yy} - S_{xy}^2 = \left(\frac{128}{x^3}\right)\left(\frac{128}{y^3}\right) - 1 > 0$ and $S_{xx} = \frac{128}{x^3} > 0$. Hence it follows that the dimensions $4 \text{ ft} \times 4 \text{ ft} \times 2 \text{ ft}$ give the minimum surface.

LAGRANGE MULTIPLIERS for MAXIMA and MINIMA

24. Consider $F(x, y, z)$ subject to the constraint condition $G(x, y, z) = 0$. Prove that a necessary condition that $F(x, y, z)$ have an extreme value is that $F_x G_y - F_y G_x = 0$.

Since $G(x, y, z) = 0$, we can consider z as a function of x and y, say $z = f(x, y)$. A necessary condition that $F[x, y, f(x, y)]$ have an extreme value is that the partial derivatives with respect to x and y be zero. This gives

    (1) $F_x + F_z z_x = 0$ *(2)* $F_y + F_z z_y = 0$

Since $G(x, y, z) = 0$, we also have

    (3) $G_x + G_z z_x = 0$ *(4)* $G_y + G_z z_y = 0$

From *(1)* and *(3)* we have *(5)* $F_x G_z - F_z G_x = 0$, and from *(2)* and *(4)* we have *(6)* $F_y G_z - F_z G_y = 0$. Then from *(5)* and *(6)* we find $F_x G_y - F_y G_x = 0$.

The above results hold only if $F_z \neq 0$, $G_z \neq 0$.

25. Referring to the preceding problem, show that the stated condition is equivalent to the conditions $\phi_x = 0$, $\phi_y = 0$ where $\phi = F + \lambda G$ and λ is a constant.

If $\phi_x = 0$, $F_x + \lambda G_x = 0$. If $\phi_y = 0$, $F_y + \lambda G_y = 0$. Elimination of λ between these equations yields $F_x G_y - F_y G_x = 0$.

The multiplier λ is the *Lagrange multiplier*. If desired we can consider equivalently $\phi = \lambda F + G$ where $\phi_x = 0$, $\phi_y = 0$.

26. Find the shortest distance from the origin to the hyperbola $x^2 + 8xy + 7y^2 = 225$, $z = 0$.

We must find the minimum value of $x^2 + y^2$ (the square of the distance from the origin to any point in the xy plane) subject to the constraint $x^2 + 8xy + 7y^2 = 225$.

According to the method of Lagrange multipliers, we consider $\phi = x^2 + 8xy + 7y^2 - 225 + \lambda(x^2 + y^2)$. Then

$$\phi_x = 2x + 8y + 2\lambda x = 0 \qquad \text{or} \qquad (1) \quad (\lambda + 1)x + 4y = 0$$
$$\phi_y = 8x + 14y + 2\lambda y = 0 \qquad \text{or} \qquad (2) \quad 4x + (\lambda + 7)y = 0$$

From (1) and (2), since $(x, y) \neq (0, 0)$, we must have

$$\begin{vmatrix} \lambda + 1 & 4 \\ 4 & \lambda + 7 \end{vmatrix} = 0, \qquad \text{i.e.} \quad \lambda^2 + 8\lambda - 9 = 0 \quad \text{or} \quad \lambda = 1, -9$$

Case 1: $\lambda = 1$. From (1) or (2), $x = -2y$ and substitution in $x^2 + 8xy + 7y^2 = 225$ yields $-5y^2 = 225$, for which no real solution exists.

Case 2: $\lambda = -9$. From (1) or (2), $y = 2x$ and substitution in $x^2 + 8xy + 7y^2 = 225$ yields $45x^2 = 225$. Then $x^2 = 5$, $y^2 = 4x^2 = 20$ and so $x^2 + y^2 = 25$. Thus the required shortest distance is $\sqrt{25} = 5$.

27. (a) Find the maximum and minimum values of $x^2 + y^2 + z^2$ subject to the constraint conditions $x^2/4 + y^2/5 + z^2/25 = 1$ and $z = x + y$. (b) Give a geometric interpretation of the result in (a).

(a) We must find the extrema of $F = x^2 + y^2 + z^2$ subject to the constraint conditions $\phi_1 = \dfrac{x^2}{4} + \dfrac{y^2}{5} + \dfrac{z^2}{25} - 1 = 0$ and $\phi_2 = x + y - z = 0$. In this case we use two Lagrange multipliers λ_1, λ_2 and consider the function

$$G = F + \lambda_1 \phi_1 + \lambda_2 \phi_2 = x^2 + y^2 + z^2 + \lambda_1 \left(\frac{x^2}{4} + \frac{y^2}{5} + \frac{z^2}{25} - 1 \right) + \lambda_2 \left(x + y - z \right)$$

Taking the partial derivatives of G with respect to x, y, z and setting them equal to zero, we find

$$G_x = 2x + \frac{\lambda_1 x}{2} + \lambda_2 = 0, \qquad G_y = 2y + \frac{2\lambda_1 y}{5} + \lambda_2 = 0, \qquad G_z = 2z + \frac{2\lambda_1 z}{25} - \lambda_2 = 0 \qquad (1)$$

Solving these equations for x, y, z, we find

$$x = \frac{-2\lambda_2}{\lambda_1 + 4}, \quad y = \frac{-5\lambda_2}{2\lambda_1 + 10}, \quad z = \frac{25\lambda_2}{2\lambda_1 + 50} \qquad (2)$$

From the second constraint condition, $x + y - z = 0$, we obtain on division by λ_2, assumed different from zero (this is justified since otherwise we would have $x = 0$, $y = 0$, $z = 0$ which would not satisfy the first constraint condition), the result

$$\frac{2}{\lambda_1 + 4} + \frac{5}{2\lambda_1 + 10} + \frac{25}{2\lambda_1 + 50} = 0$$

Multiplying both sides by $2(\lambda_1 + 4)(\lambda_1 + 5)(\lambda_1 + 25)$ and simplifying yields

$$17\lambda_1^2 + 245\lambda_1 + 750 = 0 \quad \text{or} \quad (\lambda_1 + 10)(17\lambda_1 + 75) = 0$$

from which $\lambda_1 = -10$ or $-75/17$.

Case 1: $\lambda_1 = -10.$

From (2), $x = \frac{1}{3}\lambda_2$, $y = \frac{1}{2}\lambda_2$, $z = \frac{5}{6}\lambda_2$. Substituting in the first constraint condition, $x^2/4 + y^2/5 + z^2/25 = 1$, yields $\lambda_2^2 = 180/19$ or $\lambda_2 = \pm 6\sqrt{5/19}$. This gives the two critical points

$$(2\sqrt{5/19},\ 3\sqrt{5/19},\ 5\sqrt{5/19}), \quad (-2\sqrt{5/19},\ -3\sqrt{5/19},\ -5\sqrt{5/19})$$

The value of $x^2 + y^2 + z^2$ corresponding to these critical points is $(20 + 45 + 125)/19 = 10$.

Case 2: $\lambda_1 = -75/17.$

From (2), $x = \frac{34}{7}\lambda_2$, $y = -\frac{17}{4}\lambda_2$, $z = \frac{17}{28}\lambda_2$. Substituting in the first constraint condition, $x^2/4 + y^2/5 + z^2/25 = 1$, yields $\lambda_2 = \pm 140/(17\sqrt{646})$ which gives the critical points

$$(40/\sqrt{646},\ -35/\sqrt{646},\ 5/\sqrt{646}), \quad (-40/\sqrt{646},\ 35/\sqrt{646},\ -5/\sqrt{646})$$

The value of $x^2 + y^2 + z^2$ corresponding to these is $(1600 + 1225 + 25)/646 = 75/17$.

Thus the required maximum value is 10 and the minimum value is 75/17.

(b) Since $x^2 + y^2 + z^2$ represents the square of the distance of (x, y, z) from the origin $(0, 0, 0)$, the problem is equivalent to determining the largest and smallest distances from the origin to the curve of intersection of the ellipsoid $x^2/4 + y^2/5 + z^2/25 = 1$ and the plane $z = x + y$. Since this curve is an ellipse, we have the interpretation that $\sqrt{10}$ and $\sqrt{75/17}$ are the lengths of the semi-major and semi-minor axes of this ellipse.

The fact that the maximum and minimum values happen to be given by $-\lambda_1$ in both Case 1 and Case 2 is more than a coincidence. It follows, in fact, on multiplying equations (1) by x, y and z in succession and adding, for we then obtain

$$2x^2 + \frac{\lambda_1 x^2}{2} + \lambda_2 x + 2y^2 + \frac{2\lambda_1 y^2}{5} + \lambda_2 y + 2z^2 + \frac{2\lambda_1 z^2}{25} - \lambda_2 z = 0$$

i.e.

$$x^2 + y^2 + z^2 + \lambda_1\left(\frac{x^2}{4} + \frac{y^2}{5} + \frac{z^2}{25}\right) + \lambda_2(x + y - z) = 0$$

Then using the constraint conditions, we find $x^2 + y^2 + z^2 = -\lambda_1$.

For a generalization of this problem, see Problem 76.

APPLICATIONS to ERRORS

28. The period T of a simple pendulum of length l is given by $T = 2\pi\sqrt{l/g}$. Find the (a) error and (b) percent error made in computing T by using $l = 2$ m and $g = 9.75$ m/sec^2, if the true values are $l = 1.95$ m and $g = 9.81$ m/sec^2.

(a) $T = 2\pi l^{1/2} g^{-1/2}$. Then

$$dT = (2\pi g^{-1/2})(\tfrac{1}{2}l^{-1/2}\,dl) + (2\pi l^{1/2})(-\tfrac{1}{2}g^{-3/2}\,dg) = \frac{\pi}{\sqrt{lg}}\,dl - \pi\sqrt{\frac{l}{g^3}}\,dg \tag{1}$$

Error in $g = \Delta g = dg = +0.06$; error in $l = \Delta l = dl = -0.05$

The error in T is actually ΔT, which is in this case approximately equal to dT. Thus we have from (1),

$$\text{Error in } T = dT = \frac{\pi}{\sqrt{(2)(9.75)}}(-0.05) - \pi\sqrt{\frac{2}{(9.75)^3}}(+0.06) = -0.0444 \text{ sec (approx.)}$$

The value of T for $l = 2$, $g = 9.75$ is $T = 2\pi\sqrt{\dfrac{2}{9.75}} = 2.846$ sec (approx.)

(b) Percent error (or relative error) in $T = \dfrac{dT}{T} = \dfrac{-0.0444}{2.846} = -1.56\%$.

Another method: Since $\ln T = \ln 2\pi + \tfrac{1}{2}\ln l - \tfrac{1}{2}\ln g$,

$$\frac{dT}{T} = \frac{1}{2}\frac{dl}{l} - \frac{1}{2}\frac{dg}{g} = \frac{1}{2}\left(\frac{-0.05}{2}\right) - \frac{1}{2}\left(\frac{+0.06}{9.75}\right) = -1.56\% \tag{2}$$

as before. Note that (2) can be written

$$\text{Percent error in } T = \tfrac{1}{2} \text{ Percent error in } l - \tfrac{1}{2} \text{ Percent error in } g$$

MISCELLANEOUS PROBLEMS

29. Evaluate $\displaystyle\int_0^1 \frac{x-1}{\ln x}\, dx$.

In order to evaluate this integral, we resort to the following device. Define

$$\phi(\alpha) \;=\; \int_0^1 \frac{x^\alpha - 1}{\ln x}\, dx \qquad \alpha > 0$$

Then by Leibnitz's rule

$$\phi'(\alpha) \;=\; \int_0^1 \frac{\partial}{\partial\alpha}\Big(\frac{x^\alpha - 1}{\ln x}\Big) dx \;=\; \int_0^1 \frac{x^\alpha \ln x}{\ln x}\, dx \;=\; \int_0^1 x^\alpha\, dx \;=\; \frac{1}{\alpha + 1}$$

Integrating with respect to α, $\phi(\alpha) = \ln(\alpha+1) + c$. But since $\phi(0) = 0$, $c = 0$ and so $\phi(\alpha) = \ln(\alpha + 1)$.

Then the value of the required integral is $\phi(1) = \ln 2$.

The applicability of Leibnitz's rule can be justified here, since if we define $F(x,\alpha) = (x^\alpha - 1)/\ln x$, $0 < x < 1$, $F(0,\alpha) = 0$, $F(1,\alpha) = \alpha$, then $F(x,\alpha)$ is continuous in both x and α for $0 \leqq x \leqq 1$ and all finite $\alpha > 0$.

30. Find constants a and b for which

$$F(a,b) \;=\; \int_0^\pi \{\sin x - (ax^2 + bx)\}^2\, dx$$

is a minimum.

The necessary conditions for a minimum are $\partial F/\partial a = 0$, $\partial F/\partial b = 0$. Performing these differentiations, we obtain

$$\frac{\partial F}{\partial a} \;=\; \int_0^\pi \frac{\partial}{\partial a}\{\sin x - (ax^2 + bx)\}^2\, dx \;=\; -2\int_0^\pi x^2 \{\sin x - (ax^2 + bx)\}\, dx \;=\; 0$$

$$\frac{\partial F}{\partial b} \;=\; \int_0^\pi \frac{\partial}{\partial b}\{\sin x - (ax^2 + bx)\}^2\, dx \;=\; -2\int_0^\pi x \{\sin x - (ax^2 + bx)\}\, dx \;=\; 0$$

From these we find

$$\begin{cases} a\displaystyle\int_0^\pi x^4\, dx + b\int_0^\pi x^3\, dx \;=\; \int_0^\pi x^2 \sin x\, dx \\[2mm] a\displaystyle\int_0^\pi x^3\, dx + b\int_0^\pi x^2\, dx \;=\; \int_0^\pi x \sin x\, dx \end{cases}$$

or

$$\begin{cases} \dfrac{\pi^5 a}{5} + \dfrac{\pi^4 b}{4} \;=\; \pi^2 - 4 \\[3mm] \dfrac{\pi^4 a}{4} + \dfrac{\pi^3 b}{3} \;=\; \pi \end{cases}$$

Solving for a and b, we find

$$a \;=\; \frac{20}{\pi^3} - \frac{320}{\pi^5} \;\approx\; -0.40065, \qquad b \;=\; \frac{240}{\pi^4} - \frac{12}{\pi^2} \;\approx\; 1.24798$$

We can show that for these values, $F(a,b)$ is indeed a minimum using the sufficiency conditions on Page 164.

The polynomial $ax^2 + bx$ is said to be a *least square approximation* of $\sin x$ over the interval $(0,\pi)$. The ideas involved here are of importance in many branches of mathematics and their applications.

Supplementary Problems

TANGENT PLANE and NORMAL LINE to a SURFACE

31. Find the equations of the (a) tangent plane and (b) normal line to the surface $x^2 + y^2 = 4z$ at $(2, -4, 5)$. *Ans.* (a) $x - 2y - z = 5$, (b) $\dfrac{x-2}{1} = \dfrac{y+4}{-2} = \dfrac{z-5}{-1}$.

32. If $z = f(x, y)$, prove that the equations for the tangent plane and normal line at point $P(x_0, y_0, z_0)$ are given respectively by

$$(a) \quad z - z_0 = f_x|_P\,(x - x_0) + f_y|_P\,(y - y_0) \quad \text{and} \quad (b) \quad \frac{x - x_0}{f_x|_P} = \frac{y - y_0}{f_y|_P} = \frac{z - z_0}{-1}$$

33. Prove that the acute angle γ between the z axis and the normal to the surface $F(x, y, z) = 0$ at any point is given by $\sec \gamma = \sqrt{F_x^2 + F_y^2 + F_z^2}/|F_z|$.

34. The equation of a surface is given in cylindrical coordinates by $F(\rho, \phi, z) = 0$, where F is continuously differentiable. Prove that the equations of (a) the tangent plane and (b) the normal line at the point $P(\rho_0, \phi_0, z_0)$ are given respectively by

$$A(x - x_0) + B(y - y_0) + C(z - z_0) = 0 \quad \text{and} \quad \frac{x - x_0}{A} = \frac{y - y_0}{B} = \frac{z - z_0}{C}$$

where $x_0 = \rho_0 \cos \phi_0$, $y_0 = \rho_0 \sin \phi_0$ and

$$A = F_\rho|_P \cos \phi_0 - \frac{1}{\rho} F_\phi|_P \sin \phi_0, \quad B = F_\rho|_P \sin \phi_0 + \frac{1}{\rho} F_\phi|_P \cos \phi_0, \quad C = F_z|_P$$

35. Use Problem 34 to find the equation of the tangent plane to the surface $\pi z = \rho \phi$ at the point where $\rho = 2$, $\phi = \pi/2$, $z = 1$. To check your answer work the problem using rectangular coordinates. *Ans.* $2x - \pi y + 2\pi z = 0$

TANGENT LINE and NORMAL PLANE to a CURVE

36. Find the equations of the (a) tangent line and (b) normal plane to the space curve $x = 6 \sin t$, $y = 4 \cos 3t$, $z = 2 \sin 5t$ at the point where $t = \pi/4$.
Ans. (a) $\dfrac{x - 3\sqrt{2}}{3} = \dfrac{y + 2\sqrt{2}}{-6} = \dfrac{z + \sqrt{2}}{-5}$ (b) $3x - 6y - 5z = 26\sqrt{2}$.

37. The surfaces $x + y + z = 3$ and $x^2 - y^2 + 2z^2 = 2$ intersect in a space curve. Find the equations of the (a) tangent line (b) normal plane to this space curve at the point $(1, 1, 1)$.
Ans. (a) $\dfrac{x - 1}{-3} = \dfrac{y - 1}{1} = \dfrac{z - 1}{2}$, (b) $3x - y - 2z = 0$

ENVELOPES

38. Find the envelope of each of the following families of curves in the xy plane. In each case construct a graph. (a) $y = \alpha x - \alpha^2$, (b) $\dfrac{x^2}{\alpha} + \dfrac{y^2}{1 - \alpha} = 1$.
Ans. (a) $x^2 = 4y$; (b) $x + y = \pm 1$, $x - y = \pm 1$

39. Find the envelope of a family of lines having the property that the length intercepted between the x and y axes is a constant a. *Ans.* $x^{2/3} + y^{2/3} = a^{2/3}$

40. Find the envelope of the family of circles having centers on the parabola $y = x^2$ and passing through its vertex. [Hint: Let (α, α^2) be any point on the parabola.] *Ans.* $x^2 = -y^3/(2y + 1)$

41. Find the envelope of the normals (called an *evolute*) to the parabola $y = \frac{1}{2}x^2$ and construct a graph.
Ans. $8(y - 1)^3 = 27x^2$

42. Find the envelope of the following families of surfaces:

$$(a)\ \alpha(x - y) - \alpha^2 z = 1, \qquad (b)\ (x - \alpha)^2 + y^2 = 2\alpha z$$

Ans. (a) $4z = (x - y)^2$, (b) $y^2 = z^2 + 2xz$

43. Prove that the envelope of the two parameter family of surfaces $F(x, y, z, \alpha, \beta) = 0$, if it exists, is obtained by eliminating α and β in the equations $F = 0$, $F_\alpha = 0$, $F_\beta = 0$.

44. Find the envelope of the two parameter families (a) $z = \alpha x + \beta y - \alpha^2 - \beta^2$ and (b) $x \cos \alpha + y \cos \beta + z \cos \gamma = a$ where $\cos^2\alpha + \cos^2\beta + \cos^2\gamma = 1$ and a is a constant.
Ans. (a) $4z = x^2 + y^2$, (b) $x^2 + y^2 + z^2 = a^2$

DIRECTIONAL DERIVATIVES

45. (a) Find the directional derivative of $U = 2xy - z^2$ at $(2, -1, 1)$ in a direction toward $(3, 1, -1)$. (b) In what direction is the directional derivative a maximum? (c) What is the value of this maximum?
Ans. (a) 10/3, (b) $-2\mathbf{i} + 4\mathbf{j} - 2\mathbf{k}$, (c) $2\sqrt{6}$

46. The temperature at any point (x, y) in the xy plane is given by $T = 100xy/(x^2 + y^2)$. (a) Find the directional derivative at the point $(2, 1)$ in a direction making an angle of $60°$ with the positive x axis. (b) In what direction from $(2, 1)$ would the derivative be a maximum? (c) What is the value of this maximum?
Ans. (a) $12\sqrt{3} - 6$; (b) in a direction making an angle of $\pi - \tan^{-1}2$ with the positive x axis, or in the direction $-\mathbf{i} + 2\mathbf{j}$; (c) $12\sqrt{5}$

47. Prove that if $F(\rho, \phi, z)$ is continuously differentiable, the maximum directional derivative of F at any point is given by $\sqrt{\left(\dfrac{\partial F}{\partial \rho}\right)^2 + \dfrac{1}{\rho^2}\left(\dfrac{\partial F}{\partial \phi}\right)^2 + \left(\dfrac{\partial F}{\partial z}\right)^2}$.

DIFFERENTIATION UNDER the INTEGRAL SIGN

48. If $\phi(\alpha) = \displaystyle\int_{\sqrt{\alpha}}^{1/\alpha} \cos \alpha x^2\, dx$, find $\dfrac{d\phi}{d\alpha}$.   *Ans.* $-\displaystyle\int_{\sqrt{\alpha}}^{1/\alpha} x^2 \sin \alpha x^2\, dx - \dfrac{1}{\alpha^2} \cos \dfrac{1}{\alpha} - \dfrac{1}{2\sqrt{\alpha}} \cos \alpha^2$

49. (a) If $F(\alpha) = \displaystyle\int_0^{\alpha^2} \tan^{-1}\dfrac{x}{\alpha}\, dx$, find $\dfrac{dF}{d\alpha}$ by Leibnitz's rule. (b) Check the result in (a) by direct integration.   *Ans.* (a) $2\alpha \tan^{-1}\alpha - \frac{1}{2} \ln(\alpha^2 + 1)$

50. Given $\displaystyle\int_0^1 x^p\, dx = \dfrac{1}{p + 1}$, $p > -1$. Prove that $\displaystyle\int_0^1 x^p (\ln x)^m\, dx = \dfrac{(-1)^m m!}{(p + 1)^{m+1}}$, $m = 1, 2, 3, \ldots$.

51. Prove that $\displaystyle\int_0^\pi \ln(1 + \alpha \cos x)\, dx = \pi \ln\left(\dfrac{1 + \sqrt{1 - \alpha^2}}{2}\right)$, $|\alpha| < 1$.

52. Prove that $\displaystyle\int_0^\pi \ln(1 - 2\alpha \cos x + \alpha^2)\, dx = \begin{cases} \pi \ln \alpha^2, & |\alpha| < 1 \\ 0, & |\alpha| > 1 \end{cases}$. Discuss the case $|\alpha| = 1$.

53. Show that $\displaystyle\int_0^\pi \dfrac{dx}{(5 - 3\cos x)^3} = \dfrac{59\pi}{2048}$.

INTEGRATION UNDER the INTEGRAL SIGN

54. Verify that $\displaystyle\int_0^1 \left\{\int_1^2 (\alpha^2 - x^2)\, dx\right\} d\alpha = \int_1^2 \left\{\int_0^1 (\alpha^2 - x^2)\, d\alpha\right\} dx$.

55. Starting with the result $\displaystyle\int_0^{2\pi} (\alpha - \sin x)\, dx = 2\pi\alpha$, prove that for all constants a and b,

$$\int_0^{2\pi} \{(b - \sin x)^2 - (a - \sin x)^2\}\, dx = 2\pi(b^2 - a^2)$$

56. Use the result $\displaystyle\int_0^{2\pi} \frac{dx}{\alpha + \sin x} = \frac{2\pi}{\sqrt{\alpha^2 - 1}}$, $\alpha > 1$ to prove that

$$\int_0^{2\pi} \ln\!\left(\frac{5 + 3 \sin x}{5 + 4 \sin x}\right) dx \;=\; 2\pi \ln\!\left(\frac{9}{8}\right)$$

57. (a) Use the result $\displaystyle\int_0^{\pi/2} \frac{dx}{1 + \alpha \cos x} = \frac{\cos^{-1}\alpha}{\sqrt{1 - \alpha^2}}$, $0 \leqq \alpha < 1$ to show that for $0 \leqq a < 1,\ 0 \leqq b < 1$

$$\int_0^{\pi/2} \sec x \ln\!\left(\frac{1 + b \cos x}{1 + a \cos x}\right) dx \;=\; \tfrac{1}{2}\{(\cos^{-1} a)^2 - (\cos^{-1} b)^2\}$$

    (b) Show that $\displaystyle\int_0^{\pi/2} \sec x \ln\,(1 + \tfrac{1}{2} \cos x)\, dx \;=\; \frac{5\pi^2}{72}.$

MAXIMA and MINIMA. LAGRANGE MULTIPLIERS

58. Find the maxima and minima of $F(x, y, z) = xy^2z^3$ subject to the conditions $x + y + z = 6,\ x > 0,\ y > 0,$ $z > 0$.   *Ans.* maximum value $= 108$ at $x = 1,\ y = 2,\ z = 3$

59. What is the volume of the largest rectangular parallelepiped which can be inscribed in the ellipsoid $x^2/9 + y^2/16 + z^2/36 = 1$?   *Ans.* $64\sqrt{3}$

60. (a) Find the maximum and minimum values of $x^2 + y^2$ subject to the condition $3x^2 + 4xy + 6y^2 = 140$. (b) Give a geometrical interpretation of the results in (a). *Ans.* maximum value $= 70$, minimum value $= 20$

61. Solve Problem 23 using Lagrange multipliers.

62. Prove that in any triangle ABC there is a point P such that $\overline{PA}^2 + \overline{PB}^2 + \overline{PC}^2$ is a minimum and that P is the intersection of the medians.

63. (a) Prove that the maximum and minimum values of $f(x, y) = x^2 + xy + y^2$ in the unit square $0 \leqq x \leqq 1,\ 0 \leqq y \leqq 1$ are 3 and 0 respectively. (b) Can the result of (a) be obtained by setting the partial derivatives of $f(x, y)$ with respect to x and y equal to zero. Explain.

64. Find the extreme values of z on the surface $2x^2 + 3y^2 + z^2 - 12xy + 4xz = 35$. *Ans.* maximum $= 5$, minimum $= -5$

65. Establish the method of Lagrange multipliers in the case where we wish to find the extreme values of $F(x, y, z)$ subject to the two constraint conditions $G(x, y, z) = 0,\ H(x, y, z) = 0$.

66. Prove that the shortest distance from the origin to the curve of intersection of the surfaces $xyz = a$ and $y = bx$ where $a > 0,\ b > 0$, is $3\sqrt{a(b^2 + 1)/2b}$.

67. Find the volume of the ellipsoid $11x^2 + 9y^2 + 15z^2 - 4xy + 10yz - 20xz = 80$.   *Ans.* $64\pi\sqrt{2}/3$

APPLICATIONS to ERRORS

68. The diameter of a right circular cylinder is measured as 6.0 ± 0.03 inches, while its height is measured as 4.0 ± 0.02 inches. What is the largest possible (a) error and (b) percent error made in computing the volume?   *Ans.* (a) 1.70 in³, (b) 1.5%

69. The sides of a triangle are measured to be 12.0 and 15.0 feet, and the included angle 60.0°. If the lengths can be measured to within 1% accuracy while the angle can be measured to within 2% accuracy, find the maximum error and percent error in determining the (a) area and (b) opposite side of the triangle.   *Ans.* (a) 2.501 ft², 3.21%; (b) 0.287 ft, 2.08%

MISCELLANEOUS PROBLEMS

70. If ρ and ϕ are cylindrical coordinates, a and b are any positive constants and n is a positive integer, prove that the surfaces $\rho^n \sin n\phi = a$ and $\rho^n \cos n\phi = b$ are mutually perpendicular along their curves of intersection.

71. Find an equation for the (a) tangent plane and (b) normal line to the surface $8r\theta\phi = \pi^2$ at the point where $r = 1$, $\theta = \pi/4$, $\phi = \pi/2$, (r, θ, ϕ) being spherical coordinates.

  Ans. (a) $4x - (\pi^2 + 4\pi)y + (4\pi - \pi^2)z = -\pi^2\sqrt{2}$, (b) $\dfrac{x}{-4} = \dfrac{y - \sqrt{2}/2}{\pi^2 + 4\pi} = \dfrac{z - \sqrt{2}/2}{\pi^2 - 4\pi}$

72. (a) Prove that the shortest distance from the point (a, b, c) to the plane $Ax + By + Cz + D = 0$ is

$$\left| \frac{Aa + Bb + Cc + D}{\sqrt{A^2 + B^2 + C^2}} \right|$$

  (b) Find the shortest distance from $(1, 2, -3)$ to the plane $2x - 3y + 6z = 20$.   Ans. (b) 6

73. The potential V due to a charge distribution is given in spherical coordinates (r, θ, ϕ) by

$$V = \frac{p \cos \theta}{r^2}$$

  where p is a constant. Prove that the maximum directional derivative at any point is

$$\frac{p\sqrt{\sin^2 \theta + 4 \cos^2 \theta}}{r^3}$$

74. Prove that $\displaystyle\int_0^1 \frac{x^m - x^n}{\ln x}\, dx = \ln\left(\frac{m+1}{n+1}\right)$ if $m > 0$, $n > 0$. Can you extend the result to the case $m > -1$, $n > -1$?

75. (a) If $b^2 - 4ac < 0$ and $a > 0$, $c > 0$, prove that the area of the ellipse $ax^2 + bxy + cy^2 = 1$ is $2\pi/\sqrt{4ac - b^2}$. [Hint: Find the maximum and minimum values of $x^2 + y^2$ subject to the constraint $ax^2 + bxy + cy^2 = 1$.]

76. Prove that the maximum and minimum distances from the origin to the curve of intersection defined by $x^2/a^2 + y^2/b^2 + z^2/c^2 = 1$ and $Ax + By + Cz = 0$ can be obtained by solving for d the equation

$$\frac{A^2 a^2}{a^2 - d^2} + \frac{B^2 b^2}{b^2 - d^2} + \frac{C^2 c^2}{c^2 - d^2} = 0$$

77. Prove that the last equation in the preceding problem always has two real solutions d_1^2 and d_2^2 for any real non-zero constants a, b, c and any real constants A, B, C (not all zero). Discuss the geometrical significance of this.

78. (a) Prove that $\quad I_M = \displaystyle\int_0^M \frac{dx}{(x^2 + a^2)^2} = \frac{1}{2a^3} \tan^{-1} \frac{M}{a} + \frac{M}{2a^2 (a^2 + M^2)}$

  (b) Find $\displaystyle\lim_{M \to \infty} I_M$. This can be denoted by $\displaystyle\int_0^\infty \frac{dx}{(x^2 + a^2)^2}$.

  (c) Is $\quad \displaystyle\lim_{M \to \infty} \frac{d}{da} \int_0^M \frac{dx}{(x^2 + a^2)^2} = \frac{d}{da} \lim_{M \to \infty} \int_0^M \frac{dx}{(x^2 + a^2)^2}$?

79. Find the point on the paraboloid $z = x^2 + y^2$ which is closest to the point $(3, -6, 4)$.
  Ans. $(1, -2, 5)$

80. Investigate the maxima and minima of $f(x, y) = (x^2 - 2x + 4y^2 - 8y)^2$.
  Ans. minimum value = 0

81. (a) Prove that $\quad \displaystyle\int_0^{\pi/2} \frac{\cos x \, dx}{\alpha \cos x + \sin x} = \frac{\alpha\pi}{2(\alpha^2 + 1)} - \frac{\ln \alpha}{\alpha^2 + 1}$.

  (b) Use (a) to prove that $\quad \displaystyle\int_0^{\pi/2} \frac{\cos^2 x \, dx}{(2 \cos x + \sin x)^2} = \frac{3\pi + 5 - 8 \ln 2}{50}$.

82. (a) Find sufficient conditions for a relative maximum or minimum of $w = f(x, y, z)$.
  (b) Examine $w = x^2 + y^2 + z^2 - 6xy + 8xz - 10yz$ for maxima and minima.

  [Hint: For (a) use the fact that the *quadratic form* $A\alpha^2 + B\beta^2 + C\gamma^2 + 2D\alpha\beta + 2E\alpha\gamma + 2F\beta\gamma > 0$ (i.e. is *positive definite*) if

$$A > 0, \qquad \begin{vmatrix} A & D \\ D & B \end{vmatrix} > 0, \qquad \begin{vmatrix} A & D & F \\ D & B & E \\ F & E & C \end{vmatrix} > 0 \ ]$$

Chapter 9

Multiple Integrals

DOUBLE INTEGRALS

Let $F(x, y)$ be defined in a closed region $\mathcal{R}$ of the xy plane (see Fig. 9-1). Subdivide $\mathcal{R}$ into n sub-regions $\Delta\mathcal{R}_k$ of area ΔA_k, $k = 1, 2, \ldots, n$. Let (ξ_k, η_k) be some point of $\Delta\mathcal{R}_k$. Form the sum

$$\sum_{k=1}^{n} F(\xi_k, \eta_k) \, \Delta A_k \qquad (1)$$

Consider

$$\lim_{n \to \infty} \sum_{k=1}^{n} F(\xi_k, \eta_k) \, \Delta A_k \qquad (2)$$

where the limit is taken so that the number n of subdivisions increases without limit and such that the largest linear dimension of each $\Delta\mathcal{R}_k$ approaches zero. If this limit exists it is denoted by

$$\iint\limits_{\mathcal{R}} F(x, y) \, dA \qquad (3)$$

and is called the *double integral* of $F(x, y)$ over the region $\mathcal{R}$.

It can be proved that the limit does exist if $F(x, y)$ is continuous (or sectionally continuous) in $\mathcal{R}$.

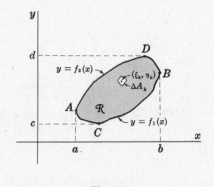

Fig. 9-1

ITERATED INTEGRALS

If $\mathcal{R}$ is such that any lines parallel to the y axis meet the boundary of $\mathcal{R}$ in at most two points (as is true in Fig. 9-1), then we can write the equations of the curves ACB and ADB bounding $\mathcal{R}$ as $y = f_1(x)$ and $y = f_2(x)$ respectively, where $f_1(x)$ and $f_2(x)$ are single-valued and continuous in $a \leqq x \leqq b$. In this case we can evaluate the double integral (3) by choosing the regions $\Delta\mathcal{R}_k$ as rectangles formed by constructing a grid of lines parallel to the x and y axes and ΔA_k as the corresponding areas. Then (3) can be written

$$\iint\limits_{\mathcal{R}} F(x, y) \, dx \, dy \;=\; \int_{x=a}^{b} \int_{y=f_1(x)}^{f_2(x)} F(x, y) \, dy \, dx \qquad (4)$$

$$=\; \int_{x=a}^{b} \left\{ \int_{y=f_1(x)}^{f_2(x)} F(x, y) \, dy \right\} dx$$

where the integral in braces is to be evaluated first (keeping x constant) and finally integrating with respect to x from a to b. The result (4) indicates how a double integral can be evaluated by expressing it in terms of two single integrals called *iterated integrals*.

180

If $\mathcal{R}$ is such that any lines parallel to the x axis meet the boundary of $\mathcal{R}$ in at most two points (as in Fig. 9-1), then the equations of curves CAD and CBD can be written $x = g_1(y)$ and $x = g_2(y)$ respectively and we find similarly

$$\iint\limits_{\mathcal{R}} F(x, y)\, dx\, dy \;=\; \int_{y=c}^{d} \int_{x=g_1(y)}^{g_2(y)} F(x, y)\, dx\, dy \tag{5}$$

$$=\; \int_{y=c}^{d} \left\{ \int_{x=g_1(y)}^{g_2(y)} F(x, y)\, dx \right\} dy$$

If the double integral exists, (4) and (5) yield the same value. (See, however, Problem 17.) In writing a double integral, either of the forms (4) or (5), whichever is appropriate, may be used. We call one form an *interchange of the order of integration* with respect to the other form.

In case $\mathcal{R}$ is not of the type shown in the above figure, it can generally be subdivided into regions $\mathcal{R}_1, \mathcal{R}_2, \ldots$ which are of this type. Then the double integral over $\mathcal{R}$ is found by taking the sum of the double integrals over $\mathcal{R}_1, \mathcal{R}_2, \ldots$.

TRIPLE INTEGRALS

The above results are easily generalized to closed regions in three dimensions. For example, consider a function $F(x, y, z)$ defined in a closed three dimensional region $\mathcal{R}$. Subdivide the region into n subregions of volume ΔV_k, $k = 1, 2, \ldots, n$. Letting (ξ_k, η_k, ζ_k) be some point in each subregion, we form

$$\lim_{n \to \infty} \sum_{k=1}^{n} F(\xi_k, \eta_k, \zeta_k)\, \Delta V_k \tag{6}$$

where the number n of subdivisions approaches infinity in such a way that the largest linear dimension of each subregion approaches zero. If this limit exists we denote it by

$$\iiint\limits_{\mathcal{R}} F(x, y, z)\, dV \tag{7}$$

called the *triple integral* of $F(x, y, z)$ over $\mathcal{R}$. The limit does exist if $F(x, y, z)$ is continuous (or sectionally continuous) in $\mathcal{R}$.

If we construct a grid consisting of planes parallel to the xy, yz and xz planes, the region $\mathcal{R}$ is subdivided into subregions which are rectangular parallelepipeds. In such case we can express the triple integral over $\mathcal{R}$ given by (7) as an *iterated integral* of the form

$$\int_{x=a}^{b} \int_{y=g_1(x)}^{g_2(x)} \int_{z=f_1(x,y)}^{f_2(x,y)} F(x, y, z)\, dx\, dy\, dz \;=\; \int_{x=a}^{b} \left[\int_{y=g_1(x)}^{g_2(x)} \left\{ \int_{z=f_1(x,y)}^{f_2(x,y)} F(x, y, z)\, dz \right\} dy \right] dx \tag{8}$$

(where the innermost integral is to be evaluated first) or the sum of such integrals. The integration can also be performed in any other order to give an equivalent result.

Extensions to higher dimensions are also possible.

TRANSFORMATIONS of MULTIPLE INTEGRALS

In evaluating a multiple integral over a region $\mathcal{R}$, it is often convenient to use coordinates other than rectangular, such as the curvilinear coordinates considered in Chapters 6 and 7.

If we let (u, v) be curvilinear coordinates of points in a plane, there will be a set of transformation equations $x = f(u, v)$, $y = g(u, v)$ mapping points (x, y) of the xy plane into points (u, v) of the uv plane. In such case the region $\mathcal{R}$ of the xy plane is mapped into a region $\mathcal{R}'$ of the uv plane. We then have

$$\iint\limits_{\mathcal{R}} F(x, y)\, dx\, dy \;=\; \iint\limits_{\mathcal{R}'} G(u, v) \left| \frac{\partial(x, y)}{\partial(u, v)} \right| du\, dv \tag{9}$$

where $G(u, v) \equiv F\{f(u, v), g(u, v)\}$ and

$$\frac{\partial(x, y)}{\partial(u, v)} \equiv \begin{vmatrix} \dfrac{\partial x}{\partial u} & \dfrac{\partial x}{\partial v} \\[2mm] \dfrac{\partial y}{\partial u} & \dfrac{\partial y}{\partial v} \end{vmatrix} \tag{10}$$

is the *Jacobian* of x and y with respect to u and v (see Chapter 6).

Similarly if (u, v, w) are curvilinear coordinates in three dimensions, there will be a set of transformation equations $x = f(u, v, w)$, $y = g(u, v, w)$, $z = h(u, v, w)$ and we can write

$$\iiint\limits_{\mathcal{R}} F(x, y, z)\, dx\, dy\, dz \;=\; \iiint\limits_{\mathcal{R}'} G(u, v, w) \left| \frac{\partial(x, y, z)}{\partial(u, v, w)} \right| du\, dv\, dw \tag{11}$$

where $G(u, v, w) \equiv F\{f(u, v, w), g(u, v, w), h(u, v, w)\}$ and

$$\frac{\partial(x, y, z)}{\partial(u, v, w)} \equiv \begin{vmatrix} \dfrac{\partial x}{\partial u} & \dfrac{\partial x}{\partial v} & \dfrac{\partial x}{\partial w} \\[2mm] \dfrac{\partial y}{\partial u} & \dfrac{\partial y}{\partial v} & \dfrac{\partial y}{\partial w} \\[2mm] \dfrac{\partial z}{\partial u} & \dfrac{\partial z}{\partial v} & \dfrac{\partial z}{\partial w} \end{vmatrix} \tag{12}$$

is the Jacobian of x, y and z with respect to u, v and w.

The results (9) and (11) correspond to change of variables for double and triple integrals.

Generalizations to higher dimensions are easily made.

Solved Problems

DOUBLE INTEGRALS

1. (a) Sketch the region $\mathcal{R}$ in the xy plane bounded by $y = x^2$, $x = 2$, $y = 1$.

   (b) Give a physical interpretation to $\iint\limits_{\mathcal{R}} (x^2 + y^2)\, dx\, dy$.

   (c) Evaluate the double integral in (b).

   (a) The required region $\mathcal{R}$ is shown shaded in Fig. 9-2 below.

   (b) Since $x^2 + y^2$ is the square of the distance from any point (x, y) to $(0, 0)$, we can consider the double integral as representing the *polar moment of inertia* (i.e. moment of inertia with respect to the origin) of the region $\mathcal{R}$ (assuming unit density).

   We can also consider the double integral as representing the *mass* of the region $\mathcal{R}$ assuming a density varying as $x^2 + y^2$.

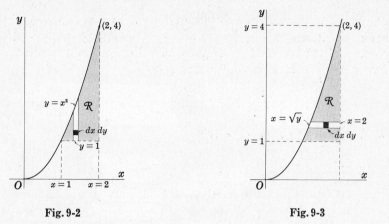

Fig. 9-2   **Fig. 9-3**

(c) **Method 1:**   The double integral can be expressed as the iterated integral

$$\int_{x=1}^{2}\int_{y=1}^{x^2}(x^2+y^2)\,dy\,dx \;=\; \int_{x=1}^{2}\left\{\int_{y=1}^{x^2}(x^2+y^2)\,dy\right\}dx \;=\; \int_{x=1}^{2}x^2y+\frac{y^3}{3}\Big|_{y=1}^{x^2}dx$$

$$=\; \int_{x=1}^{2}\left(x^4+\frac{x^6}{3}-x^2-\frac{1}{3}\right)dx \;=\; \frac{1006}{105}$$

The integration with respect to y (keeping x constant) from $y=1$ to $y=x^2$ corresponds formally to summing in a vertical column (see Fig. 9-2). The subsequent integration with respect to x from $x=1$ to $x=2$ corresponds to addition of contributions from all such vertical columns between $x=1$ and $x=2$.

Method 2:   The double integral can also be expressed as the iterated integral

$$\int_{y=1}^{4}\int_{x=\sqrt{y}}^{2}(x^2+y^2)\,dx\,dy \;=\; \int_{y=1}^{4}\left\{\int_{x=\sqrt{y}}^{2}(x^2+y^2)\,dx\right\}dy \;=\; \int_{y=1}^{4}\frac{x^3}{3}+xy^2\Big|_{x=\sqrt{y}}^{2}dy$$

$$=\; \int_{y=1}^{4}\left(\frac{8}{3}+2y^2-\frac{y^{3/2}}{3}-y^{5/2}\right)dy \;=\; \frac{1006}{105}$$

In this case the vertical column of region $\mathcal{R}$ in Fig. 9-2 above is replaced by a horizontal column as in Fig. 9-3 above. Then the integration with respect to x (keeping y constant) from $x=\sqrt{y}$ to $x=2$ corresponds to summing in this horizontal column. Subsequent integration with respect to y from $y=1$ to $y=4$ corresponds to addition of contributions for all such horizontal columns between $y=1$ and $y=4$.

2. Find the volume of the region common to the intersecting cylinders $x^2+y^2=a^2$ and $x^2+z^2=a^2$.

Required volume

= 8 times volume of region shown in Fig. 9-4

$$=\; 8\int_{x=0}^{a}\int_{y=0}^{\sqrt{a^2-x^2}}z\,dy\,dx$$

$$=\; 8\int_{x=0}^{a}\int_{y=0}^{\sqrt{a^2-x^2}}\sqrt{a^2-x^2}\,dy\,dx$$

$$=\; 8\int_{x=0}^{a}(a^2-x^2)\,dx \;=\; \frac{16a^3}{3}$$

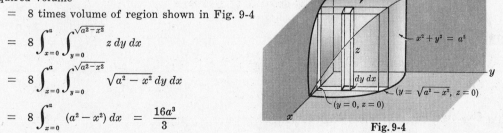

Fig. 9-4

As an aid in setting up this integral note that $z\,dy\,dx$ corresponds to the volume of a column such as shown darkly shaded in the figure. Keeping x constant and integrating with respect to y from $y=0$ to $y=\sqrt{a^2-x^2}$ corresponds to adding the volumes of all such columns in a slab parallel to the yz plane, thus giving the volume of this slab. Finally, integrating with respect to x from $x=0$ to $x=a$ corresponds to adding the volumes of all such slabs in the region, thus giving the required volume.

3. Find the volume of the region bounded by
$$z = x + y, \;\; z = 6, \;\; x = 0, \;\; y = 0, \;\; z = 0$$

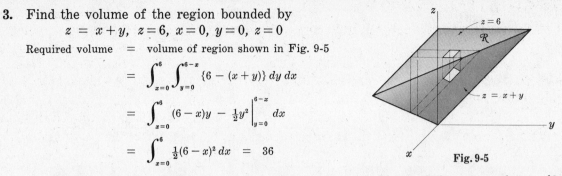

Fig. 9-5

Required volume　=　volume of region shown in Fig. 9-5

$$= \int_{x=0}^{6} \int_{y=0}^{6-x} \{6 - (x+y)\} \, dy \, dx$$

$$= \int_{x=0}^{6} (6-x)y \;-\; \tfrac{1}{2} y^2 \, \Big|_{y=0}^{6-x} \, dx$$

$$= \int_{x=0}^{6} \tfrac{1}{2}(6-x)^2 \, dx \;\; = \;\; 36$$

In this case the volume of a typical column (shown darkly shaded) corresponds to $\{6 - (x+y)\} \, dy \, dx$. The limits of integration are then obtained by integrating over the region $\mathcal{R}$ of the figure. Keeping x constant and integrating with respect to y from $y = 0$ to $y = 6 - x$ (obtained from $z = 6$ and $z = x + y$) corresponds to summing all columns in a slab parallel to the yz plane. Finally, integrating with respect to x from $x = 0$ to $x = 6$ corresponds to adding the volumes of all such slabs and gives the required volume.

TRANSFORMATION of DOUBLE INTEGRALS

4. Justify equation (9), Page 182, for changing variables in a double integral.

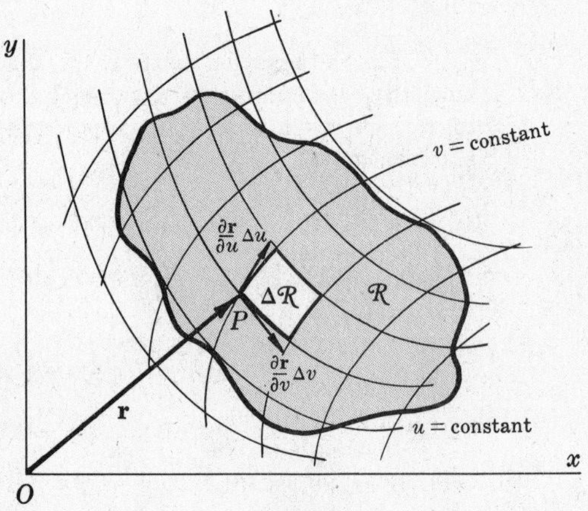

In rectangular coordinates, the double integral of $F(x, y)$ over the region $\mathcal{R}$ (shaded in Fig. 9-6) is $\iint\limits_{\mathcal{R}} F(x, y) \, dx \, dy$. We can also evaluate this double integral by considering a grid formed by a family of u and v curvilinear coordinate curves constructed on the region $\mathcal{R}$ as shown in the figure.

Let P be any point with coordinates (x, y) or (u, v), where $x = f(u, v)$ and $y = g(u, v)$. Then the vector $\mathbf{r}$ from O to P is given by $\mathbf{r} = x\mathbf{i} + y\mathbf{j} = f(u, v)\mathbf{i} + g(u, v)\mathbf{j}$. The tangent vectors to the coordinate curves $u = c_1$ and $v = c_2$, where c_1 and c_2 are constants, are $\partial\mathbf{r}/\partial v$ and $\partial\mathbf{r}/\partial u$ respectively. Then the area of region $\Delta\mathcal{R}$ of Fig. 9-6 is given approximately by $\left| \dfrac{\partial\mathbf{r}}{\partial u} \times \dfrac{\partial\mathbf{r}}{\partial v} \right| \Delta u \, \Delta v$.

Fig. 9-6

But

$$\frac{\partial\mathbf{r}}{\partial u} \times \frac{\partial\mathbf{r}}{\partial v} \;=\; \begin{vmatrix} \mathbf{i} & \mathbf{j} & \mathbf{k} \\ \dfrac{\partial x}{\partial u} & \dfrac{\partial y}{\partial u} & 0 \\ \dfrac{\partial x}{\partial v} & \dfrac{\partial y}{\partial v} & 0 \end{vmatrix} \;=\; \begin{vmatrix} \dfrac{\partial x}{\partial u} & \dfrac{\partial y}{\partial u} \\ \dfrac{\partial x}{\partial v} & \dfrac{\partial y}{\partial v} \end{vmatrix} \mathbf{k} \;=\; \frac{\partial(x, y)}{\partial(u, v)} \mathbf{k}$$

so that

$$\left| \frac{\partial\mathbf{r}}{\partial u} \times \frac{\partial\mathbf{r}}{\partial v} \right| \Delta u \, \Delta v \;=\; \left| \frac{\partial(x, y)}{\partial(u, v)} \right| \Delta u \, \Delta v$$

The double integral is the limit of the sum

$$\Sigma \, F\{f(u, v), \, g(u, v)\} \left| \frac{\partial(x, y)}{\partial(u, v)} \right| \Delta u \, \Delta v$$

taken over the entire region $\mathcal{R}$. An investigation reveals that this limit is

$$\iint\limits_{\mathcal{R}'} F\{f(u, v), \, g(u, v)\} \left| \frac{\partial(x, y)}{\partial(u, v)} \right| du \, dv$$

where $\mathcal{R}'$ is the region in the uv plane into which the region $\mathcal{R}$ is mapped under the transformation $x = f(u, v), \; y = g(u, v)$.

Another method of justifying the above method of change of variables makes use of line integrals and Green's theorem in the plane (see Chapter 10, Problem 32).

5. If $u = x^2 - y^2$ and $v = 2xy$, find $\partial(x,y)/\partial(u,v)$ in terms of u and v.

$$\frac{\partial(u,v)}{\partial(x,y)} = \begin{vmatrix} u_x & u_y \\ v_x & v_y \end{vmatrix} = \begin{vmatrix} 2x & -2y \\ 2y & 2x \end{vmatrix} = 4(x^2 + y^2)$$

From the identity $(x^2 + y^2)^2 = (x^2 - y^2)^2 + (2xy)^2$ we have

$$(x^2 + y^2)^2 = u^2 + v^2 \qquad \text{and} \qquad x^2 + y^2 = \sqrt{u^2 + v^2}$$

Then by Problem 45, Chapter 6,

$$\frac{\partial(x,y)}{\partial(u,v)} = \frac{1}{\partial(u,v)/\partial(x,y)} = \frac{1}{4(x^2 + y^2)} = \frac{1}{4\sqrt{u^2 + v^2}}$$

Another method: Solve the given equations for x and y in terms of u and v and find the Jacobian directly.

6. Find the polar moment of inertia of the region in the xy plane bounded by $x^2 - y^2 = 1$, $x^2 - y^2 = 9$, $xy = 2$, $xy = 4$ assuming unit density.

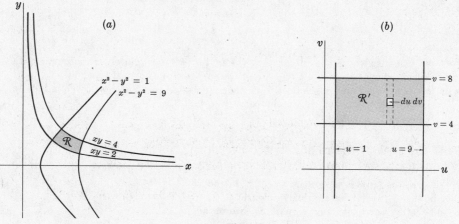

Fig. 9-7

Under the transformation $x^2 - y^2 = u$, $2xy = v$ the required region $\mathcal{R}$ in the xy plane [shaded in Fig. 9-7(a)] is mapped into region $\mathcal{R}'$ of the uv plane [shaded in Fig. 9-7(b)]. Then:

$$\text{Required polar moment of inertia} = \iint_{\mathcal{R}} (x^2 + y^2)\, dx\, dy = \iint_{\mathcal{R}'} (x^2 + y^2) \left| \frac{\partial(x,y)}{\partial(u,v)} \right| du\, dv$$

$$= \iint_{\mathcal{R}'} \sqrt{u^2 + v^2}\, \frac{du\, dv}{4\sqrt{u^2 + v^2}} = \frac{1}{4} \int_{u=1}^{9} \int_{v=4}^{8} du\, dv = 8$$

where we have used the results of Problem 5.

Note that the limits of integration for the region $\mathcal{R}'$ can be constructed directly from the region $\mathcal{R}$ in the xy plane without actually constructing the region $\mathcal{R}'$. In such case we use a grid as in Problem 4. The coordinates (u,v) are curvilinear coordinates, in this case called *hyperbolic coordinates*.

7. Evaluate $\displaystyle\iint_{\mathcal{R}} \sqrt{x^2 + y^2}\, dx\, dy$, where $\mathcal{R}$ is the region in the xy plane bounded by $x^2 + y^2 = 4$ and $x^2 + y^2 = 9$.

The presence of $x^2 + y^2$ suggests the use of polar coordinates (ρ, ϕ), where $x = \rho \cos \phi$, $y = \rho \sin \phi$ (see Problem 38, Chapter 6). Under this transformation the region $\mathcal{R}$ [Fig. 9-8(a) below] is mapped into the region $\mathcal{R}'$ [Fig. 9-8(b) below].

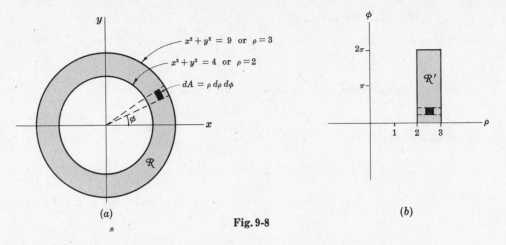

(a)

(b)

Fig. 9-8

Since $\dfrac{\partial(x, y)}{\partial(\rho, \phi)} = \rho$, it follows that

$$\iint\limits_{R} \sqrt{x^2 + y^2}\, dx\, dy \;=\; \iint\limits_{R'} \sqrt{x^2 + y^2}\left|\frac{\partial(x, y)}{\partial(\rho, \phi)}\right| d\rho\, d\phi \;=\; \iint\limits_{R'} \rho \cdot \rho\, d\rho\, d\phi$$

$$= \int_{\phi=0}^{2\pi} \int_{\rho=2}^{3} \rho^2\, d\rho\, d\phi \;=\; \int_{\phi=0}^{2\pi} \frac{\rho^3}{3}\bigg|_{2}^{3}\, d\phi \;=\; \int_{\phi=0}^{2\pi} \frac{19}{3}\, d\phi \;=\; \frac{38\pi}{3}$$

We can also write the integration limits for R' immediately on observing the region R, since for fixed ϕ, ρ varies from $\rho = 2$ to $\rho = 3$ within the sector shown dashed in Fig. 9-8(a). An integration with respect to ϕ from $\phi = 0$ to $\phi = 2\pi$ then gives the contribution from all sectors. Geometrically $\rho\, d\rho\, d\phi$ represents the area dA as shown in Fig. 9-8(a).

8. Find the area of the region in the xy plane bounded by the lemniscate $\rho^2 = a^2 \cos 2\phi$.

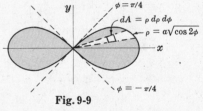

Here the curve is given directly in polar coordinates (ρ, ϕ). By assigning various values to ϕ and finding corresponding values of ρ we obtain the graph shown in Fig. 9-9. The required area (making use of symmetry) is

Fig. 9-9

$$4 \int_{\phi=0}^{\pi/4} \int_{\rho=0}^{a\sqrt{\cos 2\phi}} \rho\, d\rho\, d\phi \;=\; 4 \int_{\phi=0}^{\pi/4} \frac{\rho^2}{2}\bigg|_{\rho=0}^{a\sqrt{\cos 2\phi}}\, d\phi$$

$$= 2 \int_{\phi=0}^{\pi/4} a^2 \cos 2\phi\, d\phi \;=\; a^2 \sin 2\phi \bigg|_{\phi=0}^{\pi/4} \;=\; a^2$$

TRIPLE INTEGRALS

9. (a) Sketch the 3 dimensional region R bounded by $x + y + z = a$ ($a > 0$), $x = 0$, $y = 0$, $z = 0$.

(b) Give a physical interpretation to

$$\iiint\limits_{R} (x^2 + y^2 + z^2)\, dx\, dy\, dz$$

(c) Evaluate the triple integral in (b).

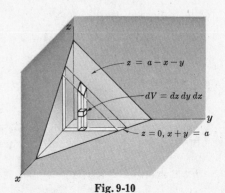

Fig. 9-10

(a)   The required region R is shown in Fig. 9-10.

(b)   Since $x^2 + y^2 + z^2$ is the square of the distance from any point (x, y, z) to $(0, 0, 0)$, we can consider the triple integral as representing the *polar moment of inertia* (i.e. moment of inertia with respect to the origin) of the region $\mathcal{R}$ (assuming unit density).

We can also consider the triple integral as representing the *mass* of the region if the density varies as $x^2 + y^2 + z^2$.

(*c*) The triple integral can be expressed as the iterated integral

$$\int_{x=0}^{a}\int_{y=0}^{a-x}\int_{z=0}^{a-x-y} (x^2 + y^2 + z^2)\, dz\, dy\, dx$$

$$= \int_{x=0}^{a}\int_{y=0}^{a-x} x^2 z + y^2 z + \frac{z^3}{3}\Big|_{z=0}^{a-x-y}\, dy\, dx$$

$$= \int_{x=0}^{a}\int_{y=0}^{a-x}\left\{ x^2(a-x) - x^2 y + (a-x)y^2 - y^3 + \frac{(a-x-y)^3}{3}\right\} dy\, dx$$

$$= \int_{x=0}^{a} x^2(a-x)y - \frac{x^2 y^2}{2} + \frac{(a-x)y^3}{3} - \frac{y^4}{4} - \frac{(a-x-y)^4}{12}\Big|_{y=0}^{a-x}\, dx$$

$$= \int_{0}^{a}\left\{ x^2(a-x)^2 - \frac{x^2(a-x)^2}{2} + \frac{(a-x)^4}{3} - \frac{(a-x)^4}{4} + \frac{(a-x)^4}{12}\right\} dx$$

$$= \int_{0}^{a}\left\{ \frac{x^2(a-x)^2}{2} + \frac{(a-x)^4}{6}\right\} dx \quad = \quad \frac{a^5}{20}$$

The integration with respect to z (keeping x and y constant) from $z = 0$ to $z = a - x - y$ corresponds to summing the polar moments of inertia (or masses) corresponding to each cube in a vertical column. The subsequent integration with respect to y from $y = 0$ to $y = a - x$ (keeping x constant) corresponds to addition of contributions from all vertical columns contained in a slab parallel to the yz plane. Finally, integration with respect to x from $x = 0$ to $x = a$ adds up contributions from all slabs parallel to the yz plane.

Although the above integration has been accomplished in the order z, y, x, any other order is clearly possible and the final answer should be the same.

10. Find the (*a*) volume and (*b*) centroid of the region $\mathcal{R}$ bounded by the parabolic cylinder $z = 4 - x^2$ and the planes $x = 0$, $y = 0$, $y = 6$, $z = 0$ assuming the density to be a constant σ.

The region $\mathcal{R}$ is shown in Fig. 9-11.

(*a*) Required volume $= \iiint\limits_{\mathcal{R}} dx\, dy\, dz$

$$= \int_{x=0}^{2}\int_{y=0}^{6}\int_{z=0}^{4-x^2} dz\, dy\, dx$$

$$= \int_{x=0}^{2}\int_{y=0}^{6} (4 - x^2)\, dy\, dx$$

$$= \int_{x=0}^{2} (4 - x^2)y\Big|_{y=0}^{6}\, dx$$

$$= \int_{x=0}^{2} (24 - 6x^2)\, dx \quad = \quad 32$$

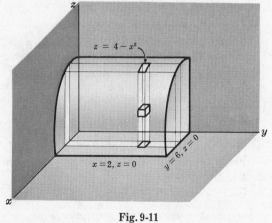

Fig. 9-11

(*b*) Total mass $= \displaystyle\int_{x=0}^{2}\int_{y=0}^{6}\int_{z=0}^{4-x^2} \sigma\, dz\, dy\, dx \quad = \quad 32\sigma$ by part (*a*), since σ is constant. Then

$$\bar{x} = \frac{\text{Total moment about } yz \text{ plane}}{\text{Total mass}} = \frac{\displaystyle\int_{x=0}^{2}\int_{y=0}^{6}\int_{z=0}^{4-x^2} \sigma x\, dz\, dy\, dx}{\text{Total mass}} = \frac{24\sigma}{32\sigma} = \frac{3}{4}$$

$$\bar{y} \;=\; \frac{\text{Total moment about } xz \text{ plane}}{\text{Total mass}} \;=\; \frac{\int_{x=0}^{2}\int_{y=0}^{6}\int_{z=0}^{4-x^2} \sigma y \, dz \, dy \, dx}{\text{Total mass}} \;=\; \frac{96\sigma}{32\sigma} \;=\; 3$$

$$\bar{z} \;=\; \frac{\text{Total moment about } xy \text{ plane}}{\text{Total mass}} \;=\; \frac{\int_{x=0}^{2}\int_{y=0}^{6}\int_{z=0}^{4-x^2} \sigma z \, dz \, dy \, dx}{\text{Total mass}} \;=\; \frac{256\sigma/5}{32\sigma} \;=\; \frac{8}{5}$$

Thus the centroid has coordinates $(3/4, 3, 8/5)$.

Note that the value for $\bar{y}$ could have been predicted because of symmetry.

TRANSFORMATION of TRIPLE INTEGRALS

11. Justify equation *(11)*, Page 182, for changing variables in a triple integral.

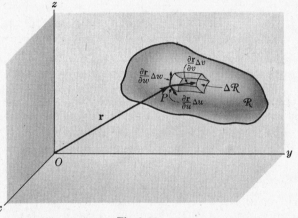

Fig. 9-12

By analogy with Problem 4, we construct a grid of curvilinear coordinate surfaces which sub-divide the region $\mathcal{R}$ into subregions, a typical one of which is $\Delta\mathcal{R}$ (see Fig. 9-12).

The vector $\mathbf{r}$ from the origin O to point P is

$$r \;=\; x\mathbf{i} + y\mathbf{j} + z\mathbf{k} \;=\; f(u,v,w)\mathbf{i} + g(u,v,w)\mathbf{j} + h(u,v,w)\mathbf{k}$$

assuming that the transformation equations are $x = f(u,v,w)$, $y = g(u,v,w)$ and $z = h(u,v,w)$.

Tangent vectors to the coordinate curves corresponding to the intersection of pairs of coordinate surfaces are given by $\partial r/\partial u$, $\partial r/\partial v$, $\partial r/\partial w$. Then the volume of the region $\Delta\mathcal{R}$ of Fig. 9-12 is given approximately by

$$\left| \frac{\partial \mathbf{r}}{\partial u} \cdot \frac{\partial \mathbf{r}}{\partial v} \times \frac{\partial \mathbf{r}}{\partial w} \right| \Delta u \, \Delta v \, \Delta w \;=\; \left| \frac{\partial(x,y,z)}{\partial(u,v,w)} \right| \Delta u \, \Delta v \, \Delta w$$

The triple integral of $F(x,y,z)$ over the region is the limit of the sum

$$\Sigma \, F\{f(u,v,w),\, g(u,v,w),\, h(u,v,w)\} \left| \frac{\partial(x,y,z)}{\partial(u,v,w)} \right| \Delta u \, \Delta v \, \Delta w$$

An investigation reveals that this limit is

$$\iiint\limits_{\mathcal{R}'} F\{f(u,v,w),\, g(u,v,w),\, h(u,v,w)\} \left| \frac{\partial(x,y,z)}{\partial(u,v,w)} \right| du \, dv \, dw$$

where $\mathcal{R}'$ is the region in the uvw space into which the region $\mathcal{R}$ is mapped under the transformation.

Another method for justifying the above change of variables in triple integrals makes use of Stokes' theorem (see Problem 84, Chapter 10).

12. Express $\displaystyle\iiint\limits_{R} F(x, y, z)\, dx\, dy\, dz$ in (a) cylindrical and (b) spherical coordinates.

(a)　The transformation equations in cylindrical coordinates are $x = \rho \cos \phi,\; y = \rho \sin \phi,\; z = z$.

As in Problem 39, Chapter 6, $\partial(x, y, z)/\partial(\rho, \phi, z) = \rho$. Then by Problem 11 the triple integral becomes

$$\iiint\limits_{R'} G(\rho, \phi, z)\, \rho\, d\rho\, d\phi\, dz$$

where R' is the region in the ρ, ϕ, z space corresponding to R and where $G(\rho, \phi, z) \equiv F(\rho \cos \phi, \rho \sin \phi, z)$.

(b)　The transformation equations in spherical coordinates are $x = r \sin \theta \cos \phi,\; y = r \sin \theta \sin \phi,\; z = r \cos \theta$.

By Problem 103, Chapter 6, $\partial(x, y, z)/\partial(r, \theta, \phi) = r^2 \sin \theta$. Then by Problem 11 the triple integral becomes

$$\iiint\limits_{R'} H(r, \theta, \phi)\, r^2 \sin \theta\, dr\, d\theta\, d\phi$$

where R' is the region in the r, θ, ϕ space corresponding to R, and where $H(r, \theta, \phi) \equiv F(r \sin \theta \cos \phi,\; r \sin \theta \sin \phi,\; r \cos \theta)$.

13. Find the volume of the region above the xy plane bounded by the paraboloid $z = x^2 + y^2$ and the cylinder $x^2 + y^2 = a^2$.

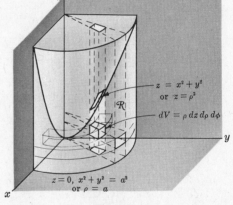

Fig. 9-13

The volume is most easily found by using cylindrical coordinates. In these coordinates the equations for the paraboloid and cylinder are respectively $z = \rho^2$ and $\rho = a$. Then

Required volume

$\quad =$　4 times volume shown in Fig. 9-13

$$= 4 \int_{\phi=0}^{\pi/2} \int_{\rho=0}^{a} \int_{z=0}^{\rho^2} \rho\, dz\, d\rho\, d\phi$$

$$= 4 \int_{\phi=0}^{\pi/2} \int_{\rho=0}^{a} \rho^3\, d\rho\, d\phi$$

$$= 4 \int_{\phi=0}^{\pi/2} \frac{\rho^4}{4}\Big|_{\rho=0}^{a}\, d\phi \;=\; \frac{\pi}{2} a^4$$

The integration with respect to z (keeping ρ and ϕ constant) from $z = 0$ to $z = \rho^2$ corresponds to summing the cubical volumes (indicated by dV) in a vertical column extending from the xy plane to the paraboloid. The subsequent integration with respect to ρ (keeping ϕ constant) from $\rho = 0$ to $\rho = a$ corresponds to addition of volumes of all columns in the wedge shaped region. Finally, integration with respect to ϕ corresponds to adding volumes of all such wedge shaped regions.

The integration can also be performed in other orders to yield the same result.

We can also set up the integral by determining the region R' in ρ, ϕ, z space into which R is mapped by the cylindrical coordinate transformation.

14. (a) Find the moment of inertia about the z axis of the region in Problem 13, assuming that the density is the constant σ. (b) Find the radius of gyration.

(a)　The moment of inertia about the z axis is

$$I_z \;=\; 4 \int_{\phi=0}^{\pi/2} \int_{\rho=0}^{a} \int_{z=0}^{\rho^2} \rho^2 \cdot \sigma \rho\, dz\, d\rho\, d\phi$$

$$= 4\sigma \int_{\phi=0}^{\pi/2} \int_{\rho=0}^{a} \rho^5\, d\rho\, d\phi \;=\; 4\sigma \int_{\phi=0}^{\pi/2} \frac{\rho^6}{6}\Big|_{\rho=0}^{a}\, d\phi \;=\; \frac{\pi a^6 \sigma}{3}$$

The result can be expressed in terms of the mass M of the region, since by Problem 13,

$$M = \text{volume} \times \text{density} = \frac{\pi}{2} a^4 \sigma \quad \text{so that} \quad I_z = \frac{\pi a^6 \sigma}{3} = \frac{\pi a^6}{3} \cdot \frac{2M}{\pi a^4} = \frac{2}{3} M a^2$$

Note that in setting up the integral for I_z we can think of $\sigma \rho \, dz \, d\rho \, d\phi$ as being the mass of the cubical volume element, $\rho^2 \cdot \sigma \rho \, dz \, d\rho \, d\phi$, as the moment of inertia of this mass with respect to the z axis and $\iiint\limits_{\mathcal{R}} \rho^2 \cdot \sigma \rho \, dz \, d\rho \, d\phi$ as the total moment of inertia about the z axis. The limits of integration are determined as in Problem 13.

(b) The radius of gyration is the value K such that $MK^2 = \frac{2}{3} M a^2$, i.e. $K^2 = \frac{2}{3} a^2$ or $K = a\sqrt{2/3}$.

The physical significance of K is that if all the mass M were concentrated in a thin cylindrical shell of radius K, then the moment of inertia of this shell about the axis of the cylinder would be I_z.

15. (a) Find the volume of the region bounded above by the sphere $x^2 + y^2 + z^2 = a^2$ and below by the cone $z^2 \sin^2 \alpha = (x^2 + y^2) \cos^2 \alpha$, where α is a constant such that $0 \leqq \alpha \leqq \pi$. (b) From the result in (a), find the volume of a sphere of radius a.

In spherical coordinates the equation of the sphere is $r = a$ and that of the cone is $\theta = \alpha$. This can be seen directly or by using the transformation equations $x = r \sin \theta \cos \phi$, $y = r \sin \theta \sin \phi$, $z = r \cos \theta$. For example, $z^2 \sin^2 \alpha = (x^2 + y^2) \cos^2 \alpha$ becomes, on using these equations,

$$r^2 \cos^2 \theta \sin^2 \alpha = (r^2 \sin^2 \theta \cos^2 \phi + r^2 \sin^2 \theta \sin^2 \phi) \cos^2 \alpha$$

i.e.,
$$r^2 \cos^2 \theta \sin^2 \alpha = r^2 \sin^2 \theta \cos^2 \alpha$$

from which $\tan \theta = \pm \tan \alpha$ and so $\theta = \alpha$ or $\theta = \pi - \alpha$. It is sufficient to consider one of these, say $\theta = \alpha$.

(a)   Required volume

= 4 times volume (shaded) in Fig. 9-14

$$= 4 \int_{\phi=0}^{\pi/2} \int_{\theta=0}^{\alpha} \int_{r=0}^{a} r^2 \sin \theta \, dr \, d\theta \, d\phi$$

$$= 4 \int_{\phi=0}^{\pi/2} \int_{\theta=0}^{\alpha} \frac{r^3}{3} \sin \theta \Big|_{r=0}^{a} d\theta \, d\phi$$

$$= \frac{4a^3}{3} \int_{\phi=0}^{\pi/2} \int_{\theta=0}^{\alpha} \sin \theta \, d\theta \, d\phi$$

$$= \frac{4a^3}{3} \int_{\phi=0}^{\pi/2} -\cos \theta \Big|_{\theta=0}^{\alpha} d\phi$$

$$= \frac{2\pi a^3}{3} (1 - \cos \alpha)$$

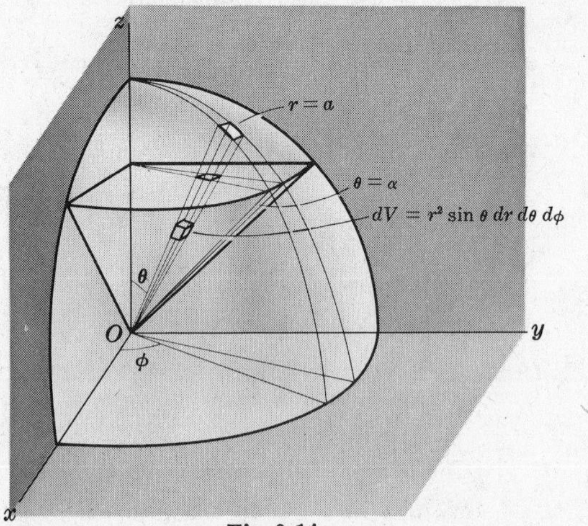

Fig. 9-14

The integration with respect to r (keeping θ and ϕ constant) from $r = 0$ to $r = a$ corresponds to summing the volumes of all cubical elements (such as indicated by dV) in a column extending from $r = 0$ to $r = a$. The subsequent integration with respect to θ (keeping ϕ constant) from $\theta = 0$ to $\theta = \pi/4$ corresponds to summing the volumes of all columns in the wedge shaped region. Finally, integration with respect to ϕ corresponds to adding volumes of all such wedge shaped regions.

(b)   Letting $\alpha = \pi$, the volume of the sphere thus obtained is

$$\frac{2\pi a^3}{3} (1 - \cos \pi) = \frac{4}{3} \pi a^3$$

16. (*a*) Find the centroid of the region in Problem 15.

(*b*) Use the result in (*a*) to find the centroid of a hemisphere.

(*a*)   The centroid $(\bar{x}, \bar{y}, \bar{z})$ is, due to symmetry, given by $\bar{x} = \bar{y} = 0$ and

$$\bar{z} \;=\; \frac{\text{Total moment about } xy \text{ plane}}{\text{Total mass}} \;=\; \frac{\iiint z\,\sigma\,dV}{\iiint \sigma\,dV}$$

Since $z = r \cos\theta$ and σ is constant the numerator is

$$4\sigma \int_{\phi=0}^{\pi/2} \int_{\theta=0}^{\alpha} \int_{r=0}^{a} r\cos\theta \cdot r^2 \sin\theta \, dr \, d\theta \, d\phi \;=\; 4\sigma \int_{\phi=0}^{\pi/2} \int_{\theta=0}^{\alpha} \left.\frac{r^4}{4}\right|_{r=0}^{a} \sin\theta\cos\theta\,d\theta\,d\phi$$

$$=\; \sigma a^4 \int_{\phi=0}^{\pi/2} \int_{\theta=0}^{\alpha} \sin\theta\cos\theta\,d\theta\,d\phi$$

$$=\; \sigma a^4 \int_{\phi=0}^{\pi/2} \left.\frac{\sin^2\theta}{2}\right|_{\theta=0}^{\alpha} d\phi \;=\; \frac{\pi\sigma a^4 \sin^2\alpha}{4}$$

The denominator, obtained by multiplying the result of Prob. 15(*a*) by σ, is $\frac{2}{3}\pi\sigma a^3(1-\cos\alpha)$. Then

$$\bar{z} \;=\; \frac{\frac{1}{4}\pi\sigma a^4 \sin^2\alpha}{\frac{2}{3}\pi\sigma a^3(1-\cos\alpha)} \;=\; \frac{3}{8}a(1+\cos\alpha).$$

(*b*)   Letting $\alpha = \pi/2$, $\bar{z} = \frac{3}{8}a$.

MISCELLANEOUS PROBLEMS

17. Prove that   (*a*) $\displaystyle\int_0^1 \left\{ \int_0^1 \frac{x-y}{(x+y)^3}\,dy \right\} dx = \frac{1}{2}$,   (*b*) $\displaystyle\int_0^1 \left\{ \int_0^1 \frac{x-y}{(x+y)^3}\,dx \right\} dy = -\frac{1}{2}$.

(*a*) $\displaystyle\int_0^1 \left\{ \int_0^1 \frac{x-y}{(x+y)^3}\,dy \right\} dx \;=\; \int_0^1 \left\{ \int_0^1 \frac{2x-(x+y)}{(x+y)^3}\,dy \right\} dx$

$$=\; \int_0^1 \left\{ \int_0^1 \left(\frac{2x}{(x+y)^3} - \frac{1}{(x+y)^2} \right) dy \right\} dx$$

$$=\; \int_0^1 \left. \left(\frac{-x}{(x+y)^2} + \frac{1}{x+y} \right) \right|_{y=0}^{1} dx$$

$$=\; \int_0^1 \frac{dx}{(x+1)^2} \;=\; \left. \frac{-1}{x+1} \right|_0^1 \;=\; \frac{1}{2}$$

(*b*)   This follows at once on formally interchanging x and y in (*a*) to obtain $\displaystyle\int_0^1 \left\{ \int_0^1 \frac{y-x}{(x+y)^3}\,dx \right\} dy = \frac{1}{2}$ and then multiplying both sides by -1.

This example shows that interchange in order of integration may not always produce equal results. A sufficient condition under which the order may be interchanged is that the double integral over the corresponding region exists. In this case $\displaystyle\iint_{\mathcal{R}} \frac{x-y}{(x+y)^3}\,dx\,dy$, where $\mathcal{R}$ is the region $0 \leqq x \leqq 1$, $0 \leqq y \leqq 1$ fails to exist because of the discontinuity of the integrand at the origin. The integral is actually an *improper* double integral (see Chapter 12).

18. Prove that $\displaystyle\int_0^x \left\{ \int_0^t F(u)\,du \right\} dt \;=\; \int_0^x (x-u)\,F(u)\,du.$

Let $\displaystyle I(x) = \int_0^x \left\{ \int_0^t F(u)\,du \right\} dt$, $\displaystyle J(x) = \int_0^x (x-u)\,F(u)\,du.$ Then

$$I'(x) \;=\; \int_0^x F(u)\,du, \qquad J'(x) \;=\; \int_0^x F(u)\,du$$

using Leibnitz's rule, Page 163. Thus $I'(x) = J'(x)$, and so $I(x) - J(x) = c$ where c is a constant. Since $I(0) = J(0) = 0$, $c = 0$ and so $I(x) = J(x)$.

The result is sometimes written in the form

$$\int_0^x \int_0^x F(x)\,dx^2 \;=\; \int_0^x (x-u)\,F(u)\,du$$

The result can be generalized to give (see Problem 54)

$$\int_0^x \int_0^x \cdots \int_0^x F(x)\,dx^n \;=\; \frac{1}{(n-1)!}\int_0^x (x-u)^{n-1}\,F(u)\,du$$

Supplementary Problems

DOUBLE INTEGRALS

19. (a) Sketch the region $\mathcal{R}$ in the xy plane bounded by $y^2 = 2x$ and $y = x$. (b) Find the area of $\mathcal{R}$. (c) Find the polar moment of inertia of $\mathcal{R}$ assuming constant density σ.

   Ans. (b) $\frac{2}{3}$; (c) $48\sigma/35 = 72M/35$, where M is the mass of $\mathcal{R}$.

20. Find the centroid of the region in the preceding problem.   Ans. $\bar{x} = \frac{4}{5}$, $\bar{y} = 1$

21. Given $\displaystyle\int_{y=0}^3 \int_{x=1}^{\sqrt{4-y}} (x+y)\,dx\,dy$. (a) Sketch the region and give a possible physical interpretation of the double integral. (b) Interchange the order of integration. (c) Evaluate the double integral.

   Ans. (b) $\displaystyle\int_{x=1}^2 \int_{y=0}^{4-x^2} (x+y)\,dy\,dx$,   (c) 241/60

22. Show that $\displaystyle\int_{x=1}^2 \int_{y=\sqrt{x}}^x \sin\frac{\pi x}{2y}\,dy\,dx \;+\; \int_{x=2}^4 \int_{y=\sqrt{x}}^2 \sin\frac{\pi x}{2y}\,dy\,dx \;=\; \frac{4(\pi+2)}{\pi^3}$.

23. Find the volume of the tetrahedron bounded by $x/a + y/b + z/c = 1$ and the coordinate planes.
   Ans. $abc/6$

24. Find the volume of the region bounded by $z = x^2 + y^2$, $z = 0$, $x = -a$, $x = a$, $y = -a$, $y = a$.
   Ans. $8a^4/3$

25. Find (a) the moment of inertia about the z axis and (b) the centroid of the region in Problem 24 assuming a constant density σ.
   Ans. (a) $\frac{112}{45}a^6\sigma = \frac{14}{15}Ma^2$, where $M =$ mass; (b) $\bar{x} = \bar{y} = 0$, $\bar{z} = \frac{7}{15}a^2$

TRANSFORMATION of DOUBLE INTEGRALS

26. Evaluate $\displaystyle\iint_{\mathcal{R}} \sqrt{x^2+y^2}\,dx\,dy$, where $\mathcal{R}$ is the region $x^2 + y^2 \leqq a^2$.   Ans. $\frac{2}{3}\pi a^3$

27. If $\mathcal{R}$ is the region of Prob. 26, evaluate $\displaystyle\iint_{\mathcal{R}} e^{-(x^2+y^2)}\,dx\,dy$.   Ans. $\pi(1 - e^{-a^2})$

28. By using the transformation $x + y = u$, $y = uv$, show that

$$\int_{x=0}^1 \int_{y=0}^{1-x} e^{y/(x+y)}\,dy\,dx \;=\; \frac{e-1}{2}$$

29. Find the area of the region bounded by $xy = 4$, $xy = 8$, $xy^3 = 5$, $xy^3 = 15$. [Hint: Let $xy = u$, $xy^3 = v$.]
    Ans. $2 \ln 3$

30. Show that the volume generated by revolving the region in the first quadrant bounded by the parabolas $y^2 = x$, $y^2 = 8x$, $x^2 = y$, $x^2 = 8y$ about the x axis is $279\pi/2$. [Hint: Let $y^2 = ux$, $x^2 = vy$.]

31. Find the area of the region in the first quadrant bounded by $y = x^3$, $y = 4x^3$, $x = y^3$, $x = 4y^3$.　　　Ans. $\frac{1}{8}$

32. Let $\mathcal{R}$ be the region bounded by $x + y = 1$, $x = 0$, $y = 0$. Show that $\iint\limits_{\mathcal{R}} \cos\left(\dfrac{x-y}{x+y}\right) dx\, dy = \dfrac{\sin 1}{2}$.
    [Hint: Let $x - y = u$, $x + y = v$.]

TRIPLE INTEGRALS

33. (a) Evaluate $\displaystyle\int_{x=0}^{1} \int_{y=0}^{1} \int_{z=\sqrt{x^2+y^2}}^{2} xyz\, dz\, dy\, dx$. (b) Give a physical interpretation to the integral in (a).
    Ans. (a) $\frac{3}{8}$

34. Find the (a) volume and (b) centroid of the region in the first octant bounded by $x/a + y/b + z/c = 1$, where a, b, c are positive.　　　Ans. (a) $abc/6$; (b) $\bar{x} = a/4$, $\bar{y} = b/4$, $\bar{z} = c/4$

35. Find the (a) moment of inertia and (b) radius of gyration about the z axis of the region in Prob. 34.
    Ans. (a) $M(a^2 + b^2)/10$, (b) $\sqrt{(a^2 + b^2)/10}$

36. Find the mass of the region corresponding to $x^2 + y^2 + z^2 \leqq 4$, $x \geqq 0$, $y \geqq 0$, $z \geqq 0$, if the density is equal to xyz.　　　Ans. $4/3$

37. Find the volume of the region bounded by $z = x^2 + y^2$ and $z = 2x$.　　　Ans. $\pi/2$

TRANSFORMATION of TRIPLE INTEGRALS

38. Find the volume of the region bounded by $z = 4 - x^2 - y^2$ and the xy plane.　　　Ans. 8π

39. Find the centroid of the region in Problem 38, assuming constant density σ.
    Ans. $\bar{x} = \bar{y} = 0$, $\bar{z} = \frac{4}{3}$

40. (a) Evaluate $\displaystyle\iiint\limits_{\mathcal{R}} \sqrt{x^2 + y^2 + z^2}\, dx\, dy\, dz$, where $\mathcal{R}$ is the region bounded by the plane $z = 3$ and the
    cone $z = \sqrt{x^2 + y^2}$. (b) Give a physical interpretation of the integral in (a). [Hint: Perform the integration in cylindrical coordinates in the order ρ, z, ϕ.]　　　Ans. $27\pi(2\sqrt{2} - 1)/2$

41. Show that the volume of the region bounded by the cone $z = \sqrt{x^2 + y^2}$ and the paraboloid $z = x^2 + y^2$ is $\pi/6$.

42. Find the moment of inertia of a right circular cylinder of radius a and height b, about its axis if the density is proportional to the distance from the axis.　　　Ans. $\frac{3}{5}Ma^2$

43. (a) Evaluate $\displaystyle\iiint\limits_{\mathcal{R}} \dfrac{dx\, dy\, dz}{(x^2 + y^2 + z^2)^{3/2}}$, where $\mathcal{R}$ is the region bounded by the spheres $x^2 + y^2 + z^2 = a^2$
    and $x^2 + y^2 + z^2 = b^2$ where $a > b > 0$. (b) Give a physical interpretation of the integral in (a).
    Ans. (a) $4\pi \ln (a/b)$

44. (a) Find the volume of the region bounded above by the sphere $r = 2a \cos \theta$, and below by the cone $\phi = \alpha$ where $0 < \alpha < \pi/2$. (b) Discuss the case $\alpha = \pi/2$.　　　Ans. $\frac{4}{3}\pi a^3(1 - \cos^4 \alpha)$

45. Find the centroid of a hemispherical shell having outer radius a and inner radius b if the density (a) is constant, (b) varies as the square of the distance from the base. Discuss the case $a = b$.
    Ans. Taking the z axis as axis of symmetry: (a) $\bar{x} = \bar{y} = 0$, $\bar{z} = \frac{3}{8}(a^4 - b^4)/(a^3 - b^3)$; (b) $\bar{x} = \bar{y} = 0$,
    $\bar{z} = \frac{5}{8}(a^6 - b^6)/(a^5 - b^5)$

MISCELLANEOUS PROBLEMS

46. Find the mass of a right circular cylinder of radius a and height b, if the density varies as the square of the distance from a point on the circumference of the base.
Ans. $\frac{1}{6}\pi a^2 bk(9a^2 + 2b^2)$, where $k = $ constant of proportionality.

47. Find the (a) volume and (b) centroid of the region bounded above by the sphere $x^2 + y^2 + z^2 = a^2$ and below by the plane $z = b$ where $a > b > 0$, assuming constant density.
Ans. (a) $\frac{1}{3}\pi(2a^3 - 3a^2b + b^3)$; (b) $\bar{x} = \bar{y} = 0$, $\bar{z} = \frac{3}{4}(a+b)^2/(2a+b)$

48. A sphere of radius a has a cylindrical hole of radius b bored from it, the axis of the cylinder coinciding with a diameter of the sphere. Show that the volume of the sphere which remains is $\frac{4}{3}\pi[a^3 - (a^2 - b^2)^{3/2}]$.

49. A simple closed curve in a plane is revolved about an axis in the plane which does not intersect the curve. Prove that the volume generated is equal to the area bounded by the curve multiplied by the distance traveled by the centroid of the area (*Pappus' theorem*).

50. Use Problem 49 to find the volume generated by revolving the circle $x^2 + (y-b)^2 = a^2$, $b > a > 0$ about the x axis. *Ans.* $2\pi^2 a^2 b$

51. Find the volume of the region bounded by the hyperbolic cylinders $xy = 1$, $xy = 9$, $xz = 4$, $xz = 36$, $yz = 25$, $yz = 49$. [Hint: Let $xy = u$, $xz = v$, $yz = w$.] *Ans.* 64

52. Evaluate $\iiint\limits_{\mathcal{R}} \sqrt{1 - (x^2/a^2 + y^2/b^2 + z^2/c^2)} \, dx \, dy \, dz$, where $\mathcal{R}$ is the region interior to the ellipsoid $x^2/a^2 + y^2/b^2 + z^2/c^2 = 1$. [Hint: Let $x = au$, $y = bv$, $z = cw$. Then use spherical coordinates.]
Ans. $\frac{1}{4}\pi^2 abc$

53. If $\mathcal{R}$ is the region $x^2 + xy + y^2 \leq 1$, prove that $\iint\limits_{\mathcal{R}} e^{-(x^2 + xy + y^2)} \, dx \, dy = \frac{2\pi}{e\sqrt{3}}(e-1)$.

[Hint: Let $x = u \cos\alpha - v \sin\alpha$, $y = u \sin\alpha + v \cos\alpha$ and choose α so as to eliminate the xy term in the integrand. Then let $u = a\rho \cos\phi$, $v = b\rho \sin\phi$ where a and b are appropriately chosen.]

54. Prove that $\displaystyle\int_0^x \int_0^x \cdots \int_0^x F(x) \, dx^n = \frac{1}{(n-1)!} \int_0^x (x-u)^{n-1} F(u) \, du$ for $n = 1, 2, 3, \ldots$ (see Prob. 18).

Chapter 10

Line Integrals, Surface Integrals and Integral Theorems

LINE INTEGRALS

Let C be a curve in the xy plane which connects points $A(a_1, b_1)$ and $B(a_2, b_2)$, (see Fig. 10-1). Let $P(x, y)$ and $Q(x, y)$ be single-valued functions defined at all points of C. Subdivide C into n parts by choosing $(n-1)$ points on it given by $(x_1, y_1), (x_2, y_2), \ldots, (x_{n-1}, y_{n-1})$. Call $\Delta x_k = x_k - x_{k-1}$ and $\Delta y_k = y_k - y_{k-1}$, $k = 1, 2, \ldots, n$ where $(a_1, b_1) \equiv (x_0, y_0)$, $(a_2, b_2) \equiv (x_n, y_n)$ and suppose that points (ξ_k, η_k) are chosen so that they are situated on C between points (x_{k-1}, y_{k-1}) and (x_k, y_k). Form the sum

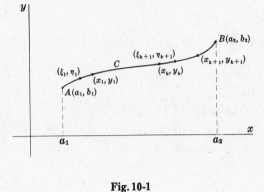

Fig. 10-1

$$\sum_{k=1}^{n} \{ P(\xi_k, \eta_k) \, \Delta x_k + Q(\xi_k, \eta_k) \, \Delta y_k \} \qquad (1)$$

The limit of this sum as $n \to \infty$ in such a way that all the quantities Δx_k, Δy_k approach zero, if such limit exists, is called a *line integral* along C and is denoted by

$$\int_C P(x, y) \, dx + Q(x, y) \, dy \qquad \text{or} \qquad \int_{(a_1, b_1)}^{(a_2, b_2)} P \, dx + Q \, dy \qquad (2)$$

The limit does exist if P and Q are continuous (or sectionally continuous) at all points of C. The value of the integral depends in general on P, Q, the particular curve C, and on the limits (a_1, b_1) and (a_2, b_2).

In an exactly analogous manner one may define a line integral along a curve C in three dimensional space as

$$\lim_{n \to \infty} \sum_{k=1}^{n} \{ A_1(\xi_k, \eta_k, \zeta_k) \, \Delta x_k + A_2(\xi_k, \eta_k, \zeta_k) \, \Delta y_k + A_3(\xi_k, \eta_k, \zeta_k) \, \Delta z_k \} \qquad (3)$$

$$= \int_C A_1 \, dx + A_2 \, dy + A_3 \, dz$$

where A_1, A_2 and A_3 are functions of x, y and z.

Other types of line integrals, depending on particular curves, can be defined. For example, if Δs_k denotes the arc length along curve C in the above figure between points (x_k, y_k) and (x_{k+1}, y_{k+1}), then

$$\lim_{n \to \infty} \sum_{k=1}^{n} U(\xi_k, \eta_k) \, \Delta s_k = \int_C U(x, y) \, ds \qquad (4)$$

is called the line integral of $U(x, y)$ along curve C. Extensions to three (or higher) dimensions are possible.

VECTOR NOTATION for LINE INTEGRALS

It is often convenient to express a line integral in vector form as an aid in physical or geometric understanding as well as for brevity of notation. For example, we can express the line integral (3) in the form

$$\int_C A_1\,dx + A_2\,dy + A_3\,dz = \int_C (A_1\mathbf{i} + A_2\mathbf{j} + A_3\mathbf{k}) \cdot (dx\,\mathbf{i} + dy\,\mathbf{j} + dz\,\mathbf{k}) \tag{5}$$

$$= \int_C \mathbf{A} \cdot d\mathbf{r}$$

where $\mathbf{A} = A_1\mathbf{i} + A_2\mathbf{j} + A_3\mathbf{k}$ and $d\mathbf{r} = dx\,\mathbf{i} + dy\,\mathbf{j} + dz\,\mathbf{k}$. The line integral (2) is a special case of this with $z = 0$.

If at each point (x, y, z) we associate a force $\mathbf{F}$ acting on an object (i.e. if a *force field* is defined), then

$$\int_C \mathbf{F} \cdot d\mathbf{r} \tag{6}$$

represents physically the total work done in moving the object along the curve C.

EVALUATION of LINE INTEGRALS

If the equation of a curve C in the plane $z = 0$ is given as $y = f(x)$, the line integral (2) is evaluated by placing $y = f(x)$, $dy = f'(x)\,dx$ in the integrand to obtain the definite integral

$$\int_{a_1}^{a_2} P\{x, f(x)\}\,dx + Q\{x, f(x)\}\,f'(x)\,dx \tag{7}$$

which is then evaluated in the usual manner.

Similarly if C is given as $x = g(y)$, then $dx = g'(y)\,dy$ and the line integral becomes

$$\int_{b_1}^{b_2} P\{g(y), y\}\,g'(y)\,dy + Q\{g(y), y\}\,dy \tag{8}$$

If C is given in parametric form $x = \phi(t)$, $y = \psi(t)$, the line integral becomes

$$\int_{t_1}^{t_2} P\{\phi(t), \psi(t)\}\,\phi'(t)\,dt + Q\{\phi(t), \psi(t)\}\,\psi'(t)\,dt \tag{9}$$

where t_1 and t_2 denote the values of t corresponding to points A and B respectively.

Combinations of the above methods may be used in the evaluation.

Similar methods are used for evaluating line integrals along space curves.

PROPERTIES of LINE INTEGRALS

Line integrals have properties which are analogous to those of ordinary integrals. For example:

1. $\displaystyle\int_C P(x, y)\,dx + Q(x, y)\,dy = \int_C P(x, y)\,dx + \int_C Q(x, y)\,dy$

2. $\displaystyle\int_{(a_1, b_1)}^{(a_2, b_2)} P\,dx + Q\,dy = -\int_{(a_2, b_2)}^{(a_1, b_1)} P\,dx + Q\,dy$

Thus reversal of the path of integration changes the sign of the line integral.

3. $\displaystyle\int_{(a_1, b_1)}^{(a_2, b_2)} P\,dx \;+\; Q\,dy \;\;=\;\; \int_{(a_1, b_1)}^{(a_3, b_3)} P\,dx \;+\; Q\,dy \;+\; \int_{(a_3, b_3)}^{(a_2, b_2)} P\,dx \;+\; Q\,dy$

       where (a_3, b_3) is another point on C.

Similar properties hold for line integrals in space.

SIMPLE CLOSED CURVES. SIMPLY and MULTIPLY-CONNECTED REGIONS

A *simple closed curve* is a closed curve which does not intersect itself anywhere. Mathematically, a curve in the xy plane is defined by the parametric equations $x = \phi(t)$, $y = \psi(t)$ where ϕ and ψ are single-valued and continuous in an interval $t_1 \leqq t \leqq t_2$. If $\phi(t_1) = \phi(t_2)$ and $\psi(t_1) = \psi(t_2)$, the curve is said to be *closed*. If $\phi(u) = \phi(v)$ and $\psi(u) = \psi(v)$ only when $u = v$ (except in the special case where $u = t_1$ and $v = t_2$), the curve is closed and does not intersect itself and so is a simple closed curve. We shall also assume, unless otherwise stated, that ϕ and ψ are piecewise differentiable in $t_1 \leqq t \leqq t_2$.

If a plane region has the property that any closed curve in it can be continuously shrunk to a point without leaving the region, then the region is called *simply-connected*, otherwise it is called *multiply-connected* (see Page 102 of Chapter 6).

As the parameter t varies from t_1 to t_2, the plane curve is described in a certain sense or direction. For curves in the xy plane, we arbitrarily describe this direction as *positive* or *negative* according as a person traversing the curve in this direction with his head pointing in the positive z direction has the region enclosed by the curve always toward his left or right respectively. If we look down upon a simple closed curve in the xy plane, this amounts to saying that traversal of the curve in the counterclockwise direction is taken as positive while traversal in the clockwise direction is taken as negative.

GREEN'S THEOREM in the PLANE

Let P, Q, $\partial P/\partial y$, $\partial Q/\partial x$ be single-valued and continuous in a simply-connected region $\mathcal{R}$ bounded by a simple closed curve C. Then

$$\oint_C P\,dx \;+\; Q\,dy \;\;=\;\; \iint_{\mathcal{R}} \left(\frac{\partial Q}{\partial x} - \frac{\partial P}{\partial y} \right) dx\,dy \qquad\qquad (10)$$

where $\displaystyle\oint_C$ is used to emphasize that C is closed and that it is described in the positive direction.

This theorem is also true for regions bounded by two or more closed curves (i.e. multiply-connected regions). See Problem 10.

CONDITIONS for a LINE INTEGRAL to be INDEPENDENT of the PATH

Theorem 1.

A necessary and sufficient condition for $\displaystyle\int_C P\,dx \;+\; Q\,dy$ to be independent of the path C joining any two given points in a region $\mathcal{R}$ is that in $\mathcal{R}$

$$\frac{\partial P}{\partial y} \;=\; \frac{\partial Q}{\partial x} \qquad\qquad (11)$$

where it is supposed that these partial derivatives are continuous in $\mathcal{R}$.

The condition (11) is also the condition that $P\,dx + Q\,dy$ is an exact differential, i.e. that there exists a function $\phi(x,y)$ such that $P\,dx + Q\,dy = d\phi$. In such case if the end points of curve C are (x_1, y_1) and (x_2, y_2), the value of the line integral is given by

$$\int_{(x_1,y_1)}^{(x_2,y_2)} P\,dx + Q\,dy = \int_{(x_1,y_1)}^{(x_2,y_2)} d\phi = \phi(x_2,y_2) - \phi(x_1,y_1) \tag{12}$$

In particular if (11) holds and C is closed, we have $x_1 = x_2$, $y_1 = y_2$ and

$$\oint_C P\,dx + Q\,dy = 0 \tag{13}$$

For proofs and related theorems, see Problems 11-13.

The results in Theorem 1 can be extended to line integrals in space. Thus we have

Theorem 2.

A necessary and sufficient condition for $\displaystyle\int_C A_1\,dx + A_2\,dy + A_3\,dz$ to be independent of the path C joining any two given points in a region $\mathcal{R}$ is that in $\mathcal{R}$

$$\frac{\partial A_1}{\partial y} = \frac{\partial A_2}{\partial x}, \quad \frac{\partial A_3}{\partial x} = \frac{\partial A_1}{\partial z}, \quad \frac{\partial A_2}{\partial z} = \frac{\partial A_3}{\partial y} \tag{14}$$

where it is supposed that these partial derivatives are continuous in $\mathcal{R}$.

The results can be expressed concisely in terms of vectors. If $\mathbf{A} = A_1\mathbf{i} + A_2\mathbf{j} + A_3\mathbf{k}$, the line integral can be written $\displaystyle\int_C \mathbf{A}\cdot d\mathbf{r}$ and condition (14) is equivalent to the condition $\nabla \times \mathbf{A} = 0$. If $\mathbf{A}$ represents a force field $\mathbf{F}$ which acts on an object, the result is equivalent to the statement that the work done in moving the object from one point to another is independent of the path joining the two points if and only if $\nabla \times \mathbf{A} = 0$. Such a force field is often called *conservative*.

The condition (14) [or the equivalent condition $\nabla \times \mathbf{A} = 0$] is also the condition that $A_1\,dx + A_2\,dy + A_3\,dz$ [or $\mathbf{A}\cdot d\mathbf{r}$] is an exact differential, i.e. that there exists a function $\phi(x,y,z)$ such that $A_1\,dx + A_2\,dy + A_3\,dz = d\phi$. In such case if the endpoints of curve C are (x_1, y_1, z_1) and (x_2, y_2, z_2), the value of the line integral is given by

$$\int_{(x_1,y_1,z_1)}^{(x_2,y_2,z_2)} \mathbf{A}\cdot d\mathbf{r} = \int_{(x_1,y_1,z_1)}^{(x_2,y_2,z_2)} d\phi = \phi(x_2,y_2,z_2) - \phi(x_1,y_1,z_1) \tag{15}$$

In particular if C is closed and $\nabla \times \mathbf{A} = 0$, we have

$$\oint_C \mathbf{A}\cdot d\mathbf{r} = 0 \tag{16}$$

SURFACE INTEGRALS

Let S be a two-sided surface having projection $\mathcal{R}$ on the xy plane as in the adjoining Fig. 10-2. Assume that an equation for S is $z = f(x,y)$, where f is single-valued and continuous for all x and y in $\mathcal{R}$. Divide $\mathcal{R}$ into n subregions of area ΔA_p, $p = 1, 2, \ldots, n$, and erect a vertical column on each of these subregions to intersect S in an area ΔS_p.

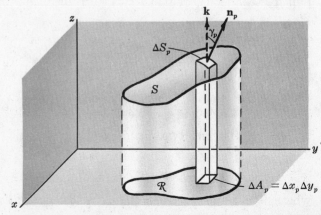

$$\Delta A_p = \Delta x_p \Delta y_p$$

Fig. 10-2

Let $\phi(x, y, z)$ be single-valued and continuous at all points of S. Form the sum

$$\sum_{p=1}^{n} \phi(\xi_p, \eta_p, \zeta_p)\, \Delta S_p \tag{17}$$

where (ξ_p, η_p, ζ_p) is some point of ΔS_p. If the limit of this sum as $n \to \infty$ in such a way that each $\Delta S_p \to 0$ exists, the resulting limit is called the *surface integral* of $\phi(x, y, z)$ over S and is designated by

$$\iint_S \phi(x, y, z)\, dS \tag{18}$$

Since $\Delta S_p = |\sec \gamma_p|\, \Delta A_p$ approximately, where γ_p is the angle between the normal line to S and the positive z axis, the limit of the sum (17) can be written

$$\iint_{\mathcal{R}} \phi(x, y, z)\, |\sec \gamma|\, dA \tag{19}$$

The quantity $|\sec \gamma|$ is given by

$$|\sec \gamma| \;=\; \frac{1}{|\mathbf{n}_p \cdot \mathbf{k}|} \;=\; \sqrt{1 + \left(\frac{\partial z}{\partial x}\right)^2 + \left(\frac{\partial z}{\partial y}\right)^2} \tag{20}$$

Then assuming that $z = f(x, y)$ has continuous (or sectionally continuous) derivatives in $\mathcal{R}$, (19) can be written in rectangular form as

$$\iint_{\mathcal{R}} \phi(x, y, z) \sqrt{1 + \left(\frac{\partial z}{\partial x}\right)^2 + \left(\frac{\partial z}{\partial y}\right)^2}\, dx\, dy \tag{21}$$

In case the equation for S is given as $F(x, y, z) = 0$, (21) can also be written

$$\iint_{\mathcal{R}} \phi(x, y, z)\, \frac{\sqrt{(F_x)^2 + (F_y)^2 + (F_z)^2}}{|F_z|}\, dx\, dy \tag{22}$$

The results (21) or (22) can be used to evaluate (18).

In the above we have assumed that S is such that any line parallel to the z axis intersects S in only one point. In case S is not of this type, we can usually subdivide S into surfaces $S_1, S_2, \ldots$ which are of this type. Then the surface integral over S is defined as the sum of the surface integrals over $S_1, S_2, \ldots$.

The results stated hold when S is projected on to a region $\mathcal{R}$ of the xy plane. In some cases it is better to project S on to the yz or xz planes. For such cases (18) can be evaluated by appropriately modifying (21) and (22).

The DIVERGENCE THEOREM

Let S be a closed surface bounding a region of volume V. Choose the outward drawn normal to the surface as the *positive normal* and assume that α, β, γ are the angles which this normal makes with the positive x, y and z axes respectively. Then if A_1, A_2 and A_3 are continuous and have continuous partial derivatives in the region

$$\iiint_V \left(\frac{\partial A_1}{\partial x} + \frac{\partial A_2}{\partial y} + \frac{\partial A_3}{\partial z}\right) dV \;=\; \iint_S (A_1 \cos \alpha + A_2 \cos \beta + A_3 \cos \gamma)\, dS \tag{23}$$

which can also be written

$$\iiint_V \left(\frac{\partial A_1}{\partial x} + \frac{\partial A_2}{\partial y} + \frac{\partial A_3}{\partial z}\right) dV \;=\; \iint_S A_1\, dy\, dz + A_2\, dz\, dx + A_3\, dx\, dy \tag{24}$$

In vector form with $\mathbf{A} = A_1\mathbf{i} + A_2\mathbf{j} + A_3\mathbf{k}$ and $\mathbf{n} = \cos\alpha\,\mathbf{i} + \cos\beta\,\mathbf{j} + \cos\gamma\,\mathbf{k}$, these can be simply written as

$$\iiint\limits_{V} \nabla \cdot \mathbf{A}\,dV \;=\; \iint\limits_{S} \mathbf{A} \cdot \mathbf{n}\,dS \tag{25}$$

In words this theorem, called the *divergence theorem* or *Green's theorem in space*, states that the surface integral of the normal component of a vector $\mathbf{A}$ taken over a closed surface is equal to the integral of the divergence of $\mathbf{A}$ taken over the volume enclosed by the surface.

STOKES' THEOREM

Let S be an open, two-sided surface bounded by a closed non-intersecting curve C (simple closed curve). Consider a directed line normal to S as positive if it is on one side of S, and negative if it is on the other side of S. The choice of which side is positive is arbitrary but should be decided upon in advance. Call the direction or sense of C positive if an observer, walking on the boundary of S with his head pointing in the direction of the positive normal, has the surface on his left. Then if A_1, A_2, A_3 are single-valued, continuous, and have continuous first partial derivatives in a region of space including S, we have

$$\int_{C} A_1\,dx + A_2\,dy + A_3\,dz \;=\; \iint\limits_{S} \left[\left(\frac{\partial A_3}{\partial y} - \frac{\partial A_2}{\partial z} \right)\cos\alpha \right. \tag{26}$$

$$\left. + \left(\frac{\partial A_1}{\partial z} - \frac{\partial A_3}{\partial x} \right)\cos\beta + \left(\frac{\partial A_2}{\partial x} - \frac{\partial A_1}{\partial y} \right)\cos\gamma \right] dS$$

In vector form with $\mathbf{A} = A_1\mathbf{i} + A_2\mathbf{j} + A_3\mathbf{k}$ and $\mathbf{n} = \cos\alpha\,\mathbf{i} + \cos\beta\,\mathbf{j} + \cos\gamma\,\mathbf{k}$, this is simply expressed as

$$\int_{C} \mathbf{A} \cdot d\mathbf{r} \;=\; \iint\limits_{S} (\nabla \times \mathbf{A}) \cdot \mathbf{n}\,dS \tag{27}$$

In words this theorem, called *Stokes' theorem*, states that the line integral of the tangential component of a vector $\mathbf{A}$ taken around a simple closed curve C is equal to the surface integral of the normal component of the curl of $\mathbf{A}$ taken over any surface S having C as a boundary. Note that if, as a special case $\nabla \times \mathbf{A} = 0$ in (27), we obtain the result (16).

Solved Problems

LINE INTEGRALS

1. Evaluate $\displaystyle\int_{(0,1)}^{(1,2)} (x^2 - y)\,dx + (y^2 + x)\,dy$ along (a) a straight line from $(0,1)$ to $(1,2)$,

(b) straight lines from $(0,1)$ to $(1,1)$ and then from $(1,1)$ to $(1,2)$, (c) the parabola $x = t,\; y = t^2 + 1$.

(a) An equation for the line joining $(0,1)$ and $(1,2)$ in the xy plane is $y = x + 1$. Then $dy = dx$ and the line integral equals

$$\int_{x=0}^{1} \{x^2 - (x+1)\}\,dx + \{(x+1)^2 + x\}\,dx \;=\; \int_{0}^{1} (2x^2 + 2x)\,dx \;=\; 5/3$$

(b) Along the straight line from $(0, 1)$ to $(1, 1)$, $y = 1$, $dy = 0$ and the line integral equals

$$\int_{x=0}^{1} (x^2 - 1)\, dx + (1 + x)(0) = \int_{0}^{1} (x^2 - 1)\, dx = -2/3$$

Along the straight line from $(1, 1)$ to $(1, 2)$, $x = 1$, $dx = 0$ and the line integral equals

$$\int_{y=1}^{2} (1 - y)(0) + (y^2 + 1)\, dy = \int_{1}^{2} (y^2 + 1)\, dy = 10/3$$

Then the required value $= -2/3 + 10/3 = 8/3$.

(c) Since $t = 0$ at $(0, 1)$ and $t = 1$ at $(1, 2)$, the line integral equals

$$\int_{t=0}^{1} \{t^2 - (t^2 + 1)\}\, dt + \{(t^2 + 1)^2 + t\}\, 2t\, dt = \int_{0}^{1} (2t^5 + 4t^3 + 2t^2 + 2t - 1)\, dt = 2$$

2. If $\mathbf{A} = (3x^2 - 6yz)\mathbf{i} + (2y + 3xz)\mathbf{j} + (1 - 4xyz^2)\mathbf{k}$, evaluate $\int_{C} \mathbf{A} \cdot d\mathbf{r}$ from $(0, 0, 0)$ to $(1, 1, 1)$ along the following paths C:

(a) $x = t,\ y = t^2,\ z = t^3$.

(b) the straight lines from $(0, 0, 0)$ to $(0, 0, 1)$, then to $(0, 1, 1)$, and then to $(1, 1, 1)$.

(c) the straight line joining $(0, 0, 0)$ and $(1, 1, 1)$.

$$\int_{C} \mathbf{A} \cdot d\mathbf{r} = \int_{C} \{(3x^2 - 6yz)\mathbf{i} + (2y + 3xz)\mathbf{j} + (1 - 4xyz^2)\mathbf{k}\} \cdot (dx\,\mathbf{i} + dy\,\mathbf{j} + dz\,\mathbf{k})$$

$$= \int_{C} (3x^2 - 6yz)\, dx + (2y + 3xz)\, dy + (1 - 4xyz^2)\, dz$$

(a) If $x = t,\ y = t^2,\ z = t^3$, points $(0, 0, 0)$ and $(1, 1, 1)$ correspond to $t = 0$ and $t = 1$ respectively. Then

$$\int_{C} \mathbf{A} \cdot d\mathbf{r} = \int_{t=0}^{1} \{3t^2 - 6(t^2)(t^3)\}\, dt + \{2t^2 + 3(t)(t^3)\}\, d(t^2) + \{1 - 4(t)(t^2)(t^3)^2\}\, d(t^3)$$

$$= \int_{t=0}^{1} (3t^2 - 6t^5)\, dt + (4t^3 + 6t^5)\, dt + (3t^2 - 12t^{11})\, dt = 2$$

Another method:

Along C, $\mathbf{A} = (3t^2 - 6t^5)\mathbf{i} + (2t^2 + 3t^4)\mathbf{j} + (1 - 4t^9)\mathbf{k}$ and $\mathbf{r} = x\mathbf{i} + y\mathbf{j} + z\mathbf{k} = t\mathbf{i} + t^2\mathbf{j} + t^3\mathbf{k}$, $d\mathbf{r} = (\mathbf{i} + 2t\mathbf{j} + 3t^2\mathbf{k})\, dt$. Then

$$\int_{C} \mathbf{A} \cdot d\mathbf{r} = \int_{0}^{1} (3t^2 - 6t^5)\, dt + (4t^3 + 6t^5)\, dt + (3t^2 - 12t^{11})\, dt = 2$$

(b) Along the straight line from $(0, 0, 0)$ to $(0, 0, 1)$, $x = 0$, $y = 0$, $dx = 0$, $dy = 0$ while z varies from 0 to 1. Then the integral over this part of the path is

$$\int_{z=0}^{1} \{3(0)^2 - 6(0)(z)\}0 + \{2(0) + 3(0)(z)\}0 + \{1 - 4(0)(0)(z^2)\}\, dz = \int_{z=0}^{1} dz = 1$$

Along the straight line from $(0, 0, 1)$ to $(0, 1, 1)$, $x = 0$, $z = 1$, $dx = 0$, $dz = 0$ while y varies from 0 to 1. Then the integral over this part of the path is

$$\int_{y=0}^{1} \{3(0)^2 - 6(y)(1)\}0 + \{2y + 3(0)(1)\}\, dy + \{1 - 4(0)(y)(1)^2\}0 = \int_{y=0}^{1} 2y\, dy = 1$$

Along the straight line from $(0, 1, 1)$ to $(1, 1, 1)$, $y = 1$, $z = 1$, $dy = 0$, $dz = 0$ while x varies from 0 to 1. Then the integral over this part of the path is

$$\int_{x=0}^{1} \{3x^2 - 6(1)(1)\}\, dx + \{2(1) + 3x(1)\}0 + \{1 - 4x(1)(1)^2\}0 = \int_{x=0}^{1} (3x^2 - 6)\, dx = -5$$

Adding, $\int_{C} \mathbf{A} \cdot d\mathbf{r} = 1 + 1 - 5 = -3$.

(c) The straight line joining $(0,0,0)$ and $(1,1,1)$ is given in parametric form by $x = t, \ y = t, \ z = t$. Then

$$\int_C \mathbf{A} \cdot d\mathbf{r} \ = \ \int_{t=0}^1 (3t^2 - 6t^2)\,dt \ + \ (2t + 3t^2)\,dt \ + \ (1 - 4t^4)\,dt \ = \ 6/5$$

3. Find the work done in moving a particle once around an ellipse C in the xy plane, if the ellipse has center at the origin with semi-major and semi-minor axes 4 and 3 respectively, as indicated in Fig. 10-3, and if the force field is given by

$$\mathbf{F} \ = \ (3x - 4y + 2z)\mathbf{i} \ + \ (4x + 2y - 3z^2)\mathbf{j} \ + \ (2xz - 4y^2 + z^3)\mathbf{k}$$

In the plane $z = 0$, $\mathbf{F} = (3x - 4y)\mathbf{i} + (4x + 2y)\mathbf{j} - 4y^2\mathbf{k}$ and $d\mathbf{r} = dx\,\mathbf{i} + dy\,\mathbf{j}$ so that the work done is

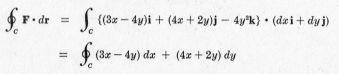

$$\oint_C \mathbf{F} \cdot d\mathbf{r} \ = \ \int_C \{(3x - 4y)\mathbf{i} + (4x + 2y)\mathbf{j} - 4y^2\mathbf{k}\} \cdot (dx\,\mathbf{i} + dy\,\mathbf{j})$$

$$= \ \oint_C (3x - 4y)\,dx \ + \ (4x + 2y)\,dy$$

Choose the parametric equations of the ellipse as $x = 4 \cos t$, $y = 3 \sin t$ where t varies from 0 to 2π (see Fig. 10-3). Then the line integral equals

Fig. 10-3

$$\int_{t=0}^{2\pi} \{3(4 \cos t) - 4(3 \sin t)\}\{-4 \sin t\}\,dt \ + \ \{4(4 \cos t) + 2(3 \sin t)\}\{3 \cos t\}\,dt$$

$$= \ \int_{t=0}^{2\pi} (48 - 30 \sin t \cos t)\,dt \ = \ (48t - 15 \sin^2 t)\big|_0^{2\pi} \ = \ 96\pi$$

In traversing C we have chosen the counterclockwise direction indicated in Fig. 10-3. We call this the *positive* direction, or say that C has been traversed in the *positive sense*. If C were traversed in the clockwise (negative) direction the value of the integral would be $- 96\pi$.

4. Evaluate $\displaystyle\int_C y\,ds$ along the curve C given by $y = 2\sqrt{x}$ from $x = 3$ to $x = 24$.

Since $ds = \sqrt{dx^2 + dy^2} = \sqrt{1 + (y')^2}\,dx = \sqrt{1 + 1/x}\,dx$, we have

$$\int_C y\,ds \ = \ \int_3^{24} 2\sqrt{x}\,\sqrt{1 + 1/x}\,dx \ = \ 2\int_3^{24} \sqrt{x + 1}\,dx \ = \ \frac{4}{3}(x+1)^{3/2}\Big|_3^{24} \ = \ 156$$

GREEN'S THEOREM in the PLANE

5. Prove Green's theorem in the plane if C is a closed curve which has the property that any straight line parallel to the coordinate axes cuts C in at most two points.

Let the equations of the curves AEB and AFB (see adjoining Fig. 10-4) be $y = Y_1(x)$ and $y = Y_2(x)$ respectively. If $\mathcal{R}$ is the region bounded by C, we have

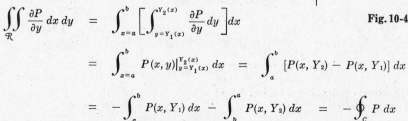

Fig. 10-4

$$\iint_{\mathcal{R}} \frac{\partial P}{\partial y}\,dx\,dy \ = \ \int_{x=a}^b \left[\int_{y=Y_1(x)}^{Y_2(x)} \frac{\partial P}{\partial y}\,dy \right] dx$$

$$= \ \int_{x=a}^b P(x,y)\big|_{y=Y_1(x)}^{Y_2(x)}\,dx \ = \ \int_a^b [P(x, Y_2) - P(x, Y_1)]\,dx$$

$$= \ -\int_a^b P(x, Y_1)\,dx \ - \ \int_b^a P(x, Y_2)\,dx \ = \ -\oint_C P\,dx$$

Then
$$(1) \qquad \oint_C P \, dx \;=\; -\iint_R \frac{\partial P}{\partial y} \, dx \, dy$$

Similarly let the equations of curves EAF and EBF be $x = X_1(y)$ and $x = X_2(y)$ respectively. Then

$$\iint_R \frac{\partial Q}{\partial x} \, dx \, dy \;=\; \int_{y=e}^{f} \left[\int_{x=X_1(y)}^{X_2(y)} \frac{\partial Q}{\partial x} \, dx \right] dy \;=\; \int_e^f \left[Q(X_2, y) - Q(X_1, y) \right] dy$$

$$=\; \int_f^e Q(X_1, y) \, dy \;+\; \int_e^f Q(X_2, y) \, dy \;=\; \oint_C Q \, dy$$

Then
$$(2) \qquad \oint_C Q \, dy \;=\; \iint_R \frac{\partial Q}{\partial x} \, dx \, dy$$

Adding (1) and (2), $\qquad \displaystyle\oint_C P \, dx + Q \, dy \;=\; \iint_R \left(\frac{\partial Q}{\partial x} - \frac{\partial P}{\partial y} \right) dx \, dy.$

6. Verify Green's theorem in the plane for
$$\oint_C (2xy - x^2) \, dx \;+\; (x + y^2) \, dy$$
where C is the closed curve of the region bounded by $y = x^2$ and $y^2 = x$.

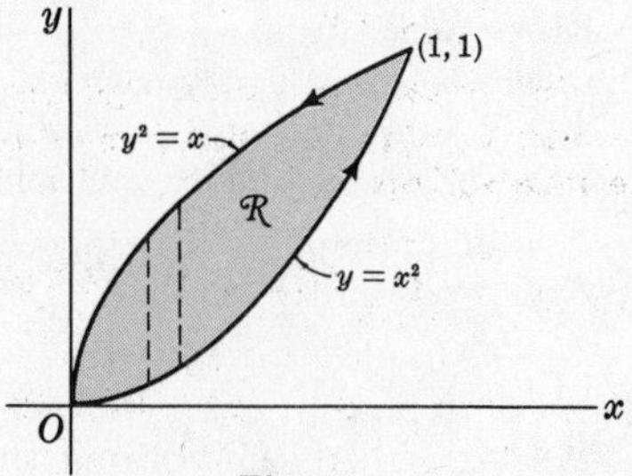

The plane curves $y = x^2$ and $y^2 = x$ intersect at $(0, 0)$ and $(1, 1)$. The positive direction in traversing C is as shown in Fig. 10-5.

Along $y = x^2$, the line integral equals

$$\int_{x=0}^{1} \{ (2x)(x^2) - x^2 \} \, dx \;+\; \{ x + (x^2)^2 \} \, d(x^2) \;=\; \int_0^1 (2x^3 + x^2 + 2x^5) \, dx \;=\; 7/6$$

Along $y^2 = x$ the line integral equals

$$\int_{y=1}^{0} \{ 2(y^2)(y) - (y^2)^2 \} \, d(y^2) \;+\; \{ y^2 + y^2 \} \, dy \;=\; \int_1^0 (4y^4 - 2y^5 + 2y^2) \, dy \;=\; -17/15$$

Then the required line integral $= 7/6 - 17/15 = 1/30$.

$$\iint_R \left(\frac{\partial Q}{\partial x} - \frac{\partial P}{\partial y} \right) dx \, dy \;=\; \iint_R \left\{ \frac{\partial}{\partial x} (x + y^2) - \frac{\partial}{\partial y} (2xy - x^2) \right\} dx \, dy$$

$$=\; \iint_R (1 - 2x) \, dx \, dy \;=\; \int_{x=0}^{1} \int_{y=x^2}^{\sqrt{x}} (1 - 2x) \, dy \, dx$$

$$=\; \int_{x=0}^{1} (y - 2xy) \big|_{y=x^2}^{\sqrt{x}} \, dx \;=\; \int_0^1 (x^{1/2} - 2x^{3/2} - x^2 + 2x^3) \, dx \;=\; 1/30$$

Hence Green's theorem is verified.

7. Extend the proof of Green's theorem in the plane given in Problem 5 to the curves C for which lines parallel to the coordinate axes may cut C in more than two points.

Consider a closed curve C such as shown in the adjoining Fig. 10-6, in which lines parallel to the axes may meet C in more than two points. By constructing line ST the region is divided into two regions $\mathcal{R}_1$ and $\mathcal{R}_2$ which are of the type considered in Problem 5 and for which Green's theorem applies, i.e.,

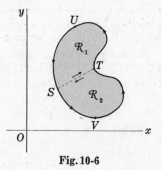

Fig. 10-6

Fig. 10-5

$$(1) \int_{STUS} P\,dx + Q\,dy \;=\; \iint_{\mathcal{R}_1}\left(\frac{\partial Q}{\partial x} - \frac{\partial P}{\partial y}\right)dx\,dy, \qquad (2) \int_{SVTS} P\,dx + Q\,dy \;=\; \iint_{\mathcal{R}_2}\left(\frac{\partial Q}{\partial x} - \frac{\partial P}{\partial y}\right)dx\,dy$$

Adding the left hand sides of (1) and (2), we have, omitting the integrand $P\,dx + Q\,dy$ in each case,

$$\int_{STUS} + \int_{SVTS} = \int_{ST} + \int_{TUS} + \int_{SVT} + \int_{TS} = \int_{TUS} + \int_{SVT} = \int_{TUSVT}$$

using the fact that $\displaystyle\int_{ST} = -\int_{TS}$.

Adding the right hand sides of (1) and (2), omitting the integrand, $\displaystyle\iint_{\mathcal{R}_1} + \iint_{\mathcal{R}_2} = \iint_{\mathcal{R}}$ where $\mathcal{R}$ consists of regions $\mathcal{R}_1$ and $\mathcal{R}_2$.

Then $\displaystyle\int_{TUSVT} P\,dx + Q\,dy = \iint_{\mathcal{R}}\left(\frac{\partial Q}{\partial x} - \frac{\partial P}{\partial y}\right)dx\,dy$ and the theorem is proved.

A region $\mathcal{R}$ such as considered here and in Problem 5, for which any closed curve lying in $\mathcal{R}$ can be continuously shrunk to a point without leaving $\mathcal{R}$, is called a *simply-connected region*. A region which is not simply-connected is called *multiply-connected*. We have shown here that Green's theorem in the plane applies to simply-connected regions bounded by closed curves. In Problem 10 the theorem is extended to multiply-connected regions.

For more complicated simply-connected regions it may be necessary to construct more lines, such as ST, to establish the theorem.

8. Show that the area bounded by a simple closed curve C is given by $\frac{1}{2}\displaystyle\oint_C x\,dy - y\,dx$.

In Green's theorem, put $P = -y$, $Q = x$. Then

$$\oint_C x\,dy - y\,dx \;=\; \iint_{\mathcal{R}}\left(\frac{\partial}{\partial x}(x) - \frac{\partial}{\partial y}(-y)\right)dx\,dy \;=\; 2\iint_{\mathcal{R}} dx\,dy \;=\; 2A$$

where A is the required area. Thus $A = \frac{1}{2}\displaystyle\oint_C x\,dy - y\,dx$.

9. Find the area of the ellipse $x = a\cos\theta$, $y = b\sin\theta$.

$$\text{Area} \;=\; \frac{1}{2}\oint_C x\,dy - y\,dx \;=\; \frac{1}{2}\int_0^{2\pi} (a\cos\theta)(b\cos\theta)\,d\theta - (b\sin\theta)(-a\sin\theta)\,d\theta$$

$$=\; \frac{1}{2}\int_0^{2\pi} ab(\cos^2\theta + \sin^2\theta)\,d\theta \;=\; \frac{1}{2}\int_0^{2\pi} ab\,d\theta \;=\; \pi ab$$

10. Show that Green's theorem in the plane is also valid for a multiply-connected region $\mathcal{R}$ such as shown in Fig. 10-7.

The shaded region $\mathcal{R}$, shown in the figure, is multiply-connected since not every closed curve lying in $\mathcal{R}$ can be shrunk to a point without leaving $\mathcal{R}$, as is observed by considering a curve surrounding $DEFGD$ for example. The boundary of $\mathcal{R}$, which consists of the exterior boundary $AHJKLA$ and the interior boundary $DEFGD$, is to be traversed in the positive direction, so that a person traveling in this direction always has the region on his left. It is seen that the positive directions are those indicated in the adjoining figure.

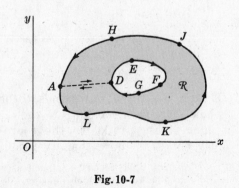

Fig. 10-7

In order to establish the theorem, construct a line, such as AD, called a *cross-cut*, connecting the exterior and interior boundaries. The region bounded by $ADEFGDALKJHA$ is simply-connected, and so Green's theorem is valid. Then

$$\oint_{ADEFGDALKJHA} P\,dx + Q\,dy = \iint_{\mathcal{R}}\left(\frac{\partial Q}{\partial x} - \frac{\partial P}{\partial y}\right)dx\,dy$$

But the integral on the left, leaving out the integrand, is equal to

$$\int_{AD} + \int_{DEFGD} + \int_{DA} + \int_{ALKJHA} = \int_{DEFGD} + \int_{ALKJHA}$$

since $\int_{AD} = -\int_{DA}$. Thus if C_1 is the curve $ALKJHA$, C_2 is the curve $DEFGD$ and C is the boundary of $\mathcal{R}$ consisting of C_1 and C_2 (traversed in the positive directions), then $\int_{C_1} + \int_{C_2} = \int_C$ and so

$$\oint_C P\,dx + Q\,dy = \iint_{\mathcal{R}}\left(\frac{\partial Q}{\partial x} - \frac{\partial P}{\partial y}\right)dx\,dy$$

INDEPENDENCE of the PATH

11. Let $P(x,y)$ and $Q(x,y)$ be continuous and have continuous first partial derivatives at each point of a simply connected region $\mathcal{R}$. Prove that a necessary and sufficient condition that $\oint_C P\,dx + Q\,dy = 0$ around every closed path C in $\mathcal{R}$ is that $\partial P/\partial y = \partial Q/\partial x$ identically in $\mathcal{R}$.

Sufficiency. Suppose $\partial P/\partial y = \partial Q/\partial x$. Then by Green's theorem,

$$\oint_C P\,dx + Q\,dy = \iint_{\mathcal{R}}\left(\frac{\partial Q}{\partial x} - \frac{\partial P}{\partial y}\right)dx\,dy = 0$$

where $\mathcal{R}$ is the region bounded by C.

Necessity.

Suppose $\oint_C P\,dx + Q\,dy = 0$ around every closed path C in $\mathcal{R}$ and that $\partial P/\partial y \neq \partial Q/\partial x$ at some point of $\mathcal{R}$. In particular suppose $\partial P/\partial y - \partial Q/\partial x > 0$ at the point (x_0, y_0).

By hypothesis $\partial P/\partial y$ and $\partial Q/\partial x$ are continuous in $\mathcal{R}$, so that there must be some region τ containing (x_0, y_0) as an interior point for which $\partial P/\partial y - \partial Q/\partial x > 0$. If Γ is the boundary of τ, then by Green's theorem

$$\oint_{\Gamma} P\,dx + Q\,dy = \iint_{\tau}\left(\frac{\partial Q}{\partial x} - \frac{\partial P}{\partial y}\right)dx\,dy > 0$$

contradicting the hypothesis that $\oint P\,dx + Q\,dy = 0$ for *all* closed curves in $\mathcal{R}$. Thus $\partial Q/\partial x - \partial P/\partial y$ cannot be positive.

Similarly we can show that $\partial Q/\partial x - \partial P/\partial y$ cannot be negative, and it follows that it must be identically zero, i.e. $\partial P/\partial y = \partial Q/\partial x$ identically in $\mathcal{R}$.

12. Let P and Q be defined as in Problem 11. Prove that a necessary and sufficient condition that $\int_A^B P\,dx + Q\,dy$ be independent of the path in $\mathcal{R}$ joining points A and B is that $\partial P/\partial y = \partial Q/\partial x$ identically in $\mathcal{R}$.

Sufficiency. If $\partial P/\partial y = \partial Q/\partial x$, then by Problem 11,

$$\int_{ADBEA} P\,dx + Q\,dy = 0$$

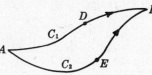

Fig. 10-8

(see Fig. 10-8). From this, omitting for brevity the integrand $P\,dx + Q\,dy$, we have

$$\int_{ADB} + \int_{BEA} = 0, \qquad \int_{ADB} = -\int_{BEA} = \int_{AEB} \qquad \text{and so} \quad \int_{C_1} = \int_{C_2}$$

i.e. the integral is independent of the path.

Necessity.

If the integral is independent of the path, then for all paths C_1 and C_2 in $\mathcal{R}$ we have

$$\int_{C_1} = \int_{C_2}, \qquad \int_{ADB} = \int_{AEB} \quad \text{and} \quad \int_{ADBEA} = 0$$

From this it follows that the line integral around any closed path in $\mathcal{R}$ is zero, and hence by Problem 11 that $\partial P/\partial y = \partial Q/\partial x$.

13. Let P and Q be as in Problem 11.

(a) Prove that a necessary and sufficient condition that $P\,dx + Q\,dy$ be an exact differential of a function $\phi(x, y)$ is that $\partial P/\partial y = \partial Q/\partial x$.

(b) Show that in such case $\displaystyle\int_A^B P\,dx + Q\,dy = \int_A^B d\phi = \phi(B) - \phi(A)$ where A and B are any two points.

(a) **Necessity.**

If $P\,dx + Q\,dy = d\phi = \dfrac{\partial \phi}{\partial x}\,dx + \dfrac{\partial \phi}{\partial y}\,dy$,

an exact differential, then (1) $\partial \phi/\partial x = P$, (2) $\partial \phi/\partial y = Q$. Thus by differentiating (1) and (2) with respect to y and x respectively, $\partial P/\partial y = \partial Q/\partial x$ since we are assuming continuity of the partial derivatives.

Sufficiency.

By Prob. 12, if $\partial P/\partial y = \partial Q/\partial x$, then $\displaystyle\int P\,dx + Q\,dy$ is independent of the path joining two points. In particular, let the two points be (a, b) and (x, y) and define

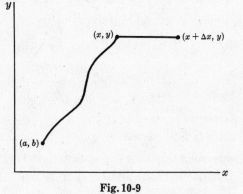

Fig. 10-9

$$\phi(x, y) = \int_{(a, b)}^{(x, y)} P\,dx + Q\,dy$$

Then

$$\phi(x + \Delta x, y) - \phi(x, y) = \int_{(a, b)}^{x + \Delta x, y} P\,dx + Q\,dy - \int_{(a, b)}^{(x, y)} P\,dx + Q\,dy$$

$$= \int_{(x, y)}^{(x + \Delta x, y)} P\,dx + Q\,dy$$

Since the last integral is independent of the path joining (x, y) and $(x + \Delta x, y)$, we can choose the path to be a straight line joining these points (see Fig. 10-9) so that $dy = 0$. Then by the mean value theorem for integrals,

$$\frac{\phi(x + \Delta x, y) - \phi(x, y)}{\Delta x} = \frac{1}{\Delta x}\int_{(x, y)}^{(x + \Delta x, y)} P\,dx = P(x + \theta\,\Delta x, y) \qquad 0 < \theta < 1$$

Taking the limit as $\Delta x \to 0$, we have $\partial \phi/\partial x = P$.

Similarly we can show that $\partial \phi/\partial y = Q$.

Thus it follows that $P\,dx + Q\,dy = \dfrac{\partial \phi}{\partial x}\,dx + \dfrac{\partial \phi}{\partial y}\,dy = d\phi$.

(b) Let $A = (x_1, y_1)$, $B = (x_2, y_2)$. From part (a),

$$\phi(x, y) = \int_{(a, b)}^{(x, y)} P\,dx + Q\,dy$$

Then omitting the integrand $P\,dx + Q\,dy$, we have

$$\int_A^B = \int_{(x_1, y_1)}^{(x_2, y_2)} = \int_{(a, b)}^{(x_2, y_2)} - \int_{(a, b)}^{(x_1, y_1)} = \phi(x_2, y_2) - \phi(x_1, y_1) = \phi(B) - \phi(A)$$

14. (a) Prove that $\displaystyle\int_{(1,2)}^{(3,4)} (6xy^2 - y^3)\,dx + (6x^2y - 3xy^2)\,dy$ is independent of the path

joining $(1,2)$ and $(3,4)$.   (b) Evaluate the integral in (a).

(a) $P = 6xy^2 - y^3$, $Q = 6x^2y - 3xy^2$.   Then $\partial P/\partial y = 12xy - 3y^2 = \partial Q/\partial x$ and by Problem 12 the line integral is independent of the path.

(b)   **Method 1:**

Since the line integral is independent of the path, choose any path joining $(1,2)$ and $(3,4)$, for example that consisting of lines from $(1,2)$ to $(3,2)$ [along which $y = 2$, $dy = 0$] and then $(3,2)$ to $(3,4)$ [along which $x = 3$, $dx = 0$].   Then the required integral equals

$$\int_{x=1}^{3} (24x - 8)\,dx + \int_{y=2}^{4} (54y - 9y^2)\,dy = 80 + 156 = 236$$

Method 2:

Since $\dfrac{\partial P}{\partial y} = \dfrac{\partial Q}{\partial x}$,   we must have   (1) $\dfrac{\partial \phi}{\partial x} = 6xy^2 - y^3$,   (2) $\dfrac{\partial \phi}{\partial y} = 6x^2y - 3xy^2$.

From (1), $\phi = 3x^2y^2 - xy^3 + f(y)$.   From (2), $\phi = 3x^2y^2 - xy^3 + g(x)$.   The only way in which these two expressions for ϕ are equal is if $f(y) = g(x) = c$,   a constant.   Hence $\phi = 3x^2y^2 - xy^3 + c$.   Then by Problem 13,

$$\int_{(1,2)}^{(3,4)} (6xy^2 - y^3)\,dx + (6x^2y - 3xy^2)\,dy = \int_{(1,2)}^{(3,4)} d(3x^2y^2 - xy^3 + c)$$

$$= 3x^2y^2 - xy^3 + c \Big|_{(1,2)}^{(3,4)} = 236$$

Note that in this evaluation the arbitrary constant c can be omitted.   See also Prob. 16, Page 115.

We could also have noted by inspection that

$$(6xy^2 - y^3)\,dx + (6x^2y - 3xy^2)\,dy = (6xy^2\,dx + 6x^2y\,dy) - (y^3\,dx + 3xy^2\,dy)$$
$$= d(3x^2y^2) - d(xy^3) = d(3x^2y^2 - xy^3)$$

from which it is clear that $\phi = 3x^2y^2 - xy^3 + c$.

15. Evaluate $\displaystyle\oint (x^2y \cos x + 2xy \sin x - y^2e^x)\,dx + (x^2 \sin x - 2ye^x)\,dy$ around the

hypocycloid $x^{2/3} + y^{2/3} = a^{2/3}$.

$P = x^2y \cos x + 2xy \sin x - y^2e^x$, $Q = x^2 \sin x - 2ye^x$.

Then $\partial P/\partial y = x^2 \cos x + 2x \sin x - 2ye^x = \partial Q/\partial x$,   so that by Problem 11 the line integral around any closed path, in particular $x^{2/3} + y^{2/3} = a^{2/3}$, is zero.

SURFACE INTEGRALS

16. If γ is the angle between the normal line to any point (x, y, z) of a surface S and the positive z axis, prove that

$$|\sec \gamma| = \sqrt{1 + z_x^2 + z_y^2} = \frac{\sqrt{F_x^2 + F_y^2 + F_z^2}}{|F_z|}$$

according as the equation for S is $z = f(x, y)$ or $F(x, y, z) = 0$.

If the equation of S is $F(x, y, z) = 0$, a normal to S at (x, y, z) is $\nabla F = F_x\mathbf{i} + F_y\mathbf{j} + F_z\mathbf{k}$.   Then

$$\nabla F \cdot \mathbf{k} = |\nabla F|\,|\mathbf{k}| \cos \gamma \qquad \text{or} \qquad F_z = \sqrt{F_x^2 + F_y^2 + F_z^2} \cos \gamma$$

from which $|\sec \gamma| = \dfrac{\sqrt{F_x^2 + F_y^2 + F_z^2}}{|F_z|}$ as required.

In case the equation is $z = f(x, y)$,   we can write $F(x, y, z) = z - f(x, y) = 0$,   from which $F_x = -z_x$, $F_y = -z_y$, $F_z = 1$ and we find $|\sec \gamma| = \sqrt{1 + z_x^2 + z_y^2}$.

17. Evaluate $\iint\limits_{S} U(x,y,z)\,dS$ where S is the surface of the paraboloid $z = 2 - (x^2 + y^2)$ above the xy plane and $U(x,y,z)$ is equal to (a) 1, (b) $x^2 + y^2$, (c) $3z$. Give a physical interpretation in each case.

The required integral is equal to

$$\iint\limits_{R} U(x,y,z)\,\sqrt{1 + z_x^2 + z_y^2}\,dx\,dy \qquad (1)$$

where R is the projection of S on the xy plane given by $x^2 + y^2 = 2$, $z = 0$.

Since $z_x = -2x$, $z_y = -2y$, (1) can be written

$$\iint\limits_{R} U(x,y,z)\,\sqrt{1 + 4x^2 + 4y^2}\,dx\,dy \qquad (2)$$

(a) If $U(x,y,z) = 1$, (2) becomes

$$\iint\limits_{R} \sqrt{1 + 4x^2 + 4y^2}\,dx\,dy$$

To evaluate this, transform to polar coordinates (ρ, ϕ). Then the integral becomes

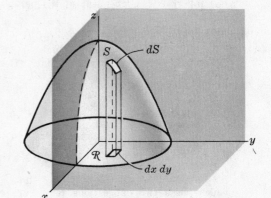

Fig. 10-10

$$\int_{\phi=0}^{2\pi} \int_{\rho=0}^{\sqrt{2}} \sqrt{1 + 4\rho^2}\,\rho\,d\rho\,d\phi \;=\; \int_{\phi=0}^{2\pi} \frac{1}{12}(1 + 4\rho^2)^{3/2}\Big|_{\rho=0}^{\sqrt{2}}\,d\phi \;=\; \frac{13\pi}{3}$$

Physically this could represent the surface area of S, or the mass of S assuming unit density.

(b) If $U(x,y,z) = x^2 + y^2$, (2) becomes $\quad \iint\limits_{R} (x^2 + y^2)\sqrt{1 + 4x^2 + 4y^2}\,dx\,dy \quad$ or in polar coordinates

$$\int_{\phi=0}^{2\pi} \int_{\rho=0}^{\sqrt{2}} \rho^3 \sqrt{1 + 4\rho^2}\,d\rho\,d\phi \;=\; \frac{149\pi}{30}$$

where the integration with respect to ρ is accomplished by the substitution $\sqrt{1 + 4\rho^2} = u$.

Physically this could represent the moment of inertia of S about the z axis assuming unit density, or the mass of S assuming a density $= x^2 + y^2$.

(c) If $U(x,y,z) = 3z$, (2) becomes

$$\iint\limits_{R} 3z\sqrt{1 + 4x^2 + 4y^2}\,dx\,dy \;=\; \iint\limits_{R} 3\{2 - (x^2 + y^2)\}\sqrt{1 + 4x^2 + 4y^2}\,dx\,dy$$

or in polar coordinates,

$$\int_{\phi=0}^{2\pi} \int_{\rho=0}^{\sqrt{2}} 3\rho(2 - \rho^2)\sqrt{1 + 4\rho^2}\,d\rho\,d\phi \;=\; \frac{111\pi}{10}$$

Physically this could represent the mass of S assuming a density $= 3z$, or three times the first moment of S about the xy plane.

18. Find the surface area of a hemisphere of radius a cut off by a cylinder having this radius as diameter.

Equations for the hemisphere and cylinder (see Fig. 10-11) are given respectively by $x^2 + y^2 + z^2 = a^2$ (or $z = \sqrt{a^2 - x^2 - y^2}$) and $(x - a/2)^2 + y^2 = a^2/4$ (or $x^2 + y^2 = ax$).

Since

$$z_x = \frac{-x}{\sqrt{a^2 - x^2 - y^2}} \quad \text{and} \quad z_y = \frac{-y}{\sqrt{a^2 - x^2 - y^2}},$$

we have

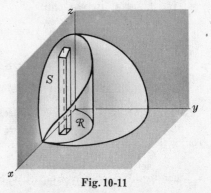

Fig. 10-11

$$\text{Required surface area} \;=\; 2\iint_{\mathcal{R}} \sqrt{1 + z_x^2 + z_y^2}\, dx\, dy \;=\; 2\iint_{\mathcal{R}} \frac{a}{\sqrt{a^2 - x^2 - y^2}}\, dx\, dy$$

Two methods of evaluation are possible.

Method 1: Using polar coordinates.

Since $x^2 + y^2 = ax$ in polar coordinates is $\rho = a\cos\phi$, the integral becomes

$$2\int_{\phi=0}^{\pi/2}\int_{\rho=0}^{a\cos\phi} \frac{a}{\sqrt{a^2 - \rho^2}}\, \rho\, d\rho\, d\phi \;=\; 2a\int_{\phi=0}^{\pi/2} -\sqrt{a^2 - \rho^2}\,\Big|_{\rho=0}^{a\cos\phi}\, d\phi$$

$$=\; 2a^2\int_0^{\pi/2}(1 - \sin\phi)\, d\phi \;=\; (\pi - 2)a^2$$

Method 2: The integral is equal to

$$2\int_{x=0}^{a}\int_{y=0}^{\sqrt{ax-x^2}} \frac{a}{\sqrt{a^2 - x^2 - y^2}}\, dy\, dx \;=\; 2a\int_{x=0}^{a} \sin^{-1}\frac{y}{\sqrt{ax-x^2}}\,\Big|_{y=0}^{\sqrt{ax-x^2}}\, dx$$

$$=\; 2a\int_0^a \sin^{-1}\sqrt{\frac{x}{a+x}}\, dx$$

Letting $x = a\tan^2\theta$, this integral becomes

$$4a^2\int_0^{\pi/4} \theta\tan\theta\sec^2\theta\, d\theta \;=\; 4a^2\left\{ \tfrac{1}{2}\theta\tan^2\theta\big|_0^{\pi/4} - \tfrac{1}{2}\int_0^{\pi/4} \tan^2\theta\, d\theta \right\}$$

$$=\; 2a^2\left\{ \theta\tan^2\theta\big|_0^{\pi/4} - \int_0^{\pi/4}(\sec^2\theta - 1)\, d\theta \right\}$$

$$=\; 2a^2\left\{ \pi/4 - (\tan\theta - \theta)\big|_0^{\pi/4} \right\} \;=\; (\pi - 2)a^2$$

Note that the above integrals are actually *improper* and should be treated by appropriate limiting procedures (see Problem 78, Chapter 5, and also Chapter 12).

19. Find the centroid of the surface in Problem 17.

By symmetry, $\bar{x} = \bar{y} = 0$ and $\quad \bar{z} = \dfrac{\iint_S z\, dS}{\iint_S dS} = \dfrac{\iint_{\mathcal{R}} z\sqrt{1 + 4x^2 + 4y^2}\, dx\, dy}{\iint_{\mathcal{R}} \sqrt{1 + 4x^2 + 4y^2}\, dx\, dy}$

The numerator and denominator can be obtained from the results of Prob. 17(c) and 17(a) respectively, and we thus have $\bar{z} = \dfrac{37\pi/10}{13\pi/3} = \dfrac{111}{130}$.

20. Evaluate $\displaystyle\iint_S \mathbf{A}\cdot\mathbf{n}\, dS$, where $\mathbf{A} = xy\,\mathbf{i} - x^2\,\mathbf{j} + (x+z)\,\mathbf{k}$, S is that portion of the plane $2x + 2y + z = 6$ included in the first octant, and $\mathbf{n}$ is a unit normal to S.

A normal to S is $\nabla(2x + 2y + z - 6) = 2\mathbf{i} + 2\mathbf{j} + \mathbf{k}$,

and so $\mathbf{n} = \dfrac{2\mathbf{i} + 2\mathbf{j} + \mathbf{k}}{\sqrt{2^2 + 2^2 + 1^2}} = \dfrac{2\mathbf{i} + 2\mathbf{j} + \mathbf{k}}{3}$. Then

$$\mathbf{A}\cdot\mathbf{n} = \{xy\,\mathbf{i} - x^2\mathbf{j} + (x+z)\mathbf{k}\}\cdot\left(\frac{2\mathbf{i} + 2\mathbf{j} + \mathbf{k}}{3}\right)$$

$$= \frac{2xy - 2x^2 + (x+z)}{3}$$

$$= \frac{2xy - 2x^2 + (x + 6 - 2x - 2y)}{3}$$

$$= \frac{2xy - 2x^2 - x - 2y + 6}{3}$$

The required surface integral is therefore

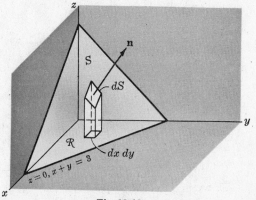

Fig. 10-12

$$\iint_S \left(\frac{2xy - 2x^2 - x - 2y + 6}{3}\right)dS = \iint_R \left(\frac{2xy - 2x^2 - x - 2y + 6}{3}\right)\sqrt{1 + z_x^2 + z_y^2}\, dx\, dy$$

$$= \iint_R \left(\frac{2xy - 2x^2 - x - 2y + 6}{3}\right)\sqrt{1^2 + 2^2 + 2^2}\, dx\, dy$$

$$= \int_{x=0}^{3} \int_{y=0}^{3-x} (2xy - 2x^2 - x - 2y + 6)\, dy\, dx$$

$$= \int_{x=0}^{3} (xy^2 - 2x^2y - xy - y^2 + 6y)\Big|_0^{3-x} dx = 27/4$$

21. In dealing with surface integrals we have restricted ourselves to surfaces which are two-sided. Give an example of a surface which is not two-sided.

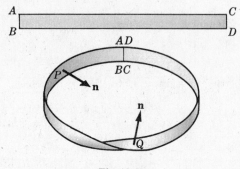

Take a strip of paper such as *ABCD* as shown in the adjoining Fig. 10-13. Twist the strip so that points *A* and *B* fall on *D* and *C* respectively, as in the adjoining figure. If **n** is the positive normal at point *P* of the surface, we find that as **n** moves around the surface it reverses its original direction when it reaches *P* again. If we tried to color only one side of the surface we would find the whole thing colored. This surface, called a *Moebius strip*, is an example of a one-sided surface. This is sometimes called a *non-orientable* surface. A two-sided surface is *orientable*.

Fig. 10-13

The DIVERGENCE THEOREM

22. Prove the divergence theorem.

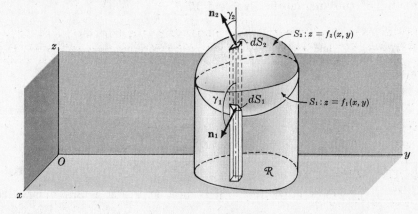

Fig. 10-14

Let *S* be a closed surface which is such that any line parallel to the coordinate axes cuts *S* in at most two points. Assume the equations of the lower and upper portions, S_1 and S_2, to be $z = f_1(x, y)$ and $z = f_2(x, y)$ respectively. Denote the projection of the surface on the *xy* plane by $\mathcal{R}$. Consider

$$\iiint_V \frac{\partial A_3}{\partial z}\, dV = \iiint_V \frac{\partial A_3}{\partial z}\, dz\, dy\, dx = \iint_R \left[\int_{z=f_1(x,y)}^{f_2(x,y)} \frac{\partial A_3}{\partial z}\, dz\right] dy\, dx$$

$$= \iint_R A_3(x, y, z)\Big|_{z=f_1}^{f_2} dy\, dx = \iint_R [A_3(x, y, f_2) - A_3(x, y, f_1)]\, dy\, dx$$

For the upper portion S_2, $dy\, dx = \cos\gamma_2\, dS_2 = \mathbf{k} \cdot \mathbf{n}_2\, dS_2$ since the normal $\mathbf{n}_2$ to S_2 makes an acute angle γ_2 with **k**.

For the lower portion S_1, $dy\,dx = -\cos\gamma_1\,dS_1 = -\mathbf{k}\cdot\mathbf{n}_1\,dS_1$ since the normal $\mathbf{n}_1$ to S_1 makes an obtuse angle γ_1 with $\mathbf{k}$.

Then

$$\iint\limits_{\mathcal{R}} A_3(x,y,f_2)\,dy\,dx \;=\; \iint\limits_{S_2} A_3\,\mathbf{k}\cdot\mathbf{n}_2\,dS_2$$

$$\iint\limits_{\mathcal{R}} A_3(x,y,f_1)\,dy\,dx \;=\; -\iint\limits_{S_1} A_3\,\mathbf{k}\cdot\mathbf{n}_1\,dS_1$$

and

$$\iint\limits_{\mathcal{R}} A_3(x,y,f_2)\,dy\,dx - \iint\limits_{\mathcal{R}} A_3(x,y,f_1)\,dy\,dx \;=\; \iint\limits_{S_2} A_3\,\mathbf{k}\cdot\mathbf{n}_2\,dS_2 + \iint\limits_{S_1} A_3\,\mathbf{k}\cdot\mathbf{n}_1\,dS_1$$

$$=\; \iint\limits_{S} A_3\,\mathbf{k}\cdot\mathbf{n}\,dS$$

so that

$$(1)\quad \iiint\limits_{V} \frac{\partial A_3}{\partial z}\,dV \;=\; \iint\limits_{S} A_3\,\mathbf{k}\cdot\mathbf{n}\,dS$$

Similarly, by projecting S on the other coordinate planes,

$$(2)\quad \iiint\limits_{V} \frac{\partial A_1}{\partial x}\,dV \;=\; \iint\limits_{S} A_1\,\mathbf{i}\cdot\mathbf{n}\,dS$$

$$(3)\quad \iiint\limits_{V} \frac{\partial A_2}{\partial y}\,dV \;=\; \iint\limits_{S} A_2\,\mathbf{j}\cdot\mathbf{n}\,dS$$

Adding (1), (2) and (3),

$$\iiint\limits_{V}\left(\frac{\partial A_1}{\partial x}+\frac{\partial A_2}{\partial y}+\frac{\partial A_3}{\partial z}\right)dV \;=\; \iint\limits_{S}(A_1\mathbf{i}+A_2\mathbf{j}+A_3\mathbf{k})\cdot\mathbf{n}\,dS$$

or

$$\iiint\limits_{V} \nabla\cdot\mathbf{A}\,dV \;=\; \iint\limits_{S} \mathbf{A}\cdot\mathbf{n}\,dS$$

The theorem can be extended to surfaces which are such that lines parallel to the coordinate axes meet them in more than two points. To establish this extension, subdivide the region bounded by S into subregions whose surfaces do satisfy this condition. The procedure is analogous to that used in Green's theorem for the plane.

23. Verify the divergence theorem for $\mathbf{A} = (2x-z)\mathbf{i} + x^2 y\,\mathbf{j} - xz^2\,\mathbf{k}$ taken over the region bounded by $x=0$, $x=1$, $y=0$, $y=1$, $z=0$, $z=1$.

We first evaluate $\displaystyle\iint\limits_{S} \mathbf{A}\cdot\mathbf{n}\,dS$ where S is the surface of the cube in Fig. 10-15.

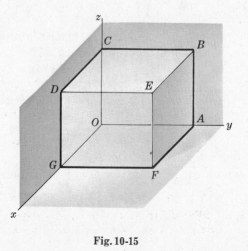

Face DEFG: $\mathbf{n}=\mathbf{i}$, $x=1$. Then

$$\iint\limits_{DEFG} \mathbf{A}\cdot\mathbf{n}\,dS \;=\; \int_0^1\int_0^1 \{(2-z)\mathbf{i}+\mathbf{j}-z^2\mathbf{k}\}\cdot\mathbf{i}\,dy\,dz$$

$$=\; \int_0^1\int_0^1 (2-z)\,dy\,dz \;=\; 3/2$$

Face ABCO: $\mathbf{n}=-\mathbf{i}$, $x=0$. Then

$$\iint\limits_{ABCO} \mathbf{A}\cdot\mathbf{n}\,dS \;=\; \int_0^1\int_0^1 (-z\mathbf{i})\cdot(-\mathbf{i})\,dy\,dz$$

$$=\; \int_0^1\int_0^1 z\,dy\,dz \;=\; 1/2$$

Fig. 10-15

Face ABEF: $\mathbf{n} = \mathbf{j}$, $y = 1$. Then

$$\iint\limits_{ABEF} \mathbf{A} \cdot \mathbf{n} \, dS \;=\; \int_0^1 \int_0^1 \{(2x-z)\mathbf{i} + x^2\mathbf{j} - xz^2\mathbf{k}\} \cdot \mathbf{j} \, dx \, dz \;=\; \int_0^1 \int_0^1 x^2 \, dx \, dz \;=\; 1/3$$

Face OGDC: $\mathbf{n} = -\mathbf{j}$, $y = 0$. Then

$$\iint\limits_{OGDC} \mathbf{A} \cdot \mathbf{n} \, dS \;=\; \int_0^1 \int_0^1 \{(2x-z)\mathbf{i} - xz^2\mathbf{k}\} \cdot (-\mathbf{j}) \, dx \, dz \;=\; 0$$

Face BCDE: $\mathbf{n} = \mathbf{k}$, $z = 1$. Then

$$\iint\limits_{BCDE} \mathbf{A} \cdot \mathbf{n} \, dS \;=\; \int_0^1 \int_0^1 \{(2x-1)\mathbf{i} + x^2 y\mathbf{j} - x\mathbf{k}\} \cdot \mathbf{k} \, dx \, dy \;=\; \int_0^1 \int_0^1 - x \, dx \, dy \;=\; -1/2$$

Face AFGO: $\mathbf{n} = -\mathbf{k}$, $z = 0$. Then

$$\iint\limits_{AFGO} \mathbf{A} \cdot \mathbf{n} \, dS \;=\; \int_0^1 \int_0^1 \{2x\mathbf{i} - x^2 y\mathbf{j}\} \cdot (-\mathbf{k}) \, dx \, dy \;=\; 0$$

Adding, $\displaystyle \iint\limits_{S} \mathbf{A} \cdot \mathbf{n} \, dS = \tfrac{3}{2} + \tfrac{1}{2} + \tfrac{1}{3} + 0 - \tfrac{1}{2} + 0 = \tfrac{11}{6}$. Since

$$\iiint\limits_{V} \nabla \cdot \mathbf{A} \, dV \;=\; \int_0^1 \int_0^1 \int_0^1 (2 + x^2 - 2xz) \, dx \, dy \, dz \;=\; \frac{11}{6}$$

the divergence theorem is verified in this case.

24. Evaluate $\displaystyle \iint\limits_{S} \mathbf{r} \cdot \mathbf{n} \, dS$, where S is a closed surface.

By the divergence theorem,

$$\iint\limits_{S} \mathbf{r} \cdot \mathbf{n} \, dS \;=\; \iiint\limits_{V} \nabla \cdot \mathbf{r} \, dV$$

$$\qquad\qquad =\; \iiint\limits_{V} \left(\frac{\partial}{\partial x}\mathbf{i} + \frac{\partial}{\partial y}\mathbf{j} + \frac{\partial}{\partial z}\mathbf{k} \right) \cdot (x\mathbf{i} + y\mathbf{j} + z\mathbf{k}) \, dV$$

$$\qquad\qquad =\; \iiint\limits_{V} \left(\frac{\partial x}{\partial x} + \frac{\partial y}{\partial y} + \frac{\partial z}{\partial z} \right) dV \;=\; 3 \iiint\limits_{V} dV \;=\; 3V$$

where V is the volume enclosed by S.

25. Evaluate $\displaystyle \iint\limits_{S} xz^2 \, dy \, dz + (x^2 y - z^3) \, dz \, dx + (2xy + y^2 z) \, dx \, dy$ where S is the entire surface of the hemispherical region bounded by $z = \sqrt{a^2 - x^2 - y^2}$ and $z = 0$ (*a*) by the divergence theorem (Green's theorem in space), (*b*) directly.

(*a*) Since $dy \, dz = dS \cos\alpha$, $dz \, dx = dS \cos\beta$, $dx \, dy = dS \cos\gamma$, the integral can be written

$$\iint\limits_{S} \{xz^2 \cos\alpha + (x^2 y - z^3) \cos\beta + (2xy + y^2 z) \cos\gamma\} \, dS \;=\; \iint\limits_{S} \mathbf{A} \cdot \mathbf{n} \, dS$$

where $\mathbf{A} = xz^2\mathbf{i} + (x^2 y - z^3)\mathbf{j} + (2xy + y^2 z)\mathbf{k}$ and $\mathbf{n} = \cos\alpha\,\mathbf{i} + \cos\beta\,\mathbf{j} + \cos\gamma\,\mathbf{k}$, the outward drawn unit normal.

Then by the divergence theorem the integral equals

$$\iiint\limits_{V} \nabla \cdot \mathbf{A} \, dV \;=\; \iiint\limits_{V} \left\{ \frac{\partial}{\partial x}(xz^2) + \frac{\partial}{\partial y}(x^2 y - z^3) + \frac{\partial}{\partial z}(2xy + y^2 z) \right\} dV \;=\; \iiint\limits_{V} (x^2 + y^2 + z^2) \, dV$$

where V is the region bounded by the hemisphere and the xy plane.

By use of spherical coordinates, as in Problem 15, Chapter 9, this integral is equal to

$$4 \int_{\phi=0}^{\pi/2} \int_{\theta=0}^{\pi/2} \int_{r=0}^{a} r^2 \cdot r^2 \sin\theta \, dr \, d\theta \, d\phi \;=\; \frac{2\pi a^5}{5}$$

(b) If S_1 is the convex surface of the hemispherical region and S_2 is the base ($z=0$), then

$$\iint_{S_1} xz^2 \, dy \, dz \;=\; \int_{y=-a}^{a} \int_{z=0}^{\sqrt{a^2-y^2}} z^2 \sqrt{a^2-y^2-z^2} \, dz \, dy \;-\; \int_{y=-a}^{a} \int_{z=0}^{\sqrt{a^2-y^2}} - z^2 \sqrt{a^2-y^2-z^2} \, dz \, dy$$

$$\iint_{S_1} (x^2 y - z^3) \, dz \, dx \;=\; \int_{x=-a}^{a} \int_{z=0}^{\sqrt{a^2-x^2}} \{x^2 \sqrt{a^2-x^2-z^2} - z^3\} \, dz \, dx$$

$$-\; \int_{x=-a}^{a} \int_{z=0}^{\sqrt{a^2-x^2}} \{-x^2 \sqrt{a^2-x^2-z^2} - z^3\} \, dz \, dx$$

$$\iint_{S_1} (2xy + y^2 z) \, dx \, dy \;=\; \int_{x=-a}^{a} \int_{y=-\sqrt{a^2-x^2}}^{\sqrt{a^2-x^2}} \{2xy + y^2 \sqrt{a^2-x^2-y^2}\} \, dy \, dx$$

$$\iint_{S_2} xz^2 \, dy \, dz \;=\; 0, \qquad \iint_{S_2} (x^2 y - z^3) \, dz \, dx \;=\; 0,$$

$$\iint_{S_2} (2xy + y^2 z) \, dx \, dy \;=\; \iint_{S_2} \{2xy + y^2(0)\} \, dx \, dy \;=\; \int_{x=-a}^{a} \int_{y=-\sqrt{a^2-x^2}}^{\sqrt{a^2-x^2}} 2xy \, dy \, dx \;=\; 0$$

By addition of the above, we obtain

$$4 \int_{y=0}^{a} \int_{z=0}^{\sqrt{a^2-y^2}} z^2 \sqrt{a^2-y^2-z^2} \, dz \, dy \;+\; 4 \int_{x=0}^{a} \int_{z=0}^{\sqrt{a^2-x^2}} x^2 \sqrt{a^2-x^2-z^2} \, dz \, dx$$

$$+\; 4 \int_{x=0}^{a} \int_{y=0}^{\sqrt{a^2-x^2}} y^2 \sqrt{a^2-x^2-y^2} \, dy \, dx$$

Since by symmetry all these integrals are equal, the result is, on using polar coordinates,

$$12 \int_{x=0}^{a} \int_{y=0}^{\sqrt{a^2-x^2}} y^2 \sqrt{a^2-x^2-y^2} \, dy \, dx \;=\; 12 \int_{\phi=0}^{\pi/2} \int_{\rho=0}^{a} \rho^2 \sin^2\phi \sqrt{a^2-\rho^2} \, \rho \, d\rho \, d\phi \;=\; \frac{2\pi a^5}{5}$$

STOKES' THEOREM

26. Prove Stokes' theorem.

Let S be a surface which is such that its projections on the xy, yz and xz planes are regions bounded by simple closed curves, as indicated in Fig. 10-16.   Assume S to have representation $z = f(x,y)$ or $x = g(y,z)$ or $y = h(x,z)$, where f, g, h are single-valued, continuous and differentiable functions.   We must show that

$$\iint_{S} (\nabla \times \mathbf{A}) \cdot \mathbf{n} \, dS$$

$$= \iint_{S} [\nabla \times (A_1 \mathbf{i} + A_2 \mathbf{j} + A_3 \mathbf{k})] \cdot \mathbf{n} \, dS$$

$$= \int_{C} \mathbf{A} \cdot d\mathbf{r}$$

where C is the boundary of S.

Consider first $\displaystyle\iint_{S} [\nabla \times (A_1 \mathbf{i})] \cdot \mathbf{n} \, dS$.

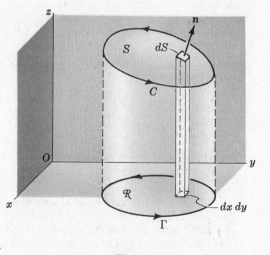

Fig. 10-16

Since $\quad \nabla \times (A_1\mathbf{i}) = \begin{vmatrix} \mathbf{i} & \mathbf{j} & \mathbf{k} \\ \frac{\partial}{\partial x} & \frac{\partial}{\partial y} & \frac{\partial}{\partial z} \\ A_1 & 0 & 0 \end{vmatrix} = \frac{\partial A_1}{\partial z}\mathbf{j} - \frac{\partial A_1}{\partial y}\mathbf{k},$

$$[\nabla \times (A_1\mathbf{i})] \cdot \mathbf{n}\, dS = \left(\frac{\partial A_1}{\partial z}\mathbf{n}\cdot\mathbf{j} - \frac{\partial A_1}{\partial y}\mathbf{n}\cdot\mathbf{k}\right) dS \tag{1}$$

If $z = f(x, y)$ is taken as the equation of S, then the position vector to any point of S is $\mathbf{r} = x\,\mathbf{i} + y\,\mathbf{j} + z\,\mathbf{k} = x\,\mathbf{i} + y\,\mathbf{j} + f(x,y)\mathbf{k}$ so that $\frac{\partial \mathbf{r}}{\partial y} = \mathbf{j} + \frac{\partial z}{\partial y}\mathbf{k} = \mathbf{j} + \frac{\partial f}{\partial y}\mathbf{k}$. But $\frac{\partial \mathbf{r}}{\partial y}$ is a vector tangent to S and thus perpendicular to $\mathbf{n}$, so that

$$\mathbf{n}\cdot\frac{\partial \mathbf{r}}{\partial y} = \mathbf{n}\cdot\mathbf{j} + \frac{\partial z}{\partial y}\mathbf{n}\cdot\mathbf{k} = 0 \qquad \text{or} \qquad \mathbf{n}\cdot\mathbf{j} = -\frac{\partial z}{\partial y}\mathbf{n}\cdot\mathbf{k}$$

Substitute in (1) to obtain

$$\left(\frac{\partial A_1}{\partial z}\mathbf{n}\cdot\mathbf{j} - \frac{\partial A_1}{\partial y}\mathbf{n}\cdot\mathbf{k}\right) dS = \left(-\frac{\partial A_1}{\partial z}\frac{\partial z}{\partial y}\mathbf{n}\cdot\mathbf{k} - \frac{\partial A_1}{\partial y}\mathbf{n}\cdot\mathbf{k}\right) dS$$

or

$$[\nabla \times (A_1\mathbf{i})]\cdot\mathbf{n}\, dS = -\left(\frac{\partial A_1}{\partial y} + \frac{\partial A_1}{\partial z}\frac{\partial z}{\partial y}\right)\mathbf{n}\cdot\mathbf{k}\, dS \tag{2}$$

Now on S, $A_1(x,y,z) = A_1[x,y,f(x,y)] = F(x,y)$; hence $\frac{\partial A_1}{\partial y} + \frac{\partial A_1}{\partial z}\frac{\partial z}{\partial y} = \frac{\partial F}{\partial y}$ and (2) becomes

$$[\nabla \times (A_1\mathbf{i})]\cdot\mathbf{n}\, dS = -\frac{\partial F}{\partial y}\mathbf{n}\cdot\mathbf{k}\, dS = -\frac{\partial F}{\partial y}\, dx\, dy$$

Then

$$\iint_S [\nabla \times (A_1\mathbf{i})]\cdot\mathbf{n}\, dS = \iint_{\mathcal{R}} -\frac{\partial F}{\partial y}\, dx\, dy$$

where $\mathcal{R}$ is the projection of S on the xy plane. By Green's theorem for the plane the last integral equals $\oint_\Gamma F\, dx$ where Γ is the boundary of $\mathcal{R}$. Since at each point (x,y) of Γ the value of F is the same as the value of A_1 at each point (x,y,z) of C, and since dx is the same for both curves, we must have

$$\oint_\Gamma F\, dx = \oint_C A_1\, dx$$

or

$$\iint_S [\nabla \times (A_1\mathbf{i})]\cdot\mathbf{n}\, dS = \oint_C A_1\, dx$$

Similarly, by projections on the other coordinate planes,

$$\iint_S [\nabla \times (A_2\mathbf{j})]\cdot\mathbf{n}\, dS = \oint_C A_2\, dy, \qquad \iint_S [\nabla \times (A_3\mathbf{k})]\cdot\mathbf{n}\, dS = \oint_C A_3\, dz$$

Thus by addition,

$$\iint_S (\nabla \times \mathbf{A})\cdot\mathbf{n}\, dS = \oint_C \mathbf{A}\cdot d\mathbf{r}$$

The theorem is also valid for surfaces S which may not satisfy the restrictions imposed above. For assume that S can be subdivided into surfaces $S_1, S_2, \ldots, S_k$ with boundaries $C_1, C_2, \ldots, C_k$ which do satisfy the restrictions. Then Stokes' theorem holds for each such surface. Adding these surface integrals, the total surface integral over S is obtained. Adding the corresponding line integrals over $C_1, C_2, \ldots, C_k$, the line integral over C is obtained.

27. Verify Stokes' theorem for $\mathbf{A} = 3y\,\mathbf{i} - xz\,\mathbf{j} + yz^2\,\mathbf{k}$, where S is the surface of the paraboloid $2z = x^2 + y^2$ bounded by $z = 2$ and C is its boundary.

The boundary C of S is a circle with equations $x^2 + y^2 = 4$, $z = 2$ and parametric equations $x = 2\cos t$, $y = 2\sin t$, $z = 2$, where $0 \le t < 2\pi$. Then

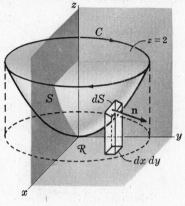

$$\oint_C \mathbf{A} \cdot d\mathbf{r} = \oint_C 3y\,dx - xz\,dy + yz^2\,dz$$

$$= \int_{2\pi}^{0} 3(2\sin t)(-2\sin t)\,dt - (2\cos t)(2)(2\cos t)\,dt$$

$$= \int_{0}^{2\pi} (12\sin^2 t + 8\cos^2 t)\,dt = 20\pi$$

Also,

$$\nabla \times \mathbf{A} = \begin{vmatrix} \mathbf{i} & \mathbf{j} & \mathbf{k} \\ \dfrac{\partial}{\partial x} & \dfrac{\partial}{\partial y} & \dfrac{\partial}{\partial z} \\ 3y & -xz & yz^2 \end{vmatrix} = (z^2 + x)\mathbf{i} - (z+3)\mathbf{k}$$

Fig. 10-17

and

$$\mathbf{n} = \frac{\nabla(x^2 + y^2 - 2z)}{|\nabla(x^2 + y^2 - 2z)|} = \frac{x\mathbf{i} + y\mathbf{j} - \mathbf{k}}{\sqrt{x^2 + y^2 + 1}}.$$

Then

$$\iint_S (\nabla \times \mathbf{A}) \cdot \mathbf{n}\,dS = \iint_{\mathcal{R}} (\nabla \times A) \cdot \mathbf{n}\frac{dx\,dy}{|\mathbf{n} \cdot \mathbf{k}|} = \iint_{\mathcal{R}} (xz^2 + x^2 + z + 3)\,dx\,dy$$

$$= \iint_{\mathcal{R}} \left\{ x\left(\frac{x^2 + y^2}{2}\right)^2 + x^2 + \frac{x^2 + y^2}{2} + 3 \right\} dx\,dy$$

In polar coordinates this becomes

$$\int_{\phi=0}^{2\pi} \int_{\rho=0}^{2} \{(\rho\cos\phi)(\rho^4/2) + \rho^2\cos^2\phi + \rho^2/2 + 3\}\,\rho\,d\rho\,d\phi = 20\pi$$

28. Prove that a necessary and sufficient condition that $\oint_C \mathbf{A} \cdot d\mathbf{r} = 0$ for every closed curve C is that $\nabla \times \mathbf{A} = 0$ identically.

Sufficiency.   Suppose $\nabla \times \mathbf{A} = 0$.   Then by Stokes' theorem

$$\oint_C \mathbf{A} \cdot d\mathbf{r} = \iint_S (\nabla \times \mathbf{A}) \cdot \mathbf{n}\,dS = 0$$

Necessity.

Suppose $\oint_C \mathbf{A} \cdot d\mathbf{r} = 0$ around every closed path C, and assume $\nabla \times \mathbf{A} \ne 0$ at some point P. Then assuming $\nabla \times \mathbf{A}$ is continuous there will be a region with P as an interior point, where $\nabla \times \mathbf{A} \ne 0$. Let S be a surface contained in this region whose normal $\mathbf{n}$ at each point has the same direction as $\nabla \times \mathbf{A}$, i.e. $\nabla \times \mathbf{A} = \alpha\mathbf{n}$ where α is a positive constant. Let C be the boundary of S. Then by Stokes' theorem

$$\oint_C \mathbf{A} \cdot d\mathbf{r} = \iint_S (\nabla \times \mathbf{A}) \cdot \mathbf{n}\,dS = \alpha \iint_S \mathbf{n} \cdot \mathbf{n}\,dS > 0$$

which contradicts the hypothesis that $\oint_C \mathbf{A} \cdot d\mathbf{r} = 0$ and shows that $\nabla \times \mathbf{A} = 0$.

It follows that $\nabla \times \mathbf{A} = 0$ is also a necessary and sufficient condition for a line integral $\displaystyle\int_{P_1}^{P_2} \mathbf{A} \cdot d\mathbf{r}$ to be independent of the path joining points P_1 and P_2.

29. Prove that a necessary and sufficient condition that $\nabla \times \mathbf{A} = 0$ is that $\mathbf{A} = \nabla \phi$.

Sufficiency. If $\mathbf{A} = \nabla \phi$, then $\nabla \times \mathbf{A} = \nabla \times \nabla \phi = 0$ by Prob. 80, Chap. 7, Page 158.

Necessity.

If $\nabla \times \mathbf{A} = 0$, then by Prob. 28, $\oint \mathbf{A} \cdot d\mathbf{r} = 0$ around every closed path and $\int_c \mathbf{A} \cdot d\mathbf{r}$ is independent of the path joining two points which we take as (a, b, c) and (x, y, z). Let us define

$$\phi(x, y, z) = \int_{(a,b,c)}^{(x,y,z)} \mathbf{A} \cdot d\mathbf{r} = \int_{(a,b,c)}^{(x,y,z)} A_1 \, dx + A_2 \, dy + A_3 \, dz$$

Then

$$\phi(x + \Delta x, y, z) - \phi(x, y, z) = \int_{(x,y,z)}^{(x+\Delta x, y, z)} A_1 \, dx + A_2 \, dy + A_3 \, dz$$

Since the last integral is independent of the path joining (x, y, z) and $(x + \Delta x, y, z)$, we can choose the path to be a straight line joining these points so that dy and dz are zero. Then

$$\frac{\phi(x + \Delta x, y, z) - \phi(x, y, z)}{\Delta x} = \frac{1}{\Delta x} \int_{(x,y,z)}^{(x+\Delta x, y, z)} A_1 \, dx = A_1(x + \theta \Delta x, y, z) \qquad 0 < \theta < 1$$

where we have applied the law of the mean for integrals.

Taking the limit of both sides as $\Delta x \to 0$ gives $\partial \phi / \partial x = A_1$.

Similarly we can show that $\partial \phi / \partial y = A_2, \quad \partial \phi / \partial z = A_3$.

Thus $\mathbf{A} = A_1 \mathbf{i} + A_2 \mathbf{j} + A_3 \mathbf{k} = \dfrac{\partial \phi}{\partial x} \mathbf{i} + \dfrac{\partial \phi}{\partial y} \mathbf{j} + \dfrac{\partial \phi}{\partial z} \mathbf{k} = \nabla \phi$.

30. (a) Prove that a necessary and sufficient condition that $A_1 \, dx + A_2 \, dy + A_3 \, dz = d\phi$, an exact differential, is that $\nabla \times \mathbf{A} = 0$ where $\mathbf{A} = A_1 \mathbf{i} + A_2 \mathbf{j} + A_3 \mathbf{k}$.

(b) Show that in such case,

$$\int_{(x_1, y_1, z_1)}^{(x_2, y_2, z_2)} A_1 \, dx + A_2 \, dy + A_3 \, dz = \int_{(x_1, y_1, z_1)}^{(x_2, y_2, z_2)} d\phi = \phi(x_2, y_2, z_2) - \phi(x_1, y_1, z_1)$$

(a) **Necessity.** If $A_1 \, dx + A_2 \, dy + A_3 \, dz = d\phi = \dfrac{\partial \phi}{\partial x} dx + \dfrac{\partial \phi}{\partial y} dy + \dfrac{\partial \phi}{\partial z} dz$, then

$$(1) \quad \frac{\partial \phi}{\partial x} = A_1 \qquad (2) \quad \frac{\partial \phi}{\partial y} = A_2 \qquad (3) \quad \frac{\partial \phi}{\partial z} = A_3$$

Then by differentiating we have, assuming continuity of the partial derivatives,

$$\frac{\partial A_1}{\partial y} = \frac{\partial A_2}{\partial x}, \qquad \frac{\partial A_2}{\partial z} = \frac{\partial A_3}{\partial y}, \qquad \frac{\partial A_1}{\partial z} = \frac{\partial A_3}{\partial x}$$

which is precisely the condition $\nabla \times \mathbf{A} = 0$.

Another method: If $A_1 \, dx + A_2 \, dy + A_3 \, dz = d\phi$, then

$$\mathbf{A} = A_1 \mathbf{i} + A_2 \mathbf{j} + A_3 \mathbf{k} = \frac{\partial \phi}{\partial x} \mathbf{i} + \frac{\partial \phi}{\partial y} \mathbf{j} + \frac{\partial \phi}{\partial z} \mathbf{k} = \nabla \phi$$

from which $\nabla \times \mathbf{A} = \nabla \times \nabla \phi = 0$.

Sufficiency. If $\nabla \times \mathbf{A} = 0$, then by Problem 29, $\mathbf{A} = \nabla \phi$ and

$$A_1 \, dx + A_2 \, dy + A_3 \, dz = \mathbf{A} \cdot d\mathbf{r} = \nabla \phi \cdot d\mathbf{r} = \frac{\partial \phi}{\partial x} dx + \frac{\partial \phi}{\partial y} dy + \frac{\partial \phi}{\partial z} dz = d\phi$$

(b) From part (a), $\phi(x, y, z) = \displaystyle\int_{(a,b,c)}^{(x,y,z)} A_1 \, dx + A_2 \, dy + A_3 \, dz.$

Then omitting the integrand $A_1 \, dx + A_2 \, dy + A_3 \, dz$, we have

$$\int_{(x_1, y_1, z_1)}^{(x_2, y_2, z_2)} = \int_{(a, b, c)}^{(x_2, y_2, z_2)} - \int_{(a, b, c)}^{(x_1, y_1, z_1)} = \phi(x_2, y_2, z_2) - \phi(x_1, y_1, z_1)$$

31. (a) Prove that $\mathbf{F} = (2xz^3 + 6y)\mathbf{i} + (6x - 2yz)\mathbf{j} + (3x^2z^2 - y^2)\mathbf{k}$ is a conservative force field. (b) Evaluate $\int_C \mathbf{F} \cdot d\mathbf{r}$ where C is any path from $(1, -1, 1)$ to $(2, 1, -1)$. (c) Give a physical interpretation of the results.

(a) A force field $\mathbf{F}$ is conservative if the line integral $\int_C \mathbf{F} \cdot d\mathbf{r}$ is independent of the path C joining any two points. A necessary and sufficient condition that $\mathbf{F}$ be conservative is that $\nabla \times \mathbf{F} = \mathbf{0}$.

Since here
$$\nabla \times \mathbf{F} = \begin{vmatrix} \mathbf{i} & \mathbf{j} & \mathbf{k} \\ \dfrac{\partial}{\partial x} & \dfrac{\partial}{\partial y} & \dfrac{\partial}{\partial z} \\ 2xz^3 + 6y & 6x - 2yz & 3x^2z^2 - y^2 \end{vmatrix} = \mathbf{0}, \quad \mathbf{F} \text{ is conservative.}$$

(b) **Method 1:**

By Problem 30, $\mathbf{F} \cdot d\mathbf{r} = (2xz^3 + 6y) \, dx + (6x - 2yz) \, dy + (3x^2z^2 - y^2) \, dz$ is an exact differential $d\phi$, where ϕ is such that

$$(1) \quad \frac{\partial \phi}{\partial x} = 2xz^3 + 6y \qquad (2) \quad \frac{\partial \phi}{\partial y} = 6x - 2yz \qquad (3) \quad \frac{\partial \phi}{\partial z} = 3x^2z^2 - y^2$$

From these we obtain respectively

$$\phi = x^2z^3 + 6xy + f_1(y, z) \qquad \phi = 6xy - y^2z + f_2(x, z) \qquad \phi = x^2z^3 - y^2z + f_3(x, y)$$

These are consistent if $f_1(y, z) = -y^2z + c$, $f_2(x, z) = x^2z^3 + c$, $f_3(x, y) = 6xy + c$ in which case $\phi = x^2z^3 + 6xy - y^2z + c$. Thus by Problem 30,

$$\int_{(1, -1, 1)}^{(2, 1, -1)} \mathbf{F} \cdot d\mathbf{r} = x^2z^3 + 6xy - y^2z + c \Big|_{(1, -1, 1)}^{(2, 1, -1)} = 15$$

Alternatively we may notice by inspection that

$$\mathbf{F} \cdot d\mathbf{r} = (2xz^3 \, dx + 3x^2z^2 \, dz) + (6y \, dx + 6x \, dy) - (2yz \, dy + y^2 \, dz)$$
$$= d(x^2z^3) + d(6xy) - d(y^2z) = d(x^2z^3 + 6xy - y^2z + c)$$

from which ϕ is determined.

Method 2:

Since the integral is independent of the path, we can choose any path to evaluate it; in particular we can choose the path consisting of straight lines from $(1, -1, 1)$ to $(2, -1, 1)$, then to $(2, 1, 1)$ and then to $(2, 1, -1)$. The result is

$$\int_{x=1}^{2} (2x - 6) \, dx + \int_{y=-1}^{1} (12 - 2y) \, dy + \int_{z=1}^{-1} (12z^2 - 1) \, dz = 15$$

where the first integral is obtained from the line integral by placing $y = -1$, $z = 1$, $dy = 0$, $dz = 0$; the second integral by placing $x = 2$, $z = 1$, $dx = 0$, $dz = 0$; and the third integral by placing $x = 2$, $y = 1$, $dx = 0$, $dy = 0$.

(c) Physically $\int_C \mathbf{F} \cdot d\mathbf{r}$ represents the work done in moving an object from $(1, -1, 1)$ to $(2, 1, -1)$ along C. In a conservative force field the work done is independent of the path C joining these points.

MISCELLANEOUS PROBLEMS

32. (a) If $x = f(u, v)$, $y = g(u, v)$ defines a transformation which maps a region $\mathcal{R}$ of the xy plane into a region $\mathcal{R}'$ of the uv plane, prove that

$$\iint\limits_{R} dx\, dy \;=\; \iint\limits_{R'} \left|\frac{\partial(x,y)}{\partial(u,v)}\right| du\, dv$$

(b) Interpret geometrically the result in (a).

(a) If C (assumed to be a simple closed curve) is the boundary of R, then by Problem 8,

$$\iint\limits_{R} dx\, dy \;=\; \frac{1}{2}\oint_C x\, dy \,-\, y\, dx \tag{1}$$

Under the given transformation the integral on the right of (1) becomes

$$\frac{1}{2}\oint_{C'} x\left(\frac{\partial y}{\partial u}du + \frac{\partial y}{\partial v}dv\right) - y\left(\frac{\partial x}{\partial u}du + \frac{\partial x}{\partial v}dv\right) \;=\; \frac{1}{2}\int_{C'}\left(x\frac{\partial y}{\partial u} - y\frac{\partial x}{\partial u}\right)du + \left(x\frac{\partial y}{\partial v} - y\frac{\partial x}{\partial v}\right)dv \tag{2}$$

where C' is the mapping of C in the uv plane (we suppose the mapping to be such that C' is a simple closed curve also).

By Green's theorem if R' is the region in the uv plane bounded by C', the right side of (2) equals

$$\frac{1}{2}\iint\limits_{R'}\left|\frac{\partial}{\partial u}\left(x\frac{\partial y}{\partial v} - y\frac{\partial x}{\partial v}\right) - \frac{\partial}{\partial v}\left(x\frac{\partial y}{\partial u} - y\frac{\partial x}{\partial u}\right)\right|du\, dv \;=\; \iint\limits_{R'}\left|\frac{\partial x}{\partial u}\frac{\partial y}{\partial v} - \frac{\partial x}{\partial v}\frac{\partial y}{\partial u}\right|du\, dv$$

$$=\; \iint\limits_{R'}\left|\frac{\partial(x,y)}{\partial(u,v)}\right|du\, dv$$

where we have inserted absolute value signs so as to ensure that the result is non-negative as is $\iint\limits_{R} dx\, dy$.

In general, we can show (see Problem 83) that

$$\iint\limits_{R} F(x,y)\, dx\, dy \;=\; \iint\limits_{R'} F\{f(u,v), g(u,v)\}\left|\frac{\partial(x,y)}{\partial(u,v)}\right|du\, dv \tag{3}$$

(b) $\iint\limits_{R} dx\, dy$ and $\iint\limits_{R'}\left|\frac{\partial(x,y)}{\partial(u,v)}\right|du\, dv$ represent the area of region R, the first expressed in rectangular coordinates, the second in curvilinear coordinates.

33. Let $\mathbf{F} = \dfrac{-y\mathbf{i} + x\mathbf{j}}{x^2 + y^2}$. (a) Calculate $\nabla \times \mathbf{F}$. (b) Evaluate $\oint \mathbf{F}\cdot d\mathbf{r}$ around any closed path and explain the results.

(a) $\nabla \times \mathbf{F} = \begin{vmatrix} \mathbf{i} & \mathbf{j} & \mathbf{k} \\ \dfrac{\partial}{\partial x} & \dfrac{\partial}{\partial y} & \dfrac{\partial}{\partial z} \\ \dfrac{-y}{x^2+y^2} & \dfrac{x}{x^2+y^2} & 0 \end{vmatrix} = \mathbf{0}$ in any region excluding $(0,0)$.

(b) $\oint \mathbf{F}\cdot d\mathbf{r} = \oint \dfrac{-y\, dx + x\, dy}{x^2 + y^2}$. Let $x = \rho\cos\phi$, $y = \rho\sin\phi$, where (ρ, ϕ) are polar coordinates.

Then

$$dx = -\rho\sin\phi\, d\phi + d\rho\cos\phi, \qquad dy = \rho\cos\phi\, d\phi + d\rho\sin\phi$$

and so

$$\frac{-y\, dx + x\, dy}{x^2 + y^2} = d\phi = d\left(\arctan\frac{y}{x}\right)$$

For a closed curve $ABCDA$ [see Fig. 10-18(a) below] surrounding the origin, $\phi = 0$ at A and $\phi = 2\pi$ after a complete circuit back to A. In this case the line integral equals $\displaystyle\int_0^{2\pi} d\phi = 2\pi$.

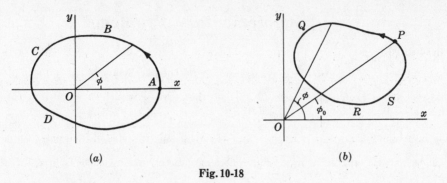

Fig. 10-18

For a closed curve $PQRSP$ [see Fig.10-18(b) above] not surrounding the origin, $\phi = \phi_0$ at P and $\phi = \phi_0$ after a complete circuit back to P. In this case the line integral equals $\displaystyle\int_{\phi_0}^{\phi_0} d\phi = 0$.

Since $\mathbf{F} = P\mathbf{i} + Q\mathbf{j}$, $\nabla \times \mathbf{F} = 0$ is equivalent to $\partial P/\partial y = \partial Q/\partial x$ and the results would seem to contradict those of Problem 11. However, no contradiction exists since $P = \dfrac{-y}{x^2 + y^2}$ and $Q = \dfrac{x}{x^2 + y^2}$ do not have continuous derivatives throughout any region including $(0, 0)$, and this was assumed in Problem 11.

34. If div $\mathbf{A}$ denotes the divergence of a vector field $\mathbf{A}$ at a point P, show that

$$\text{div } \mathbf{A} = \lim_{\Delta V \to 0} \frac{\iint_{\Delta S} \mathbf{A} \cdot \mathbf{n}\, dS}{\Delta V}$$

where ΔV is the volume enclosed by the surface ΔS and the limit is obtained by shrinking ΔV to the point P.

By the divergence theorem, $\displaystyle\iiint_{\Delta V} \text{div } \mathbf{A}\, dV = \iint_{\Delta S} \mathbf{A} \cdot \mathbf{n}\, dS$

By the mean-value theorem for integrals, the left side can be written

$$\overline{\text{div } \mathbf{A}} \iiint_{\Delta V} dV = \overline{\text{div } \mathbf{A}}\, \Delta V$$

where $\overline{\text{div } \mathbf{A}}$ is some value intermediate between the maximum and minimum of div $\mathbf{A}$ throughout ΔV. Then

$$\overline{\text{div } \mathbf{A}} = \frac{\iint_{\Delta S} \mathbf{A} \cdot \mathbf{n}\, dS}{\Delta V}$$

Taking the limit as $\Delta V \to 0$ such that P is always interior to ΔV, $\overline{\text{div } \mathbf{A}}$ approaches the value div $\mathbf{A}$ at point P; hence

$$\text{div } \mathbf{A} = \lim_{\Delta V \to 0} \frac{\iint_{\Delta S} \mathbf{A} \cdot \mathbf{n}\, dS}{\Delta V}$$

This result can be taken as a starting point for defining the divergence of $\mathbf{A}$, and from it all the properties may be derived including proof of the divergence theorem. We can also use this to extend the concept of divergence to coordinate systems other than rectangular (see Page 142).

Physically, $\left(\displaystyle\iiint_{\Delta S} \mathbf{A} \cdot \mathbf{n}\, dS\right)/\Delta V$ represents the flux or net outflow per unit volume of the vector $\mathbf{A}$ from the surface ΔS. If div $\mathbf{A}$ is positive in the neighborhood of a point P it means that the outflow from P is positive and we call P a *source*. Similarly, if div $\mathbf{A}$ is negative in the neighborhood of P the outflow is really an inflow and P is called a *sink*. If in a region there are no sources or sinks, then div $\mathbf{A} = 0$ and we call $\mathbf{A}$ a *solenoidal* vector field.

Supplementary Problems

LINE INTEGRALS

35. Evaluate $\int_{(1,1)}^{(4,2)} (x+y)\,dx + (y-x)\,dy$ along (a) the parabola $y^2 = x$, (b) a straight line, (c) straight lines from $(1,1)$ to $(1,2)$ and then to $(4,2)$, (d) the curve $x = 2t^2 + t + 1$, $y = t^2 + 1$.
    Ans. (a) 34/3, (b) 11, (c) 14, (d) 32/3

36. Evaluate $\oint (2x - y + 4)\,dx + (5y + 3x - 6)\,dy$ around a triangle in the xy plane with vertices at $(0,0)$, $(3,0)$, $(3,2)$ traversed in a counterclockwise direction.   Ans. 12

37. Evaluate the line integral in the preceding problem around a circle of radius 4 with center at $(0,0)$.
    Ans. 64π

38. (a) If $\mathbf{F} = (x^2 - y^2)\mathbf{i} + 2xy\mathbf{j}$, evaluate $\int_C \mathbf{F} \cdot d\mathbf{r}$ along the curve C in the xy plane given by $y = x^2 - x$ from the point $(1,0)$ to $(2,2)$. (b) Interpret physically the result obtained.
    Ans. (a) 124/15

39. Evaluate $\int_C (2x + y)\,ds$, where C is the curve in the xy plane given by $x^2 + y^2 = 25$ and s is the arc length parameter, from the point $(3,4)$ to $(4,3)$ along the shortest path.   Ans. 15

40. If $\mathbf{F} = (3x - 2y)\mathbf{i} + (y + 2z)\mathbf{j} - x^2\mathbf{k}$, evaluate $\int_C \mathbf{F} \cdot d\mathbf{r}$ from $(0,0,0)$ to $(1,1,1)$, where C is a path consisting of (a) the curve $x = t$, $y = t^2$, $z = t^3$, (b) a straight line joining these points, (c) the straight lines from $(0,0,0)$ to $(0,1,0)$, then to $(0,1,1)$ and then to $(1,1,1)$, (d) the curve $x = z^2$, $z = y^2$.
    Ans. (a) 23/15, (b) 5/3, (c) 0, (d) 13/15

41. If $\mathbf{T}$ is the unit tangent vector to a curve C (plane or space curve) and $\mathbf{F}$ is a given force field, prove that under appropriate conditions $\int_C \mathbf{F} \cdot d\mathbf{r} = \int_C \mathbf{F} \cdot \mathbf{T}\,ds$ where s is the arc length parameter. Interpret the result physically and geometrically.

GREEN'S THEOREM in the PLANE. INDEPENDENCE of the PATH

42. Verify Green's theorem in the plane for $\oint_C (x^2 - xy^3)\,dx + (y^2 - 2xy)\,dy$ where C is a square with vertices at $(0,0)$, $(2,0)$, $(2,2)$, $(0,2)$ and counterclockwise orientation.   Ans. common value = 8

43. Evaluate the line integrals of (a) Problem 36 and (b) Problem 37 by Green's theorem.

44. (a) Let C be any simple closed curve bounding a region having area A. Prove that if $a_1, a_2, a_3, b_1, b_2, b_3$ are constants,
    $$\oint_C (a_1 x + a_2 y + a_3)\,dx + (b_1 x + b_2 y + b_3)\,dy = (b_1 - a_2)A$$
    (b) Under what conditions will the line integral around any path C be zero?   Ans. (b) $a_2 = b_1$

45. Find the area bounded by the hypocycloid $x^{2/3} + y^{2/3} = a^{2/3}$.
    [Hint: Parametric equations are $x = a\cos^3 t$, $y = a\sin^3 t$, $0 \leqq t \leqq 2\pi$.]   Ans. $3\pi a^2/8$

46. If $x = \rho\cos\phi$, $y = \rho\sin\phi$, prove that $\frac{1}{2}\oint x\,dy - y\,dx = \frac{1}{2}\int \rho^2\,d\phi$ and interpret.

47. Verify Green's theorem in the plane for $\oint_C (x^3 - x^2 y)\,dx + xy^2\,dy$, where C is the boundary of the region enclosed by the circles $x^2 + y^2 = 4$ and $x^2 + y^2 = 16$.   Ans. common value = 120π

48. (a) Prove that $\int_{(1,0)}^{(2,1)} (2xy - y^4 + 3)\,dx + (x^2 - 4xy^3)\,dy$ is independent of the path joining $(1,0)$ and $(2,1)$. (b) Evaluate the integral in (a).   Ans. (b) 5

49. Evaluate $\int_C (2xy^3 - y^2 \cos x)\, dx + (1 - 2y \sin x + 3x^2 y^2)\, dy$ along the parabola $2x = \pi y^2$ from $(0,0)$ to $(\pi/2, 1)$.   Ans. $\pi^2/4$

50. Evaluate the line integral in the preceding problem around a parallelogram with vertices at $(0,0)$, $(3,0)$, $(5,2)$, $(2,2)$.   Ans. 0

51. (a) Prove that $G = (2x^2 + xy - 2y^2)\, dx + (3x^2 + 2xy)\, dy$ is not an exact differential.   (b) Prove that $e^{y/x} G/x$ is an exact differential of ϕ and find ϕ.   (c) Find a solution of the differential equation $(2x^2 + xy - 2y^2)\, dx + (3x^2 + 2xy)\, dy = 0$.
Ans. (b) $\phi = e^{y/x}(x^2 + 2xy) + c$, (c) $x^2 + 2xy + ce^{-y/x} = 0$

SURFACE INTEGRALS

52. (a) Evaluate $\iint_S (x^2 + y^2)\, dS$, where S is the surface of the cone $z^2 = 3(x^2 + y^2)$ bounded by $z = 0$ and $z = 3$. (b) Interpret physically the result in (a).   Ans. (a) 9π

53. Determine the surface area of the plane $2x + y + 2z = 16$ cut off by (a) $x = 0$, $y = 0$, $x = 2$, $y = 3$, (b) $x = 0$, $y = 0$ and $x^2 + y^2 = 64$.   Ans. (a) 9, (b) 24π

54. Find the surface area of the paraboloid $2z = x^2 + y^2$ which is outside the cone $z = \sqrt{x^2 + y^2}$.
Ans. $\frac{2}{3}\pi(5\sqrt{5} - 1)$

55. Find the area of the surface of the cone $z^2 = 3(x^2 + y^2)$ cut out by the paraboloid $z = x^2 + y^2$.
Ans. 6π

56. Find the surface area of the region common to the intersecting cylinders $x^2 + y^2 = a^2$ and $x^2 + z^2 = a^2$.
Ans. $16a^2$

57. (a) Obtain the surface area of the sphere $x^2 + y^2 + z^2 = a^2$ contained within the cone $z \tan \alpha = \sqrt{x^2 + y^2}$, $0 < \alpha < \pi/2$. (b) Use the result in (a) to find the surface area of a hemisphere. (c) Explain why formally placing $\alpha = \pi$ in the result of (a) yields the total surface area of a sphere.
Ans. (a) $2\pi a^2(1 - \cos \alpha)$, (b) $2\pi a^2$ (consider the limit as $\alpha \to \pi/2$)

58. Determine the moment of inertia of the surface of a sphere of radius a about a point on the surface. Assume a constant density σ.   Ans. $2Ma^2$, where mass $M = 4\pi a^2 \sigma$

59. (a) Find the centroid of the surface of the sphere $x^2 + y^2 + z^2 = a^2$ contained within the cone $z \tan \alpha = \sqrt{x^2 + y^2}$, $0 < \alpha < \pi/2$. (b) From the result in (a) obtain the centroid of the surface of a hemisphere.   Ans. (a) $\frac{1}{2}a(1 + \cos \alpha)$, (b) $a/2$

The DIVERGENCE THEOREM

60. Verify the divergence theorem for $\mathbf{A} = (2xy + z)\mathbf{i} + y^2 \mathbf{j} - (x + 3y)\mathbf{k}$ taken over the region bounded by $2x + 2y + z = 6$, $x = 0$, $y = 0$, $z = 0$.   Ans. common value $= 27$

61. Evaluate $\iint_S \mathbf{F} \cdot \mathbf{n}\, dS$, where $\mathbf{F} = (z^2 - x)\mathbf{i} - xy\mathbf{j} + 3z\mathbf{k}$ and S is the surface of the region bounded by $z = 4 - y^2$, $x = 0$, $x = 3$ and the xy plane.   Ans. 16

62. Evaluate $\iint_S \mathbf{A} \cdot \mathbf{n}\, dS$, where $\mathbf{A} = (2x + 3z)\mathbf{i} - (xz + y)\mathbf{j} + (y^2 + 2z)\mathbf{k}$ and S is the surface of the sphere having center at $(3, -1, 2)$ and radius 3.   Ans. 108π

63. Determine the value of $\iint_S x\, dy\, dz + y\, dz\, dx + z\, dx\, dy$, where S is the surface of the region bounded by the cylinder $x^2 + y^2 = 9$ and the planes $z = 0$ and $z = 3$, (a) by using the divergence theorem, (b) directly.   Ans. 81π

64. Evaluate $\iint\limits_{S} 4xz\,dy\,dz - y^2\,dz\,dx + yz\,dx\,dy$, where S is the surface of the cube bounded by $x=0$,

    $y=0$, $z=0$, $x=1$, $y=1$, $z=1$,   (a) directly,   (b) by Green's theorem in space (divergence theorem).
    Ans. 3/2

65. Prove that $\iint\limits_{S} (\nabla \times \mathbf{A}) \cdot \mathbf{n}\,dS = 0$ for any closed surface S.

66. Prove that $\iint\limits_{S} \mathbf{n}\,dS = 0$,   where n is the outward drawn normal to any closed surface S.

67. If $\mathbf{n}$ is the unit outward drawn normal to any closed surface S bounding the region V, prove that

$$\iiint\limits_{V} \operatorname{div} \mathbf{n}\,dV = S$$

STOKES' THEOREM

68. Verify Stokes' theorem for $\mathbf{A} = 2y\mathbf{i} + 3x\mathbf{j} - z^2\mathbf{k}$,   where S is the upper half surface of the sphere $x^2 + y^2 + z^2 = 9$ and C is its boundary.   Ans. common value $= 9\pi$

69. Verify Stokes' theorem for $\mathbf{A} = (y+z)\mathbf{i} - xz\mathbf{j} + y^2\mathbf{k}$,   where S is the surface of the region in the first octant bounded by $2x + z = 6$ and $y = 2$ which is not included in the   (a) xy plane,   (b) plane $y = 2$,   (c) plane $2x + z = 6$ and C is the corresponding boundary.
    Ans. The common value is   (a) -6,   (b) -9,   (c) -18

70. Evaluate $\iint\limits_{S} (\nabla \times \mathbf{A}) \cdot \mathbf{n}\,dS$, where $\mathbf{A} = (x-z)\mathbf{i} + (x^3 + yz)\mathbf{j} - 3xy^2\mathbf{k}$ and S is the surface of the

    cone $z = 2 - \sqrt{x^2 + y^2}$ above the xy plane.   Ans. 12π

71. If V is a region bounded by a closed surface S and $\mathbf{B} = \nabla \times \mathbf{A}$, prove that $\iint\limits_{S} \mathbf{B} \cdot \mathbf{n}\,dS = 0$.

72. (a) Prove that $\mathbf{F} = (2xy + 3)\mathbf{i} + (x^2 - 4z)\mathbf{j} - 4y\mathbf{k}$ is a conservative force field.   (b) Find ϕ such that

    $\mathbf{F} = \nabla \phi$.   (c) Evaluate $\int_{C} \mathbf{F} \cdot d\mathbf{r}$, where C is any path from $(3, -1, 2)$ to $(2, 1, -1)$.

    Ans. (b) $\phi = x^2 y - 4yz + 3x + \text{constant}$,   (c) 6

73. Let C be any path joining any point on the sphere $x^2 + y^2 + z^2 = a^2$ to any point on the sphere

    $x^2 + y^2 + z^2 = b^2$.   Show that if $\mathbf{F} = 5r^3\mathbf{r}$, where $\mathbf{r} = x\mathbf{i} + y\mathbf{j} + z\mathbf{k}$, then $\int_{C} \mathbf{F} \cdot d\mathbf{r} = b^5 - a^5$.

74. In Problem 73 evaluate $\int_{C} \mathbf{F} \cdot d\mathbf{r}$ if $\mathbf{F} = f(r)\mathbf{r}$, where $f(r)$ is assumed to be continuous.

    Ans. $\int_{a}^{b} r f(r)\,dr$

75. Determine whether there is a function ϕ such that $\mathbf{F} = \nabla \phi$, where:
    (a) $\mathbf{F} = (xz - y)\mathbf{i} + (x^2 y + z^3)\mathbf{j} + (3xz^2 - xy)\mathbf{k}$.
    (b) $\mathbf{F} = 2xe^{-y}\mathbf{i} + (\cos z - x^2 e^{-y})\mathbf{j} - y \sin z\,\mathbf{k}$.   If so, find it.
    Ans. (a) ϕ does not exist.   (b) $\phi = x^2 e^{-y} + y \cos z + \text{constant}$

76. Solve the differential equation $(z^3 - 4xy)\,dx + (6y - 2x^2)\,dy + (3xz^2 + 1)\,dz = 0$.
    Ans. $xz^3 - 2x^2 y + 3y^2 + z = \text{constant}$

MISCELLANEOUS PROBLEMS

77. Prove that a necessary and sufficient condition that $\oint_{C} \dfrac{\partial U}{\partial x}\,dy - \dfrac{\partial U}{\partial y}\,dx$ be zero around every simple

    closed path C in a region $\mathcal{R}$ (where U is continuous and has continuous partial derivatives of order

    two, at least) is that $\dfrac{\partial^2 U}{\partial x^2} + \dfrac{\partial^2 U}{\partial y^2} = 0$.

78. Verify Green's theorem for a multiply-connected region containing two "holes" (see Problem 10).

79. If $P\,dx + Q\,dy$ is not an exact differential but $\mu(P\,dx + Q\,dy)$ is an exact differential where μ is some function of x and y, then μ is called an *integrating factor*. (a) Prove that if F and G are functions of x alone, then $(Fy + G)\,dx + dy$ has an integrating factor μ which is a function of x alone and find μ. What must be assumed about F and G? (b) Use (a) to find solutions of the differential equation $xy' = 2x + 3y$.
   Ans. (a) $\mu = e^{\int F(x)\,dx}$ (b) $y = cx^3 - x$, where c is any constant

80. Find the surface area of the sphere $x^2 + y^2 + (z - a)^2 = a^2$ contained within the paraboloid $z = x^2 + y^2$.
   Ans. $2\pi a$

81. If $f(r)$ is a continuously differentiable function of $r = \sqrt{x^2 + y^2 + z^2}$, prove that
$$\iint_S f(r)\,\mathbf{n}\,dS \;=\; \iiint_V \frac{f'(r)}{r}\,\mathbf{r}\,dV$$

82. Prove that $\displaystyle\iint_S \nabla \times (\phi\mathbf{n})\,dS = 0$ where ϕ is any continuously differentiable scalar function of position and $\mathbf{n}$ is a unit outward drawn normal to a closed surface S.

83. Establish equation (3), Problem 32, by using Green's theorem in the plane.
   [Hint: Let the closed region $\mathcal{R}$ in the xy plane have boundary C and suppose that under the transformation $x = f(u, v)$, $y = g(u, v)$ these are transformed into $\mathcal{R}'$ and C' in the uv plane respectively.

   First prove that $\displaystyle\iint_\mathcal{R} F(x, y)\,dx\,dy = \int_C Q(x, y)\,dy$ where $\partial Q/\partial y = F(x, y)$. Then show that apart from sign this last integral is equal to $\displaystyle\int_{C'} Q\!\left[f(u, v),\, g(u, v)\right]\!\left[\frac{\partial g}{\partial u}\,du + \frac{\partial g}{\partial v}\,dv\right]$. Finally use Green's theorem to transform this into $\displaystyle\iint_{\mathcal{R}'} F[f(u, v),\, g(u, v)]\left|\frac{\partial(x, y)}{\partial(u, v)}\right| du\,dv$.

84. If $x = f(u, v, w)$, $y = g(u, v, w)$, $z = h(u, v, w)$ defines a transformation which maps a region $\mathcal{R}$ of xyz space into a region $\mathcal{R}'$ of uvw space, prove using Stokes' theorem that
$$\iiint_\mathcal{R} F(x, y, z)\,dx\,dy\,dz \;=\; \iiint_{\mathcal{R}'} G(u, v, w)\left|\frac{\partial(x, y, z)}{\partial(u, v, w)}\right| du\,dv\,dw$$
where $G(u, v, w) \equiv F[f(u, v, w),\, g(u, v, w),\, h(u, v, w)]$. State sufficient conditions under which the result is valid. See Problem 83.

85. (a) Show that in general the equation $\mathbf{r} = \mathbf{r}(u, v)$ geometrically represents a surface. (b) Discuss the geometric significance of $u = c_1$, $v = c_2$ where c_1 and c_2 are constants. (c) Prove that the element of arc length on this surface is given by
$$ds^2 \;=\; E\,du^2 + 2F\,du\,dv + G\,dv^2$$
where $E = \dfrac{\partial \mathbf{r}}{\partial u} \cdot \dfrac{\partial \mathbf{r}}{\partial u}$, $F = \dfrac{\partial \mathbf{r}}{\partial u} \cdot \dfrac{\partial \mathbf{r}}{\partial v}$, $G = \dfrac{\partial \mathbf{r}}{\partial v} \cdot \dfrac{\partial \mathbf{r}}{\partial v}$.

86. (a) Referring to Problem 85, show that the element of surface area is given by $dS = \sqrt{EG - F^2}\,du\,dv$.
   (b) Deduce from (a) that the area of a surface $\mathbf{r} = \mathbf{r}(u, v)$ is $\displaystyle\iint_S \sqrt{EG - F^2}\,du\,dv$.

   [Hint: Use the fact that $\left|\dfrac{\partial \mathbf{r}}{\partial u} \times \dfrac{\partial \mathbf{r}}{\partial v}\right| = \sqrt{\left(\dfrac{\partial \mathbf{r}}{\partial u} \times \dfrac{\partial \mathbf{r}}{\partial v}\right) \cdot \left(\dfrac{\partial \mathbf{r}}{\partial u} \times \dfrac{\partial \mathbf{r}}{\partial v}\right)}$ and then use the identity $(\mathbf{A} \times \mathbf{B}) \cdot (\mathbf{C} \times \mathbf{D}) = (\mathbf{A} \cdot \mathbf{C})(\mathbf{B} \cdot \mathbf{D}) - (\mathbf{A} \cdot \mathbf{D})(\mathbf{B} \cdot \mathbf{C})$.

87. (a) Prove that $\mathbf{r} = a \sin u \cos v\,\mathbf{i} + a \sin u \sin v\,\mathbf{j} + a \cos u\,\mathbf{k}$, $0 \leqq u \leqq \pi$, $0 \leqq v < 2\pi$ represents a sphere of radius a. (b) Use Problem 86 to show that the surface area of this sphere is $4\pi a^2$.

88. Use the result of Problem 34 to obtain div $\mathbf{A}$ in (a) cylindrical and (b) spherical coordinates. See Page 142.

Chapter 11

Infinite Series

CONVERGENCE and DIVERGENCE of INFINITE SERIES

Consider the infinite series

$$\sum_{n=1}^{\infty} u_n = u_1 + u_2 + u_3 + \cdots \qquad (1)$$

Let the sequence of *partial sums* of the series be $S_1, S_2, S_3, \ldots$ where

$$S_1 = u_1, \quad S_2 = u_1 + u_2, \quad S_3 = u_1 + u_2 + u_3, \quad \ldots, \quad S_n = u_1 + u_2 + \cdots + u_n \qquad (2)$$

If this sequence is convergent, i.e. if there exists a number S such that $\lim\limits_{n \to \infty} S_n = S$, the series (1) is called *convergent* and S is called its *sum*. If $\lim\limits_{n \to \infty} S_n$ does not exist, the series is called *divergent*. (Compare Chap. 3, Page 43). We shall sometimes abbreviate the infinite series (1) by Σu_n and shall refer to u_n as the *nth term* of the series.

Example 1: $\sum\limits_{n=1}^{\infty} \dfrac{1}{2^n} = \dfrac{1}{2} + \dfrac{1}{2^2} + \dfrac{1}{2^3} + \cdots$. Here $S_n = $ sum of first n terms $= 1 - \dfrac{1}{2^n}$ (see Prob. 25, Chap. 3). Then since $\lim\limits_{n \to \infty} \left(1 - \dfrac{1}{2^n} \right) = 1$, the series is convergent and has sum $S = 1$.

Example 2: $\sum\limits_{n=1}^{\infty} (-1)^{n-1} = 1 - 1 + 1 - 1 + \cdots$. Here $S_n = 0$ or 1 according as n is even or odd. Hence $\lim\limits_{n \to \infty} S_n$ does not exist and the series is divergent.

FUNDAMENTAL FACTS CONCERNING INFINITE SERIES

1. If Σu_n converges, then $\lim\limits_{n \to \infty} u_n = 0$ (see Prob. 26, Chap. 3). The converse however is not necessarily true, i.e. if $\lim\limits_{n \to \infty} u_n = 0$, Σu_n may or may not converge.

   It follows that if the nth term of a series does *not* approach zero the series is divergent.

2. Multiplication of each term of a series by a constant different from zero does not affect the convergence or divergence.

3. Removal (or addition) of a *finite* number of terms from (or to) a series does not affect the convergence or divergence.

SPECIAL SERIES

1. **Geometric series** $\sum\limits_{n=1}^{\infty} ar^{n-1} = a + ar + ar^2 + \cdots$, where a and r are constants, converges to $S = \dfrac{a}{1-r}$ if $|r| < 1$ and diverges if $|r| \geqq 1$. The sum of the first n terms is $S_n = \dfrac{a(1-r^n)}{1-r}$ (see Prob. 25, Chap. 3).

224

2. **The p series** $\sum\limits_{n=1}^{\infty}\dfrac{1}{n^p} = \dfrac{1}{1^p}+\dfrac{1}{2^p}+\dfrac{1}{3^p}+\cdots$ where p is a constant, converges for $p>1$ and diverges for $p\leqq 1$. The series with $p=1$ is called the *harmonic series*.

TESTS for CONVERGENCE and DIVERGENCE of SERIES of CONSTANTS

1. **Comparison test** for series of non-negative terms.
   (a) *Convergence.* Let $v_n \geqq 0$ for all $n>N$ and suppose that Σv_n converges. Then if $0 \leqq u_n \leqq v_n$ for all $n>N$, Σu_n also converges. Note that $n>N$ means *from some term onward*. Often, $N=1$.

   Example: Since $\dfrac{1}{2^n+1} \leqq \dfrac{1}{2^n}$ and $\Sigma \dfrac{1}{2^n}$ converges, $\Sigma \dfrac{1}{2^n+1}$ also converges.

   (b) *Divergence.* Let $v_n \geqq 0$ for all $n>N$ and suppose that Σv_n diverges. Then if $u_n \geqq v_n$ for all $n>N$, Σu_n also diverges.

   Example: Since $\dfrac{1}{\ln n} > \dfrac{1}{n}$ and $\sum\limits_{n=2}^{\infty}\dfrac{1}{n}$ diverges, $\sum\limits_{n=2}^{\infty}\dfrac{1}{\ln n}$ also diverges.

2. **Quotient test** for series of non-negative terms.
   (a) If $u_n \geqq 0$ and $v_n \geqq 0$ and if $\lim\limits_{n\to\infty}\dfrac{u_n}{v_n} = A \neq 0$ or ∞, then Σu_n and Σv_n either both converge or both diverge.
   (b) If $A=0$ in (a) and Σv_n converges, then Σu_n converges.
   (c) If $A=\infty$ in (a) and Σv_n diverges, then Σu_n diverges.

   This test is related to the comparison test and is often a very useful alternative to it. In particular, taking $v_n = 1/n^p$, we have from known facts about the p series the

Theorem 1. Let $\lim\limits_{n\to\infty} n^p u_n = A$. Then
   (i) Σu_n converges if $p>1$ and A is finite
   (ii) Σu_n diverges if $p\leqq 1$ and $A\neq 0$ (A may be infinite).

   Examples: 1. $\Sigma \dfrac{n}{4n^3-2}$ converges since $\lim\limits_{n\to\infty} n^2 \cdot \dfrac{n}{4n^3-2} = \dfrac{1}{4}$.

   2. $\Sigma \dfrac{\ln n}{\sqrt{n+1}}$ diverges since $\lim\limits_{n\to\infty} n^{1/2} \cdot \dfrac{\ln n}{(n+1)^{1/2}} = \infty$.

3. **Integral test** for series of non-negative terms.
   If $f(x)$ is positive, continuous and monotonic decreasing for $x \geqq N$ and is such that $f(n) = u_n$, $n = N, N+1, N+2, \ldots$, then Σu_n converges or diverges according as $\displaystyle\int_N^{\infty} f(x)\,dx = \lim\limits_{M\to\infty}\int_N^M f(x)\,dx$ converges or diverges. In particular we may have $N=1$, as is often true in practice.

   Example: $\sum\limits_{n=1}^{\infty}\dfrac{1}{n^2}$ converges since $\lim\limits_{M\to\infty}\int_1^M \dfrac{dx}{x^2} = \lim\limits_{M\to\infty}\left(1-\dfrac{1}{M}\right)$ exists.

4. **Alternating series test.** An *alternating series* is one whose successive terms are alternately positive and negative.

An alternating series converges if the following two conditions are satisfied (see Problem 15).

$$(a) \quad |u_{n+1}| \leqq |u_n| \quad \text{for } n \geqq 1$$
$$(b) \quad \lim_{n \to \infty} u_n = 0 \quad \left(\text{or } \lim_{n \to \infty} |u_n| = 0 \right)$$

Example: For the series $\quad 1 - \frac{1}{2} + \frac{1}{3} - \frac{1}{4} + \frac{1}{5} - \cdots = \sum_{n=1}^{\infty} \frac{(-1)^{n-1}}{n}$ we have $u_n = \frac{(-1)^{n-1}}{n}$, $|u_n| = \frac{1}{n}$, $|u_{n+1}| = \frac{1}{n+1}$. Then for $n \geqq 1$, $|u_{n+1}| \leqq |u_n|$. Also $\lim_{n \to \infty} |u_n| = 0$. Hence the series converges.

Theorem 2.

The numerical error made in stopping at any particular term of a convergent alternating series which satisfies conditions (a) and (b) is less than the absolute value of the next term.

Example: If we stop at the 4th term of the series $1 - \frac{1}{2} + \frac{1}{3} - \frac{1}{4} + \frac{1}{5} - \cdots$, the error made is less than $\frac{1}{5} = 0.2$.

5. **Absolute and conditional convergence.** The series Σu_n is called *absolutely convergent* if $\Sigma |u_n|$ converges. If Σu_n converges but $\Sigma |u_n|$ diverges, then Σu_n is called *conditionally convergent.*

Theorem 3.

If $\Sigma |u_n|$ converges, then Σu_n converges. In words, an absolutely convergent series is convergent (see Prob. 17).

Example 1: $\frac{1}{1^2} + \frac{1}{2^2} - \frac{1}{3^2} - \frac{1}{4^2} + \frac{1}{5^2} + \frac{1}{6^2} - \cdots$ is absolutely convergent and thus convergent, since the series of absolute values $\frac{1}{1^2} + \frac{1}{2^2} + \frac{1}{3^2} + \frac{1}{4^2} + \cdots$ converges.

Example 2: $1 - \frac{1}{2} + \frac{1}{3} - \frac{1}{4} + \cdots$ converges, but $1 + \frac{1}{2} + \frac{1}{3} + \frac{1}{4} + \cdots$ diverges. Thus $1 - \frac{1}{2} + \frac{1}{3} - \frac{1}{4} + \cdots$ is conditionally convergent.

Any of the tests used for series with non-negative terms can be used to test for absolute convergence.

6. **Ratio test.** Let $\lim_{n \to \infty} \left| \frac{u_{n+1}}{u_n} \right| = L$. Then the series Σu_n

   (a) converges (absolutely) if $L < 1$
   (b) diverges if $L > 1$.

   If $L = 1$ the test fails.

7. **The nth root test.** Let $\lim_{n \to \infty} \sqrt[n]{|u_n|} = L$. Then the series Σu_n

   (a) converges (absolutely) if $L < 1$
   (b) diverges if $L > 1$.

   If $L = 1$ the test fails.

8. **Raabe's test.** Let $\lim_{n \to \infty} n \left(1 - \left| \frac{u_{n+1}}{u_n} \right| \right) = L$. Then the series Σu_n

   (a) converges (absolutely) if $L > 1$
   (b) diverges or converges conditionally if $L < 1$.

If $L=1$ the test fails.

This test is often used when the ratio test fails.

9. **Gauss' test.** If $\left|\dfrac{u_{n+1}}{u_n}\right| = 1 - \dfrac{L}{n} + \dfrac{c_n}{n^2}$, where $|c_n| < P$ for all $n > N$, then the series Σu_n

(a) converges (absolutely) if $L > 1$

(b) diverges or converges conditionally if $L \leqq 1$.

This test is often used when Raabe's test fails.

THEOREMS on ABSOLUTELY CONVERGENT SERIES

Theorem 4.

The terms of an absolutely convergent series can be rearranged in any order, and all such rearranged series will converge to the same sum. However, if the terms of a conditionally convergent series are suitably rearranged, the resulting series may diverge or converge to *any* desired sum (see Problem 76).

Theorem 5.

The sum, difference and product of two absolutely convergent series is absolutely convergent. The operations can be performed as for finite series.

INFINITE SEQUENCES and SERIES of FUNCTIONS. UNIFORM CONVERGENCE

Let $\{u_n(x)\}$, $n = 1, 2, 3, \ldots$ be a sequence of functions defined in $[a, b]$. The sequence is said to converge to $F(x)$, or to have the limit $F(x)$ in $[a, b]$, if for each $\epsilon > 0$ and each x in $[a, b]$ we can find $N > 0$ such that $|u_n(x) - F(x)| < \epsilon$ for all $n > N$. In such case we write $\lim\limits_{n \to \infty} u_n(x) = F(x)$. The number N may depend on x as well as ϵ. If it depends *only* on ϵ and not on x, the sequence is said to converge to $F(x)$ *uniformly* in $[a, b]$ or to be *uniformly convergent* in $[a, b]$.

The infinite series of functions

$$\sum_{n=1}^{\infty} u_n(x) = u_1(x) + u_2(x) + u_3(x) + \cdots \tag{3}$$

is said to be convergent in $[a, b]$ if the sequence of partial sums $\{S_n(x)\}$, $n = 1, 2, 3, \ldots$, where $S_n(x) = u_1(x) + u_2(x) + \cdots + u_n(x)$, is convergent in $[a, b]$. In such case we write $\lim\limits_{n \to \infty} S_n(x) = S(x)$ and call $S(x)$ the *sum* of the series.

It follows that $\Sigma u_n(x)$ converges to $S(x)$ in $[a, b]$ if for each $\epsilon > 0$ and each x in $[a, b]$ we can find $N > 0$ such that $|S_n(x) - S(x)| < \epsilon$ for all $n > N$. If N depends *only* on ϵ and not on x, the series is called *uniformly convergent* in $[a, b]$.

Since $S(x) - S_n(x) = R_n(x)$, the remainder after n terms, we can equivalently say that $\Sigma u_n(x)$ is uniformly convergent in $[a, b]$ if for each $\epsilon > 0$ we can find N depending on ϵ but not on x such that $|R_n(x)| < \epsilon$ for all $n > N$ and all x in $[a, b]$.

These definitions can be modified to include other intervals besides $a \leqq x \leqq b$, such as $a < x < b$, etc.

The domain of convergence (absolute or uniform) of a series is the set of values of x for which the series of functions converges (absolutely or uniformly).

SPECIAL TESTS for UNIFORM CONVERGENCE of SERIES

1. **Weierstrass M test.** If a sequence of positive constants $M_1, M_2, M_3, \ldots$ can be found such that in some interval

    $$(a) \quad |u_n(x)| \leqq M_n \qquad n = 1, 2, 3, \ldots$$
    $$(b) \quad \Sigma M_n \text{ converges}$$

    then $\Sigma u_n(x)$ is uniformly and absolutely convergent in the interval.

    Example: $\displaystyle\sum_{n=1}^{\infty} \frac{\cos nx}{n^2}$ is uniformly and absolutely convergent in $[0, 2\pi]$ since $\left| \dfrac{\cos nx}{n^2} \right| \leqq \dfrac{1}{n^2}$ and $\Sigma \dfrac{1}{n^2}$ converges.

    This test supplies a sufficient but not a necessary condition for uniform convergence, i.e. a series may be uniformly convergent even when the test cannot be made to apply.

    One may be led because of this test to believe that uniformly convergent series must be absolutely convergent, and conversely. However, the two properties are independent, i.e. a series can be uniformly convergent without being absolutely convergent, and conversely. See Problems 30, 123.

2. **Dirichlet's test.** Suppose that

    (a) the sequence $\{a_n\}$ is a monotonic decreasing sequence of positive constants having limit zero,

    (b) there exists a constant P such that for $a \leqq x \leqq b$

    $$|u_1(x) + u_2(x) + \cdots + u_n(x)| < P \quad \text{for all } n > N.$$

    Then the series

    $$a_1 u_1(x) + a_2 u_2(x) + \cdots = \sum_{n=1}^{\infty} a_n u_n(x)$$

    is uniformly convergent in $a \leqq x \leqq b$.

THEOREMS on UNIFORMLY CONVERGENT SERIES

If an infinite series of functions is uniformly convergent, it has many of the properties possessed by sums of finite series of functions, as indicated in the following theorems.

Theorem 6.

If $\{u_n(x)\}$, $n = 1, 2, 3, \ldots$ are continuous in $[a, b]$ and if $\Sigma u_n(x)$ converges uniformly to the sum $S(x)$ in $[a, b]$, then $S(x)$ is continuous in $[a, b]$.

Briefly, this states that a uniformly convergent series of continuous functions is a continuous function. This result is often used to demonstrate that a given series is not uniformly convergent by showing that the sum function $S(x)$ is discontinuous at some point (see Problem 30).

In particular if x_0 is in $[a, b]$, then the theorem states that

$$\lim_{x \to x_0} \sum_{n=1}^{\infty} u_n(x) \;=\; \sum_{n=1}^{\infty} \lim_{x \to x_0} u_n(x) \;=\; \sum_{n=1}^{\infty} u_n(x_0)$$

where we use right or left hand limits in case x_0 is an endpoint of $[a, b]$.

Theorem 7.

If $\{u_n(x)\}$, $n = 1, 2, 3, \ldots$, are continuous in $[a, b]$ and if $\Sigma u_n(x)$ converges u̶ to the sum $S(x)$ in $[a, b]$, then

$$\int_a^b S(x)\, dx \;=\; \sum_{n=1}^{\infty} \int_a^b u_n(x)\, dx$$

or

$$\int_a^b \left\{ \sum_{n=1}^{\infty} u_n(x) \right\} dx \;=\; \sum_{n=1}^{\infty} \int_a^b u_n(x)\, dx \tag{5}$$

Briefly, a uniformly convergent series of continuous functions can be integrated term by term.

Theorem 8.

If $\{u_n(x)\}$, $n = 1, 2, 3, \ldots$, are continuous and have continuous derivatives in $[a, b]$ and if $\Sigma u_n(x)$ converges to $S(x)$ while $\Sigma u_n'(x)$ is uniformly convergent in $[a, b]$, then in $[a, b]$

$$S'(x) \;=\; \sum_{n=1}^{\infty} u_n'(x) \tag{6}$$

or

$$\frac{d}{dx} \left\{ \sum_{n=1}^{\infty} u_n(x) \right\} \;=\; \sum_{n=1}^{\infty} \frac{d}{dx}\, u_n(x) \tag{7}$$

This shows conditions under which a series can be differentiated term by term.

Theorems similar to the above can be formulated for sequences. For example, if $\{u_n(x)\}$, $n = 1, 2, 3, \ldots$ is uniformly convergent in $[a, b]$, then

$$\lim_{n \to \infty} \int_a^b u_n(x)\, dx \;=\; \int_a^b \lim_{n \to \infty} u_n(x)\, dx \tag{8}$$

which is the analog of Theorem 7.

POWER SERIES. A series having the form

$$a_0 + a_1 x + a_2 x^2 + \cdots \;=\; \sum_{n=0}^{\infty} a_n x^n \tag{9}$$

where $a_0, a_1, a_2, \ldots$ are constants, is called a *power series* in x. It is often convenient to abbreviate the series (9) as $\Sigma a_n x^n$.

In general a power series converges for $|x| < R$ and diverges for $|x| > R$, where the constant R is called the *radius of convergence* of the series. For $|x| = R$, the series may or may not converge.

The interval $|x| < R$ or $-R < x < R$, with possible inclusion of endpoints, is called the *interval of convergence* of the series. Although the ratio test is often successful in obtaining this interval, it may fail and in such cases other tests may be used (see Prob. 22).

The two special cases $R = 0$ and $R = \infty$ can arise. In the first case the series converges only for $x = 0$; in the second case it converges for all x, sometimes written $-\infty < x < \infty$ (see Problem 25). When we speak of a convergent power series we shall assume, unless otherwise indicated, that $R > 0$.

Similar remarks hold for a power series of the form (9), where x is replaced by $(x - a)$.

and absolutely in any interval which lies *entirely*

229

or integrated term by term over any interval
gence. Also, the sum of a convergent power
ntirely within its interval of convergence.

and the theorems on uniformly convergent
an be extended to include endpoints of the
orems.

series converges up to and including an endpoint of its interval of
the interval of uniform convergence also extends so far as to include this
point. See Problem 42.

Theorem 12. *Abel's limit theorem.*

If $\displaystyle\sum_{n=0}^{\infty} a_n x^n$ converges at $x = x_0$ which may be an interior point or an endpoint of
the interval of convergence, then

$$\lim_{x \to x_0}\left\{ \sum_{n=0}^{\infty} a_n x^n \right\} = \sum_{n=0}^{\infty}\left\{ \lim_{x \to x_0} a_n x^n \right\} = \sum_{n=0}^{\infty} a_n x_0^n \tag{10}$$

If x_0 is an endpoint we must use $x \to x_0+$ or $x \to x_0-$ in (10) according as x_0 is a
left or right hand endpoint.

This follows at once from Theorem 11 and Theorem 6 on the continuity of sums of
uniformly convergent series.

OPERATIONS with POWER SERIES

In the following theorems we assume that all power series are convergent in some
interval.

Theorem 13.

Two power series can be added or subtracted term by term for each value of x common to their intervals of convergence.

Theorem 14.

Two power series, for example $\displaystyle\sum_{n=0}^{\infty} a_n x^n$ and $\displaystyle\sum_{n=0}^{\infty} b_n x^n$, can be multiplied to obtain
$\displaystyle\sum_{n=0}^{\infty} c_n x^n$ where

$$c_n = a_0 b_n + a_1 b_{n-1} + a_2 b_{n-2} + \cdots + a_n b_0 \tag{11}$$

the result being valid for each x within the common interval of convergence.

Theorem 15.

If the power series $\displaystyle\sum_{n=0}^{\infty} a_n x^n$ is divided by the power series $\Sigma b_n x^n$ where $b_0 \neq 0$, the
quotient can be written as a power series which converges for sufficiently small values of x.

Theorem 16.

If $y = \sum\limits_{n=0}^{\infty} a_n x^n$, then by substituting $x = \sum\limits_{n=0}^{\infty} b_n y^n$ we can obtain the coefficients b_n in terms of a_n. This process is often called *reversion of series*.

EXPANSION of FUNCTIONS in POWER SERIES

Suppose that $f(x)$ and its derivatives $f'(x), f''(x), \ldots, f^{(n)}(x)$ exist and are continuous in the closed interval $a \leqq x \leqq b$ and that $f^{(n+1)}(x)$ exists in the open interval $a < x < b$. Then as we have seen in previous chapters (see Pages 61 and 95)

$$f(x) \;=\; f(a) + f'(a)(x-a) + \frac{f''(a)}{2!}(x-a)^2 + \cdots + \frac{f^{(n)}(a)}{n!}(x-a)^n + R_n \qquad (12)$$

where R_n, the remainder, is given in either of the forms

$$R_n \;=\; \frac{f^{(n+1)}(\xi)}{(n+1)!}(x-a)^{n+1} \qquad \text{(Lagrange's form)} \qquad (13)$$

$$R_n \;=\; \frac{f^{(n+1)}(\xi)}{n!}(x-\xi)^n (x-a) \qquad \text{(Cauchy's form)} \qquad (14)$$

where ξ, which lies between a and x, is in general different in the two forms.

As n changes, ξ also changes in general. If for all x and ξ in $[a, b]$ we have $\lim\limits_{n \to \infty} R_n = 0$, then (12) can be written

$$f(x) \;=\; f(a) + f'(a)(x-a) + \frac{f''(a)}{2!}(x-a)^2 + \frac{f'''(a)}{3!}(x-a)^3 + \cdots \qquad (15)$$

which is called the *Taylor series or expansion* of $f(x)$. In case $a = 0$, it is often called the *Maclaurin series or expansion* of $f(x)$. For problems involving such expansions see also Chapter 6.

One might be tempted to believe that if all derivatives of $f(x)$ exist at $x = a$, the expansion (15) would be valid. This however is not necessarily the case, for although one can then *formally* obtain the series on the right of (15), the resulting series may not converge to $f(x)$. For an example of this see Problem 104.

Precise conditions under which the series converges to $f(x)$ are best obtained by means of the theory of functions of a complex variable. See Chapter 17.

SOME IMPORTANT POWER SERIES

The following series, convergent to the given function in the indicated intervals, are frequently employed in practice.

1. $\sin x \quad = x - \dfrac{x^3}{3!} + \dfrac{x^5}{5!} - \dfrac{x^7}{7!} + \cdots (-1)^{n-1}\dfrac{x^{2n-1}}{(2n-1)!} + \cdots \quad -\infty < x < \infty$

2. $\cos x \quad = 1 - \dfrac{x^2}{2!} + \dfrac{x^4}{4!} - \dfrac{x^6}{6!} + \cdots (-1)^{n-1}\dfrac{x^{2n-2}}{(2n-2)!} + \cdots \quad -\infty < x < \infty$

3. $e^x \quad = 1 + x + \dfrac{x^2}{2!} + \dfrac{x^3}{3!} + \cdots + \dfrac{x^{n-1}}{(n-1)!} + \cdots \quad -\infty < x < \infty$

4. $\ln|1+x| \quad = x - \dfrac{x^2}{2} + \dfrac{x^3}{3} - \dfrac{x^4}{4} + \cdots (-1)^{n-1}\dfrac{x^n}{n} + \cdots \quad -1 < x \leqq 1$

5. $\frac{1}{2}\ln\left|\dfrac{1+x}{1-x}\right| = x + \dfrac{x^3}{3} + \dfrac{x^5}{5} + \dfrac{x^7}{7} + \cdots + \dfrac{x^{2n-1}}{2n-1} + \cdots \quad -1 < x < 1$

6. $\tan^{-1}x \quad = x - \dfrac{x^3}{3} + \dfrac{x^5}{5} - \dfrac{x^7}{7} + \cdots (-1)^{n-1}\dfrac{x^{2n-1}}{2n-1} + \cdots \quad -1 \leqq x \leqq 1$

7. $(1+x)^p \quad = 1 + px + \dfrac{p(p-1)}{2!}x^2 + \cdots + \dfrac{p(p-1)\cdots(p-n+1)}{n!}x^n + \cdots$

This is the *binomial series*.

(*a*) If p is a positive integer or zero, the series terminates.

(*b*) If $p > 0$ but is not an integer, the series converges (absolutely) for $-1 \leqq x \leqq 1$.

(*c*) If $-1 < p < 0$, the series converges for $-1 < x \leqq 1$.

(*d*) If $p \leqq -1$, the series converges for $-1 < x < 1$.

For all p the series certainly converges if $-1 < x < 1$.

SPECIAL TOPICS

1. **Functions defined by series** are often useful in applications and frequently arise as solutions of differential equations. For example, the function defined by

$$J_p(x) \; = \; \frac{x^p}{2^p p!} \left\{ 1 \, - \, \frac{x^2}{2(2p+2)} \, + \, \frac{x^4}{2 \cdot 4(2p+2)(2p+4)} \, - \, \cdots \right\} \qquad (16)$$
$$= \; \sum_{n=0}^{\infty} \frac{(-1)^n (x/2)^{p+2n}}{n!\,(n+p)!}$$

is a solution of *Bessel's differential equation* $\; x^2 y'' + xy' + (x^2 - p^2)y = 0 \;$ and is thus called a *Bessel function of order p*. See Problems 46, 106-109.

Similarly, the *hypergeometric function*

$$F(a, b; c; x) \; = \; 1 \, + \, \frac{a \cdot b}{1 \cdot c} x \, + \, \frac{a(a+1)\,b(b+1)}{1 \cdot 2 \cdot c(c+1)} x^2 \, + \, \cdots \qquad (17)$$

is a solution of *Gauss' differential equation* $\; x(1-x)y'' + \{c - (a+b+1)x\}y' - aby = 0$.

These functions have many important properties.

2. **Infinite series of complex terms**, in particular power series of the form $\displaystyle\sum_{n=0}^{\infty} a_n z^n$,

where $z = x + iy$ and a_n may be complex, can be handled in a manner similar to real series.

Such power series converge for $|z| < R$, i.e. interior to a *circle of convergence* $x^2 + y^2 = R^2$, where R is the *radius of convergence* (if the series converges only for $z = 0$, we say that the radius of convergence R is zero; if it converges for all z, we say that the radius of convergence is infinite). On the boundary of this circle, i.e. $|z| = R$, the series may or may not converge, depending on the particular z.

Note that for $y = 0$ the circle of convergence reduces to the interval of convergence for real power series. Greater insight into the behavior of power series is obtained by use of the theory of functions of a complex variable (see Chapter 17).

3. **Infinite series of functions of two (or more) variables**, such as $\displaystyle\sum_{n=1}^{\infty} u_n(x, y)$ can

be treated in a manner analogous to series in one variable. In particular, we can discuss power series in x and y having the form

$$a_{00} + (a_{10}x + a_{01}y) + (a_{20}x^2 + a_{11}xy + a_{02}y^2) + \cdots \qquad (18)$$

using double subscripts for the constants. As for one variable, we can expand suitable functions of x and y in such power series using results of Chapter 6, Page 109 and showing that the remainder $R_n \to 0$ as $n \to \infty$. In general, such power series converge inside a rectangular region $|x| < A$, $|y| < B$ and possibly on the boundary.

4. **Double Series.** Consider the array of numbers (or functions)

$$\begin{pmatrix} u_{11} & u_{12} & u_{13} & \cdots \\ u_{21} & u_{22} & u_{23} & \cdots \\ u_{31} & u_{32} & u_{33} & \cdots \\ \cdot & \cdot & \cdot \\ \cdot & \cdot & \cdot \\ \cdot & \cdot & \cdot \end{pmatrix}$$

Let $S_{mn} = \sum_{p=1}^{m} \sum_{q=1}^{n} u_{pq}$ be the sum of the numbers in the first m rows and first n columns of this array. If there exists a number S such that $\lim_{\substack{m \to \infty \\ n \to \infty}} S_{mn} = S$, we say that the double series $\sum_{p=1}^{\infty} \sum_{q=1}^{\infty} u_{pq}$ *converges* to the *sum S*; otherwise it *diverges*.

Definitions and theorems for double series are very similar to those for series already considered.

5. **Infinite Products.** Let $P_n = (1+u_1)(1+u_2)(1+u_3) \cdots (1+u_n)$ denoted by $\prod_{k=1}^{n} (1+u_k)$ where we suppose that $u_k \neq -1$, $k = 1, 2, 3, \ldots$. If there exists a number $P \neq 0$ such that $\lim_{n \to \infty} P_n = P$, we say that the *infinite product* $(1+u_1)(1+u_2)(1+u_3) \cdots \equiv \prod_{k=1}^{\infty} (1+u_k)$, or briefly $\Pi(1+u_k)$, converges to P; otherwise it diverges.

If $\Pi(1+|u_k|)$ converges, we call the infinite product $\Pi(1+u_k)$ *absolutely convergent*. It can be shown that an absolutely convergent infinite product converges and that factors can in such cases be rearranged without affecting the result.

Theorems about infinite products can (by taking logarithms), often be made to depend on theorems for infinite series. Thus, for example, we have the

Theorem. A necessary and sufficient condition that $\Pi(1+u_k)$ converge absolutely is that Σu_k converge absolutely.

6. **Summability.** Let $S_1, S_2, S_3, \ldots$ be the partial sums of a divergent series Σu_n. If the sequence $S_1, \dfrac{S_1 + S_2}{2}, \dfrac{S_1 + S_2 + S_3}{3}, \ldots$ (formed by taking arithmetic means of the first n terms of $S_1, S_2, S_3, \ldots$) converges to S, we say that the series Σu_n is *summable* in the *Césaro sense*, or *C-1 summable* to S (see Problem 51).

If Σu_n converges to S, the Césaro method also yields the result S. For this reason the Césaro method is said to be a *regular* method of summability.

In case the Césaro limit does not exist, we can apply the same technique to the sequence $S_1, \dfrac{S_1 + S_2}{2}, \dfrac{S_1 + S_2 + S_3}{3}, \ldots$. If the C-1 limit for this sequence exists and equals S, we say that Σu_k converges to S in the C-2 sense. The process can be continued indefinitely.

7. **Asymptotic series.** Consider the series

$$S(x) = a_0 + \frac{a_1}{x} + \frac{a_2}{x^2} + \cdots + \frac{a_n}{x^n} + \cdots \tag{19}$$

and suppose that

$$S_n(x) = a_0 + \frac{a_1}{x} + \frac{a_2}{x^2} + \cdots + \frac{a_n}{x^n} \qquad n = 0, 1, 2 \ldots \tag{20}$$

are the partial sums of this series.

If $R_n(x) = f(x) - S_n(x)$, where $f(x)$ is given, is such that for every n

$$\lim_{|x| \to \infty} x^n R_n(x) = 0$$

then $S(x)$ is called an *asymptotic expansion* of $f(x)$ and we denote this by writing $f(x) \sim S(x)$.

In practice, the series (*19*) diverges. However, by taking the sum of successive terms of the series, stopping *just before the terms begin to increase*, we may obtain a useful approximation for $f(x)$.

Various operations with asymptotic series are permissible. For example, asymptotic series may be multiplied or integrated term by term to yield another asymptotic series.

Solved Problems

CONVERGENCE and DIVERGENCE of SERIES of CONSTANTS

1. (*a*) Prove that $\dfrac{1}{1 \cdot 3} + \dfrac{1}{3 \cdot 5} + \dfrac{1}{5 \cdot 7} + \cdots = \displaystyle\sum_{n=1}^{\infty} \dfrac{1}{(2n-1)(2n+1)}$ converges and (*b*) find its sum.

$$u_n = \frac{1}{(2n-1)(2n+1)} = \frac{1}{2}\left(\frac{1}{2n-1} - \frac{1}{2n+1}\right). \quad \text{Then}$$

$$S_n = u_1 + u_2 + \cdots + u_n = \frac{1}{2}\left(\frac{1}{1} - \frac{1}{3}\right) + \frac{1}{2}\left(\frac{1}{3} - \frac{1}{5}\right) + \cdots + \frac{1}{2}\left(\frac{1}{2n-1} - \frac{1}{2n+1}\right)$$

$$= \frac{1}{2}\left(\frac{1}{1} - \frac{1}{3} + \frac{1}{3} - \frac{1}{5} + \frac{1}{5} - \cdots + \frac{1}{2n-1} - \frac{1}{2n+1}\right) = \frac{1}{2}\left(1 - \frac{1}{2n+1}\right)$$

Since $\displaystyle\lim_{n \to \infty} S_n = \lim_{n \to \infty} \frac{1}{2}\left(1 - \frac{1}{2n+1}\right) = \frac{1}{2}$,   the series converges and its sum is $\frac{1}{2}$.

The series is sometimes called a *telescoping series* since the terms of S_n, other than the first and last, cancel out in pairs.

2. (*a*) Prove that $\frac{2}{3} + (\frac{2}{3})^2 + (\frac{2}{3})^3 + \cdots = \displaystyle\sum_{n=1}^{\infty} (\frac{2}{3})^n$ converges and (*b*) find its sum.

$$S_n = \tfrac{2}{3} + (\tfrac{2}{3})^2 + (\tfrac{2}{3})^3 + \cdots + (\tfrac{2}{3})^n$$

$$\tfrac{2}{3}S_n = \qquad (\tfrac{2}{3})^2 + (\tfrac{2}{3})^3 + \cdots + (\tfrac{2}{3})^n + (\tfrac{2}{3})^{n+1}$$

Subtract: $\qquad \tfrac{1}{3}S_n = \tfrac{2}{3} - (\tfrac{2}{3})^{n+1} \qquad \text{or} \qquad S_n = 2\{1 - (\tfrac{2}{3})^n\}$

Since $\displaystyle\lim_{n \to \infty} S_n = \lim_{n \to \infty} 2\{1 - (\tfrac{2}{3})^n\} = 2$,   the series converges and its sum is 2.

Another method: Let $a = \frac{2}{3}$, $r = \frac{2}{3}$ in Prob. 25 of Chap. 3; then the sum is $a/(1-r) = \frac{2}{3}/(1 - \frac{2}{3}) = 2$.

3. Prove that the series $\frac{1}{2} + \frac{2}{3} + \frac{3}{4} + \frac{4}{5} + \cdots = \displaystyle\sum_{n=1}^{\infty} \frac{n}{n+1}$ diverges.

$$\lim_{n \to \infty} u_n = \lim_{n \to \infty} \frac{n}{n+1} = 1. \quad \text{Hence by Problem 26, Chapter 3, the series is divergent.}$$

4. Show that the series whose nth term is $u_n = \sqrt{n+1} - \sqrt{n}$ diverges although $\lim\limits_{n \to \infty} u_n = 0$.

   The fact that $\lim\limits_{n \to \infty} u_n = 0$ follows from Problem 14(c), Chapter 3.

   Now $S_n = u_1 + u_2 + \cdots + u_n = (\sqrt{2} - \sqrt{1}) + (\sqrt{3} - \sqrt{2}) + \cdots + (\sqrt{n+1} - \sqrt{n}) = \sqrt{n+1} - \sqrt{1}$.

   Then S_n increases without bound and the series diverges.

   This problem shows that $\lim\limits_{n \to \infty} u_n = 0$ is a *necessary* but not *sufficient* condition for the convergence of Σu_n. See also Problem 6.

COMPARISON TEST and QUOTIENT TEST

5. If $0 \leqq u_n \leqq v_n$, $n = 1, 2, 3, \ldots$ and if Σv_n converges, prove that Σu_n also converges (i.e. establish the comparison test for convergence).

   Let $S_n = u_1 + u_2 + \cdots + u_n$, $T_n = v_1 + v_2 + \cdots + v_n$.

   Since Σv_n converges, $\lim\limits_{n \to \infty} T_n$ exists and equals T, say. Also, since $v_n \geqq 0$, $T_n \leqq T$.

   Then $S_n = u_1 + u_2 + \cdots + u_n \leqq v_1 + v_2 + \cdots + v_n \leqq T$ or $0 \leqq S_n \leqq T$.

   Thus S_n is a bounded monotonic increasing sequence and must have a limit (see Chap. 3), i.e. Σu_n converges.

6. Using the comparison test prove that $1 + \frac{1}{2} + \frac{1}{3} + \cdots = \sum\limits_{n=1}^{\infty} \frac{1}{n}$ diverges.

   We have
   $$1 \geqq \tfrac{1}{2}$$
   $$\tfrac{1}{2} + \tfrac{1}{3} \geqq \tfrac{1}{4} + \tfrac{1}{4} = \tfrac{1}{2}$$
   $$\tfrac{1}{4} + \tfrac{1}{5} + \tfrac{1}{6} + \tfrac{1}{7} \geqq \tfrac{1}{8} + \tfrac{1}{8} + \tfrac{1}{8} + \tfrac{1}{8} = \tfrac{1}{2}$$
   $$\tfrac{1}{8} + \tfrac{1}{9} + \tfrac{1}{10} + \cdots + \tfrac{1}{15} \geqq \tfrac{1}{16} + \tfrac{1}{16} + \tfrac{1}{16} + \cdots + \tfrac{1}{16} \text{ (8 terms) } = \tfrac{1}{2}$$

   etc. Thus to any desired number of terms,
   $$1 + (\tfrac{1}{2} + \tfrac{1}{3}) + (\tfrac{1}{4} + \tfrac{1}{5} + \tfrac{1}{6} + \tfrac{1}{7}) + \cdots \geqq \tfrac{1}{2} + \tfrac{1}{2} + \tfrac{1}{2} + \cdots$$

   Since the right hand side can be made larger than any positive number by choosing enough terms, the given series diverges.

   By methods analogous to that used here, we can show that $\sum\limits_{n=1}^{\infty} \frac{1}{n^p}$, where p is a constant, diverges if $p \leqq 1$ and converges if $p > 1$. This can also be shown in other ways [see Problem 13(a)].

7. Test for convergence or divergence $\sum\limits_{n=1}^{\infty} \frac{\ln n}{2n^3 - 1}$.

   Since $\ln n < n$ and $\frac{1}{2n^3 - 1} \leqq \frac{1}{n^3}$, we have $\frac{\ln n}{2n^3 - 1} \leqq \frac{n}{n^3} = \frac{1}{n^2}$.

   Then the given series converges, since $\sum\limits_{n=1}^{\infty} \frac{1}{n^2}$ converges.

8. Let u_n and v_n be positive. If $\lim\limits_{n \to \infty} \frac{u_n}{v_n} = $ constant $A \neq 0$, prove that Σu_n converges or diverges according as Σv_n converges or diverges.

   By hypothesis, given $\epsilon > 0$ we can choose an integer N such that $\left| \frac{u_n}{v_n} - A \right| < \epsilon$ for all $n > N$. Then for $n = N+1, N+2, \ldots$
   $$-\epsilon < \frac{u_n}{v_n} - A < \epsilon \qquad \text{or} \qquad (A - \epsilon)v_n < u_n < (A + \epsilon)v_n \tag{1}$$

   Summing from $N+1$ to ∞ (more precisely from $N+1$ to M and then letting $M \to \infty$),
   $$(A - \epsilon) \sum\limits_{N+1}^{\infty} v_n \leqq \sum\limits_{N+1}^{\infty} u_n \leqq (A + \epsilon) \sum\limits_{N+1}^{\infty} v_n \tag{2}$$

There is no loss in generality in assuming $A - \epsilon > 0$. Then from the right hand inequality of (2), Σu_n converges when Σv_n does. From the left hand inequality of (2), Σu_n diverges when Σv_n does. For the cases $A = 0$ or $A = \infty$, see Problem 62.

9. Test for convergence: $(a) \displaystyle\sum_{n=1}^{\infty} \frac{4n^2 - n + 3}{n^3 + 2n}$, $(b) \displaystyle\sum_{n=1}^{\infty} \frac{n + \sqrt{n}}{2n^3 - 1}$, $(c) \displaystyle\sum_{n=1}^{\infty} \frac{\ln n}{n^2 + 3}$.

(a) For large n, $\dfrac{4n^2 - n + 3}{n^3 + 2n}$ is approximately $\dfrac{4n^2}{n^3} = \dfrac{4}{n}$. Taking $u_n = \dfrac{4n^2 - n + 3}{n^3 + 2n}$ and $v_n = \dfrac{4}{n}$,

we have $\displaystyle\lim_{n \to \infty} \frac{u_n}{v_n} = 1$.

Since $\Sigma v_n = 4\Sigma 1/n$ diverges, Σu_n also diverges by Problem 8.

Note that the purpose of considering the behavior of u_n for large n is to obtain an appropriate comparison series v_n. In the above we could just as well have taken $v_n = 1/n$.

Another method: $\displaystyle\lim_{n \to \infty} n\left(\frac{4n^2 - n + 3}{n^3 + 2n}\right) = 4$. Then by Theorem 1, Page 225, the series diverges.

(b) For large n, $u_n = \dfrac{n + \sqrt{n}}{2n^3 - 1}$ is approximately $v_n = \dfrac{n}{2n^3} = \dfrac{1}{2n^2}$.

Since $\displaystyle\lim_{n \to \infty} \frac{u_n}{v_n} = 1$ and $\displaystyle\sum v_n = \frac{1}{2}\sum \frac{1}{n^2}$ converges (p series with $p = 2$), the given series converges.

Another method: $\displaystyle\lim_{n \to \infty} n^2\left(\frac{n + \sqrt{n}}{2n^3 - 1}\right) = \frac{1}{2}$. Then by Theorem 1, Page 225, the series converges.

(c) $\displaystyle\lim_{n \to \infty} n^{3/2}\left(\frac{\ln n}{n^2 + 3}\right) \leq \lim_{n \to \infty} n^{3/2}\left(\frac{\ln n}{n^2}\right) = \lim_{n \to \infty} \frac{\ln n}{\sqrt{n}} = 0$ (by L'Hospital's rule or otherwise). Then by Theorem 1 with $p = 3/2$, the series converges.

Note that the method of Problem 6(a) yields $\dfrac{\ln n}{n^2 + 3} < \dfrac{n}{n^2} = \dfrac{1}{n}$, but nothing can be deduced since $\Sigma 1/n$ diverges.

10. Examine for convergence: $(a) \displaystyle\sum_{n=1}^{\infty} e^{-n^2}$, $(b) \displaystyle\sum_{n=1}^{\infty} \sin^3\left(\frac{1}{n}\right)$.

(a) $\displaystyle\lim_{n \to \infty} n^2 e^{-n^2} = 0$ (by L'Hospital's rule or otherwise). Then by Theorem 1 with $p = 2$, the series converges.

(b) For large n, $\sin(1/n)$ is approximately $1/n$. This leads to consideration of
$$\lim_{n \to \infty} n^3 \sin^3\left(\frac{1}{n}\right) = \lim_{n \to \infty} \left\{\frac{\sin(1/n)}{1/n}\right\}^3 = 1$$
from which we deduce, by Theorem 1 with $p = 3$, that the given series converges.

INTEGRAL TEST

11. Establish the integral test (see Page 225).

We perform the proof taking $N = 1$. Modifications are easily made if $N > 1$.

From the monotonicity of $f(x)$, we have
$$u_{n+1} = f(n+1) \leq f(x) \leq f(n) = u_n \qquad n = 1, 2, 3, \ldots$$

Integrating from $x = n$ to $x = n + 1$, using Property 7, Page 81,
$$u_{n+1} \leq \int_n^{n+1} f(x)\,dx \leq u_n \qquad n = 1, 2, 3, \ldots$$

Summing from $n = 1$ to $M - 1$,

$$u_2 + u_3 + \cdots + u_M \; \leqq \; \int_1^M f(x)\, dx \; \leqq \; u_1 + u_2 + \cdots + u_{M-1} \tag{1}$$

If $f(x)$ is strictly decreasing, the equality signs in (1) can be omitted.

If $\lim\limits_{M \to \infty} \int_1^M f(x)\, dx$ exists and is equal to S, we see from the left hand inequality in (1) that $u_2 + u_3 + \cdots + u_M$ is monotonic increasing and bounded above by S, so that Σu_n converges.

If $\lim\limits_{M \to \infty} \int_1^M f(x)\, dx$ is unbounded, we see from the right hand inequality in (1) that Σu_n diverges.

Thus the proof is complete.

12. Illustrate geometrically the proof in Problem 11.

Geometrically, $u_2 + u_3 + \cdots + u_M$ is the total area of the rectangles shown shaded in Fig. 11-1, while $u_1 + u_2 + \cdots + u_{M-1}$ is the total area of the rectangles which are shaded and non-shaded.

The area under the curve $y = f(x)$ from $x = 1$ to $x = M$ is intermediate in value between the two areas given above, thus illustrating the result (1) of Problem 11.

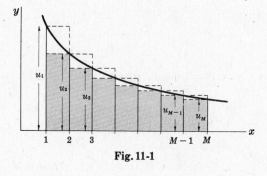

Fig. 11-1

13. Test for convergence: (a) $\displaystyle\sum_1^\infty \frac{1}{n^p}$, p = constant; (b) $\displaystyle\sum_1^\infty \frac{n}{n^2 + 1}$; (c) $\displaystyle\sum_2^\infty \frac{1}{n \ln n}$; (d) $\displaystyle\sum_1^\infty n e^{-n^2}$.

(a)　Consider　$\displaystyle\int_1^M \frac{dx}{x^p} = \int_1^M x^{-p}\, dx = \left. \frac{x^{1-p}}{1-p} \right|_1^M = \frac{M^{1-p} - 1}{1 - p}$　where $p \neq 1$.

If $p < 1$, $\displaystyle\lim_{M \to \infty} \frac{M^{1-p} - 1}{1 - p} = \infty$, so that the integral and thus the series diverges.

If $p > 1$, $\displaystyle\lim_{M \to \infty} \frac{M^{1-p} - 1}{1 - p} = \frac{1}{p - 1}$, so that the integral and thus the series converges.

If $p = 1$, $\displaystyle\int_1^M \frac{dx}{x^p} = \int_1^M \frac{dx}{x} = \ln M$ and $\displaystyle\lim_{M \to \infty} \ln M = \infty$, so that the integral and thus the series diverges.

Thus the series converges if $p > 1$ and diverges if $p \leqq 1$.

✱ (b)　$\displaystyle\lim_{M \to \infty} \int_1^M \frac{x\, dx}{x^2 + 1} = \lim_{M \to \infty} \left. \tfrac{1}{2} \ln (x^2 + 1) \right|_1^M = \lim_{M \to \infty} \left\{ \tfrac{1}{2} \ln (M^2 + 1) - \tfrac{1}{2} \ln 2 \right\} = \infty$　and the series diverges.

(c)　$\displaystyle\lim_{M \to \infty} \int_2^M \frac{dx}{x \ln x} = \lim_{M \to \infty} \left. \ln (\ln x) \right|_2^M = \lim_{M \to \infty} \left\{ \ln (\ln M) - \ln (\ln 2) \right\} = \infty$　and the series diverges.

(d)　$\displaystyle\lim_{M \to \infty} \int_1^M x e^{-x^2}\, dx = \lim_{M \to \infty} \left. -\tfrac{1}{2} e^{-x^2} \right|_1^M = \lim_{M \to \infty} \left\{ \tfrac{1}{2} e^{-1} - \tfrac{1}{2} e^{-M^2} \right\} = \tfrac{1}{2} e^{-1}$　and the series converges.

Note that when the series converges, the value of the corresponding integral is not (in general) the same as the sum of the series. However, the approximate sum of a series can often be obtained quite accurately by using integrals. See Problem 70.

14. Prove that $\dfrac{\pi}{4} < \displaystyle\sum_{n=1}^{\infty} \dfrac{1}{n^2+1} < \dfrac{1}{2} + \dfrac{\pi}{4}$.

From Problem 11 it follows that

$$\lim_{M \to \infty} \sum_{n=2}^{M} \frac{1}{n^2+1} \ < \ \lim_{M \to \infty} \int_{1}^{M} \frac{dx}{x^2+1} \ < \ \lim_{M \to \infty} \sum_{n=1}^{M-1} \frac{1}{n^2+1}$$

i.e. $\displaystyle\sum_{n=2}^{\infty} \frac{1}{n^2+1} < \frac{\pi}{4} < \sum_{n=1}^{\infty} \frac{1}{n^2+1}$, from which $\dfrac{\pi}{4} < \displaystyle\sum_{n=1}^{\infty} \frac{1}{n^2+1}$ as required.

Since $\displaystyle\sum_{n=2}^{\infty} \frac{1}{n^2+1} < \frac{\pi}{4}$, we obtain, on adding $\frac{1}{2}$ to each side, $\displaystyle\sum_{n=1}^{\infty} \frac{1}{n^2+1} < \frac{1}{2} + \frac{\pi}{4}$.

The required result is therefore proved.

ALTERNATING SERIES

15. Given the alternating series $a_1 - a_2 + a_3 - a_4 + \cdots$ where $0 \leqq a_{n+1} \leqq a_n$ and where $\lim\limits_{n \to \infty} a_n = 0$. Prove that (a) the series converges, (b) the error made in stopping at any term is not greater than the absolute value of the next term.

(a) The sum of the series to $2M$ terms is

$$\begin{aligned}
S_{2M} &= (a_1 - a_2) + (a_3 - a_4) + \cdots + (a_{2M-1} - a_{2M}) \\
&= a_1 - (a_2 - a_3) - (a_4 - a_5) - \cdots - (a_{2M-2} - a_{2M-1}) - a_{2M}
\end{aligned}$$

Since the quantities in parentheses are non-negative, we have

$$S_{2M} \geqq 0, \quad S_2 \leqq S_4 \leqq S_6 \leqq S_8 \leqq \cdots \leqq S_{2M} \leqq a_1$$

Therefore $\{S_{2M}\}$ is a bounded monotonic increasing sequence and thus has limit S.

Also, $S_{2M+1} = S_{2M} + a_{2M+1}$. Since $\lim\limits_{M \to \infty} S_{2M} = S$ and $\lim\limits_{M \to \infty} a_{2M+1} = 0$ (for, by hypothesis, $\lim\limits_{n \to \infty} a_n = 0$), it follows that $\lim\limits_{M \to \infty} S_{2M+1} = \lim\limits_{M \to \infty} S_{2M} + \lim\limits_{M \to \infty} a_{2M+1} = S + 0 = S$.

Thus the partial sums of the series approach the limit S and the series converges.

(b) The error made in stopping after $2M$ terms is

$$(a_{2M+1} - a_{2M+2}) + (a_{2M+3} - a_{2M+4}) + \cdots = a_{2M+1} - (a_{2M+2} - a_{2M+3}) - \cdots$$

and is thus non-negative and less than or equal to a_{2M+1}, the first term which is omitted.

Similarly, the error made in stopping after $2M + 1$ terms is

$$-a_{2M+2} + (a_{2M+3} - a_{2M+4}) + \cdots = -(a_{2M+2} - a_{2M+3}) - (a_{2M+4} - a_{2M+5}) - \cdots$$

which is non-positive and greater than $-a_{2M+2}$.

16. (a) Prove that the series $\displaystyle\sum_{n=1}^{\infty} \frac{(-1)^{n+1}}{2n-1}$ converges. (b) Find the maximum error made in approximating the sum by the first 8 terms and the first 9 terms of the series. (c) How many terms of the series are needed in order to obtain an error which does not exceed .001 in absolute value?

(a) The series is $1 - \frac{1}{3} + \frac{1}{5} - \frac{1}{7} + \frac{1}{9} - \cdots$. If $u_n = \dfrac{(-1)^{n+1}}{2n-1}$, then $a_n = |u_n| = \dfrac{1}{2n-1}$, $a_{n+1} = |u_{n+1}| = \dfrac{1}{2n+1}$. Since $\dfrac{1}{2n+1} \leqq \dfrac{1}{2n-1}$ and since $\lim\limits_{n \to \infty} \dfrac{1}{2n-1} = 0$, it follows by Problem 15(a) that the series converges.

(b) Use the results of Problem 15(b). Then the first 8 terms give $1 - \frac{1}{3} + \frac{1}{5} - \frac{1}{7} + \frac{1}{9} - \frac{1}{11} + \frac{1}{13} - \frac{1}{15}$ and the error is positive and does not exceed $\frac{1}{17}$.

Similarly, the first 9 terms are $1 - \frac{1}{3} + \frac{1}{5} - \frac{1}{7} + \frac{1}{9} - \frac{1}{11} + \frac{1}{13} - \frac{1}{15} + \frac{1}{17}$ and the error is negative and greater than or equal to $-\frac{1}{19}$, i.e. the error does not exceed $\frac{1}{19}$ in absolute value.

(c) The absolute value of the error made in stopping after M terms is less than $1/(2M+1)$. To obtain the desired accuracy we must have $1/(2M+1) \leq .001$, from which $M \geq 499.5$. Thus at least 500 terms are needed.

ABSOLUTE and CONDITIONAL CONVERGENCE

17. Prove that an absolutely convergent series is convergent.

Given that $\Sigma \, |u_n|$ converges, we must show that Σu_n converges.

Let $S_M = u_1 + u_2 + \cdots + u_M$ and $T_M = |u_1| + |u_2| + \cdots + |u_M|$. Then

$$
\begin{aligned}
S_M + T_M &= (u_1 + |u_1|) + (u_2 + |u_2|) + \cdots + (u_M + |u_M|) \\
&\leq 2|u_1| + 2|u_2| + \cdots + 2|u_M|
\end{aligned}
$$

Since $\Sigma \, |u_n|$ converges and since $u_n + |u_n| \geq 0$, for $n = 1, 2, 3, \ldots$, it follows that $S_M + T_M$ is a bounded monotonic increasing sequence, and so $\lim_{M \to \infty} (S_M + T_M)$ exists.

Also, since $\lim_{M \to \infty} T_M$ exists (since the series is absolutely convergent by hypothesis),

$$
\lim_{M \to \infty} S_M = \lim_{M \to \infty} (S_M + T_M - T_M) = \lim_{M \to \infty} (S_M + T_M) - \lim_{M \to \infty} T_M
$$

must also exist and the result is proved.

18. Investigate the convergence of the series $\dfrac{\sin \sqrt{1}}{1^{3/2}} - \dfrac{\sin \sqrt{2}}{2^{3/2}} + \dfrac{\sin \sqrt{3}}{3^{3/2}} - \cdots$.

Since each term is in absolute value less than or equal to the corresponding term of the series $\frac{1}{1^{3/2}} + \frac{1}{2^{3/2}} + \frac{1}{3^{3/2}} + \cdots$, which converges, it follows that the given series is absolutely convergent and hence convergent by Problem 17.

19. Examine for convergence and absolute convergence:

(a) $\displaystyle\sum_{n=1}^{\infty} \frac{(-1)^{n-1} n}{n^2 + 1}$,   (b) $\displaystyle\sum_{n=2}^{\infty} \frac{(-1)^{n-1}}{n \ln^2 n}$,   (c) $\displaystyle\sum_{n=1}^{\infty} \frac{(-1)^{n-1} 2^n}{n^2}$.

(a) The series of absolute values is $\displaystyle\sum_{n=1}^{\infty} \frac{n}{n^2 + 1}$ which is divergent by Problem 13(b). Hence the given series is not absolutely convergent.

However, if $a_n = |u_n| = \dfrac{n}{n^2+1}$ and $a_{n+1} = |u_{n+1}| = \dfrac{n+1}{(n+1)^2 + 1}$, then $a_{n+1} \leq a_n$ for all $n \geq 1$, and also $\lim_{n \to \infty} a_n = \lim_{n \to \infty} \dfrac{n}{n^2+1} = 0$. Hence by Problem 15 the series converges.

Since the series converges but is not absolutely convergent, it is *conditionally convergent*.

(b) The series of absolute values is $\displaystyle\sum_{n=2}^{\infty} \frac{1}{n \ln^2 n}$.

By the integral test, this series converges or diverges according as $\displaystyle\lim_{M \to \infty} \int_{2}^{M} \frac{dx}{x \ln^2 x}$ exists or does not exist.

If $u = \ln x$, $\int \dfrac{dx}{x \ln^2 x} = \int \dfrac{du}{u^2} = -\dfrac{1}{u} + c = -\dfrac{1}{\ln x} + c$.

Hence $\displaystyle\lim_{M \to \infty} \int_2^M \dfrac{dx}{x \ln^2 x} = \lim_{M \to \infty} \left(\dfrac{1}{\ln 2} - \dfrac{1}{\ln M} \right) = \dfrac{1}{\ln 2}$ and the integral exists. Thus the series converges.

Then $\displaystyle\sum_{n=2}^{\infty} \dfrac{(-1)^{n-1}}{n \ln^2 n}$ converges absolutely and thus converges.

Another method:

Since $\dfrac{1}{(n+1) \ln^2 (n+1)} \leq \dfrac{1}{n \ln^2 n}$ and $\displaystyle\lim_{n \to \infty} \dfrac{1}{n \ln^2 n} = 0$, it follows, by Prob. 15(a), that the given alternating series converges. To examine its absolute convergence we must proceed as above.

(c) Since $\displaystyle\lim_{n \to \infty} u_n \neq 0$ where $u_n = \dfrac{(-1)^{n-1} 2^n}{n^2}$, the given series cannot be convergent. To show that $\displaystyle\lim_{n \to \infty} u_n \neq 0$, it suffices to show that $\displaystyle\lim_{n \to \infty} |u_n| = \lim_{n \to \infty} \dfrac{2^n}{n^2} \neq 0$. This can be accomplished by L'Hospital's rule or other methods [see Prob. 21(b)].

RATIO TEST

20. Establish the ratio test for convergence.

Consider first the series $u_1 + u_2 + u_3 + \cdots$ where each term is non-negative. We must prove that if $\displaystyle\lim_{n \to \infty} \dfrac{u_{n+1}}{u_n} = L < 1$, then necessarily Σu_n converges.

By hypothesis, we can choose an integer N so large that for all $n \geq N$, $(u_{n+1}/u_n) < r$ where $L < r < 1$. Then

$$u_{N+1} < r\, u_N$$
$$u_{N+2} < r\, u_{N+1} < r^2 u_N$$
$$u_{N+3} < r\, u_{N+2} < r^3 u_N$$

etc. By addition,

$$u_{N+1} + u_{N+2} + \cdots < u_N(r + r^2 + r^3 + \cdots)$$

and so the given series converges by the comparison test, since $0 < r < 1$.

In case the series has terms with mixed signs, we consider $|u_1| + |u_2| + |u_3| + \cdots$. Then by the above proof and Problem 17, it follows that if $\displaystyle\lim_{n \to \infty} \left| \dfrac{u_{n+1}}{u_n} \right| = L < 1$, then Σu_n converges (absolutely).

Similarly we can prove that if $\displaystyle\lim_{n \to \infty} \left| \dfrac{u_{n+1}}{u_n} \right| = L > 1$ the series Σu_n diverges, while if $\displaystyle\lim_{n \to \infty} \left| \dfrac{u_{n+1}}{u_n} \right| = L = 1$ the ratio test fails [see Prob. 21(c)].

21. Investigate the convergence of $\quad(a)\ \displaystyle\sum_{n=1}^{\infty} n^4 e^{-n^2}, \quad (b)\ \sum_{n=1}^{\infty} \dfrac{(-1)^{n-1} 2^n}{n^2}, \quad (c)\ \sum_{n=1}^{\infty} \dfrac{(-1)^{n-1} n}{n^2 + 1}$.

(a) Here $u_n = n^4 e^{-n^2}$. Then

$$\lim_{n \to \infty} \left| \dfrac{u_{n+1}}{u_n} \right| = \lim_{n \to \infty} \left| \dfrac{(n+1)^4 e^{-(n+1)^2}}{n^4 e^{-n^2}} \right| = \lim_{n \to \infty} \dfrac{(n+1)^4 e^{-(n^2+2n+1)}}{n^4 e^{-n^2}}$$

$$= \lim_{n \to \infty} \left(\dfrac{n+1}{n} \right)^4 e^{-2n-1} = \lim_{n \to \infty} \left(\dfrac{n+1}{n} \right)^4 \lim_{n \to \infty} e^{-2n-1} = 1 \cdot 0 = 0$$

Since $0 < 1$, the series converges.

(b) Here $u_n = \dfrac{(-1)^{n-1} 2^n}{n^2}$. Then

$$\lim_{n \to \infty} \left| \frac{u_{n+1}}{u_n} \right| = \lim_{n \to \infty} \left| \frac{(-1)^n 2^{n+1}}{(n+1)^2} \cdot \frac{n^2}{(-1)^{n-1} 2^n} \right| = \lim_{n \to \infty} \frac{2n^2}{(n+1)^2} = 2$$

Since $2 > 1$, the series diverges. Compare Prob. 19(c).

(c) Here $u_n = \dfrac{(-1)^{n-1} n}{n^2 + 1}$. Then

$$\lim_{n \to \infty} \left| \frac{u_{n+1}}{u_n} \right| = \lim_{n \to \infty} \left| \frac{(-1)^n (n+1)}{(n+1)^2 + 1} \cdot \frac{n^2 + 1}{(-1)^{n-1} n} \right| = \lim_{n \to \infty} \frac{(n+1)(n^2+1)}{(n^2 + 2n + 2)n} = 1$$

and the ratio test fails. By using other tests [see Prob. 19(a)], the series is seen to be convergent.

MISCELLANEOUS TESTS

22. Test for convergence $1 + 2r + r^2 + 2r^3 + r^4 + 2r^5 + \cdots$ where (a) $r = 2/3$, (b) $r = -2/3$, (c) $r = 4/3$.

Here the ratio test is inapplicable, since $\left| \dfrac{u_{n+1}}{u_n} \right| = 2|r|$ or $\frac{1}{2}|r|$ depending on whether n is odd or even.

However, using the nth root test, we have

$$\sqrt[n]{|u_n|} = \begin{cases} \sqrt[n]{2|r^n|} = \sqrt[n]{2}\,|r| & \text{if } n \text{ is odd} \\ \sqrt[n]{|r^n|} = |r| & \text{if } n \text{ is even} \end{cases}$$

Then $\lim\limits_{n \to \infty} \sqrt[n]{|u_n|} = |r|$ (since $\lim\limits_{n \to \infty} 2^{1/n} = 1$).

Thus if $|r| < 1$ the series converges, and if $|r| > 1$ the series diverges.

Hence the series converges for cases (a) and (b), and diverges in case (c).

23. Test for convergence $\left(\dfrac{1}{3} \right)^2 + \left(\dfrac{1 \cdot 4}{3 \cdot 6} \right)^2 + \left(\dfrac{1 \cdot 4 \cdot 7}{3 \cdot 6 \cdot 9} \right)^2 + \cdots + \left(\dfrac{1 \cdot 4 \cdot 7 \ldots (3n-2)}{3 \cdot 6 \cdot 9 \ldots (3n)} \right)^2 + \cdots$.

The ratio test fails since $\lim\limits_{n \to \infty} \left| \dfrac{u_{n+1}}{u_n} \right| = \lim\limits_{n \to \infty} \left(\dfrac{3n+1}{3n+3} \right)^2 = 1$. However, by Raabe's test,

$$\lim_{n \to \infty} n \left(1 - \left| \frac{u_{n+1}}{u_n} \right| \right) = \lim_{n \to \infty} n \left\{ 1 - \left(\frac{3n+1}{3n+3} \right)^2 \right\} = \frac{4}{3} > 1$$

and so the series converges.

24. Test for convergence $\left(\dfrac{1}{2} \right)^2 + \left(\dfrac{1 \cdot 3}{2 \cdot 4} \right)^2 + \left(\dfrac{1 \cdot 3 \cdot 5}{2 \cdot 4 \cdot 6} \right)^2 + \cdots + \left(\dfrac{1 \cdot 3 \cdot 5 \ldots (2n-1)}{2 \cdot 4 \cdot 6 \ldots (2n)} \right)^2 + \cdots$.

The ratio test fails since $\lim\limits_{n \to \infty} \left| \dfrac{u_{n+1}}{u_n} \right| = \lim\limits_{n \to \infty} \left(\dfrac{2n+1}{2n+2} \right)^2 = 1$. Also, Raabe's test fails since

$$\lim_{n \to \infty} n \left(1 - \left| \frac{u_{n+1}}{u_n} \right| \right) = \lim_{n \to \infty} n \left\{ 1 - \left(\frac{2n+1}{2n+2} \right)^2 \right\} = 1$$

However, using long division,

$$\left| \frac{u_{n+1}}{u_n} \right| = \left(\frac{2n+1}{2n+2} \right)^2 = 1 - \frac{1}{n} + \frac{5 - 4/n}{4n^2 + 8n + 4} = 1 - \frac{1}{n} + \frac{c_n}{n^2} \quad \text{where } |c_n| < P$$

so that the series diverges by Gauss' test.

SERIES of FUNCTIONS

25. For what values of x do the following series converge?

$$(a)\ \sum_{n=1}^{\infty} \frac{x^{n-1}}{n \cdot 3^n}, \quad (b)\ \sum_{n=1}^{\infty} \frac{(-1)^{n-1} x^{2n-1}}{(2n-1)!}, \quad (c)\ \sum_{n=1}^{\infty} n!\,(x-a)^n, \quad (d)\ \sum_{n=1}^{\infty} \frac{n(x-1)^n}{2^n(3n-1)}.$$

(a) $u_n = \dfrac{x^{n-1}}{n \cdot 3^n}$.   Assuming $x \neq 0$ (if $x = 0$ the series converges), we have

$$\lim_{n \to \infty} \left| \frac{u_{n+1}}{u_n} \right| = \lim_{n \to \infty} \left| \frac{x^n}{(n+1) \cdot 3^{n+1}} \cdot \frac{n \cdot 3^n}{x^{n-1}} \right| = \lim_{n \to \infty} \frac{n}{3(n+1)} |x| = \frac{|x|}{3}$$

Then the series converges if $\dfrac{|x|}{3} < 1$, and diverges if $\dfrac{|x|}{3} > 1$. If $\dfrac{|x|}{3} = 1$, i.e. $x = \pm 3$, the test fails.

If $x = 3$ the series becomes $\displaystyle\sum_{n=1}^{\infty} \frac{1}{3n} = \frac{1}{3} \sum_{n=1}^{\infty} \frac{1}{n}$ which diverges.

If $x = -3$ the series becomes $\displaystyle\sum_{n=1}^{\infty} \frac{(-1)^{n-1}}{3n} = \frac{1}{3} \sum_{n=1}^{\infty} \frac{(-1)^{n-1}}{n}$ which converges.

Then the *interval of convergence* is $-3 \leqq x < 3$. The series diverges outside this interval.

Note that the series converges absolutely for $-3 < x < 3$. At $x = -3$ the series converges conditionally.

(b)   Proceed as in part (a) with $u_n = \dfrac{(-1)^{n-1} x^{2n-1}}{(2n-1)!}$.   Then

$$\lim_{n \to \infty} \left| \frac{u_{n+1}}{u_n} \right| = \lim_{n \to \infty} \left| \frac{(-1)^n x^{2n+1}}{(2n+1)!} \cdot \frac{(2n-1)!}{(-1)^{n-1} x^{2n-1}} \right| = \lim_{n \to \infty} \frac{(2n-1)!}{(2n+1)!} x^2$$

$$= \lim_{n \to \infty} \frac{(2n-1)!}{(2n+1)(2n)(2n-1)!} x^2 = \lim_{n \to \infty} \frac{x^2}{(2n+1)(2n)} = 0$$

Then the series converges (absolutely) for all x, i.e. the interval of (absolute) convergence is $-\infty < x < \infty$.

(c) $u_n = n!\,(x-a)^n$, $\displaystyle\lim_{n \to \infty} \left| \frac{u_{n+1}}{u_n} \right| = \lim_{n \to \infty} \left| \frac{(n+1)!\,(x-a)^{n+1}}{n!\,(x-a)^n} \right| = \lim_{n \to \infty} (n+1)\,|x-a|$.

This limit is infinite if $x \neq a$. Then the series converges only for $x = a$.

(d) $u_n = \dfrac{n(x-1)^n}{2^n(3n-1)}$, $u_{n+1} = \dfrac{(n+1)(x-1)^{n+1}}{2^{n+1}(3n+2)}$.   Then

$$\lim_{n \to \infty} \left| \frac{u_{n+1}}{u_n} \right| = \lim_{n \to \infty} \left| \frac{(n+1)(3n-1)(x-1)}{2n(3n+2)} \right| = \left| \frac{x-1}{2} \right| = \frac{|x-1|}{2}$$

Thus the series converges for $|x-1| < 2$ and diverges for $|x-1| > 2$.

The test fails for $|x-1| = 2$, i.e. $x - 1 = \pm 2$ or $x = 3$ and $x = -1$.

For $x = 3$ the series becomes $\displaystyle\sum_{n=1}^{\infty} \frac{n}{3n-1}$ which diverges since the nth term does not approach zero.

For $x = -1$ the series becomes $\displaystyle\sum_{n=1}^{\infty} \frac{(-1)^n n}{3n-1}$ which also diverges since the nth term does not approach zero.

Then the series converges only for $|x-1| < 2$, i.e. $-2 < x - 1 < 2$ or $-1 < x < 3$.

26. For what values of x does (a) $\sum_{n=1}^{\infty} \frac{1}{2n-1}\left(\frac{x+2}{x-1}\right)^n$, (b) $\sum_{n=1}^{\infty} \frac{1}{(x+n)(x+n-1)}$ converge?

(a) $u_n = \frac{1}{2n-1}\left(\frac{x+2}{x-1}\right)^n$.   Then $\lim_{n \to \infty} \left|\frac{u_{n+1}}{u_n}\right| = \lim_{n \to \infty} \frac{2n-1}{2n+1}\left|\frac{x+2}{x-1}\right| = \left|\frac{x+2}{x-1}\right|$ if $x \neq 1, -2$.

Then the series converges if $\left|\frac{x+2}{x-1}\right| < 1$, diverges if $\left|\frac{x+2}{x-1}\right| > 1$, and the test fails if $\left|\frac{x+2}{x-1}\right| = 1$, i.e. $x = -\frac{1}{2}$.

If $x = 1$ the series diverges.

If $x = -2$ the series converges.

If $x = -\frac{1}{2}$ the series is $\sum_{n=1}^{\infty} \frac{(-1)^n}{2n-1}$ which converges.

Thus the series converges for $\left|\frac{x+2}{x-1}\right| < 1$, $x = -\frac{1}{2}$ and $x = -2$, i.e. for $x \leqq -\frac{1}{2}$.

(b) The ratio test fails since $\lim_{n \to \infty} \left|\frac{u_{n+1}}{u_n}\right| = 1$, where $u_n = \frac{1}{(x+n)(x+n-1)}$. However, noting that

$$\frac{1}{(x+n)(x+n-1)} = \frac{1}{x+n-1} - \frac{1}{x+n}$$

we see that if $x \neq 0, -1, -2, \ldots, -n,$

$$S_n = u_1 + u_2 + \cdots + u_n = \left(\frac{1}{x} - \frac{1}{x+1}\right) + \left(\frac{1}{x+1} - \frac{1}{x+2}\right) + \cdots + \left(\frac{1}{x+n-1} - \frac{1}{x+n}\right)$$

$$= \frac{1}{x} - \frac{1}{x+n}$$

and $\lim_{n \to \infty} S_n = 1/x$, provided $x \neq 0, -1, -2, -3, \ldots$.

Then the series converges for all x except $x = 0, -1, -2, -3, \ldots$, and its sum is $1/x$.

UNIFORM CONVERGENCE

27. Find the domain of convergence of $(1-x) + x(1-x) + x^2(1-x) + \cdots$.

Method 1:

Sum of first n terms $= S_n(x) = (1-x) + x(1-x) + x^2(1-x) + \cdots + x^{n-1}(1-x)$
$$= 1 - x + x - x^2 + x^2 - x^3 + \cdots + x^{n-1} - x^n$$
$$= 1 - x^n$$

If $|x| < 1$, $\lim_{n \to \infty} S_n(x) = \lim_{n \to \infty} (1 - x^n) = 1$.

If $|x| > 1$, $\lim_{n \to \infty} S_n(x)$ does not exist.

If $x = 1$, $S_n(x) = 0$ and $\lim_{n \to \infty} S_n(x) = 0$.

If $x = -1$, $S_n(x) = 1 - (-1)^n$ and $\lim_{n \to \infty} S_n(x)$ does not exist.

Thus the series converges for $|x| < 1$ and $x = 1$, i.e. for $-1 < x \leqq 1$.

Method 2, using the ratio test.

The series converges if $x = 1$. If $x \neq 1$ and $u_n = x^{n-1}(1-x)$, then $\lim_{n \to \infty} \left|\frac{u_{n+1}}{u_n}\right| = \lim_{n \to \infty} |x|$.

Thus the series converges if $|x| < 1$, diverges if $|x| > 1$. The test fails if $|x| = 1$. If $x = 1$ the series converges; if $x = -1$ the series diverges. Then the series converges for $-1 < x \leqq 1$.

28. Investigate the uniform convergence of the series of Problem 27 in the interval (a) $-\frac{1}{2} < x < \frac{1}{2}$, (b) $-\frac{1}{2} \leqq x \leqq \frac{1}{2}$, (c) $-.99 \leqq x \leqq .99$, (d) $-1 < x < 1$, (e) $0 \leqq x < 2$.

(a) By Problem 27, $S_n(x) = 1 - x^n$, $S(x) = \lim_{n \to \infty} S_n(x) = 1$ if $-\frac{1}{2} < x < \frac{1}{2}$; thus the series *converges* in this interval. We have

$$\text{Remainder after } n \text{ terms} = R_n(x) = S(x) - S_n(x) = 1 - (1 - x^n) = x^n$$

The series is *uniformly convergent* in the interval if given any $\epsilon > 0$ we can find N dependent on ϵ, *but not on* x, such that $|R_n(x)| < \epsilon$ for all $n > N$. Now

$$|R_n(x)| = |x^n| = |x|^n < \epsilon \quad \text{when} \quad n \ln |x| < \ln \epsilon \quad \text{or} \quad n > \frac{\ln \epsilon}{\ln |x|}$$

since division by $\ln |x|$ (which is negative since $|x| < \frac{1}{2}$) reverses the sense of the inequality.

But if $|x| < \frac{1}{2}$, $\ln |x| < \ln (\frac{1}{2})$ and $n > \dfrac{\ln \epsilon}{\ln |x|} > \dfrac{\ln \epsilon}{\ln (\frac{1}{2})} = N$. Thus since N is independent of x, the series is uniformly convergent in the interval.

(b) In this case $|x| \leqq \frac{1}{2}$, $\ln |x| \leqq \ln (\frac{1}{2})$ and $n > \dfrac{\ln \epsilon}{\ln |x|} \geqq \dfrac{\ln \epsilon}{\ln (\frac{1}{2})} = N$, so that the series is also uniformly convergent in $-\frac{1}{2} \leqq x \leqq \frac{1}{2}$.

(c) Reasoning similar to the above, with $\frac{1}{2}$ replaced by .99, shows that the series is uniformly convergent in $-.99 \leqq x \leqq .99$.

(d) The arguments used above break down in this case, since $\dfrac{\ln \epsilon}{\ln |x|}$ can be made larger than any positive number by choosing $|x|$ sufficiently close to 1. Thus no N exists and it follows that the series is *not* uniformly convergent in $-1 < x < 1$.

(e) Since the series does not even converge at all points in this interval, it cannot converge uniformly in the interval.

29. Discuss the continuity of the sum function $S(x) = \lim_{n \to \infty} S_n(x)$ of Problem 27 for the interval $0 \leqq x \leqq 1$.

If $0 \leqq x < 1$, $S(x) = \lim_{n \to \infty} S_n(x) = \lim_{n \to \infty} (1 - x^n) = 1$.

If $x = 1$, $S_n(x) = 0$ and $S(x) = 0$.

Thus $S(x) = \begin{cases} 1 & \text{if } 0 \leqq x < 1 \\ 0 & \text{if } x = 1 \end{cases}$ and $S(x)$ is discontinuous at $x = 1$ but continuous at all other points in $0 \leqq x < 1$.

In Problem 34 it is shown that if a series is uniformly convergent in an interval, the sum function $S(x)$ must be continuous in the interval. It follows that if the sum function is not continuous in an interval, the series cannot be uniformly convergent. This fact is often used to demonstrate the non-uniform convergence of a series (or sequence).

30. Investigate the uniform convergence of $x^2 + \dfrac{x^2}{1 + x^2} + \dfrac{x^2}{(1 + x^2)^2} + \cdots + \dfrac{x^2}{(1 + x^2)^n} + \cdots$.

Suppose $x \neq 0$. Then the series is a geometric series with ratio $1/(1 + x^2)$ whose sum is (see Prob. 25, Chap. 3).

$$S(x) = \frac{x^2}{1 - 1/(1 + x^2)} = 1 + x^2$$

If $x = 0$ the sum of the first n terms is $S_n(0) = 0$; hence $S(0) = \lim\limits_{n \to \infty} S_n(0) = 0$.

Since $\lim\limits_{x \to 0} S(x) = 1 \neq S(0)$, $S(x)$ is discontinuous at $x = 0$. Then by Problem 34, the series cannot be *uniformly convergent* in any interval which includes $x = 0$, although it is (absolutely) *convergent* in any interval. However, it is uniformly convergent in any interval which excludes $x = 0$.

This can also be shown directly (see Problem 89).

WEIERSTRASS M TEST

31. Prove the Weierstrass M test, i.e. if $|u_n(x)| \leqq M_n$, $n = 1, 2, 3, \ldots$, where M_n are positive constants such that ΣM_n converges, then $\Sigma u_n(x)$ is uniformly (and absolutely) convergent.

The remainder of the series $\Sigma u_n(x)$ after n terms is $R_n(x) = u_{n+1}(x) + u_{n+2}(x) + \cdots$. Now

$$|R_n(x)| = |u_{n+1}(x) + u_{n+2}(x) + \cdots| \leqq |u_{n+1}(x)| + |u_{n+2}(x)| + \cdots \leqq M_{n+1} + M_{n+2} + \cdots$$

But $M_{n+1} + M_{n+2} + \cdots$ can be made less than ϵ by choosing $n > N$, since ΣM_n converges. Since N is clearly independent of x, we have $|R_n(x)| < \epsilon$ for $n > N$, and the series is uniformly convergent. The absolute convergence follows at once from the comparison test.

32. Test for uniform convergence:

$$(a)\ \sum_{n=1}^{\infty} \frac{\cos nx}{n^4}, \quad (b)\ \sum_{n=1}^{\infty} \frac{x^n}{n^{3/2}}, \quad (c)\ \sum_{n=1}^{\infty} \frac{\sin nx}{n}, \quad (d)\ \sum_{n=1}^{\infty} \frac{1}{n^2 + x^2}.$$

(a) $\left|\dfrac{\cos nx}{n^4}\right| \leqq \dfrac{1}{n^4} = M_n$. Then since ΣM_n converges (p series with $p = 4 > 1$), the series is uniformly (and absolutely) convergent for all x by the M test.

(b) By the ratio test, the series converges in the interval $-1 \leqq x \leqq 1$, i.e. $|x| \leqq 1$.

For all x in this interval, $\left|\dfrac{x^n}{n^{3/2}}\right| = \dfrac{|x|^n}{n^{3/2}} \leqq \dfrac{1}{n^{3/2}}$. Choosing $M_n = \dfrac{1}{n^{3/2}}$, we see that ΣM_n converges. Thus the given series converges uniformly for $-1 \leqq x \leqq 1$ by the M test.

(c) $\left|\dfrac{\sin nx}{n}\right| \leqq \dfrac{1}{n}$. However, ΣM_n, where $M_n = \dfrac{1}{n}$, does not converge. The M test cannot be used in this case and we cannot conclude anything about the uniform convergence by this test (see, however, Problem 121).

(d) $\left|\dfrac{1}{n^2 + x^2}\right| \leqq \dfrac{1}{n^2}$, and $\Sigma \dfrac{1}{n^2}$ converges. Then by the M test the given series converges uniformly for all x.

33. If a power series $\Sigma a_n x^n$ converges for $x = x_0$, prove that it converges (a) absolutely in the interval $|x| < |x_0|$, (b) uniformly in the interval $|x| \leqq |x_1|$, where $|x_1| < |x_0|$.

(a) Since $\Sigma a_n x_0^n$ converges, $\lim\limits_{n \to \infty} a_n x_0^n = 0$ and so we can make $|a_n x_0^n| < 1$ by choosing n large enough, i.e. $|a_n| < \dfrac{1}{|x_0|^n}$ for $n > N$. Then

$$\sum_{N+1}^{\infty} |a_n x^n| = \sum_{N+1}^{\infty} |a_n|\,|x|^n < \sum_{N+1}^{\infty} \frac{|x|^n}{|x_0|^n} \tag{1}$$

Since the last series in (1) converges for $|x| < |x_0|$, it follows by the comparison test that the first series converges, i.e. the given series is absolutely convergent.

(b) Let $M_n = \dfrac{|x_1|^n}{|x_0|^n}$. Then ΣM_n converges since $|x_1| < |x_0|$. As in part (a), $|a_n x^n| < M_n$ for

$|x| \leqq |x_1|$, so that by the Weierstrass M test, $\Sigma a_n x^n$ is uniformly convergent.

It follows that a power series is uniformly convergent in any interval *within* its interval of convergence.

THEOREMS on UNIFORM CONVERGENCE

34. Prove Theorem 6, Page 228.

We must show that $S(x)$ is continuous in $[a, b]$.

Now $S(x) = S_n(x) + R_n(x)$, so that $S(x+h) = S_n(x+h) + R_n(x+h)$ and thus

$$S(x+h) - S(x) \;=\; S_n(x+h) - S_n(x) + R_n(x+h) - R_n(x) \tag{1}$$

where we choose h so that both x and $x + h$ lie in $[a, b]$ (if $x = b$, for example, this will require $h < 0$).

Since $S_n(x)$ is a sum of a finite number of continuous functions, it must also be continuous. Then given $\epsilon > 0$, we can find δ so that

$$| S_n(x+h) - S_n(x) | \;<\; \epsilon/3 \quad \text{whenever } |h| < \delta \tag{2}$$

Since the series, by hypothesis, is uniformly convergent, we can choose N so that

$$|R_n(x)| \;<\; \epsilon/3 \quad \text{and} \quad | R_n(x+h) | \;<\; \epsilon/3 \quad \text{for } n > N \tag{3}$$

Then from (1), (2) and (3),

$$| S(x+h) - S(x) | \;\leqq\; | S_n(x+h) - S_n(x) | + | R_n(x+h) | + |R_n(x)| \;<\; \epsilon$$

for $|h| < \delta$, and so the continuity is established.

35. Prove Theorem 7, Page 229.

If a function is continuous in $[a, b]$, its integral exists. Then since $S(x)$, $S_n(x)$ and $R_n(x)$ are continuous,

$$\int_a^b S(x) \;=\; \int_a^b S_n(x)\,dx \;+\; \int_a^b R_n(x)\,dx$$

To prove the theorem we must show that

$$\left| \int_a^b S(x)\,dx - \int_a^b S_n(x)\,dx \right| \;=\; \left| \int_a^b R_n(x)\,dx \right|$$

can be made arbitrarily small by choosing n large enough. This, however, follows at once, since by the uniform convergence of the series we can make $|R_n(x)| < \epsilon/(b-a)$ for $n > N$ independent of x in $[a, b]$, and so

$$\left| \int_a^b R_n(x)\,dx \right| \;\leqq\; \int_a^b |R_n(x)|\,dx \;<\; \int_a^b \frac{\epsilon}{b-a}\,dx \;=\; \epsilon$$

This is equivalent to the statements

$$\int_a^b S(x)\,dx \;=\; \lim_{n \to \infty} \int_a^b S_n(x)\,dx \quad \text{or} \quad \lim_{n \to \infty} \int_a^b S_n(x)\,dx \;=\; \int_a^b \{ \lim_{n \to \infty} S_n(x)\}\,dx$$

36. Prove Theorem 8, Page 229.

Let $g(x) = \sum\limits_{n=1}^{\infty} u'_n(x)$. Since, by hypothesis, this series converges uniformly in $[a, b]$, we can integrate term by term (by Problem 35) to obtain

$$\int_a^x g(x)\,dx = \sum_{n=1}^{\infty} \int_a^x u'_n(x)\,dx = \sum_{n=1}^{\infty} \{u_n(x) - u_n(a)\}$$

$$= \sum_{n=1}^{\infty} u_n(x) - \sum_{n=1}^{\infty} u_n(a) = S(x) - S(a)$$

because, by hypothesis, $\sum\limits_{n=1}^{\infty} u_n(x)$ converges to $S(x)$ in $[a, b]$.

Differentiating both sides of $\int_a^x g(x)\,dx = S(x) - S(a)$ then shows that $g(x) = S'(x)$, which proves the theorem.

37. Let $S_n(x) = nxe^{-nx^2}$, $n = 1, 2, 3, \ldots$, $0 \leqq x \leqq 1$.

(a) Determine whether $\lim\limits_{n \to \infty} \int_0^1 S_n(x)\,dx = \int_0^1 \lim\limits_{n \to \infty} S_n(x)\,dx$.

(b) Explain the result in (a).

(a) $\int_0^1 S_n(x)\,dx = \int_0^1 nxe^{-nx^2}\,dx = -\tfrac{1}{2}e^{-nx^2}\big|_0^1 = \tfrac{1}{2}(1 - e^{-n})$. Then

$$\lim_{n \to \infty} \int_0^1 S_n(x)\,dx = \lim_{n \to \infty} \tfrac{1}{2}(1 - e^{-n}) = \tfrac{1}{2}$$

$S(x) = \lim\limits_{n \to \infty} S_n(x) = \lim\limits_{n \to \infty} nxe^{-nx^2} = 0$, whether $x = 0$ or $0 < x \leqq 1$. Then,

$$\int_0^1 S(x)\,dx = 0$$

It follows that $\lim\limits_{n \to \infty} \int_0^1 S_n(x)\,dx \neq \int_0^1 \lim\limits_{n \to \infty} S_n(x)\,dx$, i.e. the limit cannot be taken under the integral sign.

(b) The reason for the result in (a) is that although the sequence $S_n(x)$ converges to 0, it does not converge *uniformly* to 0. To show this, observe that the function nxe^{-nx^2} has a maximum at $x = 1/\sqrt{2n}$ (by the usual rules of elementary calculus), the value of this maximum being $\sqrt{\tfrac{1}{2}n}\,e^{-1/2}$. Hence *as* $n \to \infty$, $S_n(x)$ cannot be made arbitrarily small *for all x* and so cannot converge *uniformly* to 0.

38. Let $f(x) = \sum\limits_{n=1}^{\infty} \dfrac{\sin nx}{n^3}$. Prove that $\int_0^\pi f(x)\,dx = 2 \sum\limits_{n=1}^{\infty} \dfrac{1}{(2n-1)^4}$.

We have $\left| \dfrac{\sin nx}{n^3} \right| \leqq \dfrac{1}{n^3}$. Then by the Weierstrass M test the series is uniformly convergent for all x, in particular $0 \leqq x \leqq \pi$, and can be integrated term by term. Thus

$$\int_0^\pi f(x)\,dx = \int_0^\pi \left(\sum_{n=1}^{\infty} \frac{\sin nx}{n^3} \right) dx = \sum_{n=1}^{\infty} \int_0^\pi \frac{\sin nx}{n^3}\,dx$$

$$= \sum_{n=1}^{\infty} \frac{1 - \cos n\pi}{n^4} = 2\left(\frac{1}{1^4} + \frac{1}{3^4} + \frac{1}{5^4} + \cdots \right) = 2 \sum_{n=1}^{\infty} \frac{1}{(2n-1)^4}$$

POWER SERIES

39. Prove that both the power series $\sum\limits_{n=0}^{\infty} a_n x^n$ and the corresponding series of derivatives $\sum\limits_{n=0}^{\infty} n a_n x^{n-1}$ have the same radius of convergence.

Let $R > 0$ be the radius of convergence of $\Sigma a_n x^n$. Let $0 < |x_0| < R$. Then, as in Problem 33, we can choose N so that $|a_n| < \dfrac{1}{|x_0|^n}$ for $n > N$.

Thus the terms of the series $\Sigma |n a_n x^{n-1}| = \Sigma n |a_n| \, |x|^{n-1}$ can for $n > N$ be made less than corresponding terms of the series $\Sigma n \dfrac{|x|^{n-1}}{|x_0|^n}$ which converges, by the ratio test, for $|x| < |x_0| < R$.

Hence $\Sigma n a_n x^{n-1}$ converges absolutely for all points x_0 (no matter how close $|x_0|$ is to R).

If, however, $|x| > R$, $\lim\limits_{n \to \infty} a_n x^n \neq 0$ and thus $\lim\limits_{n \to \infty} n a_n x^{n-1} \neq 0$, so that $\Sigma n a_n x^{n-1}$ does not converge.

Thus R is the radius of convergence of $\Sigma n a_n x^{n-1}$.

Note that the series of derivatives may or may not converge for values of x such that $|x| = R$.

40. Illustrate Problem 39 by using the series $\sum\limits_{n=1}^{\infty} \dfrac{x^n}{n^2 \cdot 3^n}$.

$$\lim_{n \to \infty} \left| \frac{u_{n+1}}{u_n} \right| = \lim_{n \to \infty} \left| \frac{x^{n+1}}{(n+1)^2 \cdot 3^{n+1}} \cdot \frac{n^2 \cdot 3^n}{x^n} \right| = \lim_{n \to \infty} \frac{n^2}{3(n+1)^2} |x| = \frac{|x|}{3}$$

so that the series converges for $|x| < 3$. At $x = \pm 3$ the series also converges, so that the interval of convergence is $-3 \leqq x \leqq 3$.

The series of derivatives is

$$\sum_{n=1}^{\infty} \frac{n x^{n-1}}{n^2 \cdot 3^n} = \sum_{n=1}^{\infty} \frac{x^{n-1}}{n \cdot 3^n}$$

By Problem 25(a) this has the interval of convergence $-3 \leqq x < 3$.

The two series have the same radius of convergence, i.e. $R = 3$, although they do not have the same interval of convergence.

Note that the result of Problem 39 can also be proved by the ratio test *if* this test is applicable. The proof given there, however, applies even when the test is not applicable, as in the series of Problem 22.

41. Prove that in any interval *within* its interval of convergence a power series

(a) represents a continuous function, say $f(x)$,

(b) can be integrated term by term to yield the integral of $f(x)$,

(c) can be differentiated term by term to yield the derivative of $f(x)$.

We consider the power series $\Sigma a_n x^n$, although analogous results hold for $\Sigma a_n (x-a)^n$.

(a) This follows from Problems 33, 34 and the fact that each term $a_n x^n$ of the series is continuous.

(b) This follows from Problems 33, 35 and the fact that each term $a_n x^n$ of the series is continuous and thus integrable.

(c) From Problem 39, the series of derivatives of a power series always converges within the interval of convergence of the original power series and therefore is uniformly convergent within this interval. Thus the required result follows from Problems 33 and 36.

If a power series converges at one (or both) end points of the interval of convergence, it is possible to establish (a) and (b) to include the end point (or end points). See Problem 42.

42. Prove Abel's theorem that if a power series converges at an end point of its interval of convergence, then the interval of uniform convergence includes this end point.

For simplicity in the proof, we assume the power series to be $\sum\limits_{k=0}^{\infty} a_k x^k$ with the end point of its interval of convergence at $x=1$, so that the series surely converges for $0 \leqq x \leqq 1$. Then we must show that the series converges uniformly in this interval.

Let
$$R_n(x) \;=\; a_n x^n + a_{n+1} x^{n+1} + a_{n+2} x^{n+2} + \cdots, \qquad R_n = a_n + a_{n+1} + a_{n+2} + \cdots$$

To prove the required result we must show that given any $\epsilon > 0$, we can find N such that $|R_n(x)| < \epsilon$ for all $n > N$, where N is independent of the particular x in $0 \leqq x \leqq 1$.

Now
$$\begin{aligned}
R_n(x) \;&=\; (R_n - R_{n+1})x^n + (R_{n+1} - R_{n+2})x^{n+1} + (R_{n+2} - R_{n+3})x^{n+2} + \cdots \\
&=\; R_n x^n + R_{n+1}(x^{n+1} - x^n) + R_{n+2}(x^{n+2} - x^{n+1}) + \cdots \\
&=\; x^n\{R_n - (1-x)(R_{n+1} + R_{n+2}\,x + R_{n+3}\,x^2 + \cdots)\}
\end{aligned}$$

Hence for $0 \leqq x < 1$,

$$|R_n(x)| \;\leqq\; |R_n| + (1-x)(|R_{n+1}| + |R_{n+2}|\,x + |R_{n+3}|\,x^2 + \cdots) \tag{1}$$

Since Σa_k converges by hypothesis, it follows that given $\epsilon > 0$ we can choose N such that $|R_k| < \epsilon/2$ for all $k \geqq n$. Then for $n > N$ we have from (1),

$$|R_n(x)| \;\leqq\; \frac{\epsilon}{2} + (1-x)\left(\frac{\epsilon}{2} + \frac{\epsilon}{2}\,x + \frac{\epsilon}{2}\,x^2 + \cdots\right) \;=\; \frac{\epsilon}{2} + \frac{\epsilon}{2} \;=\; \epsilon \tag{2}$$

since $(1-x)(1 + x + x^2 + x^3 + \cdots) = 1$ (if $0 \leqq x < 1$)

Also, for $x = 1$, $|R_n(x)| = |R_n| < \epsilon$ for $n > N$.

Thus $|R_n(x)| < \epsilon$ for all $n > N$, where N is independent of the value of x in $0 \leqq x \leqq 1$, and the required result follows.

Extensions to other power series are easily made.

43. Prove Abel's limit theorem (see Page 230).

As in Problem 42, assume the power series to be $\sum\limits_{k=0}^{\infty} a_k x^k$, convergent for $0 \leqq x \leqq 1$.

Then we must show that $\lim\limits_{x \to 1-} \sum\limits_{k=0}^{\infty} a_k x^k = \sum\limits_{k=0}^{\infty} a_k$.

This follows at once from Problem 42 which shows that $\Sigma a_k x^k$ is uniformly convergent for $0 \leqq x \leqq 1$, and from Problem 34 which shows that $\Sigma a_k x^k$ is continuous at $x = 1$.

Extensions to other power series are easily made.

44. (a) Prove that $\quad \tan^{-1} x = x - \dfrac{x^3}{3} + \dfrac{x^5}{5} - \dfrac{x^7}{7} + \cdots \quad$ where the series is uniformly convergent in $-1 \leqq x \leqq 1$.

(b) Prove that $\dfrac{\pi}{4} = 1 - \dfrac{1}{3} + \dfrac{1}{5} - \dfrac{1}{7} + \cdots$.

(a) By Problem 25 of Chapter 3, with $r = -x^2$ and $a = 1$, we have

$$\frac{1}{1+x^2} \;=\; 1 - x^2 + x^4 - x^6 + \cdots \qquad -1 < x < 1 \tag{1}$$

Integrating from 0 to x, where $-1 < x < 1$, yields

$$\int_0^x \frac{dx}{1+x^2} \;=\; \tan^{-1} x \;=\; x - \frac{x^3}{3} + \frac{x^5}{5} - \frac{x^7}{7} + \cdots \tag{2}$$

using Problems 33 and 35.

Since the series on the right of (2) converges for $x = \pm 1$, it follows by Problem 42 that the series is uniformly convergent in $-1 \leqq x \leqq 1$ and represents $\tan^{-1} x$ in this interval.

(b) By Problem 43 and part (a), we have

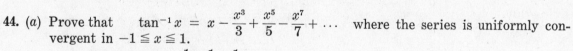

$$\lim_{x \to 1-} \tan^{-1} x \;=\; \lim_{x \to 1-} \left(x - \frac{x^3}{3} + \frac{x^5}{5} - \frac{x^7}{7} + \cdots \right) \qquad \text{or} \qquad \frac{\pi}{4} = 1 - \frac{1}{3} + \frac{1}{5} - \frac{1}{7} + \cdots$$

45. Evaluate $\int_0^1 \dfrac{1 - e^{-x^2}}{x^2}\, dx$ to 3 decimal place accuracy.

We have $\quad e^u = 1 + u + \dfrac{u^2}{2!} + \dfrac{u^3}{3!} + \dfrac{u^4}{4!} + \dfrac{u^5}{5!} + \cdots, \quad -\infty < u < \infty.$

Then if $u = -x^2, \quad e^{-x^2} = 1 - x^2 + \dfrac{x^4}{2!} - \dfrac{x^6}{3!} + \dfrac{x^8}{4!} - \dfrac{x^{10}}{5!} + \cdots, \quad -\infty < x < \infty.$

Thus $\quad \dfrac{1 - e^{-x^2}}{x^2} = 1 - \dfrac{x^2}{2!} + \dfrac{x^4}{3!} - \dfrac{x^6}{4!} + \dfrac{x^8}{5!} - \cdots.$

Since the series converges for all x and so, in particular, converges uniformly for $0 \leqq x \leqq 1$, we can integrate term by term to obtain

$$
\begin{aligned}
\int_0^1 \frac{1 - e^{-x^2}}{x^2}\, dx &= \left. x - \frac{x^3}{3 \cdot 2!} + \frac{x^5}{5 \cdot 3!} - \frac{x^7}{7 \cdot 4!} + \frac{x^9}{9 \cdot 5!} - \cdots \right|_0^1 \\
&= 1 - \frac{1}{3 \cdot 2!} + \frac{1}{5 \cdot 3!} - \frac{1}{7 \cdot 4!} + \frac{1}{9 \cdot 5!} - \cdots \\
&= 1 - 0.16666 + 0.03333 - 0.00595 + 0.00092 - \cdots = 0.862
\end{aligned}
$$

Note that the error made in adding the first four terms of the alternating series is less than the fifth term, i.e. less than 0.001 (see Problem 15).

MISCELLANEOUS PROBLEMS

46. Prove that $y = J_p(x)$ defined by (16), Page 232, satisfies Bessel's differential equation

$$x^2 y'' + xy' + (x^2 - p^2)y = 0$$

The series for $J_p(x)$ converges for all x [see Problem 106(a)]. Since a power series can be differentiated term by term within its interval of convergence, we have for all x,

$$
\begin{aligned}
y &= \sum_{n=0}^\infty \frac{(-1)^n x^{p+2n}}{2^{p+2n}\, n!\, (n+p)!} \\
y' &= \sum_{n=0}^\infty \frac{(-1)^n (p+2n)\, x^{p+2n-1}}{2^{p+2n}\, n!\, (n+p)!} \\
y'' &= \sum_{n=0}^\infty \frac{(-1)^n (p+2n)(p+2n-1)\, x^{p+2n-2}}{2^{p+2n}\, n!\, (n+p)!}
\end{aligned}
$$

Then,

$$
\begin{aligned}
(x^2 - p^2)y &= \sum_{n=0}^\infty \frac{(-1)^n x^{p+2n+2}}{2^{p+2n}\, n!\, (n+p)!} - \sum_{n=0}^\infty \frac{(-1)^n p^2\, x^{p+2n}}{2^{p+2n}\, n!\, (n+p)!} \\
xy' &= \sum_{n=0}^\infty \frac{(-1)^n (p+2n)\, x^{p+2n}}{2^{p+2n}\, n!\, (n+p)!} \\
x^2 y'' &= \sum_{n=0}^\infty \frac{(-1)^n (p+2n)(p+2n-1)\, x^{p+2n}}{2^{p+2n}\, n!\, (n+p)!}
\end{aligned}
$$

Adding,

$$
\begin{aligned}
x^2 y'' + xy' + (x^2 - p^2)y &= \sum_{n=0}^\infty \frac{(-1)^n x^{p+2n+2}}{2^{p+2n}\, n!\, (n+p)!} \\
&\quad + \sum_{n=0}^\infty \frac{(-1)^n [-p^2 + (p+2n) + (p+2n)(p+2n-1)]\, x^{p+2n}}{2^{p+2n}\, n!\, (n+p)!} \\
&= \sum_{n=0}^\infty \frac{(-1)^n x^{p+2n+2}}{2^{p+2n}\, n!\, (n+p)!} + \sum_{n=0}^\infty \frac{(-1)^n [4n(n+p)]\, x^{p+2n}}{2^{p+2n}\, n!\, (n+p)!} \\
&= \sum_{n=1}^\infty \frac{(-1)^{n-1} x^{p+2n}}{2^{p+2n-2}\, (n-1)!\, (n-1+p)!} + \sum_{n=1}^\infty \frac{(-1)^n 4 x^{p+2n}}{2^{p+2n}\, (n-1)!\, (n+p-1)!} \\
&= -\sum_{n=1}^\infty \frac{(-1)^n 4 x^{p+2n}}{2^{p+2n}\, (n-1)!\, (n+p-1)!} + \sum_{n=1}^\infty \frac{(-1)^n 4 x^{p+2n}}{2^{p+2n}\, (n-1)!\, (n+p-1)!} \\
&= 0
\end{aligned}
$$

47. Test for convergence the complex power series $\displaystyle\sum_{n=1}^{\infty} \frac{z^{n-1}}{n^3 \cdot 3^{n-1}}$.

Since $\displaystyle\lim_{n \to \infty} \left| \frac{u_{n+1}}{u_n} \right| = \lim_{n \to \infty} \left| \frac{z^n}{(n+1)^3 \cdot 3^n} \cdot \frac{n^3 \cdot 3^{n-1}}{z^{n-1}} \right| = \lim_{n \to \infty} \frac{n^3}{3(n+1)^3} |z| = \frac{|z|}{3}$, the series converges for $\dfrac{|z|}{3} < 1$, i.e. $|z| < 3$, and diverges for $|z| > 3$.

For $|z| = 3$, the series of absolute values is $\displaystyle\sum_{n=1}^{\infty} \frac{|z|^{n-1}}{n^3 \cdot 3^{n-1}} = \sum_{n=1}^{\infty} \frac{1}{n^3}$ so that the series is absolutely convergent and thus convergent for $|z| = 3$.

Thus the series converges within and on the circle $|z| = 3$.

48. Assuming the power series for e^x holds for complex numbers, show that

$$e^{ix} = \cos x + i \sin x$$

Letting $z = ix$ in $e^z = 1 + z + \dfrac{z^2}{2!} + \dfrac{z^3}{3!} + \cdots$, we have

$$e^{ix} = 1 + ix + \frac{i^2 x^2}{2!} + \frac{i^3 x^3}{3!} + \cdots = \left(1 - \frac{x^2}{2!} + \frac{x^4}{4!} - \cdots \right) + i \left(x - \frac{x^3}{3!} + \frac{x^5}{5!} - \cdots \right)$$

$$= \cos x + i \sin x$$

Similarly, $e^{-ix} = \cos x - i \sin x$. The results are called *Euler's identities*.

49. Prove that $\displaystyle\lim_{n \to \infty} \left(1 + \frac{1}{2} + \frac{1}{3} + \frac{1}{4} + \cdots + \frac{1}{n} - \ln n \right)$ exists.

Letting $f(x) = 1/x$ in (1), Problem 11, we find

$$\frac{1}{2} + \frac{1}{3} + \frac{1}{4} + \cdots + \frac{1}{M} \leqq \ln M \leqq 1 + \frac{1}{2} + \frac{1}{3} + \frac{1}{4} + \cdots + \frac{1}{M-1}$$

from which we have on replacing M by n,

$$\frac{1}{n} \leqq 1 + \frac{1}{2} + \frac{1}{3} + \frac{1}{4} + \cdots + \frac{1}{n} - \ln n \leqq 1$$

Thus the sequence $S_n = 1 + \dfrac{1}{2} + \dfrac{1}{3} + \dfrac{1}{4} + \cdots + \dfrac{1}{n} - \ln n$ is bounded by 0 and 1.

Consider $S_{n+1} - S_n = \dfrac{1}{n+1} - \ln \left(\dfrac{n+1}{n} \right)$. By integrating the inequality $\dfrac{1}{n+1} \leqq \dfrac{1}{x} \leqq \dfrac{1}{n}$ with respect to x from n to $n+1$, we have

$$\frac{1}{n+1} \leqq \ln \left(\frac{n+1}{n} \right) \leqq \frac{1}{n} \quad \text{or} \quad \frac{1}{n+1} - \frac{1}{n} \leqq \frac{1}{n+1} - \ln \left(\frac{n+1}{n} \right) \leqq 0$$

i.e. $S_{n+1} - S_n \leqq 0$, so that S_n is monotonic decreasing.

Since S_n is bounded and monotonic decreasing, it has a limit. This limit, denoted by γ, is equal to $0.577215\ldots$ and is called *Euler's constant*. It is not yet known whether γ is rational or not.

50. Prove that the infinite product $\displaystyle\prod_{k=1}^{\infty} (1 + u_k)$, where $u_k > 0$, converges if $\displaystyle\sum_{k=1}^{\infty} u_k$ converges.

By equation (1) of Problem 28, Chapter 4, $1 + x \leqq e^x$ for $x > 0$, so that

$$P_n = \prod_{k=1}^{n} (1 + u_k) = (1 + u_1)(1 + u_2) \cdots (1 + u_n) \leqq e^{u_1} \cdot e^{u_2} \cdots e^{u_n} = e^{u_1 + u_2 + \cdots + u_n}$$

Since $u_1 + u_2 + \cdots$ converges, it follows that P_n is a bounded monotonic increasing sequence and so has a limit, thus proving the required result.

51. Prove that the series $1 - 1 + 1 - 1 + 1 - 1 + \cdots$ is $C - 1$ summable to $1/2$.

The sequence of partial sums is $1, 0, 1, 0, 1, 0, \ldots$.

Then $\quad S_1 = 1, \quad \dfrac{S_1 + S_2}{2} = \dfrac{1+0}{2} = \dfrac{1}{2}, \quad \dfrac{S_1 + S_2 + S_3}{3} = \dfrac{1+0+1}{3} = \dfrac{2}{3}, \quad \ldots$.

Continuing in this manner, we obtain the sequence $1, \frac{1}{2}, \frac{2}{3}, \frac{1}{2}, \frac{3}{5}, \frac{1}{2}, \ldots$, the nth term being

$$T_n = \begin{cases} 1/2 & \text{if } n \text{ is even} \\ n/(2n-1) & \text{if } n \text{ is odd} \end{cases}. \quad \text{Thus} \quad \lim_{n \to \infty} T_n = \tfrac{1}{2} \text{ and the required result follows.}$$

52. (a) Prove that if $x > 0$ and $p > 0$,

$$f(x) = \int_x^\infty \frac{e^{-u}}{u^p}\, du$$

$$= e^{-x}\left\{ \frac{1}{x^p} - \frac{p}{x^{p+1}} + \frac{p(p+1)}{x^{p+2}} - \cdots (-1)^n \frac{p(p+1)\cdots(p+n-1)}{x^{p+n}} \right\}$$

$$(-1)^{n+1} p(p+1) \cdots (p+n) \int_x^\infty \frac{e^{-u}}{u^{p+n+1}}\, du$$

(b) Use (a) to prove that

$$f(x) = \int_x^\infty \frac{e^{-u}}{u^p}\, du \;\sim\; e^{-x}\left\{ \frac{1}{x^p} - \frac{p}{x^{p+1}} + \frac{p(p+1)}{x^{p+2}} - \cdots \right\} = S(x)$$

i.e. the series on the right is an asymptotic expansion of the function on the left.

(a) Integrating by parts, we have

$$I_p = \int_x^\infty \frac{e^{-u}}{u^p}\, du = \lim_{b \to \infty} \int_x^b e^{-u} u^{-p}\, du$$

$$= \lim_{b \to \infty} \left\{ (-e^{-u})(u^{-p}) \Big|_x^b - \int_x^b (-e^{-u})(-p u^{-p-1})\, du \right\}$$

$$= \lim_{b \to \infty} \left\{ \frac{e^{-x}}{x^p} - \frac{e^{-b}}{b^p} - p \int_x^b \frac{e^{-u}}{u^{p+1}}\, du \right\}$$

$$= \frac{e^{-x}}{x^p} - p \int_x^\infty \frac{e^{-u}}{u^{p+1}}\, du = \frac{e^{-x}}{x^p} - p I_{p+1}$$

Similarly, $\quad I_{p+1} = \dfrac{e^{-x}}{x^{p+1}} - (p+1) I_{p+2} \quad$ so that

$$I_p = \frac{e^{-x}}{x^p} - p\left\{ \frac{e^{-x}}{x^{p+1}} - (p+1) I_{p+2} \right\} = \frac{e^{-x}}{x^p} - \frac{p e^{-x}}{x^{p+1}} + p(p+1) I_{p+2}$$

By continuing in this manner the required result follows.

(b) Let $\quad S_n(x) = e^{-x}\left\{ \dfrac{1}{x^p} - \dfrac{p}{x^{p+1}} + \dfrac{p(p+1)}{x^{p+2}} - \cdots (-1)^n \dfrac{p(p+1)\cdots(p+n-1)}{x^{p+n}} \right\}.$

Then $\quad R_n(x) = f(x) - S_n(x) = (-1)^{n+1} p(p+1)\cdots(p+n) \displaystyle\int_x^\infty \frac{e^{-u}}{u^{p+n+1}}\, du.\quad$ Now

$$|R_n(x)| = p(p+1)\ldots(p+n) \int_x^\infty \frac{e^{-u}}{u^{p+n+1}}\, du \;\leqq\; p(p+1)\ldots(p+n) \int_x^\infty \frac{e^{-u}}{x^{p+n+1}}\, du$$

$$\leqq\; \frac{p(p+1)\cdots(p+n)}{x^{p+n+1}}$$

since $\displaystyle \int_x^\infty e^{-u}\, du \;\leqq\; \int_0^\infty e^{-u}\, du = 1.$ **Thus**

$$\lim_{|x| \to \infty} |x^n R_n(x)| \; \le \; \lim_{|x| \to \infty} \frac{p(p+1) \cdots (p+n)}{|x|^p} \; = \; 0$$

and it follows that $\lim_{|x| \to \infty} x^n R_n(x) = 0$. Hence the required result is proved.

Note that since $\left| \dfrac{u_{n+1}}{u_n} \right| = \left| \dfrac{p(p+1) \cdots (p+n)/x^{p+n+1}}{p(p+1) \cdots (p+n-1)/x^{p+n}} \right| = \dfrac{p+n}{|x|}$, we have for all fixed x,

$\lim_{n \to \infty} \left| \dfrac{u_{n+1}}{u_n} \right| = \infty$ and the series diverges for all x by the ratio test.

Supplementary Problems

CONVERGENCE and DIVERGENCE of SERIES of CONSTANTS

53. (a) Prove that the series $\dfrac{1}{3 \cdot 7} + \dfrac{1}{7 \cdot 11} + \dfrac{1}{11 \cdot 15} + \cdots = \sum_{n=1}^{\infty} \dfrac{1}{(4n-1)(4n+3)}$ converges and
(b) find its sum.   Ans. (b) $1/12$

54. Prove that the convergence or divergence of a series is not affected by (a) multiplying each term by the same non-zero constant, (b) removing (or adding) a finite number of terms.

55. If Σu_n and Σv_n converge to A and B respectively, prove that $\Sigma(u_n + v_n)$ converges to $A + B$.

56. Prove that the series $\tfrac{3}{2} + (\tfrac{3}{2})^2 + (\tfrac{3}{2})^3 + \cdots = \Sigma(\tfrac{3}{2})^n$ diverges.

57. Find the fallacy: Let $S = 1 - 1 + 1 - 1 + 1 - 1 + \cdots$. Then $S = 1 - (1-1) - (1-1) - \cdots = 1$ and $S = (1-1) + (1-1) + (1-1) + \cdots = 0$. Hence $1 = 0$.

COMPARISON TEST and QUOTIENT TEST

58. Test for convergence:

(a) $\sum_{n=1}^{\infty} \dfrac{1}{n^2+1}$, (b) $\sum_{n=1}^{\infty} \dfrac{n}{4n^2-3}$, (c) $\sum_{n=1}^{\infty} \dfrac{n+2}{(n+1)\sqrt{n+3}}$, (d) $\sum_{n=1}^{\infty} \dfrac{3^n}{n \cdot 5^n}$, (e) $\sum_{n=1}^{\infty} \dfrac{1}{5n-3}$, (f) $\sum_{n=1}^{\infty} \dfrac{2n-1}{(3n+2)n^{4/3}}$.

Ans. (a) conv., (b) div., (c) div., (d) conv., (e) div., (f) conv.

59. Investigate the convergence of (a) $\sum_{n=1}^{\infty} \dfrac{4n^2+5n-2}{n(n^2+1)^{3/2}}$, (b) $\sum_{n=1}^{\infty} \sqrt{\dfrac{n - \ln n}{n^2 + 10n^3}}$.   Ans. (a) conv., (b) div.

60. Establish the comparison test for divergence (see Page 225).

61. Use the comparison test to prove that

(a) $\sum_{n=1}^{\infty} \dfrac{1}{n^p}$ converges if $p > 1$ and diverges if $p \le 1$, (b) $\sum_{n=1}^{\infty} \dfrac{\tan^{-1} n}{n}$ diverges, (c) $\sum_{n=1}^{\infty} \dfrac{n^2}{2^n}$ converges.

62. Establish the results (b) and (c) of the quotient test, Page 225.

63. Test for convergence:

(a) $\sum_{n=1}^{\infty} \dfrac{(\ln n)^2}{n^2}$, (b) $\sum_{n=1}^{\infty} \sqrt{n} \tan^{-1}(1/n^3)$, (c) $\sum_{n=1}^{\infty} \dfrac{3 + \sin n}{n(1 + e^{-n})}$, (d) $\sum_{n=1}^{\infty} n \sin^2(1/n)$.

Ans. (a) conv., (b) div., (c) div., (d) div.

64. If Σu_n converges, where $u_n \geqq 0$ for $n > N$, and if $\lim\limits_{n \to \infty} nu_n$ exists, prove that $\lim\limits_{n \to \infty} nu_n = 0$.

65. (a) Test for convergence $\sum\limits_{n=1}^{\infty} \dfrac{1}{n^{1+1/n}}$. (b) Does your answer to (a) contradict the statement about the p series made on Page 225 that $\Sigma 1/n^p$ converges for $p > 1$?   Ans. (a) div.

INTEGRAL TEST

66. Test for convergence: (a) $\sum\limits_{n=1}^{\infty} \dfrac{n^2}{2n^3-1}$,   (b) $\sum\limits_{n=2}^{\infty} \dfrac{1}{n(\ln n)^3}$,   (c) $\sum\limits_{n=1}^{\infty} \dfrac{n}{2^n}$,   (d) $\sum\limits_{n=1}^{\infty} \dfrac{e^{-\sqrt{n}}}{\sqrt{n}}$,   (e) $\sum\limits_{n=2}^{\infty} \dfrac{\ln n}{n}$,

(f) $\sum\limits_{n=10}^{\infty} \dfrac{2^{\ln(\ln n)}}{n \ln n}$.   Ans. (a) div., (b) conv., (c) conv., (d) conv., (e) div., (f) div.

67. Prove that $\sum\limits_{n=2}^{\infty} \dfrac{1}{n(\ln n)^p}$, where p is a constant, (a) converges if $p > 1$ and (b) diverges if $p \leqq 1$.

68. Prove that $\dfrac{9}{8} < \sum\limits_{n=1}^{\infty} \dfrac{1}{n^3} < \dfrac{5}{4}$.

69. Investigate the convergence of $\sum\limits_{n=1}^{\infty} \dfrac{e^{\tan^{-1} n}}{n^2+1}$.
Ans. conv.

70. (a) Prove that $\frac{2}{3} n^{3/2} + \frac{1}{3} \leqq \sqrt{1} + \sqrt{2} + \sqrt{3} + \cdots + \sqrt{n} \leqq \frac{2}{3} n^{3/2} + n^{1/2} - \frac{2}{3}$.

(b) Use (a) to estimate the value of $\sqrt{1} + \sqrt{2} + \sqrt{3} + \cdots + \sqrt{100}$, giving the maximum error.

(c) Show how the accuracy in (b) can be improved by estimating, for example, $\sqrt{10} + \sqrt{11} + \cdots + \sqrt{100}$ and adding on the value of $\sqrt{1} + \sqrt{2} + \cdots + \sqrt{9}$ computed to some desired degree of accuracy.
Ans. (b) 671.5 ± 4.5

ALTERNATING SERIES

71. Test for convergence: (a) $\sum\limits_{n=1}^{\infty} \dfrac{(-1)^{n+1}}{2^n}$,   (b) $\sum\limits_{n=1}^{\infty} \dfrac{(-1)^n}{n^2+2n+2}$,   (c) $\sum\limits_{n=1}^{\infty} \dfrac{(-1)^{n+1} n}{3n-1}$, (d) $\sum\limits_{n=1}^{\infty} (-1)^n \sin^{-1} \dfrac{1}{n}$,

(e) $\sum\limits_{n=2}^{\infty} \dfrac{(-1)^n \sqrt{n}}{\ln n}$.   Ans. (a) conv., (b) conv., (c) div., (d) conv., (e) div.

72. (a) What is the largest absolute error made in approximating the sum of the series $\sum\limits_{n=1}^{\infty} \dfrac{(-1)^n}{2^n(n+1)}$ by the sum of the first 5 terms?   Ans. 1/192

(b) What is the least number of terms which must be taken in order that 3 decimal place accuracy will result?   Ans. 8 terms

73. (a) Prove that $S = \dfrac{1}{1^3} + \dfrac{1}{2^3} + \dfrac{1}{3^3} + \cdots = \dfrac{4}{3}\left(\dfrac{1}{1^3} - \dfrac{1}{2^3} + \dfrac{1}{3^3} - \cdots\right)$.

(b) How many terms of the series on the right are needed in order to calculate S to six decimal place accuracy?   Ans. (b) at least 100 terms

ABSOLUTE and CONDITIONAL CONVERGENCE

74. Test for absolute or conditional convergence:

(a) $\sum\limits_{n=1}^{\infty} \dfrac{(-1)^{n-1}}{n^2+1}$ (c) $\sum\limits_{n=2}^{\infty} \dfrac{(-1)^n}{n \ln n}$ (e) $\sum\limits_{n=1}^{\infty} \dfrac{(-1)^{n-1}}{2n-1} \sin \dfrac{1}{\sqrt{n}}$

(b) $\sum\limits_{n=1}^{\infty} \dfrac{(-1)^{n-1} n}{n^2+1}$ (d) $\sum\limits_{n=1}^{\infty} \dfrac{(-1)^n n^3}{(n^2+1)^{4/3}}$ (f) $\sum\limits_{n=1}^{\infty} \dfrac{(-1)^{n-1} n^3}{2^n-1}$

Ans. (a) abs. conv., (b) cond. conv., (c) cond. conv., (d) div., (e) abs. conv., (f) abs. conv.

75. Prove that $\sum\limits_{n=1}^{\infty} \dfrac{\cos n\pi a}{x^2+n^2}$ converges absolutely for all real x and a.

76. If $1 - \frac{1}{2} + \frac{1}{3} - \frac{1}{4} + \cdots$ converges to S, prove that the rearranged series $1 + \frac{1}{3} - \frac{1}{2} + \frac{1}{5} + \frac{1}{7} - \frac{1}{4} + \frac{1}{9} + \frac{1}{11} - \frac{1}{6} + \cdots = \frac{3}{2}S$. Explain.

[Hint: Take 1/2 of the first series and write it as $0 + \frac{1}{2} + 0 - \frac{1}{4} + 0 + \frac{1}{6} + \cdots$; then add term by term to the first series. Note that $S = \ln 2$, as shown in Problem 96.]

77. Prove that the terms of an absolutely convergent series can always be rearranged without altering the sum.

RATIO TEST

78. Test for convergence:

(a) $\sum_{n=1}^{\infty} \frac{(-1)^n n}{(n+1)e^n}$, (b) $\sum_{n=1}^{\infty} \frac{10^{2n}}{(2n-1)!}$, (c) $\sum_{n=1}^{\infty} \frac{3^n}{n^3}$, (d) $\sum_{n=1}^{\infty} \frac{(-1)^n 2^{3n}}{3^{2n}}$, (e) $\sum_{n=1}^{\infty} \frac{(\sqrt{5}-1)^n}{n^2+1}$.

Ans. (a) conv. (abs.), (b) conv., (c) div., (d) conv. (abs.), (e) div.

79. Show that the ratio test cannot be used to establish the conditional convergence of a series.

80. Prove that (a) $\sum_{n=1}^{\infty} \frac{n!}{n^n}$ converges and (b) $\lim_{n \to \infty} \frac{n!}{n^n} = 0$.

MISCELLANEOUS TESTS

81. Establish the validity of the nth root test on Page 226.

82. Apply the nth root test to work Problems 78(a), (c), (d) and (e).

83. Prove that $\frac{1}{3} + (\frac{2}{3})^2 + (\frac{1}{3})^3 + (\frac{2}{3})^4 + (\frac{1}{3})^5 + (\frac{2}{3})^6 + \cdots$ converges.

84. Test for convergence: (a) $\frac{1}{3} + \frac{1 \cdot 4}{3 \cdot 6} + \frac{1 \cdot 4 \cdot 7}{3 \cdot 6 \cdot 9} + \cdots$, (b) $\frac{2}{9} + \frac{2 \cdot 5}{9 \cdot 12} + \frac{2 \cdot 5 \cdot 8}{9 \cdot 12 \cdot 15} + \cdots$.
Ans. (a) div., (b) conv.

85. If a, b and d are positive numbers and $b > a$, prove that

$$\frac{a}{b} + \frac{a(a+d)}{b(b+d)} + \frac{a(a+d)(a+2d)}{b(b+d)(b+2d)} + \cdots$$

converges if $b - a > d$, and diverges if $b - a \leq d$.

SERIES of FUNCTIONS

86. Find the domain of convergence of the series:

(a) $\sum_{n=1}^{\infty} \frac{x^n}{n^3}$, (b) $\sum_{n=1}^{\infty} \frac{(-1)^n (x-1)^n}{2^n (3n-1)}$, (c) $\sum_{n=1}^{\infty} \frac{1}{n(1+x^2)^n}$, (d) $\sum_{n=1}^{\infty} n^2 \left(\frac{1-x}{1+x}\right)^n$, (e) $\sum_{n=1}^{\infty} \frac{e^{nx}}{n^2-n+1}$

Ans. (a) $-1 \leq x \leq 1$, (b) $-1 < x \leq 3$, (c) all $x \neq 0$, (d) $x > 0$, (e) $x \leq 0$.

87. Prove that $\sum_{n=1}^{\infty} \frac{1 \cdot 3 \cdot 5 \cdots (2n-1)}{2 \cdot 4 \cdot 6 \cdots (2n)} x^n$ converges for $-1 \leq x < 1$.

UNIFORM CONVERGENCE

88. By use of the definition, investigate the uniform convergence of the series

$$\sum_{n=1}^{\infty} \frac{x}{[1+(n-1)x][1+nx]}$$

[Hint: Resolve the nth term into partial fractions and show that the nth partial sum is $S_n(x) = 1 - \frac{1}{1+nx}$.]

Ans. Not uniformly convergent in any interval which includes $x = 0$; uniformly convergent in any other interval.

89. Work Problem 30 directly by first obtaining $S_n(x)$.

90. Investigate by any method the convergence and uniform convergence of the series:

(a) $\sum_{n=1}^{\infty} \left(\frac{x}{3}\right)^n$,   (b) $\sum_{n=1}^{\infty} \frac{\sin^2 nx}{2^n - 1}$,   (c) $\sum_{n=0}^{\infty} \frac{x}{(1+x)^n}$, $x \geqq 0$.

    Ans. (a) conv. for $|x| < 3$; unif. conv. for $|x| \leqq r < 3$.   (b) unif. conv. for all x.   (c) conv. for $x \geqq 0$;
        not unif. conv. for $x \geqq 0$, but unif. conv. for $x \geqq r > 0$.

91. If $F(x) = \sum_{n=1}^{\infty} \frac{\sin nx}{n^3}$,   prove that:

    (a) $F(x)$ is continuous for all x,   (b) $\lim_{x \to 0} F(x) = 0$,   (c) $F'(x) = \sum_{n=1}^{\infty} \frac{\cos nx}{n^2}$ is continuous everywhere.

92. Prove that $\int_0^{\pi} \left(\frac{\cos 2x}{1 \cdot 3} + \frac{\cos 4x}{3 \cdot 5} + \frac{\cos 6x}{5 \cdot 7} + \cdots\right) dx = 0$.

93. Prove that $F(x) = \sum_{n=1}^{\infty} \frac{\sin nx}{\sinh n\pi}$ has derivatives of all orders for any real x.

94. Examine the sequence $u_n(x) = \frac{1}{1 + x^{2n}}$, $n = 1, 2, 3, \ldots$,   for uniform convergence.

95. Prove that $\lim_{n \to \infty} \int_0^1 \frac{dx}{(1 + x/n)^n} = 1 - e^{-1}$.

POWER SERIES

96. (a) Prove that $\ln(1 + x) = x - \frac{x^2}{2} + \frac{x^3}{3} - \frac{x^4}{4} + \cdots$.

    (b) Prove that $\ln 2 = 1 - \frac{1}{2} + \frac{1}{3} - \frac{1}{4} + \cdots$.

    [Hint: Use the fact that $\frac{1}{1 + x} = 1 - x + x^2 - x^3 + \cdots$ and integrate.]

97. Prove that $\sin^{-1} x = x + \frac{1}{2}\frac{x^3}{3} + \frac{1 \cdot 3}{2 \cdot 4}\frac{x^5}{5} + \frac{1 \cdot 3 \cdot 5}{2 \cdot 4 \cdot 6}\frac{x^7}{7} + \cdots$, $-1 \leqq x \leqq 1$.

98. Evaluate   (a) $\int_0^{1/2} e^{-x^2} dx$,   (b) $\int_0^1 \frac{1 - \cos x}{x} dx$ to 3 decimal places, justifying all steps.

    Ans. (a) 0.461,   (b) 0.486

99. Evaluate   (a) $\sin 40°$,   (b) $\cos 65°$,   (c) $\tan 12°$ correct to 3 decimal places.
    Ans. (a) 0.643,   (b) 0.423,   (c) 0.213

100. Verify the expansions 4, 5 and 6 on Page 231.

101. By multiplying the series for $\sin x$ and $\cos x$, verify that $2 \sin x \cos x = \sin 2x$.

102. Show that $e^{\cos x} = e\left(1 - \frac{x^2}{2!} + \frac{4x^4}{4!} - \frac{31x^6}{6!} + \cdots\right)$, $-\infty < x < \infty$.

103. Obtain the expansions

        (a) $\tanh^{-1} x \quad = x + \frac{x^3}{3} + \frac{x^5}{5} + \frac{x^7}{7} + \cdots \qquad -1 < x < 1$

        (b) $\ln(x + \sqrt{x^2 + 1}) \quad = x - \frac{1}{2}\frac{x^3}{3} + \frac{1 \cdot 3}{2 \cdot 4}\frac{x^5}{5} - \frac{1 \cdot 3 \cdot 5}{2 \cdot 4 \cdot 6}\frac{x^7}{7} + \cdots \qquad -1 \leqq x \leqq 1$

104. Let $f(x) = \begin{cases} e^{-1/x^2} & x \neq 0 \\ 0 & x = 0 \end{cases}$.   Prove that the formal Taylor series about $x = 0$ corresponding to $f(x)$

    exists but that it does not converge to the given function for any $x \neq 0$.
    [Hint: See Problem 38, Chapter 4.]

105. Prove that

$$(a) \quad \frac{\ln(1+x)}{1+x} = x - \left(1 + \frac{1}{2}\right)x^2 + \left(1 + \frac{1}{2} + \frac{1}{3}\right)x^3 - \cdots \qquad \text{for } -1 < x < 1$$

$$(b) \quad \{\ln(1+x)\}^2 = x^2 - \left(1 + \frac{1}{2}\right)\frac{2x^3}{3} + \left(1 + \frac{1}{2} + \frac{1}{3}\right)\frac{2x^4}{4} - \cdots \qquad \text{for } -1 < x \leqq 1$$

MISCELLANEOUS PROBLEMS

106. Prove that the series for $J_p(x)$ converges (a) for all x, (b) absolutely and uniformly in any finite interval.

107. Prove that (a) $\dfrac{d}{dx}\{J_0(x)\} = -J_1(x)$, (b) $\dfrac{d}{dx}\{x^p J_p(x)\} = x^p J_{p-1}(x)$,

(c) $J_{p+1}(x) = \dfrac{2p}{x}J_p(x) - J_{p-1}(x)$.

108. Assuming that the result of Problem 107(c) holds for $p = 0, -1, -2, \ldots$, prove that

(a) $J_{-1}(x) = -J_1(x)$, (b) $J_{-2}(x) = J_2(x)$, (c) $J_{-n}(x) = (-1)^n J_n(x)$, $n = 1, 2, 3, \ldots$.

109. Prove that $e^{\frac{1}{2}x(t-1/t)} = \displaystyle\sum_{p=-\infty}^{\infty} J_p(x)\,t^p$.

[Hint: Write the left side as $e^{xt/2}\,e^{-x/2t}$, expand and use Problem 108.]

110. Prove that $\displaystyle\sum_{n=1}^{\infty} \frac{(n+1)z^n}{n(n+2)^2}$ is absolutely and uniformly convergent at all points within and on the circle $|z| = 1$.

111. (a) If $\displaystyle\sum_{n=1}^{\infty} a_n x^n = \sum_{n=1}^{\infty} b_n x^n$ for all x in the common interval of convergence $|x| < R$ where $R > 0$, prove that $a_n = b_n$ for $n = 0, 1, 2, \ldots$. (b) Use (a) to show that if the Taylor expansion of a function exists, the expansion is unique.

112. Suppose that $\overline{\lim} \sqrt[n]{|u_n|} = L$. Prove that Σu_n converges or diverges according as $L < 1$ or $L > 1$. If $L = 1$ the test fails.

113. Prove that the radius of convergence of the series $\Sigma a_n x^n$ can be determined by the following limits, when they exist, and give examples: (a) $\displaystyle\lim_{n\to\infty}\left|\frac{a_n}{a_{n+1}}\right|$, (b) $\displaystyle\lim_{n\to\infty}\frac{1}{\sqrt[n]{|a_n|}}$, (c) $\displaystyle\overline{\lim_{n\to\infty}}\frac{1}{\sqrt[n]{|a_n|}}$.

114. Use Problem 113 to find the radius of convergence of the series in Problem 22.

115. (a) Prove that a necessary and sufficient condition that the series Σu_n converge is that, given any $\epsilon > 0$, we can find $N > 0$ depending on ϵ such that $|S_p - S_q| < \epsilon$ whenever $p > N$ and $q > N$, where $S_k = u_1 + u_2 + \cdots + u_k$.

(b) Use (a) to prove that the series $\displaystyle\sum_{n=1}^{\infty} \frac{n}{(n+1)\,3^n}$ converges.

(c) How could you use (a) to prove that the series $\displaystyle\sum_{n=1}^{\infty} \frac{1}{n}$ diverges?

[Hint: Use the Cauchy convergence criterion, Page 43.]

116. Prove that the hypergeometric series (Page 232) (a) is absolutely convergent for $|x| < 1$, (b) is divergent for $|x| > 1$, (c) is absolutely convergent for $|x| = 1$ if $a + b - c < 0$, (d) satisfies the differential equation $x(1-x)y'' + \{c - (a+b+1)x\}y' - ab\,y = 0$.

117. If $F(a, b; c; x)$ is the hypergeometric function defined by the series on Page 232, prove that
(a) $F(-p, 1; 1; -x) = (1+x)^p$, (b) $xF(1, 1; 2; -x) = \ln(1+x)$, (c) $F(\frac{1}{2}, \frac{1}{2}; \frac{3}{2}; x^2) = (\sin^{-1}x)/x$.

118. Find the sum of the series $S(x) = x + \dfrac{x^3}{1\cdot 3} + \dfrac{x^5}{1\cdot 3\cdot 5} + \cdots$.

[Hint: Show that $S'(x) = 1 + x\,S(x)$ and solve.] $\qquad$ *Ans.* $e^{x^2/2}\displaystyle\int_0^x e^{-x^2/2}\,dx$

119. Prove that

$$1 + \frac{1}{1 \cdot 3} + \frac{1}{1 \cdot 3 \cdot 5} + \frac{1}{1 \cdot 3 \cdot 5 \cdot 7} + \cdots = \sqrt{e}\left(1 - \frac{1}{2 \cdot 3} + \frac{1}{2^2 \cdot 2! \cdot 5} - \frac{1}{2^3 \cdot 3! \cdot 7} + \frac{1}{2^4 \cdot 4! \cdot 9} - \cdots\right)$$

120. Establish the Dirichlet test on Page 228.

121. Prove that $\sum\limits_{n=1}^{\infty} \frac{\sin nx}{n}$ is uniformly convergent in any interval which does not include $0, \pm\pi, \pm 2\pi, \dots$.

[Hint: Use the Dirichlet test, Page 228, and Problem 94, Chapter 1.]

122. Establish the results on Page 232 concerning the binomial series.

[Hint: Examine the Lagrange and Cauchy forms of the remainder in Taylor's theorem.]

123. Prove that $\sum\limits_{n=1}^{\infty} \frac{(-1)^{n-1}}{n + x^2}$ converges uniformly for all x, but not absolutely.

124. Prove that $1 - \frac{1}{4} + \frac{1}{7} - \frac{1}{10} + \cdots = \frac{\pi}{3\sqrt{3}} + \frac{1}{3}\ln 2$

125. If $x = ye^y$, prove that $y = \sum\limits_{n=1}^{\infty} \frac{(-1)^{n-1} n^{n-1}}{n!} x^n$ for $-1/e < x \leqq 1/e$.

126. Prove that the equation $e^{-\lambda} = \lambda - 1$ has only one real root and show that it is given by

$$\lambda = 1 + \sum\limits_{n=1}^{\infty} \frac{(-1)^{n-1} n^{n-1} e^{-n}}{n!}$$

127. Let $\frac{x}{e^x - 1} = 1 + B_1 x + \frac{B_2 x^2}{2!} + \frac{B_3 x^3}{3!} + \cdots$. (a) Show that the numbers B_n, called the *Bernoulli numbers*, satisfy the recursion formula $(B+1)^n - B^n = 0$ where B^k is formally replaced by B_k after expanding. (b) Using (a) or otherwise, determine $B_1, \dots, B_6$.

Ans. (b) $B_1 = -\frac{1}{2}$, $B_2 = \frac{1}{6}$, $B_3 = 0$, $B_4 = -\frac{1}{30}$, $B_5 = 0$, $B_6 = \frac{1}{42}$

128. (a) Prove that $\frac{x}{e^x - 1} = \frac{x}{2}\left(\coth\frac{x}{2} - 1\right)$. (b) Use Problem 127 and part (a) to show that $B_{2k+1} = 0$ if $k = 1, 2, 3, \dots$.

129. Derive the series expansions:

(a) $\coth x = \frac{1}{x} + \frac{x}{3} - \frac{x^3}{45} + \cdots + \frac{B_{2n}(2x)^{2n}}{(2n)!\, x} + \cdots$

(b) $\cot x = \frac{1}{x} - \frac{x}{3} - \frac{x^3}{45} + \cdots (-1)^n \frac{B_{2n}(2x)^{2n}}{(2n)!\, x} + \cdots$

(c) $\tan x = x + \frac{x^3}{3} + \frac{2x^5}{15} + \cdots (-1)^{n-1} \frac{2(2^{2n} - 1) B_{2n}(2x)^{2n-1}}{(2n)!} + \cdots$

(d) $\csc x = \frac{1}{x} + \frac{x}{6} + \frac{7}{360} x^3 + \cdots (-1)^{n-1} \frac{2(2^{2n-1} - 1) B_{2n} x^{2n-1}}{(2n)!} + \cdots$

[Hint: For (a) use Problem 128; for (b) replace x by ix in (a); for (c) use $\tan x = \cot x - 2\cot 2x$; for (d) use $\csc x = \cot x + \tan x/2$.]

130. Prove that $\prod\limits_{n=1}^{\infty} \left(1 + \frac{1}{n^3}\right)$ converges.

131. Use the definition to prove that $\prod\limits_{n=1}^{\infty} \left(1 + \frac{1}{n}\right)$ diverges.

132. Prove that $\prod\limits_{n=1}^{\infty} (1 - u_n)$, where $0 < u_n < 1$, converges if and only if Σu_n converges.

133. (a) Prove that $\prod\limits_{n=2}^{\infty} \left(1 - \frac{1}{n^2}\right)$ converges to $\frac{1}{2}$. (b) Evaluate the infinite product in (a) to 2 decimal places and compare with the true value.

134. Prove that the series $1 + 0 - 1 + 1 + 0 - 1 + 1 + 0 - 1 + \cdots$ is $C - 1$ summable to zero.

135. Prove that the Césaro method of summability is regular. [Hint: See Prob. 28, Chap. 3.]

136. Prove that the series $1 + 2x + 3x^2 + 4x^3 + \cdots + nx^{n-1} + \cdots$ converges to $1/(1-x)^2$ for $|x| < 1$.

137. A series $\sum\limits_{n=0}^{\infty} a_n$ is called *Abel summable* to S if $S = \lim\limits_{x \to 1-} \sum\limits_{n=0}^{\infty} a_n x^n$ exists. Prove that

(a) $\sum\limits_{n=0}^{\infty} (-1)^n (n+1)$ is Abel summable to $1/4$ and

(b) $\sum\limits_{n=0}^{\infty} \dfrac{(-1)^n (n+1)(n+2)}{2}$ is Abel summable to $1/8$.

138. Prove that the double series $\sum\limits_{m=1}^{\infty} \sum\limits_{n=1}^{\infty} \dfrac{1}{(m^2 + n^2)^p}$, where p is a constant, converges or diverges according as $p > 1$ or $p \leqq 1$ respectively.

139. (a) Prove that $\displaystyle\int_x^\infty \frac{e^{x-u}}{u}\,du = \frac{1}{x} - \frac{1}{x^2} + \frac{2!}{x^3} - \frac{3!}{x^4} + \cdots \frac{(-1)^{n-1}(n-1)!}{x^n} + (-1)^n n! \int_x^\infty \frac{e^{x-u}}{u^{n+1}}\,du.$

(b) Use (a) to prove that $\displaystyle\int_x^\infty \frac{e^{x-u}}{u}\,du \sim \frac{1}{x} - \frac{1}{x^2} + \frac{2!}{x^3} - \frac{3!}{x^4} + \cdots$

140. Prove that

$$\int_x^\infty \frac{\sin u}{u}\,du \sim \frac{\cos x}{x}\left(1 - \frac{2!}{x^2} + \frac{4!}{x^4} - \cdots\right) + \frac{\sin x}{x}\left(\frac{1}{x} - \frac{3!}{x^3} + \frac{5!}{x^5} - \cdots\right)$$

$$\int_x^\infty \frac{\cos u}{u}\,du \sim \frac{\cos x}{x}\left(\frac{1}{x} - \frac{3!}{x^3} + \frac{5!}{x^5} - \cdots\right) - \frac{\sin x}{x}\left(1 - \frac{2!}{x^2} + \frac{4!}{x^4} - \cdots\right)$$

141. If $f(x)$ has an asymptotic expansion given by $\sum\limits_{n=2}^{\infty} a_n x^{-n}$, prove that $\displaystyle\int_x^\infty f(x)\,dx$ has the asymptotic expansion $\sum\limits_{n=2}^{\infty} \dfrac{a_n x^{1-n}}{n-1}$.

Chapter 12

Improper Integrals

DEFINITION of an IMPROPER INTEGRAL

The integral $\int_a^b f(x)\,dx$ is called an *improper integral* if

(1) $a = -\infty$ or $b = \infty$ or both, i.e. one or both integration limits is infinite,

(2) $f(x)$ is unbounded at one or more points of $a \leqq x \leqq b$. Such points are called *singularities* of $f(x)$.

Integrals corresponding to (1) and (2) are called *improper integrals of the first and second kinds* respectively. Integrals with both conditions (1) and (2) are called *improper integrals of the third kind*.

Example 1: $\int_0^\infty \sin x^2\,dx$ is an improper integral of the first kind.

Example 2: $\int_0^4 \dfrac{dx}{x - 3}$ is an improper integral of the second kind.

Example 3: $\int_0^\infty \dfrac{e^{-x}}{\sqrt{x}}\,dx$ is an improper integral of the third kind.

Example 4: $\int_0^1 \dfrac{\sin x}{x}\,dx$ is a *proper integral* since $\lim\limits_{x \to 0+} \dfrac{\sin x}{x} = 1$.

In this chapter we formulate tests for convergence or divergence of improper integrals. It will be found that such tests and proofs of theorems bear close analogy to convergence and divergence tests and corresponding theorems for infinite series (see Chapter 11).

IMPROPER INTEGRALS of the FIRST KIND

Let $f(x)$ be bounded and integrable in every finite interval $a \leqq x \leqq b$. Then we define

$$\int_a^\infty f(x)\,dx \;=\; \lim_{b \to \infty} \int_a^b f(x)\,dx \qquad\qquad (1)$$

The integral on the left is called *convergent* or *divergent* according as the limit on the right does or does not exist. Note that $\int_a^\infty f(x)\,dx$ bears close analogy to the infinite series $\sum\limits_{n=1}^\infty u_n$ where $u_n = f(n)$, while $\int_a^b f(x)\,dx$ corresponds to the partial sums of such infinite series. We often write M in place of b in (1).

Similarly, we define

$$\int_{-\infty}^b f(x)\,dx \;=\; \lim_{a \to -\infty} \int_a^b f(x)\,dx \qquad\qquad (2)$$

and call the integral on the left convergent or divergent according as the limit on the right does or does not exist.

Example 1: $\displaystyle\int_1^\infty \frac{dx}{x^2} = \lim_{b\to\infty} \int_1^b \frac{dx}{x^2} = \lim_{b\to\infty}\left(1 - \frac{1}{b}\right) = 1$ so that $\displaystyle\int_1^\infty \frac{dx}{x^2}$ converges to 1.

Example 2: $\displaystyle\int_{-\infty}^u \cos x\,dx = \lim_{a\to-\infty} \int_a^u \cos x\,dx = \lim_{a\to-\infty}(\sin u - \sin a)$. Since this limit does not exist, $\displaystyle\int_{-\infty}^u \cos x\,dx$ is divergent.

In like manner, we define

$$\int_{-\infty}^\infty f(x)\,dx = \int_{-\infty}^{x_0} f(x)\,dx + \int_{x_0}^\infty f(x)\,dx \tag{3}$$

where x_0 is a real number, and call the integral convergent or divergent according as the integrals on the right converge or not as in definitions (1) and (2).

SPECIAL IMPROPER INTEGRALS of the FIRST KIND

1. **Geometric or exponential integral** $\displaystyle\int_a^\infty e^{-tx}\,dx$, where t is a constant, converges if $t>0$ and diverges if $t\leq 0$. Note the analogy with the geometric series if $r = e^{-t}$ so that $e^{-tx} = r^x$.

2. **The p integral of the first kind** $\displaystyle\int_a^\infty \frac{dx}{x^p}$, where p is a constant and $a>0$, converges if $p>1$ and diverges if $p\leq 1$. Compare with the p series.

CONVERGENCE TESTS for IMPROPER INTEGRALS of the FIRST KIND

The following tests are given for cases where an integration limit is ∞. Similar tests exist where an integration limit is $-\infty$ (a change of variable $x = -y$ then makes the integration limit ∞). Unless otherwise specified we shall assume that $f(x)$ is continuous and thus integrable in every finite interval $a \leq x \leq b$.

1. **Comparison test** for integrals with non-negative integrands.

   (a) *Convergence.* Let $g(x) \geq 0$ for all $x \geq a$, and suppose that $\displaystyle\int_a^\infty g(x)\,dx$ converges. Then if $0 \leq f(x) \leq g(x)$ for all $x \geq a$, $\displaystyle\int_a^\infty f(x)\,dx$ also converges.

   Example: Since $\dfrac{1}{e^x+1} \leq \dfrac{1}{e^x} = e^{-x}$ and $\displaystyle\int_0^\infty e^{-x}\,dx$ converges, $\displaystyle\int_0^\infty \frac{dx}{e^x+1}$ also converges.

   (b) *Divergence.* Let $g(x) \geq 0$ for all $x \geq a$, and suppose that $\displaystyle\int_a^\infty g(x)\,dx$ diverges. Then if $f(x) \geq g(x)$ for all $x \geq a$, $\displaystyle\int_a^\infty f(x)\,dx$ also diverges.

   Example: Since $\dfrac{1}{\ln x} > \dfrac{1}{x}$ for $x \geq 2$ and $\displaystyle\int_2^\infty \frac{dx}{x}$ diverges (p integral with $p=1$), $\displaystyle\int_2^\infty \frac{dx}{\ln x}$ also diverges.

2. **Quotient test** for integrals with non-negative integrands.

(a) If $f(x) \geqq 0$ and $g(x) \geqq 0$, and if $\lim\limits_{x \to \infty} \dfrac{f(x)}{g(x)} = A \neq 0$ or ∞, then $\displaystyle\int_a^\infty f(x)\,dx$ and $\displaystyle\int_a^\infty g(x)\,dx$ either both converge or both diverge.

(b) If $A = 0$ in (a) and $\displaystyle\int_a^\infty g(x)\,dx$ converges, then $\displaystyle\int_a^\infty f(x)\,dx$ converges.

(c) If $A = \infty$ in (a) and $\displaystyle\int_a^\infty g(x)\,dx$ diverges, then $\displaystyle\int_a^\infty f(x)\,dx$ diverges.

This test is related to the comparison test and is often a very useful alternative to it. In particular, taking $g(x) = 1/x^p$, we have from known facts about the p integral, the

Theorem 1. Let $\lim\limits_{x \to \infty} x^p f(x) = A$. Then

(i) $\displaystyle\int_a^\infty f(x)\,dx$ converges if $p > 1$ and A is finite

(ii) $\displaystyle\int_a^\infty f(x)\,dx$ diverges if $p \leqq 1$ and $A \neq 0$ (A may be infinite).

Example 1: $\displaystyle\int_0^\infty \dfrac{x^2\,dx}{4x^4 + 25}$ converges since $\lim\limits_{x \to \infty} x^2 \cdot \dfrac{x^2}{4x^4 + 25} = \dfrac{1}{4}$.

Example 2: $\displaystyle\int_0^\infty \dfrac{x\,dx}{\sqrt{x^4 + x^2 + 1}}$ diverges since $\lim\limits_{x \to \infty} x \cdot \dfrac{x}{\sqrt{x^4 + x^2 + 1}} = 1$.

Similar tests can be devised using $g(x) = e^{-tx}$.

3. **Series test** for integrals with non-negative integrands. $\displaystyle\int_a^\infty f(x)\,dx$ converges or diverges according as Σu_n, where $u_n = f(n)$, converges or diverges.

4. **Absolute and conditional convergence.** $\displaystyle\int_a^\infty f(x)\,dx$ is called *absolutely convergent* if $\displaystyle\int_a^\infty |f(x)|\,dx$ converges. If $\displaystyle\int_a^\infty f(x)\,dx$ converges but $\displaystyle\int_a^\infty |f(x)|\,dx$ diverges, then $\displaystyle\int_a^\infty f(x)\,dx$ is called *conditionally convergent.*

Theorem 2. If $\displaystyle\int_a^\infty |f(x)|\,dx$ converges, then $\displaystyle\int_a^\infty f(x)\,dx$ converges. In words, an absolutely convergent integral converges.

Example 1: $\displaystyle\int_0^\infty \dfrac{\cos x}{x^2 + 1}\,dx$ is absolutely convergent and thus convergent since $\displaystyle\int_0^\infty \left| \dfrac{\cos x}{x^2 + 1} \right|\,dx \leqq$ $\displaystyle\int_0^\infty \dfrac{dx}{x^2 + 1}$ and $\displaystyle\int_0^\infty \dfrac{dx}{x^2 + 1}$ converges.

Example 2: $\displaystyle\int_0^\infty \dfrac{\sin x}{x}\,dx$ converges (see Prob. 11), but $\displaystyle\int_0^\infty \left| \dfrac{\sin x}{x} \right|\,dx$ does not converge (see Prob. 12).

Thus $\displaystyle\int_0^\infty \dfrac{\sin x}{x}\,dx$ is conditionally convergent.

Any of the tests used for integrals with non-negative integrands can be used to test for absolute convergence.

IMPROPER INTEGRALS of the SECOND KIND

If $f(x)$ becomes unbounded only at the end point $x = a$ of the interval $a \leqq x \leqq b$, then we define

$$\int_a^b f(x)\, dx = \lim_{\epsilon \to 0+} \int_{a+\epsilon}^b f(x)\, dx \tag{4}$$

If the limit on the right of (4) exists, we call the integral on the left *convergent*; otherwise it is *divergent*.

Similarly if $f(x)$ becomes unbounded only at the end point $x = b$ of the interval $a \leqq x \leqq b$, then we define

$$\int_a^b f(x)\, dx = \lim_{\epsilon \to 0+} \int_a^{b-\epsilon} f(x)\, dx \tag{5}$$

In such case the integral on the left of (5) is called convergent or divergent according as the limit on the right exists or does not exist.

If $f(x)$ becomes unbounded only at an interior point $x = x_0$ of the interval $a \leqq x \leqq b$, then we define

$$\int_a^b f(x)\, dx = \lim_{\epsilon_1 \to 0+} \int_a^{x_0 - \epsilon_1} f(x)\, dx + \lim_{\epsilon_2 \to 0+} \int_{x_0 + \epsilon_2}^b f(x)\, dx \tag{6}$$

The integral on the left of (6) converges or diverges according as the limits on the right exist or do not exist.

Extensions of these definitions can be made in case $f(x)$ becomes unbounded at two or more points of the interval $a \leqq x \leqq b$.

CAUCHY PRINCIPAL VALUE

It may happen that the limits on the right of (6) do not exist when ϵ_1 and ϵ_2 approach zero independently. In such case it is possible that by choosing $\epsilon_1 = \epsilon_2 = \epsilon$ in (6), i.e. writing

$$\int_a^b f(x)\, dx = \lim_{\epsilon \to 0+} \left\{ \int_a^{x_0 - \epsilon} f(x)\, dx + \int_{x_0 + \epsilon}^b f(x)\, dx \right\} \tag{7}$$

the limit does exist. If the limit on the right of (7) does exist, we call this limiting value the *Cauchy principal value* of the integral on the left. See Problem 14.

SPECIAL IMPROPER INTEGRALS of the SECOND KIND

1. $\displaystyle\int_a^b \frac{dx}{(x-a)^p}$ converges if $p < 1$ and diverges if $p \geqq 1$.

2. $\displaystyle\int_a^b \frac{dx}{(b-x)^p}$ converges if $p < 1$ and diverges if $p \geqq 1$.

These can be called *p integrals of the second kind*. Note that when $p \leqq 0$ the integrals are proper.

CONVERGENCE TESTS for IMPROPER INTEGRALS of the SECOND KIND

The following tests are given for the case where $f(x)$ is unbounded only at $x = a$ in the interval $a \leqq x \leqq b$. Similar tests are available if $f(x)$ is unbounded at $x = b$ or at $x = x_0$ where $a < x_0 < b$.

1. Comparison test for integrals with non-negative integrands.

(a) *Convergence.* Let $g(x) \geqq 0$ for $a < x \leqq b$, and suppose that $\int_a^b g(x)\,dx$ converges. Then if $0 \leqq f(x) \leqq g(x)$ for $a < x \leqq b$, $\int_a^b f(x)\,dx$ also converges.

Example: $\dfrac{1}{\sqrt{x^4-1}} < \dfrac{1}{\sqrt{x-1}}$ for $x > 1$. Then since $\int_1^5 \dfrac{dx}{\sqrt{x-1}}$ converges (p integral with $a = 1$,

$p = \frac{1}{2}$), $\int_1^5 \dfrac{dx}{\sqrt{x^4-1}}$ also converges.

(b) *Divergence.* Let $g(x) \geqq 0$ for $a < x \leqq b$, and suppose that $\int_a^b g(x)\,dx$ diverges. Then if $f(x) \geqq g(x)$ for $a < x \leqq b$, $\int_a^b f(x)\,dx$ also diverges.

Example: $\dfrac{\ln x}{(x-3)^4} > \dfrac{1}{(x-3)^4}$ for $x > 3$. Then since $\int_3^6 \dfrac{dx}{(x-3)^4}$ diverges (p integral with $a = 3$,

$p = 4$), $\int_3^6 \dfrac{\ln x}{(x-3)^4}\,dx$ also diverges.

2. Quotient test for integrals with non-negative integrands.

(a) If $f(x) \geqq 0$ and $g(x) \geqq 0$ for $a < x \leqq b$, and if $\lim\limits_{x \to a} \dfrac{f(x)}{g(x)} = A \neq 0$ or ∞, then $\int_a^b f(x)\,dx$ and $\int_a^b g(x)\,dx$ either both converge or both diverge.

(b) If $A = 0$ in (a), and $\int_a^b g(x)\,dx$ converges, then $\int_a^b f(x)\,dx$ converges.

(c) If $A = \infty$ in (a), and $\int_a^b g(x)\,dx$ diverges, then $\int_a^b f(x)\,dx$ diverges.

This test is related to the comparison test and is a very useful alternative to it. In particular taking $g(x) = 1/(x-a)^p$ we have from known facts about the p integral the

Theorem 3. Let $\lim\limits_{x \to a+} (x-a)^p f(x) = A$. Then

(i) $\int_a^b f(x)\,dx$ converges if $p < 1$ and A is finite

(ii) $\int_a^b f(x)\,dx$ diverges if $p \geqq 1$ and $A \neq 0$ (A may be infinite).

If $f(x)$ becomes unbounded only at the upper limit these conditions are replaced by those in

Theorem 4. Let $\lim\limits_{x \to b-} (b-x)^p f(x) = B$. Then

(i) $\int_a^b f(x)\,dx$ converges if $p < 1$ and B is finite

(ii) $\int_a^b f(x)\,dx$ diverges if $p \geqq 1$ and $B \neq 0$ (B may be infinite).

Example 1: $\int_1^5 \dfrac{dx}{\sqrt{x^4-1}}$ converges since $\lim\limits_{x \to 1+} (x-1)^{1/2} \cdot \dfrac{1}{(x^4-1)^{1/2}} = \lim\limits_{x \to 1+} \sqrt{\dfrac{x-1}{x^4-1}} = \dfrac{1}{2}$.

Example 2: $\int_0^3 \dfrac{dx}{(3-x)\sqrt{x^2+1}}$ diverges since $\lim\limits_{x \to 3-} (3-x) \cdot \dfrac{1}{(3-x)\sqrt{x^2+1}} = \dfrac{1}{\sqrt{10}}$.

3. Absolute and conditional convergence. $\int_a^b f(x)\,dx$ is called *absolutely convergent* if $\int_a^b |f(x)|\,dx$ converges. If $\int_a^b f(x)\,dx$ converges but $\int_a^b |f(x)|\,dx$ diverges, then $\int_a^b f(x)\,dx$ is called *conditionally convergent*.

Theorem 5.

If $\int_a^b |f(x)|\,dx$ converges, then $\int_a^b f(x)\,dx$ converges. In words, an absolutely convergent integral converges.

> **Example:** Since $\left| \dfrac{\sin x}{\sqrt[3]{x-\pi}} \right| \leqq \dfrac{1}{\sqrt[3]{x-\pi}}$ and $\int_\pi^{4\pi} \dfrac{dx}{\sqrt[3]{x-\pi}}$ converges (p integral with $a = \pi$, $p = \tfrac{1}{3}$),
>
> it follows that $\int_\pi^{4\pi} \left| \dfrac{\sin x}{\sqrt[3]{x-\pi}} \right| dx$ converges and thus $\int_\pi^{4\pi} \dfrac{\sin x}{\sqrt[3]{x-\pi}}\,dx$ converges (absolutely).

Any of the tests used for integrals with non-negative integrands can be used to test for absolute convergence.

IMPROPER INTEGRALS of the THIRD KIND

Improper integrals of the third kind can be expressed in terms of improper integrals of the first and second kinds, and hence the question of their convergence or divergence is answered by using results already established.

IMPROPER INTEGRALS CONTAINING a PARAMETER. UNIFORM CONVERGENCE

Let

$$\phi(\alpha) \;=\; \int_a^\infty f(x, \alpha)\,dx \qquad\qquad (8)$$

This integral is analogous to an infinite series of functions. In seeking conditions under which we may differentiate or integrate $\phi(\alpha)$ with respect to α, it is convenient to introduce the concept of *uniform convergence* for integrals by analogy with infinite series.

We shall suppose that the integral (8) converges for $\alpha_1 \leqq \alpha \leqq \alpha_2$, or briefly $[\alpha_1, \alpha_2]$.

Definition.

The integral (8) is said to be *uniformly convergent* in $[\alpha_1, \alpha_2]$ if for each $\epsilon > 0$ we can find a number N depending on ϵ but not on α such that

$$\left| \phi(\alpha) - \int_a^u f(x, \alpha)\,dx \right| < \epsilon \quad \text{for all } u > N \text{ and all } \alpha \text{ in } [\alpha_1, \alpha_2]$$

This can be restated by noting that $\left| \phi(\alpha) - \int_a^u f(x, \alpha)\,dx \right| = \left| \int_u^\infty f(x, \alpha)\,dx \right|$ which is analogous in an infinite series to the absolute value of the remainder after N terms.

The above definition and the properties of uniform convergence to be developed are formulated in terms of improper integrals of the first kind. However, analogous results can be given for improper integrals of the second and third kinds.

SPECIAL TESTS for UNIFORM CONVERGENCE of INTEGRALS

1. Weierstrass M test. If we can find a function $M(x) \geqq 0$ such that

(a) $|f(x, \alpha)| \leqq M(x)$ $\alpha_1 \leqq \alpha \leqq \alpha_2,\ x > a$

(b) $\displaystyle\int_a^\infty M(x)\, dx$ converges,

then $\displaystyle\int_a^\infty f(x, \alpha)\, dx$ is uniformly and absolutely convergent in $\alpha_1 \leqq \alpha \leqq \alpha_2$.

Example: Since $\left| \dfrac{\cos \alpha x}{x^2 + 1} \right| \leqq \dfrac{1}{x^2 + 1}$ and $\displaystyle\int_0^\infty \dfrac{dx}{x^2 + 1}$ converges, it follows that $\displaystyle\int_0^\infty \dfrac{\cos \alpha x}{x^2 + 1}\, dx$ is uniformly and absolutely convergent for all real values of α.

As in the case of infinite series, it is possible for integrals to be uniformly convergent without being absolutely convergent, and conversely.

2. Dirichlet's test.   Suppose that

(a) $\psi(x)$ is a positive monotonic decreasing function which approaches zero as $x \to \infty$

(b) $\left| \displaystyle\int_a^u f(x, \alpha)\, dx \right| < P$ for all $u > a$ and $\alpha_1 \leqq \alpha \leqq \alpha_2$.

Then the integral $\displaystyle\int_a^\infty f(x, \alpha)\, \psi(x)\, dx$

is uniformly convergent for $\alpha_1 \leqq \alpha \leqq \alpha_2$.

THEOREMS on UNIFORMLY CONVERGENT INTEGRALS

Theorem 6.

If $f(x, \alpha)$ is continuous for $x \geqq a$ and $\alpha_1 \leqq \alpha \leqq \alpha_2$, and if $\displaystyle\int_a^\infty f(x, \alpha)\, dx$ is uniformly convergent for $\alpha_1 \leqq \alpha \leqq \alpha_2$, then $\phi(\alpha) = \displaystyle\int_a^\infty f(x, \alpha)\, dx$ is continuous in $\alpha_1 \leqq \alpha \leqq \alpha_2$. In particular, if α_0 is any point of $\alpha_1 \leqq \alpha \leqq \alpha_2$, we can write

$$\lim_{\alpha \to \alpha_0} \phi(\alpha) \ = \ \lim_{\alpha \to \alpha_0} \int_a^\infty f(x, \alpha)\, dx \ = \ \int_a^\infty \lim_{\alpha \to \alpha_0} f(x, \alpha)\, dx \tag{9}$$

If α_0 is one of the end points, we use right or left hand limits.

Theorem 7.

Under the conditions of Theorem 6, we can integrate $\phi(\alpha)$ with respect to α from α_1 to α_2 to obtain

$$\int_{\alpha_1}^{\alpha_2} \phi(\alpha)\, d\alpha \ = \ \int_{\alpha_1}^{\alpha_2} \left\{ \int_a^\infty f(x, \alpha)\, dx \right\} d\alpha \ = \ \int_a^\infty \left\{ \int_{\alpha_1}^{\alpha_2} f(x, \alpha)\, d\alpha \right\} dx \tag{10}$$

which corresponds to a change of the order of integration.

Theorem 8.

If $f(x, \alpha)$ is continuous and has a continuous partial derivative with respect to α for $x \geqq a$ and $\alpha_1 \leqq \alpha \leqq \alpha_2$, and if $\displaystyle\int_a^\infty \dfrac{\partial f}{\partial \alpha}\, dx$ converges uniformly in $\alpha_1 \leqq \alpha \leqq \alpha_2$, then if a does not depend on α,

$$\frac{d\phi}{d\alpha} \ = \ \int_a^\infty \frac{\partial f}{\partial \alpha}\, dx \tag{11}$$

If a depends on α this result is easily modified (see Leibnitz's rule, Page 163).

EVALUATION of DEFINITE INTEGRALS

Evaluation of definite integrals which are improper can be achieved by a variety of techniques. One useful device consists of introducing an appropriately placed parameter in the integral and then differentiating or integrating with respect to the parameter, employing the above properties of uniform convergence.

LAPLACE TRANSFORMS

The Laplace transform of a function $F(x)$ is defined as

$$f(s) = \mathcal{L}\{F(x)\} = \int_0^\infty e^{-sx} F(x)\, dx \quad (12)$$

and is analogous to power series as seen by replacing e^{-s} by t so that $e^{-sx} = t^x$. Many properties of power series also apply to Laplace transforms. The adjacent short table of Laplace transforms is useful. In each case a is a real constant.

One useful application of Laplace transforms is to the solution of differential equations (see Problems 34-36).

$F(x)$	$\mathcal{L}\{F(x)\}$	
a	$\dfrac{a}{s}$	$s > 0$
e^{ax}	$\dfrac{1}{s-a}$	$s > a$
$\sin ax$	$\dfrac{a}{s^2 + a^2}$	$s > 0$
$\cos ax$	$\dfrac{s}{s^2 + a^2}$	$s > 0$
$x^n \quad n = 1, 2, 3, \ldots$	$\dfrac{n!}{s^{n+1}}$	$s > 0$
$Y'(x)$	$s\mathcal{L}\{Y(x)\} - Y(0)$	
$Y''(x)$	$s^2\mathcal{L}\{Y(x)\} - s\,Y(0) - Y'(0)$	

IMPROPER MULTIPLE INTEGRALS

The definitions and results for improper single integrals can be extended to improper multiple integrals.

Solved Problems

IMPROPER INTEGRALS

1. Classify according to the type of improper integral.

(a) $\displaystyle\int_{-1}^{1} \frac{dx}{\sqrt[3]{x}\,(x+1)}$ (c) $\displaystyle\int_{3}^{10} \frac{x\, dx}{(x-2)^2}$ (e) $\displaystyle\int_{0}^{\pi} \frac{1 - \cos x}{x^2}\, dx$

(b) $\displaystyle\int_{0}^{\infty} \frac{dx}{1 + \tan x}$ (d) $\displaystyle\int_{-\infty}^{\infty} \frac{x^2\, dx}{x^4 + x^2 + 1}$

(a) *Second kind* (integrand is unbounded at $x = 0$ and $x = -1$).

(b) *Third kind* (integration limit is infinite and integrand is unbounded where $\tan x = -1$).

(c) This is a *proper* integral (integrand becomes unbounded at $x = 2$, but this is *outside* the range of integration $3 \le x \le 10$).

(d) *First kind* (integration limits are infinite but integrand is bounded).

(e) This is a *proper* integral (since $\displaystyle\lim_{x \to 0+} \frac{1 - \cos x}{x^2} = \frac{1}{2}$ by applying L'Hospital's rule).

2. Show how to transform the improper integral of the second kind, $\displaystyle\int_1^2 \frac{dx}{\sqrt{x(2-x)}}$, into (a) an improper integral of the first kind, (b) a proper integral.

(a) Consider $\displaystyle\int_1^{2-\epsilon} \frac{dx}{\sqrt{x(2-x)}}$ where $0 < \epsilon < 1$, say. Let $2 - x = \dfrac{1}{y}$. Then the integral becomes $\displaystyle\int_1^{1/\epsilon} \frac{dy}{y\sqrt{2y-1}}$. As $\epsilon \to 0+$, we see that consideration of the given integral is equivalent to consideration of $\displaystyle\int_1^\infty \frac{dy}{y\sqrt{2y-1}}$ which is an improper integral of the first kind.

(b) Letting $2 - x = v^2$ in the integral of (a), it becomes $2\displaystyle\int_{\sqrt\epsilon}^1 \frac{dv}{\sqrt{v^2+2}}$. We are thus led to consideration of $2\displaystyle\int_0^1 \frac{dv}{\sqrt{v^2+2}}$ which is a proper integral.

From the above we see that an improper integral of the first kind *may* be transformed into an improper integral of the second kind, and conversely (actually this can *always* be done).

We also see that an improper integral may be transformed into a proper integral (this can only *sometimes* be done).

IMPROPER INTEGRALS of the FIRST KIND

3. Prove the comparison test (Page 261) for convergence of improper integrals of the first kind.

Since $0 \leqq f(x) \leqq g(x)$ for $x \geqq a$, we have using Property 7, Page 81,
$$0 \;\leqq\; \int_a^b f(x)\,dx \;\leqq\; \int_a^b g(x)\,dx \;\leqq\; \int_a^\infty g(x)\,dx$$
But by hypothesis the last integral exists. Thus
$$\lim_{b\to\infty} \int_a^b f(x)\,dx \text{ exists, and hence } \int_a^\infty f(x)\,dx \text{ converges}$$

4. Prove the quotient test (a) on Page 262.

By hypothesis, $\displaystyle\lim_{x\to\infty} \frac{f(x)}{g(x)} = A > 0$. Then given any $\epsilon > 0$, we can find N such that $\left|\dfrac{f(x)}{g(x)} - A\right| < \epsilon$ when $x \geqq N$. Thus for $x \geqq N$, we have
$$A - \epsilon \;\leqq\; \frac{f(x)}{g(x)} \;\leqq\; A + \epsilon \qquad \text{or} \qquad (A-\epsilon)\,g(x) \;\leqq\; f(x) \;\leqq\; (A+\epsilon)\,g(x)$$
Then
$$(A-\epsilon)\int_N^b g(x)\,dx \;\leqq\; \int_N^b f(x)\,dx \;\leqq\; (A+\epsilon)\int_N^b g(x)\,dx \tag{1}$$
There is no loss of generality in choosing $A - \epsilon > 0$.

If $\displaystyle\int_a^\infty g(x)\,dx$ converges, then by the inequality on the right of (1),
$$\lim_{b\to\infty} \int_N^b f(x)\,dx \text{ exists, and so } \int_a^\infty f(x)\,dx \text{ converges}$$
If $\displaystyle\int_a^\infty g(x)\,dx$ diverges, then by the inequality on the left of (1),
$$\lim_{b\to\infty} \int_N^b f(x)\,dx \;=\; \infty \quad \text{and so} \quad \int_a^\infty f(x)\,dx \text{ diverges}$$

For the cases where $A = 0$ and $A = \infty$, see Problem 41.

As seen in this and the preceding problem, there is in general a marked similarity between proofs for infinite series and improper integrals.

5. Test for convergence: (a) $\displaystyle\int_1^\infty \frac{x\,dx}{3x^4+5x^2+1}$, (b) $\displaystyle\int_2^\infty \frac{x^2-1}{\sqrt{x^6+16}}\,dx$.

(a) **Method 1:** For large x, the integrand is approximately $x/3x^4 = 1/3x^3$.

Since $\dfrac{x}{3x^4+5x^2+1} \leqq \dfrac{1}{3x^3}$, and $\dfrac{1}{3}\displaystyle\int_1^\infty \frac{dx}{x^3}$ converges (p integral with $p=3$), it follows

by the comparison test that $\displaystyle\int_1^\infty \frac{x\,dx}{3x^4+5x^2+1}$ also converges.

Note that the purpose of examining the integrand for large x is to obtain a suitable comparison integral.

Method 2: Let $f(x) = \dfrac{x}{3x^4+5x^2+1}$, $g(x) = \dfrac{1}{x^3}$. Since $\displaystyle\lim_{x\to\infty} \frac{f(x)}{g(x)} = \frac{1}{3}$, and $\displaystyle\int_1^\infty g(x)\,dx$ con-

verges, $\displaystyle\int_1^\infty f(x)\,dx$ also converges by the quotient test.

Note that in the comparison function $g(x)$, we have discarded the factor $\frac{1}{3}$. It could, however, just as well have been included.

Method 3: $\displaystyle\lim_{x\to\infty} x^3\left(\frac{x}{3x^4+5x^2+1}\right) = \frac{1}{3}$.   Hence by Theorem 1, Page 262, the required integral converges.

(b) **Method 1:**   For large x, the integrand is approximately $x^2/\sqrt{x^6} = 1/x$.

For $x \geqq 2$, $\dfrac{x^2-1}{\sqrt{x^6+1}} \geqq \dfrac{1}{2}\cdot\dfrac{1}{x}$. Since $\dfrac{1}{2}\displaystyle\int_2^\infty \frac{dx}{x}$ diverges, $\displaystyle\int_2^\infty \frac{x^2-1}{\sqrt{x^6+16}}\,dx$ also diverges.

Method 2: Let $f(x) = \dfrac{x^2-1}{\sqrt{x^6+16}}$, $g(x) = \dfrac{1}{x}$.   Then since $\displaystyle\lim_{x\to\infty} \frac{f(x)}{g(x)} = 1$, and $\displaystyle\int_2^\infty g(x)\,dx$

diverges, $\displaystyle\int_2^\infty f(x)\,dx$ also diverges.

Method 3: Since $\displaystyle\lim_{x\to\infty} x\left(\frac{x^2-1}{\sqrt{x^6+16}}\right) = 1$,   the required integral diverges by Theorem 1, Page 262.

Note that Method 1 may (and often does) require one to obtain a suitable inequality factor (in this case $\frac{1}{2}$, or any positive constant less than $\frac{1}{2}$) before the comparison test can be applied. Methods 2 and 3, however, do not require this.

6. Prove that $\displaystyle\int_0^\infty e^{-x^2}\,dx$ converges.

$\displaystyle\lim_{x\to\infty} x^2 e^{-x^2} = 0$ (by L'Hospital's rule or otherwise). Then by Theorem 1, with $A=0$, $p=2$, the given integral converges. Compare Problem 10(a), Chapter 11.

7. Examine for convergence:

(a) $\displaystyle\int_1^\infty \frac{\ln x}{x+a}\,dx$, where a is a positive constant; (b) $\displaystyle\int_0^\infty \frac{1-\cos x}{x^2}\,dx$.

(a) $\displaystyle\lim_{x\to\infty} x\cdot\frac{\ln x}{x+a} = \infty$. Hence by Theorem 1, Page 262, with $A=\infty$, $p=1$, the given integral diverges.

(b) $\displaystyle\int_0^\infty \frac{1-\cos x}{x^2}\,dx = \int_0^\pi \frac{1-\cos x}{x^2}\,dx + \int_\pi^\infty \frac{1-\cos x}{x^2}\,dx$

The first integral on the right converges [see Problem 1(e)].

Since $\displaystyle\lim_{x\to\infty} x^{3/2}\left(\frac{1-\cos x}{x^2}\right) = 0$, the second integral on the right converges by Theorem 1, Page 262, with $A=0$ and $p=3/2$.

Thus the given integral converges.

8. Test for convergence: $\quad (a)\ \displaystyle\int_{-\infty}^{-1} \frac{e^x}{x}\,dx, \quad (b)\ \displaystyle\int_{-\infty}^{\infty} \frac{x^3+x^2}{x^6+1}\,dx.$

(a) Let $x = -y$. Then the integral becomes $\ -\displaystyle\int_{1}^{\infty} \frac{e^{-y}}{y}\,dy.$

Method 1: $\ \dfrac{e^{-y}}{y} \leqq e^{-y}$ for $y \geqq 1$. Then since $\displaystyle\int_{1}^{\infty} e^{-y}\,dy$ converges, $\displaystyle\int_{1}^{\infty} \frac{e^{-y}}{y}\,dy$ converges; hence the given integral converges.

Method 2: $\ \displaystyle\lim_{y\to\infty} y^2\!\left(\frac{e^{-y}}{y}\right) = \lim_{y\to\infty} y e^{-y} = 0.$ Then the given integral converges by Theorem 1, Page 262, with $A = 0$ and $p = 2$.

(b) Write the given integral as $\ \displaystyle\int_{-\infty}^{0} \frac{x^3+x^2}{x^6+1}\,dx \ + \ \displaystyle\int_{0}^{\infty} \frac{x^3+x^2}{x^6+1}\,dx.$ Letting $x = -y$ in the first integral, it becomes $\ -\displaystyle\int_{0}^{\infty} \frac{y^3-y^2}{y^6+1}\,dy.$ Since $\displaystyle\lim_{y\to\infty} y^3\!\left(\frac{y^3-y^2}{y^6+1}\right) = 1,$ this integral converges.

Since $\displaystyle\lim_{x\to\infty} x^3\!\left(\frac{x^3+x^2}{x^6+1}\right) = 1,$ the second integral converges.

Thus the given integral converges.

ABSOLUTE and CONDITIONAL CONVERGENCE for IMPROPER INTEGRALS of the FIRST KIND

9. Prove that $\displaystyle\int_{a}^{\infty} f(x)\,dx$ converges if $\displaystyle\int_{a}^{\infty} |f(x)|\,dx$ converges, i.e. an absolutely convergent integral is convergent.

We have $\ -|f(x)| \leqq f(x) \leqq |f(x)|,$ i.e. $\ 0 \leqq f(x) + |f(x)| \leqq 2|f(x)|.$ Then

$$0 \ \leqq \ \int_{a}^{b} [f(x) + |f(x)|]\,dx \ \leqq \ 2\int_{a}^{b} |f(x)|\,dx$$

If $\displaystyle\int_{a}^{\infty} |f(x)|\,dx$ converges, it follows that $\displaystyle\int_{a}^{\infty} [f(x) + |f(x)|]\,dx$ converges. Hence by subtracting $\displaystyle\int_{a}^{\infty} |f(x)|\,dx$, which converges, we see that $\displaystyle\int_{a}^{\infty} f(x)\,dx$ converges.

10. Prove that $\displaystyle\int_{1}^{\infty} \frac{\cos x}{x^2}\,dx$ converges.

Method 1:

$\left|\dfrac{\cos x}{x^2}\right| \leqq \dfrac{1}{x^2}$ for $x \geqq 1$. Then by the comparison test, since $\displaystyle\int_{1}^{\infty} \frac{dx}{x^2}$ converges, it follows that $\displaystyle\int_{1}^{\infty} \left|\frac{\cos x}{x^2}\right|\,dx$ converges, i.e. $\displaystyle\int_{1}^{\infty} \frac{\cos x}{x^2}\,dx$ converges absolutely, and so converges by Problem 9.

Method 2:

Since $\displaystyle\lim_{x\to\infty} x^{3/2}\left|\frac{\cos x}{x^2}\right| = \lim_{x\to\infty} \left|\frac{\cos x}{x^{1/2}}\right| = 0,$ it follows from Theorem 1, Page 262, with $A = 0$ and $p = 3/2$, that $\displaystyle\int_{1}^{\infty} \left|\frac{\cos x}{x^2}\right|\,dx$ converges, and hence $\displaystyle\int_{1}^{\infty} \frac{\cos x}{x^2}\,dx$ converges (absolutely).

11. Prove that $\int_0^\infty \dfrac{\sin x}{x}\,dx$ converges.

Since $\int_0^1 \dfrac{\sin x}{x}\,dx$ converges $\left(\text{because } \dfrac{\sin x}{x} \text{ is continuous in } 0 < x \leqq 1 \text{ and } \lim\limits_{x \to 0+} \dfrac{\sin x}{x} = 1\right)$
we need only show that $\int_1^\infty \dfrac{\sin x}{x}\,dx$ converges.

Method 1: Integration by parts yields
$$\int_1^M \frac{\sin x}{x}\,dx \;=\; -\frac{\cos x}{x}\Big|_1^M + \int_1^M \frac{\cos x}{x^2}\,dx \;=\; \cos 1 - \frac{\cos M}{M} + \int_1^M \frac{\cos x}{x^2}\,dx \tag{1}$$

or on taking the limit on both sides of (1) as $M \to \infty$ and using the fact that $\lim\limits_{M \to \infty} \dfrac{\cos M}{M} = 0$,
$$\int_1^\infty \frac{\sin x}{x}\,dx \;=\; \cos 1 + \int_1^\infty \frac{\cos x}{x^2}\,dx \tag{2}$$

Since the integral on the right of (2) converges by Problem 10, the required result follows.

The technique of integration by parts to establish convergence is often useful in practice.

Method 2:
$$\int_0^\infty \frac{\sin x}{x}\,dx \;=\; \int_0^\pi \frac{\sin x}{x}\,dx + \int_\pi^{2\pi} \frac{\sin x}{x}\,dx + \cdots + \int_{n\pi}^{(n+1)\pi} \frac{\sin x}{x}\,dx + \cdots$$
$$=\; \sum_{n=0}^\infty \int_{n\pi}^{(n+1)\pi} \frac{\sin x}{x}\,dx$$

Letting $x = v + n\pi$, the summation becomes
$$\sum_{n=0}^\infty (-1)^n \int_0^\pi \frac{\sin v}{v + n\pi}\,dv \;=\; \int_0^\pi \frac{\sin v}{v}\,dv - \int_0^\pi \frac{\sin v}{v + \pi}\,dv + \int_0^\pi \frac{\sin v}{v + 2\pi}\,dv - \cdots$$

This is an alternating series. Since $\dfrac{1}{v + n\pi} \leqq \dfrac{1}{v + (n+1)\pi}$ and $\sin v \geqq 0$ in $[0,\pi]$, it follows that
$$\int_0^\pi \frac{\sin v}{v + n\pi}\,dv \;\leqq\; \int_0^\pi \frac{\sin v}{v + (n+1)\pi}\,dv$$

Also,
$$\lim_{n \to \infty} \int_0^\pi \frac{\sin v}{v + n\pi}\,dv \;\leqq\; \lim_{n \to \infty} \int_0^\pi \frac{dv}{n\pi} \;=\; 0$$

Thus each term of the alternating series is in absolute value less than or equal to the preceding term, and the nth term approaches zero as $n \to \infty$. Hence by the alternating series test (Page 226) the series and thus the integral converges.

12. Prove that $\int_0^\infty \dfrac{\sin x}{x}\,dx$ converges conditionally.

Since by Problem 11 the given integral converges, we must show that it is not absolutely convergent, i.e. $\int_0^\infty \left|\dfrac{\sin x}{x}\right|\,dx$ diverges.

As in Problem 11, Method 2, we have
$$\int_0^\infty \left|\frac{\sin x}{x}\right|\,dx \;=\; \sum_{n=0}^\infty \int_{n\pi}^{(n+1)\pi} \left|\frac{\sin x}{x}\right|\,dx \;=\; \sum_{n=0}^\infty \int_0^\pi \frac{\sin v}{v + n\pi}\,dv \tag{1}$$

Now $\dfrac{1}{v + n\pi} \geqq \dfrac{1}{(n+1)\pi}$ for $0 \leqq v \leqq \pi$. Hence
$$\int_0^\pi \frac{\sin v}{v + n\pi}\,dv \;\geqq\; \frac{1}{(n+1)\pi} \int_0^\pi \sin v \,dv \;=\; \frac{2}{(n+1)\pi} \tag{2}$$

Since $\sum\limits_{n=0}^\infty \dfrac{2}{(n+1)\pi}$ diverges, the series on the right of (1) diverges by the comparison test.

Hence $\int_0^\infty \left|\dfrac{\sin x}{x}\right|\,dx$ diverges and the required result follows.

IMPROPER INTEGRALS of the SECOND KIND.   CAUCHY PRINCIPAL VALUE

13. (*a*) Prove that $\displaystyle\int_{-1}^{7} \frac{dx}{\sqrt[3]{x+1}}$ converges and (*b*) find its value.

The integrand is unbounded at $x = -1$. Then we define the integral as

$$\lim_{\epsilon \to 0+} \int_{-1+\epsilon}^{7} \frac{dx}{\sqrt[3]{x+1}} \;=\; \lim_{\epsilon \to 0+} \frac{(x+1)^{2/3}}{2/3} \Big|_{-1+\epsilon}^{7} \;=\; \lim_{\epsilon \to 0+} \left(6 - \frac{3}{2}\epsilon^{2/3}\right) \;=\; 6$$

This shows that the integral converges to 6.

14. Determine whether $\displaystyle\int_{-1}^{5} \frac{dx}{(x-1)^3}$ converges (*a*) in the usual sense, (*b*) in the Cauchy principal value sense.

(*a*) By definition,

$$\int_{-1}^{5} \frac{dx}{(x-1)^3} \;=\; \lim_{\epsilon_1 \to 0+} \int_{-1}^{1-\epsilon_1} \frac{dx}{(x-1)^3} + \lim_{\epsilon_2 \to 0+} \int_{1+\epsilon_2}^{5} \frac{dx}{(x-1)^3}$$

$$=\; \lim_{\epsilon_1 \to 0+} \left(\frac{1}{8} - \frac{1}{2\epsilon_1^2}\right) + \lim_{\epsilon_2 \to 0+} \left(\frac{1}{2\epsilon_2^2} - \frac{1}{32}\right)$$

and since the limits do not exist, the integral does not converge in the usual sense.

(*b*) Since

$$\lim_{\epsilon \to 0+} \left\{ \int_{-1}^{1-\epsilon} \frac{dx}{(x-1)^3} + \int_{1+\epsilon}^{5} \frac{dx}{(x-1)^3} \right\} \;=\; \lim_{\epsilon \to 0+} \left\{ \frac{1}{8} - \frac{1}{2\epsilon^2} + \frac{1}{2\epsilon^2} - \frac{1}{32} \right\} \;=\; \frac{3}{32}$$

the integral exists in the Cauchy principal value sense. The principal value is 3/32.

15. Investigate the convergence of:

(*a*) $\displaystyle\int_{2}^{3} \frac{dx}{x^2(x^3-8)^{2/3}}$ (*c*) $\displaystyle\int_{1}^{5} \frac{dx}{\sqrt{(5-x)(x-1)}}$ (*e*) $\displaystyle\int_{0}^{\pi/2} \frac{dx}{(\cos x)^{1/n}}$, $n > 1$.

(*b*) $\displaystyle\int_{0}^{\pi} \frac{\sin x}{x^3}\, dx$ (*d*) $\displaystyle\int_{-1}^{1} \frac{2^{\sin^{-1} x}}{1-x}\, dx$

(*a*) $\displaystyle\lim_{x \to 2+} (x-2)^{2/3} \cdot \frac{1}{x^2(x^3-8)^{2/3}} = \lim_{x \to 2+} \frac{1}{x^2}\left(\frac{1}{x^2+2x+4}\right)^{2/3} = \frac{1}{8\sqrt[3]{18}}.$ Hence the integral converges by Theorem 3(*i*), Page 264.

(*b*) $\displaystyle\lim_{x \to 0+} x^2 \cdot \frac{\sin x}{x^3} = 1.$ Hence the integral diverges by Theorem 3(*ii*) on Page 264.

(*c*)   Write the integral as $\displaystyle\int_{1}^{3} \frac{dx}{\sqrt{(5-x)(x-1)}} + \int_{3}^{5} \frac{dx}{\sqrt{(5-x)(x-1)}}.$

      Since $\displaystyle\lim_{x \to 1+} (x-1)^{1/2} \cdot \frac{1}{\sqrt{(5-x)(x-1)}} = \frac{1}{2}$, the first integral converges.

      Since $\displaystyle\lim_{x \to 5-} (5-x)^{1/2} \cdot \frac{1}{\sqrt{(5-x)(x-1)}} = \frac{1}{2}$, the second integral converges.

      Thus the given integral converges.

(*d*) $\displaystyle\lim_{x \to 1-} (1-x) \cdot \frac{2^{\sin^{-1} x}}{1-x} = 2^{\pi/2}.$ Hence the integral diverges.

Another method:

$$\frac{2^{\sin^{-1} x}}{1-x} \geqq \frac{2^{-\pi/2}}{1-x}, \quad \text{and} \quad \int_{-1}^{1} \frac{dx}{1-x} \quad \text{diverges.} \quad \text{Hence the given integral diverges.}$$

(e) $\lim\limits_{x \to \frac{1}{2}\pi^-} (\pi/2 - x)^{1/n} \cdot \dfrac{1}{(\cos x)^{1/n}} = \lim\limits_{x \to \frac{1}{2}\pi^-} \left(\dfrac{\pi/2 - x}{\cos x}\right)^{1/n} = 1.$ Hence the integral converges.

16. If m and n are real numbers, prove that $\displaystyle\int_0^1 x^{m-1}(1-x)^{n-1} dx$ (a) converges if $m > 0$ and $n > 0$ simultaneously and (b) diverges otherwise.

(a) For $m \geqq 1$ and $n \geqq 1$ simultaneously, the integral converges since the integrand is continuous in $0 \leqq x \leqq 1$. Write the integral as

$$\int_0^{1/2} x^{m-1}(1-x)^{n-1} dx \;+\; \int_{1/2}^1 x^{m-1}(1-x)^{n-1} dx \tag{1}$$

If $0 < m < 1$ and $0 < n < 1$, the first integral converges since $\lim\limits_{x \to 0+} x^{1-m} \cdot x^{m-1}(1-x)^{n-1} = 1$, using Theorem 3(i), Page 264, with $p = 1 - m$ and $a = 0$.

Similarly, the second integral converges since $\lim\limits_{x \to 1-} (1-x)^{1-n} \cdot x^{m-1}(1-x)^{n-1} = 1$, using Theorem 4(i), Page 264, with $p = 1 - n$ and $b = 1$.

Thus the given integral converges if $m > 0$ and $n > 0$ simultaneously.

(b) If $m \leqq 0$, $\lim\limits_{x \to 0+} x \cdot x^{m-1}(1-x)^{n-1} = \infty$. Hence the first integral in (1) diverges, regardless of the value of n, by Theorem 3(ii), Page 264, with $p = 1$ and $a = 0$.

Similarly, the second integral diverges if $n \leqq 0$ regardless of the value of m, and the required result follows.

Some interesting properties of the given integral, called the *beta integral* or *beta function*, are considered in Chapter 13.

17. Prove that $\displaystyle\int_0^\pi \frac{1}{x} \sin \frac{1}{x} dx$ converges conditionally.

Letting $x = 1/y$, the integral becomes $\displaystyle\int_{1/\pi}^\infty \frac{\sin y}{y} dy$ and the required result follows from Prob. 12.

IMPROPER INTEGRALS of the THIRD KIND

18. If n is a real number, prove that $\displaystyle\int_0^\infty x^{n-1} e^{-x} dx$ (a) converges if $n > 0$ and (b) diverges if $n \leqq 0$.

Write the integral as

$$\int_0^1 x^{n-1} e^{-x} dx \;+\; \int_1^\infty x^{n-1} e^{-x} dx \tag{1}$$

(a) If $n \geqq 1$, the first integral in (1) converges since the integrand is continuous in $0 \leqq x \leqq 1$.

If $0 < n < 1$, the first integral in (1) is an improper integral of the second kind at $x = 0$. Since $\lim\limits_{x \to 0+} x^{1-n} \cdot x^{n-1} e^{-x} = 1$, the integral converges by Theorem 3(i), Page 264, with $p = 1 - n$ and $a = 0$.

Thus the first integral converges for $n > 0$.

If $n > 0$, the second integral in (1) is an improper integral of the first kind. Since $\lim\limits_{x \to \infty} x^2 \cdot x^{n-1} e^{-x} = 0$ (by L'Hospital's rule or otherwise), this integral converges by Theorem 1(i), Page 262, with $p = 2$.

Thus the second integral also converges for $n > 0$, and so the given integral converges for $n > 0$.

(b) If $n \leqq 0$, the first integral of (1) diverges since $\lim\limits_{x \to 0+} x \cdot x^{n-1} e^{-x} = \infty$ [Theorem 3(ii), Page 264].

If $n \leqq 0$, the second integral of (1) converges since $\lim\limits_{x \to \infty} x \cdot x^{n-1} e^{-x} = 0$ [Theorem 1(i), Page 262].

Since the first integral in (1) diverges while the second integral converges, their sum also diverges, i.e. the given integral diverges if $n \leqq 0$.

Some interesting properties of the given integral, called the *gamma function*, are considered in Chapter 13.

UNIFORM CONVERGENCE of IMPROPER INTEGRALS

19. (a) Evaluate $\phi(\alpha) = \displaystyle\int_0^\infty \alpha e^{-\alpha x}\,dx$ for $\alpha > 0$.

(b) Prove that the integral in (a) converges uniformly to 1 for $\alpha \geqq \alpha_1 > 0$.

(c) Explain why the integral does not converge uniformly to 1 for $\alpha > 0$.

(a) $\phi(\alpha) = \lim\limits_{b \to \infty} \displaystyle\int_0^b \alpha e^{-\alpha x}\,dx = \lim\limits_{b \to \infty} -e^{-\alpha x}\Big|_{x=0}^b = \lim\limits_{b \to \infty} 1 - e^{-\alpha b} = 1$ if $\alpha > 0$.

Thus the integral converges to 1 for all $\alpha > 0$.

(b) **Method 1**, using definition.

The integral converges uniformly to 1 in $\alpha \geqq \alpha_1 > 0$ if for each $\epsilon > 0$ we can find N, depending on ϵ but not on α, such that $\left| 1 - \displaystyle\int_0^u \alpha e^{-\alpha x}\,dx \right| < \epsilon$ for all $u > N$.

Since $\left| 1 - \displaystyle\int_0^u \alpha e^{-\alpha x}\,dx \right| = \left| 1 - (1 - e^{-\alpha u}) \right| = e^{-\alpha u} \leqq e^{-\alpha_1 u} < \epsilon$ for $u > \dfrac{1}{\alpha_1} \ln \dfrac{1}{\epsilon} = N$, the result follows.

Method 2, using the Weierstrass M test.

Since $\lim\limits_{x \to \infty} x^2 \cdot \alpha e^{-\alpha x} = 0$ for $\alpha \geqq \alpha_1 > 0$, we can choose $|\alpha e^{-\alpha x}| < \dfrac{1}{x^2}$ for sufficiently large x, say $x \geqq x_0$. Taking $M(x) = \dfrac{1}{x^2}$ and noting that $\displaystyle\int_{x_0}^\infty \dfrac{dx}{x^2}$ converges, it follows that the given integral is uniformly convergent to 1 for $\alpha \geqq \alpha_1 > 0$.

(c) As $\alpha_1 \to 0$, the number N in the first method of (b) increases without limit, so that the integral cannot be uniformly convergent for $\alpha > 0$.

20. If $\phi(\alpha) = \displaystyle\int_a^\infty f(x, \alpha)\,dx$ is uniformly convergent for $\alpha_1 \leqq \alpha \leqq \alpha_2$, prove that $\phi(\alpha)$ is continuous in this interval.

Let $\phi(\alpha) = \displaystyle\int_a^u f(x, \alpha)\,dx + R(u, \alpha)$, where $R(u, \alpha) = \displaystyle\int_u^\infty f(x, \alpha)\,dx$.

Then $\phi(\alpha + h) = \displaystyle\int_a^u f(x, \alpha + h)\,dx + R(u, \alpha + h)$ and so

$$\phi(\alpha + h) - \phi(\alpha) = \int_a^u \{f(x, \alpha + h) - f(x, \alpha)\}\,dx + R(u, \alpha + h) - R(u, \alpha)$$

Thus

$$|\phi(\alpha + h) - \phi(\alpha)| \leqq \int_a^u |f(x, \alpha + h) - f(x, \alpha)|\,dx + |R(u, \alpha + h)| + |R(u, \alpha)| \qquad (1)$$

Since the integral is uniformly convergent in $\alpha_1 \leqq \alpha \leqq \alpha_2$, we can, for each $\epsilon > 0$, find N independent of α such that for $u > N$,

$$|R(u, \alpha + h)| < \epsilon/3, \qquad |R(u, \alpha)| < \epsilon/3 \qquad (2)$$

Since $f(x, \alpha)$ is continuous, we can find $\delta > 0$ corresponding to each $\epsilon > 0$ such that

$$\int_a^u |f(x, \alpha + h) - f(x, \alpha)|\,dx < \epsilon/3 \qquad \text{for } |h| < \delta \qquad (3)$$

Using (2) and (3) in (1), we see that $\left| \phi(\alpha+h) - \phi(\alpha) \right| < \epsilon$ for $|h| < \delta$, so that $\phi(\alpha)$ is continuous.

Note that in this proof we assume that α and $\alpha + h$ are both in the interval $\alpha_1 \leqq \alpha \leqq \alpha_2$. Thus if $\alpha = \alpha_1$, for example, $h > 0$ and right hand continuity is assumed.

Also note the analogy of this proof with that for infinite series.

Other properties of uniformly convergent integrals can be proved similarly.

21. (a) Show that $\lim\limits_{\alpha \to 0+} \int_0^\infty \alpha e^{-\alpha x}\, dx \neq \int_0^\infty \left(\lim\limits_{\alpha \to 0+} \alpha e^{-\alpha x} \right) dx$.   (b) Explain the result in (a).

(a) $\lim\limits_{\alpha \to 0+} \int_0^\infty \alpha e^{-\alpha x}\, dx = \lim\limits_{\alpha \to 0+} 1 = 1$ by Problem 19(a).

$$\int_0^\infty \left(\lim\limits_{\alpha \to 0+} \alpha e^{-\alpha x} \right) dx = \int_0^\infty 0\, dx = 0.$$ Thus the required result follows.

(b)   Since $\phi(\alpha) = \int_0^\infty \alpha e^{-\alpha x}\, dx$ is not uniformly convergent for $\alpha \geqq 0$ (see Problem 19), there is no guarantee that $\phi(\alpha)$ will be continuous for $\alpha \geqq 0$. Thus $\lim\limits_{\alpha \to 0+} \phi(\alpha)$ may not be equal to $\phi(0)$.

22. (a) Prove that $\int_0^\infty e^{-\alpha x} \cos rx\, dx = \dfrac{\alpha}{\alpha^2 + r^2}$ for $\alpha > 0$ and any real value of r.

(b) Prove that the integral in (a) converges uniformly and absolutely for $a \leqq \alpha \leqq b$, where $0 < a < b$ and any r.

(a)   From integration formula 34, Page 84, we have

$$\lim\limits_{M \to \infty} \int_0^M e^{-\alpha x} \cos rx\, dx = \lim\limits_{M \to \infty} \frac{e^{-\alpha x}(r \sin rx - \alpha \cos rx)}{\alpha^2 + r^2}\Bigg|_0^M = \frac{\alpha}{\alpha^2 + r^2}$$

(b)   This follows at once from the Weierstrass M test for integrals, by noting that $\left| e^{-\alpha x} \cos rx \right| \leqq e^{-\alpha x}$ and $\int_0^\infty e^{-\alpha x}\, dx$ converges.

EVALUATION of DEFINITE INTEGRALS

23. Prove that $\int_0^{\pi/2} \ln \sin x\, dx = -\dfrac{\pi}{2} \ln 2$.

The given integral converges [Problem 42(f)]. Letting $x = \pi/2 - y$,

$$I = \int_0^{\pi/2} \ln \sin x\, dx = \int_0^{\pi/2} \ln \cos y\, dy = \int_0^{\pi/2} \ln \cos x\, dx$$

Then

$$2I = \int_0^{\pi/2} (\ln \sin x + \ln \cos x)\, dx = \int_0^{\pi/2} \ln \left(\frac{\sin 2x}{2} \right) dx$$

$$= \int_0^{\pi/2} \ln \sin 2x\, dx - \int_0^{\pi/2} \ln 2\, dx = \int_0^{\pi/2} \ln \sin 2x\, dx - \frac{\pi}{2} \ln 2 \qquad (1)$$

Letting $2x = v$,

$$\int_0^{\pi/2} \ln \sin 2x\, dx = \tfrac{1}{2} \int_0^\pi \ln \sin v\, dv = \tfrac{1}{2} \left\{ \int_0^{\pi/2} \ln \sin v\, dv + \int_{\pi/2}^\pi \ln \sin v\, dv \right\}$$

$$= \tfrac{1}{2}(I + I) = I \qquad \text{(letting } v = \pi - u \text{ in the last integral)}$$

Hence (1) becomes $2I = I - \dfrac{\pi}{2} \ln 2$ or $I = -\dfrac{\pi}{2} \ln 2$.

24. Prove that $\displaystyle\int_0^\pi x \ln \sin x \, dx = -\frac{\pi^2}{2} \ln 2.$

Let $x = \pi - y$. Then, using the results in the preceding problem,

$$J = \int_0^\pi x \ln \sin x \, dx = \int_0^\pi (\pi - u) \ln \sin u \, du = \int_0^\pi (\pi - x) \ln \sin x \, dx$$

$$= \pi \int_0^\pi \ln \sin x \, dx - \int_0^\pi x \ln \sin x \, dx$$

$$= -\pi^2 \ln 2 - J$$

or $J = -\dfrac{\pi^2}{2} \ln 2$.

25. (a) Prove that $\phi(\alpha) = \displaystyle\int_0^\infty \frac{dx}{x^2 + \alpha}$ is uniformly convergent for $\alpha \geqq 1$.

(b) Show that $\phi(\alpha) = \dfrac{\pi}{2\sqrt{\alpha}}$.   (c) Evaluate $\displaystyle\int_0^\infty \frac{dx}{(x^2 + 1)^2}$.

(d) Prove that $\displaystyle\int_0^\infty \frac{dx}{(x^2 + 1)^{n+1}} = \int_0^{\pi/2} \cos^{2n} \theta \, d\theta = \frac{1 \cdot 3 \cdot 5 \cdots (2n-1)}{2 \cdot 4 \cdot 6 \cdots (2n)} \frac{\pi}{2}$.

(a) The result follows from the Weierstrass test, since $\dfrac{1}{x^2 + \alpha} \leqq \dfrac{1}{x^2 + 1}$ for $\alpha \geqq 1$ and $\displaystyle\int_0^\infty \frac{dx}{x^2 + 1}$ converges.

(b) $\phi(\alpha) = \displaystyle\lim_{b \to \infty} \int_0^b \frac{dx}{x^2 + \alpha} = \lim_{b \to \infty} \frac{1}{\sqrt{\alpha}} \tan^{-1} \frac{x}{\sqrt{\alpha}} \Big|_0^b = \lim_{b \to \infty} \frac{1}{\sqrt{\alpha}} \tan^{-1} \frac{b}{\sqrt{\alpha}} = \frac{\pi}{2\sqrt{\alpha}}$.

(c) From (b), $\displaystyle\int_0^\infty \frac{dx}{x^2 + \alpha} = \frac{\pi}{2\sqrt{\alpha}}$.   Differentiating both sides with respect to α, we have

$$\int_0^\infty \frac{\partial}{\partial \alpha} \left(\frac{1}{x^2 + \alpha} \right) dx = -\int_0^\infty \frac{dx}{(x^2 + \alpha)^2} = -\frac{\pi}{4} \alpha^{-3/2}$$

the result being justified by Theorem 8, Page 266, since $\displaystyle\int_0^\infty \frac{dx}{(x^2 + \alpha)^2}$ is uniformly convergent for $\alpha \geqq 1$ (because $\dfrac{1}{(x^2 + \alpha)^2} \leqq \dfrac{1}{(x^2 + 1)^2}$ and $\displaystyle\int_0^\infty \frac{dx}{(x^2 + 1)^2}$ converges).

Taking the limit as $\alpha \to 1+$, using Theorem 6, Page 266, we find $\displaystyle\int_0^\infty \frac{dx}{(x^2 + 1)^2} = \frac{\pi}{4}$.

(d) Differentiating both sides of $\displaystyle\int_0^\infty \frac{dx}{x^2 + \alpha} = \frac{\pi}{2} \alpha^{-1/2}$ n times, we find

$$(-1)(-2)\cdots(-n) \int_0^\infty \frac{dx}{(x^2 + \alpha)^{n+1}} = \left(-\frac{1}{2}\right)\left(-\frac{3}{2}\right)\left(-\frac{5}{2}\right)\cdots\left(-\frac{2n-1}{2}\right) \frac{\pi}{2} \alpha^{-(2n+1)/2}$$

where justification proceeds as in part (c). Letting $\alpha \to 1+$, we find

$$\int_0^\infty \frac{dx}{(x^2 + 1)^{n+1}} = \frac{1 \cdot 3 \cdot 5 \cdots (2n-1)}{2^n \, n!} \frac{\pi}{2} = \frac{1 \cdot 3 \cdot 5 \cdots (2n-1)}{2 \cdot 4 \cdot 6 \cdots (2n)} \frac{\pi}{2}$$

Substituting $x = \tan \theta$, the integral becomes $\displaystyle\int_0^{\pi/2} \cos^{2n} \theta \, d\theta$ and the required result is obtained.

26. Prove that $\displaystyle\int_0^\infty \frac{e^{-ax} - e^{-bx}}{x \sec rx} \, dx = \frac{1}{2} \ln \frac{b^2 + r^2}{a^2 + r^2}$ where $a, b > 0$.

From Problem 22 and Theorem 7, Page 266, we have

$$\int_{x=0}^{\infty} \left\{ \int_{\alpha=a}^{b} e^{-\alpha x} \cos rx \, d\alpha \right\} dx = \int_{\alpha=a}^{b} \left\{ \int_{x=0}^{\infty} e^{-\alpha x} \cos rx \, dx \right\} d\alpha$$

or

$$\int_{x=0}^{\infty} \frac{e^{-\alpha x} \cos rx}{-x} \bigg|_{\alpha=a}^{b} dx = \int_{\alpha=a}^{b} \frac{\alpha}{\alpha^2 + r^2} \, d\alpha$$

i.e.

$$\int_{0}^{\infty} \frac{e^{-ax} - e^{-bx}}{x \sec rx} \, dx = \frac{1}{2} \ln \frac{b^2 + r^2}{a^2 + r^2}$$

27. Prove that $\displaystyle \int_{0}^{\infty} e^{-\alpha x} \frac{1 - \cos x}{x^2} \, dx = \tan^{-1} \frac{1}{\alpha} - \frac{\alpha}{2} \ln (\alpha^2 + 1), \quad \alpha > 0.$

By Problem 22 and Theorem 7, Page 266, we have

$$\int_{0}^{r} \left\{ \int_{0}^{\infty} e^{-\alpha x} \cos rx \, dx \right\} dr = \int_{0}^{\infty} \left\{ \int_{0}^{r} e^{-\alpha x} \cos rx \, dr \right\} dx$$

or

$$\int_{0}^{\infty} e^{-\alpha x} \frac{\sin rx}{x} \, dx = \int_{0}^{r} \frac{\alpha}{\alpha^2 + r^2} \, dr = \tan^{-1} \frac{r}{\alpha}$$

Integrating again with respect to r from 0 to r yields

$$\int_{0}^{\infty} e^{-\alpha x} \frac{1 - \cos rx}{x^2} \, dx = \int_{0}^{r} \tan^{-1} \frac{r}{\alpha} \, dr = r \tan^{-1} \frac{r}{\alpha} - \frac{\alpha}{2} \ln (\alpha^2 + r^2)$$

using integration by parts. The required result follows on letting $r = 1$.

28. Prove that $\displaystyle \int_{0}^{\infty} \frac{1 - \cos x}{x^2} \, dx = \frac{\pi}{2}.$

Since $e^{-\alpha x} \dfrac{1 - \cos x}{x^2} \leqq \dfrac{1 - \cos x}{x^2}$ for $\alpha \geqq 0, x \geqq 0$ and $\displaystyle \int_{0}^{\infty} \frac{1 - \cos x}{x^2} \, dx$ converges [see Problem

7(b)], it follows by the Weirstrass test that $\displaystyle \int_{0}^{\infty} e^{-\alpha x} \frac{1 - \cos x}{x^2} \, dx$ is uniformly convergent and repre-

sents a continuous function of α for $\alpha \geqq 0$ (Theorem 6, Page 266). Then letting $\alpha \to 0+$, using Prob. 27, we have

$$\lim_{\alpha \to 0+} \int_{0}^{\infty} e^{-\alpha x} \frac{1 - \cos x}{x^2} \, dx = \int_{0}^{\infty} \frac{1 - \cos x}{x^2} \, dx = \lim_{\alpha \to 0} \left\{ \tan^{-1} \frac{1}{\alpha} - \frac{\alpha}{2} \ln (\alpha^2 + 1) \right\} = \frac{\pi}{2}$$

29. Prove that $\displaystyle \int_{0}^{\infty} \frac{\sin x}{x} = \int_{0}^{\infty} \frac{\sin^2 x}{x^2} \, dx = \frac{\pi}{2}.$

Integrating by parts, we have

$$\int_{\epsilon}^{M} \frac{1 - \cos x}{x^2} \, dx = \left(-\frac{1}{x} \right)(1 - \cos x) \bigg|_{\epsilon}^{M} + \int_{\epsilon}^{M} \frac{\sin x}{x} \, dx = \frac{1 - \cos \epsilon}{\epsilon} - \frac{1 - \cos M}{M} + \int_{\epsilon}^{M} \frac{\sin x}{x} \, dx$$

Taking the limit as $\epsilon \to 0+$ and $M \to \infty$ shows that

$$\int_{0}^{\infty} \frac{\sin x}{x} \, dx = \int_{0}^{\infty} \frac{1 - \cos x}{x} \, dx = \frac{\pi}{2}$$

Since $\displaystyle \int_{0}^{\infty} \frac{1 - \cos x}{x^2} \, dx = 2 \int_{0}^{\infty} \frac{\sin^2 (x/2)}{x^2} \, dx = \int_{0}^{\infty} \frac{\sin^2 u}{u^2} \, du$ on letting $u = x/2$, we also have
$\displaystyle \int_{0}^{\infty} \frac{\sin^2 x}{x^2} \, dx = \frac{\pi}{2}.$

30. Prove that $\displaystyle\int_0^\infty \frac{\sin^3 x}{x}\,dx = \frac{\pi}{4}.$

$$\sin^3 x = \left(\frac{e^{ix} - e^{-ix}}{2i}\right)^3 = \frac{(e^{ix})^3 - 3(e^{ix})^2(e^{-ix}) + 3(e^{ix})(e^{-ix})^2 - (e^{-ix})^3}{(2i)^3}$$

$$= -\frac{1}{4}\left(\frac{e^{3ix} - e^{-3ix}}{2i}\right) + \frac{3}{4}\left(\frac{e^{ix} - e^{-ix}}{2i}\right) = -\frac{1}{4}\sin 3x + \frac{3}{4}\sin x$$

Then

$$\int_0^\infty \frac{\sin^3 x}{x}\,dx = \frac{3}{4}\int_0^\infty \frac{\sin x}{x}\,dx - \frac{1}{4}\int_0^\infty \frac{\sin 3x}{x}\,dx = \frac{3}{4}\int_0^\infty \frac{\sin x}{x}\,dx - \frac{1}{4}\int_0^\infty \frac{\sin u}{u}\,du$$

$$= \frac{3}{4}\left(\frac{\pi}{2}\right) - \frac{1}{4}\left(\frac{\pi}{2}\right) = \frac{\pi}{4}$$

MISCELLANEOUS PROBLEMS

31. Prove that $\displaystyle\int_0^\infty e^{-x^2}\,dx = \sqrt{\pi}/2.$

By Problem 6, the integral converges. Let
$$I_M = \int_0^M e^{-x^2}\,dx = \int_0^M e^{-y^2}\,dy \quad \text{and let} \quad \lim_{M\to\infty} I_M = I,$$
the required value of the integral. Then

$$I_M^2 = \left(\int_0^M e^{-x^2}\,dx\right)\left(\int_0^M e^{-y^2}\,dy\right)$$

$$= \int_0^M\int_0^M e^{-(x^2+y^2)}\,dx\,dy$$

$$= \iint_{\mathcal{R}_M} e^{-(x^2+y^2)}\,dx\,dy$$

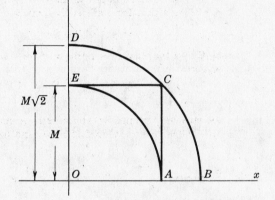

Fig. 12-1

where $\mathcal{R}_M$ is the square $OACE$ of side M (see Fig. 12-1). Since the integrand is positive, we have

$$\iint_{\mathcal{R}_1} e^{-(x^2+y^2)}\,dx\,dy \;\leqq\; I_M^2 \;\leqq\; \iint_{\mathcal{R}_2} e^{-(x^2+y^2)}\,dx\,dy \tag{1}$$

where $\mathcal{R}_1$ and $\mathcal{R}_2$ are the regions in the first quadrant bounded by the circles having radii M and $M\sqrt{2}$ respectively.

Using polar coordinates, we have from (1),

$$\int_{\phi=0}^{\pi/2}\int_{\rho=0}^{M} e^{-\rho^2}\rho\,d\rho\,d\phi \;\leqq\; I_M^2 \;\leqq\; \int_{\phi=0}^{\pi/2}\int_{\rho=0}^{M\sqrt{2}} e^{-\rho^2}\rho\,d\rho\,d\phi \tag{2}$$

or

$$\frac{\pi}{4}(1 - e^{-M^2}) \;\leqq\; I_M^2 \;\leqq\; \frac{\pi}{4}(1 - e^{-2M^2}) \tag{3}$$

Then taking the limit as $M\to\infty$ in (3), we find $\displaystyle\lim_{M\to\infty} I_M^2 = I^2 = \pi/4$ and $I = \sqrt{\pi}/2.$

32. Evaluate $\displaystyle\int_0^\infty e^{-x^2}\cos \alpha x\,dx.$

Let $\displaystyle I(\alpha) = \int_0^\infty e^{-x^2}\cos \alpha x\,dx.$ Then using integration by parts and appropriate limiting procedures,

$$\frac{dI}{d\alpha} = \int_0^\infty -xe^{-x^2}\sin \alpha x\,dx = \frac{1}{2}e^{-x^2}\sin \alpha x\Big|_0^\infty - \frac{1}{2}\alpha\int_0^\infty e^{-x^2}\cos \alpha x\,dx = -\frac{\alpha}{2}I$$

The differentiation under the integral sign is justified by Theorem 8, Page 266, and the fact that $\displaystyle\int_0^\infty xe^{-x^2}\sin \alpha x\,dx$ is uniformly convergent for all α (since by the Weierstrass test, $|xe^{-x^2}\sin \alpha x| \leqq xe^{-x^2}$ and $\displaystyle\int_0^\infty xe^{-x^2}\,dx$ converges).

From Problem 31 and the uniform convergence, and thus continuity, of the given integral (since $|e^{-x^2}\cos\alpha x| \leq e^{-x^2}$ and $\int_0^\infty e^{-x^2}\,dx$ converges, so that the Weierstrass test applies), we have $I(0) = \lim_{\alpha\to 0} I(\alpha) = \frac{1}{2}\sqrt{\pi}$.

Solving $\dfrac{dI}{d\alpha} = -\dfrac{\alpha}{2}I$ subject to $I(0) = \dfrac{\sqrt{\pi}}{2}$, we find $I(\alpha) = \dfrac{\sqrt{\pi}}{2}e^{-\alpha^2/4}$.

33. (a) Prove that $I(\alpha) = \displaystyle\int_0^\infty e^{-(x-\alpha/x)^2}\,dx = \dfrac{\sqrt{\pi}}{2}$.   (b) Evaluate $\displaystyle\int_0^\infty e^{-(x^2+x^{-2})}\,dx$.

(a) We have $I'(\alpha) = 2\displaystyle\int_0^\infty e^{-(x-\alpha/x)^2}(1-\alpha/x^2)\,dx$.

The differentiation is proved valid by observing that the integrand remains bounded as $x \to 0+$ and that for sufficiently large x,

$$e^{-(x-\alpha/x)^2}(1-\alpha/x^2) = e^{-x^2+2\alpha-\alpha^2/x^2}(1-\alpha/x^2) \leq e^{2\alpha}e^{-x^2}$$

so that $I'(\alpha)$ converges uniformly for $\alpha \geq 0$ by the Weierstrass test, since $\displaystyle\int_0^\infty e^{-x^2}\,dx$ converges. Now

$$I'(\alpha) = 2\int_0^\infty e^{-(x-\alpha/x)^2}\,dx - 2\alpha\int_0^\infty \frac{e^{-(x-\alpha/x)^2}}{x^2}\,dx = 0$$

as seen by letting $\alpha/x = y$ in the second integral. Thus $I(\alpha) = c$, a constant. To determine c, let $\alpha \to 0+$ in the required integral and use Problem 31 to obtain $c = \sqrt{\pi}/2$.

(b) From (a), $\displaystyle\int_0^\infty e^{-(x-\alpha/x)^2}\,dx = \int_0^\infty e^{-(x^2-2\alpha+\alpha^2 x^{-2})}\,dx = e^{2\alpha}\int_0^\infty e^{-(x^2+\alpha^2 x^{-2})}\,dx = \frac{\sqrt{\pi}}{2}$.

Then $\displaystyle\int_0^\infty e^{-(x^2+\alpha^2 x^{-2})}\,dx = \frac{\sqrt{\pi}}{2}e^{-2\alpha}$. Putting $\alpha = 1$, $\displaystyle\int_0^\infty e^{-(x^2+x^{-2})}\,dx = \frac{\sqrt{\pi}}{2}e^{-2}$.

34. Verify the results: (a) $\mathcal{L}\{e^{ax}\} = \dfrac{1}{s-a}$, $s > a$; (b) $\mathcal{L}\{\cos ax\} = \dfrac{s}{s^2+a^2}$, $s > 0$.

(a)
$$\mathcal{L}\{e^{ax}\} = \int_0^\infty e^{-sx}e^{ax}\,dx = \lim_{M\to\infty}\int_0^M e^{-(s-a)x}\,dx$$

$$= \lim_{M\to\infty}\frac{1-e^{-(s-a)M}}{s-a} = \frac{1}{s-a} \qquad \text{if } s > a$$

(b) $\mathcal{L}\{\cos ax\} = \displaystyle\int_0^\infty e^{-sx}\cos ax\,dx = \frac{s}{s^2+a^2}$ by Problem 22 with $\alpha = s$, $r = a$.

Another method, using complex numbers.

From part (a), $\mathcal{L}\{e^{ax}\} = \dfrac{1}{s-a}$. Replace a by ai. Then

$$\mathcal{L}\{e^{aix}\} = \mathcal{L}\{\cos ax + i\sin ax\} = \mathcal{L}\{\cos ax\} + i\,\mathcal{L}\{\sin ax\}$$

$$= \frac{1}{s-ai} = \frac{s+ai}{s^2+a^2} = \frac{s}{s^2+a^2} + i\frac{a}{s^2+a^2}$$

Equating real and imaginary parts: $\mathcal{L}\{\cos ax\} = \dfrac{s}{s^2+a^2}$, $\mathcal{L}\{\sin ax\} = \dfrac{a}{s^2+a^2}$.

The above *formal* method can be justified using methods of Chapter 17.

35. Prove that (a) $\mathcal{L}\{Y'(x)\} = s\mathcal{L}\{Y(x)\} - Y(0)$, (b) $\mathcal{L}\{Y''(x)\} = s^2\mathcal{L}\{Y(x)\} - sY(0) - Y'(0)$ under suitable conditions on $Y(x)$.

(a) By definition,
$$\mathcal{L}\{Y'(x)\} = \int_0^\infty e^{-sx}Y'(x)\,dx = \lim_{M\to\infty}\int_0^M e^{-sx}Y'(x)\,dx$$

$$= \lim_{M\to\infty}\left\{e^{-sx}Y(x)\Big|_0^M + s\int_0^M e^{-sx}Y(x)\,dx\right\}$$

$$= s\int_0^\infty e^{-sx}Y(x)\,dx - Y(0) = s\mathcal{L}\{Y(x)\} - Y(0)$$

assuming that s is such that $\lim_{M\to\infty} e^{-sM}Y(M) = 0$.

(b) Let $U(x) = Y'(x)$. Then by part (a), $\mathcal{L}\{U'(x)\} = s\mathcal{L}\{U(x)\} - U(0)$. Thus
$$\mathcal{L}\{Y''(x)\} = s\mathcal{L}\{Y'(x)\} - Y'(0) = s[s\mathcal{L}\{Y(x)\} - Y(0)] - Y'(0)$$
$$= s^2\mathcal{L}\{Y(x)\} - sY(0) - Y'(0)$$

36. Solve the differential equation $Y''(x) + Y(x) = x$, $Y(0) = 0$, $Y'(0) = 2$.

Take the Laplace transform of both sides of the given differential equation. Then by Problem 35,
$$\mathcal{L}\{Y''(x) + Y(x)\} = \mathcal{L}\{x\}, \qquad \mathcal{L}\{Y''(x)\} + \mathcal{L}\{Y(x)\} = 1/s^2$$
and so
$$s^2\mathcal{L}\{Y(x)\} - sY(0) - Y'(0) + \mathcal{L}\{Y(x)\} = 1/s^2$$

Solving for $\mathcal{L}\{Y(x)\}$ using the given conditions, we find
$$\mathcal{L}\{Y(x)\} = \frac{2s^2 + 1}{s^2(s^2 + 1)} = \frac{1}{s^2} + \frac{1}{s^2 + 1} \tag{1}$$
by methods of partial fractions.

Since $\dfrac{1}{s^2} = \mathcal{L}\{x\}$ and $\dfrac{1}{s^2+1} = \mathcal{L}\{\sin x\}$, it follows that $\dfrac{1}{s^2} + \dfrac{1}{s^2+1} = \mathcal{L}\{x + \sin x\}$.

Hence from (1), $\mathcal{L}\{Y(x)\} = \mathcal{L}\{x + \sin x\}$, from which we can conclude that $Y(x) = x + \sin x$ which is, in fact, found to be a solution.

Another method:

If $\mathcal{L}\{F(x)\} = f(s)$, we call $f(s)$ the *inverse* Laplace transform of $F(x)$ and write $f(s) = \mathcal{L}^{-1}\{F(x)\}$.

By Problem 78, $\mathcal{L}^{-1}\{f(s) + g(s)\} = \mathcal{L}^{-1}\{f(s)\} + \mathcal{L}^{-1}\{g(s)\}$. Then from (1),
$$Y(x) = \mathcal{L}^{-1}\left\{\frac{1}{s^2} + \frac{1}{s^2+1}\right\} = \mathcal{L}^{-1}\left\{\frac{1}{s^2}\right\} + \mathcal{L}^{-1}\left\{\frac{1}{s^2+1}\right\} = x + \sin x$$

Inverse Laplace transforms can be read from the Table on Page 267.

Supplementary Problems

IMPROPER INTEGRALS of the FIRST KIND

37. Test for convergence:

(a) $\displaystyle\int_0^\infty \frac{x^2+1}{x^4+1}\,dx$
(d) $\displaystyle\int_{-\infty}^\infty \frac{dx}{x^4+4}$
(g) $\displaystyle\int_{-\infty}^\infty \frac{x^2\,dx}{(x^2+x+1)^{5/2}}$

(b) $\displaystyle\int_2^\infty \frac{x\,dx}{\sqrt{x^3-1}}$
(e) $\displaystyle\int_{-\infty}^\infty \frac{2+\sin x}{x^2+1}\,dx$
(h) $\displaystyle\int_1^\infty \frac{\ln x\,dx}{x+e^{-x}}$

(c) $\displaystyle\int_1^\infty \frac{dx}{x\sqrt{3x+2}}$
(f) $\displaystyle\int_2^\infty \frac{x\,dx}{(\ln x)^3}$
(i) $\displaystyle\int_0^\infty \frac{\sin^2 x}{x^2}\,dx$

Ans. (a) conv., (b) div., (c) conv., (d) conv., (e) conv., (f) div., (g) conv., (h) div., (i) conv.

38. Prove that $\displaystyle\int_{-\infty}^{\infty} \frac{dx}{x^2 + 2ax + b^2} = \frac{\pi}{\sqrt{b^2 - a^2}}$ if $b > |a|$.

39. Test for convergence: (a) $\displaystyle\int_1^{\infty} e^{-x} \ln x \, dx$, (b) $\displaystyle\int_0^{\infty} e^{-x} \ln (1 + e^x) \, dx$, (c) $\displaystyle\int_0^{\infty} e^{-x} \cosh x^2 \, dx$.
Ans. (a) conv., (b) conv., (c) div.

40. Test for convergence, indicating absolute or conditional convergence where possible: (a) $\displaystyle\int_0^{\infty} \frac{\sin 2x}{x^3 + 1} \, dx$;

(b) $\displaystyle\int_{-\infty}^{\infty} e^{-ax^2} \cos bx \, dx$, where a, b are positive constants; (c) $\displaystyle\int_0^{\infty} \frac{\cos x}{\sqrt{x^2 + 1}} \, dx$; (d) $\displaystyle\int_0^{\infty} \frac{x \sin x}{\sqrt{x^2 + a^2}} \, dx$;

(e) $\displaystyle\int_0^{\infty} \frac{\cos x}{\cosh x} \, dx$.   Ans. (a) abs. conv., (b) abs. conv., (c) cond. conv., (d) div., (e) abs. conv.

41. Prove the quotient tests (b) and (c) on Page 262.

IMPROPER INTEGRALS of the SECOND KIND

42. Test for convergence:

(a) $\displaystyle\int_0^1 \frac{dx}{(x + 1)\sqrt{1 - x^2}}$ (d) $\displaystyle\int_0^2 \frac{\ln x}{\sqrt[3]{8 - x^3}} \, dx$ (g) $\displaystyle\int_0^3 \frac{x^2}{(3 - x)^2} \, dx$ (j) $\displaystyle\int_0^1 \frac{dx}{x^x}$

(b) $\displaystyle\int_0^1 \frac{\cos x}{x^2} \, dx$ (e) $\displaystyle\int_0^1 \frac{dx}{\sqrt{\ln (1/x)}}$ (h) $\displaystyle\int_0^{\pi/2} \frac{e^{-x} \cos x}{x} \, dx$

(c) $\displaystyle\int_{-1}^1 \frac{e^{\tan^{-1} x}}{x} \, dx$ (f) $\displaystyle\int_0^{\pi/2} \ln \sin x \, dx$ (i) $\displaystyle\int_0^1 \sqrt{\frac{1 - k^2 x^2}{1 - x^2}} \, dx, \ |k| < 1$

Ans. (a) conv., (b) div., (c) div., (d) conv., (e) conv., (f) conv., (g) div., (h) div., (i) conv., (j) conv.

43. (a) Prove that $\displaystyle\int_0^5 \frac{dx}{4 - x}$ diverges in the usual sense but converges in the Cauchy principal value sense. (b) Find the Cauchy principal value of the integral in (a) and give a geometric interpretation.
Ans. (b) $\ln 4$

44. Test for convergence, indicating absolute or conditional convergence where possible:

(a) $\displaystyle\int_0^1 \cos\left(\frac{1}{x}\right) dx$, (b) $\displaystyle\int_0^1 \frac{1}{x} \cos\left(\frac{1}{x}\right) dx$, (c) $\displaystyle\int_0^1 \frac{1}{x^2} \cos\left(\frac{1}{x}\right) dx$.
Ans. (a) abs. conv., (b) cond. conv., (c) div.

45. Prove that $\displaystyle\int_0^{4/\pi} \left(3x^2 \sin \frac{1}{x} - x \cos \frac{1}{x}\right) dx = \frac{32\sqrt{2}}{\pi^3}$.

IMPROPER INTEGRALS of the THIRD KIND

46. Test for convergence: (a) $\displaystyle\int_0^{\infty} e^{-x} \ln x \, dx$, (b) $\displaystyle\int_0^{\infty} \frac{e^{-x} \, dx}{\sqrt{x} \ln (x + 1)}$, (c) $\displaystyle\int_0^{\infty} \frac{e^{-x} \, dx}{\sqrt[3]{x} \, (3 + 2 \sin x)}$.
Ans. (a) conv., (b) div., (c) conv.

47. Test for convergence: (a) $\displaystyle\int_0^{\infty} \frac{dx}{\sqrt[3]{x^4 + x^2}}$; (b) $\displaystyle\int_0^{\infty} \frac{e^x \, dx}{\sqrt{\sinh (ax)}}, \ a > 0$.
Ans. (a) conv., (b) conv. if $a > 2$, div. if $0 < a \leqq 2$.

48. Prove that $\displaystyle\int_0^{\infty} \frac{\sinh (ax)}{\sinh (\pi x)} \, dx$ converges if $0 \leqq |a| < \pi$ and diverges if $|a| \geqq \pi$.

49. Test for convergence, indicating absolute or conditional convergence where possible:

(a) $\displaystyle\int_0^{\infty} \frac{\sin x}{\sqrt{x}} \, dx$, (b) $\displaystyle\int_0^{\infty} \frac{\sin \sqrt{x}}{\sinh \sqrt{x}} \, dx$.   Ans. (a) cond. conv., (b) abs. conv.

UNIFORM CONVERGENCE of IMPROPER INTEGRALS

50. (a) Prove that $\phi(\alpha) = \int_0^\infty \frac{\cos \alpha x}{1 + x^2}\, dx$ is uniformly convergent for all α.

(b) Prove that $\phi(\alpha)$ is continuous for all α. (c) Find $\lim_{\alpha \to 0} \phi(\alpha)$. *Ans.* (c) $\pi/2$.

51. Let $\phi(\alpha) = \int_0^\infty F(x, \alpha)\, dx$, where $F(x, \alpha) = \alpha^2 x e^{-\alpha x^2}$. (a) Show that $\phi(\alpha)$ is not continuous at $\alpha = 0$,

i.e. $\lim_{\alpha \to 0} \int_0^\infty F(x, \alpha)\, dx \neq \int_0^\infty \lim_{\alpha \to 0} F(x, \alpha)\, dx$. (b) Explain the result in (a).

52. Work Problem 51 if $F(x, \alpha) = \alpha^2 x e^{-\alpha x}$.

53. If $F(x)$ is bounded and continuous for $-\infty < x < \infty$ and

$$V(x, y) \;=\; \frac{1}{\pi} \int_{-\infty}^\infty \frac{y\, F(\lambda)\, d\lambda}{y^2 + (\lambda - x)^2}$$

prove that $\lim_{y \to 0} V(x, y) = F(x)$.

54. Prove (a) Theorem 7 and (b) Theorem 8 on Page 266.

55. Prove the Weierstrass M test for uniform convergence of integrals.

56. Prove that if $\int_0^\infty F(x)\, dx$ converges, then $\int_0^\infty e^{-\alpha x} F(x)\, dx$ converges uniformly for $\alpha \geqq 0$.

57. Prove that (a) $\phi(\alpha) = \int_0^\infty e^{-\alpha x} \frac{\sin x}{x}\, dx$ converges uniformly for $\alpha \geqq 0$, (b) $\phi(\alpha) = \frac{\pi}{2} - \tan^{-1} \alpha$,

(c) $\int_0^\infty \frac{\sin x}{x}\, dx = \frac{\pi}{2}$ (compare Problems 27-29).

58. State the definition of uniform convergence for improper integrals of the second kind.

59. State and prove a theorem corresponding to Theorem 8, Page 266, if a is a differentiable function of α.

EVALUATION of DEFINITE INTEGRALS

Establish each of the following results. Justify all steps in each case.

60. $\int_0^\infty \frac{e^{-ax} - e^{-bx}}{x}\, dx \;=\; \ln (b/a), \quad a, b > 0$

61. $\int_0^\infty \frac{e^{-ax} - e^{-bx}}{x \csc rx}\, dx \;=\; \tan^{-1}(b/r) - \tan^{-1}(a/r), \quad a, b, r > 0$

62. $\int_0^\infty \frac{\sin rx}{x(1 + x^2)}\, dx \;=\; \frac{\pi}{2}(1 - e^{-r}), \quad r \geqq 0$

63. $\int_0^\infty \frac{1 - \cos rx}{x^2}\, dx \;=\; \frac{\pi}{2}|r|$

64. $\int_0^\infty \frac{x \sin rx}{a^2 + x^2}\, dx \;=\; \frac{\pi}{2} e^{-ar}, \quad a, r \geqq 0$

65. (a) Prove that $\int_0^\infty e^{-\alpha x}\left(\frac{\cos ax - \cos bx}{x}\right) dx \;=\; \frac{1}{2}\ln\left(\frac{a^2 + b^2}{a^2 + a^2}\right), \quad \alpha \geqq 0.$

(b) Use (a) to prove that $\int_0^\infty \frac{\cos ax - \cos bx}{x}\, dx \;=\; \ln\left(\frac{b}{a}\right).$

[The results of (b) and Problem 60 are special cases of *Frullani's integral*, $\int_0^\infty \frac{F(ax) - F(bx)}{x}\, dx \;=\;$
$F(0) \ln\left(\frac{b}{a}\right)$, where $F(t)$ is continuous for $t > 0$, $F'(0)$ exists and $\int_1^\infty \frac{F(t)}{t}\, dt$ converges.]

66. Given $\int_0^\infty e^{-\alpha x^2}\, dx = \frac{1}{2}\sqrt{\pi/\alpha}, \quad \alpha > 0$. Prove that for $p = 1, 2, 3, \ldots$,

$$\int_0^\infty x^{2p} e^{-\alpha x^2}\, dx \;=\; \frac{1}{2} \cdot \frac{3}{2} \cdot \frac{5}{2} \cdots \frac{(2p - 1)}{2} \frac{\sqrt{\pi}}{2\, \alpha^{(2p+1)/2}}$$

67. If $a > 0$, $b > 0$, prove that $\displaystyle\int_0^\infty (e^{-a/x^2} - e^{-b/x^2})\, dx \;=\; \sqrt{\pi b} - \sqrt{\pi a}$.

68. Prove that $\displaystyle\int_0^\infty \frac{\tan^{-1}(x/a) - \tan^{-1}(x/b)}{x}\, dx \;=\; \frac{\pi}{2} \ln\!\left(\frac{b}{a}\right)$ where $a > 0$, $b > 0$.

69. Prove that $\displaystyle\int_{-\infty}^\infty \frac{dx}{(x^2 + x + 1)^3} \;=\; \frac{4\pi}{3\sqrt{3}}$. [Hint: Use Problem 38.]

MISCELLANEOUS PROBLEMS

70. Prove that $\displaystyle\int_0^\infty \left\{\frac{\ln(1+x)}{x}\right\}^2 dx$ converges.

71. Prove that $\displaystyle\int_0^\infty \frac{dx}{1 + x^3 \sin^2 x}$ converges. $\left[\text{Hint: Consider } \displaystyle\sum_{n=0}^\infty \int_{n\pi}^{(n+1)\pi} \frac{dx}{1 + x^3 \sin^2 x} \text{ and use the fact} \right.$

that $\displaystyle\int_{n\pi}^{(n+1)\pi} \frac{dx}{1 + x^3 \sin^2 x} \;\leqq\; \int_{n\pi}^{(n+1)\pi} \frac{dx}{1 + (n\pi)^3 \sin^2 x}.\Big]$

72. Prove that $\displaystyle\int_0^\infty \frac{x\, dx}{1 + x^3 \sin^2 x}$ diverges.

73. (a) Prove that $\displaystyle\int_0^\infty \frac{\ln(1 + \alpha^2 x^2)}{1 + x^2}\, dx \;=\; \pi \ln(1 + \alpha)$, $\alpha \geqq 0$.

(b) Use (a) to show that $\displaystyle\int_0^{\pi/2} \ln \sin \theta\, d\theta \;=\; -\frac{\pi}{2} \ln 2$.

74. Prove that $\displaystyle\int_0^\infty \frac{\sin^4 x}{x^4}\, dx \;=\; \frac{\pi}{3}$.

75. Evaluate (a) $\mathcal{L}\{1/\sqrt{x}\}$, (b) $\mathcal{L}\{\cosh ax\}$, (c) $\mathcal{L}\{(\sin x)/x\}$.

Ans. (a) $\sqrt{\pi/s}$, $s > 0$ (b) $\dfrac{s}{s^2 - a^2}$, $s > |a|$ (c) $\tan^{-1}\!\left(\dfrac{1}{s}\right)$, $s > 0$.

76. (a) If $\mathcal{L}\{F(x)\} = f(s)$, prove that $\mathcal{L}\{e^{ax} F(x)\} = f(s - a)$, (b) Evaluate $\mathcal{L}\{e^{ax} \sin bx\}$.

Ans. (b) $\dfrac{b}{(s - a)^2 + b^2}$, $s > a$

77. (a) If $\mathcal{L}\{F(x)\} = f(s)$, prove that $\mathcal{L}\{x^n F(x)\} = (-1)^n f^{(n)}(s)$, giving suitable restrictions on $F(x)$.

(b) Evaluate $\mathcal{L}\{x \cos x\}$. Ans. (b) $\dfrac{s^2 - 1}{(s^2 + 1)^2}$, $s > 0$

78. Prove that $\mathcal{L}^{-1}\{f(s) + g(s)\} \;=\; \mathcal{L}^{-1}\{f(s)\} + \mathcal{L}^{-1}\{g(s)\}$, stating any restrictions.

79. Solve using Laplace transforms, the following differential equations subject to the given conditions.

(a) $Y''(x) + 3Y'(x) + 2Y(x) = 0$; $Y(0) = 3$, $Y'(0) = 0$

(b) $Y''(x) - Y'(x) = x$; $Y(0) = 2$, $Y'(0) = -3$

(c) $Y''(x) + 2Y'(x) + 2Y(x) = 4$; $Y(0) = 0$, $Y'(0) = 0$

Ans. (a) $Y(x) = 6e^{-x} - 3e^{-2x}$, (b) $Y(x) = 4 - 2e^x - \frac{1}{2}x^2 - x$, (c) $Y(x) = 1 - e^{-x}(\sin x + \cos x)$

80. Prove that $\mathcal{L}\{F(x)\}$ exists if $F(x)$ is sectionally continuous in every finite interval $[0, b]$ where $b > 0$ and if $F(x)$ is of *exponential order* as $x \to \infty$, i.e. there exists a constant α such that $\left|e^{-\alpha x} F(x)\right| < P$ (a constant) for all $x > b$.

81. If $f(s) = \mathcal{L}\{F(x)\}$ and $g(s) = \mathcal{L}\{G(x)\}$, prove that $f(s)\,g(s) = \mathcal{L}\{H(x)\}$ where

$$H(x) \;=\; \int_0^x F(u)\,G(x-u)\,du$$

is called the *convolution* of F and G, written $F*G$.

$$\left[\text{Hint:}\quad \text{Write}\quad f(s)\,g(s) \;=\; \lim_{M \to \infty}\left\{\int_0^M e^{-su}\,F(u)\,du\right\}\left\{\int_0^M e^{-sv}\,G(v)\,dv\right\}\right.$$

$$\left.=\; \lim_{M \to \infty}\int_0^M \int_0^M e^{-s(u+v)}\,F(u)\,G(v)\,du\,dv \quad\text{and then let } u+v=t.\right]$$

82. (a) Find $\mathcal{L}^{-1}\left\{\dfrac{1}{(s^2+1)^2}\right\}$.   (b) Solve $Y''(x) + Y(x) = R(x)$, $Y(0) = Y'(0) = 0$.

   (c) Solve the integral equation $\quad Y(x) \;=\; x + \displaystyle\int_0^x Y(u)\,\sin\,(x-u)\,du$.   [Hint: Use Problem 81.]

   Ans. (a) $\tfrac{1}{2}(\sin x - x\cos x)$,   (b) $Y(x) = \displaystyle\int_0^x R(u)\,\sin\,(x-u)\,du$,   (c) $Y(x) = x + x^3/6$

83. Let $f(x)$, $g(x)$ and $g'(x)$ be continuous in every finite interval $a \leqq x \leqq b$ and suppose that $g'(x) \leqq 0$. Suppose also that $h(x) = \displaystyle\int_a^x f(x)\,dx$ is bounded for all $x \geqq a$ and $\displaystyle\lim_{x \to 0} g(x) = 0$.

   (a) Prove that $\displaystyle\int_a^\infty f(x)\,g(x)\,dx \;=\; -\int_a^\infty g'(x)\,h(x)\,dx$.

   (b) Prove that the integral on the right, and hence the integral on the left, is convergent. The result is that under the given conditions on $f(x)$ and $g(x)$, $\displaystyle\int_a^\infty f(x)\,g(x)\,dx$ converges and is sometimes called *Abel's integral test*.

   [Hint: For (a), consider $\displaystyle\lim_{b \to \infty}\int_a^b f(x)\,g(x)\,dx$ after replacing $f(x)$ by $h'(x)$ and integrating by parts. For (b), first prove that if $|h(x)| < H$ (a constant), then $\left|\displaystyle\int_a^b g'(x)\,h(x)\,dx\right| \leqq H\{g(a) - g(b)\}$; and then let $b \to \infty$.]

84. Use Problem 83 to prove that   (a) $\displaystyle\int_0^\infty \frac{\sin x}{x}\,dx$ and   (b) $\displaystyle\int_0^\infty \sin x^p\,dx$, $p > 1$, converge.

85. (a) Given that $\displaystyle\int_0^\infty \sin x^2\,dx \;=\; \int_0^\infty \cos x^2\,dx \;=\; \frac{1}{2}\sqrt{\frac{\pi}{2}}$ [see Problems 27 and 68(a), Chapter 13], evaluate

$$\int_0^\infty \int_0^\infty \sin\,(x^2+y^2)\,dx\,dy$$

   (b) Explain why the method of Problem 31 cannot be used to evaluate the multiple integral in (a). *Ans.* $\pi/4$

Chapter 13

Gamma and Beta Functions

GAMMA FUNCTION

The gamma function denoted by $\Gamma(n)$ is defined by

$$\Gamma(n) \;=\; \int_0^\infty x^{n-1} e^{-x}\, dx \tag{1}$$

which is convergent for $n > 0$ (see Problem 18, Chapter 12).

A recurrence formula for the gamma function is

$$\Gamma(n+1) \;=\; n\Gamma(n) \tag{2}$$

where $\Gamma(1) = 1$ (see Problem 1). From (2), $\Gamma(n)$ can be determined for all $n > 0$ when the values for $1 \le n < 2$ (or any other interval of unit length) are known (see table below). In particular if n is a positive integer, then

$$\Gamma(n+1) \;=\; n! \qquad n = 1, 2, 3, \ldots \tag{3}$$

For this reason $\Gamma(n)$ is sometimes called the *factorial function*.

Examples: $\Gamma(2) = 1! = 1$, $\Gamma(6) = 5! = 120$, $\dfrac{\Gamma(5)}{\Gamma(3)} = \dfrac{4!}{2!} = 12$.

It can be shown (Problem 4) that

$$\Gamma(\tfrac{1}{2}) \;=\; \sqrt{\pi} \tag{4}$$

The recurrence relation (2) is a difference equation which has (1) as a solution. By taking (1) as the definition of $\Gamma(n)$ for $n > 0$, we can generalize the gamma function to $n < 0$ by use of (2) in the form

$$\Gamma(n) \;=\; \frac{\Gamma(n+1)}{n} \tag{5}$$

See Problem 7, for example. The process is called *analytic continuation*.

TABLE of VALUES and GRAPH of the GAMMA FUNCTION

n	$\Gamma(n)$
1.00	1.0000
1.10	0.9514
1.20	0.9182
1.30	0.8975
1.40	0.8873
1.50	0.8862
1.60	0.8935
1.70	0.9086
1.80	0.9314
1.90	0.9618
2.00	1.0000

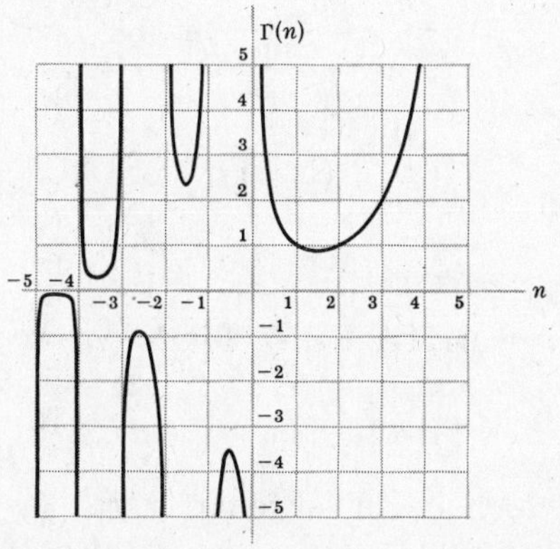

Fig. 13-1

285

ASYMPTOTIC FORMULA for $\Gamma(n)$

If n is large, the computational difficulties inherent in a calculation of $\Gamma(n)$ are apparent. A useful result in such case is supplied by the relation

$$\Gamma(n+1) \;=\; \sqrt{2\pi n}\, n^n\, e^{-n}\, e^{\theta/12(n+1)} \qquad 0 < \theta < 1 \tag{6}$$

For most practical purposes the last factor, which is very close to 1 for large n, can be omitted. If n is an integer, we can write

$$n! \;\sim\; \sqrt{2\pi n}\, n^n\, e^{-n} \tag{7}$$

where $\sim$ means "is approximately equal to for large n". This is sometimes called *Stirling's factorial approximation or asymptotic formula for* $n!$.

MISCELLANEOUS RESULTS INVOLVING the GAMMA FUNCTION

$$\Gamma(x)\,\Gamma(1-x) \;=\; \frac{\pi}{\sin x\pi} \qquad 0 < x < 1 \tag{8}$$

In particular if $x = \frac{1}{2}$, $\Gamma(\frac{1}{2}) = \sqrt{\pi}$.

$$2^{2x-1}\,\Gamma(x)\,\Gamma(x+\tfrac{1}{2}) \;=\; \sqrt{\pi}\,\Gamma(2x) \tag{9}$$

This is called the *duplication formula* for the gamma function.

$$\Gamma(x)\,\Gamma\!\left(x+\frac{1}{m}\right)\Gamma\!\left(x+\frac{2}{m}\right)\cdots\,\Gamma\!\left(x+\frac{m-1}{m}\right) \;=\; m^{\frac{1}{2}-mx}\,(2\pi)^{(m-1)/2}\,\Gamma(mx) \tag{10}$$

The result (9) is a special case of (10) with $m = 2$.

$$\frac{1}{\Gamma(x)} \;=\; xe^{\gamma x}\prod_{m=1}^{\infty}\left\{\left(1+\frac{x}{m}\right)e^{-x/m}\right\} \tag{11}$$

This is an infinite product representation for the gamma function. The constant γ is *Euler's constant* (see Problem 49, Chapter 11).

$$\Gamma(x+1) \;=\; \lim_{k\to\infty}\frac{1\cdot 2\cdot 3\ldots k}{(x+1)(x+2)\cdots(x+k)}\,k^x \;=\; \lim_{k\to\infty}\Pi(x,k) \tag{12}$$

where $\Pi(x,k)$ is sometimes called *Gauss' π function*.

$$\Gamma(x+1) \;=\; \sqrt{2\pi x}\,x^x\,e^{-x}\left\{1 + \frac{1}{12x} + \frac{1}{288x^2} - \frac{139}{51840x^3} + \cdots\right\} \tag{13}$$

This is called *Stirling's asymptotic series* for the gamma function. The series in braces is an asymptotic series (see Problem 20).

$$\Gamma'(1) \;=\; \int_0^\infty e^{-x}\ln x\,dx \;=\; -\gamma \qquad \text{where } \gamma \text{ is } \textit{Euler's constant.} \tag{14}$$

$$\frac{\Gamma'(x)}{\Gamma(x)} \;=\; -\gamma + \left(\frac{1}{1}-\frac{1}{x}\right) + \left(\frac{1}{2}-\frac{1}{x+1}\right) + \cdots + \left(\frac{1}{n}-\frac{1}{x+n-1}\right) + \cdots \tag{15}$$

The BETA FUNCTION denoted by $B(m,n)$ is defined by

$$B(m,n) \;=\; \int_0^1 x^{m-1}(1-x)^{n-1}\,dx \tag{16}$$

which is convergent for $m > 0$, $n > 0$. See Problem 16, Chapter 12.

The beta function is connected with the gamma function according to the relation

$$B(m, n) \;=\; \frac{\Gamma(m)\,\Gamma(n)}{\Gamma(m+n)} \tag{17}$$

See Problem 11.

Many integrals can be evaluated in terms of beta or gamma functions. Two useful results are

$$\int_0^{\pi/2} \sin^{2m-1}\theta \, \cos^{2n-1}\theta \, d\theta \;=\; \tfrac{1}{2}B(m, n) \;=\; \frac{\Gamma(m)\,\Gamma(n)}{2\,\Gamma(m+n)} \tag{18}$$

valid for $m > 0$ *and* $n > 0$ [see Problems (10) and (13)] and

$$\int_0^\infty \frac{x^{p-1}}{1+x}\, dx \;=\; \Gamma(p)\,\Gamma(1-p) \;=\; \frac{\pi}{\sin p\pi} \qquad 0 < p < 1 \tag{19}$$

See Problem 17.

DIRICHLET INTEGRALS

If V denotes the closed region in the first octant bounded by the surface $\left(\dfrac{x}{a}\right)^p + \left(\dfrac{y}{b}\right)^q + \left(\dfrac{z}{c}\right)^r = 1$ and the coordinate planes, then if all the constants are positive,

$$\iiint_V x^{\alpha-1}\, y^{\beta-1}\, z^{\gamma-1}\, dx\, dy\, dz \;=\; \frac{a^\alpha b^\beta c^\gamma}{pqr}\, \frac{\Gamma\!\left(\frac{\alpha}{p}\right)\Gamma\!\left(\frac{\beta}{q}\right)\Gamma\!\left(\frac{\gamma}{r}\right)}{\Gamma\!\left(1 + \frac{\alpha}{p} + \frac{\beta}{q} + \frac{\gamma}{r}\right)} \tag{20}$$

Integrals of this type are called *Dirichlet integrals* and are often useful in evaluating multiple integrals (see Problem 21).

Solved Problems

The GAMMA FUNCTION

1. Prove: $(a)\ \Gamma(n+1) = n\Gamma(n),\ n > 0;\ (b)\ \Gamma(n+1) = n!,\ n = 1, 2, 3, \dots$.

$(a)\quad \Gamma(n+1) \;=\; \int_0^\infty x^n e^{-x}\, dx \;=\; \lim_{M \to \infty} \int_0^M x^n e^{-x}\, dx$

$\qquad\qquad =\; \lim_{M \to \infty} \left\{ (x^n)(-e^{-x}) \big|_0^M - \int_0^M (-e^{-x})(nx^{n-1})\, dx \right\}$

$\qquad\qquad =\; \lim_{M \to \infty} \left\{ -M^n e^{-M} + n \int_0^M x^{n-1} e^{-x}\, dx \right\} \;=\; n\Gamma(n) \quad if\ n > 0$

$(b)\quad \Gamma(1) \;=\; \int_0^\infty e^{-x}\, dx \;=\; \lim_{M \to \infty} \int_0^M e^{-x}\, dx \;=\; \lim_{M \to \infty} (1 - e^{-M}) \;=\; 1.$

Put $n = 1, 2, 3, \dots$ in $\Gamma(n+1) = n\Gamma(n)$. Then

$$\Gamma(2) = 1\Gamma(1) = 1, \quad \Gamma(3) = 2\Gamma(2) = 2 \cdot 1 = 2!, \quad \Gamma(4) = 3\Gamma(3) = 3 \cdot 2! = 3!$$

In general, $\Gamma(n+1) = n!$ if n is a positive integer.

2. Evaluate each of the following.

(a) $\dfrac{\Gamma(6)}{2\,\Gamma(3)} = \dfrac{5!}{2\cdot 2!} = \dfrac{5\cdot 4\cdot 3\cdot 2}{2\cdot 2} = 30$

(b) $\dfrac{\Gamma(\frac{5}{2})}{\Gamma(\frac{1}{2})} = \dfrac{\frac{3}{2}\Gamma(\frac{3}{2})}{\Gamma(\frac{1}{2})} = \dfrac{\frac{3}{2}\cdot\frac{1}{2}\Gamma(\frac{1}{2})}{\Gamma(\frac{1}{2})} = \dfrac{3}{4}$

(c) $\dfrac{\Gamma(3)\,\Gamma(2.5)}{\Gamma(5.5)} = \dfrac{2!\,(1.5)(0.5)\,\Gamma(0.5)}{(4.5)(3.5)(2.5)(1.5)(0.5)\,\Gamma(0.5)} = \dfrac{16}{315}$

(d) $\dfrac{6\,\Gamma(\frac{8}{3})}{5\,\Gamma(\frac{2}{3})} = \dfrac{6(\frac{5}{3})(\frac{2}{3})\,\Gamma(\frac{2}{3})}{5\,\Gamma(\frac{2}{3})} = \dfrac{4}{3}$

3. Evaluate each integral.

(a) $\displaystyle\int_0^\infty x^3\, e^{-x}\, dx = \Gamma(4) = 3! = 6$

(b) $\displaystyle\int_0^\infty x^6\, e^{-2x}\, dx$.   Let $2x = y$. Then the integral becomes

$$\int_0^\infty \left(\frac{y}{2}\right)^6 e^{-y}\,\frac{dy}{2} = \frac{1}{2^7}\int_0^\infty y^6\, e^{-y}\, dy = \frac{\Gamma(7)}{2^7} = \frac{6!}{2^7} = \frac{45}{8}$$

4. Prove that $\Gamma(\tfrac{1}{2}) = \sqrt{\pi}$.

$\Gamma(\tfrac{1}{2}) = \displaystyle\int_0^\infty x^{-1/2}\, e^{-x}\, dx$.   Letting $x = u^2$ this integral becomes

$$2\int_0^\infty e^{-u^2}\, du = 2\left(\frac{\sqrt{\pi}}{2}\right) = \sqrt{\pi} \quad \text{using Problem 31, Chapter 12}$$

5. Evaluate each integral.

(a) $\displaystyle\int_0^\infty \sqrt{y}\, e^{-y^3}\, dy$.   Letting $y^3 = x$, the integral becomes

$$\int_0^\infty \sqrt{x^{1/3}}\, e^{-x}\cdot\frac{1}{3}\, x^{-2/3}\, dx = \frac{1}{3}\int_0^\infty x^{-1/2}\, e^{-x}\, dx = \frac{1}{3}\,\Gamma(\tfrac{1}{2}) = \frac{\sqrt{\pi}}{3}$$

(b) $\displaystyle\int_0^\infty 3^{-4z^2}\, dz = \int_0^\infty (e^{\ln 3})^{(-4z^2)}\, dz = \int_0^\infty e^{-(4\ln 3)z^2}\, dz$.   Let $(4\ln 3)z^2 = x$ and the integral becomes

$$\int_0^\infty e^{-x}\, d\left(\frac{x^{1/2}}{\sqrt{4\ln 3}}\right) = \frac{1}{2\sqrt{4\ln 3}}\int_0^\infty x^{-1/2}\, e^{-x}\, dx = \frac{\Gamma(1/2)}{2\sqrt{4\ln 3}} = \frac{\sqrt{\pi}}{4\sqrt{\ln 3}}$$

(c) $\displaystyle\int_0^1 \frac{dx}{\sqrt{-\ln x}}$.   Let $-\ln x = u$. Then $x = e^{-u}$. When $x = 1$, $u = 0$; when $x = 0$, $u = \infty$. The integral becomes

$$\int_0^\infty \frac{e^{-u}}{\sqrt{u}}\, du = \int_0^\infty u^{-1/2}\, e^{-u}\, du = \Gamma(1/2) = \sqrt{\pi}$$

6. Evaluate $\displaystyle\int_0^\infty x^m\, e^{-ax^n}\, dx$ where m, n, a are positive constants.

Letting $ax^n = y$, the integral becomes

$$\int_0^\infty \left\{\left(\frac{y}{a}\right)^{1/n}\right\}^m e^{-y}\, d\left\{\left(\frac{y}{a}\right)^{1/n}\right\} = \frac{1}{na^{(m+1)/n}}\int_0^\infty y^{(m+1)/n-1}\, e^{-y}\, dy = \frac{1}{na^{(m+1)/n}}\,\Gamma\left(\frac{m+1}{n}\right)$$

7. Evaluate (a) $\Gamma(-1/2)$, (b) $\Gamma(-5/2)$.

We use the generalization to negative values defined by $\Gamma(n) = \dfrac{\Gamma(n+1)}{n}$.

(a) Letting $n = -\frac{1}{2}$, $\Gamma(-1/2) = \dfrac{\Gamma(1/2)}{-1/2} = -2\sqrt{\pi}$.

(b) Letting $n = -3/2$, $\Gamma(-3/2) = \dfrac{\Gamma(-1/2)}{-3/2} = \dfrac{-2\sqrt{\pi}}{-3/2} = \dfrac{4\sqrt{\pi}}{3}$, using (a).

Then $\Gamma(-5/2) = \dfrac{\Gamma(-3/2)}{-5/2} = -\dfrac{8}{15}\sqrt{\pi}$.

8. Prove that $\displaystyle\int_0^1 x^m (\ln x)^n \, dx = \dfrac{(-1)^n n!}{(m+1)^{n+1}}$ where n is a positive integer and $m > -1$.

Letting $x = e^{-y}$, the integral becomes $\displaystyle(-1)^n \int_0^\infty y^n e^{-(m+1)y} \, dy$. If $(m+1)y = u$, this last integral becomes

$$(-1)^n \int_0^\infty \dfrac{u^n}{(m+1)^n} e^{-u} \dfrac{du}{m+1} \;=\; \dfrac{(-1)^n}{(m+1)^{n+1}} \int_0^\infty u^n e^{-u} \, du \;=\; \dfrac{(-1)^n}{(m+1)^{n+1}} \Gamma(n+1) \;=\; \dfrac{(-1)^n n!}{(m+1)^{n+1}}$$

Compare with Problem 50, Chapter 8, Page 177.

9. A particle is attracted toward a fixed point O with a force inversely proportional to its instantaneous distance from O. If the particle is released from rest, find the time for it to reach O.

At time $t = 0$ let the particle be located on the x axis at $x = a > 0$ and let O be the origin. Then by Newton's law

$$m \dfrac{d^2x}{dt^2} = -\dfrac{k}{x} \tag{1}$$

where m is the mass of the particle and $k > 0$ is a constant of proportionality.

Let $\dfrac{dx}{dt} = v$, the velocity of the particle. Then $\dfrac{d^2x}{dt^2} = \dfrac{dv}{dt} = \dfrac{dv}{dx} \cdot \dfrac{dx}{dt} = v \cdot \dfrac{dv}{dx}$ and (1) becomes

$$mv \dfrac{dv}{dx} = -\dfrac{k}{x} \qquad \text{or} \qquad \dfrac{mv^2}{2} = -k \ln x + c \tag{2}$$

upon integrating. Since $v = 0$ at $x = a$, we find $c = k \ln a$. Then

$$\dfrac{mv^2}{2} = k \ln \dfrac{a}{x} \qquad \text{or} \qquad v = \dfrac{dx}{dt} = -\sqrt{\dfrac{2k}{m}} \sqrt{\ln \dfrac{a}{x}} \tag{3}$$

where the negative sign is chosen since x is decreasing as t increases. We thus find that the time T taken for the particle to go from $x = a$ to $x = 0$ is given by

$$T = \sqrt{\dfrac{m}{2k}} \int_0^a \dfrac{dx}{\sqrt{\ln a/x}} \tag{4}$$

Letting $\ln a/x = u$ or $x = ae^{-u}$, this becomes

$$T = a\sqrt{\dfrac{m}{2k}} \int_0^\infty u^{-1/2} e^{-u} \, du \;=\; a\sqrt{\dfrac{m}{2k}} \Gamma(\tfrac{1}{2}) \;=\; a\sqrt{\dfrac{\pi m}{2k}}$$

The BETA FUNCTION

10. Prove that (a) $B(m, n) = B(n, m)$, (b) $B(m, n) = 2 \displaystyle\int_0^{\pi/2} \sin^{2m-1}\theta \cos^{2n-1}\theta \, d\theta$.

(a) Using the transformation $x = 1 - y$, we have

$$B(m, n) = \int_0^1 x^{m-1}(1-x)^{n-1} \, dx = \int_0^1 (1-y)^{m-1} y^{n-1} \, dy = \int_0^1 y^{n-1}(1-y)^{m-1} \, dy = B(n, m)$$

(b) Using the transformation $x = \sin^2\theta$, we have

$$B(m, n) = \int_0^1 x^{m-1}(1-x)^{n-1} \, dx = \int_0^{\pi/2} (\sin^2\theta)^{m-1}(\cos^2\theta)^{n-1} \, 2\sin\theta \cos\theta \, d\theta$$

$$= 2 \int_0^{\pi/2} \sin^{2m-1}\theta \cos^{2n-1}\theta \, d\theta$$

11. Prove that $\quad B(m,n) = \dfrac{\Gamma(m)\,\Gamma(n)}{\Gamma(m+n)} \quad m, n > 0.$

Letting $z = x^2$, we have $\quad \Gamma(m) = \displaystyle\int_0^\infty z^{m-1} e^{-z} dz = 2 \int_0^\infty x^{2m-1} e^{-x^2} dx.$

Similarly, $\quad \Gamma(n) = 2 \displaystyle\int_0^\infty y^{2n-1} e^{-y^2} dy.$ Then

$$\Gamma(m)\,\Gamma(n) = 4\left(\int_0^\infty x^{2m-1} e^{-x^2} dx\right)\left(\int_0^\infty y^{2n-1} e^{-y^2} dy\right)$$

$$= 4 \int_0^\infty \int_0^\infty x^{2m-1} y^{2n-1} e^{-(x^2+y^2)} dx\, dy$$

Transforming to polar coordinates, $x = \rho \cos\phi$, $y = \rho \sin\phi$,

$$\Gamma(m)\,\Gamma(n) = 4 \int_{\phi=0}^{\pi/2} \int_{\rho=0}^\infty \rho^{2(m+n)-1} e^{-\rho^2} \cos^{2m-1}\phi \, \sin^{2n-1}\phi \, d\rho\, d\phi$$

$$= 4\left(\int_{\rho=0}^\infty \rho^{2(m+n)-1} e^{-\rho^2} d\rho\right)\left(\int_{\phi=0}^{\pi/2} \cos^{2m-1}\phi \, \sin^{2n-1}\phi \, d\phi\right)$$

$$= 2\,\Gamma(m+n) \int_0^{\pi/2} \cos^{2m-1}\phi \, \sin^{2n-1}\phi \, d\phi \;=\; \Gamma(m+n)\,B(n,m)$$

$$= \Gamma(m+n)\,B(m,n)$$

using the results of Problem 10. Hence the required result follows.

The above argument can be made rigorous by using a limiting procedure as in Prob. 31, Chap. 12.

12. Evaluate each of the following integrals.

(a) $\displaystyle\int_0^1 x^4(1-x)^3 dx = B(5,4) = \dfrac{\Gamma(5)\,\Gamma(4)}{\Gamma(9)} = \dfrac{4!\,3!}{8!} = \dfrac{1}{280}$

(b) $\displaystyle\int_0^2 \dfrac{x^2 \, dx}{\sqrt{2-x}}.$ Letting $x = 2v$, the integral becomes

$$4\sqrt{2} \int_0^1 \dfrac{v^2}{\sqrt{1-v}} dv \;=\; 4\sqrt{2} \int_0^1 v^2(1-v)^{-1/2} dv \;=\; 4\sqrt{2}\, B(3, \tfrac{1}{2}) \;=\; \dfrac{4\sqrt{2}\,\Gamma(3)\,\Gamma(1/2)}{\Gamma(7/2)} \;=\; \dfrac{64\sqrt{2}}{15}$$

(c) $\displaystyle\int_0^a y^4 \sqrt{a^2-y^2}\, dy.$ Letting $y^2 = a^2 x$ or $y = a\sqrt{x}$, the integral becomes

$$a^6 \int_0^1 x^{3/2}(1-x)^{1/2} dx \;=\; a^6\, B(5/2, 3/2) \;=\; \dfrac{a^6\,\Gamma(5/2)\,\Gamma(3/2)}{\Gamma(4)} \;=\; \dfrac{\pi a^6}{16}$$

13. Show that $\displaystyle\int_0^{\pi/2} \sin^{2m-1}\theta \, \cos^{2n-1}\theta \, d\theta = \dfrac{\Gamma(m)\,\Gamma(n)}{2\,\Gamma(m+n)} \qquad m, n > 0.$

This follows at once from Problems 10 and 11.

14. Evaluate   (a) $\displaystyle\int_0^{\pi/2} \sin^6\theta \, d\theta,$ (b) $\displaystyle\int_0^{\pi/2} \sin^4\theta \cos^5\theta \, d\theta,$ (c) $\displaystyle\int_0^{\pi} \cos^4\theta \, d\theta.$

(a)   Let $2m-1 = 6$, $2n-1 = 0$, i.e. $m = 7/2$, $n = 1/2$, in Problem 13.

Then the required integral has the value $\dfrac{\Gamma(7/2)\,\Gamma(1/2)}{2\,\Gamma(4)} = \dfrac{5\pi}{32}.$

(b)　Letting $2m - 1 = 4$, $2n - 1 = 5$, the required integral has the value $\dfrac{\Gamma(5/2)\,\Gamma(3)}{2\,\Gamma(11/2)} = \dfrac{8}{315}$.

(c)　The given integral $= 2 \displaystyle\int_0^{\pi/2} \cos^4 \theta\, d\theta$.

　　Thus letting $2m - 1 = 0$, $2n - 1 = 4$ in Problem 13, the value is $\dfrac{2\,\Gamma(1/2)\,\Gamma(5/2)}{2\,\Gamma(3)} = \dfrac{3\pi}{8}$.

15. Prove $\displaystyle\int_0^{\pi/2} \sin^p \theta\, d\theta = \int_0^{\pi/2} \cos^p \theta\, d\theta = $ (a) $\dfrac{1 \cdot 3 \cdot 5 \cdots (p-1)}{2 \cdot 4 \cdot 6 \cdots p} \dfrac{\pi}{2}$ if p is an even positive integer,　(b) $\dfrac{2 \cdot 4 \cdot 6 \cdots (p-1)}{1 \cdot 3 \cdot 5 \cdots p}$ if p is an odd positive integer.

　　From Problem 13 with $2m - 1 = p$, $2n - 1 = 0$, we have

$$\int_0^{\pi/2} \sin^p \theta\, d\theta = \frac{\Gamma[\frac{1}{2}(p+1)]\,\Gamma(\frac{1}{2})}{2\,\Gamma[\frac{1}{2}(p+2)]}$$

(a)　If $p = 2r$, the integral equals

$$\frac{\Gamma(r + \frac{1}{2})\,\Gamma(\frac{1}{2})}{2\,\Gamma(r+1)} = \frac{(r - \frac{1}{2})(r - \frac{3}{2}) \cdots \frac{1}{2}\Gamma(\frac{1}{2}) \cdot \Gamma(\frac{1}{2})}{2r\,(r-1) \cdots 1} = \frac{(2r-1)(2r-3) \cdots 1}{2r\,(2r-2) \cdots 2} \frac{\pi}{2} = \frac{1 \cdot 3 \cdot 5 \cdots (2r-1)}{2 \cdot 4 \cdot 6 \cdots 2r} \frac{\pi}{2}$$

(b)　If $p = 2r + 1$, the integral equals

$$\frac{\Gamma(r+1)\,\Gamma(\frac{1}{2})}{2\,\Gamma(r + \frac{3}{2})} = \frac{r(r-1) \cdots 1 \cdot \sqrt{\pi}}{2(r + \frac{1}{2})(r - \frac{1}{2}) \cdots \frac{1}{2}\sqrt{\pi}} = \frac{2 \cdot 4 \cdot 6 \cdots 2r}{1 \cdot 3 \cdot 5 \cdots (2r+1)}$$

　　In both cases $\displaystyle\int_0^{\pi/2} \sin^p \theta\, d\theta = \int_0^{\pi/2} \cos^p \theta\, d\theta$, as seen by letting $\theta = \pi/2 - \phi$.

16. Evaluate　(a) $\displaystyle\int_0^{\pi/2} \cos^6 \theta\, d\theta$,　(b) $\displaystyle\int_0^{\pi/2} \sin^3 \theta \cos^2 \theta\, d\theta$,　(c) $\displaystyle\int_0^{2\pi} \sin^8 \theta\, d\theta$.

(a)　From Problem 15 the integral equals $\dfrac{1 \cdot 3 \cdot 5}{2 \cdot 4 \cdot 6} \dfrac{\pi}{2} = \dfrac{5\pi}{32}$　[compare Problem 14(a)].

(b)　The integral equals

$$\int_0^{\pi/2} \sin^3 \theta\,(1 - \sin^2 \theta)\, d\theta = \int_0^{\pi/2} \sin^3 \theta\, d\theta - \int_0^{\pi/2} \sin^5 \theta\, d\theta = \frac{2}{1 \cdot 3} - \frac{2 \cdot 4}{1 \cdot 3 \cdot 5} = \frac{2}{15}.$$
　　The method of Problem 14(b) can also be used.

(c)　The given integral equals $4 \displaystyle\int_0^{\pi/2} \sin^8 \theta\, d\theta = 4\left(\dfrac{1 \cdot 3 \cdot 5 \cdot 7}{2 \cdot 4 \cdot 6 \cdot 8} \dfrac{\pi}{2}\right) = \dfrac{35\pi}{64}$.

17. Given $\displaystyle\int_0^\infty \frac{x^{p-1}}{1 + x}\, dx = \frac{\pi}{\sin p\pi}$, show that $\Gamma(p)\,\Gamma(1 - p) = \dfrac{\pi}{\sin p\pi}$ where $0 < p < 1$.

　　Letting $\dfrac{x}{1 + x} = y$ or $x = \dfrac{y}{1 - y}$, the given integral becomes

$$\int_0^1 y^{p-1}\,(1 - y)^{-p}\, dy = B(p, 1 - p) = \Gamma(p)\,\Gamma(1 - p) \quad \text{and the result follows.}$$

18. Evaluate $\displaystyle\int_0^\infty \frac{dy}{1 + y^4}$.

　　Let $y^4 = x$. Then the integral becomes $\dfrac{1}{4} \displaystyle\int_0^\infty \frac{x^{-3/4}}{1 + x}\, dx = \dfrac{\pi}{4 \sin(\pi/4)} = \dfrac{\pi\sqrt{2}}{4}$　by Problem 17 with $p = \frac{1}{4}$.

　　The result can also be obtained by letting $y^2 = \tan \theta$.

19. Show that $\displaystyle\int_0^2 x\sqrt[3]{8-x^3}\,dx \;=\; \dfrac{16\pi}{9\sqrt{3}}$.

Letting $x^3 = 8y$ or $x = 2y^{1/3}$, the integral becomes

$$\int_0^1 2y^{1/3}\cdot\sqrt[3]{8(1-y)}\cdot\tfrac{2}{3}y^{-2/3}\,dy \;=\; \tfrac{8}{3}\int_0^1 y^{-1/3}(1-y)^{1/3}\,dy \;=\; \tfrac{8}{3}\,\mathrm{B}(\tfrac{2}{3},\tfrac{4}{3})$$

$$=\; \frac{8}{3}\,\frac{\Gamma(\tfrac{2}{3})\,\Gamma(\tfrac{4}{3})}{\Gamma(2)} \;=\; \frac{8}{9}\,\Gamma(\tfrac{1}{3})\,\Gamma(\tfrac{2}{3}) \;=\; \frac{8}{9}\cdot\frac{\pi}{\sin\pi/3} \;=\; \frac{16\pi}{9\sqrt{3}}$$

STIRLING'S FORMULA

20. Show that for large n, $\ n! = \sqrt{2\pi n}\,n^n\,e^{-n}\ $ approximately.

We have
$$\Gamma(n+1) \;=\; \int_0^\infty x^n e^{-x}\,dx \;=\; \int_0^\infty e^{n\ln x - x}\,dx \tag{1}$$

The function $n\ln x - x$ has a relative maximum for $x = n$, as is easily shown by elementary calculus. This leads us to the substitution $x = n + y$. Then (1) becomes

$$\Gamma(n+1) \;=\; e^{-n}\int_{-n}^\infty e^{n\ln(n+y)-y}\,dy \;=\; e^{-n}\int_{-n}^\infty e^{n\ln n\,+\,n\ln(1+y/n)\,-\,y}\,dy \tag{2}$$

$$=\; n^n e^{-n}\int_{-n}^\infty e^{n\ln(1+y/n)-y}\,dy$$

Up to now the analysis is rigorous. The following procedures in which we proceed formally can be made rigorous by suitable limiting procedures, but the proofs become involved and we shall omit them.

In (2) use the result
$$\ln(1+x) \;=\; x - \frac{x^2}{2} + \frac{x^3}{3} - \cdots \tag{3}$$

with $x = y/n$. Then on letting $y = \sqrt{n}\,v$, we find

$$\Gamma(n+1) \;=\; n^n e^{-n}\int_{-n}^\infty e^{-y^2/2n + y^3/3n^2 - \cdots}\,dy \;=\; n^n e^{-n}\sqrt{n}\int_{-\sqrt{n}}^\infty e^{-v^2/2 + v^3/3\sqrt{n} - \cdots}\,dv \tag{4}$$

When n is large a close approximation is

$$\Gamma(n+1) \;=\; n^n e^{-n}\sqrt{n}\int_{-\infty}^\infty e^{-v^2/2}\,dv \;=\; \sqrt{2\pi n}\,n^n e^{-n} \tag{5}$$

It is of interest that from (4) we can also obtain the result (13) on Page 286. See Problem 74.

DIRICHLET INTEGRALS

21. Evaluate $\quad I = \displaystyle\iiint_V x^{\alpha-1}\,y^{\beta-1}\,z^{\gamma-1}\,dx\,dy\,dz$

where V is the region in the first octant bounded by the sphere $x^2 + y^2 + z^2 = 1$ and the coordinate planes.

Let $x^2 = u$, $y^2 = v$, $z^2 = w$. Then

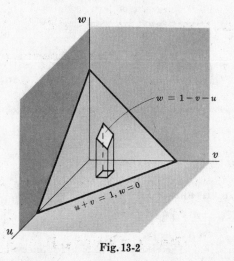

$$I \;=\; \iiint_{\mathcal{R}} u^{(\alpha-1)/2}\,v^{(\beta-1)/2}\,w^{(\gamma-1)/2}\,\frac{du}{2\sqrt{u}}\,\frac{dv}{2\sqrt{v}}\,\frac{dw}{2\sqrt{w}}$$

$$=\; \frac{1}{8}\iiint_{\mathcal{R}} u^{(\alpha/2)-1}\,v^{(\beta/2)-1}\,w^{(\gamma/2)-1}\,du\,dv\,dw \tag{1}$$

where $\mathcal{R}$ is the region in the uvw space bounded by the plane $u + v + w = 1$ and the uv, vw and uw planes as in Fig. 13-2. Thus

Fig. 13-2

$$I = \frac{1}{8} \int_{u=0}^{1} \int_{v=0}^{1-u} \int_{w=0}^{1-u-v} u^{(\alpha/2)-1} v^{(\beta/2)-1} w^{(\gamma/2)-1} \, du \, dv \, dw \tag{2}$$

$$= \frac{1}{4\gamma} \int_{u=0}^{1} \int_{v=0}^{1-u} u^{(\alpha/2)-1} v^{(\beta/2)-1} (1-u-v)^{\gamma/2} \, du \, dv$$

$$= \frac{1}{4\gamma} \int_{u=0}^{1} u^{(\alpha/2)-1} \left\{ \int_{v=0}^{1-u} v^{(\beta/2)-1} (1-u-v)^{\gamma/2} \, dv \right\} du$$

Letting $v = (1-u)t$, we have

$$\int_{v=0}^{1-u} v^{(\beta/2)-1} (1-u-v)^{\gamma/2} \, dv = (1-u)^{(\beta+\gamma)/2} \int_{t=0}^{1} t^{(\beta/2)-1} (1-t)^{\gamma/2} \, dt$$

$$= (1-u)^{(\beta+\gamma)/2} \frac{\Gamma(\beta/2) \, \Gamma(\gamma/2+1)}{\Gamma[(\beta+\gamma)/2+1]}$$

so that (2) becomes

$$I = \frac{1}{4\gamma} \frac{\Gamma(\beta/2) \, \Gamma(\gamma/2+1)}{\Gamma[(\beta+\gamma)/2+1]} \int_{u=0}^{1} u^{(\alpha/2)-1} (1-u)^{(\beta+\gamma)/2} \, du \tag{3}$$

$$= \frac{1}{4\gamma} \frac{\Gamma(\beta/2) \, \Gamma(\gamma/2+1)}{\Gamma[(\beta+\gamma)/2+1]} \cdot \frac{\Gamma(\alpha/2) \, \Gamma[(\beta+\gamma)/2+1]}{\Gamma[(\alpha+\beta+\gamma)/2+1]} = \frac{\Gamma(\alpha/2) \, \Gamma(\beta/2) \, \Gamma(\gamma/2)}{8 \, \Gamma[(\alpha+\beta+\gamma)/2+1]}$$

where we have used $(\gamma/2) \, \Gamma(\gamma/2) = \Gamma(\gamma/2+1)$.

The integral evaluated here is a special case of the Dirichlet integral (20), Page 287. The general case can be evaluated similarly.

22. Find the mass of the region bounded by $x^2 + y^2 + z^2 = a^2$ if the density is $\sigma = x^2 y^2 z^2$.

The required mass $= 8 \iiint_V x^2 y^2 z^2 \, dx \, dy \, dz$, where V is the region in the first octant bounded by the sphere $x^2 + y^2 + z^2 = a^2$ and the coordinate planes.

In the Dirichlet integral (20), Page 287, let $b = c = a$, $p = q = r = 2$ and $\alpha = \beta = \gamma = 3$. Then the required result is

$$8 \cdot \frac{a^3 \cdot a^3 \cdot a^3}{2 \cdot 2 \cdot 2} \frac{\Gamma(3/2) \, \Gamma(3/2) \, \Gamma(3/2)}{\Gamma(1+3/2+3/2+3/2)} = \frac{4\pi a^9}{945}$$

MISCELLANEOUS PROBLEMS

23. Show that $\displaystyle \int_0^1 \sqrt{1-x^4} \, dx = \frac{\{\Gamma(1/4)\}^2}{6\sqrt{2\pi}}$.

Let $x^4 = y$. Then the integral becomes

$$\frac{1}{4} \int_0^1 y^{-3/4} (1-y)^{1/2} \, dy = \frac{1}{4} \frac{\Gamma(1/4) \, \Gamma(3/2)}{\Gamma(7/4)} = \frac{\sqrt{\pi}}{6} \frac{\{\Gamma(1/4)\}^2}{\Gamma(1/4) \, \Gamma(3/4)}$$

From Problem 17 with $p = 1/4$, $\Gamma(1/4) \, \Gamma(3/4) = \pi\sqrt{2}$ so that the required result follows.

24. Prove the *duplication formula* $2^{2p-1} \Gamma(p) \, \Gamma(p + \frac{1}{2}) = \sqrt{\pi} \, \Gamma(2p)$.

Let $I = \displaystyle\int_0^{\pi/2} \sin^{2p} x \, dx$, $J = \displaystyle\int_0^{\pi/2} \sin^{2p} 2x \, dx$.

Then $I = \frac{1}{2} B(p + \frac{1}{2}, \frac{1}{2}) = \dfrac{\Gamma(p + \frac{1}{2})\sqrt{\pi}}{2 \, \Gamma(p+1)}$

Letting $2x = u$, we find

$$J = \tfrac{1}{2} \int_0^\pi \sin^{2p} u \; du = \int_0^{\pi/2} \sin^{2p} u \; du = I$$

But

$$J = \int_0^{\pi/2} (2 \sin x \cos x)^{2p} \, dx = 2^{2p} \int_0^{\pi/2} \sin^{2p} x \cos^{2p} x \; dx$$

$$= 2^{2p-1} \, B(p + \tfrac{1}{2}, \, p + \tfrac{1}{2}) = \frac{2^{2p-1} \{ \Gamma(p + \tfrac{1}{2}) \}^2}{\Gamma(2p + 1)}$$

Then since $I = J$,

$$\frac{\Gamma(p + \tfrac{1}{2}) \sqrt{\pi}}{2p \, \Gamma(p)} = \frac{2^{2p-1} \{ \Gamma(p + \tfrac{1}{2}) \}^2}{2p \, \Gamma(2p)}$$

and the required result follows.

---

25. Show that $\displaystyle \int_0^{\pi/2} \frac{d\phi}{\sqrt{1 - \tfrac{1}{2} \sin^2 \phi}} = \frac{\{ \Gamma(1/4) \}^2}{4 \sqrt{\pi}}$.

Consider

$$I = \int_0^{\pi/2} \frac{d\theta}{\sqrt{\cos \theta}} = \int_0^{\pi/2} \cos^{-1/2} \theta \; d\theta = \tfrac{1}{2} B(\tfrac{1}{4}, \tfrac{1}{2}) = \frac{\Gamma(\tfrac{1}{4}) \sqrt{\pi}}{2 \, \Gamma(\tfrac{3}{4})} = \frac{\{ \Gamma(\tfrac{1}{4}) \}^2}{2 \sqrt{2\pi}}$$

as in Problem 23.

But $\displaystyle I = \int_0^{\pi/2} \frac{d\theta}{\sqrt{\cos \theta}} = \int_0^{\pi/2} \frac{d\theta}{\sqrt{\cos^2 \theta/2 - \sin^2 \theta/2}} = \int_0^{\pi/2} \frac{d\theta}{\sqrt{1 - 2 \sin^2 \theta/2}}$.

Letting $\sqrt{2} \sin \theta/2 = \sin \phi$ in this last integral, it becomes $\displaystyle \sqrt{2} \int_0^{\pi/2} \frac{d\phi}{\sqrt{1 - \tfrac{1}{2} \sin^2 \phi}}$ from which the result follows.

---

26. Prove that $\displaystyle \int_0^\infty \frac{\cos x}{x^p} \, dx = \frac{\pi}{2 \, \Gamma(p) \cos (p\pi/2)}, \quad 0 < p < 1$.

We have $\displaystyle \frac{1}{x^p} = \frac{1}{\Gamma(p)} \int_0^\infty u^{p-1} e^{-xu} \, du$. Then

$$\int_0^\infty \frac{\cos x}{x^p} \, dx = \frac{1}{\Gamma(p)} \int_0^\infty \int_0^\infty u^{p-1} e^{-xu} \cos x \; du \, dx$$

$$= \frac{1}{\Gamma(p)} \int_0^\infty \frac{u^p}{1 + u^2} \, du \tag{1}$$

where we have reversed the order of integration and used Problem 22, Chapter 12.

Letting $u^2 = v$ in the last integral, we have by Problem 17

$$\int_0^\infty \frac{u^p}{1 + u^2} \, du = \tfrac{1}{2} \int_0^\infty \frac{v^{(p-1)/2}}{1 + v} \, dv = \frac{\pi}{2 \sin (p + 1)\pi/2} = \frac{\pi}{2 \cos p\pi/2} \tag{2}$$

Substitution of (2) in (1) yields the required result.

---

27. Evaluate $\displaystyle \int_0^\infty \cos x^2 \, dx$.

Letting $x^2 = y$, the integral becomes $\displaystyle \tfrac{1}{2} \int_0^\infty \frac{\cos y}{\sqrt{y}} \, dy = \tfrac{1}{2} \left(\frac{\pi}{2 \, \Gamma(\tfrac{1}{2}) \cos \pi/4} \right) = \tfrac{1}{2} \sqrt{\pi/2}$ by Problem 26.

This integral and the corresponding one for the sine [see Problem 68(a)] are called *Fresnel integrals*.

Supplementary Problems

The GAMMA FUNCTION

28. Evaluate (a) $\dfrac{\Gamma(7)}{2\,\Gamma(4)\,\Gamma(3)}$, (b) $\dfrac{\Gamma(3)\,\Gamma(3/2)}{\Gamma(9/2)}$, (c) $\Gamma(1/2)\,\Gamma(3/2)\,\Gamma(5/2)$.

    Ans. (a) 30, (b) 16/105, (c) $\frac{3}{8}\pi^{3/2}$

29. Evaluate (a) $\displaystyle\int_0^\infty x^4 e^{-x}\,dx$, (b) $\displaystyle\int_0^\infty x^6 e^{-3x}\,dx$, (c) $\displaystyle\int_0^\infty x^2 e^{-2x^2}\,dx$.   *Ans.* (a) 24, (b) $\dfrac{80}{243}$, (c) $\dfrac{\sqrt{2\pi}}{16}$

30. Find (a) $\displaystyle\int_0^\infty e^{-x^3}\,dx$, (b) $\displaystyle\int_0^\infty \sqrt[4]{x}\, e^{-\sqrt{x}}\,dx$, (c) $\displaystyle\int_0^\infty y^3 e^{-2y^5}\,dy$.

    Ans. (a) $\frac{1}{3}\Gamma(\frac{1}{3})$, (b) $\dfrac{3\sqrt{\pi}}{2}$, (c) $\dfrac{\Gamma(4/5)}{5\sqrt[5]{16}}$

31. Show that $\displaystyle\int_0^\infty \dfrac{e^{-st}}{\sqrt{t}}\,dt = \sqrt{\dfrac{\pi}{s}}$, $s > 0$.

32. Prove that $\Gamma(n) = \displaystyle\int_0^1 \left(\ln\dfrac{1}{x}\right)^{n-1}\,dx$, $n > 0$.

33. Evaluate (a) $\displaystyle\int_0^1 (\ln x)^4\,dx$, (b) $\displaystyle\int_0^1 (x\ln x)^3\,dx$, (c) $\displaystyle\int_0^1 \sqrt[3]{\ln(1/x)}\,dx$.

    Ans. (a) 24, (b) $-3/128$, (c) $\frac{1}{3}\Gamma(\frac{1}{3})$

34. Evaluate (a) $\Gamma(-7/2)$, (b) $\Gamma(-1/3)$.   *Ans.* (a) $(16\sqrt{\pi})/105$, (b) $-3\,\Gamma(2/3)$

35. Prove that $\displaystyle\lim_{x \to -m} \Gamma(x) = \infty$ where $m = 0, 1, 2, 3, \ldots$

36. Prove that if m is a positive integer, $\Gamma(-m + \frac{1}{2}) = \dfrac{(-1)^m\, 2^m\, \sqrt{\pi}}{1\cdot 3\cdot 5\cdots(2m-1)}$

37. Prove that $\Gamma'(1) = \displaystyle\int_0^\infty e^{-x}\ln x\,dx$ is a negative number (it is equal to $-\gamma$, where $\gamma = 0.577215\ldots$ is called *Euler's constant* as in Problem 49, Chapter 11).

The BETA FUNCTION

38. Evaluate (a) $B(3,5)$, (b) $B(3/2, 2)$, (c) $B(1/3, 2/3)$.   *Ans.* (a) 1/105, (b) 4/15, (c) $2\pi/\sqrt{3}$

39. Find (a) $\displaystyle\int_0^1 x^2(1-x)^3\,dx$, (b) $\displaystyle\int_0^1 \sqrt{(1-x)/x}\,dx$, (c) $\displaystyle\int_0^2 (4-x^2)^{3/2}\,dx$.
    Ans. (a) 1/60, (b) $\pi/2$, (c) 3π

40. Evaluate (a) $\displaystyle\int_0^4 u^{3/2}(4-u)^{5/2}\,du$, (b) $\displaystyle\int_0^3 \dfrac{dx}{\sqrt{3x-x^2}}$.   *Ans.* (a) 12π, (b) π

41. Prove that $\displaystyle\int_0^a \dfrac{dy}{\sqrt{a^4-y^4}} = \dfrac{\{\Gamma(1/4)\}^2}{4a\sqrt{2\pi}}$.

42. Evaluate (a) $\displaystyle\int_0^{\pi/2} \sin^4\theta\cos^4\theta\,d\theta$, (b) $\displaystyle\int_0^{2\pi} \cos^6\theta\,d\theta$.   *Ans.* (a) $3\pi/256$, (b) $5\pi/8$

43. Evaluate (a) $\displaystyle\int_0^\pi \sin^5\theta\,d\theta$, (b) $\displaystyle\int_0^{\pi/2} \cos^5\theta\sin^2\theta\,d\theta$.   *Ans.* (a) 16/15, (b) 8/105

44. Prove that $\displaystyle\int_0^{\pi/2} \sqrt{\tan\theta}\,d\theta = \pi/\sqrt{2}$.

45. Prove that (a) $\displaystyle\int_0^\infty \dfrac{x\,dx}{1+x^6} = \dfrac{\pi}{3\sqrt{3}}$, (b) $\displaystyle\int_0^\infty \dfrac{y^2\,dy}{1+y^4} = \dfrac{\pi}{2\sqrt{2}}$.

46. Prove that $\displaystyle\int_{-\infty}^{\infty} \frac{e^{2x}}{ae^{3x} + b}\, dx = \frac{2\pi}{3\sqrt{3}\, a^{2/3}b^{1/3}}$ where $a, b > 0$.

47. Prove that $\displaystyle\int_{-\infty}^{\infty} \frac{e^{2x}}{(e^{3x} + 1)^2}\, dx = \frac{2\pi}{9\sqrt{3}}$

[Hint: Differentiate with respect to b in Problem 46.]

48. Use the method of Problem 31, Chapter 12, to justify the procedure used in Problem 11.

DIRICHLET INTEGRALS

49. Find the mass of the region in the xy plane bounded by $x + y = 1$, $x = 0$, $y = 0$ if the density is $\sigma = \sqrt{xy}$.   Ans. $\pi/24$

50. Find the mass of the region bounded by the ellipsoid $\dfrac{x^2}{a^2} + \dfrac{y^2}{b^2} + \dfrac{z^2}{c^2} = 1$ if the density varies as the square of the distance from its center.   Ans. $\dfrac{\pi abck}{30}\,(a^2 + b^2 + c^2)$, $k = $ constant of proportionality

51. Find the volume of the region bounded by $x^{2/3} + y^{2/3} + z^{2/3} = 1$.   Ans. $4\pi/35$

52. Find the centroid of the region in the first octant bounded by $x^{2/3} + y^{2/3} + z^{2/3} = 1$.
Ans. $\bar{x} = \bar{y} = \bar{z} = 21/128$

53. Show that the volume of the region bounded by $x^m + y^m + z^m = a^m$, where $m > 0$, is given by $\dfrac{8\,\{\Gamma(1/m)\}^3}{3m^2\,\Gamma(3/m)}\, a^3$.

54. Show that the centroid of the region in the first octant bounded by $x^m + y^m + z^m = a^m$, where $m > 0$, is given by
$$\bar{x} = \bar{y} = \bar{z} = \frac{3\,\Gamma(2/m)\,\Gamma(3/m)}{4\,\Gamma(1/m)\,\Gamma(4/m)}\, a$$

MISCELLANEOUS PROBLEMS

55. Prove that $\displaystyle\int_a^b (x - a)^p (b - x)^q\, dx = (b - a)^{p+q+1}\, B(p + 1, q + 1)$ where $p > -1$, $q > -1$ and $b > a$.

[Hint: Let $x - a = (b - a)y$.]

56. Evaluate   (a) $\displaystyle\int_1^3 \frac{dx}{\sqrt{(x - 1)(3 - x)}}$,   (b) $\displaystyle\int_3^7 \sqrt[4]{(7 - x)(x - 3)}\, dx$.   Ans. (a) π,   (b) $\dfrac{2\,\{\Gamma(1/4)\}^2}{3\sqrt{\pi}}$

57. Show that $\dfrac{\{\Gamma(1/3)\}^2}{\Gamma(1/6)} = \dfrac{\sqrt{\pi}\,\sqrt[3]{2}}{\sqrt{3}}$.

58. Prove that $B(m, n) = \dfrac{1}{2}\displaystyle\int_0^1 \frac{x^{m-1} + x^{n-1}}{(1 + x)^{m+n}}\, dx$ where $m, n > 0$.

[Hint: Let $y = x/(1 + x)$.]

59. If $0 < p < 1$ prove that $\displaystyle\int_0^{\pi/2} \tan^p\theta\, d\theta = \frac{\pi}{2}\sec\frac{p\pi}{2}$.

60. Prove that $\displaystyle\int_0^1 \frac{x^{m-1}(1 - x)^{n-1}}{(x + r)^{m+n}} = \frac{B(m, n)}{r^m(1 + r)^{m+n}}$ where m, n and r are positive constants.

[Hint: Let $x = (r + 1)y/(r + y)$.]

61. Prove that $\displaystyle\int_0^{\pi/2} \frac{\sin^{2m-1}\theta\,\cos^{2n-1}\theta\, d\theta}{(a\sin^2\theta + b\cos^2\theta)^{m+n}} = \frac{B(m, n)}{2a^n b^m}$ where $m, n > 0$.

[Hint: Let $x = \sin^2\theta$ in Problem 60 and choose r appropriately.]

62. Prove that $\displaystyle\int_0^1 \frac{dx}{x^x} = \frac{1}{1^1} + \frac{1}{2^2} + \frac{1}{3^3} + \cdots$

63. Prove that for $m = 2, 3, 4, \ldots$

$$\sin\frac{\pi}{m} \sin\frac{2\pi}{m} \sin\frac{3\pi}{m} \cdots \sin\frac{(m-1)\pi}{m} = \frac{m}{2^{m-1}}$$

[Hint: Use the factored form $x^m - 1 = (x-1)(x-\alpha_1)(x-\alpha_2)\cdots(x-\alpha_{n-1})$, divide both sides by $x - 1$, and consider the limit as $x \to 1$.]

64. Prove that $\displaystyle\int_0^{\pi/2} \ln \sin x \, dx = -\pi/2 \ln 2$ using Problem 63.

[Hint: Take logarithms of the result in Prob. 63 and write the limit as $m \to \infty$ as a definite integral.]

65. Prove that $\displaystyle \Gamma\left(\frac{1}{m}\right) \Gamma\left(\frac{2}{m}\right) \Gamma\left(\frac{3}{m}\right) \cdots \Gamma\left(\frac{m-1}{m}\right) = \frac{(2\pi)^{(m-1)/2}}{\sqrt{m}}$.

[Hint: Square the left hand side and use Problem 63 and equation (8), Page 286.]

66. Prove that $\displaystyle\int_0^1 \ln \Gamma(x) \, dx = \frac{1}{2} \ln(2\pi)$.

[Hint: Take logarithms of the result in Problem 65 and let $m \to \infty$.]

67. (a) Prove that $\displaystyle\int_0^\infty \frac{\sin x}{x^p} \, dx = \frac{\pi}{2\,\Gamma(p)\sin(p\pi/2)}, \quad 0 < p < 1.$

(b) Discuss the cases $p = 0$ and $p = 1$.

68. Evaluate (a) $\displaystyle\int_0^\infty \sin x^2 \, dx$, (b) $\displaystyle\int_0^\infty x \cos x^3 \, dx$.   Ans. (a) $\frac{1}{2}\sqrt{\pi/2}$, (b) $\dfrac{\pi}{3\sqrt{3}\,\Gamma(1/3)}$

69. Prove that $\displaystyle\int_0^\infty \frac{x^{p-1} \ln x}{1+x} \, dx = -\pi^2 \csc p\pi \cot p\pi, \quad 0 < p < 1.$

70. Show that $\displaystyle\int_0^\infty \frac{\ln x}{x^4 + 1} \, dx = \frac{-\pi^2\sqrt{2}}{16}.$

71. Let $\displaystyle J_p(x) = \sum_{n=0}^\infty \frac{(-1)^n (x/2)^{p+2n}}{n!\,\Gamma(n+p+1)},$ a generalization of the Bessel function (16), Page 232, to the case where p may not be a positive integer. Prove that $J_p(x)$ satisfies the equation $x^2 y'' + xy' + (x^2 - p^2)y = 0$.

72. Referring to Problem 71, show that (a) $J_{1/2}(x) = \sqrt{2/\pi x} \sin x$, (b) $J_{-1/2}(x) = \sqrt{2/\pi x} \cos x$, (c) the results of Problem 107(b) and (c) of Page 257 are valid for all p .

73. If $a > 0$, $b > 0$ and $4ac > b^2$, prove that

$$\int_{-\infty}^\infty \int_{-\infty}^\infty e^{-(ax^2 + bxy + cy^2)} \, dx \, dy = \frac{2\pi}{\sqrt{4ac - b^2}}$$

74. Obtain (13) on Page 286 from the result (4) of Problem 20.

[Hint: Expand $e^{v^3/(3\sqrt{n})} + \cdots$ in a power series and replace the lower limit of the integral by $-\infty$.]

75. Obtain the result (15) on Page 286.

Chapter 14

Fourier Series

PERIODIC FUNCTIONS

A function $f(x)$ is said to have a *period T* or to be *periodic* with period T if for all x, $f(x + T) = f(x)$, where T is a positive constant. The least value of $T > 0$ is called the *least period* or simply *the period* of $f(x)$.

Example 1: The function $\sin x$ has periods $2\pi, 4\pi, 6\pi, \ldots$, since $\sin(x + 2\pi)$, $\sin(x + 4\pi)$, $\sin(x + 6\pi)$, $\ldots$ all equal $\sin x$. However, 2π is the *least period* or *the period* of $\sin x$.

Example 2: The period of $\sin nx$ or $\cos nx$, where n is a positive integer, is $2\pi/n$.

Example 3: The period of $\tan x$ is π.

Example 4: A constant has any positive number as period.

Other examples of periodic functions are shown in the graphs of Figures 14-1(a), (b) and (c) below.

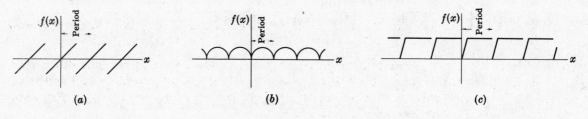

| (a) | (b) | (c) |

Fig. 14-1

FOURIER SERIES

Let $f(x)$ be defined in the interval $(-L, L)$ and outside of this interval by $f(x + 2L) = f(x)$, i.e. assume that $f(x)$ has the period $2L$. The *Fourier series* or *Fourier expansion* corresponding to $f(x)$ is given by

$$\frac{a_0}{2} + \sum_{n=1}^{\infty} \left(a_n \cos\frac{n\pi x}{L} + b_n \sin\frac{n\pi x}{L} \right) \tag{1}$$

where the *Fourier coefficients* a_n and b_n are

$$\begin{cases} a_n = \dfrac{1}{L} \displaystyle\int_{-L}^{L} f(x) \cos\dfrac{n\pi x}{L}\, dx \\[2ex] b_n = \dfrac{1}{L} \displaystyle\int_{-L}^{L} f(x) \sin\dfrac{n\pi x}{L}\, dx \end{cases} \qquad n = 0, 1, 2, \ldots \tag{2}$$

If $f(x)$ has the period $2L$, the coefficients a_n and b_n can be determined equivalently from

$$\begin{cases} a_n = \dfrac{1}{L} \displaystyle\int_{c}^{c+2L} f(x) \cos\dfrac{n\pi x}{L}\, dx \\[2ex] b_n = \dfrac{1}{L} \displaystyle\int_{c}^{c+2L} f(x) \sin\dfrac{n\pi x}{L}\, dx \end{cases} \tag{3}$$

where c is any real number. In the special case $c = -L$, (3) becomes (2).

To determine a_0 in (1), we use (2) or (3) with $n=0$. For example, from (2) we see that $a_0 = \frac{1}{L}\int_{-L}^{L} f(x)\,dx$. Note that the constant term in (1) is equal to $\frac{a_0}{2} = \frac{1}{2L}\int_{-L}^{L} f(x)\,dx$, which is the *mean* of $f(x)$ over a period.

If $L=\pi$, the series (1) and the coefficients (2) or (3) are particularly simple. The function in this case has the period 2π.

DIRICHLET CONDITIONS

Suppose that

(1) $f(x)$ is defined and single-valued except possibly at a finite number of points in $(-L, L)$

(2) $f(x)$ is periodic outside $(-L, L)$ with period $2L$

(3) $f(x)$ and $f'(x)$ are sectionally continuous in $(-L, L)$.

Then the series (1) with coefficients (2) or (3) converges to

(a) $f(x)$ if x is a point of continuity

(b) $\dfrac{f(x+0) + f(x-0)}{2}$ if x is a point of discontinuity

Here $f(x+0)$ and $f(x-0)$ are the right and left hand limits of $f(x)$ at x and represent $\lim_{\epsilon \to 0+} f(x+\epsilon)$ and $\lim_{\epsilon \to 0+} f(x-\epsilon)$ respectively. For a proof see Problems 18-23.

The conditions (1), (2) and (3) imposed on $f(x)$ are *sufficient* but not necessary, and are generally satisfied in practice. There are at present no known necessary and sufficient conditions for convergence of Fourier series. It is of interest that continuity of $f(x)$ does not *alone* insure convergence of a Fourier series.

ODD and EVEN FUNCTIONS

A function $f(x)$ is called *odd* if $f(-x) = -f(x)$. Thus x^3, $x^5 - 3x^3 + 2x$, $\sin x$, $\tan 3x$ are odd functions.

A function $f(x)$ is called *even* if $f(-x) = f(x)$. Thus x^4, $2x^6 - 4x^2 + 5$, $\cos x$, $e^x + e^{-x}$ are even functions.

The functions portrayed graphically in Figures 14-1(a) and 14-1(b) are odd and even respectively, but that of Fig. 14-1(c) is neither odd nor even.

In the Fourier series corresponding to an odd function, only sine terms can be present. In the Fourier series corresponding to an even function, only cosine terms (and possibly a constant which we shall consider a cosine term) can be present.

HALF RANGE FOURIER SINE or COSINE SERIES

A half range Fourier sine or cosine series is a series in which only sine terms or only cosine terms are present respectively. When a half range series corresponding to a given function is desired, the function is generally defined in the interval $(0, L)$ [which is half of the interval $(-L, L)$, thus accounting for the name *half range*] and then the function is specified as odd or even, so that it is clearly defined in the other half of the interval, namely $(-L, 0)$. In such case, we have

$$
\begin{cases}
a_n = 0, \quad b_n = \dfrac{2}{L} \int_0^L f(x) \sin \dfrac{n\pi x}{L}\, dx & \text{for } \textit{half range sine series} \\[2ex]
b_n = 0, \quad a_n = \dfrac{2}{L} \int_0^L f(x) \cos \dfrac{n\pi x}{L}\, dx & \text{for } \textit{half range cosine series}
\end{cases}
\tag{4}
$$

PARSEVAL'S IDENTITY states that

$$
\frac{1}{L} \int_{-L}^{L} \{f(x)\}^2\, dx \;=\; \frac{a_0^2}{2} + \sum_{n=1}^{\infty} (a_n^2 + b_n^2)
\tag{5}
$$

if a_n and b_n are the Fourier coefficients corresponding to $f(x)$ and if $f(x)$ satisfies the Dirichlet conditions.

DIFFERENTIATION and INTEGRATION of FOURIER SERIES

Differentiation and integration of Fourier series can be justified by using the theorems on Pages 228 and 229 which hold for series in general. It must be emphasized, however, that those theorems provide sufficient conditions and are not necessary. The following theorem for integration is especially useful.

Theorem:

The Fourier series corresponding to $f(x)$ may be integrated term by term from a to x, and the resulting series will converge uniformly to $\displaystyle\int_a^x f(x)\, dx$ provided that $f(x)$ is sectionally continuous in $-L \leqq x \leqq L$ and both a and x are in this interval.

COMPLEX NOTATION for FOURIER SERIES

Using Euler's identities,

$$
e^{i\theta} = \cos\theta + i\sin\theta, \quad e^{-i\theta} = \cos\theta - i\sin\theta
\tag{6}
$$

where $i = \sqrt{-1}$ (see Problem 48, Chap. 11, Page 251), the Fourier series for $f(x)$ can be written as

$$
f(x) = \sum_{n=-\infty}^{\infty} c_n e^{in\pi x/L}
\tag{7}
$$

where

$$
c_n = \frac{1}{2L} \int_{-L}^{L} f(x)\, e^{-in\pi x/L}\, dx
\tag{8}
$$

In writing the equality (7), we are supposing that the Dirichlet conditions are satisfied and further that $f(x)$ is continuous at x. If $f(x)$ is discontinuous at x, the left side of (7) should be replaced by $\dfrac{f(x+0) + f(x-0)}{2}$.

BOUNDARY-VALUE PROBLEMS

Boundary-value problems seek to determine solutions of partial differential equations satisfying certain prescribed conditions called *boundary conditions*. Some of these problems can be solved by use of Fourier series (see Problem 24).

ORTHOGONAL FUNCTIONS

Two vectors $\mathbf{A}$ and $\mathbf{B}$ are called *orthogonal* (perpendicular) if $\mathbf{A} \cdot \mathbf{B} = 0$ or $A_1 B_1 + A_2 B_2 + A_3 B_3 = 0$, where $\mathbf{A} = A_1\mathbf{i} + A_2\mathbf{j} + A_3\mathbf{k}$ and $\mathbf{B} = B_1\mathbf{i} + B_2\mathbf{j} + B_3\mathbf{k}$. Although not geometrically or physically evident, these ideas can be generalized to include vectors with more than three components. In particular we can think of a function, say $A(x)$, as being a vector with an *infinity of components* (i.e. an *infinite dimensional vector*), the value of each component being specified by substituting a particular value of x in some interval (a, b). It is natural in such case to define two functions, $A(x)$ and $B(x)$, as *orthogonal* in (a, b) if

$$\int_a^b A(x)\, B(x)\, dx \;=\; 0 \tag{9}$$

A vector $\mathbf{A}$ is called a *unit vector* or *normalized vector* if its magnitude is unity, i.e. if $\mathbf{A} \cdot \mathbf{A} = A^2 = 1$. Extending the concept, we say that the function $A(x)$ is *normal* or *normalized* in (a, b) if

$$\int_a^b \{A(x)\}^2\, dx \;=\; 1 \tag{10}$$

From the above it is clear that we can consider a set of functions $\{\phi_k(x)\}$, $k = 1, 2, 3, \ldots$, having the properties

$$\int_a^b \phi_m(x)\, \phi_n(x)\, dx \;=\; 0 \qquad m \neq n \tag{11}$$

$$\int_a^b \{\phi_m(x)\}^2\, dx \;=\; 1 \qquad m = 1, 2, 3, \ldots \tag{12}$$

In such case, each member of the set is orthogonal to every other member of the set and is also normalized. We call such a set of functions an *orthonormal set*.

The equations (11) and (12) can be summarized by writing

$$\int_a^b \phi_m(x)\, \phi_n(x)\, dx \;=\; \delta_{mn} \tag{13}$$

where δ_{mn}, called *Kronecker's symbol*, is defined as 0 if $m \neq n$ and 1 if $m = n$.

Just as any vector $\mathbf{r}$ in 3 dimensions can be expanded in a set of mutually orthogonal unit vectors $\mathbf{i}, \mathbf{j}, \mathbf{k}$ in the form $\mathbf{r} = c_1\mathbf{i} + c_2\mathbf{j} + c_3\mathbf{k}$, so we consider the possibility of expanding a function $f(x)$ in a set of orthonormal functions, i.e.,

$$f(x) \;=\; \sum_{n=1}^{\infty} c_n\, \phi_n(x) \qquad a \leq x \leq b \tag{14}$$

Such series, which are generalizations of Fourier series, are of great interest and utility both from theoretical and applied viewpoints.

Solved Problems

FOURIER SERIES

1. Graph each of the following functions.

(a) $f(x) = \begin{cases} 3 & 0 < x < 5 \\ -3 & -5 < x < 0 \end{cases}$ Period $= 10$

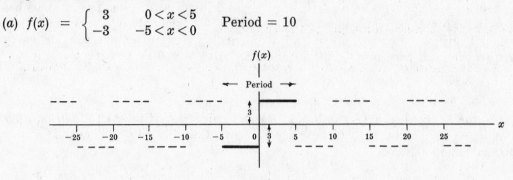

Fig. 14-2

Since the period is 10, that portion of the graph in $-5 < x < 5$ (indicated heavy in Fig. 14-2 above) is extended periodically outside this range (indicated dashed). Note that $f(x)$ is not defined at $x = 0, 5, -5, 10, -10, 15, -15$, etc. These values are the *discontinuities* of $f(x)$.

(b) $f(x) = \begin{cases} \sin x & 0 \leqq x \leqq \pi \\ 0 & \pi < x < 2\pi \end{cases}$ Period $= 2\pi$

Fig. 14-3

Refer to Fig. 14-3 above. Note that $f(x)$ is defined for all x and is continuous everywhere.

(c) $f(x) = \begin{cases} 0 & 0 \leqq x < 2 \\ 1 & 2 \leqq x < 4 \\ 0 & 4 \leqq x < 6 \end{cases}$ Period $= 6$

Fig. 14-4

Refer to Fig. 14-4 above. Note that $f(x)$ is defined for all x and is discontinuous at $x = \pm 2$, $\pm 4, \pm 8, \pm 10, \pm 14, \ldots$.

2. Prove $\displaystyle\int_{-L}^{L} \sin \frac{k\pi x}{L}\, dx = \int_{-L}^{L} \cos \frac{k\pi x}{L}\, dx = 0$ if $k = 1, 2, 3, \ldots$.

$$\int_{-L}^{L} \sin \frac{k\pi x}{L}\, dx = -\frac{L}{k\pi} \cos \frac{k\pi x}{L}\Big|_{-L}^{L} = -\frac{L}{k\pi} \cos k\pi + \frac{L}{k\pi} \cos(-k\pi) = 0$$

$$\int_{-L}^{L} \cos \frac{k\pi x}{L}\, dx = \frac{L}{k\pi} \sin \frac{k\pi x}{L}\Big|_{-L}^{L} = \frac{L}{k\pi} \sin k\pi - \frac{L}{k\pi} \sin(-k\pi) = 0$$

3. Prove   (a) $\int_{-L}^{L} \cos\frac{m\pi x}{L} \cos\frac{n\pi x}{L}\,dx = \int_{-L}^{L} \sin\frac{m\pi x}{L} \sin\frac{n\pi x}{L}\,dx = \begin{cases} 0 & m \neq n \\ L & m = n \end{cases}$

   (b) $\int_{-L}^{L} \sin\frac{m\pi x}{L} \cos\frac{n\pi x}{L}\,dx = 0$

where m and n can assume any of the values $1, 2, 3, \ldots$

(a) From trigonometry: $\cos A \cos B = \frac{1}{2}\{\cos(A - B) + \cos(A + B)\}$, $\quad \sin A \sin B = \frac{1}{2}\{\cos(A - B) - \cos(A + B)\}$.

Then, if $m \neq n$, by Problem 2,

$$\int_{-L}^{L} \cos\frac{m\pi x}{L} \cos\frac{n\pi x}{L}\,dx = \frac{1}{2}\int_{-L}^{L} \left\{ \cos\frac{(m-n)\pi x}{L} + \cos\frac{(m+n)\pi x}{L} \right\} dx = 0$$

Similarly if $m \neq n$,

$$\int_{-L}^{L} \sin\frac{m\pi x}{L} \sin\frac{n\pi x}{L}\,dx = \frac{1}{2}\int_{-L}^{L} \left\{ \cos\frac{(m-n)\pi x}{L} - \cos\frac{(m+n)\pi x}{L} \right\} dx = 0$$

If $m = n$, we have

$$\int_{-L}^{L} \cos\frac{m\pi x}{L} \cos\frac{n\pi x}{L}\,dx = \frac{1}{2}\int_{-L}^{L} \left(1 + \cos\frac{2n\pi x}{L} \right) dx = L$$

$$\int_{-L}^{L} \sin\frac{m\pi x}{L} \sin\frac{n\pi x}{L}\,dx = \frac{1}{2}\int_{-L}^{L} \left(1 - \cos\frac{2n\pi x}{L} \right) dx = L$$

Note that if $m = n = 0$ these integrals are equal to $2L$ and 0 respectively.

(b) We have $\sin A \cos B = \frac{1}{2}\{\sin(A - B) + \sin(A + B)\}$.   Then by Problem 2, if $m \neq n$,

$$\int_{-L}^{L} \sin\frac{m\pi x}{L} \cos\frac{n\pi x}{L}\,dx = \frac{1}{2}\int_{-L}^{L} \left\{ \sin\frac{(m-n)\pi x}{L} + \sin\frac{(m+n)\pi x}{L} \right\} dx = 0$$

If $m = n$,

$$\int_{-L}^{L} \sin\frac{m\pi x}{L} \cos\frac{n\pi x}{L}\,dx = \frac{1}{2}\int_{-L}^{L} \sin\frac{2n\pi x}{L}\,dx = 0$$

The results of parts (a) and (b) remain valid even when the limits of integration $-L, L$ are replaced by $c, c + 2L$ respectively.

4. If the series $A + \sum_{n=1}^{\infty} \left(a_n \cos\frac{n\pi x}{L} + b_n \sin\frac{n\pi x}{L} \right)$ converges uniformly to $f(x)$ in $(-L, L)$, show that for $n = 1, 2, 3, \ldots$,

(a) $a_n = \frac{1}{L}\int_{-L}^{L} f(x) \cos\frac{n\pi x}{L}\,dx$,   (b) $b_n = \frac{1}{L}\int_{-L}^{L} f(x) \sin\frac{n\pi x}{L}\,dx$,   (c) $A = \frac{a_0}{2}$.

(a) Multiplying

$$f(x) = A + \sum_{n=1}^{\infty} \left(a_n \cos\frac{n\pi x}{L} + b_n \sin\frac{n\pi x}{L} \right) \tag{1}$$

by $\cos\frac{m\pi x}{L}$ and integrating from $-L$ to L, using Problem 3, we have

$$\int_{-L}^{L} f(x) \cos\frac{m\pi x}{L}\,dx = A \int_{-L}^{L} \cos\frac{m\pi x}{L}\,dx \tag{2}$$
$$+ \sum_{n=1}^{\infty} \left\{ a_n \int_{-L}^{L} \cos\frac{m\pi x}{L} \cos\frac{n\pi x}{L}\,dx + b_n \int_{-L}^{L} \cos\frac{m\pi x}{L} \sin\frac{n\pi x}{L}\,dx \right\}$$
$$= a_m L \qquad \text{if } m \neq 0$$

Thus $a_m = \frac{1}{L}\int_{-L}^{L} f(x) \cos\frac{m\pi x}{L}\,dx \qquad$ if $m = 1, 2, 3, \ldots$

(b) Multiplying (1) by $\sin \dfrac{m\pi x}{L}$ and integrating from $-L$ to L, using Problem 3, we have

$$\int_{-L}^{L} f(x) \sin \frac{m\pi x}{L}\, dx \;=\; A \int_{-L}^{L} \sin \frac{m\pi x}{L}\, dx \qquad (3)$$

$$+ \; \sum_{n=1}^{\infty} \left\{ a_n \int_{-L}^{L} \sin \frac{m\pi x}{L} \cos \frac{n\pi x}{L}\, dx \;+\; b_n \int_{-L}^{L} \sin \frac{m\pi x}{L} \sin \frac{n\pi x}{L}\, dx \right\}$$

$$= \; b_m L$$

Thus $\qquad b_m \;=\; \dfrac{1}{L} \displaystyle\int_{-L}^{L} f(x) \sin \dfrac{m\pi x}{L}\, dx \qquad$ if $m = 1, 2, 3, \ldots$

(c) Integration of (1) from $-L$ to L, using Problem 2, gives

$$\int_{-L}^{L} f(x)\, dx \;=\; 2AL \qquad \text{or} \qquad A \;=\; \frac{1}{2L} \int_{-L}^{L} f(x)\, dx$$

Putting $m = 0$ in the result of part (a), we find $\quad a_0 = \dfrac{1}{L} \displaystyle\int_{-L}^{L} f(x)\, dx \quad$ and so $\quad A = \dfrac{a_0}{2}$.

The above results also hold when the integration limits $-L, L$ are replaced by $c, c + 2L$.

Note that in all parts above, interchange of summation and integration is valid because the series is *assumed* to converge uniformly to $f(x)$ in $(-L, L)$. Even when this assumption is not warranted, the coefficients a_m and b_m as obtained above are called *Fourier coefficients* corresponding to $f(x)$, and the corresponding series with these values of a_m and b_m is called the *Fourier series* corresponding to $f(x)$. An important problem in this case is to investigate conditions under which this series actually converges to $f(x)$. Sufficient conditions for this convergence are the *Dirichlet conditions* established in Problems 18 through 23 below.

5. (a) Find the Fourier coefficients corresponding to the function

$$f(x) \;=\; \begin{cases} 0 & -5 < x < 0 \\ 3 & 0 < x < 5 \end{cases} \qquad \text{Period} = 10$$

(b) Write the corresponding Fourier series.

(c) How should $f(x)$ be defined at $x = -5$, $x = 0$ and $x = 5$ in order that the Fourier series will converge to $f(x)$ for $-5 \leqq x \leqq 5$?

The graph of $f(x)$ is shown in Fig. 14-5 below.

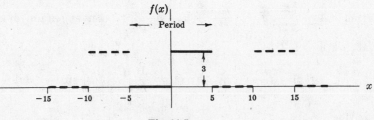

Fig. 14-5

(a) Period $= 2L = 10$ and $L = 5$. Choose the interval c to $c + 2L$ as -5 to 5, so that $c = -5$. Then

$$a_n \;=\; \frac{1}{L} \int_{c}^{c+2L} f(x) \cos \frac{n\pi x}{L}\, dx \;=\; \frac{1}{5} \int_{-5}^{5} f(x) \cos \frac{n\pi x}{5}\, dx$$

$$= \; \frac{1}{5} \left\{ \int_{-5}^{0} (0) \cos \frac{n\pi x}{5}\, dx \;+\; \int_{0}^{5} (3) \cos \frac{n\pi x}{5}\, dx \right\} \;=\; \frac{3}{5} \int_{0}^{5} \cos \frac{n\pi x}{5}\, dx$$

$$= \; \frac{3}{5} \left(\frac{5}{n\pi} \sin \frac{n\pi x}{5} \right) \Big|_{0}^{5} \;=\; 0 \qquad \text{if } n \neq 0$$

If $n = 0$, $\quad a_n \;=\; a_0 \;=\; \dfrac{3}{5} \displaystyle\int_{0}^{5} \cos \dfrac{0\pi x}{5}\, dx \;=\; \dfrac{3}{5} \int_{0}^{5} dx \;=\; 3.$

$$b_n = \frac{1}{L}\int_c^{c+2L} f(x)\sin\frac{n\pi x}{L}\,dx = \frac{1}{5}\int_{-5}^5 f(x)\sin\frac{n\pi x}{5}\,dx$$

$$= \frac{1}{5}\left\{\int_{-5}^0 (0)\sin\frac{n\pi x}{5}\,dx + \int_0^5 (3)\sin\frac{n\pi x}{5}\,dx\right\} = \frac{3}{5}\int_0^5 \sin\frac{n\pi x}{5}\,dx$$

$$= \frac{3}{5}\left(-\frac{5}{n\pi}\cos\frac{n\pi x}{5}\right)\Big|_0^5 = \frac{3(1-\cos n\pi)}{n\pi}$$

(b) The corresponding Fourier series is

$$\frac{a_0}{2} + \sum_{n=1}^\infty \left(a_n\cos\frac{n\pi x}{L} + b_n\sin\frac{n\pi x}{L}\right) = \frac{3}{2} + \sum_{n=1}^\infty \frac{3(1-\cos n\pi)}{n\pi}\sin\frac{n\pi x}{5}$$

$$= \frac{3}{2} + \frac{6}{\pi}\left(\sin\frac{\pi x}{5} + \frac{1}{3}\sin\frac{3\pi x}{5} + \frac{1}{5}\sin\frac{5\pi x}{5} + \cdots\right)$$

(c) Since $f(x)$ satisfies the Dirichlet conditions, we can say that the series converges to $f(x)$ at all points of continuity and to $\dfrac{f(x+0)+f(x-0)}{2}$ at points of discontinuity. At $x=-5,0$ and 5, which are points of discontinuity, the series converges to $(3+0)/2 = 3/2$ as seen from the graph. If we redefine $f(x)$ as follows,

$$f(x) = \begin{cases} 3/2 & x=-5 \\ 0 & -5 < x < 0 \\ 3/2 & x=0 \qquad \text{Period} = 10 \\ 3 & 0 < x < 5 \\ 3/2 & x=5 \end{cases}$$

then the series will converge to $f(x)$ for $-5 \leqq x \leqq 5$.

6. Expand $f(x) = x^2$, $0 < x < 2\pi$ in a Fourier series if (a) the period is 2π, (b) the period is not specified.

(a) The graph of $f(x)$ with period 2π is shown in Fig. 14-6 below.

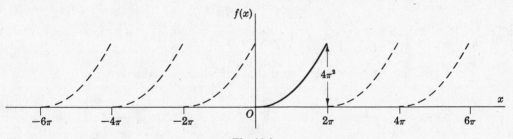

Fig. 14-6

Period $= 2L = 2\pi$ and $L = \pi$. Choosing $c = 0$, we have

$$a_n = \frac{1}{L}\int_c^{c+2L} f(x)\cos\frac{n\pi x}{L}\,dx = \frac{1}{\pi}\int_0^{2\pi} x^2\cos nx\,dx$$

$$= \frac{1}{\pi}\left\{(x^2)\left(\frac{\sin nx}{n}\right) - (2x)\left(\frac{-\cos nx}{n^2}\right) + 2\left(\frac{-\sin nx}{n^3}\right)\right\}\Big|_0^{2\pi} = \frac{4}{n^2},\quad n\neq 0$$

If $n=0$, $a_0 = \dfrac{1}{\pi}\displaystyle\int_0^{2\pi} x^2\,dx = \dfrac{8\pi^2}{3}$.

$$b_n = \frac{1}{L}\int_c^{c+2L} f(x)\sin\frac{n\pi x}{L}\,dx = \frac{1}{\pi}\int_0^{2\pi} x^2\sin nx\,dx$$

$$= \frac{1}{\pi}\left\{(x^2)\left(\frac{-\cos nx}{n}\right) - (2x)\left(\frac{-\sin nx}{n^2}\right) + (2)\left(\frac{\cos nx}{n^3}\right)\right\}\Big|_0^{2\pi} = \frac{-4\pi}{n}$$

Then $f(x) = x^2 = \dfrac{4\pi^2}{3} + \displaystyle\sum_{n=1}^\infty \left(\frac{4}{n^2}\cos nx - \frac{4\pi}{n}\sin nx\right)$.

This is valid for $0 < x < 2\pi$. At $x = 0$ and $x = 2\pi$ the series converges to $2\pi^2$.

(b)　If the period is not specified, the Fourier series cannot be determined uniquely in general.

7.　Using the results of Problem 6, prove that $\dfrac{1}{1^2} + \dfrac{1}{2^2} + \dfrac{1}{3^2} + \cdots = \dfrac{\pi^2}{6}$.

At $x = 0$ the Fourier series of Problem 6 reduces to $\dfrac{4\pi^2}{3} + \displaystyle\sum_{n=1}^{\infty} \dfrac{4}{n^2}$.

By the Dirichlet conditions, the series converges at $x = 0$ to $\frac{1}{2}(0 + 4\pi^2) = 2\pi^2$.

Then $\dfrac{4\pi^2}{3} + \displaystyle\sum_{n=1}^{\infty} \dfrac{4}{n^2} = 2\pi^2$, and so $\displaystyle\sum_{n=1}^{\infty} \dfrac{1}{n^2} = \dfrac{\pi^2}{6}$.

ODD and EVEN FUNCTIONS. HALF RANGE FOURIER SERIES

8.　Classify each of the following functions according as they are even, odd, or neither even nor odd.

(a) $f(x) = \begin{cases} 2 & 0 < x < 3 \\ -2 & -3 < x < 0 \end{cases}$　　Period $= 6$

From Fig. 14-7 below it is seen that $f(-x) = -f(x)$, so that the function is odd.

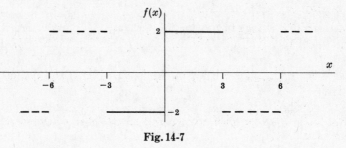

Fig. 14-7

(b) $f(x) = \begin{cases} \cos x & 0 < x < \pi \\ 0 & \pi < x < 2\pi \end{cases}$　　Period $= 2\pi$

From Fig. 14-8 below it is seen that the function is neither even nor odd.

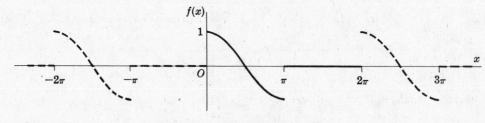

Fig. 14-8

(c) $f(x) = x(10 - x)$, $0 < x < 10$, Period $= 10$.

From Fig. 14-9 below the function is seen to be even.

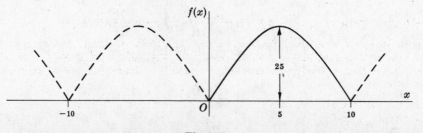

Fig. 14-9

9. Show that an even function can have no sine terms in its Fourier expansion.

Method 1:

No sine terms appear if $b_n = 0$, $n = 1, 2, 3, \ldots$. To show this, let us write

$$b_n = \frac{1}{L} \int_{-L}^{L} f(x) \sin\frac{n\pi x}{L}\, dx = \frac{1}{L} \int_{-L}^{0} f(x) \sin\frac{n\pi x}{L}\, dx + \frac{1}{L} \int_{0}^{L} f(x) \sin\frac{n\pi x}{L}\, dx \qquad (1)$$

If we make the transformation $x = -u$ in the first integral on the right of (1), we obtain

$$\frac{1}{L} \int_{-L}^{0} f(x) \sin\frac{n\pi x}{L}\, dx = \frac{1}{L} \int_{0}^{L} f(-u) \sin\left(-\frac{n\pi u}{L}\right) du = -\frac{1}{L} \int_{0}^{L} f(-u) \sin\frac{n\pi u}{L}\, du \qquad (2)$$

$$= -\frac{1}{L} \int_{0}^{L} f(u) \sin\frac{n\pi u}{L}\, du = -\frac{1}{L} \int_{0}^{L} f(x) \sin\frac{n\pi x}{L}\, dx$$

where we have used the fact that for an even function $f(-u) = f(u)$ and in the last step that the dummy variable of integration u can be replaced by any other symbol, in particular x. Thus from (1), using (2), we have

$$b_n = -\frac{1}{L} \int_{0}^{L} f(x) \sin\frac{n\pi x}{L}\, dx + \frac{1}{L} \int_{0}^{L} f(x) \sin\frac{n\pi x}{L}\, dx = 0$$

Method 2:   Assume $f(x) = \dfrac{a_0}{2} + \displaystyle\sum_{n=1}^{\infty} \left(a_n \cos\frac{n\pi x}{L} + b_n \sin\frac{n\pi x}{L}\right)$.

Then $f(-x) = \dfrac{a_0}{2} + \displaystyle\sum_{n=1}^{\infty} \left(a_n \cos\frac{n\pi x}{L} - b_n \sin\frac{n\pi x}{L}\right)$.

If $f(x)$ is even, $f(-x) = f(x)$. Hence

$$\frac{a_0}{2} + \sum_{n=1}^{\infty} \left(a_n \cos\frac{n\pi x}{L} + b_n \sin\frac{n\pi x}{L}\right) = \frac{a_0}{2} + \sum_{n=1}^{\infty} \left(a_n \cos\frac{n\pi x}{L} - b_n \sin\frac{n\pi x}{L}\right)$$

and so $\displaystyle\sum_{n=1}^{\infty} b_n \sin\frac{n\pi x}{L} = 0$,   i.e. $f(x) = \dfrac{a_0}{2} + \displaystyle\sum_{n=1}^{\infty} a_n \cos\frac{n\pi x}{L}$

and no sine terms appear.

In a similar manner we can show that an odd function has no cosine terms (or constant term) in its Fourier expansion.

10. If $f(x)$ is even, show that   (a) $a_n = \dfrac{2}{L} \displaystyle\int_{0}^{L} f(x) \cos\frac{n\pi x}{L}\, dx$,   (b) $b_n = 0$.

(a) $a_n = \dfrac{1}{L} \displaystyle\int_{-L}^{L} f(x) \cos\frac{n\pi x}{L}\, dx = \dfrac{1}{L} \displaystyle\int_{-L}^{0} f(x) \cos\frac{n\pi x}{L}\, dx + \dfrac{1}{L} \displaystyle\int_{0}^{L} f(x) \cos\frac{n\pi x}{L}\, dx$

Letting $x = -u$,

$$\frac{1}{L} \int_{-L}^{0} f(x) \cos\frac{n\pi x}{L}\, dx = \frac{1}{L} \int_{0}^{L} f(-u) \cos\left(\frac{-n\pi u}{L}\right) du = \frac{1}{L} \int_{0}^{L} f(u) \cos\frac{n\pi u}{L}\, du$$

since by definition of an even function $f(-u) = f(u)$. Then

$$a_n = \frac{1}{L} \int_{0}^{L} f(u) \cos\frac{n\pi u}{L}\, du + \frac{1}{L} \int_{0}^{L} f(x) \cos\frac{n\pi x}{L}\, dx = \frac{2}{L} \int_{0}^{L} f(x) \cos\frac{n\pi x}{L}\, dx$$

(b) This follows by Method 1 of Problem 9.

11. Expand $f(x) = \sin x$, $0 < x < \pi$, in a Fourier cosine series.

A Fourier series consisting of cosine terms alone is obtained only for an even function. Hence we extend the definition of $f(x)$ so that it becomes even (dashed part of Fig. 14-10 below). With this extension, $f(x)$ is then defined in an interval of length 2π. Taking the period as 2π, we have $2L = 2\pi$ so that $L = \pi$.

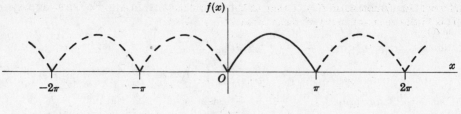

Fig. 14-10

By Problem 10, $b_n = 0$ and

$$a_n = \frac{2}{L}\int_0^L f(x)\cos\frac{n\pi x}{L}\,dx = \frac{2}{\pi}\int_0^\pi \sin x \cos nx\,dx$$

$$= \frac{1}{\pi}\int_0^\pi \{\sin(x+nx) + \sin(x-nx)\}\,dx = \frac{1}{\pi}\left\{-\frac{\cos(n+1)x}{n+1} + \frac{\cos(n-1)x}{n-1}\right\}\Big|_0^\pi$$

$$= \frac{1}{\pi}\left\{\frac{1-\cos(n+1)\pi}{n+1} + \frac{\cos(n-1)\pi - 1}{n-1}\right\} = \frac{1}{\pi}\left\{\frac{1+\cos n\pi}{n+1} - \frac{1+\cos n\pi}{n-1}\right\}$$

$$= \frac{-2(1+\cos n\pi)}{\pi(n^2-1)}\quad\text{if } n \neq 1.$$

For $n = 1$, $a_1 = \frac{2}{\pi}\int_0^\pi \sin x \cos x\,dx = \frac{2}{\pi}\frac{\sin^2 x}{2}\Big|_0^\pi = 0.$

For $n = 0$, $a_0 = \frac{2}{\pi}\int_0^\pi \sin x\,dx = \frac{2}{\pi}(-\cos x)\Big|_0^\pi = \frac{4}{\pi}.$

Then $$f(x) = \frac{2}{\pi} - \frac{2}{\pi}\sum_{n=2}^\infty \frac{(1+\cos n\pi)}{n^2-1}\cos nx$$

$$= \frac{2}{\pi} - \frac{4}{\pi}\left(\frac{\cos 2x}{2^2-1} + \frac{\cos 4x}{4^2-1} + \frac{\cos 6x}{6^2-1} + \cdots\right)$$

12. Expand $f(x) = x$, $0 < x < 2$,   in a half range   (a) sine series,   (b) cosine series.

   (a)   Extend the definition of the given function to that of the odd function of period 4 shown in Fig. 14-11 below.   This is sometimes called the *odd extension* of $f(x)$.   Then $2L = 4$, $L = 2$.

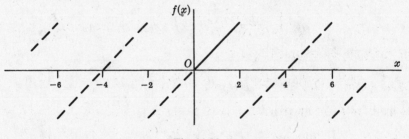

Fig. 14-11

Thus $a_n = 0$ and

$$b_n = \frac{2}{L}\int_0^L f(x)\sin\frac{n\pi x}{L}\,dx = \frac{2}{2}\int_0^2 x\sin\frac{n\pi x}{2}\,dx$$

$$= \left\{(x)\left(\frac{-2}{n\pi}\cos\frac{n\pi x}{2}\right) - (1)\left(\frac{-4}{n^2\pi^2}\sin\frac{n\pi x}{2}\right)\right\}\Big|_0^2 = \frac{-4}{n\pi}\cos n\pi$$

Then $$f(x) = \sum_{n=1}^\infty \frac{-4}{n\pi}\cos n\pi \sin\frac{n\pi x}{2}$$

$$= \frac{4}{\pi}\left(\sin\frac{\pi x}{2} - \frac{1}{2}\sin\frac{2\pi x}{2} + \frac{1}{3}\sin\frac{3\pi x}{2} - \cdots\right)$$

(b) Extend the definition of $f(x)$ to that of the even function of period 4 shown in Fig. 14-12 below. This is the *even extension* of $f(x)$.   Then $2L = 4$, $L = 2$.

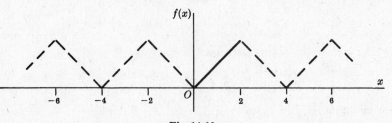

Fig. 14-12

Thus $b_n = 0$,

$$a_n = \frac{2}{L} \int_0^L f(x) \cos \frac{n\pi x}{L} \, dx = \frac{2}{2} \int_0^2 x \cos \frac{n\pi x}{2} \, dx$$

$$= \left\{ (x)\left(\frac{2}{n\pi} \sin \frac{n\pi x}{2} \right) - (1)\left(\frac{-4}{n^2 \pi^2} \cos \frac{n\pi x}{2} \right) \right\} \Big|_0^2$$

$$= \frac{4}{n^2 \pi^2} (\cos n\pi - 1) \qquad \text{if } n \neq 0$$

If $n = 0$, $a_0 = \int_0^2 x \, dx = 2.$

Then $f(x) = 1 + \sum_{n=1}^{\infty} \frac{4}{n^2 \pi^2} (\cos n\pi - 1) \cos \frac{n\pi x}{2}$

$$= 1 - \frac{8}{\pi^2} \left(\cos \frac{\pi x}{2} + \frac{1}{3^2} \cos \frac{3\pi x}{2} + \frac{1}{5^2} \cos \frac{5\pi x}{2} + \cdots \right)$$

It should be noted that the given function $f(x) = x$, $0 < x < 2$, is represented *equally well* by the two *different* series in (a) and (b).

PARSEVAL'S IDENTITY

13. Assuming that the Fourier series corresponding to $f(x)$ converges uniformly to $f(x)$ in $(-L, L)$, prove Parseval's identity

$$\frac{1}{L} \int_{-L}^{L} \{f(x)\}^2 \, dx = \frac{a_0^2}{2} + \Sigma (a_n^2 + b_n^2)$$

where the integral is assumed to exist.

If $f(x) = \frac{a_0}{2} + \sum_{n=1}^{\infty} \left(a_n \cos \frac{n\pi x}{L} + b_n \sin \frac{n\pi x}{L} \right)$,   then multiplying by $f(x)$ and integrating term

by term from $-L$ to L (which is justified since the series is uniformly convergent) we obtain

$$\int_{-L}^{L} \{f(x)\}^2 \, dx = \frac{a_0}{2} \int_{-L}^{L} f(x) \, dx + \sum_{n=1}^{\infty} \left\{ a_n \int_{-L}^{L} f(x) \cos \frac{n\pi x}{L} \, dx + b_n \int_{-L}^{L} f(x) \sin \frac{n\pi x}{L} \, dx \right\}$$

$$= \frac{a_0^2}{2} L + L \sum_{n=1}^{\infty} (a_n^2 + b_n^2) \tag{1}$$

where we have used the results

$$\int_{-L}^{L} f(x) \cos \frac{n\pi x}{L} \, dx = L a_n, \qquad \int_{-L}^{L} f(x) \sin \frac{n\pi x}{L} \, dx = L b_n, \qquad \int_{-L}^{L} f(x) \, dx = L a_0 \tag{2}$$

obtained from the Fourier coefficients.

The required result follows on dividing both sides of (1) by L. Parseval's identity is valid under less restrictive conditions than that imposed here.

14. (a) Write Parseval's identity corresponding to the Fourier series of Problem 12(b).

  (b) Determine from (a) the sum S of the series $\dfrac{1}{1^4} + \dfrac{1}{2^4} + \dfrac{1}{3^4} + \cdots + \dfrac{1}{n^4} + \cdots$.

  (a) Here $L = 2$, $a_0 = 2$, $a_n = \dfrac{4}{n^2 \pi^2}(\cos n\pi - 1)$, $n \neq 0$, $b_n = 0$.

   Then Parseval's identity becomes

$$\frac{1}{2}\int_{-2}^{2}\{f(x)\}^2\,dx \;=\; \frac{1}{2}\int_{-2}^{2} x^2\,dx \;=\; \frac{(2)^2}{2} + \sum_{n=1}^{\infty}\frac{16}{n^4\pi^4}(\cos n\pi - 1)^2$$

   or $\dfrac{8}{3} \;=\; 2 + \dfrac{64}{\pi^4}\left(\dfrac{1}{1^4} + \dfrac{1}{3^4} + \dfrac{1}{5^4} + \cdots\right)$,   i.e. $\dfrac{1}{1^4} + \dfrac{1}{3^4} + \dfrac{1}{5^4} + \cdots = \dfrac{\pi^4}{96}$.

  (b) $S \;=\; \dfrac{1}{1^4} + \dfrac{1}{2^4} + \dfrac{1}{3^4} + \cdots \;=\; \left(\dfrac{1}{1^4} + \dfrac{1}{3^4} + \dfrac{1}{5^4} + \cdots\right) + \left(\dfrac{1}{2^4} + \dfrac{1}{4^4} + \dfrac{1}{6^4} + \cdots\right)$

$$= \left(\frac{1}{1^4} + \frac{1}{3^4} + \frac{1}{5^4} + \cdots\right) + \frac{1}{2^4}\left(\frac{1}{1^4} + \frac{1}{2^4} + \frac{1}{3^4} + \cdots\right)$$

$$= \frac{\pi^4}{96} + \frac{S}{16}, \quad \text{from which } S = \frac{\pi^4}{90}$$

15. Prove that for all positive integers M,

$$\frac{a_0^2}{2} + \sum_{n=1}^{M}(a_n^2 + b_n^2) \;\leqq\; \frac{1}{L}\int_{-L}^{L}\{f(x)\}^2\,dx$$

where a_n and b_n are the Fourier coefficients corresponding to $f(x)$, and $f(x)$ is assumed sectionally continuous in $(-L, L)$.

   Let $\displaystyle S_M(x) \;=\; \frac{a_0}{2} + \sum_{n=1}^{M}\left(a_n \cos\frac{n\pi x}{L} + b_n \sin\frac{n\pi x}{L}\right)$ (1)

For $M = 1, 2, 3, \ldots$ this is the sequence of partial sums of the Fourier series corresponding to $f(x)$.

   We have

$$\int_{-L}^{L}\{f(x) - S_M(x)\}^2\,dx \;\geqq\; 0$$ (2)

since the integrand is non-negative. Expanding the integrand, we obtain

$$2\int_{-L}^{L} f(x)\,S_M(x)\,dx \;-\; \int_{-L}^{L} S_M^2(x)\,dx \;\leqq\; \int_{-L}^{L}\{f(x)\}^2\,dx$$ (3)

   Multiplying both sides of (1) by $2f(x)$ and integrating from $-L$ to L, using equations (2) of Problem 13, gives

$$2\int_{-L}^{L} f(x)\,S_M(x)\,dx \;=\; 2L\left\{\frac{a_0^2}{2} + \sum_{n=1}^{M}(a_n^2 + b_n^2)\right\}$$ (4)

   Also, squaring (1) and integrating from $-L$ to L, using Problem 3, we find

$$\int_{-L}^{L} S_M^2(x)\,dx \;=\; L\left\{\frac{a_0^2}{2} + \sum_{n=1}^{M}(a_n^2 + b_n^2)\right\}$$ (5)

   Substitution of (4) and (5) into (3) and dividing by L yields the required result.

   Taking the limit as $M \to \infty$, we obtain *Bessel's inequality*

$$\frac{a_0^2}{2} + \sum_{n=1}^{\infty}(a_n^2 + b_n^2) \;\leqq\; \frac{1}{L}\int_{-L}^{L}\{f(x)\}^2\,dx$$ (6)

   If the equality holds, we have *Parseval's identity* (Problem 13).

   We can think of $S_M(x)$ as representing an *approximation* to $f(x)$, while the left hand side of (2), divided by $2L$, represents the *mean square error* of the approximation. Parseval's identity indicates that as $M \to \infty$ the mean square error approaches zero, while Bessel's inequality indicates the possibility that this mean square error does not approach zero.

   The results are connected with the idea of *completeness* of an orthonormal set. If, for example, we were to leave out one or more terms in a Fourier series (say $\cos 4\pi x/L$, for example) we could never get the mean square error to approach zero no matter how many terms we took. For an analogy with 3 dimensional vectors, see Problem 60.

DIFFERENTIATION and INTEGRATION of FOURIER SERIES

16. (a) Find a Fourier series for $f(x) = x^2$, $0 < x < 2$, by integrating the series of Problem 12(a). (b) Use (a) to evaluate the series $\sum_{n=1}^{\infty} \dfrac{(-1)^{n-1}}{n^2}$.

(a) From Problem 12(a),

$$x = \frac{4}{\pi}\left(\sin\frac{\pi x}{2} - \frac{1}{2}\sin\frac{2\pi x}{2} + \frac{1}{3}\sin\frac{3\pi x}{2} - \cdots\right) \tag{1}$$

Integrating both sides from 0 to x (applying the theorem of Page 300) and multiplying by 2, we find

$$x^2 = C - \frac{16}{\pi^2}\left(\cos\frac{\pi x}{2} - \frac{1}{2^2}\cos\frac{2\pi x}{2} + \frac{1}{3^2}\cos\frac{3\pi x}{2} - \cdots\right) \tag{2}$$

where $C = \dfrac{16}{\pi^2}\left(1 - \dfrac{1}{2^2} + \dfrac{1}{3^2} - \dfrac{1}{4^2} + \cdots\right)$.

(b) To determine C in another way, note that (2) represents the Fourier cosine series for x^2 in $0 < x < 2$. Then since $L = 2$ in this case,

$$C = \frac{a_0}{2} = \frac{1}{L}\int_0^L f(x)\,' = \frac{1}{2}\int_0^2 x^2\,dx = \frac{4}{3}$$

Then from the value of C in (a), we have

$$\sum_{n=1}^{\infty}\frac{(-1)^{n-1}}{n^2} = 1 - \frac{1}{2^2} + \frac{1}{3^2} - \frac{1}{4^2} + \cdots = \frac{\pi^2}{16}\cdot\frac{4}{3} = \frac{\pi^2}{12}$$

17. Show that term by term differentiation of the series in Problem 12(a) is not valid.

Term by term differentiation yields $2\left(\cos\dfrac{\pi x}{2} - \cos\dfrac{2\pi x}{2} + \cos\dfrac{3\pi x}{2} - \cdots\right)$.

Since the nth term of this series does not approach 0, the series does not converge for any value of x.

CONVERGENCE of FOURIER SERIES

18. Prove that (a) $\frac{1}{2} + \cos t + \cos 2t + \cdots + \cos Mt = \dfrac{\sin(M+\frac{1}{2})t}{2\sin\frac{1}{2}t}$

(b) $\dfrac{1}{\pi}\displaystyle\int_0^{\pi}\dfrac{\sin(M+\frac{1}{2})t}{2\sin\frac{1}{2}t}\,dt = \dfrac{1}{2}$, $\dfrac{1}{\pi}\displaystyle\int_{-\pi}^0\dfrac{\sin(M+\frac{1}{2})t}{2\sin\frac{1}{2}t}\,dt = \dfrac{1}{2}$.

(a) We have $\cos nt\,\sin\frac{1}{2}t = \frac{1}{2}\{\sin(n+\frac{1}{2})t - \sin(n-\frac{1}{2})t\}$.

Then summing from $n = 1$ to M,

$$\sin\tfrac{1}{2}t\{\cos t + \cos 2t + \cdots + \cos Mt\} = (\sin\tfrac{3}{2}t - \sin\tfrac{1}{2}t) + (\sin\tfrac{5}{2}t - \sin\tfrac{3}{2}t)$$
$$+ \cdots + \left(\sin(M+\tfrac{1}{2})t - \sin(M-\tfrac{1}{2})t\right)$$
$$= \tfrac{1}{2}\{\sin(M+\tfrac{1}{2})t - \sin\tfrac{1}{2}t\}$$

On dividing by $\sin\frac{1}{2}t$ and adding $\frac{1}{2}$, the required result follows.

(b) Integrate the result in (a) from $-\pi$ to 0 and 0 to π respectively. This gives the required results, since the integrals of all the cosine terms are zero.

19. Prove that $\lim\limits_{n\to\infty}\displaystyle\int_{-\pi}^{\pi} f(x)\sin nx\,dx = \lim\limits_{n\to\infty}\displaystyle\int_{-\pi}^{\pi} f(x)\cos nx\,dx = 0$ if $f(x)$ is sectionally continuous.

This follows at once from Problem 15, since if the series $\dfrac{a_0^2}{2} + \sum\limits_{n=1}^{\infty}(a_n^2 + b_n^2)$ is convergent, $\lim\limits_{n\to\infty} a_n = \lim\limits_{n\to\infty} b_n = 0$.

The result is sometimes called *Riemann's theorem.*

20. Prove that $\displaystyle\lim_{M \to \infty} \int_{-\pi}^{\pi} f(x) \sin(M + \tfrac{1}{2})x\, dx = 0$ if $f(x)$ is sectionally continuous.

We have

$$\int_{-\pi}^{\pi} f(x) \sin(M + \tfrac{1}{2})x\, dx \;=\; \int_{-\pi}^{\pi} \{f(x) \sin \tfrac{1}{2}x\} \cos Mx\, dx \;+\; \int_{-\pi}^{\pi} \{f(x) \cos \tfrac{1}{2}x\} \sin Mx\, dx$$

Then the required result follows at once by using the result of Problem 19, with $f(x)$ replaced by $f(x) \sin \tfrac{1}{2}x$ and $f(x) \cos \tfrac{1}{2}x$ respectively which are sectionally continuous if $f(x)$ is.

The result can also be proved when the integration limits are a and b instead of $-\pi$ and π.

21. Assuming that $L = \pi$, i.e. that the Fourier series corresponding to $f(x)$ has period $2L = 2\pi$, show that

$$S_M(x) \;=\; \frac{a_0}{2} + \sum_{n=1}^{M} (a_n \cos nx + b_n \sin nx) \;=\; \frac{1}{\pi} \int_{-\pi}^{\pi} f(t+x)\, \frac{\sin(M + \tfrac{1}{2})t}{2 \sin \tfrac{1}{2}t}\, dt$$

Using the formulas for the Fourier coefficients with $L = \pi$, we have

$$a_n \cos nx + b_n \sin nx \;=\; \left(\frac{1}{\pi} \int_{-\pi}^{\pi} f(u) \cos nu\, du \right) \cos nx \;+\; \left(\frac{1}{\pi} \int_{-\pi}^{\pi} f(u) \sin nu\, du \right) \sin nx$$

$$=\; \frac{1}{\pi} \int_{-\pi}^{\pi} f(u) \left(\cos nu \cos nx + \sin nu \sin nx \right) du$$

$$=\; \frac{1}{\pi} \int_{-\pi}^{\pi} f(u) \cos n(u - x)\, du$$

Also, $$\frac{a_0}{2} \;=\; \frac{1}{2\pi} \int_{-\pi}^{\pi} f(u)\, du$$

Then $$S_M(x) \;=\; \frac{a_0}{2} + \sum_{n=1}^{M} (a_n \cos nx + b_n \sin nx)$$

$$=\; \frac{1}{2\pi} \int_{-\pi}^{\pi} f(u)\, du \;+\; \frac{1}{\pi} \sum_{n=1}^{M} \int_{-\pi}^{\pi} f(u) \cos n(u - x)\, du$$

$$=\; \frac{1}{\pi} \int_{-\pi}^{\pi} f(u) \left\{ \frac{1}{2} + \sum_{n=1}^{M} \cos n(u - x) \right\} du$$

$$=\; \frac{1}{\pi} \int_{-\pi}^{\pi} f(u)\, \frac{\sin(M + \tfrac{1}{2})(u - x)}{2 \sin \tfrac{1}{2}(u - x)}\, du$$

using Problem 18. Letting $u - x = t$, we have

$$S_M(x) \;=\; \frac{1}{\pi} \int_{-\pi - x}^{\pi - x} f(t+x)\, \frac{\sin(M + \tfrac{1}{2})t}{2 \sin \tfrac{1}{2}t}\, dt$$

Since the integrand has period 2π, we can replace the interval $-\pi - x, \pi - x$ by any other interval of length 2π, in particular $-\pi, \pi$. Thus we obtain the required result.

22. Prove that

$$S_M(x) - \left(\frac{f(x+0) + f(x-0)}{2} \right) \;=\; \frac{1}{\pi} \int_{-\pi}^{0} \left(\frac{f(t+x) - f(x-0)}{2 \sin \tfrac{1}{2}t} \right) \sin(M + \tfrac{1}{2})t\, dt$$

$$+\; \frac{1}{\pi} \int_{0}^{\pi} \left(\frac{f(t+x) - f(x+0)}{2 \sin \tfrac{1}{2}t} \right) \sin(M + \tfrac{1}{2})t\, dt$$

From Problem 21,

$$S_M(x) \;=\; \frac{1}{\pi} \int_{-\pi}^{0} f(t+x)\, \frac{\sin(M + \tfrac{1}{2})t}{2 \sin \tfrac{1}{2}t}\, dt \;+\; \frac{1}{\pi} \int_{0}^{\pi} f(t+x)\, \frac{\sin(M + \tfrac{1}{2})t}{2 \sin \tfrac{1}{2}t}\, dt \qquad (1)$$

Multiplying the integrals of Problem 18(b) by $f(x - 0)$ and $f(x + 0)$ respectively,

$$\frac{f(x+0) + f(x-0)}{2} \;=\; \frac{1}{\pi} \int_{-\pi}^{0} f(x-0)\, \frac{\sin(M + \tfrac{1}{2})t}{2 \sin \tfrac{1}{2}t}\, dt \;+\; \frac{1}{\pi} \int_{0}^{\pi} f(x+0)\, \frac{\sin(M + \tfrac{1}{2})t}{2 \sin \tfrac{1}{2}t}\, dt \qquad (2)$$

Subtracting (2) from (1) yields the required result.

23. If $f(x)$ and $f'(x)$ are sectionally continuous in $(-\pi, \pi)$, prove that

$$\lim_{M \to \infty} S_M(x) = \frac{f(x+0) + f(x-0)}{2}$$

The function $\dfrac{f(t+x) - f(x+0)}{2 \sin \frac{1}{2}t}$ is sectionally continuous in $0 < t \leq \pi$ because $f(x)$ is sectionally continuous.

Also, $\lim_{t \to 0+} \dfrac{f(t+x) - f(x+0)}{2 \sin \frac{1}{2}t} = \lim_{t \to 0+} \dfrac{f(t+x) - f(x+0)}{t} \cdot \dfrac{t}{2 \sin \frac{1}{2}t} = \lim_{t \to 0+} \dfrac{f(t+x) - f(x+0)}{t}$

exists, since by hypothesis $f'(x)$ is sectionally continuous so that the right hand derivative of $f(x)$ at each x exists.

Thus $\dfrac{f(t+x) - f(x+0)}{2 \sin \frac{1}{2}t}$ is sectionally continuous in $0 \leq t \leq \pi$.

Similarly, $\dfrac{f(t+x) - f(x-0)}{2 \sin \frac{1}{2}t}$ is sectionally continuous in $-\pi \leq t \leq 0$.

Then from Problems 20 and 22, we have

$$\lim_{M \to \infty} S_M(x) - \left\{ \frac{f(x+0) + f(x-0)}{2} \right\} = 0 \qquad \text{or} \qquad \lim_{M \to \infty} S_M(x) = \frac{f(x+0) + f(x-0)}{2}$$

BOUNDARY-VALUE PROBLEMS

24. Find a solution $U(x, t)$ of the boundary-value problem

$$\frac{\partial U}{\partial t} = 3 \frac{\partial^2 U}{\partial x^2} \qquad\qquad t > 0, \, 0 < x < 2$$

$$U(0, t) = 0, \; U(2, t) = 0 \qquad t > 0$$

$$U(x, 0) = x \qquad\qquad 0 < x < 2$$

A method commonly employed in practice is to assume the existence of a solution of the partial differential equation having the particular form $U(x, t) = X(x) \, T(t)$, where $X(x)$ and $T(t)$ are functions of x and t respectively, which we shall try to determine. For this reason the method is often called the method of *separation of variables*.

Substitution in the differential equation yields

$$(1) \quad \frac{\partial}{\partial t}(XT) = 3 \frac{\partial^2}{\partial x^2}(XT) \qquad \text{or} \qquad (2) \quad X \frac{dT}{dt} = 3T \frac{d^2X}{dx^2}$$

where we have written X and T in place of $X(x)$ and $T(t)$.

Equation (2) can be written as

$$\frac{1}{3T} \frac{dT}{dt} = \frac{1}{X} \frac{d^2X}{dx^2} \tag{3}$$

Since one side depends only on t and the other only on x, and since x and t are independent variables, it is clear that each side must be a constant c.

In Problem 47 we see that if $c \geq 0$, a solution satisfying the given boundary conditions cannot exist.

Let us thus assume that c is a negative constant which we write as $-\lambda^2$. Then from (3) we obtain two ordinary differential equations

$$\frac{dT}{dt} + 3\lambda^2 T = 0, \qquad \frac{d^2X}{dx^2} + \lambda^2 X = 0 \tag{4}$$

whose solutions are respectively

$$T = C_1 e^{-3\lambda^2 t}, \qquad X = A_1 \cos \lambda x + B_1 \sin \lambda x \tag{5}$$

A solution is given by the product of X and T which can be written

$$U(x, t) = e^{-3\lambda^2 t}(A \cos \lambda x + B \sin \lambda x) \tag{6}$$

where A and B are constants.

We now seek to determine A and B so that (6) satisfies the given boundary conditions. To satisfy the condition $U(0, t) = 0$, we must have

$$e^{-3\lambda^2 t}(A) = 0 \quad \text{or} \quad A = 0 \tag{7}$$

so that (6) becomes

$$U(x, t) = Be^{-3\lambda^2 t} \sin \lambda x \tag{8}$$

To satisfy the condition $U(2, t) = 0$, we must then have

$$Be^{-3\lambda^2 t} \sin 2\lambda = 0 \tag{9}$$

Since $B = 0$ makes the solution (8) identically zero, we avoid this choice and instead take

$$\sin 2\lambda = 0, \quad \text{i.e.} \quad 2\lambda = m\pi \text{ or } \lambda = \frac{m\pi}{2} \tag{10}$$

where $m = 0, \pm 1, \pm 2, \ldots$.

Substitution in (8) now shows that a solution satisfying the first two boundary conditions is

$$U(x, t) = B_m e^{-3m^2\pi^2 t/4} \sin \frac{m\pi x}{2} \tag{11}$$

where we have replaced B by B_m, indicating that different constants can be used for different values of m.

If we now attempt to satisfy the last boundary condition $U(x, 0) = x$, $0 < x < 2$, we find it to be impossible using (11). However, upon recognizing the fact that *sums* of solutions having the form (11) are also solutions (called the *principle of superposition*), we are led to the possible solution

$$U(x, t) = \sum_{m=1}^{\infty} B_m e^{-3m^2\pi^2 t/4} \sin \frac{m\pi x}{2} \tag{12}$$

From the condition $U(x, 0) = x$, $0 < x < 2$, we see, on placing $t = 0$, that (12) becomes

$$x = \sum_{m=1}^{\infty} B_m \sin \frac{m\pi x}{2} \qquad 0 < x < 2 \tag{13}$$

This, however, is equivalent to the problem of expanding the function $f(x) = x$ for $0 < x < 2$ into a sine series. The solution to this is given in Problem 12(a), from which we see that $B_m = \frac{-4}{m\pi} \cos m\pi$ so that (12) becomes

$$U(x, t) = \sum_{m=1}^{\infty} \left(-\frac{4}{m\pi} \cos m\pi \right) e^{-3m^2\pi^2 t/4} \sin \frac{m\pi x}{2} \tag{14}$$

which is a *formal solution*. To check that (14) is actually a solution, we must show that it satisfies the partial differential equation and the boundary conditions. The proof consists in justification of term by term differentiation and use of limiting procedures for infinite series and may be accomplished by methods of Chapter 11.

The boundary value problem considered here has an interpretation in the theory of heat conduction. The equation $\frac{\partial U}{\partial t} = k \frac{\partial^2 U}{\partial x^2}$ is the equation for heat conduction in a thin rod or wire located on the x axis between $x = 0$ and $x = L$ if the surface of the wire is insulated so that heat cannot enter or escape. $U(x, t)$ is the temperature at any place x in the rod at time t. The constant $k = K/s\rho$ (where K is the *thermal conductivity*, s is the *specific heat*, and ρ is the *density* of the conducting material) is called the *diffusivity*. The boundary conditions $U(0, t) = 0$ and $U(L, t) = 0$ indicate that the end temperatures of the rod are kept at zero units for all time $t > 0$, while $U(x, 0)$ indicates the initial temperature at any point x of the rod. In this problem the length of the rod is $L = 2$ units, while the diffusivity is $k = 3$ units.

ORTHOGONAL FUNCTIONS

25. (a) Show that the set of functions

$$1, \ \sin \frac{\pi x}{L}, \ \cos \frac{\pi x}{L}, \ \sin \frac{2\pi x}{L}, \ \cos \frac{2\pi x}{L}, \ \sin \frac{3\pi x}{L}, \ \cos \frac{3\pi x}{L}, \ldots$$

forms an orthogonal set in the interval $(-L, L)$.

(b) Determine the corresponding normalizing constants for the set in (a) so that the set is orthonormal in $(-L, L)$.

(a) This follows at once from the results of Problems 2 and 3.

(b) By Problem 3,

$$\int_{-L}^{L} \sin^2 \frac{m\pi x}{L} \, dx = L, \qquad \int_{-L}^{L} \cos^2 \frac{m\pi x}{L} \, dx = L$$

Then

$$\int_{-L}^{L} \left(\sqrt{\frac{1}{L}} \sin \frac{m\pi x}{L} \right)^2 dx = 1, \qquad \int_{-L}^{L} \left(\sqrt{\frac{1}{L}} \cos \frac{m\pi x}{L} \right)^2 dx = 1$$

Also,

$$\int_{-L}^{L} (1)^2 \, dx = 2L \quad \text{or} \quad \int_{-L}^{L} \left(\frac{1}{\sqrt{2L}} \right)^2 dx = 1$$

Thus the required orthonormal set is given by

$$\frac{1}{\sqrt{2L}}, \ \frac{1}{\sqrt{L}} \sin \frac{\pi x}{L}, \ \frac{1}{\sqrt{L}} \cos \frac{\pi x}{L}, \ \frac{1}{\sqrt{L}} \sin \frac{2\pi x}{L}, \ \frac{1}{\sqrt{L}} \cos \frac{2\pi x}{L}, \ \ldots$$

MISCELLANEOUS PROBLEMS

26. Find a Fourier series for $f(x) = \cos \alpha x$, $-\pi \leqq x \leqq \pi$, where $\alpha \neq 0, \pm 1, \pm 2, \pm 3, \ldots$.

We shall take the period as 2π so that $2L = 2\pi$, $L = \pi$. Since the function is even, $b_n = 0$ and

$$a_n = \frac{2}{L} \int_0^L f(x) \cos nx \, dx = \frac{2}{\pi} \int_0^\pi \cos \alpha x \cos nx \, dx$$

$$= \frac{1}{\pi} \int_0^\pi \{ \cos (\alpha - n)x + \cos (\alpha + n)x \} \, dx$$

$$= \frac{1}{\pi} \left\{ \frac{\sin (\alpha - n)\pi}{\alpha - n} + \frac{\sin (\alpha + n)\pi}{\alpha + n} \right\} = \frac{2\alpha \sin \alpha\pi \cos n\pi}{\pi(\alpha^2 - n^2)}$$

$$a_0 = \frac{2 \sin \alpha\pi}{\alpha\pi}$$

Then

$$\cos \alpha x = \frac{\sin \alpha\pi}{\alpha\pi} + \frac{2\alpha \sin \alpha\pi}{\pi} \sum_{n=1}^{\infty} \frac{\cos n\pi}{\alpha^2 - n^2} \cos nx$$

$$= \frac{\sin \alpha\pi}{\pi} \left(\frac{1}{\alpha} - \frac{2\alpha}{\alpha^2 - 1^2} \cos x + \frac{2\alpha}{\alpha^2 - 2^2} \cos 2x - \frac{2\alpha}{\alpha^2 - 3^2} \cos 3x + \cdots \right)$$

27. Prove that $\sin x = x \left(1 - \frac{x^2}{\pi^2} \right) \left(1 - \frac{x^2}{(2\pi)^2} \right) \left(1 - \frac{x^2}{(3\pi)^2} \right) \cdots$.

Let $x = \pi$ in the Fourier series obtained in Problem 26. Then

$$\cos \alpha\pi = \frac{\sin \alpha\pi}{\pi} \left(\frac{1}{\alpha} + \frac{2\alpha}{\alpha^2 - 1^2} + \frac{2\alpha}{\alpha^2 - 2^2} + \frac{2\alpha}{\alpha^2 - 3^2} + \cdots \right)$$

or

$$\pi \cot \alpha\pi - \frac{1}{\alpha} = \frac{2\alpha}{\alpha^2 - 1^2} + \frac{2\alpha}{\alpha^2 - 2^2} + \frac{2\alpha}{\alpha^2 - 3^2} + \cdots \qquad (1)$$

This result is of interest since it represents an expansion of the cotangent into partial fractions.

By the Weierstrass M test, the series on the right of (1) converges uniformly for $0 \leq |\alpha| \leq |x| < 1$ and the left hand side of (1) approaches zero as $\alpha \to 0$, as is seen by using L'Hospital's rule. Thus we can integrate both sides of (1) from 0 to x to obtain

$$\int_0^x \left(\pi \cot \alpha \pi - \frac{1}{\alpha} \right) d\alpha \;=\; \int_0^x \frac{2\alpha}{\alpha^2 - 1} \, d\alpha \;+\; \int_0^x \frac{2\alpha}{\alpha^2 - 2^2} \, d\alpha \;+\; \cdots$$

or

$$\ln \left(\frac{\sin \alpha \pi}{\alpha \pi} \right) \Big|_0^x \;=\; \ln \left(1 - \frac{x^2}{1^2} \right) + \ln \left(1 - \frac{x^2}{2^2} \right) + \cdots$$

i.e.,

$$\ln \left(\frac{\sin \pi x}{\pi x} \right) \;=\; \lim_{n \to \infty} \; \ln \left(1 - \frac{x^2}{1^2} \right) + \ln \left(1 - \frac{x^2}{2^2} \right) + \cdots + \ln \left(1 - \frac{x^2}{n^2} \right)$$

$$= \lim_{n \to \infty} \; \ln \left\{ \left(1 - \frac{x^2}{1^2} \right) \left(1 - \frac{x^2}{2^2} \right) \cdots \left(1 - \frac{x^2}{n^2} \right) \right\}$$

$$= \ln \left\{ \lim_{n \to \infty} \left(1 - \frac{x^2}{1^2} \right) \left(1 - \frac{x^2}{2^2} \right) \cdots \left(1 - \frac{x^2}{n^2} \right) \right\}$$

so that

$$\frac{\sin \pi x}{\pi x} \;=\; \lim_{n \to \infty} \left(1 - \frac{x^2}{1^2} \right) \left(1 - \frac{x^2}{2^2} \right) \cdots \left(1 - \frac{x^2}{n^2} \right) \;=\; \left(1 - \frac{x^2}{1^2} \right) \left(1 - \frac{x^2}{2^2} \right) \cdots \qquad (2)$$

Replacing x by x/π, we obtain

$$\sin x \;=\; x \left(1 - \frac{x^2}{\pi^2} \right) \left(1 - \frac{x^2}{(2\pi)^2} \right) \cdots \qquad (3)$$

called the *infinite product for sin x*, which can be shown valid for all x. The result is of interest since it corresponds to a factorization of $\sin x$ in a manner analogous to factorization of a polynomial.

28. Prove that $\quad \dfrac{\pi}{2} \;=\; \dfrac{2 \cdot 2 \cdot 4 \cdot 4 \cdot 6 \cdot 6 \cdot 8 \cdot 8 \cdots}{1 \cdot 3 \cdot 3 \cdot 5 \cdot 5 \cdot 7 \cdot 7 \cdot 9 \cdots}$.

Let $x = 1/2$ in equation (2) of Problem 27. Then,

$$\frac{2}{\pi} \;=\; \left(1 - \frac{1}{2^2} \right) \left(1 - \frac{1}{4^2} \right) \left(1 - \frac{1}{6^2} \right) \cdots \;=\; \left(\frac{1}{2} \cdot \frac{3}{2} \right) \left(\frac{3}{4} \cdot \frac{5}{4} \right) \left(\frac{5}{6} \cdot \frac{7}{6} \right) \cdots$$

Taking reciprocals of both sides, we obtain the required result which is often called *Wallis' product*.

Supplementary Problems

FOURIER SERIES

29. Graph each of the following functions and find their corresponding Fourier series using properties of even and odd functions wherever applicable.

(a) $f(x) = \begin{cases} 8 & 0 < x < 2 \\ -8 & 2 < x < 4 \end{cases}$ Period 4

(b) $f(x) = \begin{cases} -x & -4 \leqq x \leqq 0 \\ x & 0 \leqq x \leqq 4 \end{cases}$ Period 8

(c) $f(x) = 4x, \; 0 < x < 10,$ Period 10

(d) $f(x) = \begin{cases} 2x & 0 \leqq x < 3 \\ 0 & -3 < x < 0 \end{cases}$ Period 6

Ans. (a) $\dfrac{16}{\pi} \displaystyle\sum_{n=1}^{\infty} \dfrac{(1 - \cos n\pi)}{n} \sin \dfrac{n\pi x}{2}$

(b) $2 - \dfrac{8}{\pi^2} \displaystyle\sum_{n=1}^{\infty} \dfrac{(1 - \cos n\pi)}{n^2} \cos \dfrac{n\pi x}{4}$

(c) $20 - \dfrac{40}{\pi} \displaystyle\sum_{n=1}^{\infty} \dfrac{1}{n} \sin \dfrac{n\pi x}{5}$

(d) $\dfrac{3}{2} + \displaystyle\sum_{n=1}^{\infty} \left\{ \dfrac{6(\cos n\pi - 1)}{n^2 \pi^2} \cos \dfrac{n\pi x}{3} - \dfrac{6 \cos n\pi}{n\pi} \sin \dfrac{n\pi x}{3} \right\}$

30. In each part of Problem 29, tell where the discontinuities of $f(x)$ are located and to what value the series converges at these discontinuities.

Ans. (a) $x = 0, \pm 2, \pm 4, \ldots; \; 0$ (b) no discontinuities

(c) $x = 0, \pm 10, \pm 20, \ldots; \; 20$ (d) $x = \pm 3, \pm 9, \pm 15, \ldots; \; 3$

31. Expand $f(x) = \begin{cases} 2 - x & 0 < x < 4 \\ x - 6 & 4 < x < 8 \end{cases}$ in a Fourier series of period 8.

Ans. $\dfrac{16}{\pi^2} \left\{ \cos \dfrac{\pi x}{4} + \dfrac{1}{3^2} \cos \dfrac{3\pi x}{4} + \dfrac{1}{5^2} \cos \dfrac{5\pi x}{4} + \cdots \right\}$

32. (a) Expand $f(x) = \cos x, \; 0 < x < \pi,$ in a Fourier sine series.

(b) How should $f(x)$ be defined at $x = 0$ and $x = \pi$ so that the series will converge to $f(x)$ for $0 \leqq x \leqq \pi$?

Ans. (a) $\dfrac{8}{\pi} \displaystyle\sum_{n=1}^{\infty} \dfrac{n \sin 2nx}{4n^2 - 1}$ (b) $f(0) = f(\pi) = 0$

33. (a) Expand in a Fourier series $f(x) = \cos x, \; 0 < x < \pi$ if the period is π; and (b) compare with the result of Problem 32, explaining the similarities and differences if any.

Ans. Answer is the same as in Problem 32.

34. Expand $f(x) = \begin{cases} x & 0 < x < 4 \\ 8 - x & 4 < x < 8 \end{cases}$ in a series of (a) sines, (b) cosines.

Ans. (a) $\dfrac{32}{\pi^2} \displaystyle\sum_{n=1}^{\infty} \dfrac{1}{n^2} \sin \dfrac{n\pi}{2} \sin \dfrac{n\pi x}{8}$

(b) $\dfrac{16}{\pi^2} \displaystyle\sum_{n=1}^{\infty} \left(\dfrac{2 \cos n\pi/2 - \cos n\pi - 1}{n^2} \right) \cos \dfrac{n\pi x}{8}$

35. Prove that for $0 \leqq x \leqq \pi$,

(a) $x(\pi - x) = \dfrac{\pi^2}{6} - \left(\dfrac{\cos 2x}{1^2} + \dfrac{\cos 4x}{2^2} + \dfrac{\cos 6x}{3^2} + \cdots \right)$

(b) $x(\pi - x) = \dfrac{8}{\pi} \left(\dfrac{\sin x}{1^3} + \dfrac{\sin 3x}{3^3} + \dfrac{\sin 5x}{5^3} + \cdots \right)$

36. Use the preceding problem to show that

(a) $\displaystyle\sum_{n=1}^{\infty} \dfrac{1}{n^2} = \dfrac{\pi^2}{6},$ (b) $\displaystyle\sum_{n=1}^{\infty} \dfrac{(-1)^{n-1}}{n^2} = \dfrac{\pi^2}{12},$ (c) $\displaystyle\sum_{n=1}^{\infty} \dfrac{(-1)^{n-1}}{(2n-1)^3} = \dfrac{\pi^3}{32}.$

37. Show that $\dfrac{1}{1^3} + \dfrac{1}{3^3} - \dfrac{1}{5^3} - \dfrac{1}{7^3} + \dfrac{1}{9^3} + \dfrac{1}{11^3} - \cdots = \dfrac{3\pi^2 \sqrt{2}}{16}.$

DIFFERENTIATION and INTEGRATION of FOURIER SERIES

38. (a) Show that for $-\pi < x < \pi$,

$$x = 2\left(\frac{\sin x}{1} - \frac{\sin 2x}{2} + \frac{\sin 3x}{3} - \cdots\right)$$

(b) By integrating the result of (a), show that for $-\pi \leq x \leq \pi$,

$$x^2 = \frac{\pi^2}{3} - 4\left(\frac{\cos x}{1^2} - \frac{\cos 2x}{2^2} + \frac{\cos 3x}{3^2} - \cdots\right)$$

(c) By integrating the result of (b), show that for $-\pi \leq x \leq \pi$,

$$x(\pi - x)(\pi + x) = 12\left(\frac{\sin x}{1^3} - \frac{\sin 2x}{2^3} + \frac{\sin 3x}{3^3} - \cdots\right)$$

39. (a) Show that for $-\pi < x < \pi$,

$$x \cos x = -\tfrac{1}{2}\sin x + 2\left(\frac{2}{1\cdot 3}\sin 2x - \frac{3}{2\cdot 4}\sin 3x + \frac{4}{3\cdot 5}\sin 4x - \cdots\right)$$

(b) Use (a) to show that for $-\pi \leq x \leq \pi$,

$$x \sin x = 1 - \tfrac{1}{2}\cos x - 2\left(\frac{\cos 2x}{1\cdot 3} - \frac{\cos 3x}{2\cdot 4} + \frac{\cos 4x}{3\cdot 5} - \cdots\right)$$

40. By differentiating the result of Problem 35(b), prove that for $0 \leq x \leq \pi$,

$$x = \frac{\pi}{2} - \frac{4}{\pi}\left(\frac{\cos x}{1^2} + \frac{\cos 3x}{3^2} + \frac{\cos 5x}{5^2} + \cdots\right)$$

PARSEVAL'S IDENTITY

41. By using Problem 35 and Parseval's identity, show that

$$(a)\ \sum_{n=1}^{\infty}\frac{1}{n^4} = \frac{\pi^4}{90} \qquad (b)\ \sum_{n=1}^{\infty}\frac{1}{n^6} = \frac{\pi^6}{945}$$

42. Show that $\dfrac{1}{1^2\cdot 3^2} + \dfrac{1}{3^2\cdot 5^2} + \dfrac{1}{5^2\cdot 7^2} + \cdots = \dfrac{\pi^2-8}{16}$. [Hint: Use Problem 11.]

43. Show that $(a)\ \sum_{n=1}^{\infty}\dfrac{1}{(2n-1)^4} = \dfrac{\pi^4}{96}$, $(b)\ \sum_{n=1}^{\infty}\dfrac{1}{(2n-1)^6} = \dfrac{\pi^6}{960}$.

44. Show that $\dfrac{1}{1^2\cdot 2^2\cdot 3^2} + \dfrac{1}{2^2\cdot 3^2\cdot 4^2} + \dfrac{1}{3^2\cdot 4^2\cdot 5^2} + \cdots = \dfrac{4\pi^2-39}{16}$.

BOUNDARY-VALUE PROBLEMS

45. (a) Solve $\dfrac{\partial U}{\partial t} = 2\dfrac{\partial^2 U}{\partial x^2}$ subject to the conditions $U(0,t)=0,\ U(4,t)=0,\ U(x,0)=3\sin\pi x - 2\sin 5\pi x$, where $0 < x < 4,\ t > 0$.
(b) Give a possible physical interpretation of the problem and solution.
Ans. (a) $U(x,t) = 3e^{-2\pi^2 t}\sin\pi x - 2e^{-50\pi^2 t}\sin 5\pi x$.

46. Solve $\dfrac{\partial U}{\partial t} = \dfrac{\partial^2 U}{\partial x^2}$ subject to the conditions $U(0,t)=0,\ U(6,t)=0,\ U(x,0)=\begin{cases}1 & 0<x<3\\0 & 3<x<6\end{cases}$
and interpret physically.
Ans. $U(x,t) = \sum_{m=1}^{\infty} 2\left[\dfrac{1-\cos(m\pi/3)}{m\pi}\right]e^{-m^2\pi^2 t/36}\sin\dfrac{m\pi x}{6}$

47. Show that if each side of equation (3), Page 313, is a constant c where $c \geq 0$, then there is no solution satisfying the boundary-value problem.

48. A flexible string of length π is tightly stretched between points $x = 0$ and $x = \pi$ on the x axis, its ends fixed at these points. When set into small transverse vibration the displacement $Y(x, t)$ from the x axis of any point x at time t is given by $\dfrac{\partial^2 Y}{\partial t^2} = a^2 \dfrac{\partial^2 Y}{\partial x^2}$ where $a^2 = T/\rho$, $T = $ tension, $\rho = $ mass per unit length.

(a) Find a solution of this equation (sometimes called the *wave equation*) with $a^2 = 4$ which satisfies the conditions $Y(0, t) = 0$, $Y(\pi, t) = 0$, $Y(x, 0) = 0.1 \sin x + 0.01 \sin 4x$, $Y_t(x, 0) = 0$ for $0 < x < \pi$, $t > 0$.

(b) Interpret physically the boundary conditions in (a) and the solution.

Ans. (a) $Y(x, t) = 0.1 \sin x \cos 2t + 0.01 \sin 4x \cos 8t$

49. (a) Solve the boundary-value problem $\dfrac{\partial^2 Y}{\partial t^2} = 9 \dfrac{\partial^2 Y}{\partial x^2}$ subject to the conditions $Y(0, t) = 0$, $Y(2, t) = 0$, $Y(x, 0) = 0.05x(2 - x)$, $Y_t(x, 0) = 0$, where $0 < x < 2$, $t > 0$. (b) Interpret physically.

Ans. (a) $Y(x, t) = \dfrac{1.6}{\pi^3} \sum\limits_{n=1}^{\infty} \dfrac{1}{(2n-1)^3} \sin \dfrac{(2n-1)\pi x}{2} \cos \dfrac{3(2n-1)\pi t}{2}$

50. Solve the boundary-value problem $\dfrac{\partial U}{\partial t} = \dfrac{\partial^2 U}{\partial x^2}$, $U(0, t) = 1$, $U(\pi, t) = 3$, $U(x, 0) = 2$.

[Hint: Let $U(x, t) = V(x, t) + F(x)$ and choose $F(x)$ so as to simplify the differential equation and boundary conditions for $V(x, t)$.]

Ans. $U(x, t) = 1 + \dfrac{2x}{\pi} + \sum\limits_{m=1}^{\infty} \dfrac{4 \cos m\pi}{m\pi} e^{-m^2 t} \sin mx$

51. Give a physical interpretation to Problem 50.

52. Solve Problem 49 with the boundary conditions for $Y(x, 0)$ and $Y_t(x, 0)$ interchanged, i.e. $Y(x, 0) = 0$, $Y_t(x, 0) = 0.05x(2 - x)$, and give a physical interpretation.

Ans. $Y(x, t) = \dfrac{3.2}{3\pi^4} \sum\limits_{n=1}^{\infty} \dfrac{1}{(2n-1)^4} \sin \dfrac{(2n-1)\pi x}{2} \sin \dfrac{3(2n-1)\pi t}{2}$

53. Verify that the boundary-value problem of Problem 24 actually has the solution (*14*), Page 314.

MISCELLANEOUS PROBLEMS

54. If $-\pi < x < \pi$ and $\alpha \neq 0, \pm 1, \pm 2, \ldots$, prove that

$$\frac{\pi}{2} \frac{\sin \alpha x}{\sin \alpha \pi} = \frac{\sin x}{1^2 - \alpha^2} - \frac{2 \sin 2x}{2^2 - \alpha^2} + \frac{3 \sin 3x}{3^2 - \alpha^2} - \cdots$$

55. If $-\pi < x < \pi$, prove that

(a) $$\frac{\pi}{2} \frac{\sinh \alpha x}{\sinh \alpha \pi} = \frac{\sin x}{\alpha^2 + 1^2} - \frac{2 \sin 2x}{\alpha^2 + 2^2} + \frac{3 \sin 3x}{\alpha^2 + 3^2} - \cdots$$

(b) $$\frac{\pi}{2} \frac{\cosh \alpha x}{\sinh \alpha \pi} = \frac{1}{2\alpha} - \frac{\alpha \cos x}{\alpha^2 + 1^2} + \frac{\alpha \cos 2x}{\alpha^2 + 2^2} - \cdots$$

56. Prove that $\sinh x = x \left(1 + \dfrac{x^2}{\pi^2}\right) \left(1 + \dfrac{x^2}{(2\pi)^2}\right) \left(1 + \dfrac{x^2}{(3\pi)^2}\right) \cdots$

57. Prove that $\cos x = \left(1 - \dfrac{4x^2}{\pi^2}\right) \left(1 - \dfrac{4x^2}{(3\pi)^2}\right) \left(1 - \dfrac{4x^2}{(5\pi)^2}\right) \cdots$

[Hint: $\cos x = (\sin 2x)/(2 \sin x)$.]

58. Show that (a) $\dfrac{\sqrt{2}}{2} = \dfrac{1 \cdot 3 \cdot 5 \cdot 7 \cdot 9 \cdot 11 \cdot 13 \cdot 15 \cdots}{2 \cdot 2 \cdot 6 \cdot 6 \cdot 10 \cdot 10 \cdot 14 \cdot 14 \cdots}$

(b) $\pi\sqrt{2} = 4 \left(\dfrac{4 \cdot 4 \cdot 8 \cdot 8 \cdot 12 \cdot 12 \cdot 16 \cdot 16 \cdots}{3 \cdot 5 \cdot 7 \cdot 9 \cdot 11 \cdot 13 \cdot 15 \cdot 17 \cdots}\right)$

59. (a) Prove that if $\alpha \neq 0, \pm 1, \pm 2, \ldots$, then

$$\frac{\pi}{\sin \alpha\pi} = \frac{1}{\alpha} - \frac{2\alpha}{\alpha^2 - 1^2} + \frac{2\alpha}{\alpha^2 - 2^2} - \frac{2\alpha}{\alpha^2 - 3^2} + \cdots$$

(b) Prove that if $0 < \alpha < 1$,

$$\int_0^\infty \frac{x^{\alpha-1}}{1+x}\, dx = \int_0^1 \frac{x^{\alpha-1} - x^{-\alpha}}{1+x}\, dx$$

$$= \frac{1}{\alpha} - \frac{2\alpha}{\alpha^2 - 1^2} + \frac{2\alpha}{\alpha^2 - 2^2} - \frac{2\alpha}{\alpha^2 - 3^2} + \cdots$$

(c) From (a), (b) and Problem 17, Chapter 13, complete the proof of the fact that

$$\Gamma(\alpha)\,\Gamma(1-\alpha) = \frac{\pi}{\sin \alpha\pi}, \qquad 0 < \alpha < 1$$

[Hint: For (a), use Problem 26. For (b), write the given integral as the sum of the integrals from 0 to 1 and 1 to ∞, and let $x = 1/y$ in the last integral thus obtained. Then use the fact that $\frac{1}{1+x} = 1 - x + x^2 - x^3 + \cdots$.]

60. Let **r** be any three dimensional vector. Show that

$$(a)\ (\mathbf{r} \cdot \mathbf{i})^2 + (\mathbf{r} \cdot \mathbf{j})^2 \leqq (\mathbf{r})^2, \qquad (b)\ (\mathbf{r} \cdot \mathbf{i})^2 + (\mathbf{r} \cdot \mathbf{j})^2 + (\mathbf{r} \cdot \mathbf{k})^2 = \mathbf{r}^2$$

and discuss these with reference to Bessel's inequality and Parseval's identity.

61. If $\{\phi_n(x)\}$, $n = 1, 2, 3, \ldots$ is orthonormal in (a, b), prove that $\displaystyle\int_a^b \{f(x) - \sum_{n=1}^\infty c_n \phi_n(x)\}^2\, dx$ is a minimum when

$$c_n = \int_a^b f(x)\, \phi_n(x)\, dx$$

Discuss the relevance of this result to Fourier series.

Chapter 15

Fourier Integrals

The FOURIER INTEGRAL

Let us assume the following conditions on $f(x)$:

1. $f(x)$ satisfies the Dirichlet conditions (Page 299) in every finite interval $(-L, L)$.

2. $\int_{-\infty}^{\infty} |f(x)|\, dx$ converges, i.e. $f(x)$ is absolutely integrable in $(-\infty, \infty)$.

Then *Fourier's integral theorem* states that

$$f(x) = \int_0^\infty \{A(\alpha) \cos \alpha x + B(\alpha) \sin \alpha x\}\, d\alpha \tag{1}$$

where

$$\begin{cases} A(\alpha) = \dfrac{1}{\pi} \int_{-\infty}^{\infty} f(x) \cos \alpha x \, dx \\[2mm] B(\alpha) = \dfrac{1}{\pi} \int_{-\infty}^{\infty} f(x) \sin \alpha x \, dx \end{cases} \tag{2}$$

The result (1) holds if x is a point of continuity of $f(x)$. If x is a point of discontinuity, we must replace $f(x)$ by $\dfrac{f(x+0) + f(x-0)}{2}$ as in the case of Fourier series. Note that the above conditions are sufficient but not necessary.

The similarity of (1) and (2) with corresponding results for Fourier series is apparent. The right hand side of (1) is sometimes called a *Fourier integral expansion* of $f(x)$.

EQUIVALENT FORMS of FOURIER'S INTEGRAL THEOREM

Fourier's integral theorem can also be written in the forms

$$f(x) = \frac{1}{\pi} \int_{\alpha=0}^{\infty} \int_{u=-\infty}^{\infty} f(u) \cos \alpha(x-u) \, du \, d\alpha \tag{3}$$

$$f(x) = \frac{1}{2\pi} \int_{-\infty}^{\infty} e^{-i\alpha x} \, d\alpha \int_{-\infty}^{\infty} f(u) \, e^{i\alpha u} \, du \tag{4}$$

$$= \frac{1}{2\pi} \int_{-\infty}^{\infty} \int_{-\infty}^{\infty} f(u) \, e^{i\alpha(u-x)} \, du \, d\alpha$$

where it is understood that if $f(x)$ is not continuous at x the left side must be replaced by $\dfrac{f(x+0) + f(x-0)}{2}$.

These results can be simplified somewhat if $f(x)$ is either an odd or an even function, and we have

$$f(x) = \frac{2}{\pi} \int_0^\infty \cos \alpha x \, d\alpha \int_0^\infty f(u) \cos \alpha u \, du \qquad \text{if } f(x) \text{ is even} \tag{5}$$

$$f(x) = \frac{2}{\pi} \int_0^\infty \sin \alpha x \, d\alpha \int_0^\infty f(u) \sin \alpha u \, du \qquad \text{if } f(x) \text{ is odd} \tag{6}$$

FOURIER TRANSFORMS

From (4) it follows that if

$$F(\alpha) \;=\; \frac{1}{\sqrt{2\pi}} \int_{-\infty}^{\infty} f(u)\, e^{i\alpha u}\, du \tag{7}$$

then

$$f(x) \;=\; \frac{1}{\sqrt{2\pi}} \int_{-\infty}^{\infty} F(\alpha)\, e^{-i\alpha x}\, d\alpha \tag{8}$$

The function $F(\alpha)$ is called the *Fourier transform* of $f(x)$ and is sometimes written $F(\alpha) = \mathcal{F}\{f(x)\}$. The function $f(x)$ is the *inverse Fourier transform* of $F(\alpha)$ and is written $f(x) = \mathcal{F}^{-1}\{F(\alpha)\}$.

Note: The constants preceding the integral signs in (7) and (8) were here taken as equal to $1/\sqrt{2\pi}$. However, they can be any constants different from zero so long as their product is $1/2\pi$. The above is called the *symmetric form*.

If $f(x)$ is an even function, equation (5) yields

$$\begin{cases} F_c(\alpha) \;=\; \sqrt{\dfrac{2}{\pi}} \displaystyle\int_{0}^{\infty} f(u)\, \cos \alpha u \, du \\[2ex] f(x) \;=\; \sqrt{\dfrac{2}{\pi}} \displaystyle\int_{0}^{\infty} F_c(\alpha)\, \cos \alpha x \, d\alpha \end{cases} \tag{9}$$

and we call $F_c(\alpha)$ and $f(x)$ *Fourier cosine transforms* of each other.

If $f(x)$ is an odd function, equation (6) yields

$$\begin{cases} F_s(\alpha) \;=\; \sqrt{\dfrac{2}{\pi}} \displaystyle\int_{0}^{\infty} f(u)\, \sin \alpha u \, du \\[2ex] f(x) \;=\; \sqrt{\dfrac{2}{\pi}} \displaystyle\int_{0}^{\infty} F_s(\alpha)\, \sin \alpha x \, d\alpha \end{cases} \tag{10}$$

and we call $F_s(\alpha)$ and $f(x)$ *Fourier sine transforms* of each other.

PARSEVAL'S IDENTITIES for FOURIER INTEGRALS

If $F_s(\alpha)$ and $G_s(\alpha)$ are Fourier sine transforms of $f(x)$ and $g(x)$ respectively, then

$$\int_{0}^{\infty} F_s(\alpha)\, G_s(\alpha)\, d\alpha \;=\; \int_{0}^{\infty} f(x)\, g(x)\, dx \tag{11}$$

Similarly if $F_c(\alpha)$ and $G_c(\alpha)$ are Fourier cosine transforms of $f(x)$ and $g(x)$, then

$$\int_{0}^{\infty} F_c(\alpha)\, G_c(\alpha)\, d\alpha \;=\; \int_{0}^{\infty} f(x)\, g(x)\, dx \tag{12}$$

In the special case where $f(x) = g(x)$, (11) and (12) become respectively

$$\int_{0}^{\infty} \{F_s(\alpha)\}^2\, d\alpha \;=\; \int_{0}^{\infty} \{f(x)\}^2\, dx \tag{13}$$

$$\int_{0}^{\infty} \{F_c(\alpha)\}^2\, d\alpha \;=\; \int_{0}^{\infty} \{f(x)\}^2\, dx \tag{14}$$

The above relations are known as *Parseval's identities* for integrals. Similar relations hold for general Fourier transforms. Thus if $F(\alpha)$ and $G(\alpha)$ are Fourier transforms of $f(x)$ and $g(x)$ respectively, we can prove that

$$\int_{-\infty}^{\infty} F(\alpha)\,\overline{G(\alpha)}\,d\alpha \;=\; \int_{-\infty}^{\infty} f(x)\,\overline{g(x)}\,dx \tag{15}$$

where the bar signifies the complex conjugate obtained by replacing i by $-i$. See Prob. 30.

The CONVOLUTION THEOREM

If $F(\alpha)$ and $G(\alpha)$ are the Fourier transforms of $f(x)$ and $g(x)$ respectively, then

$$\int_{-\infty}^{\infty} F(\alpha)\,G(\alpha)\,e^{-i\alpha x}\,d\alpha \;=\; \int_{-\infty}^{\infty} f(u)\,g(x-u)\,du \tag{16}$$

If we define the *convolution*, denoted by $f * g$, of the functions f and g to be

$$f * g \;=\; \frac{1}{\sqrt{2\pi}} \int_{-\infty}^{\infty} f(u)\,g(x-u)\,du \tag{17}$$

then (16) can be written

$$\mathcal{F}\{f * g\} \;=\; \mathcal{F}\{f\}\,\mathcal{F}\{g\} \tag{18}$$

or in words, the Fourier transform of the convolution of two functions is equal to the product of their Fourier transforms. This is called the *convolution theorem for Fourier transforms*.

Solved Problems

The FOURIER INTEGRAL and FOURIER TRANSFORMS

1. (a) Find the Fourier transform of $f(x) = \begin{cases} 1 & |x| < a \\ 0 & |x| > a \end{cases}$.

   (b) Graph $f(x)$ and its Fourier transform for $a = 3$.

   (a) The Fourier transform of $f(x)$ is

$$F(\alpha) \;=\; \frac{1}{\sqrt{2\pi}} \int_{-\infty}^{\infty} f(u)\,e^{i\alpha u}\,du \;=\; \frac{1}{\sqrt{2\pi}} \int_{-a}^{a} (1)\,e^{i\alpha u}\,du \;=\; \frac{1}{\sqrt{2\pi}} \frac{e^{i\alpha u}}{i\alpha}\bigg|_{-a}^{a}$$

$$=\; \frac{1}{\sqrt{2\pi}}\left(\frac{e^{i\alpha a} - e^{-i\alpha a}}{i\alpha}\right) \;=\; \sqrt{\frac{2}{\pi}}\,\frac{\sin\alpha a}{\alpha}\,, \quad \alpha \neq 0$$

   For $\alpha = 0$, we obtain $F(\alpha) = \sqrt{2/\pi}\,a$.

   (b) The graphs of $f(x)$ and $F(\alpha)$ for $a = 3$ are shown in Figures 15-1 and 15-2 respectively.

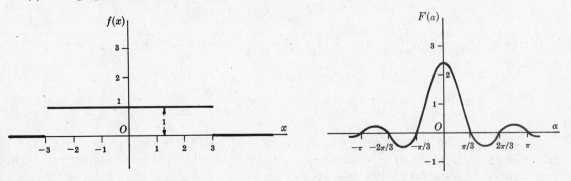

Fig. 15-1   Fig. 15-2

2. *(a)* Use the result of Problem 1 to evaluate $\displaystyle\int_{-\infty}^{\infty} \frac{\sin \alpha a \cos \alpha x}{\alpha}\, d\alpha$

(b) Deduce the value of $\displaystyle\int_{0}^{\infty} \frac{\sin u}{u}\, du$.

(a)   From Fourier's integral theorem, if

$$F(\alpha) = \frac{1}{\sqrt{2\pi}} \int_{-\infty}^{\infty} f(u)\, e^{i\alpha u}\, du \qquad \text{then} \qquad f(x) = \frac{1}{\sqrt{2\pi}} \int_{-\infty}^{\infty} F(\alpha)\, e^{-i\alpha x}\, d\alpha$$

Then from Problem 1,

$$\frac{1}{\sqrt{2\pi}} \int_{-\infty}^{\infty} \sqrt{\frac{2}{\pi}} \frac{\sin \alpha a}{\alpha}\, e^{-i\alpha x}\, d\alpha = \begin{cases} 1 & |x| < a \\ 1/2 & |x| = a \\ 0 & |x| > a \end{cases} \tag{1}$$

The left side of *(1)* is equal to

$$\frac{1}{\pi} \int_{-\infty}^{\infty} \frac{\sin \alpha a \cos \alpha x}{\alpha}\, d\alpha \; - \; \frac{i}{\pi} \int_{-\infty}^{\infty} \frac{\sin \alpha a \sin \alpha x}{\alpha}\, d\alpha \tag{2}$$

The integrand in the second integral of *(2)* is odd and so the integral is zero. Then from *(1)* and *(2)*, we have

$$\int_{-\infty}^{\infty} \frac{\sin \alpha a \cos \alpha x}{\alpha}\, d\alpha = \begin{cases} \pi & |x| < a \\ \pi/2 & |x| = a \\ 0 & |x| > a \end{cases} \tag{3}$$

(b)   If $x = 0$ and $a = 1$ in the result of *(a)*, we have

$$\int_{-\infty}^{\infty} \frac{\sin \alpha}{\alpha}\, d\alpha = \pi \qquad \text{or} \qquad \int_{0}^{\infty} \frac{\sin \alpha}{\alpha}\, d\alpha = \frac{\pi}{2}$$

since the integrand is even.

3.   If $f(x)$ is an even function show that:

(a) $F(\alpha) = \sqrt{\dfrac{2}{\pi}} \displaystyle\int_{0}^{\infty} f(u) \cos \alpha u\, du,$ *(b)* $f(x) = \sqrt{\dfrac{2}{\pi}} \displaystyle\int_{0}^{\infty} F(\alpha) \cos \alpha x\, d\alpha.$

We have

$$F(\alpha) = \frac{1}{\sqrt{2\pi}} \int_{-\infty}^{\infty} f(u)\, e^{i\alpha u}\, du = \frac{1}{\sqrt{2\pi}} \int_{-\infty}^{\infty} f(u) \cos \alpha u\, du + \frac{i}{\sqrt{2\pi}} \int_{-\infty}^{\infty} f(u) \sin \alpha u\, du \tag{1}$$

(a)   If $f(u)$ is even, $f(u) \cos \lambda u$ is even and $f(u) \sin \lambda u$ is odd. Then the second integral on the right of *(1)* is zero and the result can be written

$$F(\alpha) = \frac{2}{\sqrt{2\pi}} \int_{0}^{\infty} f(u) \cos \alpha u\, du = \sqrt{\frac{2}{\pi}} \int_{0}^{\infty} f(u) \cos \alpha u\, du$$

(b)   From *(a)*, $F(-\alpha) = F(\alpha)$ so that $F(\alpha)$ is an even function. Then by using a proof exactly analogous to that in *(a)*, the required result follows.

A similar result holds for odd functions and can be obtained by replacing the cosine by the sine.

4.   Solve the integral equation $\displaystyle\int_{0}^{\infty} f(x) \cos \alpha x\, dx = \begin{cases} 1 - \alpha & 0 \le \alpha \le 1 \\ 0 & \alpha > 1 \end{cases}$

Let $\sqrt{\dfrac{2}{\pi}} \displaystyle\int_{0}^{\infty} f(x) \cos \alpha x\, dx = F(\alpha)$ and choose $F(\alpha) = \begin{cases} \sqrt{2/\pi}\,(1 - \alpha) & 0 \le \alpha \le 1 \\ 0 & \alpha > 1 \end{cases}$. Then by Prob. 3,

$$f(x) = \sqrt{\frac{2}{\pi}} \int_{0}^{\infty} F(\alpha) \cos \alpha x\, d\alpha = \sqrt{\frac{2}{\pi}} \int_{0}^{1} \sqrt{\frac{2}{\pi}} (1 - \alpha) \cos \alpha x\, d\alpha$$

$$= \frac{2}{\pi} \int_{0}^{1} (1 - \alpha) \cos \alpha x\, d\alpha = \frac{2(1 - \cos x)}{\pi x^2}$$

5. Use Problem 4 to show that $\displaystyle\int_0^\infty \frac{\sin^2 u}{u^2}\,du = \frac{\pi}{2}$.

As obtained in Problem 4,

$$\frac{2}{\pi}\int_0^\infty \frac{1-\cos x}{x^2}\cos\alpha x\,dx = \begin{cases} 1-\alpha & 0\le\alpha\le 1 \\ 0 & \alpha>1 \end{cases}$$

Taking the limit as $\alpha\to 0+$, we find

$$\int_0^\infty \frac{1-\cos x}{x^2}\,dx = \frac{\pi}{2}$$

But this integral can be written as $\displaystyle\int_0^\infty \frac{2\sin^2(x/2)}{x^2}\,dx$ which becomes $\displaystyle\int_0^\infty \frac{\sin^2 u}{u^2}\,du$ on letting $x = 2u$, so that the required result follows.

6. Show that $\displaystyle\int_0^\infty \frac{\cos\alpha x}{\alpha^2+1}\,d\alpha = \frac{\pi}{2}e^{-x}, \quad x\ge 0$.

Let $f(x) = e^{-x}$ in the Fourier integral theorem

$$f(x) = \frac{2}{\pi}\int_0^\infty \cos\alpha x\,d\alpha \int_0^\infty f(u)\cos\lambda u\,du$$

Then

$$\frac{2}{\pi}\int_0^\infty \cos\alpha x\,d\alpha \int_0^\infty e^{-u}\cos\alpha u\,du = e^{-x}$$

But by Problem 22, Chapter 12, we have $\displaystyle\int_0^\infty e^{-u}\cos\alpha u\,du = \frac{1}{\alpha^2+1}$. Then

$$\frac{2}{\pi}\int_0^\infty \frac{\cos\alpha x}{\alpha^2+1}\,d\alpha = e^{-x} \quad \text{or} \quad \int_0^\infty \frac{\cos\alpha x}{\alpha^2+1}\,d\alpha = \frac{\pi}{2}e^{-x}$$

PARSEVAL'S IDENTITY

7. Verify Parseval's identity for Fourier integrals for the Fourier transforms of Prob. 1.

We must show that

$$\int_{-\infty}^\infty \{f(x)\}^2\,dx = \int_{-\infty}^\infty \{F(\alpha)\}^2\,d\alpha$$

where $f(x) = \begin{cases} 1 & |x| < a \\ 0 & |x| > a \end{cases}$ and $F(\alpha) = \sqrt{\dfrac{2}{\pi}}\,\dfrac{\sin\alpha a}{\alpha}$.

This is equivalent to

$$\int_{-a}^a (1)^2\,dx = \int_{-\infty}^\infty \frac{2}{\pi}\frac{\sin^2\alpha a}{\alpha^2}\,d\alpha$$

or

$$\int_{-\infty}^\infty \frac{\sin^2\alpha a}{\alpha^2}\,d\alpha = 2\int_0^\infty \frac{\sin^2\alpha a}{\alpha^2}\,d\alpha = \pi a$$

i.e.,

$$\int_0^\infty \frac{\sin^2\alpha a}{\alpha^2}\,d\alpha = \frac{\pi a}{2}$$

By letting $\alpha a = u$ and using Problem 5, it is seen that this is correct. The method can also be used to find $\displaystyle\int_0^\infty \frac{\sin^2 u}{u^2}\,du$ directly.

PROOF of the FOURIER INTEGRAL THEOREM

8. Present a heuristic demonstration of Fourier's integral theorem by use of a limiting form of Fourier series.

Let

$$f(x) = \frac{a_0}{2} + \sum_{n=1}^{\infty} \left(a_n \cos \frac{n\pi x}{L} + b_n \sin \frac{n\pi x}{L} \right) \tag{1}$$

where $a_n = \frac{1}{L} \int_{-L}^{L} f(u) \cos \frac{n\pi u}{L} \, du$ and $b_n = \frac{1}{L} \int_{-L}^{L} f(u) \sin \frac{n\pi u}{L} \, du.$

Then by substitution (see Problem 21, Chapter 14),

$$f(x) = \frac{1}{2L} \int_{-L}^{L} f(u) \, du + \frac{1}{L} \sum_{n=1}^{\infty} \int_{-L}^{L} f(u) \cos \frac{n\pi}{L} (u - x) \, du \tag{2}$$

If we assume that $\int_{-\infty}^{\infty} |f(u)| \, du$ converges, the first term on the right of (2) approaches zero as $L \to \infty$, while the remaining part appears to approach

$$\lim_{L \to \infty} \frac{1}{L} \sum_{n=1}^{\infty} \int_{-\infty}^{\infty} f(u) \cos \frac{n\pi}{L} (u - x) \, du \tag{3}$$

This last step is not rigorous and makes the demonstration heuristic.

Calling $\Delta \alpha = \pi/L$, (3) can be written

$$f(x) = \lim_{\Delta \alpha \to 0} \sum_{n=1}^{\infty} \Delta \alpha \, F(n \, \Delta \alpha) \tag{4}$$

where we have written

$$F(\alpha) = \frac{1}{\pi} \int_{-\infty}^{\infty} f(u) \cos \alpha(u - x) \, du \tag{5}$$

But the limit (4) is equal to

$$f(x) = \int_{0}^{\infty} F(\alpha) \, d\alpha = \frac{1}{\pi} \int_{0}^{\infty} d\alpha \int_{-\infty}^{\infty} f(u) \cos \alpha(u - x) \, du$$

which is Fourier's integral formula.

This demonstration serves only to provide a possible result. To be rigorous, we start with the integral

$$\frac{1}{\pi} \int_{0}^{\infty} d\alpha \int_{-\infty}^{\infty} f(u) \cos \alpha(u - x) \, dx$$

and examine the convergence. This method is considered in Problems 9-12.

9. Prove that: (a) $\lim_{\alpha \to \infty} \int_{0}^{L} \frac{\sin \alpha v}{v} \, dv = \frac{\pi}{2},$ (b) $\lim_{\alpha \to \infty} \int_{-L}^{0} \frac{\sin \alpha v}{v} \, dv = \frac{\pi}{2}.$

(a) Let $\alpha v = y$. Then $\lim_{\alpha \to \infty} \int_{0}^{L} \frac{\sin \alpha v}{v} \, dv = \lim_{\alpha \to \infty} \int_{0}^{\alpha L} \frac{\sin y}{y} \, dy = \int_{0}^{\infty} \frac{\sin y}{y} \, dy = \frac{\pi}{2}$ by Prob. 29, Chap. 12.

(b) Let $\alpha v = -y$. Then $\lim_{\alpha \to \infty} \int_{-L}^{0} \frac{\sin \alpha v}{v} \, dv = \lim_{\alpha \to \infty} \int_{0}^{\alpha L} \frac{\sin y}{y} \, dy = \frac{\pi}{2}.$

10. Riemann's theorem states that if $F(x)$ is sectionally continuous in (a, b), then

$$\lim_{\alpha \to \infty} \int_{a}^{b} F(x) \sin \alpha x \, dx = 0$$

with a similar result for the cosine (see Problem 31). Use this to prove that

$$\text{(a)}\quad \lim_{\alpha \to \infty} \int_0^L f(x+v)\frac{\sin \alpha v}{v}\,dv \;=\; \frac{\pi}{2}f(x+0)$$

$$\text{(b)}\quad \lim_{\alpha \to \infty} \int_{-L}^0 f(x+v)\frac{\sin \alpha v}{v}\,dv \;=\; \frac{\pi}{2}f(x-0)$$

where $f(x)$ and $f'(x)$ are assumed sectionally continuous in $(0,L)$ and $(-L,0)$ respectively.

(a) Using Problem 9(a), it is seen that a proof of the given result amounts to proving that

$$\lim_{\alpha \to \infty} \int_0^L \{f(x+v) - f(x+0)\}\frac{\sin \alpha v}{v}\,dv \;=\; 0$$

This follows at once from Riemann's theorem, because $F(v) = \dfrac{f(x+v) - f(x+0)}{v}$ is sectionally continuous in $(0,L)$ since $\lim\limits_{v \to 0+} F(v)$ exists and $f(x)$ is sectionally continuous.

(b) A proof of this is analogous to that in part (a) if we make use of Problem 9(b).

11. If $f(x)$ satisfies the additional condition that $\displaystyle\int_{-\infty}^{\infty} |f(x)|\,dx$ converges, prove that

(a) $\displaystyle\lim_{\alpha \to \infty} \int_0^{\infty} f(x+v)\frac{\sin \alpha v}{v}\,dv = \frac{\pi}{2}f(x+0)$, (b) $\displaystyle\lim_{\alpha \to \infty} \int_{-\infty}^0 f(x+v)\frac{\sin \alpha v}{v}\,dv = \frac{\pi}{2}f(x-0)$.

We have

$$\int_0^{\infty} f(x+v)\frac{\sin \alpha v}{v}\,dv \;=\; \int_0^L f(x+v)\frac{\sin \alpha v}{v}\,dv \;+\; \int_L^{\infty} f(x+v)\frac{\sin \alpha v}{v}\,dv \tag{1}$$

$$\int_0^{\infty} f(x+0)\frac{\sin \alpha v}{v}\,dv \;=\; \int_0^L f(x+0)\frac{\sin \alpha v}{v}\,dv \;+\; \int_L^{\infty} f(x+0)\frac{\sin \alpha v}{v}\,dv \tag{2}$$

Subtracting,

$$\int_0^{\infty} \{f(x+v) - f(x+0)\}\frac{\sin \alpha v}{v}\,dv \tag{3}$$

$$=\; \int_0^L \{f(x+v) - f(x+0)\}\frac{\sin \alpha v}{v}\,dv \;+\; \int_L^{\infty} f(x+v)\frac{\sin \alpha v}{v}\,dv \;-\; \int_L^{\infty} f(x+0)\frac{\sin \alpha v}{v}\,dv$$

Denoting the integrals in (3) by I, I_1, I_2 and I_3 respectively, we have $I = I_1 + I_2 + I_3$ so that

$$|I| \;\leqq\; |I_1| \;+\; |I_2| \;+\; |I_3| \tag{4}$$

Now

$$|I_2| \;\leqq\; \int_L^{\infty} \left| f(x+v)\frac{\sin \alpha v}{v} \right|\,dv \;\leqq\; \frac{1}{L}\int_L^{\infty} |f(x+v)|\,dv$$

Also

$$|I_3| \;\leqq\; |f(x+0)| \left| \int_L^{\infty} \frac{\sin \alpha v}{v}\,dv \right|$$

Since $\displaystyle\int_0^{\infty} |f(x)|\,dx$ and $\displaystyle\int_0^{\infty} \frac{\sin \alpha v}{v}\,dv$ both converge, we can choose L so large that $|I_2| \leqq \epsilon/3$, $|I_3| \leqq \epsilon/3$. Also, we can choose α so large that $|I_1| \leqq \epsilon/3$. Then from (4) we have $|I| < \epsilon$ for α and L sufficiently large, so that the required result follows.

This result follows by reasoning exactly analogous to that in part (a).

12. Prove Fourier's integral formula where $f(x)$ satisfies the conditions stated on Page 321.

We must prove that $\displaystyle\lim_{L \to \infty} \frac{1}{\pi}\int_{\alpha=0}^L \int_{u=-\infty}^{\infty} f(u)\cos \alpha(x-u)\,du\,d\alpha \;=\; \frac{f(x+0) + f(x-0)}{2}$

Since $\left| \displaystyle\int_{-\infty}^{\infty} f(u)\cos \alpha(x-u)\,du \right| \leqq \displaystyle\int_{-\infty}^{\infty} |f(u)|\,du$ which converges, it follows by the Weirstrass test that $\displaystyle\int_{-\infty}^{\infty} f(u)\cos \alpha(x-u)\,du$ converges absolutely and uniformly for all α. Thus we can reverse the order of integration to obtain

$$\frac{1}{\pi} \int_{\alpha=0}^{L} d\alpha \int_{u=-\infty}^{\infty} f(u) \cos \alpha(x-u) \, du \;=\; \frac{1}{\pi} \int_{u=-\infty}^{\infty} f(u) \, du \int_{\alpha=0}^{L} \cos \alpha(x-u) \, du$$

$$=\; \frac{1}{\pi} \int_{u=-\infty}^{\infty} f(u) \, \frac{\sin L(u-x)}{u-x} \, du$$

$$=\; \frac{1}{\pi} \int_{v=-\infty}^{\infty} f(x+v) \, \frac{\sin Lv}{v} \, dv$$

$$=\; \frac{1}{\pi} \int_{-\infty}^{0} f(x+v) \, \frac{\sin Lv}{v} \, dv \;+\; \frac{1}{\pi} \int_{0}^{\infty} f(x+v) \, \frac{\sin Lv}{v} \, dv$$

where we have let $u = x + v$.

Letting $L \to \infty$, we see by Problem 11 that the given integral converges to $\dfrac{f(x+0) + f(x-0)}{2}$ as required.

MISCELLANEOUS PROBLEMS

13. Solve $\dfrac{\partial U}{\partial t} = \dfrac{\partial^2 U}{\partial x^2}$ subject to the conditions $\quad U(0,t) = 0, \quad U(x,0) = \begin{cases} 1 & 0 < x < 1 \\ 0 & x \geqq 1 \end{cases}$, $U(x,t)$ is bounded where $x > 0$, $t > 0$.

We proceed as in Problem 24, Chapter 14. A solution satisfying the partial differential equation and the first boundary condition is given by $Be^{-\lambda^2 t} \sin \lambda x$. Unlike Problem 24, Chapter 14, the boundary conditions do not prescribe specific values for λ, so we must assume that all values of λ are possible. By analogy with that problem we sum over all possible values of λ, which corresponds to an integration in this case, and are led to the possible solution

$$U(x,t) \;=\; \int_0^\infty B(\lambda) \, e^{-\lambda^2 t} \sin \lambda x \, d\lambda \tag{1}$$

where $B(\lambda)$ is undetermined. By the second condition, we have

$$\int_0^\infty B(\lambda) \sin \lambda x \, d\lambda \;=\; \begin{cases} 1 & 0 < x < 1 \\ 0 & x \geqq 1 \end{cases} \;=\; f(x) \tag{2}$$

from which we have by Fourier's integral formula

$$B(\lambda) \;=\; \frac{2}{\pi} \int_0^\infty f(x) \sin \lambda x \, dx \;=\; \frac{2}{\pi} \int_0^1 \sin \lambda x \, dx \;=\; \frac{2(1 - \cos \lambda)}{\pi \lambda} \tag{3}$$

so that, at least formally, the solution is given by

$$U(x,t) \;=\; \frac{2}{\pi} \int_0^\infty \left(\frac{1 - \cos \lambda}{\lambda} \right) e^{-\lambda^2 t} \sin \lambda x \, dx \tag{4}$$

See Problem 26.

14. Show that $e^{-x^2/2}$ is its own Fourier transform.

Since $e^{-x^2/2}$ is even, its Fourier transform is given by $\quad \sqrt{2/\pi} \displaystyle\int_0^\infty e^{-x^2/2} \cos x\alpha \, dx.$

Letting $x = \sqrt{2}\, u$ and using Problem 32, Chapter 12, the integral becomes

$$\frac{2}{\sqrt{\pi}} \int_0^\infty e^{-u^2} \cos(\alpha \sqrt{2}\, u) \, du \;=\; \frac{2}{\sqrt{\pi}} \cdot \frac{\sqrt{\pi}}{2} e^{-\alpha^2/2} \;=\; e^{-\alpha^2/2}$$

which proves the required result.

15. Solve the integral equation

$$y(x) \;=\; g(x) \;+\; \int_{-\infty}^{\infty} y(u) \, r(x-u) \, du$$

where $g(x)$ and $r(x)$ are given.

Suppose that the Fourier transforms of $y(x)$, $g(x)$ and $r(x)$ exist, and denote them by $Y(\alpha)$, $G(\alpha)$ and $R(\alpha)$ respectively. Then taking the Fourier transform of both sides of the given integral equation, we have by the convolution theorem

$$Y(\alpha) \;=\; G(\alpha) \;+\; \sqrt{2\pi}\, Y(\alpha) \, R(\alpha) \qquad \text{or} \qquad Y(\alpha) \;=\; \frac{G(\alpha)}{1 - \sqrt{2\pi}\, R(\alpha)}$$

Then $\qquad y(x) \;=\; \mathcal{J}^{-1}\left\{\dfrac{G(\alpha)}{1 - \sqrt{2\pi}\,R(\alpha)}\right\} \;=\; \dfrac{1}{\sqrt{2\pi}} \int_{-\infty}^{\infty} \dfrac{G(\alpha)}{1 - \sqrt{2\pi}\,R(\alpha)}\, e^{-i\alpha x}\, d\alpha$

assuming this integral exists.

Supplementary Problems

The FOURIER INTEGRAL and FOURIER TRANSFORMS

16. (a) Find the Fourier transform of $\quad f(x) \;=\; \begin{cases} 1/2\epsilon & |x| \leqq \epsilon \\ 0 & |x| > \epsilon \end{cases}$

(b) Determine the limit of this transform as $\epsilon \to 0+$ and discuss the result.

Ans. (a) $\dfrac{1}{\sqrt{2\pi}} \dfrac{\sin \alpha\epsilon}{\alpha\epsilon}$, (b) $\dfrac{1}{\sqrt{2\pi}}$

17. (a) Find the Fourier transform of $\quad f(x) \;=\; \begin{cases} 1 - x^2 & |x| < 1 \\ 0 & |x| > 1 \end{cases}$

(b) Evaluate $\quad \displaystyle\int_0^\infty \left(\dfrac{x \cos x - \sin x}{x^3}\right) \cos \dfrac{x}{2}\, dx.$

Ans. (a) $2\sqrt{\dfrac{2}{\pi}}\left(\dfrac{\alpha \cos \alpha - \sin \alpha}{\alpha^3}\right)$, (b) $\dfrac{3\pi}{16}$

18. If $\quad f(x) \;=\; \begin{cases} 1 & 0 \leqq x < 1 \\ 0 & x \geqq 1 \end{cases}\quad$ find the (a) Fourier sine transform, (b) Fourier cosine transform of $f(x)$.

In each case obtain the graph of $f(x)$ and its transform.

Ans. (a) $\sqrt{\dfrac{2}{\pi}}\left(\dfrac{1 - \cos \alpha}{\alpha}\right)$, (b) $\sqrt{\dfrac{2}{\pi}} \dfrac{\sin \alpha}{\alpha}$.

19. (a) Find the Fourier sine transform of $e^{-x},\ x \geqq 0$.

(b) Show that $\displaystyle\int_0^\infty \dfrac{x \sin mx}{x^2 + 1}\, dx \;=\; \dfrac{\pi}{2} e^{-m},\ m > 0\quad$ by using the result in (a).

(c) Explain from the viewpoint of Fourier's integral theorem why the result in (b) does not hold for $m = 0$.

Ans. (a) $\sqrt{2/\pi}\,[\alpha/(1 + \alpha^2)]$

20. Solve for $Y(x)$ the integral equation

$$\int_0^\infty Y(x) \sin xt\, dx \;=\; \begin{cases} 1 & 0 \leqq t < 1 \\ 2 & 1 \leqq t < 2 \\ 0 & t \geqq 2 \end{cases}$$

and verify the solution by direct substitution.

Ans. $Y(x) \;=\; (2 + 2\cos x - 4\cos 2x)/\pi x$

PARSEVAL'S IDENTITY

21. Evaluate (a) $\displaystyle\int_0^\infty \dfrac{dx}{(x^2 + 1)^2}$, (b) $\displaystyle\int_0^\infty \dfrac{x^2\, dx}{(x^2 + 1)^2}$ by use of Parseval's identity.

[Hint: Use the Fourier sine and cosine transforms of $e^{-x},\ x > 0$.] Ans. (a) $\pi/4$, (b) $\pi/4$

22. Use Problem 18 to show that   (a) $\int_0^\infty \left(\dfrac{1 - \cos x}{x}\right)^2 dx = \dfrac{\pi}{2}$,   (b) $\int_0^\infty \dfrac{\sin^4 x}{x^2} dx = \dfrac{\pi}{2}$.

23. Show that $\int_0^\infty \dfrac{(x \cos x - \sin x)^2}{x^6} dx = \dfrac{\pi}{15}$.

MISCELLANEOUS PROBLEMS

24. (a) Solve $\dfrac{\partial U}{\partial t} = 2 \dfrac{\partial^2 U}{\partial x^2}$, $U(0, t) = 0$, $U(x, 0) = e^{-x}$, $x > 0$, $U(x, t)$ is bounded where $x > 0$, $t > 0$.

(b) Give a physical interpretation.

Ans. $U(x, t) = \dfrac{2}{\pi} \int_0^\infty \dfrac{\lambda e^{-2\lambda^2 t} \sin \lambda x}{\lambda^2 + 1} d\lambda$

25. Solve $\dfrac{\partial U}{\partial t} = \dfrac{\partial^2 U}{\partial x^2}$, $U_x(0, t) = 0$, $U(x, 0) = \begin{cases} x & 0 \leq x \leq 1 \\ 0 & x > 1 \end{cases}$, $U(x, t)$ is bounded where $x > 0$, $t > 0$.

Ans. $U(x, t) = \dfrac{2}{\pi} \int_0^\infty \left(\dfrac{\sin \lambda}{\lambda} + \dfrac{\cos \lambda - 1}{\lambda^2}\right) e^{-\lambda^2 t} \cos \lambda x \, d\lambda$

26. (a) Show that the solution to Problem 13 can be written

$$U(x, t) = \frac{2}{\sqrt{\pi}} \int_0^{x/2\sqrt{t}} e^{-v^2} dv - \frac{1}{\sqrt{\pi}} \int_{(1-x)/2\sqrt{t}}^{(1+x)/2\sqrt{t}} e^{-v^2} dv$$

(b) Prove directly that the function in (a) satisfies $\dfrac{\partial U}{\partial t} = \dfrac{\partial^2 U}{\partial x^2}$ and the conditions of Problem 13.

27. Verify the convolution theorem for the functions $f(x) = g(x) = \begin{cases} 1 & |x| < 1 \\ 0 & |x| > 1 \end{cases}$.

28. Establish equation (4), Page 321, from equation (3), Page 321.

29. Prove the result (18), Page 323.

[Hint: If $F(\alpha) = \dfrac{1}{\sqrt{2\pi}} \int_{-\infty}^\infty f(u) \, e^{i\alpha u} \, du$ and $G(\alpha) = \dfrac{1}{\sqrt{2\pi}} \int_{-\infty}^\infty g(v) \, e^{i\alpha v} \, dv$,   then

$$F(\alpha) \, G(\alpha) = \frac{1}{2\pi} \int_{-\infty}^\infty \int_{-\infty}^\infty e^{i\alpha(u+v)} f(u) \, g(v) \, du \, dv$$

Now make the transformation $u + v = x$.]

30. (a) If $F(\alpha)$ and $G(\alpha)$ are the Fourier transforms of $f(x)$ and $g(x)$ respectively, prove that

$$\int_{-\infty}^\infty F(\alpha) \, \overline{G(\alpha)} \, d\alpha = \int_{-\infty}^\infty f(x) \, \overline{g(x)} \, dx$$

where the bar signifies the complex conjugate.

(b) From (a) obtain the results $(11) - (14)$, Page 322.

31. Prove Riemann's theorem (see Problem 10).

Chapter 16

Elliptic Integrals

The INCOMPLETE ELLIPTIC INTEGRAL of the FIRST KIND is defined as

$$u = F(k, \phi) = \int_0^\phi \frac{d\theta}{\sqrt{1 - k^2 \sin^2 \theta}} \qquad 0 < k < 1 \qquad (1)$$

where ϕ is the *amplitude* of $F(k, \phi)$ or u, written $\phi = \text{am } u$, and k is its *modulus*, written $k = \text{mod } u$. The integral is also called *Legendre's form for the elliptic integral of the first kind*.

If $\phi = \pi/2$ the integral is called the *complete integral of the first kind* and is denoted by $K(k)$ or simply K. For all purposes it will be assumed that k is a given constant.

The INCOMPLETE ELLIPTIC INTEGRAL of the SECOND KIND is defined by

$$E(k, \phi) = \int_0^\phi \sqrt{1 - k^2 \sin^2 \theta} \, d\theta \qquad 0 < k < 1 \qquad (2)$$

also called *Legendre's form for the elliptic integral of the second kind*.

If $\phi = \pi/2$ the integral is called the *complete elliptic integral of the second kind* and is denoted by $E(k)$ or simply E. This integral arises in the determination of the length of arc of an ellipse and supplies a reason for use of the term elliptic integral.

The INCOMPLETE ELLIPTIC INTEGRAL of the THIRD KIND is defined by

$$\Pi(k, n, \phi) = \int_0^\phi \frac{d\theta}{(1 + n \sin^2 \theta) \sqrt{1 - k^2 \sin^2 \theta}} \qquad 0 < k < 1 \qquad (3)$$

also called *Legendre's form for the elliptic integral of the third kind*. Here n is a constant assumed different from zero since if $n = 0$, (3) reduces to (1).

If $\phi = \pi/2$ the integral is called the *complete elliptic integral of the third kind*.

JACOBI'S FORMS for the ELLIPTIC INTEGRALS

If the transformation $v = \sin \theta$ is made in the Legendre forms of the elliptic integrals above, we obtain the following integrals with $x = \sin \phi$.

$$F_1(k, x) = \int_0^x \frac{dv}{\sqrt{(1 - v^2)(1 - k^2 v^2)}} \qquad (4)$$

$$E_1(k, x) = \int_0^x \sqrt{\frac{1 - k^2 v^2}{1 - v^2}} \, dv \qquad (5)$$

331

$$\Pi_1(k, n, x) \;=\; \int_0^x \frac{dv}{(1 + nv^2)\sqrt{(1 - v^2)(1 - k^2 v^2)}} \tag{6}$$

called *Jacobi's forms for the elliptic integrals of* first, second and third kinds respectively. These are complete elliptic integrals if $x = 1$.

INTEGRALS REDUCIBLE to ELLIPTIC TYPE

If $R(x, y)$ is a rational algebraic function of x and y, i.e. the quotient of two polynomials in x and y, then

$$\int R(x, y)\, dx \tag{7}$$

can be evaluated in terms of the usual elementary functions (algebraic, trigonometric, inverse trigonometric, exponential and logarithmic) if $y = \sqrt{ax + b}$ or $y = \sqrt{ax^2 + bx + c}$, where a, b, c are given constants.

If $y = \sqrt{ax^3 + bx^2 + cx + d}$ or $y = \sqrt{ax^4 + bx^3 + cx^2 + dx + e}$ where a, b, c, d, e are given constants, (7) can be evaluated in terms of elliptic integrals of first, second or third kinds or for special cases in terms of elementary functions.

If $y = \sqrt{P(x)}$ where $P(x)$ is a polynomial of degree greater than four, (7) may be integrated with the aid of *hyper-elliptic functions*.

JACOBI'S ELLIPTIC FUNCTIONS

The upper limit x in the Jacobi form for the elliptic integral of the first kind is related to the upper limit ϕ in the Legendre form by $x = \sin \phi$. Since $\phi = \operatorname{am} u$, it follows that $x = \sin(\operatorname{am} u)$. Thus we are led to define the elliptic functions

$$x \;=\; \sin(\operatorname{am} u) \qquad \equiv \quad \operatorname{sn} u \tag{8}$$

$$\sqrt{1 - x^2} \;=\; \cos(\operatorname{am} u) \qquad \equiv \quad \operatorname{cn} u \tag{9}$$

$$\sqrt{1 - k^2 x^2} \;=\; \sqrt{1 - k^2 \operatorname{sn}^2 u} \quad \equiv \quad \operatorname{dn} u \tag{10}$$

$$\frac{x}{\sqrt{1 - x^2}} \;=\; \frac{\operatorname{sn} u}{\operatorname{cn} u} \qquad \equiv \quad \operatorname{tn} u \tag{11}$$

which have many important properties analogous to those of trigonometric functions as presented in the problems.

It is also possible to define *inverse elliptic functions*; for example, if $x = \operatorname{sn} u$ then $u = \operatorname{sn}^{-1} x$. Note that u depends on k. To emphasize this dependence we sometimes write $u = \operatorname{sn}^{-1}(x, k)$ or $u = \operatorname{sn}^{-1} x$, mod k.

LANDEN'S TRANSFORMATION

By use of the transformation

$$\tan \phi \;=\; \frac{\sin 2\phi_1}{k + \cos 2\phi_1} \qquad \text{or} \qquad k \sin \phi \;=\; \sin(2\phi_1 - \phi) \tag{12}$$

called *Landen's transformation*, we can show that

$$\int_0^\phi \frac{d\phi}{\sqrt{1 - k^2 \sin^2 \phi}} \;=\; \frac{2}{1 + k} \int_0^{\phi_1} \frac{d\phi_1}{\sqrt{1 - k_1^2 \sin^2 \phi_1}} \tag{13}$$

where $k_1 = \dfrac{2\sqrt{k}}{1 + k}$ (see Problem 61). This can be written

$$F(k, \phi) \;=\; \frac{2}{1 + k} F(k_1, \phi_1) \tag{14}$$

It is seen that $k < k_1 < 1$. By successive applications of Landen's transformation, a sequence of moduli k_n, $n = 1, 2, 3, \ldots$ is obtained such that $k < k_1 < k_2 < \cdots < 1$ and we can prove that $\lim_{n \to \infty} k_n = 1$. From this it follows that

$$F(k, \phi) \;=\; \sqrt{\frac{k_1 k_2 k_3 \cdots}{k}} \int_0^\Phi \frac{d\theta}{\sqrt{1 - \sin^2 \theta}} \;=\; \sqrt{\frac{k_1 k_2 k_3 \cdots}{k}} \, \ln \tan\left(\frac{\pi}{4} + \frac{\Phi}{2}\right) \quad (15)$$

where

$$k_1 = \frac{2\sqrt{k}}{1+k}, \quad k_2 = \frac{2\sqrt{k_1}}{1+k_1}, \quad k_3 = \frac{2\sqrt{k_2}}{1+k_2}, \quad \ldots \quad \text{and} \quad \Phi = \lim_{n \to \infty} \phi_n \quad (16)$$

By using this result, $F(k, \phi)$ can be computed. In practice it is possible to achieve accurate results after only a few applications of the transformation.

Solved Problems

ELLIPTIC INTEGRALS

1. Prove that if $0 < k < 1$,

$$K(k) \;=\; \int_0^{\pi/2} \frac{d\theta}{\sqrt{1 - k^2 \sin^2 \theta}} \;=\; \frac{\pi}{2}\left\{1 + \left(\frac{1}{2}\right)^2 k^2 + \left(\frac{1 \cdot 3}{2 \cdot 4}\right)^2 k^4 + \left(\frac{1 \cdot 3 \cdot 5}{2 \cdot 4 \cdot 6}\right)^2 k^6 + \cdots\right\}$$

By the binomial theorem,

$$(1 - x)^{-1/2} \;=\; 1 + (-1/2)(-x) + \frac{(-1/2)(-3/2)}{2!}(-x)^2 + \frac{(-1/2)(-3/2)(-5/2)}{3!}(-x)^3 + \cdots$$

$$= \; 1 + \frac{1}{2}x + \frac{1 \cdot 3}{2 \cdot 4}x^2 + \frac{1 \cdot 3 \cdot 5}{2 \cdot 4 \cdot 6}x^3 + \cdots$$

Let $x = k^2 \sin^2 \theta$. Then by the uniform convergence of the series, we can integrate term by term from 0 to $\pi/2$ to obtain

$$\int_0^{\pi/2} \frac{d\theta}{\sqrt{1 - k^2 \sin^2 \theta}} \;=\; \int_0^{\pi/2}\left\{1 + \frac{1}{2}k^2 \sin^2 \theta + \frac{1 \cdot 3}{2 \cdot 4}k^4 \sin^4 \theta + \frac{1 \cdot 3 \cdot 5}{2 \cdot 4 \cdot 6}k^6 \sin^6 \theta + \cdots\right\} d\theta$$

$$= \; \frac{\pi}{2}\left\{1 + \left(\frac{1}{2}\right)^2 k^2 + \left(\frac{1 \cdot 3}{2 \cdot 4}\right)^2 k^4 + \left(\frac{1 \cdot 3 \cdot 5}{2 \cdot 4 \cdot 6}\right)^2 k^6 + \cdots\right\}$$

using Problem 15, Chapter 13.

2. Evaluate $\displaystyle\int_0^{\pi/2} \frac{dx}{\sqrt{\sin x}}$ to 3 decimal places by first expressing the integral as an elliptic integral.

Let $x = \pi/2 - y$ and the integral becomes $\displaystyle\int_0^{\pi/2} \frac{dy}{\sqrt{\cos y}}$.

Let $\cos y = \cos^2 u$. Then $-\sin y \, dy = -2 \cos u \sin u \, du$ and

$$dy \;=\; \frac{2 \cos u \sin u \, du}{\sqrt{1 - \cos^4 u}} \;=\; \frac{2 \cos u \sin u \, du}{\sqrt{1 - \cos^2 u}\,\sqrt{1 + \cos^2 u}} \;=\; \frac{2 \cos u \, du}{\sqrt{1 + \cos^2 u}}$$

Hence

$$\int_0^{\pi/2} \frac{dy}{\sqrt{\cos y}} \;=\; 2 \int_0^{\pi/2} \frac{du}{\sqrt{1 + \cos^2 u}} \;=\; 2 \int_0^{\pi/2} \frac{du}{\sqrt{2 - \sin^2 u}} \;=\; \sqrt{2} \int_0^{\pi/2} \frac{du}{\sqrt{1 - \frac{1}{2} \sin^2 u}}$$

$$= \; \sqrt{2}\, F(\sqrt{\tfrac{1}{2}}, \pi/2) \quad \text{or} \quad \sqrt{2}\, K(\sqrt{\tfrac{1}{2}}).$$

Substituting $k = \sqrt{\tfrac{1}{2}}$ in the result of Problem 1 yields the value 2.622 for the integral.

Another method:

$$\int_0^{\pi/2} \frac{dx}{\sqrt{\sin x}} = \int_0^{\pi/2} \frac{dy}{\sqrt{\cos y}} = \int_0^{\pi/2} \frac{dy}{\sqrt{\cos^2 y/2 - \sin^2 y/2}} = \int_0^{\pi/2} \frac{dy}{\sqrt{1 - 2 \sin^2 y/2}}$$

Let $\sqrt{2} \sin y/2 = \sin \phi$. Then $\sqrt{2}/2 \cos y/2\, dy = \cos \phi\, d\phi$,

$$dy = \frac{\sqrt{2} \cos \phi\, d\phi}{\sqrt{1 - \tfrac{1}{2} \sin^2 \phi}} \quad \text{and the integral becomes} \quad \sqrt{2} \int_0^{\pi/2} \frac{d\phi}{\sqrt{1 - \tfrac{1}{2} \sin^2 \phi}} = \sqrt{2}\, K(\sqrt{\tfrac{1}{2}})$$

3. Evaluate $\displaystyle\int_0^{\pi/2} \sqrt{\cos x}\, dx$ in terms of elliptic integrals.

Let $\cos x = \cos^2 u$ as in Problem 2. Then

$$\int_0^{\pi/2} \sqrt{\cos x}\, dx = \int_0^{\pi/2} \cos u \left(\frac{2 \cos u}{\sqrt{1 + \cos^2 u}} \right) du = 2 \int_0^{\pi/2} \frac{\cos^2 u}{\sqrt{1 + \cos^2 u}}\, du$$

$$= 2 \int_0^{\pi/2} \frac{(1 + \cos^2 u) - 1}{\sqrt{1 + \cos^2 u}}\, du = 2 \int_0^{\pi/2} \sqrt{1 + \cos^2 u}\, du - 2 \int_0^{\pi/2} \frac{du}{\sqrt{1 + \cos^2 u}}$$

$$= 2 \int_0^{\pi/2} \sqrt{2 - \sin^2 u}\, du - 2 \int_0^{\pi/2} \frac{du}{\sqrt{2 - \sin^2 u}}$$

$$= 2\sqrt{2} \int_0^{\pi/2} \sqrt{1 - \tfrac{1}{2} \sin^2 u}\, du - \frac{2}{\sqrt{2}} \int_0^{\pi/2} \frac{du}{\sqrt{1 - \tfrac{1}{2} \sin^2 u}}$$

$$= 2\sqrt{2}\, E(\sqrt{\tfrac{1}{2}}) - \sqrt{2}\, K(\sqrt{\tfrac{1}{2}})$$

4. Evaluate $\displaystyle\int_0^{\pi/2} \sqrt{1 + 4 \sin^2 x}\, dx$.

The integral equals $\displaystyle\int_0^{\pi/2} \sqrt{1 + 4(1 - \cos^2 x)}\, dx = \int_0^{\pi/2} \sqrt{5 - 4 \cos^2 x}\, dx$.

Letting $x = \pi/2 - y$, the integral becomes

$$\int_0^{\pi/2} \sqrt{5 - 4 \sin^2 y}\, dy = \sqrt{5} \int_0^{\pi/2} \sqrt{1 - \tfrac{4}{5} \sin^2 y}\, dy = \sqrt{5}\, E(\sqrt{4/5})$$

5. Express $\displaystyle\int_0^x \sqrt{1 - 4 \sin^2 u}\, du$ in terms of incomplete elliptic integrals where $0 \leqq x \leqq \pi/6$.

Let $\sqrt{4 \sin^2 u} = 2 \sin u = \sin \phi$.

Then $2 \cos u\, du = \cos \phi\, d\phi$, $du = \dfrac{\cos \phi\, d\phi}{2 \cos u} = \dfrac{\cos \phi\, d\phi}{2\sqrt{1 - \tfrac{1}{4} \sin^2 \phi}}$ and so

$$\int_0^x \sqrt{1 - 4 \sin^2 u}\, du = \int_0^\phi \cos \phi \cdot \frac{\cos \phi}{2\sqrt{1 - \tfrac{1}{4} \sin^2 \phi}}\, d\phi = \frac{1}{2} \int_0^\phi \frac{\cos^2 \phi\, d\phi}{\sqrt{1 - \tfrac{1}{4} \sin^2 \phi}}$$

$$= \frac{1}{2} \int_0^\phi \frac{-3 + 4(1 - \tfrac{1}{4} \sin^2 \phi)}{\sqrt{1 - \tfrac{1}{4} \sin^2 \phi}}\, d\phi$$

$$= -\frac{3}{2} \int_0^\phi \frac{d\phi}{\sqrt{1 - \tfrac{1}{4} \sin^2 \phi}} + 2 \int_0^\phi \sqrt{1 - \tfrac{1}{4} \sin^2 \phi}\, d\phi$$

$$= -\tfrac{3}{2} F(\tfrac{1}{2}, \phi) + 2 E(\tfrac{1}{2}, \phi)$$

where $\phi = \sin^{-1}(2 \sin x)$.

6. Show that $\int \dfrac{dx}{\sqrt{2 - \cos x}}$ is reducible to integrals of elliptic type.

$2 - \cos x = 2 - (\cos^2 x/2 - \sin^2 x/2) = 3 - 2 \cos^2 x/2$ and the given integral can be written

$$\int \frac{dx}{\sqrt{3 - 2 \cos^2 x/2}} \;=\; \frac{1}{\sqrt{3}} \int \frac{dx}{\sqrt{1 - \tfrac{2}{3} \cos^2 x/2}}$$

Letting $x/2 = \pi/2 - u$ in this last integral, it becomes $\; -\dfrac{2}{\sqrt{3}} \displaystyle\int \frac{du}{\sqrt{1 - \tfrac{2}{3}\sin^2 u}} \;$ which is of elliptic type.

7. Prove that $\displaystyle\int_0^{\pi/2} \frac{dx}{\sqrt{2 - \cos x}} = \frac{2}{\sqrt{3}} \{ F(\sqrt{2/3},\, \pi/2) - F(\sqrt{2/3},\, \pi/4) \}.$

From Problem 6, $\; I \;=\; \displaystyle\int_0^{\pi/2} \frac{dx}{\sqrt{2 - \cos x}} \;=\; -\frac{2}{\sqrt{3}} \int_{\pi/2}^{\pi/4} \frac{du}{\sqrt{1 - \tfrac{2}{3}\sin^2 u}}$

since from $x/2 = \pi/2 - u$ it is seen that $u = \pi/2$ and $\pi/4$ respectively when $x = 0$ and $\pi/2$. Then

$$I \;=\; \frac{2}{\sqrt{3}} \int_{\pi/4}^{\pi/2} \frac{du}{\sqrt{1 - \tfrac{2}{3}\sin^2 u}} \;=\; \frac{2}{\sqrt{3}} \left\{ \int_0^{\pi/2} \frac{du}{\sqrt{1 - \tfrac{2}{3}\sin^2 u}} - \int_0^{\pi/4} \frac{du}{\sqrt{1 - \tfrac{2}{3}\sin^2 u}} \right\}$$

$$=\; \frac{2}{\sqrt{3}} \left\{ F\!\left(\sqrt{\frac{2}{3}}, \frac{\pi}{2}\right) - F\!\left(\sqrt{\frac{2}{3}}, \frac{\pi}{4}\right) \right\}.$$

8. Find the length of arc of the curve $y = \sin x$, $0 \le x \le \pi$.

$$\text{Arc length} \;=\; \int_0^\pi \sqrt{1 + (dy/dx)^2}\, dx \;=\; \int_0^\pi \sqrt{1 + \cos^2 x}\, dx \;=\; 2 \int_0^{\pi/2} \sqrt{1 + \cos^2 x}\, dx$$

$$=\; 2 \int_0^{\pi/2} \sqrt{2 - \sin^2 x}\, dx \;=\; 2\sqrt{2} \int_0^{\pi/2} \sqrt{1 - \tfrac{1}{2}\sin^2 x}\, dx \;=\; 2\sqrt{2}\, E(\sqrt{\tfrac{1}{2}})$$

9. Find the length of arc of the ellipse $x = a \sin \phi$, $y = b \cos \phi$, $a > b > 0$.

$$\text{Arc length} \;=\; 4 \int_0^{\pi/2} \sqrt{(dx)^2 + (dy)^2} \;=\; 4 \int_0^{\pi/2} \sqrt{a^2 \cos^2 \phi + b^2 \sin^2 \phi}\, d\phi$$

$$=\; 4 \int_0^{\pi/2} \sqrt{a^2 - (a^2 - b^2) \sin^2 \phi}\, d\phi \;=\; 4a \int_0^{\pi/2} \sqrt{1 - e^2 \sin^2 \phi}\, d\phi$$

where $e^2 = (a^2 - b^2)/a^2 = c^2/a^2$ is the square of the eccentricity of the ellipse.

The result can be written as $4a\, E(e, \pi/2)$ or $4a\, E(e)$.

10. Express $\displaystyle\int_0^\phi \frac{d\phi}{\sqrt{1 + k^2 \sin^2 \phi}}$ in terms of elliptic integrals.

Let $k \sin \phi = \tan u$. Then $k \cos \phi\, d\phi = \sec^2 u\, du$ and

$$\int_0^\phi \frac{d\phi}{\sqrt{1 + k^2 \sin^2 \phi}} \;=\; \int_0^u \frac{\sec u}{k \cos \phi}\, du \;=\; \int_0^u \frac{\sec u}{\sqrt{k^2 - \tan^2 u}}\, du \;=\; \int_0^u \frac{du}{\sqrt{k^2 \cos^2 u - \sin^2 u}}$$

$$=\; \int_0^u \frac{du}{\sqrt{k^2 - (k^2 + 1) \sin^2 u}}$$

Now proceed as in Problem 5. Let $\sqrt{k^2 + 1}\, \sin u = k \sin x$. Then $\sqrt{k^2 + 1}\, \cos u\, du = k \cos x\, dx$ and

$$\int_0^u \frac{du}{\sqrt{k^2 - (k^2 + 1) \sin^2 u}} \;=\; \frac{1}{\sqrt{k^2 + 1}} \int_0^x \frac{dx}{\sqrt{1 - [k^2/(k^2 + 1)] \sin^2 x}} \;=\; \frac{1}{\sqrt{k^2 + 1}}\, F\!\left(\frac{k}{\sqrt{k^2 + 1}},\, x\right)$$

Upon retracing steps it is observed that the upper limit x in the last integral is related to the upper limit ϕ in the original integral by $\; x = \sin^{-1}\!\left(\dfrac{\sqrt{k^2 + 1}\, \sin \phi}{\sqrt{1 + k^2 \sin^2 \phi}}\right).$

11. Evaluate each of the following in terms of elliptic integrals.

(a) $\displaystyle\int_0^2 \frac{dx}{\sqrt{(4-x^2)(9-x^2)}}$.   Letting $x = 2\sin\theta$, the integral becomes

$$\int_0^{\pi/2} \frac{d\theta}{\sqrt{9-4\sin^2\theta}} \;=\; \frac{1}{3}\int_0^{\pi/2}\frac{d\theta}{\sqrt{1-\frac{4}{9}\sin^2\theta}} \;=\; \frac{1}{3}F(2/3,\,\pi/2)$$

(b) $\displaystyle\int_0^1 \frac{dx}{\sqrt{(1+x^2)(1+2x^2)}}$.   Letting $x = \tan\theta$, the integral becomes

$$\int_0^{\pi/4}\frac{\sec^2\theta\,d\theta}{\sqrt{1+\tan^2\theta}\,\sqrt{1+2\tan^2\theta}} \;=\; \int_0^{\pi/4}\frac{d\theta}{\sqrt{\cos^2\theta+2\sin^2\theta}} \;=\; \int_0^{\pi/4}\frac{d\theta}{\sqrt{2-\cos^2\theta}}$$

$$=\; \frac{1}{\sqrt{2}}\int_0^{\pi/4}\frac{d\theta}{\sqrt{1-\frac{1}{2}\cos^2\theta}}$$

Letting $\theta = \pi/2 - \phi$, this last integral becomes

$$\frac{1}{\sqrt{2}}\int_{\pi/4}^{\pi/2}\frac{d\phi}{\sqrt{1-\frac{1}{2}\sin^2\phi}} \;=\; \frac{1}{\sqrt{2}}\{F(\sqrt{1/2},\,\pi/2) - F(\sqrt{1/2},\,\pi/4)\}$$

(c) $\displaystyle\int_4^6 \frac{dx}{\sqrt{(x-1)(x-2)(x-3)}}$.   Let $\sqrt{x-3} = u$ or $x = 3 + u^2$. Then the integral becomes

$$2\int_1^{\sqrt{3}}\frac{du}{\sqrt{(u^2+2)(u^2+1)}}$$

Letting $u = \tan\theta$, this integral becomes

$$2\int_{\pi/4}^{\pi/3}\frac{d\theta}{\sqrt{2\cos^2\theta+\sin^2\theta}} \;=\; 2\int_{\pi/4}^{\pi/3}\frac{d\theta}{\sqrt{2-\sin^2\theta}} \;=\; \sqrt{2}\int_{\pi/4}^{\pi/3}\frac{d\theta}{\sqrt{1-\frac{1}{2}\sin^2\theta}}$$

$$=\; \sqrt{2}\,\{F(\sqrt{1/2},\,\pi/3) - F(\sqrt{1/2},\,\pi/4)\}$$

In general $\displaystyle\int\frac{dx}{\sqrt{P_3(x)}}$ where $P_3(x)$ is a third degree polynomial with real zeros, can be transformed into an elliptic integral by the method indicated.

12. Evaluate $\displaystyle\int_1^\infty \frac{du}{\sqrt{(u^2-1)(u^2+3)}}$.

Let $u = \sec\theta$ and the integral becomes

$$\int_0^{\pi/2}\frac{\sec\theta\tan\theta\,d\theta}{\sqrt{(\sec^2\theta-1)(\sec^2\theta+3)}} \;=\; \int_0^{\pi/2}\frac{d\theta}{\sqrt{1+3\cos^2\theta}} \;=\; \int_0^{\pi/2}\frac{d\theta}{\sqrt{4-3\sin^2\theta}}$$

$$=\; \frac{1}{2}\int_0^{\pi/2}\frac{d\theta}{\sqrt{1-\frac{3}{4}\sin^2\theta}} \;=\; \frac{1}{2}F(\sqrt{3}/2,\,\pi/2) \;=\; \frac{1}{2}K(\sqrt{3}/2)$$

13. Show how to evaluate $\displaystyle\int\frac{dx}{\sqrt{(x-1)(x-2)(x-3)(x-4)}}$ in terms of elliptic integrals.

Make the *fractional linear transformation* $x = \dfrac{at+b}{ct+d}$ and choose a, b, c, d so that $x = 1, 2, 3$ correspond to $t = 0, 1, \infty$ respectively.

This leads to $1 = \dfrac{b}{d}$, $2 = \dfrac{a+b}{c+d}$, $3 = \dfrac{a}{c}$ from which $a = 3d$, $b = d$, $c = d$ so that $x = \dfrac{3t+1}{t+1}$.

Using this transformation the given integral becomes $\displaystyle\int \dfrac{dt}{\sqrt{t(t-1)(t+3)}}$.

Then if $t = u^2$, we obtain $2 \displaystyle\int \dfrac{du}{\sqrt{(u^2-1)(u^2+3)}}$ and the method of Problem 12 is applicable.

In general $\displaystyle\int \dfrac{dx}{\sqrt{P_4(x)}}$ where $P_4(x)$ is a fourth degree polynomial having real zeros, can be transformed into an elliptic integral by the method indicated. Similar procedures are available if some or all zeros are complex (see Problem 14; the method employed there can also be used in this problem).

14. Evaluate $\displaystyle\int \dfrac{dx}{\sqrt{(x^2-2x+10)(x^2+x+7)}}$ in terms of elliptic integrals.

Here the polynomial under the radical has no real zeros, so that the method of Problem 13 is inapplicable. We proceed as follows. Let $x = y + \alpha$, where α is a constant. Then the integral becomes

$$I = \int \dfrac{dy}{\sqrt{\{y^2 + (2\alpha - 2)y + \alpha^2 - 2\alpha + 10\}\{y^2 + (2\alpha + 1)y + \alpha^2 + \alpha + 7\}}}$$

Choose α so that the constant terms of each quadratic are equal, i.e. $\alpha^2 - 2\alpha + 10 = \alpha^2 + \alpha + 7$ or $\alpha = 1$. Then

$$I = \int \dfrac{dy}{\sqrt{(y^2+9)(y^2+3y+9)}}$$

Let $y = \beta u$, where β is a positive constant. Then

$$I = \beta \int \dfrac{du}{\sqrt{(\beta^2 u^2 + 9)(\beta^2 u^2 + 3\beta u + 9)}}$$

Choose β so that the coefficient of u^2 in each quadratic is equal to the constant term, i.e. $\beta^2 = 9$ or $\beta = 3$. Then

$$I = \int \dfrac{du}{\sqrt{(u^2+1)(u^2+u+1)}}$$

Let $u = \dfrac{1+t}{1-t}$, $du = \dfrac{2\,dt}{(1-t)^2}$. Then

$$I = \sqrt{2} \int \dfrac{dt}{\sqrt{(t^2+1)(t^2+3)}}$$

which is reducible to elliptic type by substituting $t = \tan\theta$ as in Problem 11(b).

15. Express $\displaystyle\int \dfrac{dx}{\sqrt{(x^2-2x+10)(x^2-2x+7)}}$ as an elliptic integral.

If we let $x = y + \alpha$ as in Problem 14, we are led to the condition $\alpha^2 - 2\alpha + 10 = \alpha^2 - 2\alpha + 7$ which is impossible. In this case however, completing the square in each quadratic suggests the transformation $x - 1 = y$ and the integral becomes

$$\int \dfrac{dy}{\sqrt{(y^2+9)(y^2+6)}}$$

which is reducible to standard form by letting $y = \sqrt{6} \tan\theta$.

16. Evaluate $\displaystyle\int_1^\infty \frac{du}{(3u^2+1)\sqrt{(u^2-1)(u^2+3)}}$.

Letting $u = \sec\theta$, the integral becomes

$$
\int_0^{\pi/2} \frac{\cos^2\theta\,d\theta}{(3+\cos^2\theta)\sqrt{1+3\cos^2\theta}} \;=\; \int_0^{\pi/2} \frac{(3+\cos^2\theta)-3}{(3+\cos^2\theta)\sqrt{1+3\cos^2\theta}}\,d\theta
$$

$$
=\; \int_0^{\pi/2} \frac{d\theta}{\sqrt{1+3\cos^2\theta}} \;-\; 3\int_0^{\pi/2}\frac{d\theta}{(3+\cos^2\theta)\sqrt{1+3\cos^2\theta}}
$$

$$
=\; \frac{1}{2}\int_0^{\pi/2}\frac{d\theta}{\sqrt{1-\frac{3}{4}\sin^2\theta}} \;-\; \frac{3}{8}\int_0^{\pi/2}\frac{d\theta}{(1-\frac{1}{4}\sin^2\theta)\sqrt{1-\frac{3}{4}\sin^2\theta}}
$$

$$
=\; \tfrac{1}{2}F(\sqrt{3}/2,\,\pi/2) \;-\; \tfrac{3}{8}\Pi(\sqrt{3}/2,\,-1/4,\,\pi/2)
$$

where the second integral is the complete elliptic integral of the third kind.

17. Show how to evaluate $\displaystyle\int \frac{dx}{x\sqrt{(x-1)(x-2)(x-3)(x-4)}}$ in terms of elliptic integrals.

Using the same transformation as in Problem 13, the integral becomes $\displaystyle\int \frac{(t+1)\,dt}{(3t+1)\sqrt{t(t-1)(t+3)}}$.

Now let $t = u^2$ to obtain

$$
2\int \frac{(u^2+1)\,du}{(3u^2+1)\sqrt{(u^2-1)(u^2+3)}} \;=\; 2\int \frac{\frac{1}{3}(3u^2+1)+\frac{2}{3}}{(3u^2+1)\sqrt{(u^2-1)(u^2+3)}}\,du
$$

$$
=\; \frac{2}{3}\int \frac{du}{\sqrt{(u^2-1)(u^2+3)}} \;+\; \frac{4}{3}\int \frac{du}{(3u^2+1)\sqrt{(u^2-1)(u^2+3)}}
$$

Then proceed as in Problems 12 and 16.

Another method: Letting $x - 1 = 1/y$, the integral becomes

$$
-\int \frac{dy}{\sqrt{(1-y)(1-2y)(1-3y)}}
$$

for which the method of Problem 11(c) is applicable.

ELLIPTIC FUNCTIONS

18. Prove $(a)\ \dfrac{d}{du}(\operatorname{sn}u) = \operatorname{cn}u\,\operatorname{dn}u$, $(b)\ \dfrac{d}{du}(\operatorname{cn}u) = -\operatorname{sn}u\,\operatorname{dn}u$.

By definition, if $u = \displaystyle\int_0^\phi \frac{d\theta}{\sqrt{1-k^2\sin^2\theta}}$ then

$$
\operatorname{sn}u = \sin\phi = \sin(\operatorname{am}u), \qquad \operatorname{cn}u = \cos\phi = \cos(\operatorname{am}u)
$$

Thus

(a) $\dfrac{d}{du}(\operatorname{sn}u) = \dfrac{d}{du}(\sin\phi) = \cos\phi\,\dfrac{d\phi}{du} = \operatorname{cn}u\sqrt{1-k^2\sin^2\phi} = \operatorname{cn}u\,\operatorname{dn}u$

(b) $\dfrac{d}{du}(\operatorname{cn}u) = \dfrac{d}{du}(\cos\phi) = -\sin\phi\,\dfrac{d\phi}{du} = -\operatorname{sn}u\sqrt{1-k^2\sin^2\phi} = -\operatorname{sn}u\,\operatorname{dn}u$

since $\dfrac{du}{d\phi} = \dfrac{1}{\sqrt{1-k^2\sin^2\phi}}$ and so $\dfrac{d\phi}{du} = \sqrt{1-k^2\sin^2\phi} = \operatorname{dn}u$.

19. Prove $\dfrac{d}{du}(\operatorname{dn} u) = -k^2 \operatorname{sn} u \operatorname{cn} u$.

$$\frac{d}{du}(\operatorname{dn} u) = \frac{d}{du}(\sqrt{1 - k^2 \sin^2 \phi}) = \frac{d}{du}(\sqrt{1 - k^2 \operatorname{sn}^2 u})$$

$$= \frac{1}{2}(1 - k^2 \operatorname{sn}^2 u)^{-1/2} \frac{d}{du}(-k^2 \operatorname{sn}^2 u)$$

$$= \frac{-k^2}{2 \operatorname{dn} u} \cdot 2(\operatorname{sn} u) \frac{d}{du}(\operatorname{sn} u) = \frac{-k^2}{2 \operatorname{dn} u} \cdot 2 \operatorname{sn} u \operatorname{cn} u \operatorname{dn} u$$

$$= -k^2 \operatorname{sn} u \operatorname{cn} u$$

20. Prove (a) $\operatorname{sn}(-u) = -\operatorname{sn} u$, (b) $\operatorname{cn}(-u) = \operatorname{cn} u$, (c) $\operatorname{dn}(-u) = \operatorname{dn} u$, (d) $\operatorname{tn}(-u) = -\operatorname{tn} u$.

(a) Let $u = \displaystyle\int_0^\phi \frac{d\theta}{\sqrt{1 - k^2 \sin^2 \theta}}$, $v = \displaystyle\int_0^{-\phi} \frac{d\theta}{\sqrt{1 - k^2 \sin^2 \theta}}$.

Then $\operatorname{sn} u = \sin \phi$, $\operatorname{sn} v = \sin(-\phi) = -\sin \phi = -\operatorname{sn} u$.

Now letting $\theta = -\psi$ in the second integral, it becomes $v = -\displaystyle\int_0^\phi \frac{d\psi}{\sqrt{1 - k^2 \sin^2 \psi}} = -u$.

Thus $\operatorname{sn}(-u) = -\operatorname{sn} u$.

(b) Since $\operatorname{cn} u = \sqrt{1 - \operatorname{sn}^2 u}$, $\operatorname{cn}(-u) = \sqrt{1 - \operatorname{sn}^2(-u)} = \sqrt{1 - \operatorname{sn}^2 u} = \operatorname{cn} u$.

(c) Since $\operatorname{dn} u = \sqrt{1 - k^2 \operatorname{sn}^2 u}$, $\operatorname{dn}(-u) = \sqrt{1 - k^2 \operatorname{sn}^2(-u)} = \sqrt{1 - k^2 \operatorname{sn}^2 u} = \operatorname{dn} u$.

(d) $\operatorname{tn}(-u) = \dfrac{\operatorname{sn}(-u)}{\operatorname{cn}(-u)} = -\dfrac{\operatorname{sn} u}{\operatorname{cn} u} = -\operatorname{tn} u$.

21. Show that if $K = \displaystyle\int_0^{\pi/2} \frac{d\theta}{\sqrt{1 - k^2 \sin^2 \theta}}$, then

(a) $\operatorname{sn}(u + 2K) = -\operatorname{sn} u$, (b) $\operatorname{cn}(u + 2K) = -\operatorname{cn} u$, (c) $\operatorname{dn}(u + 2K) = \operatorname{dn} u$, (d) $\operatorname{tn}(u + 2K) = \operatorname{tn} u$.

(a) Consider $\displaystyle\int_0^{\phi + \pi} \frac{d\theta}{\sqrt{1 - k^2 \sin^2 \theta}} = \int_0^\pi \frac{d\theta}{\sqrt{1 - k^2 \sin^2 \theta}} + \int_\pi^{\phi + \pi} \frac{d\theta}{\sqrt{1 - k^2 \sin^2 \theta}}$.

The first integral on the right $= 2\displaystyle\int_0^{\pi/2} \frac{d\theta}{\sqrt{1 - k^2 \sin^2 \theta}} = 2K$.

Letting $\theta = \pi + \psi$, the second integral on the right becomes $\displaystyle\int_0^\phi \frac{d\psi}{\sqrt{1 - k^2 \sin^2 \psi}} = u$.

Then $\displaystyle\int_0^{\phi + \pi} \frac{d\theta}{\sqrt{1 - k^2 \sin^2 \theta}} = u + 2K$ or $\operatorname{am}(u + 2K) = \phi + \pi$

and $\operatorname{sn}(u + 2K) = \sin(\phi + \pi) = -\sin \phi = -\operatorname{sn} u$

(b) $\operatorname{cn}(u + 2K) = \cos(\phi + \pi) = -\cos \phi = -\operatorname{cn} u$

(c) $\operatorname{dn}(u + 2K) = \sqrt{1 - k^2 \operatorname{sn}^2(u + 2K)} = \sqrt{1 - k^2 \operatorname{sn}^2 u} = \operatorname{dn} u$

(d) $\operatorname{tn}(u + 2K) = \dfrac{\operatorname{sn}(u + 2K)}{\operatorname{cn}(u + 2K)} = \dfrac{\operatorname{sn} u}{\operatorname{cn} u} = \operatorname{tn} u$

22. Prove that $\operatorname{sn} u$ and $\operatorname{cn} u$ have periods equal to $4K$, while $\operatorname{dn} u$ and $\operatorname{tn} u$ have periods $2K$.

Replacing u by $u + 2K$ in Problem 21,

$$\operatorname{sn}(u + 4K) = -\operatorname{sn}(u + 2K) = -(-\operatorname{sn} u) = \operatorname{sn} u$$
$$\operatorname{cn}(u + 4K) = -\operatorname{cn}(u + 2K) = -(-\operatorname{cn} u) = \operatorname{cn} u$$

and the proof is complete.

From Problem 21(c) it is seen that dn u and tn u have period $2K$.

It can be shown that the elliptic functions have other periods which are complex. For example, sn u has periods $4K$ and $2iK'$, cn u has periods $4K$ and $2K + 2iK'$, dn u has periods $2K$ and $4iK'$, where

$$K' = \int_0^{\pi/2} \frac{d\theta}{\sqrt{1 - k'^2 \sin^2\theta}}, \quad k' = \sqrt{1 - k^2}$$

For this reason elliptic functions are sometimes known as *doubly-periodic functions*.

23. Prove that (a) $\dfrac{d}{dx} \, \mathrm{sn}^{-1}(x, k) = \dfrac{1}{\sqrt{(1 - x^2)(1 - k^2 x^2)}}$

(b) $\displaystyle\int_0^x \frac{dv}{\sqrt{(1 - v^2)(1 - k^2 v^2)}} = \mathrm{sn}^{-1}(x, k) = F(k, \sin^{-1} x).$

(a) We shall write $\mathrm{sn}^{-1} x$ in place of $\mathrm{sn}^{-1}(x, k)$, the dependence on the modulus k being understood. By Problem 18, if $x = \mathrm{sn}\, u$, then

$$\frac{dx}{du} = \frac{d}{du}(\mathrm{sn}\, u) = \mathrm{cn}\, u \, \mathrm{dn}\, u = \sqrt{1 - x^2}\,\sqrt{1 - k^2 x^2} = \sqrt{(1 - x^2)(1 - k^2 x^2)}$$

Hence $\dfrac{du}{dx} = \dfrac{d}{dx}(\mathrm{sn}^{-1} x) = \dfrac{1}{\sqrt{(1 - x^2)(1 - k^2 x^2)}}$

(b) Integrating the result in (a) from 0 to x. we have, since $x = \sin\phi$ where $\phi = \mathrm{am}\, u$,

$$\int_0^x \frac{dv}{\sqrt{(1 - v^2)(1 - k^2 v^2)}} = \mathrm{sn}^{-1} x = F(k, \phi) = F(k, \sin^{-1} x)$$

Note the similarity with the result for trigonometric functions (corresponding to $k = 0$):

$\displaystyle\int_0^x \frac{dv}{\sqrt{1 - v^2}} = \sin^{-1} x$. The elliptic functions are generalizations of trigonometric functions.

MISCELLANEOUS PROBLEMS

24. Prove that $F(\sqrt{1/2}, \pi/2) = \dfrac{1}{2}\sqrt{\dfrac{\pi}{2} \dfrac{\Gamma(\tfrac{1}{4})}{\Gamma(\tfrac{3}{4})}}.$

From the properties of the gamma function,

$$\int_0^\pi \frac{d\theta}{\sqrt{\sin\theta}} = \int_0^{\pi/2} \sin^{-1/2}\theta \, d\theta = \frac{\Gamma(\tfrac{1}{4})\,\Gamma(\tfrac{1}{2})}{2\,\Gamma(\tfrac{3}{4})} = \frac{\sqrt{\pi}}{2}\frac{\Gamma(\tfrac{1}{4})}{\Gamma(\tfrac{3}{4})}$$

But from Problem 2, this integral equals $\sqrt{2}\, F(\sqrt{1/2}, \pi/2)$ and so the required result follows at once.

25. Determine the period T of a simple pendulum of length l.

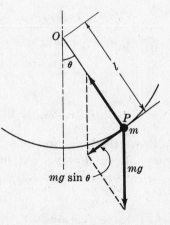

A simple pendulum consists of a mass m attached to a rigid rod OP of negligible mass and length l (see Fig. 16-1). Assuming the rod to be suspended from a fixed point O, the differential equation of motion of mass m is given by

$$ml\frac{d^2\theta}{dt^2} = -mg\sin\theta \quad \text{or} \quad \frac{d^2\theta}{dt^2} = -\frac{g}{l}\sin\theta$$

Letting $\dfrac{d\theta}{dt} = p, \; \dfrac{d^2\theta}{dt^2} = \dfrac{dp}{dt} = \dfrac{dp}{d\theta}\dfrac{d\theta}{dt} = p\dfrac{dp}{d\theta}$, the

equation becomes $p\dfrac{dp}{d\theta} = -\dfrac{g}{l}\sin\theta$. Integrating,

$\dfrac{p^2}{2} = \dfrac{g}{l}\cos\theta + c$.

Fig. 16-1

If the pendulum makes an angle $\theta = \theta_0 > 0$ at time $t = 0$ and is released from rest, i.e. $p = d\theta/dt = 0$ when $\theta = \theta_0$, then $c = -(g/l) \cos \theta_0$. We thus have

$$d\theta/dt = \pm \sqrt{2g/l} \sqrt{\cos \theta - \cos \theta_0}$$

When the pendulum goes from $\theta = \theta_0$ to $\theta = 0$ (which corresponds to one quarter of a period, or $T/4$), $d\theta/dt$ is negative; hence choose the $-$ sign. Then

$$\frac{T}{4} = -\sqrt{\frac{l}{2g}} \int_{\theta_0}^{0} \frac{d\theta}{\sqrt{\cos \theta - \cos \theta_0}} = \sqrt{\frac{l}{2g}} \int_{0}^{\theta_0} \frac{d\theta}{\sqrt{\cos \theta - \cos \theta_0}}$$

Thus

$$T = 4\sqrt{\frac{l}{2g}} \int_{0}^{\theta_0} \frac{d\theta}{\sqrt{\cos \theta - \cos \theta_0}} = 4\sqrt{\frac{l}{2g}} \int_{0}^{\theta_0} \frac{d\theta}{\sqrt{1 - 2\sin^2 \theta/2 - (1 - 2\sin^2 \theta_0/2)}}$$

$$= 2\sqrt{\frac{l}{g}} \int_{0}^{\theta_0} \frac{d\theta}{\sqrt{\sin^2 \theta_0/2 - \sin^2 \theta/2}}$$

Letting $\sin \theta/2 = \sin \theta_0/2 \sin u$, this integral becomes

$$T = 4\sqrt{\frac{l}{g}} \int_{0}^{\pi/2} \frac{du}{\sqrt{1 - k^2 \sin^2 u}}, \qquad k = \sin \theta_0/2$$

If $k = 0$, $T = 2\pi\sqrt{l/g}$ which is the approximate period for small oscillations.

26. Prove that $\quad \operatorname{sn}(u+v) = \dfrac{\operatorname{sn} u \operatorname{cn} v \operatorname{dn} v + \operatorname{cn} u \operatorname{sn} v \operatorname{dn} u}{1 - k^2 \operatorname{sn}^2 u \operatorname{sn}^2 v}$.

Let $u + v = \alpha$, a constant. Then $dv/du = -1$. Let us define $U = \operatorname{sn} u$, $V = \operatorname{sn} v$. It follows that

$$\frac{dU}{du} = \dot{U} = \operatorname{cn} u \operatorname{dn} u, \qquad \frac{dV}{du} = \dot{V} = \frac{dV}{dv} \frac{dv}{du} = -\operatorname{cn} v \operatorname{dn} v$$

where dots denote differentiation with respect to u. Then

$$\dot{U}^2 = (1 - U^2)(1 - k^2 U^2) \quad \text{and} \quad \dot{V}^2 = (1 - V^2)(1 - k^2 V^2)$$

Differentiating and simplifying, we find

$$(1) \quad \ddot{U} = 2k^2 U^3 - (1 + k^2)U \qquad (2) \quad \ddot{V} = 2k^2 V^3 - (1 + k^2)V$$

Multiplying (1) by V, (2) by U, and subtracting, we have

$$\ddot{U}V - U\ddot{V} = 2k^2 UV(U^2 - V^2) \tag{3}$$

It can be verified that (see Problem 58)

$$\dot{U}^2 V^2 - U^2 \dot{V}^2 = (1 - k^2 U^2 V^2)(V^2 - U^2) \tag{4}$$

or

$$\dot{U}V - U\dot{V} = \frac{(1 - k^2 U^2 V^2)(V^2 - U^2)}{\dot{U}V + U\dot{V}} \tag{5}$$

Dividing equations (3) and (5) we have

$$\frac{\ddot{U}V - U\ddot{V}}{\dot{U}V - U\dot{V}} = -\frac{2k^2 UV(\dot{U}V + U\dot{V})}{1 - k^2 U^2 V^2} \tag{6}$$

But $\ddot{U}V - U\ddot{V} = \dfrac{d}{du}(\dot{U}V - U\dot{V})$ and $-2k^2 UV(\dot{U}V + U\dot{V}) = \dfrac{d}{du}(1 - k^2 U^2 V^2)$, so that (6) becomes

$$\frac{d(\dot{U}V - U\dot{V})}{\dot{U}V - U\dot{V}} = \frac{d(1 - k^2 U^2 V^2)}{1 - k^2 U^2 V^2}$$

An integration yields $\dfrac{\dot{U}V - U\dot{V}}{1 - k^2 U^2 V^2} = c$ (a constant), i.e.,

$$\frac{\operatorname{sn} u \operatorname{cn} v \operatorname{dn} v + \operatorname{cn} u \operatorname{sn} v \operatorname{dn} u}{1 - k^2 \operatorname{sn}^2 u \operatorname{sn}^2 v} = c$$

is a solution of the differential equation. It is also clear that $u + v = \alpha$ is a solution. The two solutions must be related as follows:

$$\frac{\operatorname{sn} u \operatorname{cn} v \operatorname{dn} v + \operatorname{cn} u \operatorname{sn} v \operatorname{dn} u}{1 - k^2 \operatorname{sn}^2 u \operatorname{sn}^2 v} = f(u + v)$$

Putting $v = 0$, we see that $f(u) = \operatorname{sn} u$.

Then $f(u + v) = \operatorname{sn}(u + v)$ and the required result follows.

Supplementary Problems

ELLIPTIC INTEGRALS

27. (a) Use the binomial theorem to show that if $|x| < 1$,

$$\sqrt{1 - x} = 1 - \frac{x}{2} - \left(\frac{1}{2}\right)\frac{x^2}{4} - \left(\frac{1 \cdot 3}{2 \cdot 4}\right)\frac{x^3}{6} - \left(\frac{1 \cdot 3 \cdot 5}{2 \cdot 4 \cdot 6}\right)\frac{x^4}{8} + \cdots$$

(b) If $|k| < 1$, prove that

$$E(k, \pi/2) = \int_0^{\pi/2} \sqrt{1 - k^2 \sin^2 \theta}\, d\theta$$

$$= \frac{\pi}{2}\left\{1 - \left(\frac{1}{2}\right)^2 k^2 - \left(\frac{1 \cdot 3}{2 \cdot 4}\right)^2 \frac{k^4}{3} - \left(\frac{1 \cdot 3 \cdot 5}{2 \cdot 4 \cdot 6}\right)^2 \frac{k^6}{5} - \cdots\right\}$$

28. Evaluate (a) $E(\sqrt{2}/2, \pi/2)$, (b) $F(\sqrt{2}/2, \pi/2)$, (c) $E(0.5)$, (d) $K(\sqrt{3}/2)$.

   Ans. (a) 1.3506, (b) 1.8541, (c) 1.4675, (d) 2.1565

29. Show that (a) $E(1, \phi) = \sin \phi$, (b) $F(1, \phi) = \ln(\sec \phi + \tan \phi) = \ln \tan(\pi/4 + \phi/2)$.

30. Find the perimeter of the ellipse $x^2/9 + y^2/4 = 1$. [Hint: Let $x = 3 \sin \phi$, $y = 2 \cos \phi$ be the parametric equations.]   Ans. 15.865

31. Evaluate $\displaystyle\int_0^{\pi/2} \frac{dx}{\sqrt{\sin^2 x + 2 \cos^2 x}}$ in terms of elliptic integrals.   Ans. $\dfrac{1}{\sqrt{2}} K\left(\dfrac{1}{\sqrt{2}}\right)$

32. Express $\displaystyle\int_0^t \frac{dx}{\sqrt{1 - 3 \sin^2 x}}$ in terms of elliptic integrals.

   Ans. $\dfrac{1}{\sqrt{3}} F\left(\dfrac{1}{\sqrt{3}}, \phi\right)$, where $\phi = \sin^{-1}(\sqrt{3} \sin t)$

33. Show that $\displaystyle\int_0^{\pi/2} \frac{d\theta}{\sqrt{1 + 2 \sin \theta}} = F\left(\dfrac{\sqrt{3}}{2}, \sin^{-1} \sqrt{\dfrac{2}{3}}\right)$.

34. Evaluate $\displaystyle\int_0^1 \frac{dx}{\sqrt{x(1-x)(1+x)}}$.   *Ans.* $\sqrt{2}\,K\left(\dfrac{1}{\sqrt{2}}\right)$

35. Evaluate $\displaystyle\int_1^3 \frac{dx}{\sqrt{x^4+4x^2+3}}$.   *Ans.* $\dfrac{1}{\sqrt{3}}\left\{F\left(\sqrt{\dfrac{2}{3}},\dfrac{\pi}{3}\right) - F\left(\sqrt{\dfrac{2}{3}},\dfrac{\pi}{4}\right)\right\}$

36. Show that $\displaystyle\int_1^\infty \frac{dx}{\sqrt{x^4-1}} = \dfrac{1}{\sqrt{2}}K\left(\dfrac{1}{\sqrt{2}}\right)$

37. Express each of the following in terms of elliptic integrals.

(a) $\displaystyle\int_0^2 \frac{dx}{\sqrt{(16-x^2)(25-x^2)}}$ (c) $\displaystyle\int_0^1 \sqrt{\frac{3-x^2}{1-x^2}}\,dx$

(b) $\displaystyle\int_3^5 \frac{dx}{\sqrt{(x^2-9)(25-x^2)}}$ (d) $\displaystyle\int_0^\infty \frac{x\,dx}{\sqrt{(x^4+16)(x^4+25)}}$

Ans. (a) $\frac{1}{5}K(\frac{4}{5})$, (b) $\frac{1}{5}K(\frac{4}{5})$, (c) $\sqrt{3}\,E(\sqrt{\frac{1}{3}})$, (d) $\frac{1}{10}K(\frac{3}{5})$

38. Evaluate (a) $\displaystyle\int_0^1 \frac{dx}{\sqrt{(5-x)(3-x)(1-x)}}$, (b) $\displaystyle\int_6^\infty \frac{dx}{x\sqrt{(x-2)(x-4)(x-6)}}$.

Ans. (a) $F(\sqrt{2}/2, \tan^{-1}\sqrt{2}/2)$, (b) $\frac{1}{4}F(\sqrt{2}/2, \pi/2) - \frac{1}{12}\Pi(\sqrt{2}/2, -2/3, \pi/2)$

39. Show that $\displaystyle\int_0^\infty \frac{dx}{\sqrt{(x+1)(x+3)(x+5)(x+7)}} = \dfrac{1}{2}\left\{F\left(\dfrac{\sqrt{3}}{2}, \cos^{-1}\dfrac{1}{\sqrt{15}}\right) - F\left(\dfrac{\sqrt{3}}{2}, \cos^{-1}\dfrac{1}{\sqrt{3}}\right)\right\}$

40. Show that $\displaystyle\int_0^\infty \frac{dx}{\sqrt{x^4+x^2+1}} = \dfrac{4}{3}F\left(\dfrac{2\sqrt{2}}{3}, \dfrac{\pi}{3}\right)$.

[Hint: $x^4 + x^2 + 1 = (x^2+x+1)(x^2-x+1)$.]

41. Evaluate (a) $\displaystyle\int_2^{10} \frac{dx}{\sqrt{(x^2-2x+4)(x^2-4x+8)}}$, (b) $\displaystyle\int_1^\infty \frac{dx}{\sqrt{(x^2-4x+5)(x^2-4x+10)}}$.

Ans. (a) $\dfrac{1}{\sqrt{6}}\left\{F\left(\sqrt{\dfrac{2}{3}}, \dfrac{\pi}{4}\right) + F\left(\sqrt{\dfrac{2}{3}}, \tan^{-1}\dfrac{3}{5}\right)\right\}$ (b) $\dfrac{1}{\sqrt{6}}\left\{F\left(\sqrt{\dfrac{5}{6}}, \dfrac{\pi}{4}\right) + F\left(\sqrt{\dfrac{5}{6}}, \dfrac{\pi}{2}\right)\right\}$

ELLIPTIC FUNCTIONS

42. Show that (a) $\operatorname{sn}0 = 0$, (b) $\operatorname{cn}0 = 1$, (c) $\operatorname{dn}0 = 1$, (d) $\operatorname{tn}0 = 0$, (e) $\operatorname{am}0 = 0$.

43. Prove that (a) $\operatorname{sn}^2 u + \operatorname{cn}^2 u = 1$, (b) $\operatorname{dn}^2 u + k^2\operatorname{sn}^2 u = 1$.

44. Prove that $\operatorname{dn}^2 u - k^2\operatorname{cn}^2 u = k'^2$ where $k' = \sqrt{1-k^2}$.

45. Prove that (a) $\operatorname{sn}^2 u = \dfrac{1 - \operatorname{cn}2u}{1 + \operatorname{dn}2u}$, (b) $\sqrt{\dfrac{1 - \operatorname{cn}2u}{1 + \operatorname{cn}2u}} = \dfrac{\operatorname{sn}u\,\operatorname{dn}u}{\operatorname{cn}u}$.

46. Find (a) $\dfrac{d}{du}(\operatorname{sn}u\,\operatorname{cn}u)$, (b) $\dfrac{d}{du}(\operatorname{dn}u)^3$.

Ans. (a) $(\operatorname{cn}^2 u - \operatorname{sn}^2 u)\operatorname{dn}u$, (b) $-3k^2\operatorname{dn}^2 u\,\operatorname{sn}u\,\operatorname{cn}u$

47. Find (a) $\dfrac{d}{du}(\operatorname{tn}u)$, (b) $\dfrac{d}{du}\operatorname{sech}^{-1}(k\,\operatorname{sn}u)$.   *Ans.* (a) $\dfrac{\operatorname{dn}u}{\operatorname{cn}^2 u}$, (b) $\dfrac{\operatorname{cn}u}{\operatorname{sn}u}$

48. Verify the results (a) $\displaystyle\int \operatorname{cn}u\,du = \dfrac{1}{k}\cos^{-1}(\operatorname{dn}u) + c$, (b) $\displaystyle\int \frac{du}{\operatorname{sn}u} = \ln\left(\dfrac{\operatorname{sn}u}{\operatorname{cn}u + \operatorname{dn}u}\right) + c$.

49. Show that $\displaystyle\int_0^u \operatorname{sn}^2 x\,dx = \dfrac{1}{k^2}\{u - E(k, \operatorname{am}u)\}$.

50. Prove that (a) $\operatorname{sn}(u+K) = \dfrac{\operatorname{cn}u}{\operatorname{dn}u}$, (b) $\operatorname{cn}(u+K) = -\dfrac{k'\operatorname{sn}u}{\operatorname{dn}u}$, (c) $\operatorname{dn}(u+K) = \dfrac{k'}{\operatorname{dn}u}$ where $k' = \sqrt{1-k^2}$ is the complementary modulus.

51. Prove that $\displaystyle\int_0^x \frac{dv}{\sqrt{(1+v^2)(1+k'^2v^2)}} = \operatorname{tn}^{-1}(x,k) = F(k, \tan^{-1} x)$ where $k' = \sqrt{1-k^2}$. We often write $\operatorname{tn}^{-1}(x,k)$ briefly as $\operatorname{tn}^{-1} x$, the modulus k being understood.

52. Prove that $\displaystyle\int_x^\infty \frac{dv}{\sqrt{v^4 + v^2 + 1}} = \frac{1}{2}\operatorname{cn}^{-1}\left(\frac{x^2-1}{x^2+1}, \frac{\sqrt{3}}{2}\right)$.

MISCELLANEOUS PROBLEMS

53. Prove that $\displaystyle\int_0^1 \sin^{-1} x^2\, dx = \frac{\pi}{2} - 2\sqrt{2}\, E\left(\frac{1}{\sqrt{2}}, \frac{\pi}{2}\right) + \sqrt{2}\, F\left(\frac{1}{\sqrt{2}}, \frac{\pi}{2}\right)$.

54. A pendulum of length 2 ft is released at a position in which it makes an angle of 60° with the vertical. Determine its period of oscillation assuming the acceleration due to gravity $= g = 32$ ft/sec². *Ans.* 1.686 seconds

55. Show that at any time t the angle θ for the pendulum of Problem 25 is given by

$$\sin \theta/2 = \sin \theta_0/2 \,\operatorname{sn}\left(\sqrt{g/l}\, t\right)$$

where t is measured from the instant where the pendulum rod is vertical.

56. Show that $\displaystyle\int_0^{\pi/2} \frac{d\theta}{\sqrt{\sin\theta + \cos\theta}} = 2\sqrt[4]{2}\, F\left(\frac{1}{\sqrt{2}}, \cos^{-1}\frac{1}{\sqrt[4]{2}}\right)$.

[Hint: $\sin\theta + \cos\theta = \sqrt{2}\sin(\theta + \pi/4)$.]

57. Obtain the expansion

$$\operatorname{sn} u = u - (1+k^2)\frac{u^3}{3!} + (1 + 14k^2 + k^4)\frac{u^5}{5!} - (1 + 135k^2 + 135k^4 + k^6)\frac{u^7}{7!} + \cdots$$

58. Verify equation (4) of Problem 26.

59. Prove that

(a) $\operatorname{cn}(u+v) = \dfrac{\operatorname{cn} u \operatorname{cn} v - \operatorname{sn} u \operatorname{sn} v \operatorname{dn} u \operatorname{dn} v}{1 - k^2 \operatorname{sn}^2 u \operatorname{sn}^2 v}$

(b) $\operatorname{dn}(u+v) = \dfrac{\operatorname{dn} u \operatorname{dn} v - k^2 \operatorname{sn} u \operatorname{sn} v \operatorname{cn} u \operatorname{cn} v}{1 - k^2 \operatorname{sn}^2 u \operatorname{sn}^2 v}$

60. Evaluate (a) $F(\sqrt{2}/2, \pi/3)$, (b) $F(0.5, \pi/4)$ by using Landen's transformation. *Ans.* (a) 1.1424, (b) 0.8044

61. Verify the result (13), Page 332, using Landen's transformation.

62. Prove that if $k_n = \dfrac{2\sqrt{k_{n-1}}}{1 + k_{n-1}}$, then $\displaystyle\lim_{n\to\infty} k_n = 1$. Hence verify (15), Page 333.

Chapter 17

Functions of a Complex Variable

FUNCTIONS

If to each of a set of complex numbers which a variable z may assume there corresponds one or more values of a variable w, then w is called a *function of the complex variable* z, written $w = f(z)$. The fundamental operations with complex numbers have already been considered in Chapter 1.

A function is *single-valued* if for each value of z there corresponds only one value of w; otherwise it is *multiple-valued* or *many-valued*. In general we can write $w = f(z) = u(x, y) + i v(x, y)$, where u and v are real functions of x and y.

> **Example:** $w = z^2 = (x + iy)^2 = x^2 - y^2 + 2ixy = u + iv$ so that $u(x, y) = x^2 - y^2$, $v(x, y) = 2xy$. These are called the *real* and *imaginary parts* of $w = z^2$ respectively.

Unless otherwise specified we shall assume that $f(z)$ is single-valued. A function which is multiple-valued can be considered as a collection of single-valued functions.

LIMITS and CONTINUITY

Definitions of limits and continuity for functions of a complex variable are analogous to those for a real variable. Thus $f(z)$ is said to have the *limit* l as z approaches z_0 if, given any $\epsilon > 0$, there exists a $\delta > 0$ such that $|f(z) - l| < \epsilon$ whenever $0 < |z - z_0| < \delta$.

Similarly, $f(z)$ is said to be *continuous* at z_0 if, given any $\epsilon > 0$, there exists a $\delta > 0$ such that $|f(z) - f(z_0)| < \epsilon$ whenever $|z - z_0| < \delta$. Alternatively, $f(z)$ is continuous at z_0 if $\lim_{z \to z_0} f(z) = f(z_0)$.

DERIVATIVES

If $f(z)$ is single-valued in some region of the z plane the *derivative* of $f(z)$, denoted by $f'(z)$, is defined as

$$\lim_{\Delta z \to 0} \frac{f(z + \Delta z) - f(z)}{\Delta z} \tag{1}$$

provided the limit exists independent of the manner in which $\Delta z \to 0$. If the limit (1) exists for $z = z_0$, then $f(z)$ is called *analytic* at z_0. If the limit exists for all z in a region $\mathcal{R}$, then $f(z)$ is called *analytic in* $\mathcal{R}$. In order to be analytic, $f(z)$ must be single-valued and continuous. The converse, however, is not necessarily true.

We define elementary functions of a complex variable by a natural extension of the corresponding functions of a real variable. Where series expansions for real functions $f(x)$ exist, we can use as definition the series with x replaced by z. The convergence of such complex series has already been considered in Chapter 11.

> **Example 1:** We define $e^z = 1 + z + \dfrac{z^2}{2!} + \dfrac{z^3}{3!} + \cdots$, $\sin z = z - \dfrac{z^3}{3!} + \dfrac{z^5}{5!} - \dfrac{z^7}{7!} + \cdots$, $\cos z = 1 - \dfrac{z^2}{2!} + \dfrac{z^4}{4!} - \dfrac{z^6}{6!} + \cdots$ From these we can show that $e^z = e^{x+iy} = e^x (\cos y + i \sin y)$, as well as numerous other relations.

Example 2:   We define a^b as $e^{b \ln a}$ even when a and b are complex numbers. Since $e^{2k\pi i} = 1$, it follows that $e^{i\phi} = e^{i(\phi + 2k\pi)}$ and we define $\ln z = \ln(\rho e^{i\phi}) = \ln \rho + i(\phi + 2k\pi)$. Thus $\ln z$ is a many-valued function. The various single-valued functions of which this many-valued function is composed are called its *branches*.

Rules for differentiating functions of a complex variable are much the same as for those of real variables. Thus $\dfrac{d}{dz}(z^n) = nz^{n-1}$, $\dfrac{d}{dz}(\sin z) = \cos z$, etc.

CAUCHY-RIEMANN EQUATIONS

A necessary condition that $w = f(z) = u(x,y) + i\,v(x,y)$ be analytic in a region $\mathcal{R}$ is that u and v satisfy the *Cauchy-Riemann equations*

$$\frac{\partial u}{\partial x} = \frac{\partial v}{\partial y}, \qquad \frac{\partial u}{\partial y} = -\frac{\partial v}{\partial x} \tag{2}$$

(see Problem 7). If the partial derivatives in (2) are continuous in $\mathcal{R}$, the equations are sufficient conditions that $f(z)$ be analytic in $\mathcal{R}$.

If the second derivatives of u and v with respect to x and y exist and are continuous, we find by differentiating (2) that

$$\frac{\partial^2 u}{\partial x^2} + \frac{\partial^2 u}{\partial y^2} = 0, \qquad \frac{\partial^2 v}{\partial x^2} + \frac{\partial^2 v}{\partial y^2} = 0 \tag{3}$$

Thus the real and imaginary parts satisfy Laplace's equation in two dimensions. Functions satisfying Laplace's equation are called *harmonic functions*.

INTEGRALS

If $f(z)$ is defined, single-valued and continuous in a region $\mathcal{R}$, we define the *integral* of $f(z)$ along some path C in $\mathcal{R}$ from point z_1 to point z_2, where $z_1 = x_1 + iy_1$, $z_2 = x_2 + iy_2$, as

$$\int_C f(z)\,dz \;=\; \int_{(x_1,y_1)}^{(x_2,y_2)} (u+iv)(dx + i\,dy) \;=\; \int_{(x_1,y_1)}^{(x_2,y_2)} u\,dx - v\,dy \;+\; i\int_{(x_1,y_1)}^{(x_2,y_2)} v\,dx + u\,dy$$

with this definition the integral of a function of a complex variable can be made to depend on line integrals for real functions already considered in Chapter 10. An alternative definition based on the limit of a sum, as for functions of a real variable, can also be formulated and turns out to be equivalent to the one above.

The rules for complex integration are similar to those for real integrals. An important result is

$$\left| \int_C f(z)\,dz \right| \;\leqq\; \int_C |f(z)|\,|dz| \;\leqq\; M\int_C ds \;=\; ML \tag{4}$$

where M is an upper bound of $|f(z)|$ on C, i.e. $|f(z)| \leqq M$, and L is the length of the path C.

CAUCHY'S THEOREM

Let C be a simple closed curve. If $f(z)$ is analytic within the region bounded by C as well as on C, then we have *Cauchy's theorem* that

$$\int_C f(z)\,dz \;=\; \oint_C f(z)\,dz \;=\; 0 \tag{5}$$

where the second integral emphasizes the fact that C is a simple closed curve.

Expressed in another way, (5) is equivalent to the statement that $\int_{z_1}^{z_2} f(z)\,dz$ has a value *independent of the path* joining z_1 and z_2. Such integrals can be evaluated as $F(z_2) - F(z_1)$ where $F'(z) = f(z)$. These results are similar to corresponding results for line integrals developed in Chapter 10.

> **Example:** Since $f(z) = 2z$ is analytic everywhere, we have for any simple closed curve C
>
> $$\oint_C 2z\,dz = 0$$
>
> Also,
> $$\int_{2i}^{1+i} 2z\,dz = z^2 \Big|_{2i}^{1+i} = (1+i)^2 - (2i)^2 = 2i + 4$$

CAUCHY'S INTEGRAL FORMULAS

If $f(z)$ is analytic within and on a simple closed curve C and a is any point interior to C, then

$$f(a) = \frac{1}{2\pi i} \oint_C \frac{f(z)}{z-a}\,dz \tag{6}$$

where C is traversed in the positive (counterclockwise) sense.

Also, the nth derivative of $f(z)$ at $z = a$ is given by

$$f^{(n)}(a) = \frac{n!}{2\pi i} \oint_C \frac{f(z)}{(z-a)^{n+1}}\,dz \tag{7}$$

These are called *Cauchy's integral formulas*. They are quite remarkable because they show that if the function $f(z)$ is known *on* the closed curve C then it is also known *within* C, and the various derivatives at points within C can be calculated. Thus if a function of a complex variable has a first derivative, it has all higher derivatives as well. This of course is not necessarily true for functions of real variables.

TAYLOR'S SERIES

Let $f(z)$ be analytic inside and on a circle having its center at $z = a$. Then for all points z in the circle we have the *Taylor series* representation of $f(z)$ given by

$$f(z) = f(a) + f'(a)(z-a) + \frac{f''(a)}{2!}(z-a)^2 + \frac{f'''(a)}{3!}(z-a)^3 + \cdots \tag{8}$$

See Problem 21.

SINGULAR POINTS

A singular point of a function $f(z)$ is a value of z at which $f(z)$ fails to be analytic. If $f(z)$ is analytic everywhere in some region except at an interior point $z = a$, we call $z = a$ an *isolated singularity* of $f(z)$.

> **Example:** If $f(z) = \dfrac{1}{(z-3)^2}$, then $z = 3$ is an isolated singularity of $f(z)$.

POLES

If $f(z) = \dfrac{\phi(z)}{(z-a)^n}$, $\phi(a) \neq 0$, where $\phi(z)$ is analytic everywhere in a region including $z = a$, and if n is a positive integer, then $f(z)$ has an isolated singularity at $z = a$ which is called a *pole of order n*. If $n = 1$, the pole is often called a *simple pole*; if $n = 2$ it is called a *double pole*, etc.

Example 1: $f(z) = \dfrac{z}{(z-3)^2(z+1)}$ has two singularities: a pole of order 2 or double pole at $z = 3$, and a pole of order 1 or simple pole at $z = -1$.

Example 2: $f(z) = \dfrac{3z-1}{z^2+4} = \dfrac{3z-1}{(z+2i)(z-2i)}$ has two simple poles at $z = \pm 2i$.

A function can have other types of singularities besides poles. For example, $f(z) = \sqrt{z}$ has a *branch point* at $z = 0$ (see Problem 37). The function $f(z) = \dfrac{\sin z}{z}$ has a singularity at $z = 0$. However, due to the fact that $\lim\limits_{z \to 0} \dfrac{\sin z}{z}$ is finite, we call such a singularity a *removable singularity*.

LAURENT'S SERIES

If $f(z)$ has a pole of order n at $z = a$ but is analytic at every other point inside and on a circle C with center at a, then $(z-a)^n f(z)$ is analytic at all points inside and on C and has a Taylor series about $z = a$ so that

$$f(z) = \frac{a_{-n}}{(z-a)^n} + \frac{a_{-n+1}}{(z-a)^{n-1}} + \cdots + \frac{a_{-1}}{z-a} + a_0 + a_1(z-a) + a_2(z-a)^2 + \cdots \quad (9)$$

This is called a *Laurent series* for $f(z)$. The part $a_0 + a_1(z-a) + a_2(z-a)^2 + \cdots$ is called the *analytic part*, while the remainder consisting of inverse powers of $z-a$ is called the *principal part*. More generally, we refer to the series $\sum\limits_{k=-\infty}^{\infty} a_k(z-a)^k$ as a Laurent series where the terms with $k < 0$ constitute the principal part. A function which is analytic in a region bounded by two concentric circles having center at $z = a$ can always be expanded into such a Laurent series (see Problem 92).

It is possible to define various types of singularities of a function $f(z)$ from its Laurent series. For example, when the principal part of a Laurent series has a finite number of terms and $a_{-n} \neq 0$ while $a_{-n-1}, a_{-n-2}, \ldots$ are all zero, then $z = a$ is a pole of order n. If the principal part has infinitely many terms, $z = a$ is called an *essential singularity* or sometimes a *pole of infinite order*.

Example: The function $e^{1/z} = 1 + \dfrac{1}{z} + \dfrac{1}{2!\,z^2} + \cdots$ has an essential singularity at $z = 0$.

RESIDUES

The coefficients in (9) can be obtained in the customary manner by writing the coefficients for the Taylor series corresponding to $(z-a)^n f(z)$. In further developments, the coefficient a_{-1}, called the *residue* of $f(z)$ at the pole $z = a$, is of considerable importance. It can be found from the formula

$$a_{-1} = \lim_{z \to a} \frac{1}{(n-1)!} \frac{d^{n-1}}{dz^{n-1}} \{(z-a)^n f(z)\} \quad (10)$$

where n is the order of the pole. For simple poles the calculation of the residue is of particular simplicity since it reduces to

$$a_{-1} = \lim_{z \to a} (z-a) f(z) \quad (11)$$

RESIDUE THEOREM

If $f(z)$ is analytic in a region $\mathcal{R}$ except for a pole of order n at $z = a$ and if C is any simple closed curve in $\mathcal{R}$ containing $z = a$, then $f(z)$ has the form (9). Integrating (9), using the fact that

$$\oint_C \frac{dz}{(z-a)^n} \;=\; \begin{cases} 0 & \text{if } n \neq 1 \\ 2\pi i & \text{if } n = 1 \end{cases} \tag{12}$$

(see Problem 13), it follows that

$$\oint_C f(z)\,dz \;=\; 2\pi i a_{-1} \tag{13}$$

i.e. the integral of $f(z)$ around a closed path enclosing a single pole of $f(z)$ is $2\pi i$ times the residue at the pole.

More generally, we have the following important

Theorem. If $f(z)$ is analytic within and on the boundary C of a region $\mathcal{R}$ except at a finite number of poles $a, b, c, \ldots$ within $\mathcal{R}$, having residues $a_{-1}, b_{-1}, c_{-1}, \ldots$ respectively, then

$$\oint_C f(z)\,dz \;=\; 2\pi i(a_{-1} + b_{-1} + c_{-1} + \cdots) \tag{14}$$

i.e. the integral of $f(z)$ is $2\pi i$ times the sum of the residues of $f(z)$ at the poles enclosed by C. Cauchy's theorem and integral formulas are special cases of this result which we call the *residue theorem*.

EVALUATION of DEFINITE INTEGRALS

The evaluation of various definite integrals can often be achieved by using the residue theorem together with a suitable function $f(z)$ and a suitable path or *contour C*, the choice of which may require great ingenuity. The following types are most common in practice.

1. $\displaystyle\int_0^\infty F(x)\,dx,$ $F(x)$ is an even function.

   Consider $\displaystyle\oint_C F(z)\,dz$ along a contour C consisting of the line along the x axis from $-R$ to $+R$ and the semi-circle above the x axis having this line as diameter. Then let $R \to \infty$. See Problems 29, 30.

2. $\displaystyle\int_0^{2\pi} G(\sin\theta, \cos\theta)\,d\theta,$ G is a rational function of $\sin\theta$ and $\cos\theta$.

   Let $z = e^{i\theta}$. Then $\sin\theta = \dfrac{z - z^{-1}}{2i}$, $\cos\theta = \dfrac{z + z^{-1}}{2}$ and $dz = ie^{i\theta}\,d\theta$ or $d\theta = dz/iz$. The given integral is equivalent to $\displaystyle\oint_C F(z)\,dz$ where C is the unit circle with center at the origin. See Problems 31, 32.

3. $\displaystyle\int_{-\infty}^\infty F(x) \begin{Bmatrix} \cos mx \\ \sin mx \end{Bmatrix} dx,$ $F(x)$ is a rational function.

   Here we consider $\displaystyle\oint_C F(z)\,e^{imz}\,dz$ where C is the same contour as that in Type 1. See Problem 34.

4. Miscellaneous integrals involving particular contours. See Problems 35, 38.

Solved Problems

FUNCTIONS, LIMITS, CONTINUITY

1. Determine the locus represented by

    (a) $|z-2| = 3$,   (b) $|z-2| = |z+4|$,   (c) $|z-3| + |z+3| = 10$.

    (a) **Method 1:** $|z-2| = |x+iy-2| = |x-2+iy| = \sqrt{(x-2)^2+y^2} = 3$ or $(x-2)^2 + y^2 = 9$, a circle with center at $(2,0)$ and radius 3.

    Method 2: $|z-2|$ is the distance between the complex numbers $z = x+iy$ and $2+0i$. If this distance is always 3, the locus is a circle of radius 3 with center at $2+0i$ or $(2,0)$.

    (b) **Method 1:** $|x+iy-2| = |x+iy+4|$ or $\sqrt{(x-2)^2+y^2} = \sqrt{(x+4)^2+y^2}$. Squaring, we find $x = -1$, a straight line.

    Method 2: The locus is such that the distances from any point on it to $(2,0)$ and $(-4,0)$ are equal. Thus the locus is the perpendicular bisector of the line joining $(2,0)$ and $(-4,0)$, or $x=-1$.

    (c) **Method 1:** The locus is given by $\sqrt{(x-3)^2+y^2} + \sqrt{(x+3)^2+y^2} = 10$ or $\sqrt{(x-3)^2+y^2} = 10 - \sqrt{(x+3)^2+y^2}$. Squaring and simplifying, $25 + 3x = 5\sqrt{(x+3)^2+y^2}$. Squaring and simplifying again yields $\dfrac{x^2}{25} + \dfrac{y^2}{16} = 1$, an ellipse with semi-major and semi-minor axes of lengths 5 and 4 respectively.

    Method 2: The locus is such that the sum of the distances from any point on it to $(3,0)$ and $(-3,0)$ is 10. Thus the locus is an ellipse whose foci are at $(-3,0)$ and $(3,0)$ and whose major axis has length 10.

2. Determine the region in the z plane represented by each of the following.

    (a) $|z| < 1$.

    Interior of a circle of radius 1. See Fig. 17-1(a) below.

    (b) $1 < |z+2i| \le 2$.

    $|z+2i|$ is the distance from z to $-2i$, so that $|z+2i| = 1$ is a circle of radius 1 with center at $-2i$, i.e. $(0,-2)$; and $|z+2i| = 2$ is a circle of radius 2 with center at $-2i$. Then $1 < |z+2i| \le 2$ represents the region *exterior* to $|z+2i| = 1$ but *interior* to or *on* $|z+2i| = 2$. See Fig. 17-1(b) below.

    (c) $\pi/3 \le \arg z \le \pi/2$.

    Note that $\arg z = \phi$, where $z = \rho e^{i\phi}$. The required region is the infinite region bounded by the lines $\phi = \pi/3$ and $\phi = \pi/2$, including these lines. See Fig. 17-1(c) below.

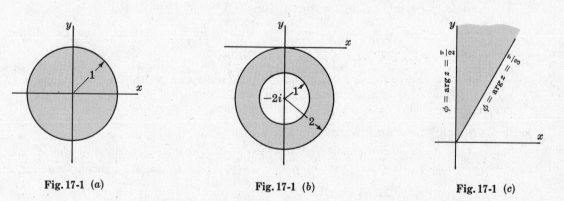

Fig. 17-1 (a)   Fig. 17-1 (b)   Fig. 17-1 (c)

3. Express each function in the form $u(x,y) + iv(x,y)$, where u and v are real:

    (a) z^3,   (b) $1/(1-z)$,   (c) e^{3z},   (d) $\ln z$.

    (a) $w = z^3 = (x+iy)^3 = x^3 + 3x^2(iy) + 3x(iy)^2 + (iy)^3 = x^3 + 3ix^2y - 3xy^2 - iy^3$
    $= x^3 - 3xy^2 + i(3x^2y - y^3)$

    Then $u(x,y) = x^3 - 3xy^2,\;\; v(x,y) = 3x^2y - y^3$.

(b) $\quad w \;=\; \dfrac{1}{1-z} \;=\; \dfrac{1}{1-(x+iy)} \;=\; \dfrac{1}{1-x-iy}\cdot\dfrac{1-x+iy}{1-x+iy} \;=\; \dfrac{1-x+iy}{(1-x)^2+y^2}$

Then $\quad u(x,y) \;=\; \dfrac{1-x}{(1-x)^2+y^2}, \quad v(x,y) \;=\; \dfrac{y}{(1-x)^2+y^2}.$

(c) $\quad e^{3z} \;=\; e^{3(x+iy)} \;=\; e^{3x}e^{3iy} \;=\; e^{3x}(\cos 3y + i\sin 3y) \quad$ and $\quad u = e^{3x}\cos 3y, \quad v = e^{3x}\sin 3y$

(d) $\quad \ln z \;=\; \ln(\rho e^{i\phi}) \;=\; \ln\rho + i\phi \;=\; \ln\sqrt{x^2+y^2} + i\tan^{-1}y/x \quad$ and

$$u \;=\; \tfrac{1}{2}\ln(x^2+y^2), \quad v \;=\; \tan^{-1}y/x$$

Note that $\ln z$ is a multiple-valued function (in this case it is *infinitely* many-valued) since ϕ can be increased by any multiple of 2π. The *principal value* of the logarithm is defined as that value for which $0 \leqq \phi < 2\pi$ and is called the *principal branch* of $\ln z$.

4. Prove (a) $\quad \sin(x+iy) = \sin x\cosh y + i\cos x\sinh y$

(b) $\quad \cos(x+iy) = \cos x\cosh y - i\sin x\sinh y.$

We use the relations $\quad e^{iz} = \cos z + i\sin z, \quad e^{-iz} = \cos z - i\sin z$, from which

$$\sin z \;=\; \frac{e^{iz}-e^{-iz}}{2i}, \quad \cos z \;=\; \frac{e^{iz}+e^{-iz}}{2}$$

Then $\quad \sin z \;=\; \sin(x+iy) \;=\; \dfrac{e^{i(x+iy)}-e^{-i(x+iy)}}{2i} \;=\; \dfrac{e^{ix-y}-e^{-ix+y}}{2i}$

$\qquad =\; \dfrac{1}{2i}\{e^{-y}(\cos x + i\sin x) - e^{y}(\cos x - i\sin x)\} \;=\; (\sin x)\left(\dfrac{e^y+e^{-y}}{2}\right) + i(\cos x)\left(\dfrac{e^y-e^{-y}}{2}\right)$

$\qquad =\; \sin x\cosh y + i\cos x\sinh y$

Similarly, $\cos z \;=\; \cos(x+iy) \;=\; \dfrac{e^{i(x+iy)}+e^{-i(x+iy)}}{2}$

$\qquad =\; \tfrac{1}{2}\{e^{ix-y}+e^{-ix+y}\} \;=\; \tfrac{1}{2}\{e^{-y}(\cos x + i\sin x) + e^{y}(\cos x - i\sin x)\}$

$\qquad =\; (\cos x)\left(\dfrac{e^y+e^{-y}}{2}\right) - i(\sin x)\left(\dfrac{e^y-e^{-y}}{2}\right) \;=\; \cos x\cosh y - i\sin x\sinh y$

DERIVATIVES. CAUCHY-RIEMANN EQUATIONS

5. Prove that $\dfrac{d}{dz}\bar{z}$, where $\bar{z}$ is the conjugate of z, does not exist anywhere.

By definition, $\dfrac{d}{dz}f(z) = \lim\limits_{\Delta z\to 0}\dfrac{f(z+\Delta z)-f(z)}{\Delta z}$ if this limit exists independent of the manner in which $\Delta z = \Delta x + i\Delta y$ approaches zero. Then

$$\frac{d}{dz}\bar{z} \;=\; \lim_{\Delta z\to 0}\frac{\overline{z+\Delta z}-\bar{z}}{\Delta z} \;=\; \lim_{\substack{\Delta x\to 0 \\ \Delta y\to 0}}\frac{\overline{x+iy+\Delta x+i\Delta y}-\overline{x+iy}}{\Delta x+i\Delta y}$$

$$=\; \lim_{\substack{\Delta x\to 0 \\ \Delta y\to 0}}\frac{x-iy+\Delta x-i\Delta y-(x-iy)}{\Delta x+i\Delta y} \;=\; \lim_{\substack{\Delta x\to 0 \\ \Delta y\to 0}}\frac{\Delta x-i\Delta y}{\Delta x+i\Delta y}$$

If $\Delta y = 0$, the required limit is $\quad \lim\limits_{\Delta x\to 0}\dfrac{\Delta x}{\Delta x} = 1.$

If $\Delta x = 0$, the required limit is $\quad \lim\limits_{\Delta y\to 0}\dfrac{-i\Delta y}{i\Delta y} = -1.$

These two possible approaches show that the limit depends on the manner in which $\Delta z \to 0$, so that the derivative does not exist; i.e. $\bar{z}$ is *non-analytic* anywhere.

6. (a) If $w = f(z) = \dfrac{1+z}{1-z}$, find $\dfrac{dw}{dz}$. (b) Determine where w is non-analytic.

(a) **Method 1:** $\dfrac{dw}{dz} = \lim\limits_{\Delta z \to 0} \dfrac{\dfrac{1 + (z + \Delta z)}{1 - (z + \Delta z)} - \dfrac{1+z}{1-z}}{\Delta z} = \lim\limits_{\Delta z \to 0} \dfrac{2}{(1 - z - \Delta z)(1 - z)}$

$= \dfrac{2}{(1-z)^2}$ provided $z \ne 1$, independent of the manner in which $\Delta z \to 0$.

Method 2: The usual rules of differentiation apply provided $z \ne 1$. Thus by the quotient rule for differentiation,

$$\frac{d}{dz}\left(\frac{1+z}{1-z}\right) = \frac{(1-z)\dfrac{d}{dz}(1+z) - (1+z)\dfrac{d}{dz}(1-z)}{(1-z)^2} = \frac{(1-z)(1) - (1+z)(-1)}{(1-z)^2} = \frac{2}{(1-z)^2}$$

(b) The function is analytic everywhere except at $z = 1$, where the derivative does not exist; i.e. the function is non-analytic at $z = 1$.

7. Prove that a necessary condition for $w = f(z) = u(x, y) + i\, v(x, y)$ to be analytic in a region is that the Cauchy-Riemann equations $\dfrac{\partial u}{\partial x} = \dfrac{\partial v}{\partial y},\ \dfrac{\partial u}{\partial y} = -\dfrac{\partial v}{\partial x}$ be satisfied in the region.

Since $f(z) = f(x + iy) = u(x, y) + i\, v(x, y)$, we have

$$f(z + \Delta z) = f[x + \Delta x + i(y + \Delta y)] = u(x + \Delta x, y + \Delta y) + i\, v(x + \Delta x, y + \Delta y)$$

Then

$$\lim_{\Delta z \to 0} \frac{f(z + \Delta z) - f(z)}{\Delta z} = \lim_{\substack{\Delta x \to 0 \\ \Delta y \to 0}} \frac{u(x + \Delta x, y + \Delta y) - u(x, y) + i\{v(x + \Delta x, y + \Delta y) - v(x, y)\}}{\Delta x + i\Delta y}$$

If $\Delta y = 0$, the required limit is

$$\lim_{\Delta x \to 0} \frac{u(x + \Delta x, y) - u(x, y)}{\Delta x} + i\left\{\frac{v(x + \Delta x, y) - v(x, y)}{\Delta x}\right\} = \frac{\partial u}{\partial x} + i\frac{\partial v}{\partial x}$$

If $\Delta x = 0$, the required limit is

$$\lim_{\Delta y \to 0} \frac{u(x, y + \Delta y) - u(x, y)}{i\Delta y} + \left\{\frac{v(x, y + \Delta y) - v(x, y)}{\Delta y}\right\} = \frac{1}{i}\frac{\partial u}{\partial y} + \frac{\partial v}{\partial y}$$

If the derivative is to exist, these two special limits must be equal, i.e.,

$$\frac{\partial u}{\partial x} + i\frac{\partial v}{\partial x} = \frac{1}{i}\frac{\partial u}{\partial y} + \frac{\partial v}{\partial y} = -i\frac{\partial u}{\partial y} + \frac{\partial v}{\partial y}$$

so that we must have $\dfrac{\partial u}{\partial x} = \dfrac{\partial v}{\partial y}$ and $\dfrac{\partial v}{\partial x} = -\dfrac{\partial u}{\partial y}$.

Conversely, we can prove that if the first partial derivatives of u and v with respect to x and y are continuous in a region, then the Cauchy-Riemann equations provide sufficient conditions for $f(z)$ to be analytic.

8. (a) If $f(z) = u(x, y) + i\, v(x, y)$ is analytic in a region $\mathcal{R}$, prove that the one parameter families of curves $u(x, y) = C_1$ and $v(x, y) = C_2$ are orthogonal families. (b) Illustrate by using $f(z) = z^2$.

(a) Consider any two particular members of these families $u(x, y) = u_0,\ v(x, y) = v_0$ which intersect at the point (x_0, y_0).

Since $du = u_x\, dx + u_y\, dy = 0$, we have $\dfrac{dy}{dx} = -\dfrac{u_x}{u_y}$.

Also since $dv = v_x\, dx + v_y\, dy = 0$, $\dfrac{dy}{dx} = -\dfrac{v_x}{v_y}$.

When evaluated at (x_0, y_0), these represent respectively the slopes of the two curves at this point of intersection.

By the Cauchy-Riemann equations, $u_x = v_y$, $u_y = -v_x$, we have the product of the slopes at the point (x_0, y_0) equal to

$$\left(-\frac{u_x}{u_y}\right)\left(-\frac{v_x}{v_y}\right) = -1$$

so that any two members of the respective families are orthogonal, and thus the two families are orthogonal.

(b) If $f(z) = z^2$, then $u = x^2 - y^2$, $v = 2xy$. The graphs of several members of $x^2 - y^2 = C_1$, $2xy = C_2$ are shown in Fig. 17-2.

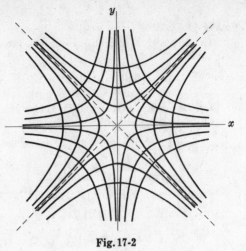

Fig. 17-2

9. In aerodynamics and fluid mechanics, the functions ϕ and ψ in $f(z) = \phi + i\psi$, where $f(z)$ is analytic, are called the *velocity potential* and *stream function* respectively. If $\phi = x^2 + 4x - y^2 + 2y$, (a) find ψ and (b) find $f(z)$.

(a) By the Cauchy-Riemann equations, $\dfrac{\partial \phi}{\partial x} = \dfrac{\partial \psi}{\partial y}$, $\dfrac{\partial \psi}{\partial x} = -\dfrac{\partial \phi}{\partial y}$. Then

$$(1) \quad \frac{\partial \psi}{\partial y} = 2x + 4 \qquad (2) \quad \frac{\partial \psi}{\partial x} = 2y - 2$$

Method 1: Integrating (1), $\psi = 2xy + 4y + F(x)$.
Integrating (2), $\psi = 2xy - 2x + G(y)$.

These are identical if $F(x) = -2x + c$, $G(y) = 4y + c$ where c is any real constant. Thus $\psi = 2xy + 4y - 2x + c$.

Method 2:

Integrating (1), $\psi = 2xy + 4y + F(x)$. Then substituting in (2), $2y + F'(x) = 2y - 2$ or $F'(x) = -2$ and $F(x) = -2x + c$. Hence $\psi = 2xy + 4y - 2x + c$.

(b) From (a), $f(z) = \phi + i\psi = x^2 + 4x - y^2 + 2y + i(2xy + 4y - 2x + c)$
$= (x^2 - y^2 + 2ixy) + 4(x + iy) - 2i(x + iy) + ic = z^2 + 4z - 2iz + c_1$

where c_1 is a pure imaginary constant.

This can also be accomplished by noting that $z = x + iy$, $\bar{z} = x - iy$ so that $x = \dfrac{z + \bar{z}}{2}$, $y = \dfrac{z - \bar{z}}{2i}$. The result is then obtained by substitution; the terms involving $\bar{z}$ drop out.

INTEGRALS, CAUCHY'S THEOREM, CAUCHY'S INTEGRAL FORMULAS

10. Evaluate $\displaystyle\int_{1+i}^{2+4i} z^2\, dz$

(a) along the parabola $x = t$, $y = t^2$ where $1 \le t \le 2$,

(b) along the straight line joining $1 + i$ and $2 + 4i$,

(c) along straight lines from $1 + i$ to $2 + i$ and then to $2 + 4i$.

We have

$$\int_{1+i}^{2+4i} z^2\, dz = \int_{(1,1)}^{(2,4)} (x + iy)^2 (dx + i\, dy) = \int_{(1,1)}^{(2,4)} (x^2 - y^2 + 2ixy)(dx + i\, dy)$$

$$= \int_{(1,1)}^{(2,4)} (x^2 - y^2)\, dx - 2xy\, dy + i \int_{(1,1)}^{(2,4)} 2xy\, dx + (x^2 - y^2)\, dy$$

Method 1:

(a) The points $(1,1)$ and $(2,4)$ correspond to $t=1$ and $t=2$ respectively. Then the above line integrals become

$$\int_{t=1}^{2} \{(t^2 - t^4)\, dt - 2(t)(t^2)2t\, dt\} \;+\; i\int_{t=1}^{2} \{2(t)(t^2)\, dt + (t^2 - t^4)(2t)\, dt\} \;=\; -\frac{86}{3} - 6i$$

(b) The line joining $(1,1)$ and $(2,4)$ has the equation $y - 1 = \dfrac{4-1}{2-1}(x-1)$ or $y = 3x - 2$. Then we find

$$\int_{x=1}^{2} \{[x^2 - (3x-2)^2]\, dx - 2x(3x-2)3\, dx\}$$
$$+ \; i\int_{x=1}^{2} \{2x(3x-2)\, dx + [x^2 - (3x-2)^2]3\, dx\} \;=\; -\frac{86}{3} - 6i$$

(c) From $1+i$ to $2+i$ [or $(1,1)$ to $(2,1)$], $y=1$, $dy=0$ and we have

$$\int_{x=1}^{2} (x^2 - 1)\, dx \;+\; i\int_{x=1}^{2} 2x\, dx \;=\; \frac{4}{3} + 3i$$

From $2+i$ to $2+4i$ [or $(2,1)$ to $(2,4)$], $x=2$, $dx=0$ and we have

$$\int_{y=1}^{4} -4y\, dy \;+\; i\int_{y=1}^{4} (4 - y^2)\, dy \;=\; -30 - 9i$$

Adding, $(\frac{4}{3} + 3i) + (-30 - 9i) = -\frac{86}{3} - 6i$.

Method 2:

By the methods of Chapter 10 it is seen that the line integrals are independent of the path, thus accounting for the same values obtained in (a), (b) and (c) above. In such case the integral can be evaluated directly, as for real variables, as follows:

$$\int_{1+i}^{2+4i} z^2\, dz \;=\; \frac{z^3}{3}\Big|_{1+i}^{2+4i} \;=\; \frac{(2+4i)^3}{3} - \frac{(1+i)^3}{3} \;=\; -\frac{86}{3} - 6i$$

11. (a) Prove Cauchy's theorem: If $f(z)$ is analytic inside and on a simple closed curve C, then $\displaystyle\oint_C f(z)\, dz = 0$.

(b) Under these conditions prove that $\displaystyle\int_{P_1}^{P_2} f(z)\, dz$ is independent of the path joining P_1 and P_2.

(a) $$\oint_C f(z)\, dz \;=\; \oint_C (u + iv)(dx + i\, dy) \;=\; \oint_C u\, dx - v\, dy \;+\; i\oint_C v\, dx + u\, dy$$

By Green's theorem (Chapter 10),

$$\oint_C u\, dx - v\, dy \;=\; \iint_{\mathcal{R}} \left(-\frac{\partial v}{\partial x} - \frac{\partial u}{\partial y}\right) dx\, dy, \qquad \oint_C v\, dx + u\, dy \;=\; \iint_{\mathcal{R}} \left(\frac{\partial u}{\partial x} - \frac{\partial v}{\partial y}\right) dx\, dy$$

where $\mathcal{R}$ is the region (simply-connected) bounded by C.

Since $f(z)$ is analytic, $\dfrac{\partial u}{\partial x} = \dfrac{\partial v}{\partial y}$, $\dfrac{\partial v}{\partial x} = -\dfrac{\partial u}{\partial y}$ (Problem 7), and so the above integrals are zero. Then $\displaystyle\oint_C f(z)\, dz = 0$, assuming $f'(z)$ [and thus the partial derivatives] to be continuous.

(b) Consider any two paths joining points P_1 and P_2 (see Fig. 17-3). By Cauchy's theorem,

$$\int_{P_1 A P_2 B P_1} f(z)\, dz \;=\; 0$$

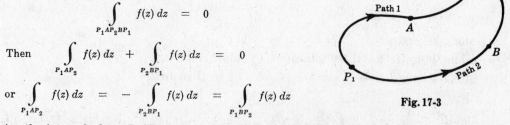

Then $$\int_{P_1 A P_2} f(z)\, dz \;+\; \int_{P_2 B P_1} f(z)\, dz \;=\; 0$$

or $$\int_{P_1 A P_2} f(z)\, dz \;=\; -\int_{P_2 B P_1} f(z)\, dz \;=\; \int_{P_1 B P_2} f(z)\, dz$$

Fig. 17-3

i.e. the integral along $P_1 A P_2$ (path 1) = integral along $P_1 B P_2$ (path 2), and so the integral is independent of the path joining P_1 and P_2.

This explains the results of Problem 10, since $f(z) = z^2$ is analytic.

12. If $f(z)$ is analytic within and on the boundary of a region bounded by two closed curves C_1 and C_2 (see Fig. 17-4), prove that

$$\oint_{C_1} f(z)\, dz \;=\; \oint_{C_2} f(z)\, dz$$

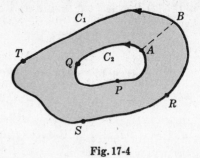

Fig. 17-4

As in Fig. 17-4, construct line AB (called a *cross-cut*) connecting any point on C_2 and a point on C_1. By Cauchy's theorem (Problem 11),

$$\int_{AQPABRSTBA} f(z)\, dz \;=\; 0$$

since $f(z)$ is analytic within the region shaded and also on the boundary. Then

$$\int_{AQPA} f(z)\, dz \;+\; \int_{AB} f(z)\, dz \;+\; \int_{BRSTB} f(z)\, dz \;+\; \int_{BA} f(z)\, dz \;=\; 0 \qquad (1)$$

But $\displaystyle\int_{AB} f(z)\, dz \;=\; -\int_{BA} f(z)\, dz$.   Hence (1) gives

$$\int_{AQPA} f(z)\, dz \;=\; -\int_{BRSTB} f(z)\, dz \;=\; \int_{BTSRB} f(z)\, dz$$

i.e.

$$\oint_{C_1} f(z)\, dz \;=\; \oint_{C_2} f(z)\, dz$$

Note that $f(z)$ need not be analytic *within* curve C_2.

13. (a) Prove that $\displaystyle\oint_C \frac{dz}{(z-a)^n} \;=\; \begin{cases} 2\pi i & \text{if } n=1 \\ 0 & \text{if } n=2,3,4,\ldots \end{cases}$ where C is a simple closed curve bounding a region having $z=a$ as interior point.

(b) What is the value of the integral if $n = 0, -1, -2, -3, \ldots$?

(a) Let C_1 be a circle of radius ϵ having center at $z = a$ (see Fig. 17-5). Since $(z-a)^{-n}$ is analytic within and on the boundary of the region bounded by C and C_1, we have by Problem 12,

$$\oint_C \frac{dz}{(z-a)^n} \;=\; \oint_{C_1} \frac{dz}{(z-a)^n}$$

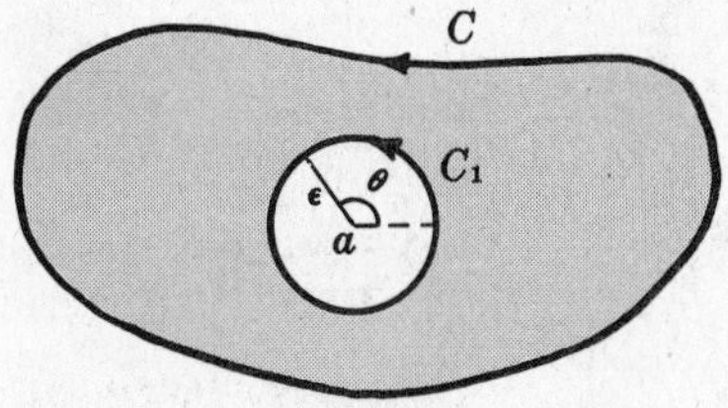

To evaluate this last integral, note that on C_1, $|z-a| = \epsilon$ or $z - a = \epsilon e^{i\theta}$ and $dz = i\epsilon e^{i\theta}\, d\theta$. The integral equals

Fig. 17-5

$$\int_0^{2\pi} \frac{i\epsilon e^{i\theta}\, d\theta}{\epsilon^n e^{in\theta}} \;=\; \frac{i}{\epsilon^{n-1}} \int_0^{2\pi} e^{(1-n)i\theta}\, d\theta \;=\; \frac{i}{\epsilon^{n-1}} \left. \frac{e^{(1-n)i\theta}}{(1-n)i} \right|_0^{2\pi} \;=\; 0 \qquad \text{if } n \ne 1$$

If $n = 1$, the integral equals $\displaystyle i\int_0^{2\pi} d\theta \;=\; 2\pi i$.

(b) For $n = 0, -1, -2, \ldots$ the integrand is $1, (z-a), (z-a)^2, \ldots$ and is analytic everywhere inside C_1, including $z = a$. Hence by Cauchy's theorem the integral is zero.

14. Evaluate $\displaystyle\oint_C \frac{dz}{z-3}$ where C is (a) the circle $|z| = 1$, (b) the circle $|z+i| = 4$.

(a) Since $z = 3$ is not interior to $|z| = 1$, the integral equals zero (Problem 11).

(b) Since $z = 3$ is interior to $|z+i| = 4$, the integral equals $2\pi i$ (Problem 13).

15. If $f(z)$ is analytic inside and on a simple closed curve C, and a is any point within C, prove that

$$f(a) = \frac{1}{2\pi i} \oint_C \frac{f(z)}{z - a} \, dz$$

Referring to Problem 12 and the figure of Problem 13, we have

$$\oint_C \frac{f(z)}{z - a} \, dz = \oint_{C_1} \frac{f(z)}{z - a} \, dz$$

Letting $z - a = \epsilon e^{i\theta}$, the last integral becomes $i \int_0^{2\pi} f(a + \epsilon e^{i\theta}) \, d\theta$. But since $f(z)$ is analytic, it is continuous. Hence

$$\lim_{\epsilon \to 0} i \int_0^{2\pi} f(a + \epsilon e^{i\theta}) \, d\theta = i \int_0^{2\pi} \lim_{\epsilon \to 0} f(a + \epsilon e^{i\theta}) \, d\theta = i \int_0^{2\pi} f(a) \, d\theta = 2\pi i f(a)$$

and the required result follows.

16. Evaluate (a) $\oint_C \dfrac{\cos z}{z - \pi} \, dz$, (b) $\oint_C \dfrac{e^z}{z(z + 1)} \, dz$ where C is the circle $|z - 1| = 3$.

(a) Since $z = \pi$ lies within C, $\dfrac{1}{2\pi i} \oint_C \dfrac{\cos z}{z - \pi} \, dz = \cos \pi = -1$ by Problem 15 with $f(z) = \cos z$,

$a = \pi$. Then $\oint_C \dfrac{\cos z}{z - \pi} \, dz = -2\pi i$.

(b)
$$\oint_C \frac{e^z}{z(z + 1)} \, dz = \oint_C e^z \left(\frac{1}{z} - \frac{1}{z + 1} \right) dz = \oint_C \frac{e^z}{z} \, dz - \oint_C \frac{e^z}{z + 1} \, dz$$
$$= 2\pi i e^0 - 2\pi i e^{-1} = 2\pi i (1 - e^{-1})$$

by Problem 15, since $z = 0$ and $z = -1$ are both interior to C.

17. Evaluate $\oint_C \dfrac{5z^2 - 3z + 2}{(z - 1)^3} \, dz$ where C is any simple closed curve enclosing $z = 1$.

Method 1: By Cauchy's integral formula, $f^{(n)}(a) = \dfrac{n!}{2\pi i} \oint_C \dfrac{f(z)}{(z - a)^{n+1}} \, dz$.

If $n = 2$ and $f(z) = 5z^2 - 3z + 2$, then $f''(1) = 10$. Hence

$$10 = \frac{2!}{2\pi i} \oint_C \frac{5z^2 - 3z + 2}{(z - 1)^3} \, dz \quad \text{or} \quad \oint_C \frac{5z^2 - 3z + 2}{(z - 1)^3} \, dz = 10\pi i$$

Method 2: $5z^2 - 3z + 2 = 5(z - 1)^2 + 7(z - 1) + 4$. Then

$$\oint_C \frac{5z^2 - 3z + 2}{(z - 1)^3} \, dz = \oint_C \frac{5(z - 1)^2 + 7(z - 1) + 4}{(z - 1)^3} \, dz$$
$$= 5 \oint_C \frac{dz}{z - 1} + 7 \oint_C \frac{dz}{(z - 1)^2} + 4 \oint_C \frac{dz}{(z - 1)^3} = 5(2\pi i) + 7(0) + 4(0)$$
$$= 10\pi i$$

by Problem 13.

SERIES and SINGULARITIES

18. For what values of z does each series converge?

(a) $\sum_{n=1}^{\infty} \dfrac{z^n}{n^2 2^n}$.   The nth term $= u_n = \dfrac{z^n}{n^2 2^n}$.   Then

$$\lim_{n \to \infty} \left| \frac{u_{n+1}}{u_n} \right| \;=\; \lim_{n \to \infty} \left| \frac{z^{n+1}}{(n+1)^2 2^{n+1}} \cdot \frac{n^2 2^n}{z^n} \right| \;=\; \frac{|z|}{2}$$

By the ratio test the series converges if $|z| < 2$ and diverges if $|z| > 2$. If $|z| = 2$ the ratio test fails.

However, the series of absolute values $\sum_{n=1}^{\infty} \left| \dfrac{z^n}{n^2 2^n} \right| = \sum_{n=1}^{\infty} \dfrac{|z|^n}{n^2 2^n}$ converges if $|z| = 2$, since $\sum_{n=1}^{\infty} \dfrac{1}{n^2}$ converges.

Thus the series converges (absolutely) for $|z| \leqq 2$, i.e. at all points inside and on the circle $|z| = 2$.

(b) $\sum_{n=1}^{\infty} \dfrac{(-1)^{n-1} z^{2n-1}}{(2n-1)!} \;=\; z - \dfrac{z^3}{3!} + \dfrac{z^5}{5!} - \cdots$.   We have

$$\lim_{n \to \infty} \left| \frac{u_{n+1}}{u_n} \right| \;=\; \lim_{n \to \infty} \left| \frac{(-1)^n z^{2n+1}}{(2n+1)!} \cdot \frac{(2n-1)!}{(-1)^{n-1} z^{2n-1}} \right| \;=\; \lim_{n \to \infty} \left| \frac{-z^2}{2n(2n+1)} \right| \;=\; 0$$

Then the series, which represents $\sin z$, converges for all values of z.

(c) $\sum_{n=1}^{\infty} \dfrac{(z-i)^n}{3^n}$.   We have $\lim_{n \to \infty} \left| \dfrac{u_{n+1}}{u_n} \right| = \lim_{n \to \infty} \left| \dfrac{(z-i)^{n+1}}{3^{n+1}} \cdot \dfrac{3^n}{(z-i)^n} \right| = \dfrac{|z-i|}{3}$.

The series converges if $|z - i| < 3$, and diverges if $|z - i| > 3$.

If $|z - i| = 3$, then $z - i = 3e^{i\theta}$ and the series becomes $\sum_{n=1}^{\infty} e^{in\theta}$. This series diverges since the nth term does not approach zero as $n \to \infty$.

Thus the series converges within the circle $|z - i| = 3$ but not on the boundary.

19. If $\sum_{n=0}^{\infty} a_n z^n$ is absolutely convergent for $|z| \leqq R$, show that it is uniformly convergent for these values of z.

The definitions, theorems and proofs for series of complex numbers are analogous to those for real series.

In this case we have $|a_n z^n| \leqq |a_n| R^n = M_n$. Since by hypothesis $\sum_{n=1}^{\infty} M_n$ converges, it follows by the Weirstrass M test that $\sum_{n=0}^{\infty} a_n z^n$ converges uniformly for $|z| \leqq R$.

20. Locate in the finite z plane all the singularities, if any, of each function and name them.

(a) $\dfrac{z^2}{(z+1)^3}$. $z = -1$ is a pole of order 3.

(b) $\dfrac{2z^3 - z + 1}{(z-4)^2 (z-i)(z-1+2i)}$. $z = 4$ is a pole of order 2 (double pole); $z = i$ and $z = 1 - 2i$ are poles of order 1 (simple poles).

(c) $\dfrac{\sin mz}{z^2 + 2z + 2}$, $m \neq 0$.   Since $z^2 + 2z + 2 = 0$ when $z = \dfrac{-2 \pm \sqrt{4-8}}{2} = \dfrac{-2 \pm 2i}{2} = -1 \pm i$, we can write $z^2 + 2z + 2 = \{z - (-1+i)\}\{z - (-1-i)\} = (z+1-i)(z+1+i)$.

The function has the two simple poles: $z = -1 + i$ and $z = -1 - i$.

(d) $\dfrac{1-\cos z}{z}$. $\quad z = 0$ appears to be a singularity. However, since $\lim\limits_{z \to 0} \dfrac{1-\cos z}{z} = 0$, it is a removable singularity.

Another method:

Since $\dfrac{1-\cos z}{z} = \dfrac{1}{z}\left\{1 - \left(1 - \dfrac{z^2}{2!} + \dfrac{z^4}{4!} - \dfrac{z^6}{6!} + \cdots\right)\right\} = \dfrac{z}{2!} - \dfrac{z^3}{4!} + \cdots$, we see that $z = 0$ is a removable singularity.

(e) $e^{-1/(z-1)^2} = 1 - \dfrac{1}{(z-1)^2} + \dfrac{1}{2!\,(z-1)^4} - \cdots$.

This is a Laurent series where the principal part has an infinite number of non-zero terms. Then $z = 1$ is an *essential singularity*.

(f) e^z.

This function has no finite singularity. However, letting $z = 1/u$, we obtain $e^{1/u}$ which has an essential singularity at $u = 0$. We conclude that $z = \infty$ is an essential singularity of e^z.

In general, to determine the nature of a possible singularity of $f(z)$ at $z = \infty$, we let $z = 1/u$ and then examine the behavior of the new function at $u = 0$.

21. If $f(z)$ is analytic at all points inside and on a circle of radius R with center at a, and if $a + h$ is any point inside C, prove *Taylor's theorem* that

$$f(a+h) = f(a) + h\,f'(a) + \dfrac{h^2}{2!}\,f''(a) + \dfrac{h^3}{3!}\,f'''(a) + \cdots$$

By Cauchy's integral formula (Problem 15), we have

$$f(a+h) = \frac{1}{2\pi i}\oint_C \frac{f(z)\,dz}{z-a-h} \tag{1}$$

By division,

$$\frac{1}{z-a-h} = \frac{1}{(z-a)[1-h/(z-a)]}$$

$$= \frac{1}{(z-a)}\left\{1 + \frac{h}{(z-a)} + \frac{h^2}{(z-a)^2} + \cdots + \frac{h^n}{(z-a)^n} + \frac{h^{n+1}}{(z-a)^n\,(z-a-h)}\right\} \tag{2}$$

Substituting (2) in (1) and using Cauchy's integral formulas, we have

$$f(a+h) = \frac{1}{2\pi i}\oint_C \frac{f(z)\,dz}{z-a} + \frac{h}{2\pi i}\oint_C \frac{f(z)\,dz}{(z-a)^2} + \cdots + \frac{h^n}{2\pi i}\oint_C \frac{f(z)\,dz}{(z-a)^{n+1}} + R_n$$

$$= f(a) + h\,f'(a) + \frac{h^2}{2!}\,f''(a) + \cdots + \frac{h^n}{n!}\,f^{(n)}(a) + R_n$$

where

$$R_n = \frac{h^{n+1}}{2\pi i}\oint_C \frac{f(z)\,dz}{(z-a)^{n+1}\,(z-a-h)}$$

Now when z is on C, $\left|\dfrac{f(z)}{z-a-h}\right| \le M$ and $|z-a| = R$, so that by (4), Page 346, we have, since $2\pi R$ is the length of C,

$$|R_n| \le \frac{|h|^{n+1}M}{2\pi\,R^{n+1}} \cdot 2\pi R$$

As $n \to \infty$, $|R_n| \to 0$. Then $R_n \to 0$ and the required result follows.

If $f(z)$ is analytic in an annular region $r_1 \le |z-a| \le r_2$, we can generalize the Taylor series to a Laurent series (see Problem 92). In some cases, as shown in Problem 22, the Laurent series can be obtained by use of known Taylor series.

22. Find Laurent series about the indicated singularity for each of the following functions. Name the singularity in each case and give the region of convergence of each series.

(a) $\dfrac{e^z}{(z-1)^2}$; $z=1$.   Let $z-1=u$. Then $z=1+u$ and

$$\frac{e^z}{(z-1)^2} = \frac{e^{1+u}}{u^2} = e \cdot \frac{e^u}{u^2} = \frac{e}{u^2}\left\{1 + u + \frac{u^2}{2!} + \frac{u^3}{3!} + \frac{u^4}{4!} + \cdots\right\}$$

$$= \frac{e}{(z-1)^2} + \frac{e}{z-1} + \frac{e}{2!} + \frac{e(z-1)}{3!} + \frac{e(z-1)^2}{4!} + \cdots$$

$z=1$ is a *pole of order 2*, or *double pole*.

The series converges for all values of $z \neq 1$.

(b) $z \cos \dfrac{1}{z}$; $z=0$.

$$z \cos \frac{1}{z} = z\left(1 - \frac{1}{2!\,z^2} + \frac{1}{4!\,z^4} - \frac{1}{6!\,z^6} + \cdots\right) = z - \frac{1}{2!\,z} + \frac{1}{4!\,z^3} - \frac{1}{6!\,z^5} + \cdots$$

$z=0$ is an *essential singularity*.

The series converges for all values of $z \neq 0$.

(c) $\dfrac{\sin z}{z-\pi}$; $z=\pi$.   Let $z-\pi=u$. Then $z=\pi+u$ and

$$\frac{\sin z}{z-\pi} = \frac{\sin(u+\pi)}{u} = -\frac{\sin u}{u} = -\frac{1}{u}\left(u - \frac{u^3}{3!} + \frac{u^5}{5!} - \cdots\right)$$

$$= -1 + \frac{u^2}{3!} - \frac{u^4}{5!} + \cdots = -1 + \frac{(z-\pi)^2}{3!} - \frac{(z-\pi)^4}{5!} + \cdots$$

$z=\pi$ is a *removable singularity*.

The series converges for all values of z.

(d) $\dfrac{z}{(z+1)(z+2)}$; $z=-1$.   Let $z+1=u$. Then

$$\frac{z}{(z+1)(z+2)} = \frac{u-1}{u(u+1)} = \frac{u-1}{u}(1 - u + u^2 - u^3 + u^4 - \cdots)$$

$$= -\frac{1}{u} + 2 - 2u + 2u^2 - 2u^3 + \cdots$$

$$= -\frac{1}{z+1} + 2 - 2(z+1) + 2(z+1)^2 - \cdots$$

$z=-1$ is a *pole of order 1*, or *simple pole*.

The series converges for values of z such that $0 < |z+1| < 1$.

(e) $\dfrac{1}{z(z+2)^3}$; $z=0,-2$.

Case 1, $z=0$. Using the binomial theorem,

$$\frac{1}{z(z+2)^3} = \frac{1}{8z(1+z/2)^3} = \frac{1}{8z}\left\{1 + (-3)\left(\frac{z}{2}\right) + \frac{(-3)(-4)}{2!}\left(\frac{z}{2}\right)^2 + \frac{(-3)(-4)(-5)}{3!}\left(\frac{z}{2}\right)^3 + \cdots\right\}$$

$$= \frac{1}{8z} - \frac{3}{16} + \frac{3}{16}z - \frac{5}{32}z^2 + \cdots$$

$z=0$ is a *pole of order 1*, or *simple pole*.

The series converges for $0 < |z| < 2$.

Case 2, z = −2. Let $z + 2 = u$. Then

$$\frac{1}{z(z+2)^3} = \frac{1}{(u-2)u^3} = \frac{1}{-2u^3(1-u/2)} = -\frac{1}{2u^3}\left\{1 + \frac{u}{2} + \left(\frac{u}{2}\right)^2 + \left(\frac{u}{2}\right)^3 + \left(\frac{u}{2}\right)^4 + \cdots\right\}$$

$$= -\frac{1}{2u^3} - \frac{1}{4u^2} - \frac{1}{8u} - \frac{1}{16} - \frac{1}{32}u - \cdots$$

$$= -\frac{1}{2(z+2)^3} - \frac{1}{4(z+2)^2} - \frac{1}{8(z+2)} - \frac{1}{16} - \frac{1}{32}(z+2) - \cdots$$

$z = −2$ is a *pole of order 3.*

The series converges for $0 < |z+2| < 2$.

RESIDUES and the RESIDUE THEOREM

23. If $f(z)$ is analytic everywhere inside and on a simple closed curve C except at $z = a$ which is a pole of order n so that

$$f(z) = \frac{a_{-n}}{(z-a)^n} + \frac{a_{-n+1}}{(z-a)^{n-1}} + \cdots + a_0 + a_1(z-a) + a_2(z-a)^2 + \cdots$$

where $a_{-n} \neq 0$, prove that

(a) $\displaystyle\oint_C f(z)\,dz = 2\pi i\, a_{-1}$

(b) $\displaystyle a_{-1} = \lim_{z \to a} \frac{1}{(n-1)!} \frac{d^{n-1}}{dz^{n-1}}\{(z-a)^n f(z)\}.$

(a) By integration, we have on using Problem 13

$$\oint_C f(z)\,dz = \oint_C \frac{a_{-n}}{(z-a)^n}\,dz + \cdots + \oint_C \frac{a_{-1}}{z-a}\,dz + \oint_C \{a_0 + a_1(z-a) + a_2(z-a)^2 + \cdots\}\,dz$$
$$= 2\pi i\, a_{-1}$$

Since only the term involving a_{-1} remains, we call a_{-1} the *residue* of $f(z)$ at the pole $z = a$.

(b) Multiplication by $(z-a)^n$ gives the Taylor series

$$(z-a)^n f(z) = a_{-n} + a_{-n+1}(z-a) + \cdots + a_{-1}(z-a)^{n-1} + \cdots$$

Taking the $(n-1)$st derivative of both sides and letting $z \to a$, we find

$$(n-1)!\, a_{-1} = \lim_{z \to a} \frac{d^{n-1}}{dz^{n-1}}\{(z-a)^n f(z)\}$$

from which the required result follows.

24. Determine the residues of each function at the indicated poles.

(a) $\dfrac{z^2}{(z-2)(z^2+1)}$; $z = 2, i, -i$.   These are simple poles. Then:

Residue at $z = 2$ is $\displaystyle\lim_{z \to 2}(z-2)\left\{\frac{z^2}{(z-2)(z^2+1)}\right\} = \frac{4}{5}$.

Residue at $z = i$ is $\displaystyle\lim_{z \to i}(z-i)\left\{\frac{z^2}{(z-2)(z-i)(z+i)}\right\} = \frac{i^2}{(i-2)(2i)} = \frac{1-2i}{10}$.

Residue at $z = -i$ is $\displaystyle\lim_{z \to -i}(z+i)\left\{\frac{z^2}{(z-2)(z-i)(z+i)}\right\} = \frac{i^2}{(-i-2)(-2i)} = \frac{1+2i}{10}$.

(b) $\dfrac{1}{z(z+2)^3}$; $z = 0, -2$. $z = 0$ is a simple pole, $z = -2$ is a pole of order 3. Then:

Residue at $z = 0$ is $\lim\limits_{z \to 0} z \cdot \dfrac{1}{z(z+2)^3} = \dfrac{1}{8}$.

Residue at $z = -2$ is $\lim\limits_{z \to -2} \dfrac{1}{2!} \dfrac{d^2}{dz^2} \left\{ (z+2)^3 \cdot \dfrac{1}{z(z+2)^3} \right\}$

$$= \lim\limits_{z \to -2} \dfrac{1}{2} \dfrac{d^2}{dz^2} \left(\dfrac{1}{z} \right) = \lim\limits_{z \to -2} \dfrac{1}{2} \left(\dfrac{2}{z^3} \right) = -\dfrac{1}{8}.$$

Note that these residues can also be obtained from the coefficients of $1/z$ and $1/(z+2)$ in the respective Laurent series [see Problem 22(e)].

(c) $\dfrac{ze^{zt}}{(z-3)^2}$; $z = 3$, a pole of order 2 or double pole. Then:

Residue is $\lim\limits_{z \to 3} \dfrac{d}{dz} \left\{ (z-3)^2 \cdot \dfrac{ze^{zt}}{(z-3)^2} \right\} = \lim\limits_{z \to 3} \dfrac{d}{dz} (ze^{zt}) = \lim\limits_{z \to 3} (e^{zt} + zte^{zt})$

$$= e^{3t} + 3te^{3t}$$

(d) $\cot z$; $z = 5\pi$, a pole of order 1. Then:

Residue is $\lim\limits_{z \to 5\pi} (z - 5\pi) \cdot \dfrac{\cos z}{\sin z} = \left(\lim\limits_{z \to 5\pi} \dfrac{z - 5\pi}{\sin z} \right)\left(\lim\limits_{z \to 5\pi} \cos z \right) = \left(\lim\limits_{z \to 5\pi} \dfrac{1}{\cos z} \right)(-1)$

$$= (-1)(-1) = 1$$

where we have used L'Hospital's rule, which can be shown applicable for functions of a complex variable.

25. If $f(z)$ is analytic within and on a simple closed curve C except at a number of poles $a, b, c, \ldots$ interior to C, prove that

$$\oint_C f(z)\, dz = 2\pi i \,\{\text{sum of residues of } f(z) \text{ at poles } a, b, c, \text{ etc.}\}$$

Refer to Fig. 17-6.

By reasoning similar to that of Problem 12 (i.e. by constructing cross cuts from C to C_1, C_2, C_3, etc.), we have

$$\oint_c f(z)\, dz = \oint_{c_1} f(z)\, dz + \oint_{c_2} f(z)\, dz + \cdots$$

For pole a,

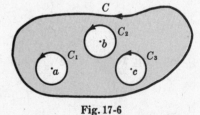

Fig. 17-6

$$f(z) = \dfrac{a_{-m}}{(z-a)^m} + \cdots + \dfrac{a_{-1}}{(z-a)} + a_0 + a_1(z-a) + \cdots$$

hence, as in Problem 23, $\oint_{c_1} f(z)\, dz = 2\pi i\, a_{-1}$.

Similarly for pole b, $f(z) = \dfrac{b_{-n}}{(z-b)^n} + \cdots + \dfrac{b_{-1}}{(z-b)} + b_0 + b_1(z-b) + \cdots$

so that $$\oint_{c_2} f(z)\, dz = 2\pi i\, b_{-1}$$

Continuing in this manner, we see that

$$\oint_c f(z)\, dz = 2\pi i (a_{-1} + b_{-1} + \cdots) = 2\pi i (\text{sum of residues})$$

26. Evaluate $\displaystyle\oint_C \frac{e^z\,dz}{(z-1)(z+3)^2}$ where C is given by (a) $|z|=3/2$, (b) $|z|=10$.

Residue at simple pole $z=1$ is $\displaystyle\lim_{z\to 1}\left\{(z-1)\,\frac{e^z}{(z-1)(z+3)^2}\right\}\ =\ \frac{e}{16}$

Residue at double pole $z=-3$ is

$$\lim_{z\to-3}\frac{d}{dz}\left\{(z+3)^2\,\frac{e^z}{(z-1)(z+3)^2}\right\}\ =\ \lim_{z\to-3}\frac{(z-1)e^z-e^z}{(z-1)^2}\ =\ \frac{-5e^{-3}}{16}$$

(a) Since $|z|=3/2$ encloses only the pole $z=1$,

$$\text{the required integral}\ =\ 2\pi i\left(\frac{e}{16}\right)\ =\ \frac{\pi i e}{8}$$

(b) Since $|z|=10$ encloses both poles $z=1$ and $z=-3$,

$$\text{the required integral}\ =\ 2\pi i\left(\frac{e}{16}-\frac{5e^{-3}}{16}\right)\ =\ \frac{\pi i(e-5e^{-3})}{8}$$

EVALUATION of DEFINITE INTEGRALS

27. If $|f(z)|\le\dfrac{M}{R^k}$ for $z=Re^{i\theta}$, where $k>1$ and M are constants, prove that $\displaystyle\lim_{R\to\infty}\int_\Gamma f(z)\,dz\ =\ 0$ where Γ is the semi-circular arc of radius R shown in Fig. 17-7.

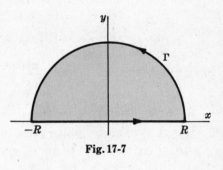

Fig. 17-7

By the result (4), Page 346, we have

$$\left|\int_\Gamma f(z)\,dz\right|\ \le\ \int_\Gamma|f(z)|\,|dz|\ \le\ \frac{M}{R^k}\cdot\pi R\ =\ \frac{\pi M}{R^{k-1}}$$

since the length of arc $L=\pi R$. Then

$$\lim_{R\to\infty}\left|\int_\Gamma f(z)\,dz\right|\ =\ 0\qquad\text{and so}\qquad\lim_{R\to\infty}\int_\Gamma f(z)\,dz\ =\ 0$$

28. Show that for $z=Re^{i\theta}$, $|f(z)|\le\dfrac{M}{R^k}$, $k>1$ if $f(z)=\dfrac{1}{1+z^4}$.

If $z=Re^{i\theta}$, $|f(z)|=\left|\dfrac{1}{1+R^4e^{4i\theta}}\right|\le\dfrac{1}{|R^4e^{4i\theta}|-1}=\dfrac{1}{R^4-1}\le\dfrac{2}{R^4}$ if R is large enough (say $R>2$, for example) so that $M=2,\ k=4$.

Note that we have made use of the inequality $|z_1+z_2|\ge|z_1|-|z_2|$ with $z_1=R^4e^{4i\theta}$ and $z_2=1$.

29. Evaluate $\displaystyle\int_0^\infty \frac{dx}{x^4+1}$.

Consider $\displaystyle\oint_C \frac{dz}{z^4+1}$, where C is the closed contour of Problem 27 consisting of the line from $-R$ to R and the semi-circle Γ, traversed in the positive (counterclockwise) sense.

Since $z^4+1=0$ when $z=e^{\pi i/4},\,e^{3\pi i/4},\,e^{5\pi i/4},\,e^{7\pi i/4}$, these are simple poles of $1/(z^4+1)$. Only the poles $e^{\pi i/4}$ and $e^{3\pi i/4}$ lie within C. Then using L'Hospital's rule,

$$\text{Residue at } e^{\pi i/4} \;=\; \lim_{z \to e^{\pi i/4}} \left\{ (z - e^{\pi i/4}) \frac{1}{z^4 + 1} \right\}$$

$$\;=\; \lim_{z \to e^{\pi i/4}} \frac{1}{4z^3} \;=\; \frac{1}{4} e^{-3\pi i/4}$$

$$\text{Residue at } e^{3\pi i/4} \;=\; \lim_{z \to e^{3\pi i/4}} \left\{ (z - e^{3\pi i/4}) \frac{1}{z^4 + 1} \right\}$$

$$\;=\; \lim_{z \to e^{3\pi i/4}} \frac{1}{4z^3} \;=\; \frac{1}{4} e^{-9\pi i/4}$$

Thus

$$\oint_C \frac{dz}{z^4 + 1} \;=\; 2\pi i \{ \tfrac{1}{4} e^{-3\pi i/4} + \tfrac{1}{4} e^{-9\pi i/4} \} \;=\; \frac{\pi \sqrt{2}}{2} \tag{1}$$

i.e.

$$\int_{-R}^{R} \frac{dx}{x^4 + 1} \;+\; \int_{\Gamma} \frac{dz}{z^4 + 1} \;=\; \frac{\pi \sqrt{2}}{2} \tag{2}$$

Taking the limit of both sides of (2) as $R \to \infty$ and using the results of Problem 28, we have

$$\lim_{R \to \infty} \int_{-R}^{R} \frac{dx}{x^4 + 1} \;=\; \int_{-\infty}^{\infty} \frac{dx}{x^4 + 1} \;=\; \frac{\pi \sqrt{2}}{2}$$

Since $\displaystyle \int_{-\infty}^{\infty} \frac{dx}{x^4 + 1} \;=\; 2 \int_{0}^{\infty} \frac{dx}{x^4 + 1}$, the required integral has the value $\dfrac{\pi \sqrt{2}}{4}$.

30. **Show that** $\displaystyle \int_{-\infty}^{\infty} \frac{x^2 \, dx}{(x^2 + 1)^2 (x^2 + 2x + 2)} \;=\; \frac{7\pi}{50}$.

The poles of $\dfrac{z^2}{(z^2 + 1)^2 (z^2 + 2z + 2)}$ enclosed by the contour C of Problem 27 are $z = i$ of order 2 and $z = -1 + i$ of order 1.

Residue at $z = i$ is $\displaystyle \lim_{z \to i} \frac{d}{dz} \left\{ (z - i)^2 \frac{z^2}{(z + i)^2 (z - i)^2 (z^2 + 2z + 2)} \right\} \;=\; \frac{9i - 12}{100}$.

Residue at $z = -1 + i$ is $\displaystyle \lim_{z \to -1 + i} (z + 1 - i) \frac{z^2}{(z^2 + 1)^2 (z + 1 - i)(z + 1 + i)} \;=\; \frac{3 - 4i}{25}$

Then $\displaystyle \oint_C \frac{z^2 \, dz}{(z^2 + 1)^2 (z^2 + 2z + 2)} \;=\; 2\pi i \left\{ \frac{9i - 12}{100} + \frac{3 - 4i}{25} \right\} \;=\; \frac{7\pi}{50}$

or $\displaystyle \int_{-R}^{R} \frac{x^2 \, dx}{(x^2 + 1)^2 (x^2 + 2x + 2)} \;+\; \int_{\Gamma} \frac{z^2 \, dz}{(z^2 + 1)^2 (z^2 + 2z + 2)} \;=\; \frac{7\pi}{50}$

Taking the limit as $R \to \infty$ and noting that the second integral approaches zero by Problem 27, we obtain the required result.

31. **Evaluate** $\displaystyle \int_{0}^{2\pi} \frac{d\theta}{5 + 3 \sin \theta}$.

Let $z = e^{i\theta}$. Then $\sin \theta = \dfrac{e^{i\theta} - e^{-i\theta}}{2i} = \dfrac{z - z^{-1}}{2i}$, $dz = i e^{i\theta} \, d\theta = iz \, d\theta$ so that

$$\int_{0}^{2\pi} \frac{d\theta}{5 + 3 \sin \theta} \;=\; \oint_C \frac{dz/iz}{5 + 3 \left(\dfrac{z - z^{-1}}{2i} \right)} \;=\; \oint_C \frac{2 \, dz}{3z^2 + 10iz - 3}$$

where C is the circle of unit radius with center at the origin, as shown in Fig. 17-8 below.

The poles of $\dfrac{2}{3z^2 + 10iz - 3}$ are the simple poles

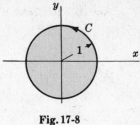

Fig. 17-8

$$z = \frac{-10i \pm \sqrt{-100 + 36}}{6}$$

$$\cdot = \frac{-10i \pm 8i}{6}$$

$$= -3i, \; -i/3.$$

Only $-i/3$ lies inside C.

Residue at $-i/3 \;=\; \lim_{z \to -i/3} \left(z + \dfrac{i}{3}\right)\!\left(\dfrac{2}{3z^2 + 10iz - 3}\right) \;=\; \lim_{z \to -i/3} \dfrac{2}{6z + 10i} \;=\; \dfrac{1}{4i}$ by L'Hospital's rule.

Then $\displaystyle\oint_C \dfrac{2\,dz}{3z^2 + 10iz - 3} \;=\; 2\pi i\!\left(\dfrac{1}{4i}\right) \;=\; \dfrac{\pi}{2},$ the required value.

32. Show that $\displaystyle\int_0^{2\pi} \dfrac{\cos 3\theta}{5 - 4\cos\theta}\,d\theta \;=\; \dfrac{\pi}{12}.$

If $z = e^{i\theta}$, $\cos\theta = \dfrac{z + z^{-1}}{2}$, $\cos 3\theta = \dfrac{e^{3i\theta} + e^{-3i\theta}}{2} = \dfrac{z^3 + z^{-3}}{2}$, $dz = iz\,d\theta$.

Then $\displaystyle\int_0^{2\pi} \dfrac{\cos 3\theta}{5 - 4\cos\theta}\,d\theta \;=\; \oint_C \dfrac{(z^3 + z^{-3})/2}{5 - 4\!\left(\dfrac{z + z^{-1}}{2}\right)}\,\dfrac{dz}{iz}$

$$= \; -\frac{1}{2i} \oint_C \frac{z^6 + 1}{z^3(2z - 1)(z - 2)}\,dz$$

where C is the contour of Problem 31.

The integrand has a pole of order 3 at $z = 0$ and a simple pole $z = \tfrac{1}{2}$ within C.

Residue at $z = 0$ is $\displaystyle\lim_{z \to 0} \dfrac{1}{2!}\dfrac{d^2}{dz^2}\!\left\{z^3 \cdot \dfrac{z^6 + 1}{z^3(2z - 1)(z - 2)}\right\} \;=\; \dfrac{21}{8}.$

Residue at $z = \tfrac{1}{2}$ is $\displaystyle\lim_{z \to 1/2}\left\{(z - \tfrac{1}{2}) \cdot \dfrac{z^6 + 1}{z^3(2z - 1)(z - 2)}\right\} \;=\; -\dfrac{65}{24}.$

Then $-\dfrac{1}{2i}\displaystyle\oint_C \dfrac{z^6 + 1}{z^3(2z - 1)(z - 2)}\,dz \;=\; -\dfrac{1}{2i}\,(2\pi i)\!\left\{\dfrac{21}{8} - \dfrac{65}{24}\right\} \;=\; \dfrac{\pi}{12}$ as required.

33. If $|f(z)| \leqq \dfrac{M}{R^k}$ for $z = Re^{i\theta}$, where $k > 0$ and M are constants, prove that

$$\lim_{R \to \infty} \int_\Gamma e^{imz} f(z)\,dz \;=\; 0$$

where Γ is the semi-circular arc of the contour in Problem 27 and m is a positive constant.

If $z = Re^{i\theta}$, $\displaystyle\int_\Gamma e^{imz} f(z)\,dz \;=\; \int_0^\pi e^{imRe^{i\theta}} f(Re^{i\theta})\, iRe^{i\theta}\,d\theta.$

Then $\left|\displaystyle\int_0^\pi e^{imRe^{i\theta}} f(Re^{i\theta})\, iRe^{i\theta}\,d\theta\right| \;\leqq\; \int_0^\pi \left|e^{imRe^{i\theta}} f(Re^{i\theta})\, iRe^{i\theta}\right|\,d\theta$

$$= \; \int_0^\pi \left|e^{imR\cos\theta - mR\sin\theta} f(Re^{i\theta})\, iRe^{i\theta}\right|\,d\theta$$

$$= \; \int_0^\pi e^{-mR\sin\theta}\, |f(Re^{i\theta})|\, R\,d\theta$$

$$\leqq \; \frac{M}{R^{k-1}} \int_0^\pi e^{-mR\sin\theta}\,d\theta \;=\; \frac{2M}{R^{k-1}} \int_0^{\pi/2} e^{-mR\sin\theta}\,d\theta$$

Now $\sin\theta \geqq 2\theta/\pi$ for $0 \leqq \theta \leqq \pi/2$ (see Problem 77, Chapter 4). Then the last integral is less than or equal to

$$\frac{2M}{R^{k-1}} \int_0^{\pi/2} e^{-2mR\theta/\pi}\, d\theta \;=\; \frac{\pi M}{mR^k}(1 - e^{-mR})$$

As $R \to \infty$ this approaches zero, since m and k are positive, and the required result is proved.

34. Show that $\displaystyle\int_0^{\infty} \frac{\cos mx}{x^2+1}\, dx \;=\; \frac{\pi}{2}e^{-m}, \quad m > 0.$

Consider $\displaystyle\oint_C \frac{e^{imz}}{z^2+1}\, dz$ where C is the contour of Problem 27.

The integrand has simple poles at $z = \pm i$, but only $z = i$ lies within C.

Residue at $z = i$ is $\displaystyle\lim_{z \to i}\left\{(z-i)\,\frac{e^{imz}}{(z-i)(z+i)}\right\} \;=\; \frac{e^{-m}}{2i}.$

Then

$$\oint_C \frac{e^{imz}}{z^2+1}\, dz \;=\; 2\pi i\left(\frac{e^{-m}}{2i}\right) \;=\; \pi e^{-m}$$

or

$$\int_{-R}^{R} \frac{e^{imx}}{x^2+1}\, dx \;+\; \int_{\Gamma} \frac{e^{imz}}{z^2+1}\, dz \;=\; \pi e^{-m}$$

i.e.

$$\int_{-R}^{R} \frac{\cos mx}{x^2+1}\, dx \;+\; i\int_{-R}^{R} \frac{\sin mx}{x^2+1}\, dx \;+\; \int_{\Gamma} \frac{e^{imz}}{z^2+1}\, dz \;=\; \pi e^{-m}$$

and so

$$2\int_0^{R} \frac{\cos mx}{x^2+1}\, dx \;+\; \int_{\Gamma} \frac{e^{imz}}{z^2+1}\, dz \;=\; \pi e^{-m}$$

Taking the limit as $R \to \infty$ and using Problem 33 to show that the integral around Γ approaches zero, we obtain the required result.

35. Show that $\displaystyle\int_0^{\infty} \frac{\sin x}{x}\, dx \;=\; \frac{\pi}{2}.$

The method of Problem 34 leads us to consider the integral of e^{iz}/z around the contour of Problem 27. However, since $z = 0$ lies on this path of integration and since we cannot integrate through a singularity, we modify that contour by indenting the path at $z = 0$, as shown in Fig. 17-9, which we call contour C' or $ABDEFGHJA$.

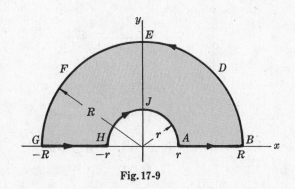

Fig. 17-9

Since $z = 0$ is outside C', we have

$$\int_{C'} \frac{e^{iz}}{z}\, dz \;=\; 0$$

or

$$\int_{-R}^{-r} \frac{e^{iz}}{x}\, dx \;+\; \int_{HJA} \frac{e^{iz}}{z}\, dz \;+\; \int_{r}^{R} \frac{e^{iz}}{x}\, dx \;+\; \int_{BDEFG} \frac{e^{iz}}{z}\, dz \;=\; 0$$

Replacing x by $-x$ in the first integral and combining with the third integral, we find,

$$\int_{r}^{R} \frac{e^{ix} - e^{-ix}}{x}\, dx \;+\; \int_{HJA} \frac{e^{iz}}{z}\, dz \;+\; \int_{BDEFG} \frac{e^{iz}}{z}\, dz \;=\; 0$$

or

$$2i\int_{r}^{R} \frac{\sin x}{x}\, dx \;=\; -\int_{HJA} \frac{e^{iz}}{z}\, dz \;-\; \int_{BDEFG} \frac{e^{iz}}{z}\, dz$$

Let $r \to 0$ and $R \to \infty$. By Problem 33, the second integral on the right approaches zero. The first integral on the right approaches

$$-\lim_{r \to 0} \int_\pi^0 \frac{e^{ire^{i\theta}}}{re^{i\theta}} ire^{i\theta}\, d\theta \;=\; -\lim_{r \to 0} \int_\pi^0 ie^{ire^{i\theta}}\, d\theta \;=\; \pi i$$

since the limit can be taken under the integral sign.

Then we have

$$\lim_{\substack{R \to \infty \\ r \to 0}} 2i \int_r^R \frac{\sin x}{x}\, dx \;=\; \pi i \quad \text{or} \quad \int_0^\infty \frac{\sin x}{x}\, dx \;=\; \frac{\pi}{2}$$

MISCELLANEOUS PROBLEMS

36. Let $w = z^2$ define a transformation from the z plane (xy plane) to the w plane (uv plane). Consider a triangle in the z plane with vertices at $A(2,1)$, $B(4,1)$, $C(4,3)$. (a) Show that the *image* or *mapping* of this triangle is a curvilinear triangle in the uv plane. (b) Find the angles of this curvilinear triangle and compare with those of the original triangle.

(a)   Since $w = z^2$, we have $u = x^2 - y^2$, $v = 2xy$ as the transformation equations. Then point $A(2,1)$ in the xy plane maps into point $A'(3,4)$ of the uv plane (see figures below). Similarly, points B and C map into points B' and C' respectively. The line segments AC, BC, AB of triangle ABC map respectively into parabolic segments $A'C', B'C', A'B'$ of curvilinear triangle $A'B'C'$ with equations as shown in Figures 17-10(a) and (b).

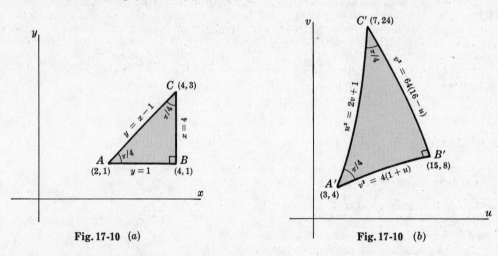

Fig. 17-10 (a)   Fig. 17-10 (b)

(b)   The slope of the tangent to the curve $v^2 = 4(1 + u)$ at $(3,4)$ is $m_1 = \dfrac{dv}{du}\Big|_{(3,4)} = \dfrac{2}{v}\Big|_{(3,4)} = \dfrac{1}{2}$.

The slope of the tangent to the curve $u^2 = 2v + 1$ at $(3,4)$ is $m_2 = \dfrac{dv}{du}\Big|_{(3,4)} = u = 3$.

Then the angle θ between the two curves at A' is given by

$$\tan \theta = \frac{m_2 - m_1}{1 + m_1 m_2} = \frac{3 - \frac{1}{2}}{1 + (3)(\frac{1}{2})} = 1, \text{ and } \theta = \pi/4$$

Similarly we can show that the angle between $A'C'$ and $B'C'$ is $\pi/4$, while the angle between $A'B'$ and $B'C'$ is $\pi/2$. Therefore the angles of the curvilinear triangle are equal to the corresponding ones of the given triangle. In general, if $w = f(z)$ is a transformation where $f(z)$ is analytic, the angle between two curves in the z plane intersecting at $z = z_0$ has the same magnitude and sense (orientation) as the angle between the images of the two curves, so long as $f'(z_0) \neq 0$. This property is called the conformal property of analytic functions and for this reason the transformation $w = f(z)$ is often called a *conformal transformation* or *conformal mapping function*.

37. Let $w = \sqrt{z}$ define a transformation from the z plane to the w plane. A point moves counterclockwise along the circle $|z| = 1$. Show that when it has returned to its starting position for the first time its image point has not yet returned, but that when it has returned for the second time its image point returns for the first time.

Let $z = e^{i\theta}$. Then $w = \sqrt{z} = e^{i\theta/2}$. Let $\theta = 0$ correspond to the starting position. Then $z = 1$ and $w = 1$ [corresponding to A and P in Figures 17-11(a) and (b)].

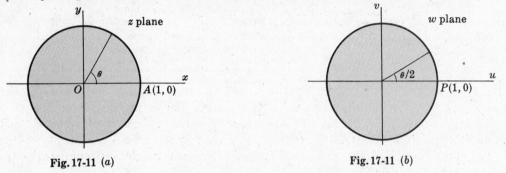

Fig. 17-11 (a)   **Fig. 17-11 (b)**

When one complete revolution in the z plane has been made, $\theta = 2\pi$, $z = 1$ but $w = e^{i\theta/2} = e^{i\pi} = -1$ so the image point has not yet returned to its starting position.

However, after two complete revolutions in the z plane have been made, $\theta = 4\pi$, $z = 1$ and $w = e^{i\theta/2} = e^{2\pi i} = 1$ so the image point has returned for the first time.

It follows from the above that w is not a single-valued function of z but is a *double-valued function of z*; i.e. given z, there are two values of w. If we wish to consider it a single-valued function, we must restrict θ. We can, for example, choose $0 \leqq \theta < 2\pi$, although other possibilities exist. This represents one branch of the double-valued function $w = \sqrt{z}$. In continuing beyond this interval we are on the second branch, e.g. $2\pi \leqq \theta < 4\pi$. The point $z = 0$ about which the rotation is taking place is called a *branch point*. Equivalently, we can insure that $f(z) = \sqrt{z}$ will be single-valued by agreeing not to cross the line Ox, called a *branch line*.

38. Show that $\displaystyle\int_0^\infty \frac{x^{p-1}}{1+x}\,dx = \frac{\pi}{\sin p\pi}$, $\quad 0 < p < 1$.

Consider $\displaystyle\oint_C \frac{z^{p-1}}{1+z}\,dz$. Since $z = 0$ is a branch point, choose C as the contour of Fig. 17-12 where AB and GH are actually coincident with the x axis but are shown separated for visual purposes.

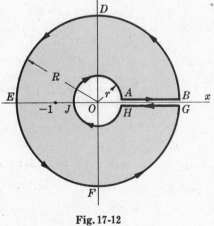

The integrand has the pole $z = -1$ lying within C.

Residue at $z = -1 = e^{\pi i}$ is

$$\lim_{z \to -1} (z+1)\frac{z^{p-1}}{1+z} = (e^{\pi i})^{p-1} = e^{(p-1)\pi i}$$

Then $\displaystyle\oint_C \frac{z^{p-1}}{1+z}\,dz = 2\pi i\, e^{(p-1)\pi i}$

or, omitting the integrand,

$$\int_{AB} + \int_{BDEFG} + \int_{GH} + \int_{HJA} = 2\pi i\, e^{(p-1)\pi i}$$

Fig. 17-12

We thus have

$$\int_r^R \frac{x^{p-1}}{1+x}\,dx + \int_0^{2\pi} \frac{(Re^{i\theta})^{p-1} iRe^{i\theta}\,d\theta}{1+Re^{i\theta}} + \int_R^r \frac{(xe^{2\pi i})^{p-1}}{1+xe^{2\pi i}}\,dx + \int_{2\pi}^0 \frac{(re^{i\theta})^{p-1} ire^{i\theta}\,d\theta}{1+re^{i\theta}} = 2\pi i\, e^{(p-1)\pi i}$$

where we have to use $z = xe^{2\pi i}$ for the integral along GH, since the argument of z is increased by 2π in going around the circle $BDEFG$.

Taking the limit as $r \to 0$ and $R \to \infty$ and noting that the second and fourth integrals approach zero, we find

$$\int_0^\infty \frac{x^{p-1}}{1+x}\,dx + \int_\infty^0 \frac{e^{2\pi i(p-1)}\,x^{p-1}}{1+x}\,dx = 2\pi\,e^{(p-1)\pi i}$$

or

$$(1 - e^{2\pi i(p-1)}) \int_0^\infty \frac{x^{p-1}}{1+x}\,dx = 2\pi i\,e^{(p-1)\pi i}$$

so that

$$\int_0^\infty \frac{x^{p-1}}{1+x}\,dx = \frac{2\pi i\,e^{(p-1)\pi i}}{1 - e^{2\pi i(p-1)}} = \frac{2\pi i}{e^{p\pi i} - e^{-p\pi i}} = \frac{\pi}{\sin p\pi}$$

Supplementary Problems

FUNCTIONS, LIMITS, CONTINUITY

39. Describe the locus represented by (a) $|z + 2 - 3i| = 5$, (b) $|z + 2| = 2|z - 1|$, (c) $|z + 5| - |z - 5| = 6$. Construct a figure in each case.

 Ans. (a) Circle $(x + 2)^2 + (y - 3)^2 = 25$, center $(-2, 3)$, radius 5.

 (b) Circle $(x - 2)^2 + y^2 = 4$, center $(2, 0)$, radius 2.

 (c) Branch of hyperbola $x^2/9 - y^2/16 = 1$, where $x \geqq 3$.

40. Determine the region in the z plane represented by each of the following:

 (a) $|z - 2 + i| \geqq 4$, (b) $|z| \leqq 3$, $0 \leqq \arg z \leqq \dfrac{\pi}{4}$, (c) $|z - 3| + |z + 3| < 10$. Construct a figure in each case.

 Ans. (a) Boundary and exterior of circle $(x - 2)^2 + (y + 1)^2 = 16$.

 (b) Region in the first quadrant bounded by $x^2 + y^2 = 9$, the x axis and the line $y = x$.

 (c) Interior of ellipse $x^2/25 + y^2/16 = 1$.

41. Express each function in the form $u(x, y) + iv(x, y)$, where u and v are real.

 (a) $z^3 + 2iz$, (b) $z/(3 + z)$, (c) e^{z^2}, (d) $\ln(1 + z)$.

 Ans. (a) $u = x^3 - 3xy^2 - 2y$, $v = 3x^2y - y^3 + 2x$

 (b) $u = \dfrac{x^2 + 3x + y^2}{x^2 + 6x + y^2 + 9}$, $v = \dfrac{3y}{x^2 + 6x + y^2 + 9}$

 (c) $u = e^{x^2 - y^2} \cos 2xy$, $v = e^{x^2 - y^2} \sin 2xy$

 (d) $u = \frac{1}{2}\ln\{(1 + x)^2 + y^2\}$, $v = \tan^{-1}\dfrac{y}{1 + x} + 2k\pi$, $k = 0, \pm1, \pm2, \ldots$

42. Prove that (a) $\lim\limits_{z \to z_0} z^2 = z_0^2$, (b) $f(z) = z^2$ is continuous at $z = z_0$ directly from the definition.

43. (a) If $z = \omega$ is any root of $z^5 = 1$ different from 1, prove that all the roots are $1, \omega, \omega^2, \omega^3, \omega^4$.

 (b) Show that $1 + \omega + \omega^2 + \omega^3 + \omega^4 = 0$.

 (c) Generalize the results in (a) and (b) to the equation $z^n = 1$.

DERIVATIVES, CAUCHY-RIEMANN EQUATIONS

44. (a) If $w = f(z) = z + \dfrac{1}{z}$, find $\dfrac{dw}{dz}$ directly from the definition.

 (b) For what finite values of z is $f(z)$ non-analytic?

 Ans. (a) $1 - 1/z^2$, (b) $z = 0$

45. Given the function $w = z^4$. (a) Find real functions u and v such that $w = u + iv$. (b) Show that the Cauchy-Riemann equations hold at all points in the finite z plane. (c) Prove that u and v are harmonic functions. (d) Determine dw/dz.   Ans. (a) $u = x^4 - 6x^2y^2 + y^4$, $v = 4x^3y - 4xy^3$ (d) $4z^3$

46. Prove that $f(z) = z |z|$ is not analytic anywhere.

47. Prove that $f(z) = \dfrac{1}{z - 2}$ is analytic in any region not including $z = 2$.

48. If the imaginary part of an analytic function is $2x(1 - y)$, determine (a) the real part, (b) the function. Ans. (a) $y^2 - x^2 - 2y + c$, (b) $2iz - z^2 + c$, where c is real

49. Construct an analytic function $f(z)$ whose real part is $e^{-x}(x \cos y + y \sin y)$ and for which $f(0) = 1$. Ans. $ze^{-z} + 1$

50. Prove that there is no analytic function whose imaginary part is $x^2 - 2y$.

51. Find $f(z)$ such that $f'(z) = 4z - 3$ and $f(1 + i) = -3i$.   Ans. $f(z) = 2z^2 - 3z + 3 - 4i$

INTEGRALS, CAUCHY'S THEOREM, CAUCHY'S INTEGRAL FORMULAS

52. Evaluate $\displaystyle\int_{1 - 2i}^{3 + i} (2z + 3)\, dz$:
    (a) along the path $x = 2t + 1$, $y = 4t^2 - t - 2$ $0 \leq t \leq 1$.
    (b) along the straight line joining $1 - 2i$ and $3 + i$.
    (c) along straight lines from $1 - 2i$ to $1 + i$ and then to $3 + i$.
    Ans. $17 + 19i$ in all cases

53. Evaluate $\displaystyle\int_C (z^2 - z + 2)\, dz$, where C is the upper half of the circle $|z| = 1$ traversed in the positive sense.   Ans. $-14/3$

54. Evaluate $\displaystyle\oint_C \frac{z\, dz}{2z - 5}$, where C is the circle (a) $|z| = 2$, (b) $|z - 3| = 2$.   Ans. (a) 0, (b) $5\pi i/2$

55. Evaluate $\displaystyle\oint_C \frac{z^2}{(z + 2)(z - 1)}\, dz$, where C is: (a) a square with vertices at $-1 - i, -1 + i, -3 + i, -3 - i$; (b) the circle $|z + i| = 3$; (c) the circle $|z| = \sqrt{2}$.   Ans. (a) $-8\pi i/3$ (b) $-2\pi i$ (c) $2\pi i/3$

56. Evaluate (a) $\displaystyle\oint_C \frac{\cos \pi z}{z - 1}\, dz$, (b) $\displaystyle\oint_C \frac{e^z + z}{(z - 1)^4}\, dz$ where C is any simple closed curve enclosing $z = 1$. Ans. (a) $-2\pi i$ (b) $\pi i e/3$

57. Prove Cauchy's integral formulas.
    [Hint: Use the definition of derivative and then apply mathematical induction.]

SERIES and SINGULARITIES

58. For what values of z does each series converge?

    (a) $\displaystyle\sum_{n=1}^{\infty} \frac{(z + 2)^n}{n!}$,   (b) $\displaystyle\sum_{n=1}^{\infty} \frac{n(z - i)^n}{n + 1}$,   (c) $\displaystyle\sum_{n=1}^{\infty} (-1)^n (z^2 + 2z + 2)^{2n}$.

    Ans. (a) all z (b) $|z - i| < 1$ (c) $z = -1 \pm i$

59. Prove that the series $\displaystyle\sum_{n=1}^{\infty} \frac{z^n}{n(n + 1)}$ is (a) absolutely convergent, (b) uniformly convergent for $|z| \leq 1$.

60. Prove that the series $\displaystyle\sum_{n=0}^{\infty} \frac{(z + i)^n}{2^n}$ converges uniformly within any circle of radius R such that $|z + i| < R < 2$.

61. Locate in the finite z plane all the singularities, if any, of each function and name them:

    (a) $\dfrac{z-2}{(2z+1)^4}$,   (b) $\dfrac{z}{(z-1)(z+2)^2}$,   (c) $\dfrac{z^2+1}{z^2+2z+2}$,   (d) $\cos\dfrac{1}{z}$,   (e) $\dfrac{\sin(z-\pi/3)}{3z-\pi}$,   (f) $\dfrac{\cos z}{(z^2+4)^2}$.

    Ans.   (a) $z=-\frac{1}{2}$, pole of order 4   (d) $z=0$, essential singularity

           (b) $z=1$, simple pole; $z=-2$, double pole   (e) $z=\pi/3$, removable singularity

           (c) Simple poles $z=-1\pm i$ (f) $z=\pm 2i$, double poles

62. Find Laurent series about the indicated singularity for each of the following functions, naming the singularity in each case. Indicate the region of convergence of each series.

    (a) $\dfrac{\cos z}{z-\pi}$; $z=\pi$ (b) $z^2 e^{-1/z}$; $z=0$ (c) $\dfrac{z^2}{(z-1)^2(z+3)}$; $z=1$

    Ans.   (a) $-\dfrac{1}{z-\pi}+\dfrac{z-\pi}{2!}-\dfrac{(z-\pi)^3}{4!}+\dfrac{(z-\pi)^5}{6!}-\cdots$,   simple pole,   all $z\ne\pi$

          (b) $z^2-z+\dfrac{1}{2!}-\dfrac{1}{3!\,z}+\dfrac{1}{4!\,z^2}-\dfrac{1}{5!\,z^3}+\cdots$,   essential singularity,   all $z\ne 0$

          (c) $\dfrac{1}{4(z-1)^2}+\dfrac{7}{16(z-1)}+\dfrac{9}{64}-\dfrac{9(z-1)}{256}+\cdots$,   double pole, $0<|z-1|<4$

RESIDUES and the RESIDUE THEOREM

63. Determine the residues of each function at its poles:

    (a) $\dfrac{2z+3}{z^2-4}$,   (b) $\dfrac{z-3}{z^3+5z^2}$,   (c) $\dfrac{e^{zt}}{(z-2)^3}$,   (d) $\dfrac{z}{(z^2+1)^2}$.

    Ans.   (a) $z=2$; $7/4$, $z=-2$; $1/4$ (c) $z=2$; $\frac{1}{2}t^2 e^{2t}$

          (b) $z=0$; $8/25$, $z=-5$; $-8/25$ (d) $z=i$; 0, $z=-i$; 0

64. Find the residue of $e^{zt}\tan z$ at the simple pole $z=3\pi/2$.   *Ans.* $-e^{3\pi t/2}$

65. Evaluate $\displaystyle\oint_C \frac{z^2\,dz}{(z+1)(z+3)}$, where C is a simple closed curve enclosing all the poles.   *Ans.* $-8\pi i$

66. If C is a simple closed curve enclosing $z=\pm i$, show that

$$\oint_C \frac{z e^{zt}}{(z^2+1)^2}\,dz = \tfrac{1}{2}t\sin t$$

67. If $f(z)=P(z)/Q(z)$, where $P(z)$ and $Q(z)$ are polynomials such that the degree of $P(z)$ is at least two less than the degree of $Q(z)$, prove that $\displaystyle\oint_C f(z)\,dz = 0$, where C encloses all the poles of $f(z)$.

EVALUATION of DEFINITE INTEGRALS

Use contour integration to verify each of the following

68. $\displaystyle\int_0^\infty \frac{x^2\,dx}{x^4+1} = \frac{\pi}{2\sqrt{2}}$ **73.** $\displaystyle\int_{-\infty}^\infty \frac{dx}{(x^2+1)^2(x^2+4)} = \frac{\pi}{9}$

69. $\displaystyle\int_{-\infty}^\infty \frac{dx}{x^6+a^6} = \frac{2\pi}{3a^5}$, $a>0$ **74.** $\displaystyle\int_0^{2\pi} \frac{d\theta}{2-\cos\theta} = \frac{2\pi}{\sqrt{3}}$

70. $\displaystyle\int_0^\infty \frac{dx}{(x^2+4)^2} = \frac{\pi}{32}$ **75.** $\displaystyle\int_0^{2\pi} \frac{d\theta}{(2+\cos\theta)^2} = \frac{4\pi\sqrt{3}}{9}$

71. $\displaystyle\int_0^\infty \frac{\sqrt{x}}{x^3+1}\,dx = \frac{\pi}{3}$ **76.** $\displaystyle\int_0^\pi \frac{\sin^2\theta}{5-4\cos\theta}\,d\theta = \frac{\pi}{8}$

72. $\displaystyle\int_0^\infty \frac{dx}{(x^4+a^4)^2} = \frac{3\pi}{8\sqrt{2}}a^{-7}$, $a>0$ **77.** $\displaystyle\int_0^{2\pi} \frac{d\theta}{(1+\sin^2\theta)^2} = \frac{3\pi}{2\sqrt{2}}$

78. $\displaystyle\int_0^{2\pi} \frac{\cos n\theta\ d\theta}{1 - 2a \cos \theta + a^2} = \frac{2\pi a^n}{1 - a^2},\quad n = 0, 1, 2, 3, \ldots,\quad 0 < a < 1$

79. $\displaystyle\int_0^{2\pi} \frac{d\theta}{(a + b \cos \theta)^3} = \frac{(2a^2 + b^2)\pi}{(a^2 - b^2)^{5/2}},\quad a > |b|$

80. $\displaystyle\int_0^\infty \frac{x \sin 2x}{x^2 + 4}\, dx = \frac{\pi e^{-4}}{4}$

83. $\displaystyle\int_0^\infty \frac{\sin x}{x(x^2 + 1)^2}\, dx = \frac{\pi(2e - 3)}{4e}$

81. $\displaystyle\int_0^\infty \frac{\cos 2\pi x}{x^4 + 4}\, dx = \frac{\pi e^{-\pi}}{8}$

84. $\displaystyle\int_0^\infty \frac{\sin^2 x}{x^2}\, dx = \frac{\pi}{2}$

82. $\displaystyle\int_0^\infty \frac{x \sin \pi x}{(x^2 + 1)^2}\, dx = \frac{\pi^2 e^{-\pi}}{4}$

85. $\displaystyle\int_0^\infty \frac{\sin^3 x}{x^3}\, dx = \frac{3\pi}{8}$

86. $\displaystyle\int_0^\infty \frac{\cos x}{\cosh x}\, dx = \frac{\pi}{2 \cosh (\pi/2)}$. [Hint: Consider $\displaystyle\oint_C \frac{e^{iz}}{\cosh z}\, dz$, where C is a rectangle with vertices at $(-R, 0),\ (R, 0),\ (R, \pi),\ (-R, \pi)$. Then let $R \to \infty$.]

MISCELLANEOUS PROBLEMS

87. If $z = \rho e^{i\phi}$ and $f(z) = u(\rho, \phi) + i\, v(\rho, \phi)$, where ρ and ϕ are polar coordinates, show that the Cauchy-Riemann equations are

$$\frac{\partial u}{\partial \rho} = \frac{1}{\rho} \frac{\partial v}{\partial \phi}, \qquad \frac{\partial v}{\partial \rho} = -\frac{1}{\rho} \frac{\partial u}{\partial \phi}$$

88. If $w = f(z)$, where $f(z)$ is analytic, defines a transformation from the z plane to the w plane where $z = x + iy$ and $w = u + iv$, prove that the Jacobian of the transformation is given by

$$\frac{\partial(u, v)}{\partial(x, y)} = |f'(z)|^2$$

89. Let $F(x, y)$ be transformed to $G(u, v)$ by the transformation $w = f(z)$. Show that if $\dfrac{\partial^2 F}{\partial x^2} + \dfrac{\partial^2 F}{\partial y^2} = 0$, then at all points where $f'(z) \neq 0$, $\dfrac{\partial^2 G}{\partial u^2} + \dfrac{\partial^2 G}{\partial v^2} = 0$.

90. Show that by the *bilinear transformation* $w = \dfrac{az + b}{cz + d}$, where $ad - bc \neq 0$, circles in the z plane are transformed into circles of the w plane.

91. If $f(z)$ is analytic inside and on the circle $|z - a| = R$, prove *Cauchy's inequality*, namely,

$$|f^{(n)}(a)| \leq \frac{n!\, M}{R^n}$$

where $|f(z)| \leq M$ on the circle. [Hint: Use Cauchy's integral formulas.]

92. Let C_1 and C_2 be concentric circles having center a and radii r_1 and r_2 respectively, where $r_1 < r_2$. If $a + h$ is any point in the annular region bounded by C_1 and C_2, and $f(z)$ is analytic in this region, prove *Laurent's theorem* that

$$f(a + h) = \sum_{-\infty}^{\infty} a_n h^n$$

where

$$a_n = \frac{1}{2\pi i} \oint_C \frac{f(z)\, dz}{(z - a)^{n+1}}$$

C being any closed curve in the angular region surrounding C_1.

$\left[\text{Hint: Write}\quad f(a + h) = \dfrac{1}{2\pi i} \oint_{C_2} \dfrac{f(z)\, dz}{z - (a + h)} - \dfrac{1}{2\pi i} \oint_{C_1} \dfrac{f(z)\, dz}{z - (a + h)}\quad \text{and expand}\ \dfrac{1}{z - a - h}\ \text{in}\right.$
$\left.\text{two different ways.}\right]$

93. Find a Laurent series expansion for the function $f(z) = \dfrac{z}{(z+1)(z+2)}$ which converges for $1 < |z| < 2$ and diverges elsewhere.

$$\left[\text{Hint: Write }\quad \frac{z}{(z+1)(z+2)} \;=\; \frac{-1}{z+1} + \frac{2}{z+2} \;=\; \frac{-1}{z(1+1/z)} + \frac{1}{1+z/2}.\right]$$

Ans. $\cdots - \dfrac{1}{z^5} + \dfrac{1}{z^4} - \dfrac{1}{z^3} + \dfrac{1}{z^2} - \dfrac{1}{z} + 1 - \dfrac{z}{2} + \dfrac{z^2}{4} - \dfrac{z^3}{8} + \cdots$

94. Let $\displaystyle\int_0^\infty e^{-st} F(t)\, dt = f(s)$ where $f(s)$ is a given rational function with numerator of degree less than that of the denominator. If C is a simple closed curve enclosing all the poles of $f(s)$, we can show that

$$F(t) \;=\; \frac{1}{2\pi i} \oint_C e^{zt} f(z)\, dz \;=\; \text{sum of residues of } e^{zt} f(z) \text{ at its poles}$$

Use this result to find $F(t)$ if $f(s)$ is   (a) $\dfrac{s}{s^2+1}$, (b) $\dfrac{1}{s^2+2s+5}$, (c) $\dfrac{s^2+1}{s(s-2)^2}$, (d) $\dfrac{1}{(s^2+1)^2}$ and check results in each case.

[Note that $f(s)$ is the *Laplace transform* of $F(t)$, and $F(t)$ is the *inverse Laplace transform* of $f(s)$ (see Chapter 12). Extensions to other functions $f(s)$ are possible.]

Ans. (a) $\cos t$, (b) $\frac{1}{2}e^{-t} \sin 2t$, (c) $\frac{1}{4} + \frac{5}{2}te^{2t} + \frac{3}{4}e^{2t}$, (d) $\frac{1}{2}(\sin t - t \cos t)$

INDEX